Common Units

	BRITISH		SI (INTERNATIONAL SYSTEM OF UNITS)		
	Name	*Symbol*	*Name*	*Symbol*	
Length	foot	ft	meter	m	
Time	second	s	second	s	
Velocity		ft/s		m/s	
Acceleration		ft/s^2		m/s^2	
Mass	slug		kilogram	kg	
Force	pound	lb	newton	N	$kg \cdot m/s^2$
Capacity	gallon	gal	liter	L	$1\ L = 1000\ cm^3$
Pressure		lb/ft^2	pascal	Pa	N/m^2
Density		lb/ft^3		kg/m^3	
Work, Energy	foot-pound	$ft \cdot lb$	joule	J	$N \cdot m$
Power	horsepower	hp	watt	W	J/s
Current	ampere	A	ampere	A	
Charge	coulomb	C	coulomb	C	
Electric potential	volt	V	volt	V	
Capacitance	farad	F	farad	F	
Inductance	henry	H	henry	H	
Resistance	ohm	Ω	ohm	Ω	
Quantity of heat	British thermal unit	Btu	joule	J	
Temperature	degree Fahrenheit	°F	degree Celsius	°C	
Absolute temperature			kelvin	K	

Introduction to Technical Mathematics

Books in the Kuhfittig series in technical mathematics:

Introduction to Technical Mathematics
Basic Technical Mathematics
Basic Technical Mathematics with Calculus
Technical Calculus with Analytic Geometry

Peter K. F. Kuhfittig
Milwaukee School of Engineering

Introduction to Technical Mathematics

Brooks/Cole Publishing Company
Monterey, California

Brooks/Cole Publishing Company
A Division of Wadsworth, Inc.

Printed in the United States of America

10 9 8 7 6 5 4

Library of Congress Cataloging-in-Publication Data
Kuhfittig, Peter K. F.
Introduction to technical mathematics.
Includes index.
1. Mathematics—1961— I. Title.
QA39.2.K84 1986 512'.1 85-17437

ISBN 0-534-05802-7

Sponsoring Editor: Craig Barth, Jeremy Hayhurst
Editorial Assistant: Eileen Galligan, Amy Mayfield
Production Management: Miller/Scheier Associates
Manuscript Editor: Carol Dondrea
Permissions Editor: Carline Haga
Interior and Cover Design: Rick Chafian
Cover Photo: West Light, © Chuck O'Rear, Lawrence Livermore
Art Coordinator: Miller/Scheier Associates
Interior Illustrations: John Foster
Typesetting: Bi-Comp, Inc.
Printing and Binding: R. R. Donnelley & Sons

Preface

Purpose of This Book

This book provides an introduction to technical mathematics for students in various technical programs. It is designed for students whose academic background does not include algebra and geometry. The topics covered include a review of arithmetic, basic algebra, graphing, introduction to trigonometry, vectors, and logarithms. Since the discussion begins at a very basic level and emphasizes technical applications, the book is suitable for students in terminal courses and various special programs. At the same time, the topic selection provides a solid foundation for more advanced mathematical and technical subjects.

Main Features

This book does more than teach mathematical skills. It shows the student how mathematics is used in technology. This approach, together with a clear and informal presentation, gives the student the motivation to master the material. This book fills the need of today's technology student in many ways:

1. The discussion is concrete and intuitive.
2. Most sections contain problems that illustrate how mathematics is applied to technical situations.
3. Whenever feasible, sections end with a technical application.
4. Drill exercises make use of notations commonly encountered in technical fields.
5. Calculator operations are explained throughout the book.

6. There is an appendix on the BASIC programming language. BASIC notations are introduced in several places in the text.

The book's presentation is exceptionally student-oriented:

1. The most important concepts are boxed and labeled for easy reference; many other concepts are identified by marginal labels.
2. The use of a second color is functional, rather than decorative. Its main purpose is to help explain difficult steps.
3. Important procedures are summarized; step-by-step procedures are provided whenever appropriate.
4. Common pitfalls are pointed out in special segments called *Common Errors*.
5. Examples are worked out in great detail. Liberal use of marginal notes helps explain steps.
6. A large amount of drill material reinforces basic concepts.
7. All the answers to the chapter review exercises and cumulative review exercises are supplied. This feature helps the student in reviewing the material from time to time.

The use of realistic notation is particularly important at this level, since beginning students often have great difficulty in transferring mathematical skills to technical problems that use a different notation.

The contemporary design, as exemplified by the use of a second color, the boxing and labeling of important concepts and procedures, and the clear separation of the examples from the text material, greatly helps the student in reading the book.

Some other features are:

1. Both SI and English units are used throughout the book.
2. Every chapter ends with a set of review exercises. Cumulative review exercises appear at the end of every third chapter.
3. The only background assumed is arithmetic. However, a review of arithmetic is presented in the first chapter.

Flexibility

In addition to the review of arithmetic, the book includes many topics normally covered in intermediate algebra. These topics, together with intuitive geometry, graphing, basic trigonometry, vectors, graphs of sinusoidal functions, and logarithms make the book suitable for programs that are terminal in nature. However, the topics are covered thoroughly enough to provide a solid background for further work in mathematics.

Coverage and Scope

Chapter 1 is a review of arithmetic, including fractions, decimals, and percent. Calculator operations are introduced at the end of the chapter.

Chapter 2 contains a detailed discussion of approximate numbers, SI and English units of measurement, and reduction and conversion of units.

Chapter 3 covers operations with signed numbers. Chapters 4 and 5 introduce the basic algebraic concepts, expressions, formulas, and the four fundamental operations.

Chapter 6 is an introduction to geometry. Emphasis is placed on the properties of geometric figures and on applications of geometry rather than on formal proof.

Chapter 7 deals with the solution of linear equations. Special attention is paid to formula rearrangement and word problems.

Chapter 8 on factoring is organized to allow a gradual mastery of factoring by alternating special products with the corresponding factoring cases. Chapter 9 is devoted to reduction of fractions, operations with fractions, and fractional equations.

Chapter 10 includes a detailed treatment of exponents and radicals: zero, negative, and fractional exponents, simplification of radicals, and operations with radicals.

Chapter 11 covers quadratic equations and their applications. The methods of solution discussed are factoring and the use of the quadratic formula.

Chapter 12 discusses functions and graphs, ratio and proportion, and functional variation.

Chapter 13 is devoted to the solution of systems of linear equations in two unknowns. Both graphical and algebraic methods of solution are discussed.

Chapter 14 continues the discussion of informal geometry begun in Chapter 6. Some of the topics covered are the Pythagorean theorem, congruence and similarity, and solid geometric figures.

Chapter 15 is an introduction to trigonometry. Topics covered are the definitions of trigonometric functions, right-triangle applications, special angles, and values of trigonometric functions. Emphasis is on the use of calculators, rather than tables.

Chapter 16 discusses vectors, vector addition by components, and applications of vectors. The sections on the sine and cosine laws contain additional vector applications.

Chapter 17 introduces radian measure and applications of radian measure and continues with a discussion of graphs of sinusoidal functions.

Chapter 18 covers logarithms and exponential equations. Appendix A is an introduction to BASIC.

Supplements

The answers to odd-numbered exercises in each section are given in the answer section; the answers to even-numbered exercises are published in a separate Instructor's Manual. For the chapter review exercises and the cumulative review exercises, all the answers are given in the answer section to help the student review the material.

I would like to thank the staff of Brooks/Cole Publishing Company and the following reviewers for their cooperation and help in the preparation of this book: Darrell Abney, Nashville State Technical Institute; Barbara Bates, Trident Technical College; Dale Boyle, Schoolcraft College; Tom Divver, Spartanburg Technical College; Charles Ferrell, Delaware Technical Community College—Stanton Campus; Harold Hackett, SUNY-Agricultural and Technical College; Patricia Hirschey, Delaware Technical Community College; Frank Iacobucci, Cincinnati Technical College; Mary Ann Justinger, Erie Community College—South Campus; Don Osborne, Sumter Technical College.

Peter Kuhfittig

Contents

Chapter 12 Functions, Graphs, and Functional Variation 367

Chapter 13 Simultaneous Linear Equations 403

Chapter 14 Additional Topics from Geometry 430

Chapter 15 Trigonometry 479

Introduction to Technical Mathematics

CHAPTER 1

Review of Arithmetic

Mathematics is the language of science and technology. A good grasp of the fundamentals of mathematics is essential for anyone aspiring to a scientific or technical career.

Most technical problems involve numerical computations, and today such computations are often done by computers. However, since arithmetic skills form the basis of more advanced studies in mathematics, they deserve their share of attention. This chapter gives you a chance to review the fundamentals of arithmetic. You should know these concepts well enough to interpret results, identify errors, and determine the accuracy of measurements. In the process of reviewing you will be building a solid foundation for some of the other topics we discuss, particularly algebra and geometry. For this reason, all calculations are to be done on paper—calculator operations will be introduced at the end of the chapter.

1.1 Addition and Subtraction of Whole Numbers

We begin our study of mathematics by reviewing the four fundamental operations with whole numbers. In this section we discuss addition and subtraction.

Whole Numbers

A number is an abstract concept that can be represented by many symbols. For example, the number 5 can be represented by the Roman numeral V, by 𝍸, or by $3 + 2$. A symbol that represents a number is called a **numeral.**

The decimal or Hindu–Arabic system of numeration uses the **digits** 0, 1, 2, 3, 4, 5, 6, 7, 8, and 9 for its basic numerals. Larger numbers are repre-

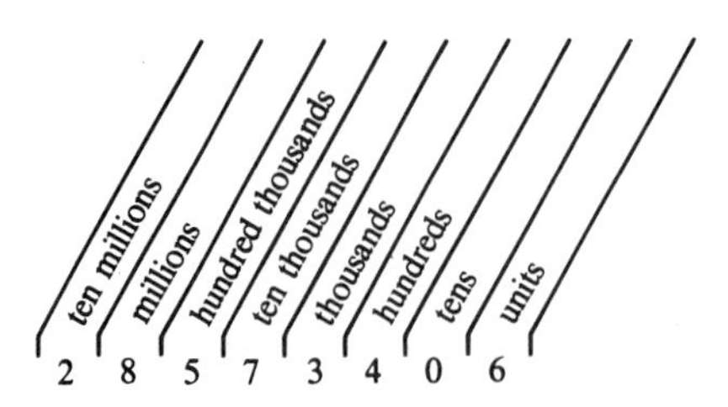

Figure 1.1

sented by using different arrangements of these digits, as shown in Figure 1.1. For example, 432 has 2 units, 3 tens, and 4 hundreds, representing four hundred thirty-two. The numeral 0 represents an empty place. For example, 302 has 2 units, zero tens, and 3 hundreds, and represents three hundred two. This *place-value* numeration, in which each numeral represents a number whose size depends on the position, greatly simplifies arithmetic operations such as addition and subtraction.

Natural numbers
Whole numbers

The numbers 1, 2, 3, and so on, are called the **natural numbers.** The natural numbers, together with the number 0, make up the **whole numbers.** The set of whole numbers is infinitely large in the sense that for every whole number N there exists a successor to that whole number: $N + 1$. Therefore, there is no largest number.

Properties of Whole Numbers

Numbers have certain properties that we know and use automatically. For example,

$$3 + 4 = 4 + 3 \qquad \text{and} \qquad 2 \times 3 = 3 \times 2$$

Commutative law

In other words, the order in which we add or multiply numbers doesn't matter. This property is known as the **commutative law,** which says that addition and multiplication are independent of order.

Associative law

The **associative law** says that addition and multiplication do not depend on the order of grouping. Thus

$$3 + (5 + 7) = (3 + 5) + 7$$

and

$$3 \times (5 \times 7) = (3 \times 5) \times 7$$

Distributive law

Finally, a property that equates a basic product with a basic sum is the **distributive law.** The distributive law says that multiplying a first number by the sum of two other numbers is equivalent to multiplying the first number by each of the two other numbers and then adding the products. For example,

$$2 \times (3 + 5) = (2 \times 3) + (2 \times 5) = 6 + 10 = 16$$

Note: The parentheses around 3 + 5 indicate that the addition operation is to be carried out before multiplication. This operation yields

$$2 \times (3 + 5) = 2 \times 8 = 16$$

in agreement with the result obtained using the distributive law.

Addition of Whole Numbers

To add 57 and 69 we arrange our work as follows:

$$\begin{array}{r} 57 \\ +69 \\ \hline \end{array}$$

Written in this form, the addition problem means

$$\begin{array}{r} 5 \text{ tens} + 7 \text{ ones} \\ + \ 6 \text{ tens} + 9 \text{ ones} \\ \hline 11 \text{ tens} + 16 \text{ ones} \end{array} \quad \text{Total}$$

Now, 16 ones = 1 ten + 6 ones. So the total is

$$\begin{aligned} 11 \text{ tens} + 1 \text{ ten} + 6 \text{ ones} &= 12 \text{ tens} + 6 \text{ ones} \\ &= 10 \text{ tens} + 2 \text{ tens} + 6 \text{ ones} \\ &= 1 \text{ hundred} + 2 \text{ tens} + 6 \text{ ones} \qquad 10 \text{ tens} = 1 \text{ hundred} \\ &= 126 \end{aligned}$$

In the column arrangement, the operation can be performed as follows:

$$\begin{array}{r} 1 \\ 57 \\ +\ 69 \\ \hline 126 \end{array} \quad \text{1 carried to tens column}$$

Addends
Sum

In **addition**, the numbers that are added are called the **addends**, and the result of the addition is called the **sum**.

Example 1 Add: 376 + 485 + 193 + 249

Solution:

$$\begin{array}{r} 32 \\ 376 \\ 485 \\ 193 \\ +\ 249 \\ \hline 1303 \end{array}$$

$6 + 5 + 3 + 9 = 23 = 3 + 20$ (units column)
$2 + 7 + 8 + 9 + 4 = 30 = 0 + 30$ (tens column)
$3 + 3 + 4 + 1 + 2 = 13$ (hundreds column)

Note that the digits in the units column add to 23. We write only the 3 and carry the 2 to the tens column. *Carrying a digit to the next column is necessary whenever the sum exceeds 9.*

It is good practice to check your work. A simple check can be performed by using the commutative law and adding the numbers in reverse order. If you add the numbers by coming down each column, check the addition by going up.

Example 2

$$\begin{array}{r} {}^{11}\\ 608 \\ 396 \\ +\ 454 \\ \hline 1458 \end{array} \qquad \begin{array}{r} 8+6+4=18=8+10 \\ 1+0+9+5=15=5+10 \\ 1+6+3+4=14 \end{array}$$

Check:

$$\begin{array}{r} 454 \\ 396 \\ +\ 608 \\ \hline 1458 \end{array}$$

Subtraction of Whole Numbers

Subtraction is what we do when we want to know the difference between two numbers. For example, if the voltage across a resistor is 80 volts and increases to 100 volts, then the difference is 100 volts − 80 volts = 20 volts. The first number (100) is called the **minuend,** the number that is subtracted (80, in this case) is called the **subtrahend,** and the result (20) is called the **difference.**

Minuend
Subtrahend
Difference

The method of subtraction is based on the same principles as addition. For example, to subtract 27 from 62, we arrange our work in columns as follows:

$$\begin{array}{l} 62 = 6 \text{ tens} + 2 \text{ ones} \\ \underline{27 = 2 \text{ tens} + 7 \text{ ones}} \end{array}$$

The problem is that we cannot subtract 7 from 2. However, if we "borrow" 1 ten from the tens column, we can increase the unit column by 10 units. Note that

$$\begin{aligned} 6 \text{ tens} + 2 \text{ ones} &= 5 \text{ tens} + \mathbf{1\ ten} + 2 \text{ ones} \\ &= 5 \text{ tens} + \mathbf{10\ ones} + 2 \text{ ones} \\ &= 5 \text{ tens} + 12 \text{ ones} \end{aligned}$$

We now have

$$\begin{array}{r} 62 = 5 \text{ tens} + 12 \text{ ones} \\ \underline{27 = 2 \text{ tens} + \ \ 7 \text{ ones}} \\ 3 \text{ tens} + \ \ 5 \text{ ones} = 35 \end{array} \qquad \text{Subtracting}$$

This operation can be performed with the original numbers by the following scheme:

Borrowing a ten (10 ones)

```
   5
   1
  6̸2     12 − 7 = 5   (units column)
 −27      5 − 2 = 3   (tens column)
 ───
  35
```

Check: If we add the difference to the subtrahend, we get the minuend:

$35 + 27 = 62$ ✓

Consider another example.

Example 3 Subtract:

```
  342
−  98
─────
```

Solution. Here we need to borrow from the tens column. Since the tens column is also short, we first borrow from the hundreds column:

```
  23
  11
  3̸4̸2    12 − 8 = 4
−  98    13 − 9 = 4
─────
  244     2 − 0 = 2
```

Check: If we add the difference to the subtrahend, we get the minuend:

$244 + 98 = 342$ ✓

Example 4 Subtract:

```
  1404
−  275
──────
```

Solution. While it is necessary to borrow from the tens column, there are no tens to borrow from. So we first borrow from the hundreds column to increase the tens column:

```
  39
    1      14 − 5 = 9
  14̸0̸4      9 − 7 = 2
−  275      3 − 2 = 1
──────
  1129      1 − 0 = 1
```

Check:

$$\begin{array}{r} 1129 \\ +\ 275 \\ \hline 1404 \end{array} \checkmark \qquad \begin{array}{l} \text{Difference} \\ +\ \text{Subtrahend} \\ \hline =\ \text{Minuend} \end{array}$$

Remark. We will discuss calculator operations at the end of this chapter with the intent of using calculators throughout the book. Until then, you should perform all arithmetic operations on paper.

In technology it is often necessary to add or subtract measurements that have units assigned. In such a case, only numbers with like units can be added or subtracted. For example, 4 ft + 10 ft = 14 ft. However, 4 ft and 10 in. cannot be added directly since the sum 4 + 10 = 14 has no meaning here.

Table 1.1 summarizes some familiar units and their relationships.

Table 1.1

Unit	Symbol	
inch	in.	
foot	ft	1 ft = 12 in.
yard	yd	1 yd = 3 ft
mile	mi	1 mi = 5280 ft
pound	lb	1 lb = 16 oz
ounce	oz	
gallon	gal	1 gal = 4 qt
quart	qt	
minute	min	

Other, less familiar, symbols will be defined as needed. A complete discussion of units can be found in Chapter 2.

Example 5 Perform the following subtraction:

$$\begin{array}{r} 7 \text{ ft } 2 \text{ in.} \\ -4 \text{ ft } 8 \text{ in.} \\ \hline \end{array}$$

Solution. Since we cannot subtract 8 in. from 2 in., we need to change 7 ft 2 in. as follows:

$$\begin{aligned} 7 \text{ ft } 2 \text{ in.} &= 7 \text{ ft} + 2 \text{ in.} \\ &= 6 \text{ ft} + 1 \text{ ft} + 2 \text{ in.} \\ &= 6 \text{ ft} + 12 \text{ in.} + 2 \text{ in.} \qquad 1 \text{ ft} = 12 \text{ in.} \\ &= 6 \text{ ft } 14 \text{ in.} \end{aligned}$$

Since we borrowed 1 ft as 12 in., the subtraction can now be performed:

$$\begin{array}{r} 6 \text{ ft } 14 \text{ in.} \\ -4 \text{ ft } 8 \text{ in.} \\ \hline 2 \text{ ft } 6 \text{ in.} \end{array}$$

Example 6 Consider the layout of the machine shop shown in Figure 1.2. Find the length of the shop.

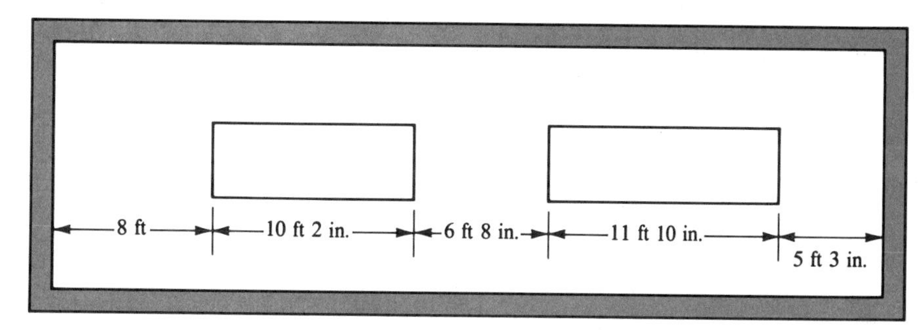

Figure 1.2

Solution. Recall that 12 in. = 1 ft. We can calculate the total length of the shop by arranging our work in columns as follows:

$$\begin{array}{rr} 8 \text{ ft} & \\ 10 \text{ ft} & 2 \text{ in.} \\ 6 \text{ ft} & 8 \text{ in.} \\ 11 \text{ ft} & 10 \text{ in.} \\ 5 \text{ ft} & 3 \text{ in.} \\ \hline 40 \text{ ft} & 23 \text{ in.} \end{array} \quad \text{Total of each column}$$

Now, 23 in. = 12 in. + 11 in. = 1 ft + 11 in. So the total is

$$\begin{aligned} 40 \text{ ft } 23 \text{ in.} &= 40 \text{ ft} + 1 \text{ ft} + 11 \text{ in.} \\ &= 41 \text{ ft} + 11 \text{ in.} = 41 \text{ ft } 11 \text{ in.} \end{aligned}$$

Example 7 Find the perimeter of the machine part in Figure 1.3. (The *perimeter* of a geometric figure is the distance around the figure.)

Solution. Since the perimeter is the distance around the figure, we get

$$\text{perimeter} = 4 \text{ cm} + 3 \text{ cm} + 5 \text{ cm} = 12 \text{ cm}$$

5 cm
4 cm
3 cm

Figure 1.3

Exercises / Section 1.1

In Exercises 1–24, add the given numbers.

1. 868 398	**2.** 165 276	**3.** 572 474	**4.** 995 239
5. 407 816 119	**6.** 321 708 117	**7.** 658 917 282	**8.** 385 237 814
9. 291 323 973	**10.** 370 659 357 264	**11.** 715 371 642 287	**12.** 165 152 830 467
13. 3964 7526 9181 7609	**14.** 2876 1077 8546 8308	**15.** 6424 2879 2237 1825	**16.** 3515 3406 9944 1093
17. 15,201 96,705 64,398	**18.** 29,414 34,145 89,236	**19.** 34,733 5,082 7,340	**20.** 69,876 9,548 2,459
21. 4514 926 9913 705 1866	**22.** 3473 5968 476 884 147	**23.** 1479 643 215 3627 6964	**24.** 9703 749 867 8570 2114

In Exercises 25–40, perform the indicated subtractions.

25. 3027 −1758	**26.** 4198 −2687	**27.** 6993 −6844	**28.** 9033 −5668
29. 8172 −1635	**30.** 9811 −1417	**31.** 1692 − 994	**32.** 6568 − 689
33. 48,145 − 5,656	**34.** 43,984 −23,869	**35.** 17,653 −14,994	**36.** 41,445 −33,482
37. 30,143 − 9,932	**38.** 24,301 − 3,825	**39.** 34,013 −19,076	**40.** 90,051 −19,673

In Exercises 41 and 42 find the perimeter of the patterns in Figures 1.4 and 1.5. (See Example 7.)

41.

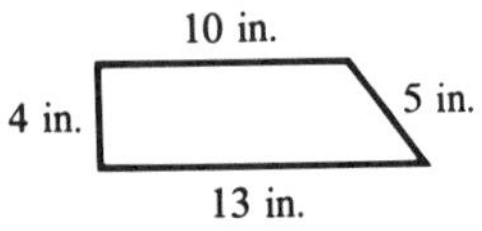

Figure 1.4

42.

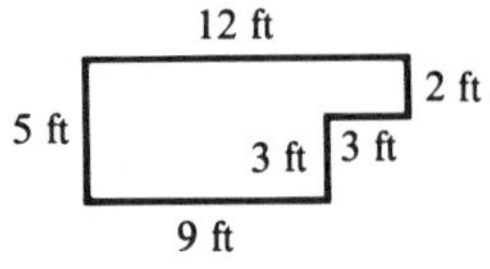

Figure 1.5

43. Five castings weigh 89 lb, 93 lb, 85 lb, 100 lb, and 120 lb, respectively. Find the total weight of the castings.

44. A technician orders four tools costing \$49, \$59, \$45, and \$29, respectively. What is the total cost of the order?

45. An electronics manufacturing company employs 829 people full-time and 237 part-time. How many people does the company employ?

46. A foundry shipped a number of castings with the following weights: flywheel, 210 lb; gear blanks, 325 lb; bearing caps, 540 lb; and foundry flasks, 2100 lb. Find the total weight of the shipment.

47. A variable resistor is set at 215 Ω (ohms) and then increased to 233 Ω. What is the increase in the resistance?

48. The voltage across a resistor increased from 98 V (volts) to 220 V. Determine the increase in the voltage.

49. The price of replacement parts for a computer increased to \$476. The original price was \$391. Determine the price increase.

50. The area of Alaska is 586,000 square miles, which includes 15,340 square miles of water. What is the land area of Alaska?

51. A technician earns \$2230 per month. His federal tax deduction is \$393, his state tax deduction is \$92, and other deductions come to \$75. What is his take-home pay?

52. The director of a computer center has \$1047 in her budget for miscellaneous expenses. If \$349 is taken out for emergency repairs, how much money is left?

53. Find the length of the center portion cut from the metal plate in Figure 1.6.

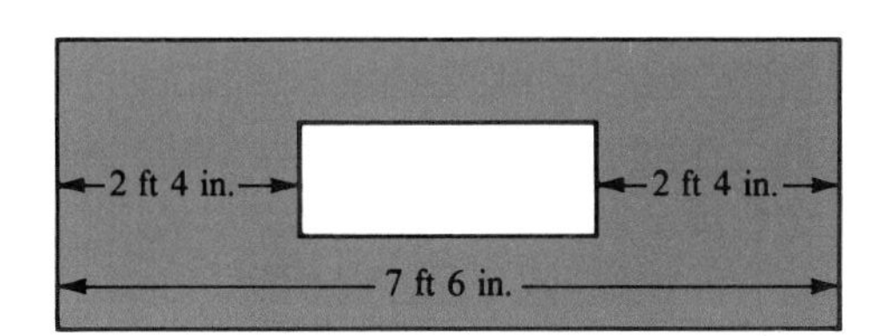

Figure 1.6

54. A time indicator on an engine reaches 1436 hours. The engine needs to be serviced every 80 hours and was last serviced when the indicator read 1390 hours. In how many hours will it need to be serviced again?

55. A chemical tank is filled to a level of 7 ft 3 in. After a certain amount is withdrawn, the level reads 2 ft 9 in. By how much has the level been reduced? (See Example 5.)

56. Three gears weigh a total of 25 kg (kilograms). If one gear weighs 6 kg and another 11 kg, how much does the remaining gear weigh?

57. A wire 87 cm (centimeters) long is cut into three parts. If one part is 19 cm long and another 32 cm long, how long is the remaining part?

1.2 Multiplication of Whole Numbers

In this section we discuss multiplication of whole numbers. Multiplication is defined as repeated addition. For example,

$$3 \times 4 \quad \text{means} \quad 4 + 4 + 4 \text{ (that is, 4 three times)}$$

and

$$4 \times 6 \quad \text{means} \quad 6 + 6 + 6 + 6 \text{ (that is, 6 four times)}$$

Thus $3 \times 4 = 12$ and $4 \times 6 = 24$.

Multiplicand
Multiplier
Product
Factors

In a multiplication such as $3 \times 4 = 12$, the first number, 3, is called the **multiplicand** and the second number, 4, is called the **multiplier.** The resulting number, 12, is called the **product.** Together, the multiplicand and multiplier are called the **factors.**

Notation for multiplication

Notation. Other common symbols for multiplication are a dot or parentheses. So, 3×4 can also be written $3 \cdot 4$ or $(3)(4)$.

To perform multiplication of whole numbers, we use the commutative and distributive laws mentioned in Section 1.1:

$$3 \times 7 = 7 \times 3 \qquad \text{Commutative law}$$
$$3 \times (4 + 10) = (3 \times 4) + (3 \times 10) \qquad \text{Distributive law}$$

Note: To perform multiplication of larger numbers readily, you should know all multiplications up to $9 \times 9 = 81$ from memory.

Example 1 Multiply 34×2.

Solution. Let's first perform the multiplication in expanded form:

$$\begin{aligned} 34 \times 2 &= 2 \times 34 && \text{Commutative law} \\ &= 2 \times (30 + 4) \\ &= (2 \times 30) + (2 \times 4) && \text{Distributive law} \\ &= 60 + 8 \\ &= 68 \end{aligned}$$

Using the column arrangement, the multiplication is done as follows:

$$\begin{array}{r} 34 \\ \times\ 2 \\ \hline 68 \end{array}$$

Multiply 2×4 and bring down the 8.
Multiply 2×3 and bring down the 6.

When multiplying larger numbers, certain digits may have to be "carried" to the next column. Consider the next example.

Example 2 Multiply 68 by 4.

Solution. Since $68 = 60 + 8$,

$$68 \times 4 = (60 + 8) \times 4 = (60 \times 4) + (8 \times 4)$$

by the distributive law. In column form the expanded multiplication is performed as follows:

$$\begin{array}{rl} 68 = 60 + 8 & \quad 4 \times 8 = 32 \\ \times \quad 4 & \quad 4 \times 60 = 240 \\ \hline 240 + 32 = 240 + 30 + 2 & \\ = (240 + 30) + 2 & \\ = 270 + 2 = 272 & \end{array}$$

This result shows that the 3 in 32 has to be carried over to the tens column:

$$\begin{array}{r} 3 \\ 68 \\ \underline{4} \\ 272 \end{array}$$

Multiply $4 \times 8 = 32$; bring down the 2 and carry the 3. Now multiply $4 \times 6 = 24$, add 3, and bring down the 27.

The next multiplication involves two 2-digit numbers.

Example 3 Multiply 79 by 36.

Solution. Since both factors are 2-digit numbers, the multiplication uses the distributive law twice:

$$\begin{aligned} 79 \times 36 &= 79 \times (30 + 6) \\ &= (79 \times 30) + (79 \times 6) && \text{Distributive law} \\ &= (30 \times 79) + (6 \times 79) && \text{Commutative law} \\ &= 30 \times (70 + 9) + 6 \times (70 + 9) \\ &= (30 \times 70) + (30 \times 9) + (6 \times 70) + (6 \times 9) && \text{Distributive law} \\ &= 2100 + 270 + 420 + 54 \\ &= 2844 \end{aligned}$$

In column form, the multiplication is performed as follows:

$$\begin{array}{rl} 79 = \quad 70 + 9 & \\ 36 = \quad \underline{30 + 6} & \\ 420 + 54 & \quad (70 + 9) \times 6 \\ \underline{2100 + 270} & \quad (70 + 9) \times 30 \\ 2100 + 690 + 54 = 2844 & \quad \text{Adding} \end{array}$$

The expanded form suggests the following procedure:

$$\begin{array}{r} 79 \\ \underline{36} \\ 474 \\ \underline{237} \\ 2844 \end{array}$$

Multiply $6 \times 79 = 474$ and bring down 474. Now multiply $3 \times 9 = 27$, place the 7 in the next row in the second column from the right, and carry the 2. Multiply $3 \times 7 = 21$ and add the carried 2 to yield 23. Write 23 in the second row in front of the 7. Adding the columns, we get 2844.

In this multiplication, 3×9 is actually $30 \times 9 = 270$. This is why we place the 7 in the tens column while the 0 may be omitted.

Example 4

$$\begin{array}{r} 369 \\ \times\ 285 \\ \hline 1845 \\ 2952 \\ 738 \\ \hline 105165 \end{array} \qquad \begin{array}{r} 8258 \\ \times\ 974 \\ \hline 33032 \\ 57806 \\ 74322 \\ \hline 8043292 \end{array}$$

If a whole number is multiplied by 0, we get 0 for the product. For example, since multiplication is repeated addition,

$$3 \times 0 = 0 + 0 + 0 = 0$$

In summary,

$N \times 0 = 0 \times N = 0$, for any whole number N.

Example 5

$$\begin{array}{r} 607 \\ \times\ 43 \\ \hline 1821 \\ 2428 \\ \hline 26101 \end{array} \qquad \begin{aligned} 607 \times 3 &= (600 + 0 + 7) \times 3 \\ &= 1800 + 0 + 21 \\ &= 1821 \end{aligned}$$

Example 6 A molder makes a casting with the dimensions shown in Figure 1.7. Find the volume. (*Note:* cm means "centimeter.")

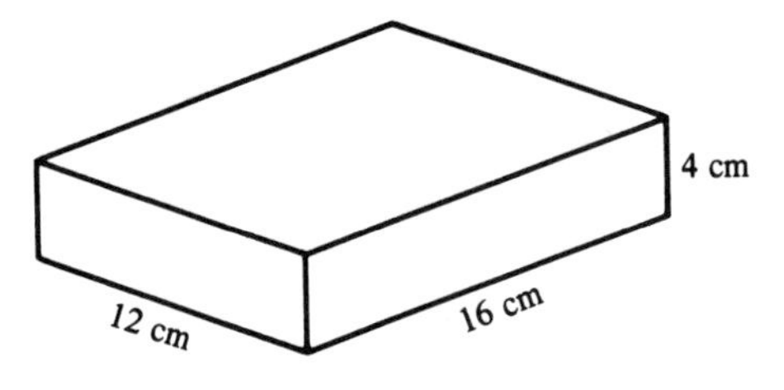

Figure 1.7

Solution. The volume V of a rectangular solid (box) is

$$V = \text{length} \times \text{width} \times \text{height}$$

For the casting in the figure,

$$V = (16\text{ cm}) \times (12\text{ cm}) \times (4\text{ cm}) = 768\text{ cm}^3$$

where cm^3 means "cubic centimeters."

Exercises / Section 1.2

In Exercises 1–20, perform the indicated multiplications.

1. 256×38 **2.** 927×46 **3.** 341×63 **4.** 682×75
5. 751×446 **6.** 665×128 **7.** 988×810 **8.** 276×407
9. 7304×32 **10.** 4610×59 **11.** 2360×102 **12.** 5003×602
13. $4496 \cdot 372$ **14.** $7666 \cdot 376$ **15.** $3062 \cdot 6704$ **16.** $1233 \cdot 474$
17. $(1812)(3055)$ **18.** $(7228)(5106)$ **19.** $(9379)(5731)$ **20.** $(7654)(6244)$

In Exercises 21 and 22, find the areas of the rectangular patterns in Figures 1.8 and 1.9 (area = length $\times$ width).

21.

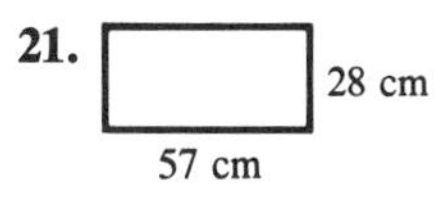

Figure 1.8

Area is expressed in square centimeters (cm^2).

22.

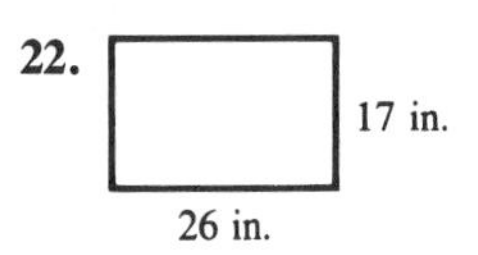

Figure 1.9

Area is expressed in square inches ($in.^2$).

23. A wrench costs \$12. What is the cost of 29 wrenches?

24. A chemical tank is 8 ft long, 7 ft wide, and 4 ft high. Find the volume. (See Example 6.)

25. An engine requires 21 gal of gasoline per week. How many gallons are required to run the engine for a year?

26. A shipment of washers arrives in a large case containing 36 boxes of 75 washers each. How many washers are there in the shipment?

27. A rocket travels from the earth to the moon at the rate of 26,560 miles per hour. If the trip takes about 9 days, determine the approximate distance to the moon. (*Note:* distance = speed $\times$ time.)

28. The current through a 146-Ω resistor (Ω = ohm) is 2 A (amperes). Find the voltage (in volts) across the resistor (voltage = current $\times$ resistance).

29. The current through a 217-Ω resistor is 2 A. Find the voltage across the resistor. (Refer to Exercise 28.)

30. To help determine its energy requirements for the year, a company estimates its average daily consumption of diesel fuel to be 129 gal. How many gallons are required for the year?

31. A certain type of motor requires 125 ft of wire. A manufacturer receives an order for 98 motors. How many feet of wiring will he need?

32. Each student in a class of 25 is allotted \$102 worth of computer time. How much is allotted for the whole class?

1.3 Division of Whole Numbers; Order of Operations

Division of Whole Numbers

Our next operation with whole numbers is **division.** Division is the inverse of multiplication, just as subtraction is the inverse of addition. For example, since $3 \times 5 = 15$, then $15 \div 5 = 3$, and we say that 5 goes into 15 three times.

A common symbol for division is $\div$. However, in algebra a preferred symbol is the bar, —, which is also used to write fractions. Thus

$$\frac{15}{5} = 3 \quad \text{means} \quad 15 \div 5 = 3$$

Dividend
Divisor
Quotient

(Occasionally we use a slash; thus 5/8 means $5 \div 8$.) Here 15 is the **dividend,** 5 the **divisor,** and 3 the **quotient.**

Division of larger whole numbers is not as straightforward as multiplication of such numbers. The scheme depends partly on finding successive estimates. Consider, for example, the division

$$546 \div 26$$

We set up the division problem as follows:

```
26)546
```

Now estimate the number of times that 26 goes into 54. Since $26 \times 2 = 52$, the number of times is 2. This number is the leftmost digit in the quotient (written directly above the 4 of the 54).

```
     2
26)546
   52
   --
    2
```

Next, multiply $2 \times 26 = 52$; place the product under the 54 and subtract.

```
     2
26)546
   52
   --
    26
```

Then bring down the next digit (6) of the dividend.

```
    21
26)546
   52
   --
    26
```

Divide 26 into the number in the last row. Since $26 \div 26 = 1$, place 1 in the ones column of the quotient.

```
    21
26)546
   52
   --
    26
    26
    --
     0
```

Multiply $1 \times 26 = 26$, place the 26 in the next row, and subtract. The remainder is 0.

We have shown that $546 \div 26 = 21$. As a check, we multiply the quotient by the divisor and get

$21 \times 26 = 546$, the dividend

Example 1 Divide 3275 by 119.

Solution. $119\overline{)3275}$

We estimate the number of times that 119 goes into 327. Since $2 \times 11 = 22$, while $3 \times 11 = 33$ (over 32), the correct number must be 2. Now proceed as before:

$$\begin{array}{r} 27 \\ 119\overline{)3275} \\ \underline{238} \\ 895 \\ \underline{833} \\ 62 \end{array} \quad \text{Remainder}$$

So 27 is the quotient and 62 is the remainder. *Check:*

$27 \times 119 + 62 = 3213 + 62 = 3275$ ✓

We can see from Example 1 that a division problem is checked as follows: Multiply the quotient by the divisor and add the remainder to the result. The final value should equal the dividend.

Example 2

$$\begin{array}{r} 379 \\ 256\overline{)97165} \\ \underline{768} \\ 2036 \\ \underline{1792} \\ 2445 \\ \underline{2304} \\ 141 \end{array} \quad \text{Remainder}$$

Check:

$379 \times 256 + 141 = 97165$ ✓

Operations with Zero

We saw in the last section that for any whole number N, N multiplied by 0 results in 0, or $N \times 0 = 0$. We may also add 0 to a number or subtract 0. For example,

$$3 + 0 = 3$$

and

$$7 - 0 = 7$$

Only division by 0 is not allowed. To see why, recall that

$$\frac{12}{4} = 3 \quad \text{because} \quad 12 = 3 \times 4$$

Suppose we try dividing 1 by 0. Let's call the quotient a for now. Then

$$\frac{1}{0} = a \quad \text{leads to} \quad 1 = a \times 0$$

but $a \times 0 = 0$, no matter what a is. So $1 = a \times 0$ says that $1 = 0$, which is not true. So $1 \div 0$ cannot equal any number whatever.

On the other hand, the quotient

$$\frac{0}{0} = a \quad \text{leads to} \quad 0 = a \times 0$$

This statement is actually correct for any number a. This means that $0 \div 0$ is equal to *any* number and not a specific number. Either way, we conclude: Since division by 0 is undefined,

Never divide by 0.

Order of Operations

If a problem involves more than one operation, we use the following rule: *In order, from left to right, perform all multiplications and divisions first; then perform the additions and subtractions.*

Example 3 Evaluate $10 - 8 + 3$.

Solution.

$$\begin{aligned} 10 - 8 + 3 &= 2 + 3 \qquad 10 - 8 = 2 \text{ (left to right)} \\ &= 5 \end{aligned}$$

Example 4 Evaluate $10 \times 3 \div 5$.

Solution.

$$\begin{aligned} 10 \times 3 \div 5 &= 30 \div 5 \qquad 10 \times 3 = 30 \text{ (left to right)} \\ &= 6 \end{aligned}$$

Example 5 Evaluate $3 \times 4 - 10 \div 5$.

Solution. Since multiplication and division are performed first, we first evaluate $3 \times 4 = 12$ and $10 \div 5 = 2$. We now get

$$3 \times 4 - 10 \div 5 = 12 - 2 = 10$$

Example 6 Evaluate $7 - 20 \div 4 + 7 \times 6$.

Solution. Performing the division and multiplication operations first, we have

$$7 - \mathbf{20 \div 4} + 7 \times 6 = 7 - \mathbf{5} + 42 = 44$$

Example 7 A technician determines the tensile strengths of five iron bars: 628 lb, 624 lb, 623 lb, 629 lb, and 621 lb. Determine the average tensile strength of these bars.

Solution. The average of n numbers is found by adding the numbers and dividing by n. For the five iron bars we get

$$\text{average} = \frac{628 + 624 + 623 + 629 + 621}{5} = \frac{3125}{5}$$
$$= 625 \text{ lb}$$

Exercises / Section 1.3

In Exercises 1–20, perform each of the indicated divisions.

1. $126 \div 3$ **2.** $284 \div 4$ **3.** $4636 \div 76$

4. $5292 \div 36$ **5.** $15{,}936 \div 32$ **6.** $34{,}276 \div 41$

7. $25{,}921 \div 49$ **8.** $45{,}900 \div 68$ **9.** $13{,}175 \div 25$

10. $15{,}795 \div 65$ **11.** $103{,}664 \div 124$ **12.** $91{,}948 \div 724$

13. $42{,}586 \div 398$ **14.** $192{,}404 \div 206$ **15.** $\frac{590{,}410}{34}$

16. $\frac{2{,}103{,}006}{74}$ **17.** $\frac{197{,}247}{16}$ **18.** $\frac{261{,}704}{19}$

19. $\frac{79{,}085}{77}$ **20.** $\frac{207{,}651}{74}$

In Exercises 21–32, perform the indicated operations. (See Examples 5 and 6.)

21. $4 + 2 \times 6$ **22.** $20 - 4 \times 3$ **23.** $16 \div 4 + 7$

24. $20 \div 10 + 6$ **25.** $3 \times 5 - \frac{25}{5} + 6$ **26.** $\frac{28}{7} - 4 + 7 \times 6$

27. $\frac{40}{8} + \frac{27}{9} - 4 \times 2$

28. $4 \times 3 + 2 - \frac{50}{10}$

29. $3 \times 6 + 2 \times 10 - \frac{30}{10}$

30. $4 \times 7 - 3 \times 6 - \frac{18}{9}$

31. $\frac{100}{10} + 4 \times 6 - 5 \times 2$

32. $\frac{80}{20} + \frac{20}{5} - 3 \times 2$

In Exercises 33–34, find the length of the metal plates pictured in Figures 1.10 and 1.11 (length = area ÷ width).

33. 348 in.² | 12 in.

Figure 1.10

34. 833 cm² | 17 cm

Figure 1.11

35. A woman earns $28,008 per year. What is her monthly salary?

36. The production of a foundry for one week was 105,600 lb at a cost of $1,267,200. What is the cost per pound?

37. An order for 120 identical frames requires 2160 rivets. How many rivets does each frame require?

38. The voltage across a 139-Ω resistor (Ω = ohms) is 417 V (volts). Find the current in amperes (current = voltage ÷ resistance).

39. A technician takes the following readings of a pressure gauge: 45 lb, 51 lb, 46 lb, 44 lb, 53 lb, and 49 lb. Find the average pressure. (See Example 7.)

40. A flywheel makes 75,600 revolutions per minute. What is the rate of rotation in revolutions per second?

41. A shipment of windshields passes inspection if there are no more than four blemishes (bubbles, scratches, and so on) per windshield on the average. The quality control technician found 602 blemishes in a shipment of 152 windshields. Does the lot pass the inspection?

42. A machine is transported by truck to a factory 408 mi away. If the truck averages 51 miles per hour, how long does the trip take?

1.4 Prime Factors

The next several sections are devoted to operations with fractions. To make our work easier, we need to examine a useful property of whole numbers involving **prime numbers.**

Prime number

A whole number is said to be **prime** if it cannot be divided evenly by any number except itself and 1. (The number 1 itself is not considered prime.) For example, the number 5 cannot be divided evenly by any whole number other than itself and 1 and is therefore a prime number. The first few primes are

2, 3, 5, 7, 11, 13, 17, and so on

A number that is not prime is called *composite*. For example, 15 is composite, since $15 = 3 \times 5$.

A fundamental principle of whole numbers is the following:

Every whole number can be written as a product of primes.

Writing a number as a product of primes can be difficult since the procedure is basically one of trial and error. For example, the product

$$779 = 19 \times 41$$

can be obtained only by trying out every prime up to 19. Writing a number as a product of primes is called **factoring** the number, or writing the number in **prime factored form.** The numbers in the product are called the **factors.** The prime factors of 779 are 19 and 41.

Factoring

Factors

Factoring can often be done step by step. For example, it is not difficult to see that $24 = 4 \times 6$. Since neither 4 nor 6 are primes, however, we continue by writing $4 = 2 \times 2$ and $6 = 2 \times 3$. So

$$24 = 4 \times 6 = 2 \times 2 \times 2 \times 3$$

Consider another example.

Example 1 Factor 84 into prime numbers.

Solution. Since 84 is divisible by 4, we write

$$84 = 4 \times 21$$

Now observe that $4 = 2 \times 2$ and $21 = 3 \times 7$. It thus follows that

$$84 = 4 \times 21 = 2 \times 2 \times 3 \times 7$$

Regardless of the factors used in the immediate steps we always get the same prime factors in the end:

$$84 = 2 \times 42 = 2 \times 6 \times 7 = 2 \times 2 \times 3 \times 7$$

Example 2 Write 630 as a product of primes.

Solution. Note that 630 is divisible by 10:

$$630 = 10 \times 63$$

Since $10 = 2 \times 5$ and $63 = 9 \times 7$, we get

$$630 = 10 \times 63 = (2 \times 5) \times (9 \times 7)$$
$$= 2 \times 5 \times 3 \times 3 \times 7$$

Since $9 = 3 \times 3$

Exercises / Section 1.4

Write each number as a product of primes.

1. 12	**2.** 15	**3.** 48	**4.** 72	**5.** 36
6. 30	**7.** 28	**8.** 90	**9.** 84	**10.** 105
11. 126	**12.** 168	**13.** 99	**14.** 525	**15.** 550

16. 441 **17.** 189 **18.** 1925 **19.** 363 **20.** 308
21. 2457 **22.** 750 **23.** 3432 **24.** 1287 **25.** 2548
26. 442 **27.** 117 **28.** 114 **29.** 204 **30.** 391
31. 551 **32.** 259

1.5 Fractions

So far we have worked only with whole numbers. In this section we discuss fractions, which are parts of whole numbers.

The Meaning of Fractions

A fraction is obtained by dividing one whole number by another. For example, $\frac{3}{8}$ is a fraction, obtained by dividing 3 by 8. (Since $3 \div 8 = \frac{3}{8} = 3/8$, the bar and slash are often used to denote division.) The fraction $\frac{3}{8}$ is pictured in Figure 1.12. According to the figure, $\frac{3}{8}$ may be viewed as 3 parts out of 8. For example, if 3 out of 8 of your friends got A's on the last test, then you could also say that $\frac{3}{8}$ of your friends got A's.

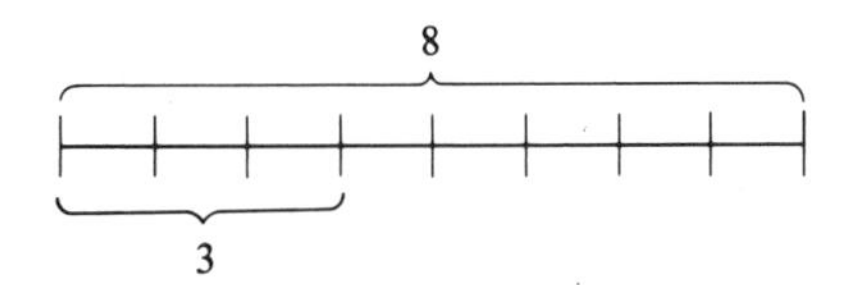

Figure 1.12

Numerator
Denominator

In the fraction $\frac{3}{8}$, 3 is called the **numerator** and 8 is called the **denominator**.

Example 1

a. The fraction $\frac{11}{23}$ has numerator 11 and denominator 23.
b. In the fraction $\frac{15}{32}$, the numerator is 15 and the denominator is 32.
c. A whole number is a fraction with a 1 for the denominator:

$$15 = \frac{15}{1}, \; 25 = \frac{25}{1}, \text{ and } 100 = \frac{100}{1}$$

Proper fraction
Improper fraction

A fraction in which the numerator is less than the denominator is called a **proper fraction.** If the numerator is greater than the denominator, it is called an **improper fraction.** For example, $\frac{4}{7}$ is a proper fraction and $\frac{8}{5}$ is an improper fraction.

Mixed number

An improper fraction can also be written as a **mixed number,** which consists of a whole number part and a fractional part. For example, $\frac{8}{5}$ can be written $1\frac{3}{5}$, which means $1 + \frac{3}{5}$. (We will study mixed numbers in Section 1.8.)

Equivalent Fractions

Equivalent fractions

Two fractions are said to be **equivalent** if they have the same value. To see how two fractions can be equivalent, let us return to the fraction $\frac{3}{8}$, shown in Figure 1.13. We have seen that $\frac{3}{8}$ can be viewed as *3 out of 8*. We can see from Figure 1.13 that "3 out of 8" means the same as "6 out of 16." In other words,

$$\frac{3}{8} = \frac{6}{16}$$

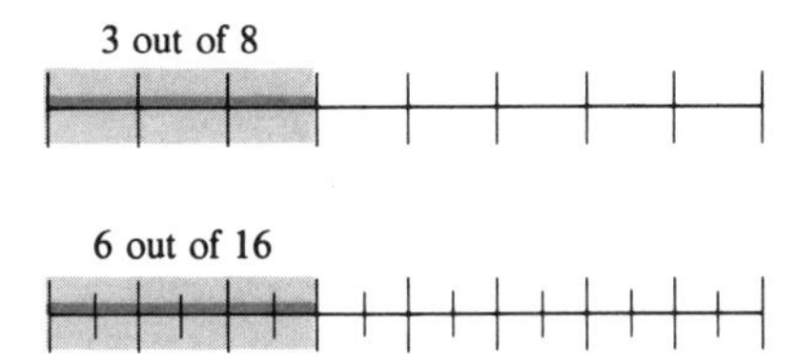

Figure 1.13

Similarly,

$$\frac{3}{8} = \frac{6}{16} = \frac{9}{24} = \frac{12}{32}, \text{ and so on}$$

This example suggests the following rule:

> **Equivalent fractions**
>
> Two fractions are equivalent if multiplying the numerator and denominator of one fraction by a nonzero number results in the other fraction.

Consider the following example.

Example 2 Show that $\frac{2}{3}$ is equivalent to $\frac{16}{24}$.

Solution. Multiplying the numerator and denominator of $\frac{2}{3}$ by 8, we get

$$\frac{2 \times 8}{3 \times 8} = \frac{16}{24}$$

So $\frac{2}{3}$ and $\frac{16}{24}$ are equivalent.

Every fraction has many equivalent fractions. In this section we are mainly interested in obtaining the "simplest" equivalent fraction. We want the numerator and denominator to be as small as possible. Finding such an equivalent fraction is called *reducing the fraction to lowest terms.*

> A fraction is said to be **reduced to lowest terms** if the numerator and denominator have no common factor except 1.

As an example, the numerator and denominator of $\frac{4}{10}$ have the common factor 2. If we now divide the 4 and 10 by 2, we get

$$\frac{4}{10} = \frac{2}{5}$$

Thus $\frac{2}{5}$ is the reduced fraction.

Example 3 **a.** Reducing the fraction $\frac{7}{21}$ to lowest terms yields the following:

$$\frac{7}{21} = \frac{1}{3}$$

which is obtained by dividing the numerator and denominator by 7.
b. Dividing the numerator and denominator by 3, we see that

$$\frac{9}{12} = \frac{3}{4}$$

To reduce more complicated fractions, we return to the technique studied in Section 1.4: factoring. If both numerator and denominator are factored, then any common factor can be divided out by canceling. For example,

$$\frac{33}{39} = \frac{11}{13}$$

which is obtained by dividing the numerator and denominator by 3. If we write

$$\frac{33}{39} = \frac{3 \times 11}{3 \times 13}$$

the same division can be carried out more simply by crossing out the 3s:

$$\frac{33}{39} = \frac{\overset{1}{\cancel{3}} \times 11}{\underset{1}{\cancel{3}} \times 13} = \frac{11}{13}$$

Canceling

Crossing out a common factor is called **canceling** the factor. The 1s next to the canceled factors are often omitted:

$$\frac{\not{3} \times 11}{\not{3} \times 13} = \frac{11}{13}$$

However, the 1s should be retained if all the factors in the numerator or denominator cancel. For example,

$$\frac{42}{210} = \frac{\overset{1}{\not{3}} \times \overset{1}{\not{2}} \times \overset{1}{\not{7}}}{\underset{1}{\not{3}} \times \underset{1}{\not{2}} \times \underset{1}{\not{7}} \times 5} = \frac{1}{5}$$

Example 4

Reduce the fraction $\frac{35}{45}$ to lowest terms.

Solution. Factoring the numerator and denominator, we get

$$\begin{aligned}\frac{35}{45} &= \frac{5 \times 7}{5 \times 3 \times 3} \\ &= \frac{\not{5} \times 7}{\not{5} \times 3 \times 3} \qquad \text{Canceling 5} \\ &= \frac{7}{9}\end{aligned}$$

Example 5

Reduce the following fraction to lowest terms:

$$\frac{1155}{1330}$$

Solution.

$$\begin{aligned}\frac{1155}{1330} &= \frac{7 \times 5 \times 3 \times 11}{7 \times 5 \times 2 \times 19} \qquad \text{Factoring} \\ &= \frac{\not{7} \times \not{5} \times 3 \times 11}{\not{7} \times \not{5} \times 2 \times 19} \qquad \text{Canceling} \\ &= \frac{3 \times 11}{2 \times 19} = \frac{33}{38}\end{aligned}$$

Alternate solution. The reduction can also be accomplished by successive cancellation. For example, if you see that 1155 and 1330 are both divisible by 5, then you can divide by 5 first:

$$\frac{\overset{231}{\not{1155}}}{\underset{266}{\not{1330}}} = \frac{231}{266} \qquad \text{Dividing by 5}$$

You can then further reduce the resulting fraction by dividing the numerator and denominator by 7:

$$\frac{\overset{\overset{33}{\cancel{231}}}{\cancel{1155}}}{\underset{\underset{38}{\cancel{266}}}{\cancel{1330}}} = \frac{33}{38}$$

Example 6 Reduce to lowest terms:

$$\frac{168}{456}$$

Solution. Factoring the numerator and denominator, we obtain

$$\frac{168}{456} = \frac{\cancel{2} \times \cancel{2} \times \cancel{2} \times \cancel{3} \times 7}{\cancel{2} \times \cancel{2} \times \cancel{2} \times \cancel{3} \times 19} = \frac{7}{19}$$

Alternate solution. Dividing the numerator and denominator by 4 and then by 6, we get

$$\frac{\overset{\overset{7}{\cancel{42}}}{\cancel{168}}}{\underset{\underset{19}{\cancel{114}}}{\cancel{456}}} = \frac{7}{19}$$

Example 7 Noise is usually measured in *decibels*. The noise made by rustling leaves is 10 decibels and the sound of riveting is 95 decibels. Suppose a new muffler is developed that reduces the noise of a car from 30 to 25 decibels. Then

$$\frac{25}{30} = \frac{5}{6}$$

indicates that a car with the new muffler is only $\frac{5}{6}$ as noisy as the same car with the old muffler.

Exercises / Section 1.5

In Exercises 1–6, identify the numerator and denominator in each case. (See Example 1.)

1. $\frac{9}{16}$

2. $\frac{2}{11}$

3. $\frac{1}{10}$

4. $\frac{1}{6}$

5. $\frac{25}{31}$ **6.** $17 \left(17 = \frac{17}{1}\right)$

In Exercises 7–34, reduce the given fractions to lowest terms.

7. $\frac{6}{14}$ **8.** $\frac{18}{22}$ **9.** $\frac{21}{18}$ **10.** $\frac{20}{28}$

11. $\frac{42}{54}$ **12.** $\frac{60}{72}$ **13.** $\frac{30}{45}$ **14.** $\frac{40}{90}$

15. $\frac{30}{72}$ **16.** $\frac{36}{54}$ **17.** $\frac{42}{98}$ **18.** $\frac{198}{234}$

19. $\frac{63}{140}$ **20.** $\frac{84}{60}$ **21.** $\frac{198}{78}$ **22.** $\frac{132}{88}$

23. $\frac{165}{198}$ **24.** $\frac{147}{294}$ **25.** $\frac{196}{189}$ **26.** $\frac{132}{297}$

27. $\frac{156}{351}$ **28.** $\frac{306}{765}$ **29.** $\frac{306}{408}$ **30.** $\frac{315}{825}$

31. $\frac{765}{570}$ **32.** $\frac{630}{2850}$ **33.** $\frac{348}{372}$ **34.** $\frac{255}{435}$

In each of the following exercises, reduce the fractions to lowest terms.

35. A survey showed that 48 out of 64 small business operations in a certain district are using computers. What fraction of the operations is using computers?

36. The weekly fuel bill of a computer consultant firm was $108. After installing more efficient insulation, the bill was reduced to $84 per week. Determine what fraction of the original cost is the new cost.

37. A transformer reduces the voltage in a transmission line from 220 V to 80 V. The reduced voltage is what fraction of the original voltage?

38. If 368 people in a town of 1472 voted in the last election, what fraction of the population voted?

39. During a recent snow emergency, 35 out of 63 employees were absent. What fraction reported for work?

40. The quality control engineer of a production line determined that 15 out of 385 flashbulbs sampled were defective. What fractional part was satisfactory?

41. A technician mixes 90 mL (milliliters) of hydrochloric acid with 27 mL of water. What fraction of the mixture is water?

42. If 110 lb of a certain alloy contains 35 lb of nickel and 55 lb of copper, what fraction is nickel and what fraction is copper?

43. A company required 1200 ft^2 (square feet) of storage capacity before reducing its inventory. After that, only 800 ft^2 were required. By what fractional part was the storage area reduced?

44. A technical supervisor read 160 pages of a 220-page technical manual. What fraction of the manual does he still have to read?

1.6 Multiplication and Division of Fractions

The topic of this section is multiplication and division of fractions. These operations are carried out by the following rules:

Multiplication and division of fractions

$$\frac{a}{b} \times \frac{c}{d} = \frac{ac}{bd} \tag{1.1}$$

$$\frac{a}{b} \div \frac{c}{d} = \frac{a}{b} \times \frac{d}{c} = \frac{ad}{bc} \tag{1.2}$$

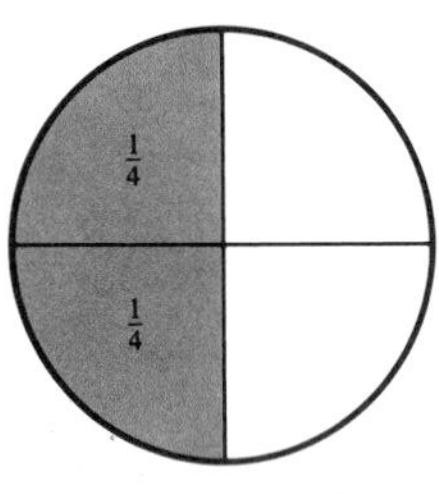

Shaded area $= \frac{1}{2}$

Figure 1.14

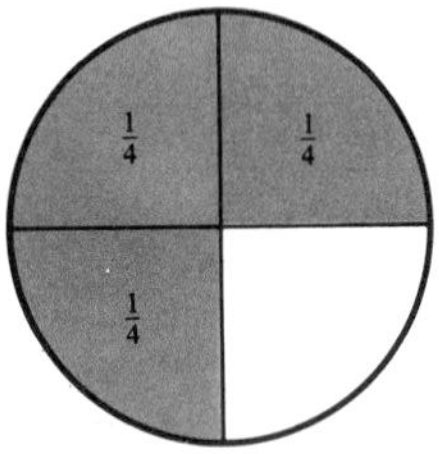

Shaded area $= \frac{3}{4}$

Figure 1.15

Multiplication

According to rule (1.1), we multiply two fractions by multiplying the numerators and the denominators and then simplifying. To see why, let us multiply $\frac{1}{4}$ by 2. Writing 2 as $\frac{2}{1}$, we get, by rule (1.1):

$$\frac{1}{4} \times \frac{2}{1} = \frac{1 \times 2}{4 \times 1} = \frac{2}{4} = \frac{1}{2}$$

This result is reasonable since $\frac{1}{2}$ is twice as large as $\frac{1}{4}$. (See Figure 1.14.) Next, let's multiply $\frac{3}{4}$ by $\frac{1}{3}$. By rule (1.1)

$$\frac{3}{4} \times \frac{1}{3} = \frac{3 \times 1}{4 \times 3} = \frac{3}{12} = \frac{1}{4}$$

This result is also reasonable since one-third of $\frac{3}{4}$ is $\frac{1}{4}$. (See Figure 1.15.)

Multiplication by $\frac{2}{3}$ can be thought of as a combination of the two preceding examples since

$$\frac{2}{3} = \frac{2}{1} \times \frac{1}{3}$$

So, multiplying a number by $\frac{2}{3}$ doubles the number and, at the same time, takes one-third of the number.

Example 1

$$\frac{3}{8} \times \frac{4}{15} = \frac{12}{120} = \frac{1}{10} \qquad \text{By rule (1.1)}$$

A closer look at Example 1 shows that a direct application of rule (1.1) is not very efficient. Consider the way in which the product is simplified:

$$\frac{3}{8} \times \frac{4}{15} = \frac{12}{120} = \frac{3 \times 4}{8 \times 15} = \frac{\overset{1}{\cancel{3}} \times 4}{8 \times \underset{5}{\cancel{15}}} = \frac{1 \times 4}{8 \times 5}$$

$$= \frac{1 \times \overset{1}{\cancel{4}}}{\underset{2}{\cancel{8}} \times 5} = \frac{1 \times 1}{2 \times 5} = \frac{1}{10}$$

Now, it is poor practice to multiply the numerators and denominators knowing that on the very next step the numbers have to be factored again to permit cancellation. Why not cancel *before* performing the multiplication? If we do, we get quite simply

$$\frac{\overset{1}{\cancel{3}}}{\underset{2}{\cancel{8}}} \times \frac{\overset{1}{\cancel{4}}}{\underset{5}{\cancel{15}}} = \frac{1 \times 1}{2 \times 5} = \frac{1}{10}$$

In summary, *whenever we multiply two or more fractions, we perform all cancellations prior to multiplying.*

Example 2 Perform the following multiplication:

$$\frac{2}{9} \times \frac{14}{15} \times \frac{3}{7}$$

Solution. Following the preceding plan, we cancel first and then multiply:

$$\frac{2}{\underset{3}{\cancel{9}}} \times \frac{\overset{2}{\cancel{14}}}{15} \times \frac{\cancel{3}}{\cancel{7}} = \frac{2 \times 2}{3 \times 15} = \frac{4}{45}$$

The factors do not have to be next to each other in order to cancel. Thus 9 and 3 are both divided by 3, even though $\frac{2}{9}$ and $\frac{3}{7}$ are not next to each other.

Sometimes canceled factors can be canceled again, as shown in the next example.

Example 3 Multiply:

$$\frac{15}{16} \times \frac{7}{9} \times \frac{12}{25} \times \frac{7}{9}$$

Solution. As before, we perform all cancellations prior to multiplying. Let us first divide 15 and 25 by 5, and **12** and **16** by 4:

$$\frac{\overset{3}{\cancel{15}}}{\underset{4}{\cancel{16}}} \times \frac{7}{9} \times \frac{\overset{3}{\cancel{12}}}{\underset{5}{\cancel{25}}} \times \frac{7}{9} = \frac{\overset{3}{\cancel{15}}}{\underset{4}{\cancel{16}}} \times \frac{7}{9} \times \frac{\overset{3}{\cancel{12}}}{\underset{5}{\cancel{25}}} \times \frac{7}{9}$$

Even though both 3s are the result of a cancellation, both can be canceled again:

$$\frac{\overset{3}{\cancel{15}}}{\underset{4}{\cancel{16}}} \times \frac{7}{\underset{3}{\cancel{9}}} \times \frac{\overset{3}{\cancel{12}}}{\underset{5}{\cancel{25}}} \times \frac{7}{\underset{3}{\cancel{9}}} = \frac{1 \times 7 \times 1 \times 7}{4 \times 3 \times 5 \times 3} = \frac{49}{180}$$

Common error Multiplying numerator and denominator directly and not reducing the result. For example,

$$\frac{13}{17} \times \frac{51}{26} = \frac{663}{442}$$

by rule (1.1). The resulting fraction is difficult to reduce. Avoid this problem by canceling first:

$$\frac{\cancel{13}}{\cancel{17}} \times \frac{\overset{3}{\cancel{51}}}{\underset{2}{\cancel{26}}} = \frac{3}{2}$$

Example 4 A machine performs 12 operations in $\frac{2}{7}$ s (second). How long will it take to perform 36 operations?

Solution. Since $36 = 12 \times 3$, it will take three times as long to perform 36 operations. So we need to multiply $\frac{2}{7}$ s by 3:

$$\frac{2}{7} \times 3 = \frac{2}{7} \times \frac{3}{1} \qquad 3 = \frac{3}{1}$$

$$= \frac{6}{7} \text{ s}$$

Division

By rule (1.2), we divide one fraction by another by inverting the divisor and proceeding as in multiplication. For example,

$$3 \div \frac{1}{2} = 3 \times \frac{2}{1} = 6$$

This result is reasonable, since $\frac{1}{2}$ goes into 3 six times in the sense that

$$\frac{1}{2} + \frac{1}{2} + \frac{1}{2} + \frac{1}{2} + \frac{1}{2} + \frac{1}{2} = 3$$

Example 5 Perform the following division:

$$\frac{13}{15} \div \frac{39}{50}$$

Solution. We invert the divisor $\frac{39}{50}$ and multiply:

$$\frac{13}{15} \div \frac{39}{50} = \frac{13}{15} \times \frac{50}{39} = \frac{\overset{1}{\cancel{13}}}{\underset{3}{\cancel{15}}} \times \frac{\overset{10}{\cancel{50}}}{\underset{3}{\cancel{39}}} = \frac{10}{9}$$

The rule for division can also be stated in terms of reciprocals: The **reciprocal** of a number is 1 divided by the number. For example,

The reciprocal of 5 is $\frac{1}{5}$

The reciprocal of $\frac{1}{4}$ is $\frac{1}{\frac{1}{4}} = \frac{1}{1} \times \frac{4}{1} = 4$

and

The reciprocal of $\frac{2}{7}$ is $\frac{1}{\frac{2}{7}} = \frac{1}{1} \times \frac{7}{2} = \frac{7}{2}$

The general case is given next.

> The **reciprocal** of $\frac{a}{b}$ is $\frac{b}{a}$.

Divison

Division: To divide a number by a fraction, multiply by the reciprocal of the fraction.

Example 6 Divide:

$$\frac{55}{49} \div \frac{11}{7}$$

Solution. The reciprocal of $\frac{11}{7}$ is $\frac{7}{11}$. Multiplying by the reciprocal, we get

$$\frac{55}{49} \div \frac{11}{7} = \frac{\overset{5}{\cancel{55}}}{\underset{7}{\cancel{49}}} \times \frac{\cancel{7}}{\cancel{11}} = \frac{5}{7}$$

The next example illustrates division by a whole number.

Example 7

$$\frac{3}{4} \div 5 = \frac{3}{4} \div \frac{5}{1} \qquad 5 = \frac{5}{1}$$

$$= \frac{3}{4} \times \frac{1}{5} = \frac{3}{20}$$

Example 8 The final stage in the manufacture of a certain type of calculator requires two operations: electronic assembly and case assembly. If the case assembly takes $\frac{4}{3}$ min per calculator, how long does it take to assemble 75 calculators?

Solution. The total time is found by multiplying the time required to assemble each calculator by the number of calculators. Thus

$$\left(\frac{4}{3}\text{ min}\right)(75) = \frac{4 \times \overset{25}{\cancel{75}}}{\underset{1}{\cancel{3}}} = 100 \text{ min}$$

Exercises / Section 1.6

In Exercises 1–32, perform the indicated operations.

1. $\frac{2}{3} \times \frac{3}{5}$
2. $\frac{3}{4} \times \frac{5}{9}$
3. $\frac{4}{5} \times \frac{15}{16}$
4. $\frac{2}{7} \times \frac{21}{26}$
5. $\frac{3}{7} \div \frac{9}{14}$
6. $\frac{2}{5} \div \frac{4}{15}$
7. $\frac{21}{2} \div \frac{7}{2}$
8. $\frac{25}{3} \div \frac{5}{3}$
9. $2 \div \frac{1}{6}$
10. $\frac{1}{2} \div 3$
11. $\frac{1}{3} \times \frac{5}{7} \times \frac{3}{5}$
12. $\frac{1}{2} \times \frac{14}{19} \times \frac{4}{7}$
13. $\frac{2}{3} \times \frac{4}{5} \times \frac{15}{16}$
14. $\frac{45}{49} \div \frac{63}{14}$
15. $\frac{16}{39} \div \frac{8}{13}$
16. $\frac{55}{13} \times \frac{39}{77}$
17. $\frac{22}{13} \times \frac{39}{44} \times \frac{4}{77}$
18. $\frac{10}{49} \times \frac{3}{16} \times \frac{14}{25}$
19. $\left(\frac{2}{3} \times \frac{6}{11}\right) \div \frac{5}{22}$
20. $\left(\frac{4}{7} \times \frac{28}{29}\right) \div \frac{4}{5}$
21. $\left(\frac{9}{11} \div \frac{4}{5}\right) \div \frac{35}{24}$
22. $\left(\frac{4}{13} \div \frac{16}{9}\right) \div \frac{3}{26}$
23. $4 \times \frac{19}{20} \times \frac{13}{38}$
24. $7 \times \frac{5}{12} \times \frac{48}{49}$
25. $\frac{3}{2} \times \frac{11}{16} \times \frac{1}{5} \times \frac{32}{33}$
26. $\frac{7}{8} \times \frac{2}{3} \times \frac{6}{7} \times \frac{4}{5}$
27. $\frac{3}{16} \times \frac{38}{25} \times \frac{5}{9} \times \frac{8}{19}$
28. $\frac{4}{17} \times \frac{7}{13} \times \frac{2}{21} \times \frac{51}{28}$
29. $\frac{19}{21} \times \frac{5}{14} \times \frac{3}{35} \times \frac{28}{57}$
30. $\frac{11}{12} \times \frac{34}{3} \times \frac{7}{51} \times \frac{2}{55}$
31. $\frac{5}{3} \times \frac{117}{125} \times \frac{25}{13}$
32. $\frac{11}{12} \times \frac{14}{121} \times \frac{48}{49} \times \frac{11}{16}$

33. A box contains 35 small castings weighing $\frac{2}{7}$ lb each. What is the total weight of the contents of the box?

34. A rectangular computer chip measures $\frac{3}{8}$ mm (millimeter) by $\frac{16}{9}$ mm. Find the area (area = length × width).

35. The tip of a vibrating tuning fork moves back and forth in $\frac{3}{100}$ s (second). How long does it take to complete 15 cycles? (A cycle is one movement back and forth.)

36. Find the volume of the machine part in Figure 1.16 (volume = length × width × height).

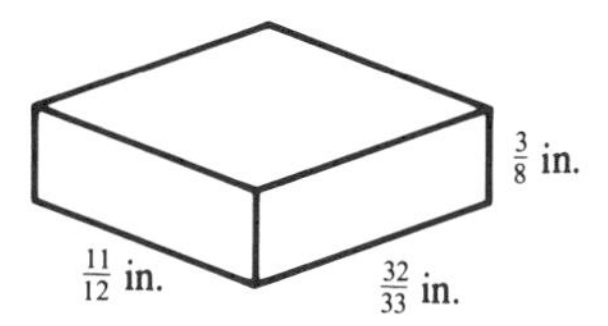

Figure 1.16

37. A card sorter can sort 10 cards in $\frac{2}{3}$ s (second). How long does it take to sort 25 cards? (See Example 4.)

38. The current in a circuit reverses directions 12 times in $\frac{1}{5}$ s (second). How many times does the current reverse directions in 1 s?

39. An electron moves in a magnetic field at the rate of $\frac{4}{3}$ ft/ms (feet per millisecond). How long does it take to travel $\frac{8}{11}$ ft? (*Note:* time = distance ÷ rate.)

40. How far does the electron in Exercise 39 travel in $\frac{9}{16}$ ms?

41. Diesel fuel flows into a tank at the rate of $\frac{17}{15}$ gal/min. How many gallons will be in the tank at the end of half an hour?

42. Steel plates $\frac{3}{8}$ in. thick are stacked on a pallet to a height of 2 ft. How many plates are stacked on the pallet?

43. Six identical machines cover $\frac{36}{5}$ ft^2 (square feet) of floor space. How much floor space does each machine cover?

44. How long can an engine run on 5 qt of oil if $\frac{5}{3}$ qt are used every $\frac{9}{4}$ hour?

45. An engine uses $\frac{5}{6}$ gal of gasoline every $\frac{3}{2}$ hour. How long will $\frac{10}{3}$ gal last?

1.7 Addition and Subtraction of Fractions

In this section we study the remaining two operations with fractions: **addition** and **subtraction.**

Adding and subtracting fractions is not as straightforward as multiplying fractions. To get an overview of the problem, let us add

$$\frac{1}{2} + \frac{1}{6}$$

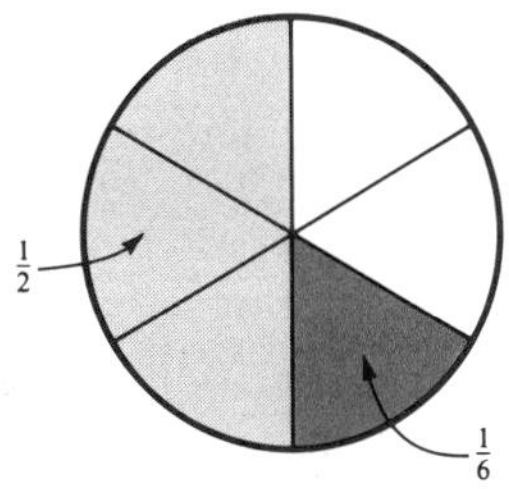

by thinking of $\frac{1}{2}$ as half a pie and $\frac{1}{6}$ as $\frac{1}{6}$ of a pie. (See Figure 1.17.) Note that by cutting the pie into 6 pieces, half a pie is really 3 pieces, or 3 pieces out of 6. This can be expressed as $\frac{3}{6}$. We now have

$$\frac{3}{6} + \frac{1}{6} = \frac{4}{6} = \frac{2}{3}$$

Figure 1.17

The diagram confirms that the total is $\frac{2}{3}$ of a pie.

Like Fractions

Fractions with the same denominator are called **like fractions.** The preceding example tells us that like fractions can be added by simply adding the numerators.

> **Combining like fractions:** To add (or subtract) like fractions, we add (or subtract) the numerators and place the sum (or difference) over the denominator common to them.

Example 1

a. $\frac{1}{7} + \frac{2}{7} = \frac{1+2}{7} = \frac{3}{7}$ Adding numerators

b. $\frac{3}{10} + \frac{5}{10} - \frac{1}{10} = \frac{3+5-1}{10}$ Combining numerators

$= \frac{7}{10}$

c. $\frac{11}{12} - \frac{5}{12} = \frac{11-5}{12} = \frac{6}{12} = \frac{1}{2}$

Fractions with Different Denominators

To combine fractions with different denominators, we change them to equivalent fractions with a common denominator. For the fractions shown in Figure 1.17, $\frac{1}{2}$ is changed to $\frac{3}{6}$, so that

$$\frac{1}{2} + \frac{1}{6} = \frac{3}{6} + \frac{1}{6} = \frac{4}{6} = \frac{2}{3}$$

The number 6 is called a *common denominator.* There exist many common denominators. For example, 12 is a common denominator, since 12 is divisible by both 2 and 6. Using 12, we get

$$\frac{1}{2} + \frac{1}{6} = \frac{6}{12} + \frac{2}{12} = \frac{8}{12} = \frac{2}{3}$$

However, the addition is simpler if the denominator has the smallest possible value, in this case 6. The number 6 is called the **lowest common denominator,** abbreviated LCD.

> **Lowest common denominator (LCD)**
>
> The **lowest common denominator** of a set of fractions is the smallest number that is divisible by all the denominators.

For simple fractions, the lowest common denominator can be obtained by observation. For example, it is not difficult to see that

$$\text{The LCD of } \frac{1}{6} \text{ and } \frac{1}{8} \text{ is } 24$$

(24 is the smallest number divisible by both 6 and 8.) Similarly,

$$\text{The LCD of } \frac{1}{3} \text{ and } \frac{1}{4} \text{ is } 12$$

since 12 is the smallest number divisible by both 3 and 4.

Example 2

a. $\frac{3}{4} - \frac{1}{6} = \frac{3 \times 3}{4 \times 3} - \frac{1 \times 2}{6 \times 2}$ LCD is 12, since 12 is the smallest number divisible by both 4 and 6

$= \frac{9}{12} - \frac{2}{12} = \frac{7}{12}$

b. $\frac{1}{4} - \frac{1}{8} = \frac{1 \times 2}{4 \times 2} - \frac{1}{8}$ LCD = 8

$= \frac{2}{8} - \frac{1}{8} = \frac{1}{8}$

c. $2 + \frac{1}{3} - \frac{1}{5} = \frac{2}{1} + \frac{1}{3} - \frac{1}{5}$ $2 = \frac{2}{1}$

$= \frac{30}{15} + \frac{5}{15} - \frac{3}{15}$ LCD = 15

$= \frac{32}{15}$

For more complicated fractions we need a procedure by which the LCD can be found systematically.

> To find the lowest common denominator (LCD):
>
> **1.** Factor the denominator of each fraction into a product of primes.
> **2.** Write the LCD as a product of all the different factors, each taken the greatest number of times that it appears in any of the denominators.

As an example, to find the LCD of

$$\frac{1}{6} \quad \text{and} \quad \frac{1}{10}$$

we write

$$6 = 2 \times 3 \qquad \text{and} \qquad 10 = 2 \times 5$$

The factor 2 occurs only one time in each denominator. So it is included only once in the LCD. The factors 3 and 5 likewise occur only once. So by the rule we get

$$\text{LCD} = 2 \times 3 \times 5 = 30$$

Example 3 Find the lowest common denominator of the fractions

$$\frac{1}{20} \quad \text{and} \quad \frac{1}{75}$$

Solution. To find the LCD, we write each denominator as a product of primes:

$$\begin{aligned} 20 &= 4 \times 5 = 2 \times 2 \times 5 \\ 75 &= 25 \times 3 = \underline{\phantom{2 \times 2 \times{}} 5 \times 5 \times 3} \\ &\text{LCD} = 2 \times 2 \times 5 \times 5 \times 3 \end{aligned}$$

The factor 2 occurs two times in 20 (and not at all in 75). So 2 × 2 is included in the LCD. The factor 5 occurs two times in 75 (and only once in 20). So 5 × 5 is included in the LCD. Finally, 3 occurs just once and must be included once in the LCD. It follows that

$$\text{LCD} = 2 \times 2 \times 5 \times 5 \times 3 = 300$$

Using this procedure, we can combine more complicated fractions by finding the LCD and changing each fraction (if necessary) to an equivalent fraction with the LCD as the denominator. The resulting fractions can be combined using the procedure for combining like fractions.

Example 4 Combine the following fractions:

$$\frac{5}{18} - \frac{1}{50}$$

Solution. First we find the LCD by factoring each of the denominators:

$$\begin{aligned} 18 &= 9 \times 2 = 3 \times 3 \times 2 \\ 50 &= 25 \times 2 = \underline{\phantom{3 \times 3 \times{}} 2 \times 5 \times 5} \\ &\text{LCD} = 3 \times 3 \times 2 \times 5 \times 5 \end{aligned}$$

The factor 3 occurs two times in 18 (and not at all in 50). So 3 × 3 is included in the LCD. For the same reason, 5 × 5 is included. The factor 2 occurs just once in both 18 and 50. So 2 is included once in the LCD. Thus

$$\text{LCD} = 3 \times 3 \times 2 \times 5 \times 5 = 450$$

Next we change each fraction to an equivalent fraction having the LCD for a denominator. Note that

$$450 \div 18 = 25$$

So

$$\frac{5}{18} = \frac{5 \times 25}{18 \times 25} = \frac{125}{450}$$

Similarly,

$$450 \div 50 = 9$$

so that

$$\frac{1}{50} = \frac{1 \times 9}{50 \times 9} = \frac{9}{450}$$

We now have

$$\frac{5}{18} - \frac{1}{50} = \frac{125}{450} - \frac{9}{450} = \frac{116}{450}$$

The best way to reduce the resulting fraction is by rewriting the denominator in its factored form. Since

$$\text{LCD} = 3 \times 3 \times 5 \times 5 \times 2$$

we have

$$\frac{116}{450} = \frac{116}{3 \times 3 \times 5 \times 5 \times 2}$$

$$= \frac{58 \times \cancel{2}}{3 \times 3 \times 5 \times 5 \times \cancel{2}} \qquad \text{Canceling 2}$$

$$= \frac{58}{3 \times 3 \times 5 \times 5}$$

Since 58 is not divisible by 5 or 3, no other cancellations are possible. So the reduced form is

$$\frac{58}{225} \qquad \text{Reduced form}$$

Example 5 Combine the following fractions:

$$\frac{9}{14} - \frac{10}{63} + \frac{5}{12}$$

Solution. To find the LCD, we write each denominator as a product of primes:

$$14 = 2 \times 7$$
$$63 = 9 \times 7 = 3 \times 3 \times 7$$
$$12 = 2 \times 2 \times 3$$

The factor 2 occurs at most two times (in 12), the factor 3 at most two times (in 63), and the factor 7 at most once. So

$$\text{LCD} = 2 \times 2 \times 3 \times 3 \times 7 = 252$$

We now change each fraction to an equivalent fraction with 252 for a denominator. Since

$$252 \div 14 = 18$$

the first fraction becomes

$$\frac{9}{14} = \frac{9 \times 18}{14 \times 18} = \frac{162}{252}$$

Similarly, since $252 \div 63 = 4$,

$$\frac{10}{63} = \frac{10 \times 4}{63 \times 4} = \frac{40}{252}$$

Finally, since $252 \div 12 = 21$,

$$\frac{5}{12} = \frac{5 \times 21}{12 \times 21} = \frac{105}{252}$$

We now have

$$\frac{9}{14} - \frac{10}{63} + \frac{5}{12} = \frac{162}{252} - \frac{40}{252} + \frac{105}{252}$$
$$= \frac{162 - 40 + 105}{252} = \frac{227}{252} \qquad \text{Adding numerators}$$

To check if the resulting fraction can be reduced, we write the LCD in its factored form:

$$\frac{227}{252} = \frac{227}{2 \times 2 \times 3 \times 3 \times 7}$$

Since 227 is not divisible by any of the factors in the denominator, we conclude that $\frac{227}{252}$ is in lowest form.

Example 6 Combine the following fractions:

$$\frac{8}{45} + \frac{9}{10} - \frac{7}{40}$$

Solution. To find the LCD, we write each of the denominators in factored form:

$$45 = 9 \times 5 = 3 \times 3 \times 5$$
$$10 = 2 \times 5$$
$$40 = 8 \times 5 = 2 \times 2 \times 2 \times 5$$

Observe that 5 occurs just once in each denominator, while 3 occurs at most twice, and 2 at most three times. So

$$\text{LCD} = 5 \times 3 \times 3 \times 2 \times 2 \times 2 = 360$$

Now we change each fraction to an equivalent fraction each with denominator 360:

$$\frac{8}{45} = \frac{8 \times 8}{45 \times 8} = \frac{64}{360} \qquad 360 \div 45 = 8$$

$$\frac{9}{10} = \frac{9 \times 36}{10 \times 36} = \frac{324}{360} \qquad 360 \div 10 = 36$$

$$\frac{7}{40} = \frac{7 \times 9}{40 \times 9} = \frac{63}{360} \qquad 360 \div 40 = 9$$

Combining the resulting fractions, we get

$$\frac{64}{360} + \frac{324}{360} - \frac{63}{360} = \frac{64 + 324 - 63}{360} = \frac{325}{360}$$
$$= \frac{325}{5 \times 3 \times 3 \times 2 \times 2 \times 2}$$
$$= \frac{65}{3 \times 3 \times 2 \times 2 \times 2} \qquad \text{Dividing by 5}$$
$$= \frac{65}{72}$$

Common error Confusing the procedure for multiplication and addition. While

$$\frac{1}{2} \times \frac{1}{3} = \frac{1 \times 1}{2 \times 3} = \frac{1}{6}$$

it is **not correct** to write

$$\frac{1}{2} + \frac{1}{3} \quad \text{as} \quad \frac{1 + 1}{2 + 3}$$

The correct procedure is

$$\frac{1}{2} + \frac{1}{3} = \frac{3}{6} + \frac{2}{6} = \frac{3 + 2}{6} = \frac{5}{6}$$

Complex fraction

A fraction in which the numerator or denominator (or both) is the sum or difference of two fractions is called a **complex fraction.** Complex fractions can be simplified by combining the fractions in the numerator and denominator into single fractions and then dividing the resulting fractions.

Example 7 Reduce the following complex fraction:

$$\frac{\frac{1}{2}-\frac{1}{3}}{\frac{1}{4}+\frac{1}{9}}$$

Solution. We combine the fractions in the numerator and denominator separately and then divide the resulting fractions:

$$\frac{\frac{1}{2}-\frac{1}{3}}{\frac{1}{4}+\frac{1}{9}} = \frac{\frac{3}{6}-\frac{2}{6}}{\frac{9}{36}+\frac{4}{36}} \qquad \begin{matrix}\text{LCD} = 6 \\ \text{LCD} = 36\end{matrix}$$

$$= \frac{\frac{1}{6}}{\frac{13}{36}} = \frac{1}{6} \div \frac{13}{36}$$

$$= \frac{1}{6} \times \frac{36}{13} = \frac{6}{13}$$

Example 8 The total resistance of two or more resistors connected in series is the sum of the individual resistances. Find the combined resistance of the resistors in Figure 1.18. (Ω is the symbol for ohms.)

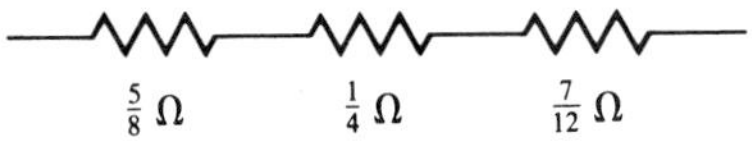

$\frac{5}{8}\,\Omega$ $\frac{1}{4}\,\Omega$ $\frac{7}{12}\,\Omega$

Figure 1.18

Solution. The total resistance is

$$\frac{5}{8}\,\Omega + \frac{1}{4}\,\Omega + \frac{7}{12}\,\Omega = \left(\frac{15}{24} + \frac{6}{24} + \frac{14}{24}\right)\Omega = \frac{35}{24}\,\Omega$$

Exercises / Section 1.7

In Exercises 1–42, combine the given fractions and simplify.

1. $\frac{1}{3}+\frac{2}{3}$ **2.** $\frac{3}{11}+\frac{4}{11}$ **3.** $\frac{7}{15}-\frac{2}{15}$ **4.** $\frac{3}{20}-\frac{1}{20}$

5. $\frac{1}{2}+\frac{1}{4}$
6. $\frac{1}{4}+\frac{1}{8}$
7. $\frac{1}{2}-\frac{1}{4}$
8. $\frac{1}{4}-\frac{1}{8}$
9. $\frac{1}{3}+\frac{1}{5}$
10. $\frac{1}{2}-\frac{1}{5}$
11. $\frac{1}{2}-\frac{1}{6}$
12. $\frac{1}{2}+\frac{1}{8}$
13. $\frac{1}{3}-\frac{1}{4}$
14. $\frac{2}{3}-\frac{1}{2}$
15. $\frac{1}{7}+\frac{1}{14}$
16. $\frac{1}{5}+\frac{5}{6}$
17. $5-\frac{1}{4}$
18. $6-\frac{4}{3}$
19. $\frac{5}{12}-\frac{1}{24}$
20. $\frac{5}{8}+\frac{3}{16}$
21. $\frac{5}{6}-\frac{3}{8}$
22. $\frac{7}{15}-\frac{3}{20}$
23. $\frac{3}{8}+\frac{5}{14}$
24. $\frac{7}{40}-\frac{2}{25}$
25. $\frac{1}{15}+\frac{2}{21}$
26. $\frac{9}{24}-\frac{5}{18}$
27. $\frac{15}{54}+\frac{5}{36}$
28. $\frac{31}{54}-\frac{7}{30}$
29. $\frac{1}{3}-\frac{1}{12}+\frac{5}{16}$
30. $\frac{3}{8}+\frac{2}{9}+\frac{5}{18}$
31. $\frac{23}{28}-\frac{1}{12}+\frac{5}{6}$
32. $\frac{5}{16}+\frac{3}{20}-\frac{1}{14}$
33. $\frac{22}{35}-\frac{1}{28}-\frac{3}{14}$
34. $\frac{11}{42}-\frac{3}{98}-\frac{1}{28}$
35. $\frac{9}{28}+\frac{5}{56}-\frac{1}{84}$
36. $\frac{2}{15}-\frac{3}{25}+\frac{1}{70}$
37. $\dfrac{\frac{1}{2}-\frac{1}{3}}{\frac{1}{12}}$
38. $\dfrac{\frac{1}{3}-\frac{1}{4}}{\frac{2}{3}}$
39. $\dfrac{\frac{5}{4}}{\frac{1}{4}-\frac{1}{8}}$
40. $\dfrac{\frac{15}{49}}{\frac{4}{7}-\frac{1}{28}}$
41. $\dfrac{\frac{1}{2}+\frac{1}{3}}{\frac{5}{12}-\frac{1}{6}}$
42. $\dfrac{\frac{3}{4}+\frac{1}{6}}{\frac{3}{10}+\frac{1}{15}}$

43. A plate is $\frac{11}{8}$ in. thick. A planer is used to reduce the thickness by $\frac{5}{24}$ in. What is the final thickness?
44. A bar is to be $\frac{51}{64}$ in. thick. If the original thickness is $\frac{7}{8}$ in., by how much must it be reduced?
45. The current in a wire increases from $\frac{1}{5}$ A (ampere) to $\frac{4}{7}$ A. What is the increase in the current?
46. In Figure 1.18, if the respective resistances are $\frac{1}{8}\ \Omega$, $\frac{7}{15}\ \Omega$, and $\frac{3}{20}\ \Omega$, what is the combined resistance?
47. The manager of a small business figures that $\frac{4}{9}$ of her profit goes for taxes, $\frac{9}{40}$ for rent, and $\frac{1}{15}$ for insurance. What fraction is left?
48. A wire 2 ft long is cut into three pieces. If one piece is $\frac{3}{10}$ ft long and another $\frac{3}{14}$ ft long, how long is the third?
49. The reading on a pressure gauge increases by $\frac{3}{4}$ lb, then by $\frac{5}{6}$ lb, and then decreases by $\frac{11}{10}$ lb. What is the net increase?
50. Find the perimeter of the pattern in Figure 1.19. (The *perimeter* is the distance around the figure.)

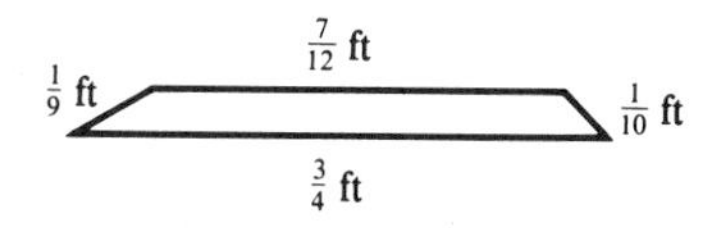

Figure 1.19

51. Find the average of the following weights: $\frac{5}{6}$ lb, $\frac{11}{10}$ lb, $\frac{5}{12}$ lb, and $\frac{2}{15}$ lb.
52. The setting on a variable resistor increases by $\frac{1}{6}\ \Omega$ (ohm), then by $\frac{7}{9}\ \Omega$, and then decreases by $\frac{1}{20}\ \Omega$. What is the net increase?

53. On a certain engine, oil needs to be added whenever the level drops by 1 in. or more. Readings taken at regular intervals have shown the level to decrease $\frac{1}{7}$ in., $\frac{3}{16}$ in., and $\frac{5}{28}$ in. How much farther can the level drop before oil has to be added?

1.8 Operations with Mixed Numbers

In this section we will make a brief study of mixed numbers.

Mixed number

Recall that a **mixed number** is a combination of a whole number and a fraction. For example,

$$3\frac{1}{4} \quad \text{means} \quad 3 + \frac{1}{4}$$

The mixed number can be changed to an improper fraction (numerator larger than denominator) by adding 3 and $\frac{1}{4}$:

$$3\frac{1}{4} = 3 + \frac{1}{4} = \frac{12}{4} + \frac{1}{4} = \frac{13}{4}$$

To change a mixed number to an improper fraction:

1. Write the mixed number as a whole number plus a fraction.
2. Convert the whole number to an equivalent fraction with denominator equal to the denominator of the given fraction.
3. Add the fractions.

Example 1 Change $7\frac{2}{9}$ to an improper fraction.

Solution. Adding the whole number part to the fractional part, we get

$$7\frac{2}{9} = 7 + \frac{2}{9} = \frac{63}{9} + \frac{2}{9} = \frac{65}{9}$$

To write an improper fraction as a mixed number:

1. Divide the numerator by the denominator.
2. Write the quotient (whole number part) plus the fraction whose numerator is the remainder and whose denominator is that of the given fraction.

Example 2 Change $\frac{76}{8}$ to a mixed number.

Solution. First divide the numerator by the denominator:

$$\begin{array}{r} 9 \\ 8\overline{)76} \\ \underline{72} \\ 4 \end{array}$$

The quotient is 9 and the remainder 4, so we get

$$\frac{76}{8} = 9 + \frac{4}{8} = 9 + \frac{1}{2} = 9\frac{1}{2}$$

Since mixed numbers can be changed to improper fractions, all operations with mixed numbers can be performed by first changing the mixed numbers to improper fractions. For multiplication and division, this procedure should always be used, as shown in the next example.

Example 3 Multiply: $(4\frac{2}{7}) \times (3\frac{1}{4})$.

Solution. Changing the mixed numbers to improper fractions, we get

$$\left(4\frac{2}{7}\right) \times \left(3\frac{1}{4}\right) = \frac{30}{7} \times \frac{13}{4}$$

$$= \frac{\overset{15}{\cancel{30}}}{7} \times \frac{13}{\underset{2}{\cancel{4}}} = \frac{195}{14} = 13\frac{13}{14}$$

To perform addition with mixed numbers, we could change all the mixed numbers to improper fractions first and then add. Another method is the following:

1. Add the whole numbers.
2. Add the fractions.
3. Add the results and simplify.

Example 4

a. $3\frac{1}{4} + 5\frac{3}{8} = (3 + 5) + \left(\frac{1}{4} + \frac{3}{8}\right) = 8 + \left(\frac{2}{8} + \frac{3}{8}\right)$

$$= 8 + \frac{5}{8} = 8\frac{5}{8}$$

b. $5\frac{5}{6} + 4\frac{7}{12} = (5 + 4) + \left(\frac{5}{6} + \frac{7}{12}\right) = 9 + \left(\frac{10}{12} + \frac{7}{12}\right)$

$$= 9 + \left(\frac{17}{12}\right) = 9 + \left(1 + \frac{5}{12}\right) = 10 + \frac{5}{12} = 10\frac{5}{12}$$

The procedure for subtraction is similar to that for addition. However, if the fractional part of the minuend is too small, we need to borrow. See the next example.

Example 5 Subtract: $5\frac{1}{3} - 2\frac{5}{9}$.

Solution. Since $\frac{1}{3}$ is less than $\frac{5}{9}$, we cannot subtract $\frac{5}{9}$ from $\frac{1}{3}$. For the number $5\frac{1}{3}$, if we borrow 1 from the whole number, we get

$$5\frac{1}{3} = 5 + \frac{1}{3} = (4 + 1) + \frac{1}{3} = 4 + \left(1 + \frac{1}{3}\right)$$

$$= 4 + \left(\frac{3}{3} + \frac{1}{3}\right) = 4 + \frac{4}{3} = 4\frac{4}{3}$$

Now the subtraction can be carried out:

$$4\frac{4}{3} - 2\frac{5}{9} = (4 - 2) + \left(\frac{4}{3} - \frac{5}{9}\right) = (4 - 2) + \left(\frac{12}{9} - \frac{5}{9}\right)$$

$$= 2 + \frac{7}{9} = 2\frac{7}{9}$$

Example 6 Find the total distance from the top of the tank to the end of the pipe in Figure 1.20.

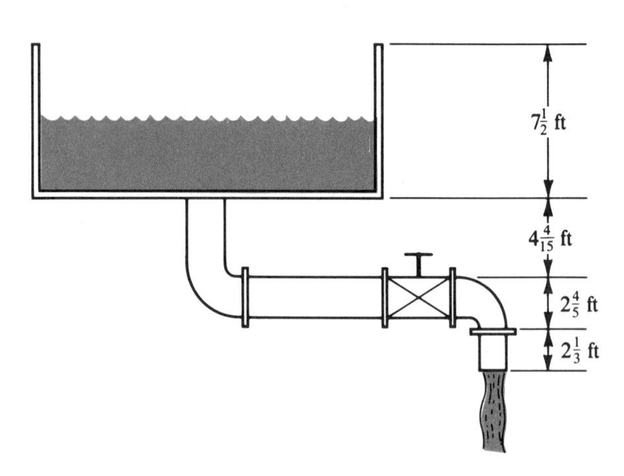

Figure 1.20

Solution. The total distance is

$$7\frac{1}{2}\text{ ft} + 4\frac{4}{15}\text{ ft} + 2\frac{4}{5}\text{ ft} + 2\frac{1}{3}\text{ ft}$$

The sum, in feet, is

$$7\frac{15}{30} + 4\frac{8}{30} + 2\frac{24}{30} + 2\frac{10}{30} \qquad \text{LCD} = 30$$

$$= (7 + 4 + 2 + 2) + \left(\frac{15}{30} + \frac{8}{30} + \frac{24}{30} + \frac{10}{30}\right)$$

$$= 15 + \frac{57}{30} = 15 + \left(1 + \frac{27}{30}\right)$$

$$= 16 + \frac{27}{30} = 16 + \frac{9}{10} = 16\frac{9}{10}\text{ ft}$$

Exercises / Section 1.8

In Exercises 1–8, write each mixed number as an improper fraction.

1. $4\frac{1}{5}$ **2.** $5\frac{3}{7}$ **3.** $2\frac{3}{11}$

4. $3\frac{4}{5}$ **5.** $10\frac{1}{8}$ **6.** $11\frac{1}{10}$

7. $9\frac{3}{13}$ **8.** $5\frac{2}{17}$

In Exercises 9–16, write each improper fraction as a mixed number.

9. $\frac{20}{3}$ **10.** $\frac{21}{4}$ **11.** $\frac{40}{13}$

12. $\frac{52}{17}$ **13.** $\frac{30}{7}$ **14.** $\frac{28}{5}$

15. $\frac{41}{6}$ **16.** $\frac{77}{8}$

In Exercises 17–32, perform the indicated operations. Write each answer as a mixed number.

17. $2\frac{1}{3} + 1\frac{1}{6}$ **18.** $5\frac{1}{7} + 2\frac{1}{3}$ **19.** $5\frac{5}{9} - 2\frac{2}{9}$

20. $7\frac{9}{14} - 3\frac{1}{14}$ **21.** $10\frac{1}{4} - 3\frac{1}{2}$ **22.** $12\frac{1}{6} - 2\frac{1}{3}$

23. $2\frac{1}{3} + 4\frac{3}{4} + 5\frac{1}{2}$ **24.** $4\frac{1}{6} + 3\frac{1}{5} + 5\frac{1}{15}$ **25.** $\left(4\frac{1}{2}\right) \times \left(3\frac{1}{4}\right)$

26. $\left(7\frac{4}{5}\right) \times \left(3\frac{1}{9}\right)$ **27.** $\left(4\frac{5}{6}\right) \times \left(2\frac{1}{8}\right)$ **28.** $\left(3\frac{1}{3}\right) \div \left(4\frac{4}{5}\right)$

29. $\left(5\frac{3}{8}\right) \div \left(3\frac{1}{4}\right)$

30. $\left(5\frac{1}{2}\right) \div \left(6\frac{1}{4}\right)$

31. $\left(1\frac{1}{2}\right) \times \left(3\frac{1}{4}\right) \div \left(1\frac{1}{8}\right)$

32. $\left(2\frac{1}{4}\right) \times \left(1\frac{1}{8}\right) \div \left(2\frac{1}{8}\right)$

33. A cam shaft is $\frac{38}{9}$ in. long. Write the length as a mixed number.

34. An electric motor is $\frac{51}{11}$ ft in height. Express the height as a mixed number.

35. Find the area of a rectangular plate $5\frac{1}{3}$ cm by $\frac{4}{15}$ cm.

36. A shop requires $2\frac{5}{7}$ lb of sweeping compound daily. How much is needed during a 5-day workweek?

37. Find the perimeter of the triangle (three sides) whose sides have lengths $6\frac{2}{9}$ cm, $9\frac{3}{7}$ cm, and $5\frac{2}{3}$ cm, respectively.

38. Find the length of the rectangular hole in the metal plate in Figure 1.21.

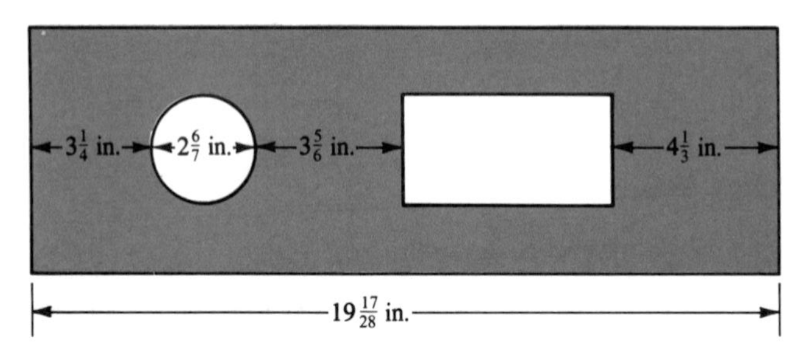

Figure 1.21

39. A technician needs to cut three pieces of molding with the following lengths: $10\frac{1}{4}$ in., $2\frac{1}{6}$ in., and $3\frac{1}{8}$ in. If she has to allow $\frac{1}{4}$ in. for each cut, what length piece of molding does she need?

40. An office building maintenance man uses $1\frac{1}{4}$ lb of solder a week. How long does his 20-lb supply last?

41. Find the total weight of the following shipment: 12 castings weighing $32\frac{1}{2}$ lb each, 4 castings weighing $69\frac{1}{4}$ lb each, and 9 castings weighing $93\frac{3}{4}$ lb each.

42. The current in a wire increases from $1\frac{1}{15}$ A to $2\frac{2}{5}$ A (amperes). What is the increase?

43. How many pieces of wire $2\frac{3}{5}$ in. long can be cut from a wire 1 ft long? How much will be left over?

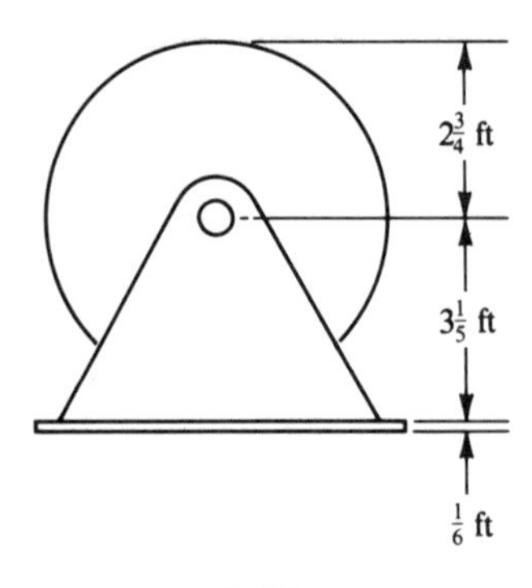

Figure 1.22

44. What is the total length of lumber needed to make 5 shelves each $3\frac{2}{9}$ ft long?

45. If you walk at the rate of $4\frac{4}{5}$ km/h (kilometers per hour), how long will it take you to walk 8 km?

46. A truck delivering heavy machinery uses $6\frac{3}{4}$ gal of diesel to travel 40 mi. What is the fuel consumption in miles per gallon?

47. Find the height of the motor in Figure 1.22.

1.9 Decimals

In this section we discuss fractions represented in decimal form. Before we can do so, however, we need to extend our place-value system for whole numbers to include fractions. Consider the chart in Figure 1.23. As with whole numbers, every digit has a meaning that depends on its position.

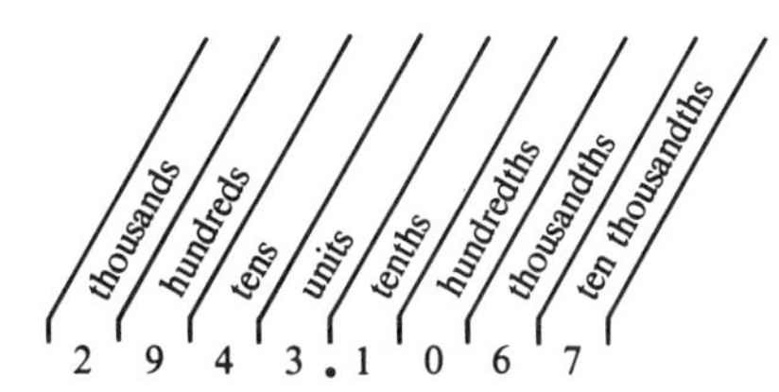

Figure 1.23

Consider the following number:

36.937

- 9 means $9 \times \frac{1}{10} = \frac{9}{10}$
- 3 means $3 \times \frac{1}{100} = \frac{3}{100}$
- 7 means $7 \times \frac{1}{1000} = \frac{7}{1000}$

It is read "36 and 937 thousandths." The following illustrates the reason for this:

$$
\begin{aligned}
36.937 &= 36 + (0.937) \\
&= 36 + \left(\frac{9}{10} + \frac{3}{100} + \frac{7}{1000}\right) \\
&= 36 + \left(\frac{900}{1000} + \frac{30}{1000} + \frac{7}{1000}\right) \\
&= 36 + \frac{937}{1000}
\end{aligned}
$$

Example 1 Read the following numbers: 3.14, 2.718, and 92.3041.

Solution.

3.14	means	3 and 14 hundredths
2.718	means	2 and 718 thousandths
92.3041	means	92 and 3041 ten thousandths

The use of place values enables us to perform the fundamental operations with decimals in much the same way as with whole numbers. However, careful attention must be paid to the placement of the decimal point.

Addition and Subtraction

In adding whole numbers we set up the problem so that only like units are added; that is, we add tens to tens, hundreds to hundreds, and so on. When adding decimals, we do the same thing: We add tenths to tenths, hundredths to hundredths, and so on. To do so, we need to line up the decimal points in the same column.

> **Addition and subtraction of decimals**
>
> When adding or subtracting decimals, arrange the numbers in a column so that the decimal points are in the same column.

Example 2 Add the following numbers: 324.6, 79.4, 3.786, and 0.567.

Solution. We arrange the numbers in a column with the decimal points lined up vertically:

$$\begin{array}{r@{.}l}
324&6\\
79&4\\
3&786\\
+\quad 0&567\\
\hline
408&353
\end{array}\quad \text{Adding}$$

Note that the decimal point is "brought down" when writing the sum.

Example 3 Add:

$$89 + 0.76 + 1.503$$

Solution. Since 89 is a whole number, the decimal point is understood to be on the right ($89 = 89.0$).

$$\begin{array}{r} 89 \\ 0.76 \\ +\ 1.503 \\ \hline 91.263 \end{array}$$

We saw in Example 3 that 89 = 89.0. For decimals, we can always insert additional zeros in this way. For example, 0.6 = 0.60 = 0.600. When subtracting one decimal from another, it is often necessary to insert additional zeros, as shown in the next example.

Example 4 Subtract 961.16 from 7625.3.

Solution. To obtain the same number of places to the right of the decimal point in each case, we write the minuend 7625.3 as 7625.30:

$$\begin{array}{rl} 7625.30 & \text{Inserting a 0} \\ -\ 961.16 & \\ \hline 6664.14 & \text{Subtracting} \end{array}$$

Multiplication

To see how we obtain the rule for multiplication, consider the product

$$(0.4) \times (0.3)$$

Changing to common fractions, we get

$$(0.4) \times (0.3) = \frac{4}{10} \times \frac{3}{10} = \frac{12}{100} = 0.12$$

We get the same answer if we multiply the decimals as if they were whole numbers and then count the number of places to the right of the decimal point:

$$\begin{array}{rl} 0.4 & \text{One decimal place} \\ \times\ 0.3 & \text{One decimal place} \\ \hline 0.12 & \text{Two decimal places} \end{array}$$

The general rule is given next.

> **Multiplication of decimals**
>
> To multiply two decimals:
>
> 1. Multiply the numbers as if they were whole numbers.
> 2. Place the decimal point in the product so that the number of places to the right of the decimal point is equal to the sum of the number of places to the right of the decimal point in the factors.

Example 5 Perform the following multiplication:

$$\begin{array}{rl} 7.63 & \text{2 decimal places} \\ \times \quad 40.8 & \text{1 decimal place} \\ \hline 6\,104 & \\ 00\,00 & \\ 305\,2 & \\ \hline 311.304 & \text{Total: 3 decimal places} \end{array}$$

Division

Division of decimals is similar to division of whole numbers. However, before performing the division, we need to move the decimal point in the divisor by enough places so that the divisor becomes a whole number.

> **Division of decimals**
>
> To divide two decimals:
>
> 1. Move the decimal point in the divisor to the right to make the divisor a whole number.
> 2. Move the decimal point in the dividend the same number of places. Add zeros, if necessary.
> 3. Place the decimal point in the quotient directly above the newly placed decimal point in the dividend.
> 4. Perform the division the same way as for whole numbers, but place the first digit of the quotient directly above the rightmost digit of that part of the dividend used to determine the first digit of the quotient.

To justify this rule, note that

$$\frac{4.8}{1.6} = \frac{4.8}{1.6} \times \frac{10}{10} = \frac{48}{16}$$

In other words, an equivalent fraction may be obtained by moving the decimal point in the numerator and denominator the same number of places.

Example 6 Perform the following division:

$$0.2\overline{)1.8}$$

Solution. To obtain a whole number divisor, we move the decimal point one place to the right:

$0.2\overline{)1.8}$

We now have

$$\begin{array}{r} 9. \\ 2\overline{)18.} \\ \underline{18} \\ 0 \end{array}$$

The quotient is 9. Note that the decimal point in the quotient is placed directly above the newly placed decimal point in the dividend.

Example 7 Adjust the decimal point in each of the given divisions:

a. $0.43\overline{)1.832}$ $\quad 0.43\overline{)1.832}$ $\quad 43\overline{)183.2}$

b. $0.78\overline{)84}$ $\quad 0.78\overline{)84.00}$ $\quad 78\overline{)8400.}$

Note that two zeros have to be added to the dividend before the decimal point can be moved.

c. $109\overline{)1.367}$

Since the divisor is already a whole number, no adjustment is necessary.

Example 8 Perform the following division:

$31.92 \div 7.6$

Solution. We set up the division problem as follows:

$7.6\overline{)31.92}$

Now move the decimal point one place:

$7.6\overline{)31.92}$

$$\begin{array}{r} 4.2 \\ 76\overline{)319.2} \\ \underline{304} \\ 152 \\ \underline{152} \end{array}$$

Since the first digit 4 of the quotient is obtained by dividing 76 into 319, the 4 is placed directly above the rightmost digit of 319.

Example 9 Perform the indicated divisions:

a. $10.8625 \div 1.25$

```
1.25)10.8625                 8.69
                       125)1086.25
                           1000
                             862
                             750
                             1125
                             1125
```

Since the first digit, 8, of the quotient is estimated by dividing 125 into 1086, the 8 is placed above the 6.

b. $0.017856 \div 0.48$

```
0.48)0.017856               .0372
                        48)1.7856
                           1 44
                             345
                             336
                              96
                              96
```

Since the digit 3 of the quotient must be placed above the 8, it is necessary to insert a zero after the decimal point.

c. $18.2 \div 0.013$

```
0.013)18.200               1400
(add zeros)             13)18200
                           13
                            52
                            52
```

Common error Forgetting that for a whole number the decimal point is understood to be on the right. For example, to perform the division

```
   1.3
3)3.90
```

no adjustment of the decimal point is required. In particular, don't write the division problem as

```
3)39.0
```

Similarly, to perform the division

```
0.13)39
```

it is incorrect to write

$$0.13\overline{)\,39}$$ Incorrect

Since 39 = 39.0, the correct procedure is

$$0.13\overline{)\,39.00}$$ Correct

Rounding Off the Quotient

If the division does not come out evenly, the quotient has to be expressed in approximate form by **rounding off.** (The question of how many decimal places to carry will be discussed in detail in the next chapter.) To round off a decimal to a specified number of digits, we examine the digit immediately to the right of the specified number of places. If this digit is 5 or more, increase the last digit by one. If this same digit is less than 5, leave the last digit as it is. Consider the next example.

Example 10

$1.372 = 1.37$ To two decimal places

$23.358 = 23.4$ To one decimal place

$9.7637 = 9.764$ To three decimal places

Example 11 Write $\frac{23}{49}$ in decimal form. Round off the quotient to two decimal places.

Solution. Placing zeros after the 23, we get

$$\begin{array}{r} 0.469 \\ 49\overline{)\,23.000} \\ \underline{19\,6} \\ 3\,40 \\ \underline{2\,94} \\ 460 \\ \underline{441} \\ 19 \end{array}$$

Rounding off the quotient 0.469 to two decimal places, we get 0.47.

Example 11 illustrates the conversion of a fraction to a decimal: To *convert a fraction to a decimal,* divide the numerator by the denominator and round off the quotient, if necessary.

To *change a decimal to a fraction,* write the decimal in its equivalent fractional form and simplify, as shown in the next example.

Example 12

$$0.6 = \frac{6}{10} = \frac{3}{5}$$

$$0.25 = \frac{25}{100} = \frac{1}{4}$$

$$0.125 = \frac{125}{1000} = \frac{1}{8}$$

$$0.0075 = \frac{75}{10,000} = \frac{3}{400}$$

Example 13 The distance a screw will move during one complete turn is called the *pitch* of the thread (Figure 1.24). If the pitch is 0.13 in. for a screw, how far will it move in 8.5 turns?

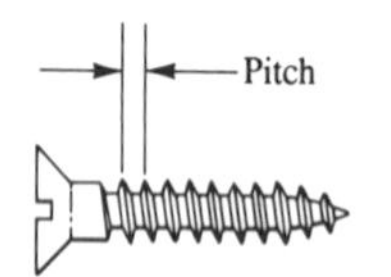

Figure 1.24

Solution. Since one turn moves the screw 0.13 in., it follows that 8.5 turns will move the screw 8.5 times as far, or

$$\text{distance} = (0.13 \text{ in.})(8.5) = 1.1 \text{ in.}$$

Exercises / Section 1.9

In Exercises 1–16, change the given fractions to decimals. Round off to two decimal places, if necessary. (See Example 11.)

1. $\frac{4}{5}$ **2.** $\frac{3}{4}$ **3.** $\frac{4}{25}$

4. $\frac{29}{50}$ **5.** $\frac{131}{250}$ **6.** $\frac{81}{125}$

7. $\frac{1}{11}$ **8.** $\frac{5}{6}$ **9.** $\frac{3}{7}$

10. $\frac{13}{29}$ **11.** $\frac{15}{32}$ **12.** $\frac{3}{7}$

13. $\frac{17}{26}$ **14.** $\frac{14}{61}$ **15.** $\frac{16}{49}$

16. $\frac{12}{23}$

In Exercises 17–24, change the given decimals to fractions in simplest form. (See Example 12.)

17. 0.75 **18.** 0.22 **19.** 0.08

20. 0.06 **21.** 0.0025 **22.** 0.0175

23. 0.375 **24.** 0.075

In Exercises 25–48, perform the indicated operations.

25. 6.3 + 2.75 + 7.605
26. 0.12 + 1.789 + 12.6
27. 26.4 + 7.26 + 5
28. 0.37 + 0.126 + 21.4
29. 3.721 − 1.25
30. 21.26 − 7
31. 34 − 19.84
32. 9.82 − 0.776
33. 5.32 − 2.709
34. 96.1 − 69.86
35. 25.2 × 3.7
36. 5.22 × 4.8
37. 8.64 × 7.6
38. (0.321)(0.95)
39. (2.7)(0.927)
40. (0.05)(3.2003)
41. 0.0025 × 0.0026
42. 2.3 × 5.369
43. 18 ÷ 2.5
44. 3.1154 ÷ 0.37
45. 0.31234 ÷ 3.22
46. 6.642 ÷ 92.25
47. 0.044826 ÷ 0.062
48. 0.4646 ÷ 0.092

In Exercises 49–54, perform the indicated divisions. Round off the quotient to the number of decimal places indicated.

49. 5.382 ÷ 6.28 (3)
50. 0.596 ÷ 3.5 (2)
51. 1.3764 ÷ 9.072 (3)
52. 0.129 ÷ 22.31 (4)
53. 0.1208 ÷ 6.75 (4)
54. 0.0962 ÷ 2.39 (2)

55. A rectangular computer chip measures 2.71 mm (millimeters) by 3.62 mm. Find its area (area = length × width).

56. The number π is defined as the circumference of a circle divided by the diameter. If a circle has a circumference of 18.1 cm and a diameter of 5.76 cm, find π to two decimal places.

57. Students in a laboratory measured the acceleration due to gravity by means of a pendulum. The following measurements were taken: 9.61 m/s^2, 9.52 m/s^2, 9.91 m/s^2, 10.1 m/s^2, and 9.82 m/s^2 (m/s^2 means "meters per second, per second"). What is the average of these measurements?

58. A casting weighs 29.8 lb. Find the cost at 30.6¢ per pound.

59. The respective resistances of four resistors connected in series are 12.4 Ω (ohms), 15.7 Ω, 18.9 Ω, and 16.4 Ω. Find the total resistance.

60. A step-up transformer changes the voltage in a transmission line from 112.4 V (volts) to 214.1 V. What is the increase in the voltage?

61. Number 5 gauge plate steel has a thickness of 0.21875 in. What is the approximate height of a stack of 56 sheets?

62. The mass density of aluminum is 2.7 g/cm^3 (grams per cubic centimeter). Determine the mass of 3.2 cm^3 of aluminum.

63. Cast iron has a weight density of 0.26 $lb/in.^3$ (pounds per cubic inch). Find the weight of a cast-iron wheel containing 96 $in.^3$.

64. If your car gets 28.5 mi to the gallon, how far can you travel on a full tank of 14.2 gal?

65. A foundry shipped a number of castings with the following weights: flywheel, 140.4 lb; gear blanks, 160.6 lb; and foundry flasks, 327.4 lb. Find the total weight of the shipment.

66. In the construction of a new office building, the following amounts of concrete were used during a week: 226.4 ft^3 (cubic feet), 293.1 ft^3, 252.3 ft^3, 269.6 ft^3, and 245.3 ft^3. What is the total amount of concrete used?

67. Light travels at the rate of 670,000,000 mi/h (miles per hour). The average distance from the earth to the sun is 93,000,000 mi. How long does it take for light to reach the earth? (Recall: time = distance ÷ rate.)

68. A motor shaft is to have a length of 9.175 in., with a *tolerance* of 0.002 in. What are the allowable dimensions?

1.10 Percent

Problems involving percents are common in technology, as well as in everyday life. The literal meaning of *percent* is "per hundred," denoted by the symbol %, read "percent." Thus 10% is read "ten percent" and means 10 out of 100 or $\frac{10}{100}$.

In a sense, percents are fractions in disguise. Since they are based on 100, we can convert percents to decimals in one step. For example,

$$10\% = 0.10$$

Both 10% and 0.10 mean 10 out of 100, or $10 \div 100$.

Division by 100 moves the decimal point two places to the left, so we obtain the following rule:

1. To change a **percent to a decimal,** move the decimal point two places to the left and omit the percent symbol.
2. To change a **decimal to a percent,** move the decimal point two places to the right and attach the percent symbol.

Consider the following example:

Example 1

	Decimal point moved
a. 22.7% = 0.227	2 places to left
b. 0.073% = 0.00073	2 places to left
c. 0.12 = 12%	2 places to right
d. 0.0137 = 1.37%	2 places to right

A common application of percent is to find the percent of a given number. For example, if your hourly wage is \$8.50 and you get an 8% raise, then your raise is 8% of \$8.50. The value is obtained by changing 8% to 0.08 and multiplying:

$$8\% \text{ of } \$8.50 \text{ is } (0.08) \times (8.50) = \$0.68$$

or 68¢.

To find the percent of a given number:

1. Write the percent as a decimal.
2. Multiply the decimal by the given number.

Example 2 A manufacturing process for microchips produces 0.6% defectives. Out of a sample of 2500 chips, about how many can be expected to be defective?

Solution. We change 0.6% to 0.006 and multiply by 2500:

$$(0.006) \times (2500) = 15 \text{ chips}$$

So about 15 out of the 2500 are defective. (The actual number will vary since 2500 is just a typical sample.)

Sometimes we need to know what percent one number is of another. For example, if 5 students out of a class of 25 are absent, what percent is absent? Since 5 out of 25 is $\frac{5}{25} = \frac{1}{5}$, we divide 1 by 5 to obtain a decimal. This decimal is then changed to a percent:

$$\frac{1}{5} = 1 \div 5 = 0.20 = 20\%$$

To determine what percent a first number is of a second number:

1. Write a fraction with the first number as the numerator and the second number (after the "of") as the denominator.
2. Write the fraction as a decimal.
3. Change the decimal to a percent.

Example 3 If 45 out of 425 shims did not pass inspection, what percent failed?

Solution. Dividing 45 by 425, we get

$$\frac{45}{425} = 0.106 = 10.6\%$$

So 10.6% of the shims did not pass inspection.

Another application of percent is finding a number when a percent of this number is known. For example, if 8 is 20% of some number, what is this number? If we denote this number by N, then we can say that

$$20\% \text{ of } N \text{ is } 8$$

or

$$0.20 \times N = 8$$

If we divide both sides of the last equality by 0.20, we get

$$\frac{0.20 \times N}{0.20} = \frac{8}{0.20}$$

or

$$\frac{\cancel{0.20}N}{\cancel{0.20}} = \frac{8}{0.20} \qquad \text{Cancellation}$$

$$N = \frac{8}{0.20} = 40$$

So $N = 40$. As a check, 20% of 40 is $0.20 \times 40 = 8$, as required.

> **To determine a number, given a percent of this number:**
>
> 1. Write the percent as a decimal.
> 2. Divide the given number by this decimal.

Example 4 The power delivered to a resistor is 32.1 W (watts). This is 80% of the maximum possible power. Find the maximum power.

Solution. Following the rule just given, we write 80% as 0.80 and divide this number into 32.1. So the maximum power is

$$32.1 \div 0.80 = 40.1 \text{ W}$$

Exercises / Section 1.10

In Exercises 1–8, change the given percents to decimals.

1. 29% **2.** 35% **3.** 86.2% **4.** 91.9%

5. 1.34% **6.** 7.69% **7.** 57.36% **8.** 42.92%

In Exercises 9–14, change the given decimals to percents.

9. 0.36 **10.** 0.72 **11.** 0.4219 **12.** 0.6629

13. 1.62 **14.** 2.36

In Exercises 15–20, find the indicated quantities.

15. 10% of 60 **16.** 12% of 84 **17.** 25% of 60 **18.** 35% of 80

19. 12.8% of 75 **20.** 70.3% of 140

21. 20 is what percent of 45?

22. 30 is what percent of 94?

23. Determine what percent 60 is of 75.

24. Determine what percent 95 is of 120.

25. What percent is 25 of 65?

26. What percent is 35 of 45?

27. 25% of a number is 40. What is the number?

28. 58% of a number is 83. What is the number?

29. 65% of a number is 92.6. What is the number?

30. 72% of a number is 25.2. What is the number?

31. 53% of a number is 90.1. Find the number.

32. 28% of a number is 63.7. Find the number.

33. A certain calculator costs $35.50. If the store offers a 12% discount, how much do you have to pay?

34. The voltage across a capacitor is 120 V (volts). If the voltage increases by 15%, what is the new voltage?

35. A technician adds 15 mL (milliliters) of water to 60 mL of sulfuric acid. What percent of the mixture is acid?

36. A molder makes 150 molds of which 8 are defective. What percent is defective?

37. In a certain town, 10,510 people voted in the last election. This number represents 40% of all the people eligible to vote. How many eligible voters are there?

38. A type of breakfast cereal contains 0.153 mg (milligrams) of vitamin C, which is 6% of the daily adult requirement. What is the daily requirement for an adult?

39. A type of brass contains 60% copper, 6% lead, and 34% zinc. How many pounds of each are there in 90 lb of brass?

40. The list price for a small home computer is $2500. If the sales tax is 4%, how much do you have to pay?

41. An alloy for making motor shafts contains 85% tin (by volume). A shaft weighing 44.5 lb contains 32.5 lb of tin. How much lighter would the shaft be if it were made entirely of tin?

42. A farmer plants 80 acres of corn. If 75% of his land is used for other crops, how big is his farm?

43. The *efficiency* of a motor is 80%; that is, the output is 80% of the input. Determine the output of a 150-hp (horsepower) motor.

44. Referring to Exercise 43, if a motor delivers 21 hp (output) and has an efficiency of 84%, what is the input?

45. Suppose your hourly wage is $8.75. If, due to a decline in business, your wage is reduced by 12% and if you receive a 12% pay raise shortly after, will your wage be back to $8.75 per hour?

46. A 6.5-yd^3 (cubic yard) mixture of concrete contains 1.5 yd^3 of gravel. What percent of the mixture is gravel?

47. Refer to Exercise 46. If 23% of the mixture is gravel, and if 2.3 yd^3 of gravel are available for mixing, how large a mixture will you get?

1.11 Power

In this section we briefly discuss the definition of **power.**

If a number is multiplied by itself repeatedly, the product can be written by use of *exponents*. For example,

$$2 \times 2 \times 2 \quad \text{is written} \quad 2^3$$

Base
Exponent

The number 2 is called the **base** and the number 3 the **exponent.** Similarly,

$$3^5 = 3 \times 3 \times 3 \times 3 \times 3$$

Here 3 is the base and 5 the exponent. In general,

$a \cdot a = a^2$ is read "a squared" or "a to the second power"

$a \cdot a \cdot a = a^3$ is read "a cubed" or "a to the third power"

$a \cdot a \cdot a \cdot a = a^4$ is read "a to the fourth power"

or simply "a to the second," "a to the third," "a to the fourth," and so on.

The *n*th power

$$a^n = a \cdot a \cdot \cdots \cdot a \ (n \text{ factors})$$

is read "a to the nth power"; a is called the **base** and n the **exponent.**

Example 1 Evaluate 5^3, 4^4, and 10^2.

Solution.

$$5^3 = 5 \times 5 \times 5 = 125$$
$$4^4 = 4 \times 4 \times 4 \times 4 = 256$$
$$10^2 = 10 \times 10 = 100$$

Example 2

$$2^3 \times 3^2 = (2 \times 2 \times 2) \times (3 \times 3) = 8 \times 9 = 72$$
$$4^2 \times 5^3 = (4 \times 4) \times (5 \times 5 \times 5) = 16 \times 125 = 2000$$

Exercises / Section 1.11

In Exercises 1–8, evaluate the given powers.

1. 3^2 **2.** 4^2 **3.** 5^3 **4.** 5^4

5. 7^2 **6.** 8^2 **7.** 4^3 **8.** 3^4

In Exercises 9–20, evaluate the given products.

9. $2^4 \times 3^2$ **10.** $3^3 \times 5^2$ **11.** $4^3 \times 3^2$ **12.** $5^3 \times 2^3$

13. $2^3 \times 6^2$ **14.** $3^2 \times 4^2 \times 5^3$ **15.** $7^2 \times 4^3 \times 2^2$ **16.** $6^3 \times 5^2 \times 2^4$

17. $9^2 \times 2^3 \times 6^2$ **18.** $3^4 \times 4^3 \times 5^2$ **19.** $4^3 \times 5^3 \times 3^2$ **20.** $8^2 \times 5^3 \times 3^4$

1.12 Arithmetic Operations with Calculators

We now turn to arithmetic (pronounced arith·*met*·ic) operations carried out with a calculator.

In this first part of the book, a calculator with only arithmetic operations (including a square root key) is sufficient. In the latter part of the book, and in more advanced courses you may plan to take, a scientific calculator is essential.

Since many types of calculators are available, we cannot discuss all of them here. So we confine ourselves to the types used by most students. These include most scientific calculators using *algebraic logic*.

General Remarks

A calculator cannot accept numbers that are too large or too small for the display. Also, if the number resulting from a calculation is too large or too small, the calculator will give an error indication (such as an E).

Other operations resulting in an error designation are division by zero, taking the square root of a negative number, and (for many calculators) raising a negative number to a power.

The Control Keys

Certain keys, often referred to as **control keys,** are used to "erase" data from the calculator. Most calculators have two types: the clear key, [C], is used to clear the displayed value, as well as all information being calculated. The clear key *will not* erase information stored in memory.

Another key, designated by [CE] or [CD], will erase only the last entry. Its main function is to erase a wrong entry.

Example 1 Suppose we wish to enter 2 + 6, but enter 2 + 9 by mistake. The [CD] key can be used to erase the 9, so that 6 can be entered instead:

[2] [+] [9]	9 entered by mistake
[CD]	Clear last entry (9)
[6]	6 replaces 9

The equal sign key, [=], also clears all information being calculated. However, the display and the information stored in memory will not be erased.

Data Entry Keys

The data entry keys are used to read in numbers in order to perform calculations with them. The data entry keys are the digits [0], [1], [2], . . . , [9]; the decimal point key, [·]; and the pi key, [π]. The change of sign key [+/−] will be discussed in Chapter 3 and the key [EE] in Chapter 10. (This key is used to enter numbers in scientific notation.)

Example 2 To enter the number 703, use the sequence

[7] [0] [3]

The display will now read 703.

Example 3 To enter the number 23.76, use the sequence

[2][3][·][7][6]

The display will now read 23.76.

Example 4 When pressing the pi key, [π], the display will read

3.1415927

(The displayed number is an approximation for π.) The number π is used frequently in geometry (Chapter 6).

Arithmetic Functions

The basic arithmetic functions are used to perform addition, subtraction, multiplication, and division of whole numbers and decimals. The keys are

[+], [−], [×], and [÷]

The key [=] is used to tell the calculator to complete the calculation with the data entered to that point. This key is used both for intermediate and final results. For example, to multiply 236×125, enter 236, press [×], enter 125, and press [=]. The result is 29,500.

The calculator sequence is abbreviated as follows:

236 [×] 125 [=] → 29500

(From now on we will omit the boxes around the individual digits.)

Example 5 Perform the following addition:

$3.76 + 9.82$

Solution. The sequence is

3 [·] 76 [+] 9 [·] 82 [=] → 13.58

Example 6 Perform the following subtraction:

$12.0765 - \pi$

Solution. The sequence is

12 [·] 0765 [−] [π] [=] → 8.9349073

From now on we will omit the decimal point key in describing a sequence.

Example 7 Multiply: 72.4×20.3.

Solution. The sequence is

72.4 [×] 20.3 [=] → 1469.72

Example 8 Divide: $783 \div 12.5$.

Solution. The sequence is

783 [÷] 12.5 [=] → 62.64

Order of Operations

Whenever several arithmetic operations are involved in a calculation, the following order of operations should be followed:

1. Evaluate any expression enclosed in parentheses (or other grouping symbols).
2. Evaluate powers.
3. Perform all multiplications and divisions from left to right.
4. Perform all additions and subtractions from left to right.

Scientific calculators are programmed to perform these operations in the correct order.

Example 9 Evaluate: $7.4 + 2.8 \times 4.9$.

Solution. The correct order of operations is: multiplication first and then addition. Since scientific calculators perform the operations in the correct order automatically, a proper sequence is

7.4 [+] 2.8 [×] 4.9 [=] → 21.12

For calculators that perform the operations in the order entered, the correct sequence is

2.8 [×] 4.9 [=] [+] 7.4 [=] → 21.12

Memory

Storage key

Recall key

For a longer string of calculations, the memory feature may be helpful. The **memory** works like an electronic scratch pad, saving us the trouble of writing down intermediate steps. Any value in the register can be stored by using the **storage key,** [STO] or [x→M]. The stored number can be recovered and used directly in a calculation by pressing the **recall key,** [RCL] or [MR].

Example 10 Evaluate $12.4 \times 7 - 7.6 \times 5$.

Solution. For scientific calculators, which perform the operations in the correct order automatically, a correct sequence is

12.4 [×] 7 [−] 7.6 [×] 5 [=] → 48.8

For nonscientific calculators, the memory feature can be used to ensure that the operations will be done in the correct order: obtain the second product, 7.6×5, and store the value in the memory. This value can be subtracted directly from the first product by pressing [−] [MR]. The sequence is

7.6 [×] 5 [=] [STO] 12.4 [×] 7 [=] [−] [MR] [=] → 48.8

Problems arise when the order of operations is altered. For example, if 3 is to be multiplied by the sum of 4 and 5, we need to use parentheses to indicate that the addition operation is to be performed first:

$$3 \times (4 + 5) = 3 \times 9 = 27$$

Without parentheses we have

$$3 \times 4 + 5 = 12 + 5 = 17$$

which is a totally different value.

Parentheses are also called *grouping symbols*. Expressions containing grouping symbols can be evaluated with a calculator in two ways:

1. Using parentheses
2. Using the memory feature

Consider the next example.

Example 11 Evaluate:

$$(4.6 + 6.9) \times 10.4 - (2.1 + 7.3) \times 4.2$$

Solution.

1. *Using parentheses.* Since scientific calculators perform arithmetic operations in the correct order, the numbers and parentheses can be entered in the order in which they occur:

   [(] 4.6 [+] 6.9 [)] [×] 10.4 [−] [(] 2.1 [+] 7.3 [)] [×] 4.2 [=] → 80.12

2. *Using the memory feature.* Use of the memory feature is a good alternative to the preceding sequence, especially for nonscientific calculators that do not have parentheses. We find the value of the second quantity,

(2.1 + 7.3) × 4.2, and store this value in memory. We then evaluate the first quantity and subtract the contents of the memory by pressing [−] [MR]. The complete sequence is

2.1 [+] 7.3 [=] [×] 4.2 [=] [STO] 4.6 [+] 6.9 [=] [×] 10.4 [=] [−] [MR] [=] → 80.12

Example 12 Evaluate:

$$(1307 + 728) \div (732 - 325)$$

Solution. Using parentheses, the sequence is

[(] 1307 [+] 728 [)] [÷] [(] 732 [−] 325 [)] [=] → 5

Using the memory feature, the sequence is

732 [−] 325 [=] [STO] 1307 [+] 728 [=] [÷] [MR] [=] → 5

Special Function Keys

Every scientific calculator has a number of special function keys. These special functions are particularly useful because they perform operations on the quantity displayed without your having to press the equal sign key. As a result, the special function keys can be used in a chain of operations.

> The special functions operate only on the quantity displayed.

In this section we consider only three special functions: the reciprocal, square, and square root keys.

Reciprocal key

The **reciprocal key** [1/x] is used to find the reciprocal of a number. For example, to find $\frac{1}{50}$, we may use the sequence

50 [1/x] → 0.02

There is no need to press [=]. (Note that the same result can be obtained by using the sequence 1 [÷] 50 [=].)

Example 13 To find the reciprocal of 1.25, use the sequence

1.25 [1/x] → 0.8

As already noted, the special function keys are particularly useful when several operations are involved, as shown in the next example.

Example 14 Evaluate:

$$\left(1 + \frac{1}{3.4}\right) \times \frac{1}{7.6}$$

Solution. Using the reciprocal key [1/x] we get the following sequence:

[(] 1 [+] 3.4 [1/x] [)] [×] 7.6 [1/x] [=] → 0.1702786

So

$$\left(1 + \frac{1}{3.4}\right) \times \frac{1}{7.6} \approx 0.17$$

The symbol ≈ means "approximately equal to." Rounding off an answer to the proper number of digits will be considered in detail in the next chapter.

Square key

The **square key** [x^2] is used to find the square of a number displayed. For example, to evaluate 16^2, we use the sequence

16 [x^2] → 256

As with all special function keys, there is no need to press [=]. Also, the square key is particularly useful in a chain of operations.

Example 15 Evaluate $(2 + 17^2)^2 + 15^2$.

Solution. The sequence is

[(] 2 [+] 17 [x^2] [)] [x^2] [+] 15 [x^2] [=] → 84906

The square root. Our final operation in this section is evaluating the square root of a given number. Square root is defined next.

> **Square root**
>
> The **square root** of a given number is the number that, when multiplied by itself, is equal to the given number.

For example, we know that $5^2 = 25$. It follows that 5 is the square root of 25. The symbol for square root is $\sqrt{\ }$. So

$$\sqrt{25} = 5$$

Example 16

$$\sqrt{36} = 6 \quad \text{since} \quad 6^2 = 36$$
$$\sqrt{121} = 11 \quad \text{since} \quad 11^2 = 121$$

A whole number whose square root is also a whole number is called a *perfect square*. For example, 49 is a perfect square since $\sqrt{49} = 7$. The square root of a number that is not a perfect square can only be approximated. A method exists for finding square roots that is somewhat similar to long division. However, with calculators so readily available, it is no longer necessary to do such calculations by hand.

Square root key

To find the square root of a given number with a calculator, we use the special **square root key** [√]. As with the other special function keys, there is no need to press [=].

Example 17 Use a calculator to find an approximation of $\sqrt{50}$.

Solution. The sequence is

50 [√] → 7.0710678

Thus $\sqrt{50} \approx 7.07$ to two decimal places.

In the next example the square root key is used in a chain of operations.

Example 18 Evaluate:

$$\sqrt{\sqrt{361} - \sqrt{126}}$$

Solution. The sequence is

361 [√] [−] 126 [√] [=] [√] → 2.7883737

Example 19 The *period P* of a pendulum is the time required to complete one cycle (Figure 1.25) and is given by

$$P = 2\pi\sqrt{\frac{l}{32}}$$

The period is measured in seconds and the length l of the pendulum in feet. Find the period of a pendulum 2.6 ft in length.

Solution. Since $l = 2.6$ ft, we need to evaluate

$$P = 2\pi\sqrt{\frac{2.6}{32}}$$

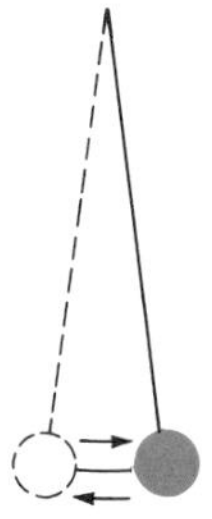

Figure 1.25

(Recall that the absence of any operational symbols indicates multiplication; also recall that $\pi \approx 3.14159$.) A possible sequence is

$$2 \boxed{\times} \boxed{\pi} \boxed{=} \boxed{\text{STO}} 2.6 \boxed{\div} 32 \boxed{=} \boxed{\sqrt{\ }} \boxed{\times} \boxed{\text{MR}} \boxed{=} \rightarrow 1.7909834$$

A simpler sequence is

$$2.6 \boxed{\div} 32 \boxed{=} \boxed{\sqrt{\ }} \boxed{\times} 2 \boxed{\times} \boxed{\pi} \boxed{=}$$

So $P = 1.8$ seconds (to the nearest tenth of a second).

Exercises / Section 1.12

In Exercises 1–40, perform the indicated operations with a calculator. (Round off answers to the nearest hundredth, if necessary.)

1. $76 + 94$
2. $123 + 68$
3. $293 - 142$
4. $646 - 381$
5. 83×92
6. 21×109
7. $1643 \div 31$
8. $1363 \div 47$
9. $\dfrac{1}{0.25}$ (use $\boxed{1/\text{x}}$)
10. $\dfrac{1}{0.125}$ (use $\boxed{1/\text{x}}$)
11. $(97)^2$
12. $(68)^2$
13. $(12.4)^2$
14. $(23.7)^2$
15. $\sqrt{676}$
16. $\sqrt{961}$
17. $\sqrt{2143.69}$
18. $\sqrt{571.21}$
19. $5.2 + 3.6 \times 4.7$
20. $98 - 3.7 \times 8.6$
21. $(78 - 20) \times (29 + 6)$
22. $(150 - 90) \times (50 - 15)$
23. $(9.7)^2 - 2\pi$ (Recall that 2π means $2 \times \pi$.)
24. $8\pi - (3.6)^2$
25. $\pi^2 - (2.1)^2$
26. $\pi^2 - (2.9)^2$
27. $(1.4 + 3.6) \times 8.2 - (6.3 - 4.2) \times 2.1$
28. $(29.3 - 21.6) \times 6.4 - (1.7 + 2.4) \times 3.9$
29. $(36.4 + 12.3) \div (76.3 - 29.6)$
30. $(100 - 7.6) \div (15.2 + 8.3)$
31. $\left(\dfrac{1}{2.1} + \dfrac{1}{6.4}\right) \times 3.7$
32. $\left(\dfrac{1}{4.6} - \dfrac{1}{8.7}\right) \times 5.2$
33. $\left(\dfrac{1}{3.6} + 2.4\right)^2 \times \left(\dfrac{1}{7.2} + 1\right)$
34. $\left(6.6 - \dfrac{1}{3.7}\right)^2 \times \left(\dfrac{1}{4.8} + 1\right)$
35. $(2 + 15^2)^2$
36. $(16^2 - 13^2)^2$
37. $(\sqrt{729} - 4)^2$
38. $(\sqrt{1156} - 6)^2$
39. $(\sqrt{1764} - \sqrt{121})^2$
40. $(\sqrt{2916} - \sqrt{1225})^2$
41. Find the period of a pendulum 12 ft in length. (See Example 19.)
42. Find the length of the support in Figure 1.26, which is given by $l = \sqrt{(6.2)^2 + (2.5)^2}$ (in feet).

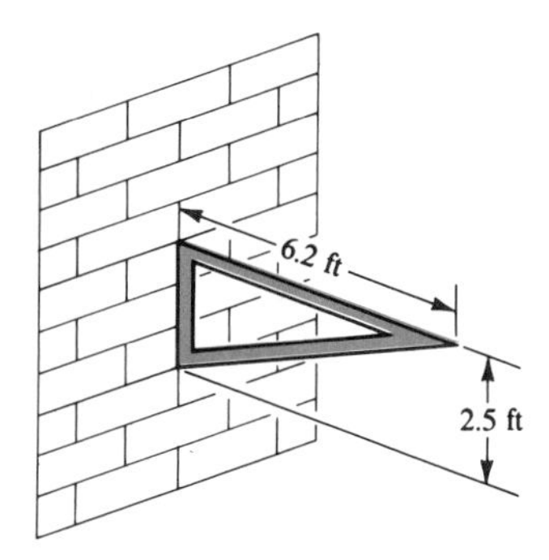

Figure 1.26

43. The combined resistance R (in ohms) of two parallel resistors, R_1 and R_2, is given by

$$R = \frac{R_1 R_2}{R_1 + R_2}$$

Find R if $R_1 = 31.4\ \Omega$ (ohms) and $R_2 = 45.7\ \Omega$.

44. The velocity v (in meters per second) of a body falling from rest is given by $v = \sqrt{19.6x}$, where x is the distance traveled (in meters). Find the velocity of a body that has fallen 35.1 m.

45. The impedance (in ohms) of a certain circuit is

$$Z = \sqrt{(12.2)^2 + \left(\frac{1}{0.015}\right)^2}$$

Evaluate Z.

46. The density d (in grams per cubic centimeter) of a certain gas is given by

$$d = 0.13 \times \left(1 + \frac{500}{273}\right) \times \frac{520}{3.2}$$

Evaluate d.

Review Exercises / Chapter 1*

In Exercises 1–67, do not use a calculator.

1. Add:

$$\begin{array}{r} 3798 \\ 9347 \\ 2109 \\ \underline{8046} \end{array}$$

2. Add:

$$\begin{array}{r} 537 \\ 403 \\ 719 \\ 164 \\ \underline{672} \end{array}$$

In Exercises 3 and 4, perform the indicated subtractions.

3.

$$\begin{array}{r} 9712 \\ \underline{-3978} \end{array}$$

4.

$$\begin{array}{r} 8002 \\ \underline{-2799} \end{array}$$

* All the answers to the Review Exercises are given in the Answers Section.

In Exercises 5–8, perform the indicated operations.

5. 8806×512

6. 2316×5109

7. $398{,}112 \div 928$

8. $106{,}680 \div 35$

In Exercises 9 and 10, write each number as a product of prime factors.

9. 252

10. 1350

In Exercises 11 and 12, reduce each fraction to lowest terms.

11. $\frac{126}{420}$

12. $\frac{315}{330}$

In Exercises 13–36, perform the indicated operations.

13. $\frac{2}{3} \times \frac{7}{8} \times \frac{6}{11} \times \frac{4}{21}$

14. $\frac{5}{9} \times \frac{11}{14} \times \frac{7}{44} \times \frac{3}{10}$

15. $\frac{55}{63} \div \frac{66}{7}$

16. $\left(\frac{2}{3} \times \frac{4}{7}\right) \div \frac{16}{7}$

17. $\frac{5}{12} - \frac{1}{6}$

18. $1 - \frac{7}{20} - \frac{2}{25}$

19. $\frac{13}{14} + \frac{1}{16} - \frac{7}{20}$

20. $\frac{19}{28} - \frac{3}{14} + \frac{2}{35}$

21. $3\frac{3}{4} + 5\frac{1}{8}$

22. $7\frac{1}{6} - 2\frac{5}{12}$

23. $\left(2\frac{2}{3}\right) \times \left(4\frac{5}{6}\right)$

24. $\left(4\frac{2}{7}\right) \div \left(2\frac{1}{2}\right)$

25. $\left(5\frac{1}{2} - 1\frac{3}{4}\right) \div 2\frac{1}{5}$

26. $\left(6\frac{1}{8} - 2\frac{3}{4}\right) \div 1\frac{1}{2}$

27. $83.1 + 5.98 - 2.883$

28. $5.71 - 2.73 + 21$

29. $(1.84)(3.65)$

30. $(6.18)(72.3)$

31. $(0.035)(0.0017)$

32. $(0.012)(0.91)$

33. $36 \div 2.5$

34. $0.288 \div 0.3$

35. $0.00312 \div 0.012$

36. $0.00864 \div 0.036$

In Exercises 37–40, change the given fractions to decimals.

37. $\frac{3}{75}$

38. $\frac{7}{25}$

39. $\frac{9}{125}$

40. $\frac{13}{250}$

In Exercises 41–44, change the given decimals to fractions.

41. 0.025

42. 0.0095

43. 0.12

44. 0.024

45. Change 46.3% to a decimal.

46. Change 0.072 to a percent.

47. Find 0.125% of 60.

48. Find 30% of 75.

49. 25 is what percent of 80?

50. 30 is what percent of 75?

51. 52% of a number is 98. What is the number?

52. 46% of a number is 84. What is the number?

53. Evaluate $2^5 \times 3^2$.

54. Evaluate $3^3 \times 4^3 \times 5^2$.

55. A wooden box containing computer parts weighs 35 lb 10 oz. The box and packing weigh a total of 8 lb 12 oz. How much do the computer parts weigh?

56. The current through a 120-Ω (ohm) resistor is 1.9 A (amperes). Find the voltage across the resistor (voltage = current × resistance).

57. An order of 26 motors requires 3361.8 ft of wire. How many feet of wire does each motor require?

58. A technician mixes 124 mL (milliliters) of sulfuric acid with 28 mL of water. What fractional part of the mixture is water?

59. The bob on a pendulum moves back and forth every $2\frac{2}{9}$ s (seconds). How long does it take to complete 18 cycles?

60. How long does a 110-lb supply of sweeping compound last if $2\frac{3}{4}$ lb are used daily?

61. A transformer increases the voltage from $100\frac{2}{3}$ V (volts) to $120\frac{1}{12}$ V. What is the increase in the voltage?

62. A technician making an hourly check on the temperature of an engine obtains the following readings (in degrees Celsius): 40.2, 41.3, 39.2, 42.5, 41.6, 40.5, 42.3, and 40.8. What is the average temperature during his shift?

63. A tank full of water weighs 3200 lb. The tank has a capacity of 50 ft^3 (cubic feet) and weighs 80 lb when empty. Determine the density of water (weight per unit volume) in pounds per cubic foot.

64. A 120-lb block of an alloy contains 30 lb of nickel. What percent of the alloy is nickel?

65. Twenty out of 130 bushings did not meet specifications. What percent did not meet specifications?

66. A farmer plants corn on 37.7% of his farm. Given that he plants 92 acres of corn, how large is his farm?

67. The manager of the finance department figures that computerizing her operation will reduce operating costs by 34.6%. If the monthly cost is now $2360 per month, what is the reduced cost to the nearest dollar?

In Exercises 68–74, carry out the indicated operations with a calculator.

68. $(7.6 + 4.3)(6.4) - (10.7 - 8.6)(2.8)$

69. $(43.7 - 7.6) \div (16.4 - 11.9)$

70. $\left(\frac{1}{5.4} - \frac{1}{12.2}\right) \times \frac{1}{7.6}$

71. $\left(1 - \frac{1}{8.6}\right)^2 \times \left(2.4 + \frac{1}{3.6}\right)$

72. $(\sqrt{784} - 12)^2$

73. $(\sqrt{1024} - \sqrt{225})^2$

74. $\left(\frac{1}{3.6} + \frac{1}{2.7} - \frac{1}{8.4}\right) \times (2.4 + \sqrt{196})$

CHAPTER 2

Approximation and Measurement

2.1 Approximate Numbers

In mathematics we generally assume that numbers are exact. Thus "10" means "exactly 10." In technology the situation is different in that numbers are often obtained through measurement. Measured quantities are necessarily **approximate** since no measurement can be perfect. For example, if the length of a steel beam is given as 16.2 ft, it is understood that the number 16.2 is only approximate.

The importance of measurement does not, however, eliminate exact numbers in technology. For example, 1 ft = 12 in. by definition. There is no measurement involved. So the number 12 is exact. Similarly, there are *exactly* 60 minutes in an hour.

Numbers obtained through a simple counting process may also be exact. For example, if you buy a carton labeled "one dozen eggs," you would expect to find exactly 12 eggs in the carton. On the other hand, if an inventory sheet contains the entry "15,000 bolts," chances are that the number is only approximate. Monetary values, such as prices, are usually exact, but very large sums, such as the gross national product, are only approximate. So in many cases, to determine whether a number is exact or approximate, we have to know how the number was obtained. A number is ordinarily exact if it is obtained (1) by counting, (2) from a definition, or (3) as a result of a computation with numbers that are also exact.

Example 1 State whether the numbers in the following statements are exact or approximate:

a. There are 24 hours in a day.
b. A pipe wrench costs $12.90.
c. The acceleration due to gravity is 32.2 ft/s^2 (feet per second per second).

Solution.

a. Since there are 24 hours in a day by definition, the number 24 is exact.
b. Since $12.90 is the price of a single item, the number 12.90 is exact.
c. The acceleration due to gravity is known experimentally (by measurement). So 32.2 is approximate.

Accuracy

Accuracy

The **accuracy** of a measurement refers to the number of digits used to express the measurement. Suppose the length of a steel beam is given as 16.2 ft. What do we know about its actual length? If the measurement is accurate to the nearest tenth of a foot, then its true length lies between 16.15 ft and 16.25 ft. Since this information is conveyed by "16.2 ft," we say that 16.2 has three **significant digits.**

Now consider the measurement 120 in. As written, we know only that the exact length is closer to 120 in. than to 110 in. or 130 in. We say that 120 has *two* significant digits. Similarly, the number 34,000 has two significant digits.

In general, any nonzero digit is significant. The only questionable digits are final zeros. For example, if we are told that the measurement of 120 in. is accurate to the nearest inch, the final zero would be significant. If this information is not available, then we assume that 120 has two significant digits. The rules for significant digits are summarized next.

Significant digits

1. The digits 1 through 9 are always significant.
2. Zeros are not always significant:
   a. Zeros are significant if they lie between two significant digits.
   b. Unless otherwise specified, final zeros are significant only if they lie to the right of the decimal point.
   c. Initial (beginning) zeros are not significant.

These rules are illustrated in the next example.

Example 2

a. 3.742 has four significant digits (no zeros).
b. 3.024 has four significant digits since the zero lies between two significant digits.
c. 4.0007 and 7.0509 each have five significant digits since all zeros lie between significant digits.
d. 3200 and 780,000 each have two significant digits. (The final zeros lie to the *left* of the decimal point.)

e. 3.200 has four significant digits. (The final zeros lie to the *right* of the decimal point.)
f. 78.0000 has six significant digits, since the final zeros lie to the right of the decimal point.
g. 0.0034 and 0.00065 each have two significant digits. (Initial zeros are not significant.)

The rule for determining whether zeros are significant may also be stated as follows: *Zeros are not significant if their only function is to place the decimal point.* For example, the zeros in the numbers 0.0032 and 32,000 serve only to place the decimal point. By the last rule, the zeros, therefore, are not significant. So each number has two significant digits. This conclusion agrees with our first rule.

Remark. Why initial zeros are not considered significant can be seen from the following example: Given that 10 mm (millimeters) = 1 cm (centimeter), it follows that

$$0.1 \text{ mm} = 0.01 \text{ cm}$$

In each case the zeros serve only to place the decimal point and are therefore not significant. So each number has one significant digit. This conclusion is reasonable since 0.1 mm and 0.01 cm describe exactly the same measurement, but with different units of measure.

Precision

Precision

When dealing with approximate numbers, we sometimes need to consider the precision with which a given measurement is made. The **precision** of a number refers to the position of the rightmost significant digit. For example, 10.23 ft is more precise than 10.2 ft since 10.23 ft was measured to the nearest hundredth of a foot but 10.2 ft only to the nearest tenth of a foot. Similarly, 0.0124 volt is more precise than 0.012 volt.

Example 3 Suppose a rectangular metal plate is measured to be 4.6 cm (centimeters) long and 2.91 cm wide. Then the width was measured with greater precision than the length. (Note that the measure of the width is also more accurate.)

Example 4 Suppose a metal bar with a circular cross section has a diameter of 5.6 in. and a coating 0.12 in. thick (Figure 2.1). Then the coating has been measured with greater precision than the diameter. (However, note that the accuracy is the same, two significant digits.)

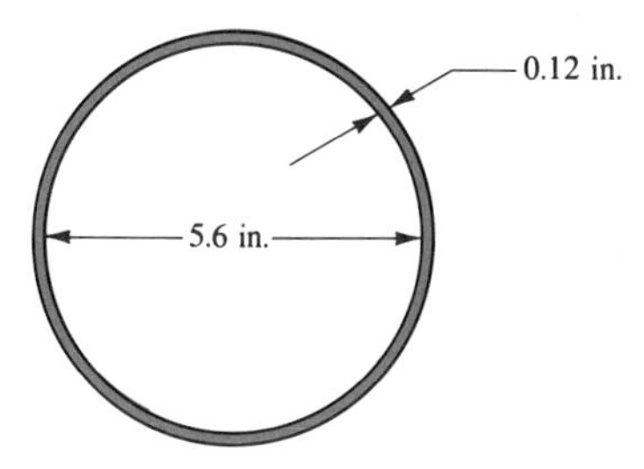

Figure 2.1

Note that if one measurement is more precise than another, it is not necessarily more accurate, as we saw in Example 4. Consider another example.

Example 5

a. 7.12 is more precise than 6.2, and
7.12 is also more accurate than 6.2.

b. 0.139 is more precise than 48.6, but the accuracy is the same (three significant digits).

c. 2.72 is more precise than 286.7, but
286.7 is more accurate.

Rounding Off

Numbers resulting from a calculation with approximate numbers often have to be rounded off. (Calculations with approximate numbers will be studied in the next section.)

To round off a number to a specific number of digits, examine the next digit to the right. If this digit is 5 or more, increase the last digit by 1. This digit then becomes the last significant digit. If the next digit is less than 5, leave the last digit as it is. (Again, this digit becomes the last significant digit.)

Example 6 Round off 3.146 to three significant digits.

Solution. The first three digits of the number 3.146 are 3.14. Since the next digit is 6, which is greater than 5, we increase the last digit by 1 and write

3.15

Example 7 Round off 192.14 to four significant digits.

Solution. The first four digits in the number 192.14 are 192.1. The next digit, 4, is less than 5. So we leave the last digit as it is and write

192.1

When rounding off numbers, it is sometimes necessary to replace final digits by zeros, as shown in the next example.

Example 8

a. 8063 rounded to three significant digits is 8060.
b. 8067 rounded to three significant digits is 8070.
c. 2940 rounded to two significant digits is 2900.
d. 46,960 rounded to two significant digits is 47,000.
e. 1.55 rounded to two significant digits is 1.6.
f. 3.347 rounded to three significant digits is 3.35.
g. 8.104 rounded to three significant digits is 8.10.
h. 71.98 rounded to three significant digits is 72.0.
i. 2.096 rounded to three significant digits is 2.10.

Example 9 Suppose that the length of a metal bar is measured to be 25.7 in. (three significant digits). Then the actual length must lie between 25.65 in. and 25.75 in. since any number between 25.65 and 25.75 is rounded off to 25.7 in.

Exercises / Section 2.1

In Exercises 1–8, state whether the numbers in the given statements are exact or approximate. (See Example 1.)

1. The floor of a laboratory has an area of 1580 ft^2.

2. A programmable calculator costs $110.95.

3. The voltage across a resistor is 30.6 V (volts).

4. The distance from the earth to the moon is 240,000 mi.

5. There are 100 centimeters in a meter.

6. There are 60 seconds in a minute.

7. A certain computer has 17,000 bytes of memory.

8. There are 21 students in the class.

In Exercises 9–28, determine the number of significant digits in each case.

9. 1.041 **10.** 59.1 **11.** 8.86 **12.** 749.7

13. 36,200 **14.** 50,400 **15.** 817,000 **16.** 99,540

17. 1.8500
18. 3.9900
19. 9.2020
20. 70.5000
21. 45.7000
22. 8.4700
23. 0.051
24. 0.0910
25. 0.08200
26. 0.005070
27. 0.00640
28. 0.0018400

In Exercises 29–40, round off each number to the number of significant figures indicated.

29. 60,536 (3)
30. 47,370 (3)
31. 28,490 (2)
32. 21,467 (2)
33. 18,916 (2)
34. 22,594 (3)
35. 2.5893 (3)
36. 59.221 (3)
37. 3.5527 (3)
38. 0.01564 (3)
39. 0.005496 (3)
40. 0.0040596 (4)

In Exercises 41–48, state which of each pair of numbers is (a) more accurate; (b) more precise.

41. 1.23, 5.230
42. 12.2, 0.736
43. 0.76, 38.7
44. 0.390, 64.23
45. 0.0042, 3.764
46. 0.0073, 4.61
47. 0.0960, 541
48. 0.0470, 860

49. A solar panel was measured to be 8.6 ft in length. Between what values must the actual length lie? (See Example 9.)

50. A communications satellite is 22,300 mi above the surface of the earth. What are the largest and smallest possible altitudes of the satellite? (See Example 9.)

51. The length of a steel rod is known to lie between 5.368 in. and 5.372 in. How should the length be expressed using three significant digits?

52. The length of a bolt is at least 9.26 cm but no more than 9.34 cm. Express the length using two significant digits.

53. Consider the washer in Figure 2.2. Which measurement is more accurate? More precise?

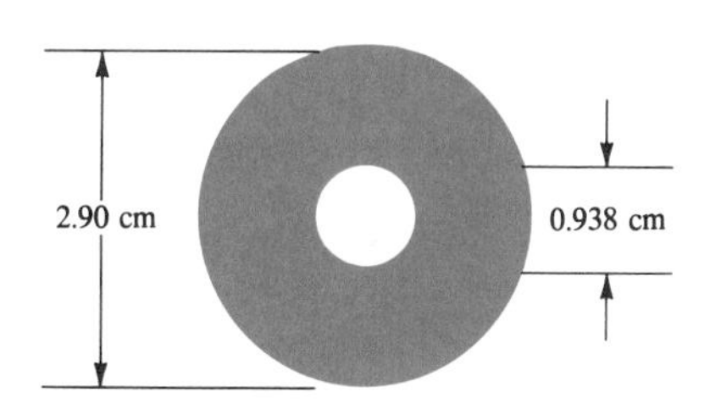

Figure 2.2

54. Figure 2.3 shows a cross section of a tube. Which measurement is more accurate? More precise?

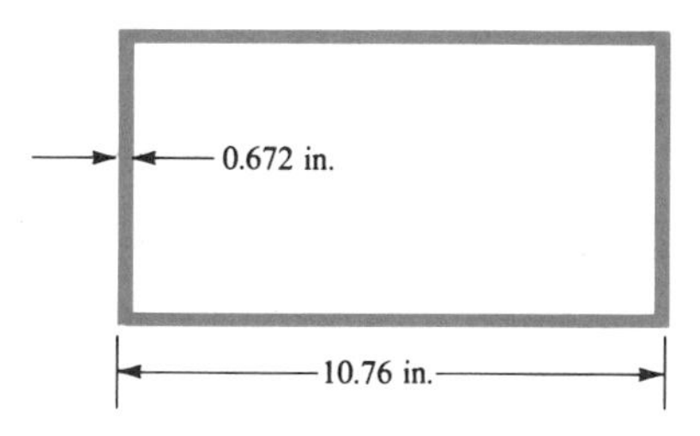

Figure 2.3

2.2 Calculations with Approximate Numbers

When calculations are performed with approximate numbers, the calculated values are also approximate. For example, suppose a rectangular metal plate has a length of 18.3 in. and a width of 10.8 in. Then the area A, which is equal to the length multiplied by the width, is

$$A = (18.3 \text{ in.}) \times (10.8 \text{ in.}) = 197.64 \text{ in.}^2$$

(The symbol in.^{2} stands for "square inches.") Can the value of the area really be expressed to this level of accuracy? No, since the measurements given contain only three significant digits. Because of this, the value of the area should be rounded off to three significant digits to yield

$$A = 198 \text{ in.}^2$$

Greater accuracy cannot be justified. Of course, you may feel that the area should be rounded off to the nearest tenth, or 197.6 in.^{2}, expressed with the same precision as the given dimensions. However, note that

18.3 in. lies between 18.25 in. and 18.35 in.

while

10.8 in. lies between 10.75 in. and 10.85 in.

So the correct value of the area is at least

$$(18.25 \text{ in.}) \times (10.75 \text{ in.}) = 196.2 \text{ in.}^2$$

but no more than

$$(18.35 \text{ in.}) \times (10.85 \text{ in.}) = 199.1 \text{ in.}^2$$

So 197.6 in.^{2} is not correct. The general rule is given next.

1. When approximate numbers are multiplied or divided, the result is no more accurate than the least accurate number.
2. The square root of an approximate number is as accurate as the number.

These rules are illustrated in the first three examples.

Example 1 Perform the indicated division and round off the quotient to the proper number of significant digits:

$$23.87 \div 6.41$$

Solution. The calculator sequence is

23.87 [÷] 6.41 [=] → 3.723869

Note that 23.87 has four significant digits and 6.41 has three significant digits. Since 6.41 is the less accurate number, we round off the answer to three significant digits to obtain 3.72.

Example 2 A box is 4.20 ft long, 2.67 ft wide, and 0.98 ft high. Determine the volume.

Solution. The volume, V, in cubic feet (ft^3), is equal to length $\times$ width $\times$ height. The sequence is

$$4.20 \boxed{\times} 2.67 \boxed{\times} 0.98 \boxed{=} \rightarrow 10.98972$$

Since the least accurate number is 0.98 (two significant digits), we need to round off 10.98972 to two significant digits to obtain 11 ft^3.

Example 3 Find $\sqrt{4.32}$.

Solution. The sequence is

$$4.32 \boxed{\sqrt{\ }} \rightarrow 2.078461$$

By the rule, the square root of a number is as accurate as the number. Rounding off to three significant digits, we obtain

$$\sqrt{4.32} = 2.08$$

The rule for addition and subtraction is stated in terms of precision.

> When approximate numbers are added or subtracted, the result is only as precise as the least precise number.

To see the reason for this rule, let us add 2.4 ft and 1.12 ft:

$$\begin{array}{l} 2.4\ \text{ft} \\ \underline{1.12\ \text{ft}} \\ 3.52\ \text{ft} \end{array}$$

Knowing that the less precise number, 2.4 ft, lies between 2.35 ft and 2.45 ft, consider the smallest and largest possible sums:

$$\begin{array}{lcl} 2.35\ \text{ft} & \qquad & 2.45\ \text{ft} \\ \underline{1.12\ \text{ft}} & & \underline{1.12\ \text{ft}} \\ 3.47\ \text{ft} & & 3.57\ \text{ft} \end{array}$$

Thus the sum may be as low as 3.47 ft or as high as 3.57 ft. These results show that we are not justified in carrying any digits beyond tenths.

All the rules in this section apply only to approximate numbers. Exact numbers have unlimited accuracy and do not affect the accuracy of the result. Calculations involving both exact and approximate numbers will be shown in Section 2.4.

Example 4 A shipping crate contains four castings weighing 10.1 lb, 9.28 lb, 12.0 lb, and 11.73 lb, respectively. Find the total weight of the contents.

Solution. The calculator sequence is

10.1 [+] 9.28 [+] 12.0 [+] 11.73 [=] → 43.11

The least precise numbers are 10.1 and 12.0. So the sum is rounded off to the nearest tenth, or 43.1 lb.

Exercises / Section 2.2

In Exercises 1–26, carry out the indicated operations with a calculator. Round off the results to the proper number of significant digits. (The numbers are approximate.)

1. $3.52 + 9.8 - 0.91$

2. $4.231 + 5.38 - 4.14$

3. $99.28 - 0.312 + 5.67$

4. $0.0665 + 0.456 - 0.09$

5. $0.5621 + 88.65 - 0.237$

6. $23.1 - 3.98 - 0.429$

7. $38.095 - 6.619 + 42.02$

8. $0.379 + 0.5149 + 0.942$

9. $(0.769)(0.027)$

10. $(0.916)(0.071)$

11. $(2.903)(0.223)$

12. $(0.0446)(0.166)$

13. $(0.037)(0.515)$

14. $(9.17)(24.26)$

15. $27{,}000 \div 16{,}000$

16. $3400 \div 4610$

17. $52.87 \div 1.46$

18. $0.7030 \div 1.216$

19. $271.22 \div 7.817$

20. $51.39 \div 90.2$

21. $\sqrt{8.548}$

22. $\sqrt{7.2319}$

23. $\sqrt{47.784}$

24. $\sqrt{23.340}$

25. $\sqrt{0.01328}$

26. $\sqrt{0.0561}$

Use a calculator in the remaining problems.

27. A wooden box weighing 6.8 lb when empty contains gears with the following weights: 0.73 lb, 1.28 lb, 6.9 lb, 9.6 lb, and 11.3 lb. Find the weight of the full box.

28. A box delivered to a retail store weighs 30.4 lb. The label says that the contents weigh 25.25 lb. What is the combined weight of the empty box and packaging materials?

29. Find the length of the machine part in Figure 2.4.

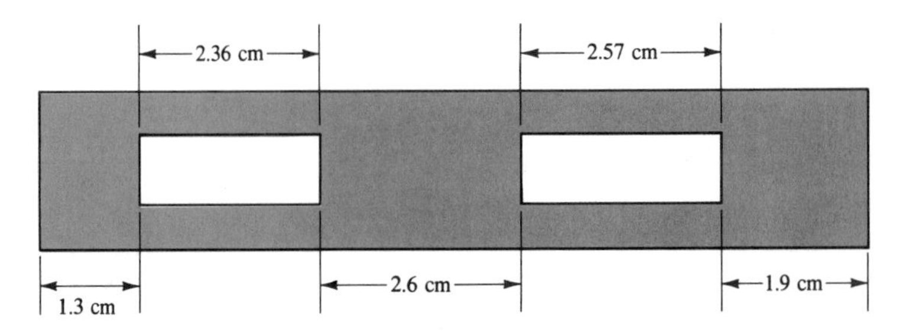

Figure 2.4

30. Four resistances of 5.25 Ω (ohms), 7.3 Ω, 10.36 Ω, and 15.5 Ω, respectively, are connected in series. Find the resistance of the combination, which is the sum of the individual resistances.

31. According to Ohm's law, the voltage drop across a resistor is the product of the current I and the resistance R. If $I = 2.5$ A (amperes) and $R = 15.3\ \Omega$ (ohms), find the voltage drop.

32. If the current in a circuit is 1.8 A (amperes), then the power P delivered to a 10.1-ohm resistor is $P = (1.8)^2(10.1)$ W (watts). Find the value of P.

33. A car travels to a town 174 mi away at an average speed of 51.3 miles per hour. How long does it take to get there?

34. Find the area of a rectangular plate whose length is 5.80 cm and whose width is 6.72 cm.

35. The velocity v (in feet per second) of an object dropped from rest is given by $v = \sqrt{64.4d}$, where d is the distance from the starting point. Find the velocity after the object has fallen 95.7 ft.

36. The period P (in seconds) of a pendulum is the time required to complete one full swing. It is given by $P = 1.1\sqrt{l}$, where l is the length of the pendulum (in feet). Find the period of a pendulum 16 ft long.

2.3 The Metric System (SI)

Two systems of measurement are in use in the United States today: the English system and the metric system. While the metric system is expected to become the standard system eventually, a complete changeover will take many more years. Consequently, the technologist and engineer has to be familiar with both systems.

SI system

As is the case with the English system, the metric system uses certain base units. To make the notation for the base units uniform, a standard system of units was adopted in 1960. This system is called the **International System of Units,** abbreviated **SI.** In technology, the system that is referred to as the metric system is usually understood to be the SI system.

Common Units

Base units

Both the English and SI systems have certain **base units** for expressing length, time, capacity, and so on. Other units, such as the units for area and volume, are expressed in terms of the base units.

Length: The base unit of length is the **meter** in the SI system and the **foot** in the English system.

Time: The base unit for time is the **second** in both systems.

Temperature: Temperature is measured in **degrees Fahrenheit** in the English system and in **degrees Celsius** in the SI system. Also, *absolute temperature* is measured in **kelvin** in the SI system.

Capacity: The base unit of capacity (volume) is the **liter** in the SI system and the **gallon** in the English system.

Mass and Weight: In everyday life we tend to use the terms *weight* and *mass* interchangeably. But in physics and technology we carefully distin-

Weight
Mass

guish between the two. **Weight** is the measure of the force exerted on the body by gravity. **Mass** is the measure of the resistance of the body to motion (the body's inertia). In other words, the greater the mass of the body, the greater the force needed to move it. While the weight of a body decreases as it moves away from the center of the earth, the mass of the body remains the same.

In the English system the base unit of weight (or force) is the **pound** and the base unit of mass, the **slug.** In the SI system the base unit of weight is the **newton** and the base unit of mass, the **kilogram.**

Table 2.1
Common units

	British		SI (International system of units)		
	Name	Symbol	Name	Symbol	
Length	foot	ft	meter	m	
Mass	slug		kilogram	kg	
Force	pound	lb	newton	N	kg · m/s^2
Capacity	gallon	gal	liter	L or ℓ	1 L = 1000 cm^3
Pressure		lb/ft^2	pascal	Pa	N/m^2
Work, Energy	foot-pound	ft · lb	joule	J	N · m
Power	horsepower	hp	watt	W	J/s
Current	ampere	A	ampere	A	
Charge	coulomb	C	coulomb	C	
Electric potential	volt	V	volt	V	
Capacitance	farad	F	farad	F	
Inductance	henry	H	henry	H	
Resistance	ohm	Ω	ohm	Ω	
Quantity of heat	British thermal unit	Btu	joule	J	
Temperature	degree Fahrenheit	°F	degree Celsius	°C	
Absolute temperature			kelvin	K	
Time	second minute hour	s min h	second minute hour	s min h	

Although the weight of an object varies with altitude, this variation is of no significance if the object remains near the surface of the earth. For this reason **the kilogram is often used as a measure of weight.** This explains the relation 1 lb = 0.454 kg.

Electrical units: The base unit for electric current is the ampere in both systems. This and other electrical units are listed in Table 2.1.

Table 2.1 lists the various base units in the English and SI systems. The table also shows the symbols used for these units. For example, m is used for *meter,* Ω for *ohm,* A for *ampere,* and so on.

Also note that the SI system includes standard notations for time measurements. Thus h is used for "hour," min for "minute," and s for "second." You may be used to seeing *second* abbreviated "sec." The single-letter abbreviation is used for consistency and to allow the use of metric prefixes, discussed next.

Metric Prefixes

In the metric system, units of different magnitudes are obtained by multiplying certain basic units by powers of 10. These multiples are denoted by prefixes, some of which are listed in Table 2.2.

Table 2.2

Prefix	Symbol	Base unit multiplied by	Meaning
mega-	M	$10^6 = 1{,}000{,}000$	million
kilo-	k	$10^3 = 1{,}000$	thousand
centi-	c	$\frac{1}{10^2} = 0.01$	hundredth
milli-	m	$\frac{1}{10^3} = 0.001$	thousandth
micro-	μ	$\frac{1}{10^6} = 0.000001$	millionth

The prefixes in the table are used with many units. For example, the prefix *kilo-,* meaning "thousand," is combined with *gram* to form *kilogram.* Some of these combinations are shown in the next example.

Example 1 Prefix: *kilo-*

Unit	*Symbol*	*Name*	*Meaning*
g	kg	kilogram	1000 g
m	km	kilometer	1000 m
W	kW	kilowatt	1000 W

The next example shows some combinations with the prefix *milli-.*

Example 2 Prefix: *milli-*

Unit	*Symbol*	*Name*	*Meaning*
m	mm	millimeter	0.001 m
L	mL	milliliter	0.001 L
A	mA	milliampere	0.001 A
Ω	mΩ	milliohm	0.001 Ω

Table 2.2 also lists the number that the unit has to be multiplied by to yield a prefixed unit.

Example 3

μF = 0.000001 F (microfarad)
mΩ = 0.001 Ω (milliohm)
MC = 1,000,000 C (megacoulomb)
cL = 0.01 L (centiliter)
μs = 0.000001 s (microsecond)

Reduction of Units

Reduction
Conversion

Changing a unit of measurement within a system is called a **reduction.** A change from one system to another is called a **conversion.** Conversions will be taken up in the next section.

We have seen that reduction within the SI system can be accomplished by multiplying or dividing a given measurement by a multiple of 10. Since multiplication and division by a multiple of 10 moves the decimal point, a reduction can be performed readily with the help of the diagram in Figure 2.5.

Reduction in the SI system

To change from one unit to another, move the decimal point as many places as you move along the scale.

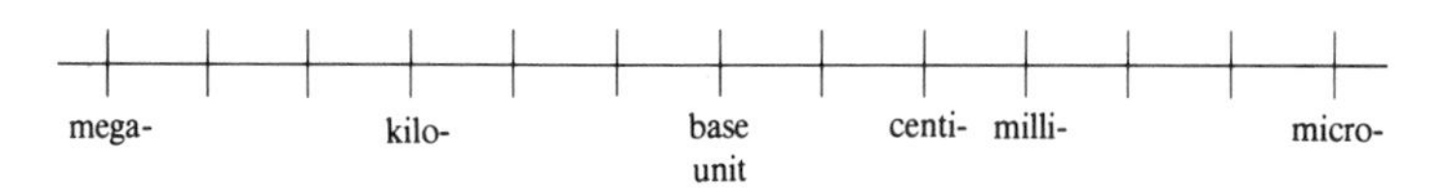

Figure 2.5

1. To change to a unit farther to the right, move the decimal point to the right.
2. To change to a unit farther to the left, move the decimal point to the left.

Example 4 Change 3.2 μs to milliseconds.

Solution.

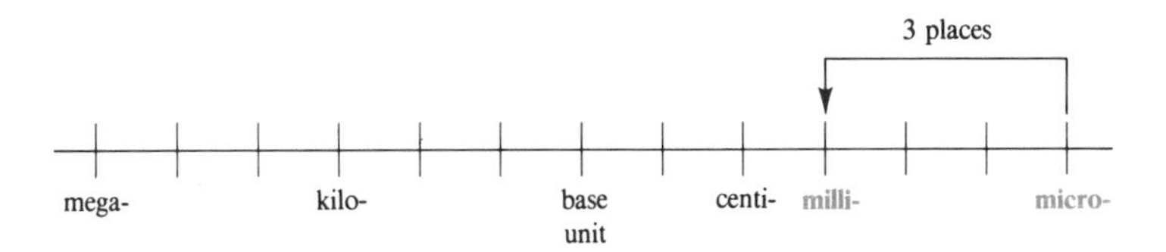

Figure 2.6

We see from Figure 2.6 that the decimal point is moved 3 units to the left. In order to do this, we need to insert zeros:

$$3.2\ \mu\text{s} = 0.0032\ \text{ms}$$

Example 5 Change 0.7634 L to milliliters.

Solution.

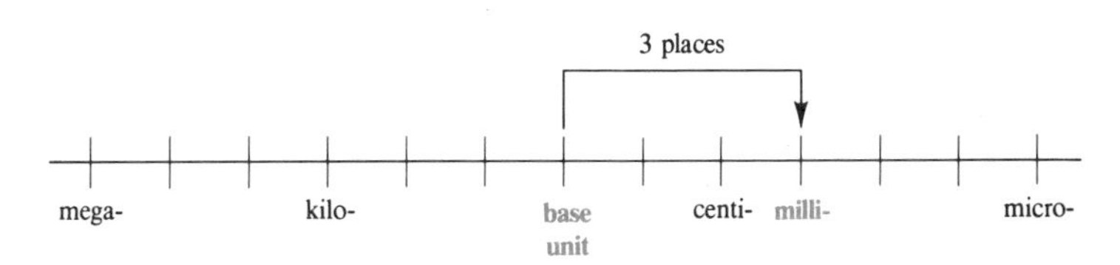

Figure 2.7

Figure 2.7 shows that the decimal point must be moved 3 places to the right:

$$0.7634\ \text{L} = 763.4\ \text{mL}$$

Example 6 Change 0.0034 kΩ to milliohms.

Solution.

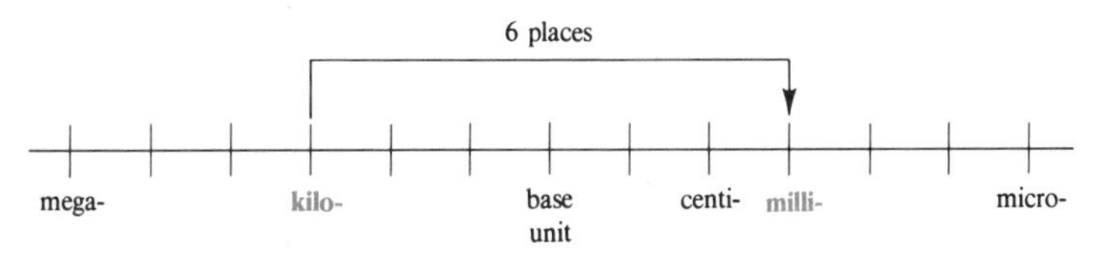

Figure 2.8

By Figure 2.8, we have to move the decimal point 6 places to the right. Inserting zeros, we get

$$0.0034 \text{ k}\Omega = 3400 \text{ m}\Omega$$

Exercises / Section 2.3

In Exercises 1–16, write the proper symbol for each unit.

1. millimeter **2.** milliliter **3.** milligram **4.** kilogram
5. centiliter **6.** megawatt **7.** kilowatt **8.** centimeter
9. microfarad **10.** kilometer **11.** milliohm **12.** megaohm
13. microsecond **14.** millisecond **15.** milliampere **16.** microcoulomb

In Exercises 17–28, give the meaning of each of the given units.

17. ms **18.** μF **19.** MΩ **20.** mΩ
21. cL **22.** km **23.** mg **24.** $\mu\Omega$
25. cm **26.** mA **27.** μs **28.** MC

In Exercises 29–50, use the diagram in Figure 2.9 to reduce the given units to the units indicated.

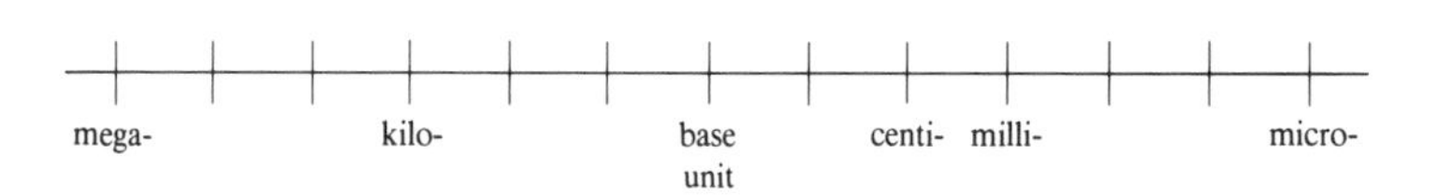

Figure 2.9

29. 0.34 m to millimeters
30. 9.84 cm to millimeters
31. 0.264 L to milliliters
32. 7.404 Ω to milliohms
33. 23 mg to grams
34. 1.24 kg to grams
35. 4.06 kW to watts
36. 72 ms to seconds
37. 102 ms to seconds
38. 23 ms to microseconds
39. 0.89 ms to microseconds
40. 26 μs to milliseconds
41. 25 $\mu\Omega$ to milliohms
42. 7.6 mΩ to microohms
43. 36.4 μF to millifarads
44. 29.3 μF to millifarads
45. 0.076 MΩ to milliohms
46. 0.00042 MΩ to milliohms
47. 0.000726 kW to centiwatts
48. 0.0000030 kW to milliwatts
49. 0.0000096 MW to milliwatts
50. 0.00000000234 MW to microwatts

51. The capacitance of a capacitor is 230 μF. What is the capacitance in farads?

52. The current in a certain microcircuit is 360 mA. Express the current in amperes.

53. An electron moves across a vacuum tube in 20 ms. What is the time in seconds?

54. The voltage across a certain filament is 50 μV. Express this voltage in volts.

55. A certain computer can perform an addition in 0.036 μs. Express this time in milliseconds.

56. The wavelength of red light is 0.00065 mm. Express the wavelength in micrometers.

2.4 Conversion of Units

In this section we study a method by which we can convert units systematically. To do this, we need a list of basic conversion factors relating the English and SI systems. See Table 2.3.

Table 2.3 Basic conversion units

1 in. = 2.54 cm (exact)	1 kg = 2.205 lb
1 km = 0.6214 mi	1 lb = 4.448 N
1 lb = 454 g	1 ft^3 = 28.32 L
	1 cm^3 = 1 mL

The unit ft^3 stands for *cubic foot* and is a measure of volume. Note especially the relationship 1 cm^3 = 1 mL, which is actually the definition of milliliter. (An equivalent relationship is 1 L = 1000 cm^3.)

Also note that the basic units of weight (or force), pounds and newtons, are related by the equality 1 lb = 4.448 N. However, as mentioned in the last section, the mass units in the metric system, grams and kilograms, are also used as weight units. For this reason, the relationships 1 lb = 454 g and 1 kg = 2.205 lb are included in Table 2.3.

Our method of converting units is similar to the method of multiplying fractions. For example, to multiply $\frac{2}{3}$ by $\frac{3}{7}$, we cancel the 3s and obtain the reduced form

$$\frac{2}{\not{3}} \times \frac{\not{3}}{7} = \frac{2}{7}$$

Similarly, a velocity of 1 ft/s can be converted to feet per minute by the operation

$$1\,\frac{\text{ft}}{\text{s}} = 1\,\frac{\text{ft}}{\not{\text{s}}} \times \frac{60\,\not{\text{s}}}{1\text{ min}} = 60\,\frac{\text{ft}}{\text{min}} \qquad 60\text{ s} = 1\text{ min}$$

Note that the cancellation of the time unit s was carried out as if s were a number.

The object of any such cancellation is to eliminate one unit and introduce another. This can be accomplished by multiplying the given quantity by 1. More precisely, we multiply the given quantity by a fraction that has a value of 1. These fractions are obtained from Table 2.3. Since all values are given as equal pairs, we have

$$\frac{1\text{ in.}}{2.54\text{ cm}} = 1, \qquad \frac{1\text{ km}}{0.6214\text{ mi}} = 1, \qquad \frac{454\text{ g}}{1\text{ lb}} = 1$$

and so on. Multiplying a given quantity by any of these fractions does not change the value. Consider the following example.

Example 1 Convert 2.36 mi to kilometers.

Solution. To eliminate mi and introduce km, we multiply 2.36 mi by

$$\frac{1\text{ km}}{0.6214\text{ mi}} = 1 \qquad 1\text{ km} = 0.6214\text{ mi}$$

This multiplication yields

$$\frac{2.36\text{ mi}}{1} \times \frac{1\text{ km}}{0.6214\text{ mi}} = \frac{2.36}{0.6214}\text{ km}$$
$$= 3.80\text{ km}$$

Note that the cancellation of the unit mi leaves only km, the unit desired. (Since 0.6214 has four significant digits, but 2.36 only three, the answer is rounded off to three significant digits.)

The general rule for converting units is given next.

Rule for converting units

1. Multiply the given quantity by one or more fractions (from Table 2.3), each of which has a value of 1.
2. Cancel units in the resulting product as if they were numbers.

Example 2 Convert 20.0 km/min to miles per minute.

Solution. This example is similar to Example 1: We multiply the given quantity by

$$\frac{0.6214\text{ mi}}{1\text{ km}} \qquad 1\text{ km} = 0.6214\text{ mi}$$

Rounding off to three significant digits, we get

$$20.0\,\frac{\text{km}}{\text{min}} \times \frac{0.6214\text{ mi}}{1\text{ km}} = 12.4\,\frac{\text{mi}}{\text{min}}$$

It was pointed out in Section 2.1 that exact numbers have an unlimited number of significant digits. In the next example we use the relationships

$$60\text{ min} = 1\text{ h} \qquad \text{and} \qquad 60\text{ s} = 1\text{ min}$$

These numbers are exact and have no bearing on the accuracy of the result.

Example 3 Convert 25 mi/h to kilometers per second.

Solution. Since 1 km = 0.6214 mi, we multiply by

$$\frac{1\ \text{km}}{0.6214\ \text{mi}}$$

in order to cancel mi and introduce km. Similarly, we multiply by

$$\frac{1\ \text{h}}{60\ \text{min}}$$

to cancel h and introduce 1/min. Finally, we multiply by

$$\frac{1\ \text{min}}{60\ \text{s}}$$

to cancel min and introduce 1/s. Successive multiplication yields

$$25\ \frac{\cancel{\text{mi}}}{\cancel{\text{h}}} \times \frac{1\ \text{km}}{0.6214\ \cancel{\text{mi}}} \times \frac{1\ \cancel{\text{h}}}{60\ \cancel{\text{min}}} \times \frac{1\ \cancel{\text{min}}}{60\ \text{s}} = \frac{25}{(0.6214)(60)(60)}\ \frac{\text{km}}{\text{s}}$$

The calculator sequence is

25 [÷] 0.6214 [÷] 60 [÷] 60 [=] → 0.0111754

Since 60 is exact, the answer is rounded off to the accuracy of 25 mi/h (two significant digits). So

$$25\ \frac{\text{mi}}{\text{h}} = 0.011\ \frac{\text{km}}{\text{s}}$$

The next example involves the exact numbers 1 lb = 16 oz.

Example 4 Convert 25.4 oz to newtons.

Solution. According to Table 2.3, 1 lb = 4.448 N, the weight units. However, to use this relationship, we must first change ounces to pounds by multiplying the given quantity by

$$\frac{1\ \text{lb}}{16\ \text{oz}}$$

Next, we multiply by

$$\frac{4.448\ \text{N}}{1\ \text{lb}}$$

to cancel lb and introduce N. Successive multiplication yields

$$\frac{25.4\ \cancel{\text{oz}}}{1} \times \frac{1\ \cancel{\text{lb}}}{16\ \cancel{\text{oz}}} \times \frac{4.448\ \text{N}}{1\ \cancel{\text{lb}}} = \frac{(25.4)(4.448)}{16}\ \text{N}$$

The sequence is

25.4 [×] 4.448 [÷] 16 [=] → 7.0612

Rounding off to three significant digits, we have

25.4 oz = 7.06 N

The next example involves the relationship 1 in. = 2.54 cm, which is exact by definition.

Example 5

$$76.340 \text{ in.} = 76.340 \cancel{\text{in.}} \left(\frac{2.54 \text{ cm}}{1 \cancel{\text{in.}}}\right)$$
$$= 193.90 \text{ cm}$$

Since 2.54 is exact, the answer is rounded off to the accuracy of 76.340 in. (five significant digits).

If the sides of a rectangle are measured in inches, then the area is expressed in *square inches,* denoted by in.^{2}. (The area of a rectangle is equal to length × width.) Similarly, if the lengths of the sides of a box are measured in inches, then the volume is expressed in *cubic inches,* denoted by in.^{3}. (The volume is equal to length × width × height.) Other units of measure are ft^2, cm^2, and m^2 (for area) and ft^3, cm^3, and m^3 (for volume). The use of exponents enables us to cancel these units in the same way as we cancel base units.

Example 6 Convert

$$23.7 \frac{\text{lb}}{\text{in.}^2}$$

to newtons per square meter.

Solution. Since 1 in. = 2.54 cm, we multiply the given quantity by

$$\left(\frac{1 \text{ in.}}{2.54 \text{ cm}}\right)^2 = \frac{1 \text{ in.}}{2.54 \text{ cm}} \times \frac{1 \text{ in.}}{2.54 \text{ cm}} = \frac{1^2\text{in.}^2}{(2.54)^2\text{cm}^2}$$

in order to cancel in.^{2} and introduce 1/cm^2. To cancel cm^2, we multiply by

$$\left(\frac{100 \text{ cm}}{1 \text{ m}}\right)^2 = \frac{100 \text{ cm}}{1 \text{ m}} \times \frac{100 \text{ cm}}{1 \text{ m}} = \frac{(100)^2\text{cm}^2}{1^2\text{m}^2}$$

The squared units may now be canceled in the same way as base units:

$$23.7\ \frac{\text{lb}}{\text{in.}^2} = \left(23.7\ \frac{\text{lb}}{\text{in.}^2}\right)\left(\frac{1\ \text{in.}}{2.54\ \text{cm}}\right)^2\left(\frac{100\ \text{cm}}{1\ \text{m}}\right)^2\left(\frac{4.448\ \text{N}}{1\ \text{lb}}\right)$$

$$= 23.7\ \frac{\cancel{\text{lb}}}{\cancel{\text{in.}^2}} \times \frac{1^2\cancel{\text{in.}^2}}{(2.54)^2\cancel{\text{cm}^2}} \times \frac{(100)^2\cancel{\text{cm}^2}}{1^2\text{m}^2} \times \frac{4.448\ \text{N}}{1\ \cancel{\text{lb}}}$$

$$= \frac{(23.7)(100)^2(4.448)}{(2.54)^2}\ \frac{\text{N}}{\text{m}^2} = 163{,}000\ \frac{\text{N}}{\text{m}^2}$$

The sequence is

23.7 [×] 100 [x^2] [×] 4.448 [÷] 2.54 [x^2] [=]

Common error Squaring the units, but forgetting to square the number. Thus

$(10\ \text{m})^2$ is not the same as $10\ \text{m}^2$

Instead,

$(10\ \text{m})^2 = 10^2\text{m}^2 = 100\ \text{m}^2$

Changing units within the SI system (reduction) can also be carried out by the methods of this section.

Example 7

a. $300\ \text{ms} = \dfrac{300\ \cancel{\text{ms}}}{1} \times \dfrac{1\ \text{s}}{1000\ \cancel{\text{ms}}} = 0.3\ \text{s}$

b. $0.004\ \Omega = \dfrac{0.004\ \cancel{\Omega}}{1} \times \dfrac{1\ \mu\Omega}{\dfrac{1}{10^6}\ \cancel{\Omega}} = \left(\dfrac{0.004}{1}\right) \times (1{,}000{,}000)\mu\Omega = 4000\ \mu\Omega$

c. $0.52\ \text{A} = \dfrac{0.52\ \cancel{\text{A}}}{1} \times \dfrac{1\ \text{mA}}{\dfrac{1}{10^3}\ \cancel{\text{A}}} = (0.52) \times (1000)\text{mA} = 520\ \text{mA}$

Since the liter is a measure of volume, cubic inches (and similar volume measures) can be converted to liters, as shown in the next example.

Example 8 Express 75.62 in.³ in liters.

Solution. We can use the relationship $1\ \text{ft}^3 = 28.32\ \text{L}$:

$$\frac{75.62\ \text{in.}^3}{1} \times \left(\frac{1\ \text{ft}}{12\ \text{in.}}\right)^3 \times \frac{28.32\ \text{L}}{1\ \text{ft}^3} = \frac{75.62\ \cancel{\text{in.}^3}}{1} \times \frac{1^3\ \cancel{\text{ft}^3}}{12^3\ \cancel{\text{in.}^3}} \times \frac{28.32\ \text{L}}{1\ \cancel{\text{ft}^3}}$$

$$= 1.239\ \text{L}$$

We can also use the relationship 1 cm³ = 1 mL:

$$\frac{75.62 \not{\text{in.}^3}}{1} \times \frac{(2.54)^3 \not{\text{cm}^3}}{1^3 \not{\text{in.}^3}} \times \frac{1 \not{\text{mL}}}{1 \not{\text{cm}^3}} \times \frac{1 \text{ L}}{1000 \not{\text{mL}}} = 1.239 \text{ L}$$

Example 9 The density of lead is 11.3 g/cm³. Determine the density in pounds per cubic foot.

Solution. Here the gram is treated as a weight unit. So we use the relationship 1 lb = 454 g to get

$$11.3 \frac{\text{g}}{\text{cm}^3} \times \frac{1 \text{ lb}}{454 \text{ g}} \times \left(\frac{2.54 \text{ cm}}{1 \text{ in.}}\right)^3 \times \left(\frac{12 \text{ in.}}{1 \text{ ft}}\right)^3$$

$$= 11.3 \frac{\not{\text{g}}}{\not{\text{cm}^3}} \times \frac{1 \text{ lb}}{454 \not{\text{g}}} \times \frac{(2.54)^3 \not{\text{cm}^3}}{1^3 \not{\text{in.}^3}} \times \frac{12^3 \not{\text{in.}^3}}{1^3 \text{ ft}^3}$$

$$= 705 \frac{\text{lb}}{\text{ft}^3}$$

Exercises / Section 2.4

In Exercises 1–12, express

1. 19.2 in. in feet

2. 35.7 oz in pounds

3. 2.30 yd in inches

4. 112 in. in yards

5. 0.131 s in milliseconds

6. 0.00768 s in microseconds

7. 0.000238 A in microamperes

8. 0.000296 F in microfarads

9. 28 mA in microamperes

10. 36 $\mu\Omega$ in milliohms

11. 47 μF in millifarads

12. 0.0724 mC in microcoulombs

In Exercises 13–44, convert the given units to the units indicated. (Use a calculator.)

13. 2.0 in. to centimeters

14. 3.0 in. to centimeters

15. 10.0 cm to inches

16. 12.2 cm to inches

17. 4.0 in.² to square centimeters

18. 5.0 in.² to square centimeters

19. 2.5 ft to centimeters

20. 3.6 ft to centimeters

21. 10.7 mi to kilometers

22. 20.07 mi to kilometers

23. 3.76 lb to newtons

24. 10.0 N to pounds

25. 902 g to pounds

26. 3.7 lb to grams

27. $0.25 \frac{\text{ft}}{\text{s}}$ to centimeters per second

28. $10.5 \frac{\text{ft}}{\text{min}}$ to centimeters per minute

29. $2.76 \frac{\text{mi}}{\text{h}}$ to kilometers per minute

30. $73{,}000 \frac{\text{km}}{\text{s}}$ to miles per minute

31. $3.92 \frac{\text{lb}}{\text{in.}^2}$ to newtons per square centimeter

32. $10.12 \frac{\text{lb}}{\text{ft}^2}$ to newtons per square meter

33. $72.40 \frac{N}{m^2}$ to pounds per square inch

34. $8.34 \frac{kg}{m^2}$ to pounds per square foot

35. $6.00 \frac{kg}{m^3}$ to pounds per cubic inch

36. $43.1 \frac{lb}{in.^2}$ to kilograms per square meter

37. $0.274 \frac{lb}{in.^2}$ to kilograms per square meter

38. $22.3 \frac{lb}{ft^2}$ to kilograms per square meter

39. $10.34 \frac{lb}{in.^3}$ to newtons per cubic meter

40. $2096 \frac{N}{m^3}$ to pounds per cubic inch

41. 48 in.^{3} to liters

42. 92 in.^{3} to liters

43. 251 ft^3 to liters

44. 16.0 ft^3 to liters

45. The distance from the earth to the moon is about 239,000 miles. Express this distance in kilometers.

46. The speed of light is approximately 186,000 mi/s. Express the speed in kilometers per second.

47. The density of water is 62.4 lb/ft^3. Express the density in newtons per cubic meter.

48. Write the density of water in kilograms per cubic meter. (Refer to Exercise 47.)

49. The speed of sound is approximately 1130 ft/s. Change the units to kilometers per hour.

50. The acceleration due to gravity is 32 ft/s^2 (feet per second per second). Convert the units to meters per second per second (m/s^2).

51. Aluminum weighs 2.7 g/cm^3. Convert this weight to pounds per cubic foot.

52. A chemical tank is 4.56 ft long, 3.50 ft wide, and 2.85 ft high. What is its capacity in liters?

53. In the manufacture of liquid oxygen, the incoming air is compressed and passed through an expansion valve. The air pressure is 2200 lb/in.^{2}. Write the pressure in kilograms per square centimeter.

54. The velocity of a communications satellite is 6900 mi/h. Find its velocity in kilometers per second.

Review Exercises / Chapter 2

In Exercises 1 and 2, state whether the given numbers are exact or approximate.

1. The velocity of light is $186{,}000 \frac{mi}{s}$.

2. One inch measures 2.54 cm.

In Exercises 3–6, determine the number of significant digits.

3. 470

4. 3940

5. 0.0560

6. 0.07200

In Exercises 7–10, round off each number to the number of significant digits indicated.

7. 27,663 (3)

8. 99,627 (4)

9. 0.05763 (3)

10. 0.002984 (2)

In Exercises 11–18, carry out the indicated operations with a calculator and round off the answers to the proper number of significant digits. (The numbers are approximate.)

11. $9.021 + 5.9706 - 0.31$

12. $48.89 - 7.7093 + 3.626$

13. $(54.51)(8.35)$

14. $(5.142)(0.0904)$

15. $59.68 \div 2.39$

16. $6.544 \div 23.7$

17. $\sqrt{1.236}$

18. $\sqrt{0.25529}$

In Exercises 19–24, write the meaning of each of the given symbols.

19. μF

20. ms

21. MΩ

22. mΩ

23. mL

24. μs

In Exercises 25–28, express

25. 0.0372 s in milliseconds

26. 0.0000279 A in microamperes

27. 2,700,000 Ω in megaohms

28. 17 μF in farads

In Exercises 29–36, convert each of the given units to the units indicated. Use a calculator.

29. 8.43 in.^{2} to square centimeters

30. 115 cm^2 to square inches

31. $0.8400 \frac{\text{ft}}{\text{s}}$ to centimeters per minute

32. $1280 \frac{\text{mi}}{\text{min}}$ to kilometers per second

33. $275 \frac{\text{N}}{\text{m}^2}$ to pounds per square inch

34. $0.25 \frac{\text{lb}}{\text{ft}^2}$ to newtons per square meter

35. $1.75 \frac{\text{lb}}{\text{ft}^2}$ to kilograms per square meter

36. $83.0 \frac{\text{kg}}{\text{m}^2}$ to pounds per square inch

37. Consider the A-frame in Figure 2.10. Which measurement is (a) more accurate? (b) More precise?

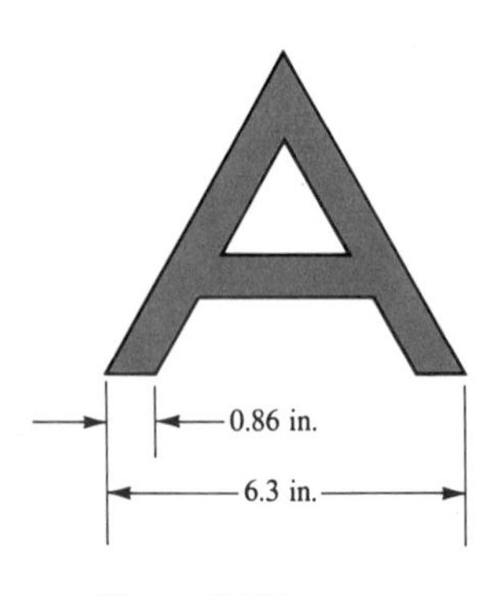

Figure 2.10

38. Three resistors are connected in series. If their respective resistances are 50.5 Ω, 70.37 Ω, and 60.9 Ω, find the resistance of the combination (the sum).

39. The voltage drop across a 10.5-Ω resistor is 25.0 V. Find the current, obtained by dividing voltage by resistance.

40. The current in a certain transistor is 51 μA. What is the current in amperes?

41. The diameter of a piston was measured with an error no greater than 2.3 μm. What is the error in millimeters?

42. Gold weighs 19.3 g/cm^3. Convert to pounds per cubic foot.

43. The speed of the earth relative to the sun is 18.5 mi/s. Write the speed in kilometers per hour.

44. A box containing office supplies has a capacity of 33.9 ft^3. What is the capacity in liters?

CHAPTER 3

Signed Numbers

3.1 The Meaning of Signed Numbers

The main purpose of this section is to introduce the concept of a **signed number.** Inequality and absolute value are also discussed.

To see the need for signed numbers, compare the following subtractions:

$$9 - 4 \qquad \text{and} \qquad 4 - 9$$

We know that $9 - 4 = 5$, but $4 - 9$ does not seem to have an answer, since 4 is smaller than 9. Yet there are many situations in technology in which a larger number must be subtracted from a smaller number. To be able to do this, we need to introduce numbers that are **less than zero.** In our example, $4 - 9$ is a subtraction that leaves a "deficit" of 5, a number less than zero. This number is called "negative 5." Similarly, $2 - 6$ equals "negative 4," and $1 - 7$ equals "negative 6."

The simplest way to picture negative numbers is to display them on a line. (See Figure 3.1.) The numbers to the left of zero are called **negative**

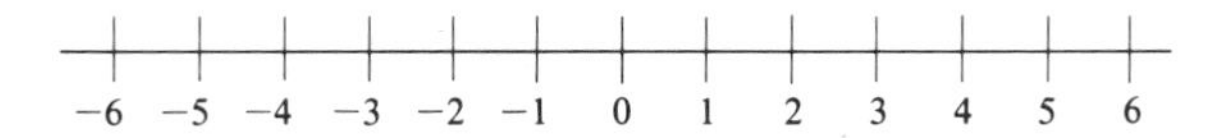

Figure 3.1

numbers. The numbers to the right of zero are called **positive numbers.** (The positive numbers are the numbers we dealt with in our earlier chapters.) The number 0 is neither positive nor negative. Collectively, these numbers are called **signed numbers.**

To indicate whether a number is negative or positive, we use the same symbols that we used for addition and subtraction: A negative number is preceded by a **negative sign** (−); a positive number is preceded by a **positive sign** (+). If there is no danger of confusion, the positive sign for a positive number may be left out. (See Figure 3.1.) For example, the positive number five can be written as either 5 or +5. However, we never omit the sign of a negative number.

Negative sign
Positive sign

Example 1

+6 and +3 are positive numbers.
6 and 3 are positive numbers.
−7 and −4 are negative numbers.
−9.6 and −10.2 are negative numbers.
0 is neither positive nor negative.

Negative numbers are not unfamiliar to you. People giving weather reports use negative numbers to denote temperatures below 0°F (or below 0°C). Similarly, gains or losses in the stock market are designated by + or −, as a look at the financial page of your newspaper will confirm.

Negative numbers are also common in technology. For example, positive current indicates current in one direction and negative current indicates current in the opposite direction.

The Integers

As already noted, positive and negative numbers can be placed on a number line. To do so, we first pick a point, called the **origin,** to which we assign the number 0. The positive whole numbers are placed equally far apart on the right side of the origin. The negative numbers are then placed on the left side of the origin in such a way that every positive number has a mirror image on the left side of the origin. For example, −1 is the mirror image of +1; −2 is the mirror image of +2, and so on. (See Figure 3.2.)

Origin

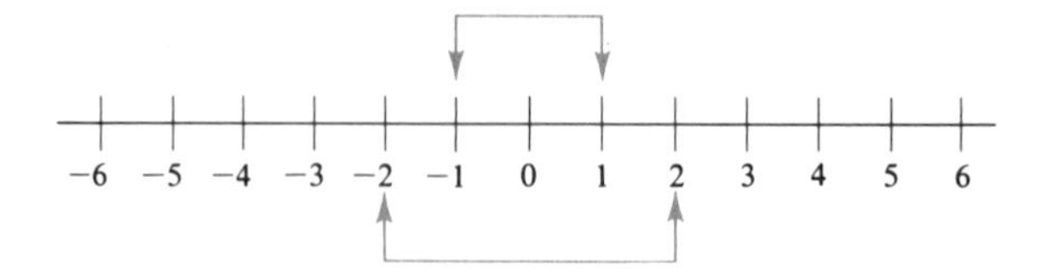

Figure 3.2

A natural number (nonzero whole number) is called a **positive integer.** The mirror image of a positive integer is called a **negative integer.** The number 0 is neither positive nor negative.

Integers

An **integer** is a positive integer, a negative integer, or zero.

Other numbers are placed on the number line between the integers. For example, the number $\frac{3}{2}$ is located midway between 1 and 2, and the number $-\frac{5}{2}$ is placed midway between -2 and -3. These and some other numbers are shown in Figure 3.3.

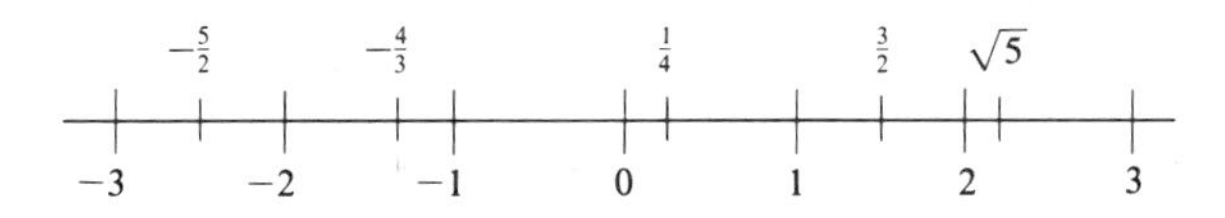

Figure 3.3

Example 2

When describing the motion of an object hurled vertically upward, it is convenient to assign a direction to the motion. If the upward direction is designated positive, then the downward direction is designated negative (Figure 3.4). The origin may be placed anywhere along the path, but it is frequently taken to be ground level.

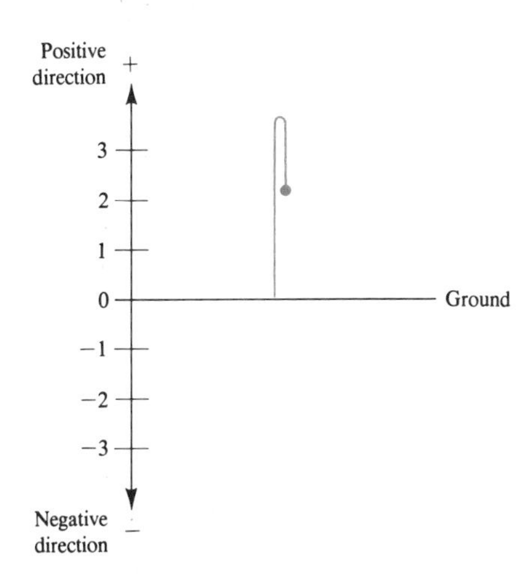

Figure 3.4

Equality and Inequality

A statement that says two quantities are equal is called an equality or an **equation.** Examples of equalities are

$$\frac{1}{2} = \frac{2}{4} \qquad \text{and} \qquad 2.54 \text{ cm} = 1 \text{ in.}$$

Inequality

A statement that says that two quantities are not equal is called an **inequality.** The symbol for inequality is $\neq$. For example,

$$7 \neq 4$$

means "7 is not equal to 4."

Whenever two numbers are not equal, then one has to be greater than the other. To indicate which number is greater or less, we use the following symbols:

Less than
Greater than

1. The symbol $<$ means "is less than."
2. The symbol $>$ means "is greater than."

For example,

$2 < 5$ means "2 is less than 5"

and

$7 > 4$ means "7 is greater than 4"

A convenient way to think of inequality is by means of the number line. For example, since $2 < 5$, 2 is to the *left* of 5; since $7 > 4$, 7 is to the *right* of 4.

Using the idea of position on a number line, inequality can be defined for signed numbers. For example, since -3 is to the left of -1, we have $\mathbf{-3 < -1}$. Similarly, since 2 is to the right of -1, we have $\mathbf{2 > -1}$. (See Figure 3.5.)

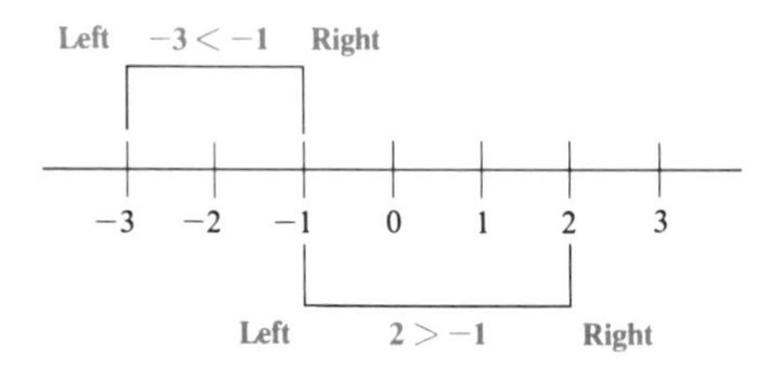

Figure 3.5

Example 3 Referring to Figure 3.5, other examples of inequalities are $-3 < -2$, $-1 > -2$, $-1 < 0$, $3 > 0$, and $-1 < 3$.

In general, we define inequality as follows:

1. If a is to the left of b on the number line, then $a < b$.
2. If a is to the right of b on the number line, then $a > b$.

Absolute Value

In technology, situations occasionally arise in which we are interested in a number, but not in the sign. For example, suppose two objects are moving along a line in opposite directions with velocities equal to 10 ft/s and −15 ft/s, respectively. Then the *faster* object moves with a speed of 15 ft/s, even though −15 is less than 10. In other words, the speed with which an object moves depends on the *numerical value* of the velocity, not on the sign—the sign indicates only the direction. The number 15 is said to be the *absolute value* of −15.

Absolute value

The **absolute value** of a number is the numerical value without any sign. For example, the absolute value of 6 (or +6) is 6. The absolute value of −4 is 4.

The absolute value of a number is denoted by the symbol | |, which is placed around the number. Thus

$$|6| = 6, \qquad |+6| = 6, \qquad \text{and } |-4| = 4$$

Example 4

a. The absolute value of +10 is 10, or

$$|+10| = 10$$

b. The absolute value of 8.13 is 8.13, or

$$|8.13| = 8.13$$

c. The absolute value of −3 is 3, or

$$|-3| = 3$$

d. The absolute value of $-\frac{3}{2}$ is $\frac{3}{2}$, or

$$\left|-\frac{3}{2}\right| = \frac{3}{2}$$

The concept of absolute value will be used in the next section to define addition of signed numbers.

Example 5 An alternating current i changes direction regularly. Suppose that $i = 1.2$ A at one instant and that $i = -1.2$ A at another instant. The signs tell us that the current is moving in opposite directions at these instants. To indicate that the magnitude is the same (1.2 A), we write

$$|i| = 1.2 \text{ A}$$

Exercises / Section 3.1

In Exercises 1–16, place < or > between the given numbers.

1. 2, 4 **2.** $7, \frac{1}{2}$ **3.** −4, −2 **4.** −3, −7

5. 4, −4 **6.** −7, 7 **7.** $0, -\frac{1}{2}$ **8.** $0, \frac{1}{3}$

9. $-\frac{3}{4}, -\frac{1}{4}$ **10.** $-\frac{1}{3}, -\frac{2}{3}$ **11.** $-\frac{1}{3}, 4$ **12.** $2, -\frac{1}{2}$

13. $-2.5, -5.4$ **14.** $-6.3, -2.1$ **15.** $0, -2.7$ **16.** $0, -1.3$

In Exercises 17–30, find the absolute value of each number.

17. -2 **18.** $-\frac{1}{2}$ **19.** $+4$ **20.** $+10$

21. -7 **22.** -12 **23.** 15 **24.** 26

25. $-\sqrt{4}$ **26.** $-\sqrt{9}$ **27.** $\sqrt{49}$ **28.** $\sqrt{64}$

29. -1.8 **30.** -2.3

31. For a particular stock, $+3.4$ indicates an increase in the dollar value. What does -1.6 represent?

32. The current i in a circuit is 1.0 A at some instant. What is i at the instant when the current is 0.8 A in the opposite direction?

33. Let the distance d (in feet) above the ground be positive, and let d be zero on the ground. What is d 2 feet below the ground?

34. A spring is hanging from the ceiling. Assume that the lower end of the spring is at the origin and that -2 cm represents a distance of 2 cm above the origin. The end of the spring is pulled 3.5 cm below the origin. Write this distance as a signed number.

35. Two forces are pulling in opposite directions. If one force is 20 lb, write the 40-lb opposite force as a signed number.

36. If the respective forces in Exercise 35 are 30 lb and -50 lb, which is the stronger force?

37. If the countdown gives the (positive) time in seconds before a rocket takes off, what does -6 s represent?

38. The consumer price index is based on the year 1967. If 1967 is designated as year 0 and if 1977 is year 10, what does year -2 represent?

39. If -10, -20, -30, and -50 denote financial losses (in dollars), what does $|-50|$ represent?

40. Two objects are moving along a line in opposite directions. The respective velocities are 20 m/s and -30 m/s. What does $|-30|$ m/s represent?

3.2 Addition

In the last section we discussed the meaning of signed numbers. In this section we study the first of the four fundamental operations: **addition** of signed numbers. We need to consider two cases:

1. Addition of two numbers with like signs.
2. Addition of two numbers with unlike signs.

Like Signs

If two numbers have like signs, they are either both positive or both negative.

Addition of two positive numbers is ordinary arithmetic; the sum can be found in the usual way. For example,

$$(+3) + (+4) = +7 \qquad \text{or} \qquad 3 + 4 = 7$$

If both numbers are negative, we simply think of each number as a debt. Suppose you owe your brother \$4 and your friend \$10. Then your total debt is \$4 + \$10 = \$14. If we represent the debts as negative numbers, then the addition looks like this:

$$(-4) + (-10) = -14$$

The addition $(-4) + (-10) = -14$ can be shown on the number line. Consider the positive and negative directions shown in Figure 3.6. Note that we have 4 units to the left plus 10 units to the left equals 14 units to the left.

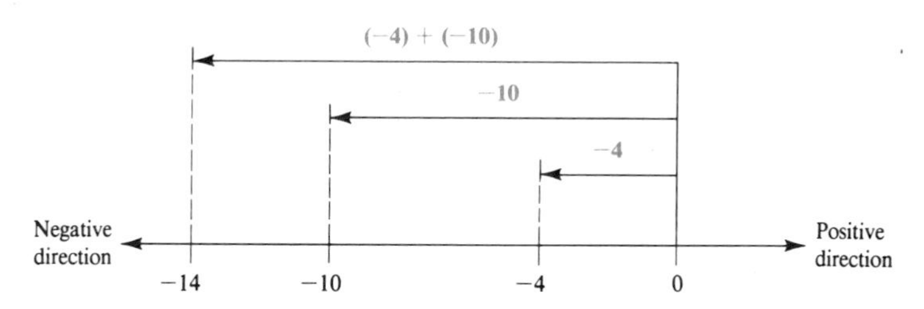

Figure 3.6

To add two numbers with like signs, then, we first ignore the signs and add the numerical values. Then we affix the common sign to the result.

Unlike Signs

To add two numbers with unlike signs, think of the negative number as a loss and the positive number as a gain. Thus $(-10) + (+4)$ combines a loss of 10 with a gain of 4. The loss of 10 is partially offset by a gain of 4, resulting in a loss of 6, represented by -6. So

$$(-10) + (+4) = -6 \qquad \text{or} \qquad (-10) + (4) = -6$$

On the number line (Figure 3.7), we have 10 units to the left plus 4 units to the right equals 6 units to the left.

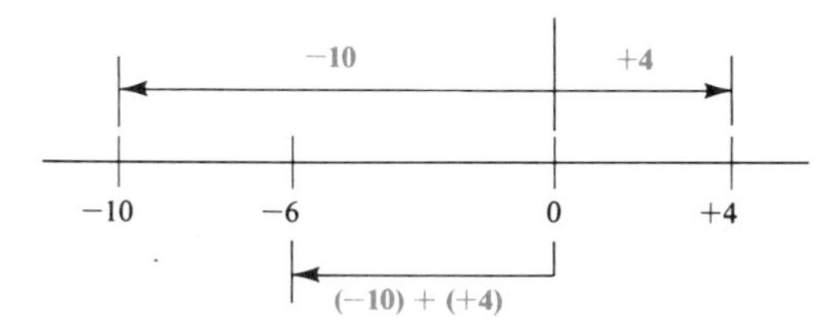

Figure 3.7

Now consider the sum

$$(+12) + (-5)$$

A gain of 12 and a loss of 5 results in a gain of 7. So

$$(+12) + (-5) = +7 \qquad \text{or} \qquad (12) + (-5) = 7$$

On the number line (Figure 3.8) we have 12 units to the right plus 5 units to the left equals 7 units to the right.

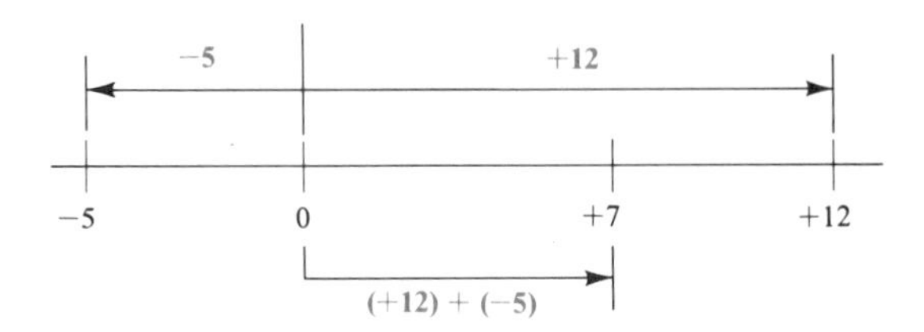

Figure 3.8

In either case, to add two numbers with unlike signs, we subtract the smaller absolute value from the larger absolute value and affix the sign of the larger absolute value to the result.

The General Rule

Let us now summarize the rules for addition of signed numbers.

> **Addition of signed numbers**
>
> 1. To add two numbers with like signs, add their absolute values and affix their common sign to the result.
> 2. To add two numbers with unlike signs, subtract the smaller absolute value from the larger absolute value and affix to the result the sign of the number having the larger absolute value.

We will now illustrate the rule for addition with several examples.

Example 1 Perform the following additions:

a. $(+6) + (+10)$ **b.** $(-3) + (-9)$

Solution. a. Since both numbers are positive, the sum is merely the arithmetic sum. In fact, $(+6) + (+10)$ can be written simply as

$$6 + 10 = 16$$

b. To find the sum $(-3) + (-9)$, we first add the absolute values of the numbers:

$$3 + 9 = 12$$

Affixing the common negative sign, we get

$$(-3) + (-9) = -12$$

The steps in this addition can be carried out more quickly as follows:

$$(-3) + (-9) = \text{the negative of } (3 + 9)$$

or

$$(-3) + (-9) = -(3 + 9) \\ = -(12) = -12$$

The parentheses in $-(3 + 9)$ indicate that the addition inside the parentheses must be carried out first.

Example 2

$$(-5) + (-8) + (-14) = -(5 + 8 + 14) = -27$$

$$(-15) + (-1) + (-3) + (-7) = -(15 + 1 + 3 + 7) = -26$$

The next example illustrates addition of two numbers with unlike signs.

Example 3 Add: $(-17) + 9$.

Solution. Note that $|-17| = 17$ and $|9| = 9$. By the rule for addition of numbers with unlike signs, we subtract 9 from 17 and affix the sign of the number with the larger absolute value, -17. Since $17 - 9 = 8$, we get

$$(-17) + 9 = -8$$

Example 4

$(-20) + (5) = -15$ Since $20 > 5$

$(30) + (-10) = 20$ Since $30 > 10$

$(-50) + (20) = -30$ Since $50 > 20$

$(-22.5) + (10.2) = -12.3$ Since $22.5 > 10.2$

If several signed numbers are to be added, proceed as follows:

1. Add all positive numbers.
2. Add all negative numbers.
3. Add the resulting sums.

To justify this rule, as well as the other rules in this chapter, it is important to note that the commutative, associative, and distributive laws are valid for signed numbers. For example, $(-3) + (-4) = (-4) + (-3)$ by the commutative law.

Example 5 Perform the following addition:

$$(-3) + (-4) + (3) + (-6) + (4) + (-5)$$

Solution. By the commutative law, we may group the positive and negative numbers together:

$$\begin{aligned}(-3) + (-4) + (-6) + (-5) + (3) + (4) \\ &= -(3 + 4 + 6 + 5) + (3 + 4) \\ &= -(18) + (7) \\ &= -11\end{aligned}$$

So far we have placed parentheses around every number in a sum. However, for positive numbers parentheses are really unnecessary and are usually left out. For example,

$$(-4) + (6) \quad \text{is usually written} \quad (-4) + 6$$

We may also leave out the parentheses around the first number in a sum. Thus $(-4) + 6$ can be written

$$-4 + 6$$

Similarly, the sum

$$(-3) + (9) + (-6) + (2)$$

can be written as

$$-3 + 9 + (-6) + 2$$

Note, however, that the parentheses around -6 must be kept. Otherwise we get two signs in succession: $+ \ -6$. This is not a customary notation.

Common error Leaving out parentheses when adding a negative number. The sum of 2 and -3 should *not* be written as $2 + -3$. The correct notation is

$$2 + (-3)$$

Example 6 Perform the following addition:

$$-6 + 4 + (-8) + 7 + 6 + (-10) + 0$$

Solution. As before, we group the positive and negative numbers together to get

$$\begin{aligned}-6 + (-8) + (-10) + 4 + 7 + 6 + 0 \\ &= -(6 + 8 + 10) + (4 + 7 + 6) + 0 \\ &= -24 + 17 + 0 = -7 + 0 = -7 \qquad a + 0 = a\end{aligned}$$

Example 7 Perform the following addition:

$$(-2.5) + 1.7 + 3.9 + (-6.4) + (-5.3) + 9.9$$

Solution. Grouping positive and negative numbers together, we get

$$\begin{aligned}(-2.5) + (-6.4) + (-5.3) + 1.7 + 3.9 + 9.9 \\ = -(2.5 + 6.4 + 5.3) + (1.7 + 3.9 + 9.9) \\ = -14.2 + 15.5 = 1.3\end{aligned}$$

Example 8 The forces acting on the weight in Figure 3.9 are considered positive if they act to the right and negative if they act to the left. Determine the net force on the weight.

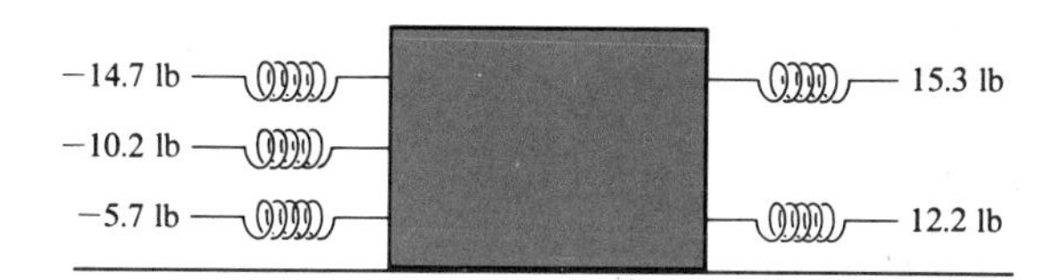

Figure 3.9

Solution. The net force (with proper sign) is the algebraic sum of all the forces:

$$\begin{aligned}\text{net force} &= (-14.7) + (-10.2) + (-5.7) + 15.3 + 12.2 \\ &= -(14.7 + 10.2 + 5.7) + (15.3 + 12.2) \\ &= -30.6 + 27.5 = -3.1 \text{ lb}\end{aligned}$$

We conclude that the net force is 3.1 lb to the left.

Exercises / Section 3.2

In Exercises 1–32, perform the indicated additions.

1. $(5) + (-2)$

2. $(-7) + (4)$

3. $(-10) + 7$

4. $6 + (-3)$

5. $(-3) + 4 + (-6)$

6. $(-7) + 11 + (-5)$

7. $(-4) + 6 + (-2)$

8. $(-8) + 5 + (3)$

9. $(-12) + (-3) + 0 + 4$

10. $3 + (-15) + 0 + 6$

11. $9 + (-7) + (-6) + 13 - 0$

12. $(-2) + 1 + (-13) + 12 - 0$

13. $-4 + 6 + (-7) + (-20) + 13$

14. $-11 + 5 + (-2) + (-18) + 20$

15. $-21 + (-11) + 14 + 25 + (-2)$

16. $(-25) + (-4) + 18 + (-2) + 20$

17. $25 + (-50) + 4 + (-8) + (-3) + 0$

18. $-45 + (-5) + 25 + (-1) + 7$

19. $(-22) + 2 + (-5) + 27 + (-3) + 1$

20. $-17 + (-13) + 20 + (-2) + 40 + (-15)$

21. $3 + (-2) + (-7) + 25 + 2 + (-30)$

22. $7 + (-5) + (-12) + 10 + 3 + (-1) + (-7)$

23. $(-1) + 2 + (-4) + 3 + (-15) + 2 + (-3)$

24. $17 + (-4) + (-3) + (-19) + 8 + 3$

25. $-6 + (-16) + 14 + (-1) + 23 + 4$

26. $-28 + (-3) + 35 + 2 + (-8) + (-6)$

27. $(-14) + (-3) + 26 + (-4) + (-19) + 2$

28. $36 + (-26) + (-4) + (-4) + 17 + (-6)$

29. $(-2.5) + (-7.9) + 1.4 + (-1.1) + 0.2$

30. $(-10.1) + (-8.3) + 4.7 + 1.0 + (-0.4)$

31. $4.86 + (-2.31) + (-5.37) + 0.46$

32. $3.52 + (-2.16) + (-1.94) + 2.40 + (-0.70)$

33. If the temperature outside is -10°C, and it increases by 4°C, what is the final temperature?

34. If the temperature inside a freezer is -5°F, and it increases by 12°F, what is the new temperature?

35. Bill owes his brother \$20.00, but Jim and George owe Bill \$5.25 and \$8.75, respectively. Use addition of signed numbers to determine how much Bill owes.

36. Four forces are acting on an object in two opposite directions. If the forces are 10.46 lb, -7.62 lb, 8.37 lb, and -9.92 lb, determine the net force on the object. (See Example 8.)

37. The value of a certain stock shows the following gains (or losses) in cents during a 5-day period: -23, -12, $+25$, $+19$, and -4. Determine the gain (or loss) during this period.

38. Figure 3.10 shows part of a circuit. By *Kirchhoff's current law,* the algebraic sum of the currents through A is 0. Assume that the currents toward A are positive and away from A negative. Given that $|I_1| = 1.5$ A and $|I_2| = 0.7$ A, find I_3.

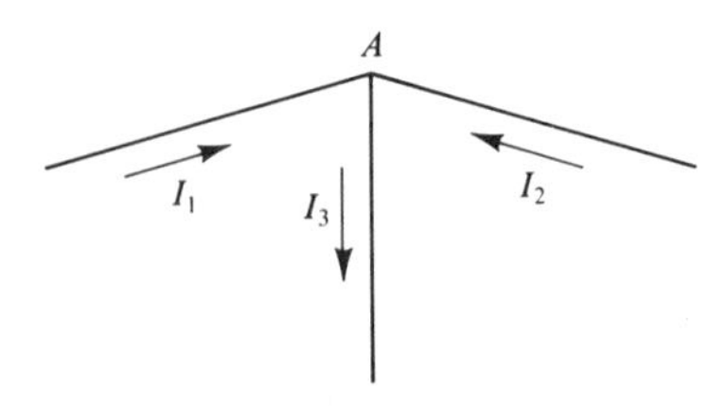

Figure 3.10

39. In Figure 3.10, find I_2, given that $|I_1| = 1.3$ A and $|I_3| = 2.7$ A. (Refer to Exercise 38.)

3.3 Subtraction

In the last section we studied addition of signed numbers. In this section we discuss **subtraction.**

Subtraction is the opposite of addition in this sense: If we subtract one number from another, the result is a number which, when added to the subtrahend (the number after the subtraction symbol), gives us the first number. For example,

$$8 - 3 = 5 \quad \text{because} \quad 5 + 3 = 8$$

The same principle holds for signed numbers. For example,

$$8 - (-3) = 11 \quad \text{because} \quad 11 + (-3) = 8$$

Similarly

$$-6 - (-10) = 4 \quad \text{because} \quad 4 + (-10) = -6$$

Note that in all cases the subtraction is carried out by changing the sign of the subtrahend and adding. In particular

Change sign

$$8 - 3 = 8 - (+3) = 8 + (-3) = 5$$

Add

Change sign

$$8 - (-3) = 8 + (+3) = 11$$

Add

and

Change sign

$$-6 - (-10) = -6 + (+10) = 4$$

Add

The general rule is given next.

Subtraction

To subtract one number from another, change the sign of the subtrahend (the number after the subtraction symbol), and proceed as in addition. In symbols:

$$a - b = a + (-b)$$

Example 1 Perform the indicated subtractions:

a. $4 - (-3)$ **b.** $-10 - (-15)$

Solution. a. Note that the subtrahend is -3. If we change the sign of the subtrahend, we get $+3$. This number is added to 4, for a total of 7:

$$4 - (-3) = 4 + (+3) = 4 + 3 = 7$$

b. The subtrahend is -15. After changing the sign, we obtain $+15$. Proceeding as in addition, we have

$$-10 - (-15) = -10 + (+15) = -10 + 15 = 5$$

We noted in the last section that whenever we add a positive number, the parentheses around the number should be left out. Similarly, whenever we subtract a positive number, we leave out the parentheses. Thus $-4 - (3)$ is written $-4 - 3$.

Example 2 **a.** $20 - 6 = 20 + (-6) = 14$

b. $-7 + 4 = -3$ By the rule for addition

c. $-7 - 4 = -7 + (-4) = -11$

d. $-10 + 5 = -5$ By the rule for addition

Suppose we look at the subtractions in Example 2 in another way. By the rule for subtraction,

$$20 - 6 = 20 + (-6) = 14$$

and

$$-7 - 4 = -7 + (-4) = -11$$

Observe that in both cases, **subtracting a positive number is the same as adding the corresponding negative number.**

This rule is particularly useful if several numbers are combined by addition or subtraction. Since subtracting a positive number is the same as adding the corresponding negative number, the numbers can be rearranged so that all positive and all negative numbers may be combined separately. For example,

$$\begin{aligned} &-4 + 8 - 20 + 25 - 10 \\ &\quad = -4 + 8 + (-20) + 25 + (-10) && \text{Rule for subtraction} \\ &\quad = -4 + (-20) + (-10) + (8 + 25) && \text{Rearranging} \\ &\quad = -34 + 33 \\ &\quad = -1 \end{aligned}$$

These operations are easier to perform if the numbers are rearranged right away:

$$\begin{aligned} \mathbf{-4} + 8 \mathbf{- 20} + 25 \mathbf{- 10} &= \mathbf{-4 - 20 - 10} + 8 + 25 \\ &= -34 + 33 = -1 \end{aligned}$$

Here we used the fact that $-4 - 20 - 10$ is equal to the sum of -4, -20, and -10.

Example 3 Perform the indicated operations:

$$5 - 10 + 6 + 2 - 20 - 3$$

Solution. We first arrange the numbers so that all negative and all positive numbers are grouped together. Thus

$$\begin{aligned} &5 - 10 + 6 + 2 - 20 - 3 \\ &\quad = 5 + 6 + 2 - 10 - 20 - 3 && \text{Rearranging} \\ &\quad = 13 - 33 && \text{Combining all positive} \\ &\quad = -20 && \text{and all negative numbers} \end{aligned}$$

(We used the fact that $-10 - 20 - 3$ is equal to the sum of -10, -20, and -3.)

Example 4 Combine the following numbers:

$$-8 + 3 - 15 + 9 - 20 + 5$$

Solution.

$$\begin{aligned} &-8 + 3 - 15 + 9 - 20 + 5 \\ &\quad = -8 - 15 - 20 + 3 + 9 + 5 && \text{Rearranging} \\ &\quad = -43 + 17 && \text{Combining all positive} \\ &\quad = -26 && \text{and all negative numbers} \end{aligned}$$

Subtraction of a negative number leads to addition of the corresponding positive number. For example,

$$\begin{aligned} 4 - (-3) &= 4 + (+3) && \text{Rule for subtraction} \\ &= 4 + 3 = 7 \end{aligned}$$

In other words, $-(-3)$ becomes $+3$. In general,

$$-(-a) = +a$$

Example 5 Combine the following numbers:

$$5 - 4 - (-6) + 7 - 2 + 8 + 2 - (-10)$$

Solution. This problem involves the subtraction of two negative numbers. So we change $-(-6)$ to $+6$ and $-(-10)$ to $+10$ and proceed as in Examples 3 and 4:

$$\begin{aligned} &5 - 4 - (-6) + 7 - 2 + 8 + 2 - (-10) \\ &\quad = 5 - 4 + 6 + 7 - 2 + 8 + 2 + 10 && -(-6) = 6,\ -(-10) = 10 \\ &\quad = 5 + 6 + 7 + 8 + 2 + 10 - 4 - 2 && \text{Rearranging} \\ &\quad = 38 - 6 && \text{Combining all positive} \\ &\quad = 32 && \text{and all negative numbers} \end{aligned}$$

Example 6 Combine the following numbers:

$$-20 - (-11) + 4 - (-7) - 0 - (-15)$$

Solution.

$$\begin{aligned} & -20 - (\mathbf{-11}) + 4 - (\mathbf{-7}) - 0 - (\mathbf{-15}) \\ &= -20 + \mathbf{11} + 4 + \mathbf{7} + \mathbf{15} \qquad a - 0 = a \\ &= -20 + 37 = 17 \end{aligned}$$

Example 7 If the temperature outside is $-10°C$ and the temperature inside is $20°C$, then the difference in temperatures is

$$20°C - (-10°C) = 20°C + 10°C = 30°C$$

Example 8 Determine the difference between the boiling point of oxygen, $-297.4°F$, and the melting point of oxygen, $-361.1°F$.

Solution. As in ordinary arithmetic, the difference is found by subtracting the smaller number from the larger:

$$-297.4 - (-361.1) = -297.4 + 361.1 = 63.7°F$$

Exercises / Section 3.3

In Exercises 1–26, perform the indicated operations.

1. $-10 - 4$
2. $-16 - 12$
3. $5 - (-4)$
4. $9 - (-6)$
5. $17 - 20$
6. $21 - 30$
7. $-2 - 3 + 4 - 6 + 0$
8. $4 - 7 - 11 + 15 - 0$
9. $-3 - 8 - 15 + 2$
10. $4 - 0 - 3 - 20 + 9$
11. $25 - 16 - 12 + 14 - 5 - 0$
12. $-17 + 20 - 4 - 7 - 3$
13. $5 - (-3) - 3 - (-7)$
14. $6 - (-4) - (-10) - 2$
15. $-12 + (-3) - (-4) + 4 - 0$
16. $-20 + (-16) - (-5) + 25 - 0$
17. $-6 - (-3) - 4 - 7 + 6$
18. $19 - (-10) - 12 - 14 + 12$
19. $17 - (-5) - 5 + 0 - 17$
20. $14 - 10 - (-4) - (-3) - 6 + 11$
21. $2.3 - (-6.4) - 1.4 + 3.7$
22. $-7.9 - 8.6 - (-0.7) + 4.2$
23. $3.4 - (-7.2) - 5.9 + 1.7$
24. $10.2 - (-0.4) - 4.7 - 8.4$
25. $\frac{1}{2} - \frac{1}{4} - \left(-\frac{1}{3}\right)$
26. $\frac{1}{6} - \left(-\frac{1}{12}\right) + \frac{3}{4} - \frac{1}{4}$
27. If the voltage across an element with respect to the ground is -25.0 V and it changes to -11.0 V, what is the absolute value of the change in the voltage?

28. If the temperature outside is $-15°C$ and it decreases by $6°C$, what is the final temperature?

29. The pressure in a tank increases by 10 lb/in.^{2}, decreases by 12 lb/in.^{2}, then increases again by 5 lb/in.^{2}, and finally decreases by 8 lb/in.^{2} Determine the net increase or decrease in the pressure.

30. The top of Mt. McKinley, 20,320 ft above sea level, is the highest point in the United States, while Death Valley is the lowest point, 282 ft below sea level. Write the altitudes as signed numbers and find the difference between them.

31. If an object is moving along a line, the velocity in one direction is commonly designated as positive and that in the opposite direction is designated as negative. If two objects move along a line with respective velocities of 40 ft/s and -20 ft/s, find the absolute value of the difference between the velocities.

32. Referring to Exercise 31, if the respective velocities are -40 ft/s and -20 ft/s, find the absolute value of the difference between the velocities. Explain the results.

33. If the distances above the ground are considered positive and the distances below the ground negative, use signed numbers to write an expression for the distance from a point 100 ft above the ground to a point 50 ft below the ground. (The points are in the same vertical line.)

3.4 Multiplication and Division

In this section we discuss the remaining operations with signed numbers: multiplication and division.

Multiplication

Recall that besides the symbol $\times$ for multiplication, we may also use a dot or parentheses. For example, both $4 \cdot 5$ and $(4)(5)$ mean "4 times 5." These conventions will be used most of the time from now on.

To understand the rules for multiplication of signed numbers, let us recall that multiplication is actually repeated addition. For example,

$$3 \cdot 5 = 5 + 5 + 5 = 15$$

Similarly,

$$3 \cdot (-5) = (-5) + (-5) + (-5) = -15$$

Since multiplication of signed numbers is commutative, we also have

$$(-5)(3) = 3(-5) = -15$$

Based on these observations, the product of a negative number and a positive number is negative. The product of two negative numbers (as well as two positive numbers) is positive, as we will see shortly. First let us state the rule for multiplication.

Multiplication

1. The product of two numbers with **like signs** is positive.
2. The product of two numbers with **unlike signs** is negative.

Example 1

$5(-7) = -35$ Since 5 and -7 have unlike signs

$-8(10) = -80$ Since -8 and 10 have unlike signs

$-12 \cdot 4 = -48$ Unlike signs

$6 \cdot (-7) = -42$ Unlike signs

$(-3)(-9) = 27$ Since -3 and -9 have like signs

$3 \cdot 9 = 27$ Since 3 and 9 have like signs

To see why the product of two negative numbers is positive, we need to recall that the laws of positive numbers are also valid for signed numbers. For example,

$$-3(-4 + 4) = (-3)(-4) + (-3)(4)$$

by the distributive law.

Now consider the product of two negative numbers, say

$$(-3)(-4)$$

First observe that

$$(-3)(-4) + (-3)(4) = -3(-4 + 4) \qquad \text{Distributive law}$$
$$= -3 \cdot 0 = 0 \qquad a \cdot 0 = 0 \text{ for all } a$$

So

$$(-3)(-4) + (-3)(4) = 0$$

and

$$(-3)(-4) + (-12) = 0 \qquad (-3)(4) = -12$$

This is possible only if

$$(-3)(-4) = 12 \qquad 12 + (-12) = 0$$

We conclude that the product of two negative numbers is positive.

Example 2

$(-4)(-5) = 20$ -4 and -5 have like signs

$(-10)(-15) = 150$

$(3)(8) = 24$

$(-20)(0) = 0$

If several numbers are multiplied, we perform the multiplication in pairs, as shown in the next example.

Example 3

a. $(-4)(-5)(6) = (20)(6) = 120$ — $(-4)(-5) = 20$

b. $(-7)(2)(-1)(-4) = (-7)(-1)(-4)(2)$ — Rearranging

$= (-7)(-1)(-4)(2)$

$= (7)(-4)(2)$ — $(-7)(-1) = 7$

$= (7)(-4)(2)$

$= (-28)(2)$ — $(7)(-4) = -28$

$= -56$

Example 3 shows that if a product contains two negative factors, the result is positive. If the product contains three negative factors, the result is negative. The general rule is given next.

1. If a product contains an even number of negative factors, then the product is positive.
2. If a product contains an odd number of negative factors, then the product is negative.
3. The product of positive numbers is positive.

Example 4

a. $(-4)(-5)(6)(-10) = -(4)(5)(6)(10)$ — Odd number of negatives

$= -1200$

b. $(-6)(-3)(-4)(-5) = +(6)(3)(4)(5)$ — Even number of negatives

$= 360$

c. $(5)(4)(20) = 400$ — Positive numbers

The principle illustrated in Example 4 also applies to powers. Recall from Section 1.11 that an exponent indicates the number of times that a given number is a factor. For example,

$$3^3 = 3 \cdot 3 \cdot 3 = 27$$

A negative number can also be raised to a power. For example,

$$(-2)^4 = (-2)(-2)(-2)(-2) = 16 \quad \text{Even number of negatives}$$

and

$$(-2)^3 = (-2)(-2)(-2) = -8 \quad \text{Odd number of negatives}$$

We obtain the following rule:

A negative number raised to an even power is positive.
A negative number raised to an odd power is negative.

Example 5

$(-3)^4 = 81$ Even power
$(-5)^3 = -125$ Odd power
$(-1)^6 = 1$ Even power
$(-1)^7 = -1$ Odd power

Division

As far as the signs are concerned, the rule for division is the same as the rule for multiplication. The reason is that

$$\frac{a}{b} \text{ is the same as } a \cdot \frac{1}{b}$$

Division

1. The quotient of two numbers with like signs is positive.
2. The quotient of two numbers with unlike signs is negative.

Example 6

$\frac{-25}{5} = -5$ Unlike signs

$\frac{-30}{-6} = 5$ Like signs

$\frac{15}{5} = 3$ Like signs

Since multiplication and division follow the same rule, combinations of multiplication and division can be performed by first determining the sign and then the numerical value.

Example 7 Perform the indicated operations:

$$\frac{(-7)(-4)(3)}{-12}$$

Solution. This problem is a combination of multiplication and division. Since we have three negatives (an odd number), we know that the answer is negative. By placing the minus sign in front, we can obtain the numerical value by cancellation:

$$\frac{(-7)(-4)(3)}{-12} = -\frac{(7)(\not{4})(\not{3})}{\not{12}} = -7$$

So the answer is -7.

Example 8 Perform the indicated operations:

$$\frac{(-8)(-4)(3)(5)}{(-28)(24)(-15)}$$

Solution. Since the total number of negatives is even (four), we know that the answer is positive. So we place the plus sign in front and obtain the numerical value by cancellation:

$$\frac{(-8)(-4)(3)(5)}{(-28)(24)(-15)} = +\frac{(\overset{1}{\not{8}})(\overset{1}{\not{4}})(\overset{1}{\not{3}})(\overset{1}{\not{5}})}{(\underset{7}{\not{28}})(\underset{\not{8}}{\not{24}})(\underset{3}{\not{15}})} = \frac{1}{7 \cdot 3} = \frac{1}{21}$$

Example 9 A laboratory technician takes a number of measurements to determine the freezing point of a saturated solution of common salt in water. He obtains the following measurements: $-6°F$, $-4°F$, $0°F$, $1°F$, $-5°F$, $-5°F$, and $-9°F$. What is the average value of these measurements?

Solution. The average of a set of signed numbers is found the same way as the average of a set of positive numbers: We add all the measurements and divide by the total number of measurements. So the average temperature in degrees Fahrenheit is

$$\frac{(-6) + (-4) + 0 + 1 + (-5) + (-5) + (-9)}{7} = -4°F$$

Exercises / Section 3.4

In Exercises 1–54, perform the indicated operations.

1. $(-8)(-9)$
2. $(-7)(4)$
3. $(-15)(3)$
4. $(6)(-11)$
5. $(3)(-1)(0)$
6. $(-4)(0)(-6)$
7. $(4)(-2)(-8)$
8. $(3)(-1)(7)(10)$
9. $(-4)(5)(-2)(6)$
10. $(-10)(3)(4)(-12)$
11. $(-3)(-6)(2)(-1)$
12. $(-6)(7)(-2)(3)$
13. $(-9)(4)(-6)(2)$
14. $(-5)(4)(-6)(-7)$
15. $(-2)(-4)(-8)(-3)$

16. $(-4)(-6)(-10)(-2)$

17. $(-5)(-7)(4)(-12)(3)$

18. $(-13)(4)(-9)(-8)(6)$

19. $(-20)(5)(-6)(-4)(-2)$

20. $(-3)(-15)(-7)(-8)(-1)$

21. $\frac{-9}{3}$

22. $\frac{-25}{5}$

23. $\frac{45}{-15}$

24. $\frac{-60}{-12}$

25. $\frac{-9}{-15}$

26. $\frac{-14}{-21}$

27. $\frac{(-27)(2)}{-9}$

28. $\frac{(-3)(36)}{-12}$

29. $\frac{(16)(-3)}{(6)(-4)}$

30. $\frac{(45)(-3)}{(-9)(-15)}$

31. $\frac{(-8)(-12)(15)}{-10}$

32. $\frac{(-9)(6)(-15)}{(5)(2)(-21)}$

33. $\frac{(-7)(77)(-45)}{(14)(-9)(-11)}$

34. $\frac{(-63)(60)(16)}{(-56)(-7)(-12)}$

35. $\frac{(-48)(33)(-7)}{(22)(21)(-36)}$

36. $\frac{(-13)(4)(-5)(-8)}{(-32)(-39)(16)}$

37. $\frac{(12)(-7)(6)(-2)}{(15)(18)(-21)(-16)}$

38. $\frac{(22)(-5)(-24)(7)}{(-16)(4)(-6)(55)}$

39. $(-2)^2$

40. $(-2)^5$

41. $(-2)^3(-3)^2$

42. $(-9)^2(-2)^2$

43. $(-5)^3(-2)^2$

44. $(-4)^2(-2)^2$

45. $(-4)^2(-2)^4$

46. $(-3)^3(-2)^2$

47. $(-3)^2(-2)^4$

48. $(-2)^4(-5)^2$

49. $(-1)^5$

50. $(-1)^9$

51. $(-1)^8$

52. $(-1)^4$

53. $(-1)^{11}$

54. $(-1)^{15}$

55. The freezing point of mercury is about $-38°C$ and that of ethyl alcohol is three times as low. What is the freezing point of ethyl alcohol?

56. The current in a certain circuit is -2.73 A. This is three times the current measured earlier. What was the current earlier?

57. In determining the boiling point of oxygen, the following measurements were taken: $-183.4°C$, $-183.1°C$, $-179.9°C$, $-179.8°C$, and $-179.8°C$. What is the average of these measurements?

58. Forces acting in opposite directions are assumed to have opposite signs. What force is equal to one-half of the combined forces of -73.4 lb and $+36.8$ lb?

59. The resistance in a wire increases by 0.1 Ω for every increase in temperature of 5°F. What is the increase in the resistance if the temperature increases from $-80°F$ to $-50°F$?

60. If distances above the ground are considered positive and below the ground negative, how is a distance of 40 ft below the ground represented? What about $\frac{5}{8}$ of this distance?

61. The voltage across a resistor is obtained by multiplying the current by the resistance. Referring to Figure 3.11, consider the clockwise current positive and the counterclockwise current negative. If the absolute value of the current is 1.5 A, what are the two voltages?

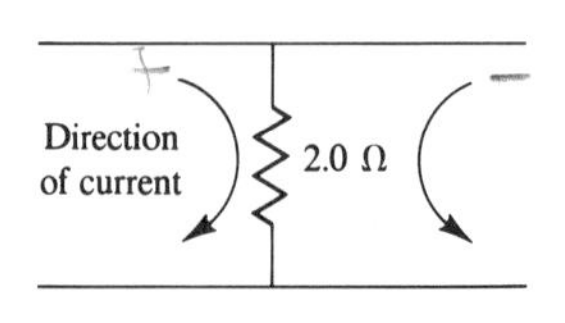

Figure 3.11

3.5 Order of Operations

We saw in Chapter 1 that whenever a problem involves several operations, the quantities must be evaluated in the following order:

1. Evaluate all quantities enclosed in parentheses (or other grouping symbols).
2. Evaluate powers.
3. Perform all multiplications and divisions in order, from left to right.
4. Perform all additions and subtractions in order, from left to right.

The same order of operations holds for signed numbers, as illustrated by several examples.

Example 1 Evaluate $-4 + 3(-2)$.

Solution. The multiplication must be performed before the addition. So

$$\begin{aligned} -4 + 3(-2) &= -4 + (-6) && \text{Multiplication performed first} \\ &= -10 \end{aligned}$$

Example 2 Evaluate $\frac{-4}{-2} - (-2)(5)$.

Solution.

$$\begin{aligned} \frac{-4}{-2} - (-2)(5) &= 2 - (-10) && \text{Division and multiplication performed first} \\ &= 2 + (+10) = 12 \end{aligned}$$

Examples 1 and 2 involve only the four fundamental operations. The next example involves a power.

Example 3 Evaluate $(-2)(6) - 3(-4)^2$.

Solution. In this problem the power, $(-4)^2$, must be evaluated first:

$$\begin{aligned} (-2)(6) - 3(-4)^2 &= (-2)(6) - 3(16) && (-4)^2 = (-4)(-4) = 16 \\ &= -12 - 48 \\ &= -12 + (-48) = -60 \end{aligned}$$

The next example involves the subtraction of two numbers enclosed in parentheses.

Example 4 Evaluate $(2 - 4)(6) + \frac{8}{-2}$.

Solution. The quantity enclosed in parentheses must be evaluated first:

$$\begin{aligned} (2-4)(6) + \frac{8}{-2} &= (-2)(6) + \frac{8}{-2} && 2 - 4 = -2 \\ &= -12 + (-4) && (-2)(6) = -12;\ \frac{8}{-2} = -4 \\ &= -16 \end{aligned}$$

The fraction bar can also be a symbol of grouping, as shown in the next example.

Example 5 Evaluate

$$\frac{25-7}{9} + (-3)^2$$

Solution. Here the subtraction operation above the bar must be performed first:

$$\begin{aligned} \frac{25-7}{9} + (-3)^2 &= \frac{18}{9} + (-3)^2 \\ &= 2 + 9 && \frac{18}{9} = 2;\ (-3)^2 = 9 \\ &= 11 \end{aligned}$$

Exercises / Section 3.5

Evaluate the given quantities.

1. $2 + (-3)(4)$

2. $-4 - (3)(-2)$

3. $7 - \frac{12}{-4}$

4. $5 + \frac{-6}{-2}$

5. $(-2)(4) + (-3)(4)$

6. $(-4)(-2) + (-2)(6)$

7. $\frac{-10}{5} - (3)(-2)$

8. $\frac{-20}{-4} + (-2)(6)$

9. $\frac{15}{-5} + \frac{-8}{2}$

10. $\frac{-18}{-9} - \frac{12}{3}$

11. $1 + 2(-3)^2$

12. $2 - 4(2)^3$

13. $-3 - 4(-2)^3$

14. $-4 - 2(-3)^2$

15. $2 - \frac{-18}{6} - 8$

16. $4 + \frac{25}{-5} - 7$

17. $(-3)(-6) - 4(-2)^2$

18. $(-2)(-8) + 2(-4)^2$

19. $(-1)(-4) + 2 - (3)(0)$

20. $4 - (-1)(6) + 0(-7)$

21. $3(-2 - 1) + 2(-7)$

22. $2(7 - 2) + (4)(-1)$

23. $7 - 2(6 - 10)$

24. $8 + (-3)(-1 + 4)$

25. $3^2 + 2(-1 + 10)$

26. $(-2)^2 - 3(2 + 4)$

27. $\dfrac{9 + 3}{-4} + (-2)^3$

28. $\dfrac{10 + 5}{-3} + (-3)^2$

29. $2(-4)^2 - \dfrac{2 + 6}{-2}$

30. $3(-2)^2 + \dfrac{-2 - 8}{-2}$

3.6 Calculator Operations

Scientific calculators are programmed to perform the four fundamental operations with signed numbers.

To enter a negative number, enter the absolute value of the number and press the **change of sign key** [+/−]. The change of sign key changes the sign of the number displayed. (Every number, when entered, is assumed to be positive.)

Example 1 Multiply $(-3)(-7)(-9)(4)$.

Solution. The sequence is

3 [+/−] [×] 7 [+/−] [×] 9 [+/−] [×] 4 [=] → −756

As we saw in Chapter 1, combinations of the four fundamental operations can be performed easily since scientific calculators perform multiplication and division before addition and subtraction.

Example 2 Evaluate

$$8.36 - \frac{-7.21}{(3.63)(-1.02)} - (-5.00)$$

Solution. The sequence is

8.36 [−] 7.21 [+/−] [÷] 3.63 [÷] 1.02 [+/−] [−] 5.00 [+/−] [=]
→ 11.41272

Rounding off to three significant digits, we get 11.4.

To find the power of a number, use the key [y^x] or [x^y]. For example, to find 2^4, use the sequence

2 [y^x] 4 [=] → 16

Example 3 Evaluate $3^7 - 2000$.

Solution. The sequence is

$$3 \boxed{y^x} \boxed{7} \boxed{-} 2000 \boxed{=} \rightarrow 187$$

Caution: When using the key $\boxed{y^x}$, some calculators may display an error if the base is negative. It is therefore necessary to determine the sign of the result and then use the key $\boxed{y^x}$ with the corresponding positive base to obtain the numerical value.

Example 4 Evaluate $(-2)^9$.

Solution. For most calculators the sequence

$$2 \boxed{+/-} \boxed{y^x} 9 \boxed{=}$$

results in an error designation since the base is negative.

In such a case we use the sequence

$$2 \boxed{y^x} 9 \boxed{=} \rightarrow 512$$

to obtain the numerical value. Then, since $(-2)^9$ has an odd exponent, the result is negative, and it follows that

$$(-2)^9 = -512$$

Exercises / Section 3.6

In Exercises 1–28, use a calculator to perform the indicated operations. Round off the answers to the proper number of significant digits.

1. $2.1 - (-4.6)$

2. $-7.6 - (-1.3)$

3. $(-2.76)(4.83)$

4. $(-10.41)(-2.006)$

5. $\dfrac{-1.975}{4.694}$

6. $\dfrac{-11.6}{8.58}$

7. $67.6 - (-1.62)(9.35)$

8. $126.9 - (55.41)(-5.70)$

9. $0.9872 - \dfrac{-0.9354}{0.1059}$

10. $-9.49 + \dfrac{-2.391}{1.19}$

11. $-0.606 - \dfrac{0.1654}{-0.0184} + 0.3867$

12. $8.091 - \dfrac{9.866}{1.957} + 2.446$

13. $\dfrac{(-3.40)(0.0393)}{(-8.86)(-7.43)}$

14. $\dfrac{(46.47)(-17.60)}{0.8223}$

15. $\dfrac{92.34}{(-9.718)(0.7370)} - 6.383$

16. $\dfrac{(-0.834)(0.0963)}{0.0357} - 0.869$

17. $(-4.53)(9.139) - \frac{0.807}{-0.5312}$

18. $(2.674)(-6.092) - \frac{-0.500}{7.62}$

19. $\frac{2.27 - 1.669}{0.0384 - 0.7688}$

20. $\frac{4.80 - (-5.96)}{2.27 - 9.35}$

21. $\frac{(0.0970)(-0.673)}{0.2855 - 0.631}$

22. $\frac{0.1013 - 8.690}{(0.837)(-2.93)}$

23. $(-2.84)^6$

24. $(-1.94)^8$

25. $(-0.9470)^5$

26. $(-1.034)^7$

27. $(-0.736)^9 + 2$

28. $(-0.826)^7 - 1$

29. The current (in amperes) in a certain circuit is given by

$$\frac{4.32 - (2.67)(-6.25)}{(5.25)(7.62)}$$

Find the current.

30. Find the average of the following temperature readings (in degrees Celsius): $-2.76°$, $-1.84°$, $5.24°$, $1.37°$, $-3.04°$, and $-1.27°$.

31. The velocity (in centimeters per second) of an object moving along a line is $-5.36 - \sqrt{7.29}$. Find the velocity.

32. The tensile strength (in pounds) of a wire at a certain temperature is $473.1 + (-0.1320\sqrt{8.620})$. Determine the tensile strength.

Review Exercises / Chapter 3

In Exercises 1–34, perform the indicated operations. Do not use a calculator.

1. $(-3) + (-7) + 6 - (-2)$

2. $(-4) + (-8) + 5 + 12 - (-4)$

3. $7 + (-6) + (-10) + (-4)$

4. $4 + (-6) + (-12) + 9 + (-7)$

5. $8 - (-5) - (-10) - 4$

6. $-25 + (-15) - (-7) + 20 - 0$

7. $19 - (-7) - 6 - 7 + 10 + 0$

8. $14 - 9 - (-5) - (-4) - 13 + 3$

9. $-2.3 - (-6.1) - 5.2$

10. $-10.3 - 6.2 - (-0.4)$

11. $6.4 - (-1.7) + 6.0 - 8.3$

12. $-2.1 - (-0.4) + 0.9 - 6.4$

13. $\frac{1}{2} - \left(-\frac{1}{3}\right) - \frac{1}{12}$

14. $\frac{1}{5} - \left(-\frac{1}{15}\right) - \frac{1}{10}$

15. $(4)(-1)(-6)(3)$

16. $(16)(-2)(-3)(-4)$

17. $(-7)(-3)(-5)(6)$

18. $(-10)(-6)(-3)(0)$

19. $2 + \frac{(4)(-8)}{-2}$

20. $-4 + \frac{(-25)(4)}{10}$

21. $\frac{-16}{(4)(-8)} - 2$

22. $\frac{-12}{(-4)(-9)} + 3$

23. $2(-3) - 4(-2)$

24. $1 + 2(-4) - 3(-7)$

25. $5(-3) - 6(-7) + 1$

26. $(-6)(-2) - (-3)^2$

27. $\frac{(6)(-2)(-14)}{(-21)(-5)(18)(-8)}$

28. $\frac{(-16)(-33)(9)}{(-15)(48)(-44)}$

29. $\frac{(-39)(25)(-12)}{(-125)(-8)(13)}$

30. $\frac{(26)(-18)(49)(-2)}{(-27)(5)(-7)(-39)}$

31. $(-3)^3(-2)^2$

32. $(-2)^4(-3)^2$

33. $(-4)^2(-3)^2$

34. $(-2)^2 - 3(-1 + 4)$

35. The respective voltages across two resistors, with respect to the ground, are -10.3 V and -27.8 V. What is the absolute value of the difference?

36. Which is larger: -3.4 or 2.1? Which has the larger absolute value: -3.4 A or 2.1 A?

37. If a loss is denoted by a negative number, how must a loss of \$2150 be written? What about $2\frac{1}{2}$ times this loss?

38. The current in a certain circuit changes from -3.05 A to 2.47 A. What is the change in the current?

39. Find the sum of the following currents: -1.0 A, 3.0 A, -2.4 A, -3.8 A, and 4.1 A.

40. If the outside temperature is -10°C, and it decreases by 4°C, what is the final temperature?

41. The highest point on earth is the top of Mount Everest: 29,028 ft above sea level. The lowest point is the Mariana Trench: 36,201 ft below sea level. Write the altitudes as signed numbers and find the difference between them.

42. Four forces are acting on an object, two forces from one direction and two from the opposite direction. The respective forces are -10.25 N, 4.12 N, 6.21 N, and -2.36 N. What single force will have the same effect?

43. The boiling point of hydrogen is 56.97°C higher than the boiling point of nitrogen, which is -195.81°C. What is the boiling point of hydrogen?

44. Suppose the velocity of an object is positive when moving to the right and negative when moving to the left. If the velocity of the object is -5.3 m/s and if the absolute value of the velocity increases at the rate of 2.0 m/s every second, determine the velocity 8 s later. (Assume that the object continues to move to the left.)

45. In Figure 3.12, assume that the currents toward A are positive and away from A negative. The algebraic sum of the currents is 0. Given that $|I_1| = 2.0$ A, $|I_2| = 2.4$ A, and $|I_3| = 1.5$ A, find I_4.

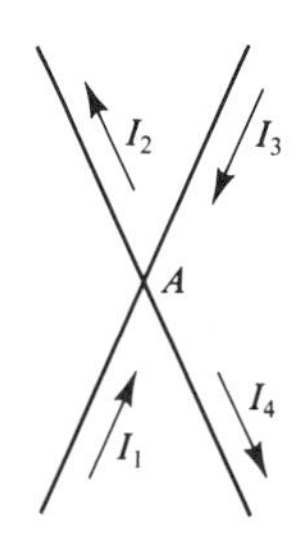

Figure 3.12

46. Referring to Figure 3.13, assume that a clockwise current is positive and a counterclockwise current negative. Given that $|I_1| = 0.80$ A and $|I_2| = 1.3$ A, find the voltage across each resistor (voltage = current × resistance).

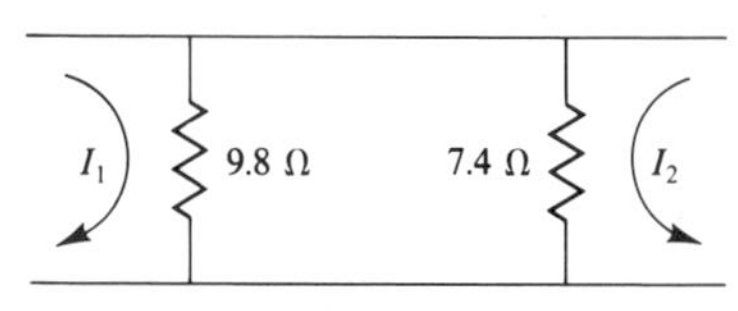

Figure 3.13

Cumulative Review Exercises / Chapters 1–3

In Exercises 1–16, do not use a calculator.

1. Add:

$$\begin{array}{r} 7984 \\ 4706 \\ 1941 \\ +2809 \\ \hline \end{array}$$

2. Subtract:

$$\begin{array}{r} 8202 \\ -2987 \\ \hline \end{array}$$

3. Multiply:

a. 394×738
b. 3.94×73.8

4. Divide:

a. $54{,}416 \div 152$
b. $54.416 \div 1.52$

5. Add:

$10.14 + 2.7394$

In Exercises 6–16, perform the indicated operations.

6. $320 - 5 \times 46$

7. $\frac{168}{252}$ (reduce to lowest terms)

8. $\frac{7}{4} \times \frac{15}{26} \times \frac{32}{21} \times \frac{13}{45}$

9. $5\frac{5}{8} \times 2\frac{6}{25}$

10. $\left(5\frac{5}{9}\right) \div \left(7\frac{1}{3}\right)$

11. $\frac{5}{18} - \frac{7}{12} + \frac{13}{42}$

12. $-10 + 14 - (-6) - (-7)$

13. $(-4)(2)(-1)(-6)$

14. $-2 - (-6) + \frac{(-8)(-2)}{-4}$

15. $2 - \frac{(-16)(21)}{(-3)(-8)} - (-4)$

16. $(-2)^4 \cdot (-3)^3 + (-2)^2$

In the remaining exercises use a calculator whenever convenient.

17. Given that 1 kg = 2.205 lb, change 14.2 oz to kilograms.

18. Given that 1 km = 0.6214 mi, change 50.3 mi/min to kilometers per hour.

19. A type of brass contains 35.2% zinc. How many pounds of zinc are there in a 76.5-pound brass rail?

20. The efficiency of an engine is 78.4% (output is 78.4% of input). If the output is 184 horsepower, what is the input?

21. The impedance Z (in ohms) of an alternating current circuit is $Z = \sqrt{R^2 + X^2}$, where R is the resistance and X the reactance. Find Z if $R = 19.8\ \Omega$ and $X = 15.2\ \Omega$.

22. The distance from the earth to the moon is about 239,000 mi. A radio signal travels at 670,000,000 mi/h. How many seconds does a radio signal from the moon take to reach the earth? (time = distance ÷ rate)

23. A shipment contains gear blanks with the following weights: 2.1 lb, 10.3 lb, 5.27 lb, and 3.74 lb. Find the total weight of the shipment.

24. A computer is shipped in a crate 4.32 ft long, 3.26 ft wide, and 3.15 ft long. Find the capacity in liters (1 ft^3 = 28.32 L).

25. Express the capacity of the crate in Exercise 24 in cubic meters (1 in. = 2.54 cm).

26. The current in a circuit is 2.3 A at some instant. At another instant the current is 1.2 A in the opposite direction. Write the latter current as a signed number.

27. Two objects are moving along a line at −20 ft/s and −35 ft/s. What is the absolute value of the difference in their velocities?

CHAPTER 4

Algebraic Expressions and Operations

One reason for studying algebra is its usefulness in problem solving. Algebra also provides a shortcut for representing certain mathematical relationships. As a result, algebra is a major tool in science and technology, as well as in more advanced mathematical areas. In this chapter we introduce some of the fundamental concepts of algebra.

4.1 Literal Symbols and Formulas

Algebra is essentially a generalization of arithmetic. Arithmetic deals with numbers; algebra deals with letters that represent numbers. These ideas are not new to you. For example, the area of a rectangle is found by multiplying the length of the rectangle by the width. (See Figure 4.1.) If we let the letter

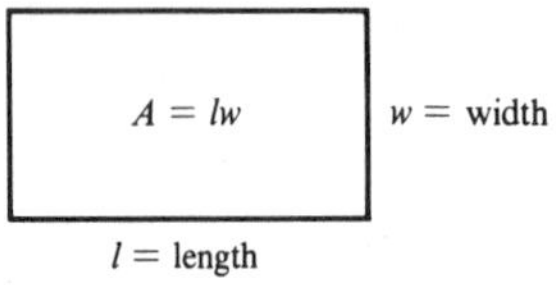

$A = lw$

w = width

l = length

Figure 4.1

A represent the area, l the length, and w the width, we obtain a shorthand notation for the expression for the area:

$$A = l \cdot w$$

The equality $A = l \cdot w$ is called a *formula*. A formula is useful because of its generality: $A = l \cdot w$ shows how to find the area of *any* rectangle, regardless of its dimensions. For example, if $l = 4.0$ in. and $w = 2.0$ in., then

$$A = l \cdot w$$
$$A = (4.0 \text{ in.})(2.0 \text{ in.}) = 8.0 \text{ in.}^2$$

Literal symbols
Variable
Constant

The letters used to represent numbers are called **literal symbols.** If a literal symbol can take on various values, it is called a **variable.** A number or a literal symbol that does not vary is called a **constant.** For example, suppose R (in ohms) is the resistance of a variable resistor. If the current through the resistor is 2 A, then the voltage V (in volts) is $V = 2 \cdot R$. Here 2 is a constant (since the current does not change) and R and V are variables (since the resistance and voltage vary).

Notation. To indicate multiplication of literal symbols or of a number and a literal symbol, we simply place the symbols together. For example,

$$A = lw \quad \text{means} \quad A = l \cdot w$$
$$abc \quad \text{means} \quad a \cdot b \cdot c$$

and

$$2xy \quad \text{means} \quad 2 \cdot x \cdot y$$

To indicate division, we use a fraction bar or a slash. Thus

$$\frac{a}{b} \quad \text{means} \quad a \div b$$

and

$$3/x \quad \text{means} \quad 3 \div x$$

In other words, in algebra a division is usually expressed in the form of a fraction.

Formula

As already noted, the algebraic statement $A = lw$ is an example of a formula. A **formula** expresses a relationship between constants and variables. Quite often this relationship is a geometric or physical law. Consider the following example.

Example 1 The distance that a car travels is equal to the average rate multiplied by the time traveled. If d is the distance, r the average rate, and t the time, this statement can be abbreviated

$$d = rt$$

Note that in the formula $d = rt$ (Example 1), all the literal symbols are variables. In the next example the formula contains literal symbols that represent constants.

Example 2 The velocity V (in feet per second or in meters per second) of a falling object is given by

$$V = gt + v$$

where t is the time in seconds. Here g represents the acceleration due to gravity (32 ft/s^2 or 9.8 m/s^2) and v the *initial velocity* (the velocity when the motion begins). In any particular problem, g and v are constants, while V and t are variables.

It was mentioned earlier that variables can assume different values. Assigning a particular value to a variable is called *substituting* the value for the variable.

Example 3 A woman tosses a ball downward from the top of a building at the rate of 12 ft/s. What is the velocity 2.5 s later? (Refer to Example 2.)

Solution. Since the initial velocity is 12 ft/s, $v = 12$ ft/s in the formula $V = gt + v$. Since $g = 32$ ft/s^2, a constant, we get

$$V = 32t + 12$$

If $t = 2.5$ s, the velocity is

$$V = 32\,\frac{\text{ft}}{\text{s}^2}\,(2.5\text{ s}) + 12\,\frac{\text{ft}}{\text{s}} = 92\,\frac{\text{ft}}{\text{s}}$$

Common error Switching capital and lowercase letters. Do not write

$$a - 2b \quad \text{as} \quad A - 2B$$

or

$$4R + C \quad \text{as} \quad 4r + c$$

In a technical problem, R and r will represent entirely different variables, as, for example, in the formula $T = R + r$.

Example 4 Write the formula for the area A of a rectangle x ft long and 5 ft wide.

Solution. The area A of a rectangle is found by multiplying length by width. The length is x ft and the width is 5 ft. So $A = x \cdot 5$, or

$$A = 5x$$

Example 5 If one gear costs \$2, how much do 10 gears cost? If the cost of one gear is \$$N$, write the formula for the cost C of x gears.

Solution. 10 gears cost $\$2 \cdot 10 = \$20$. In other words, the total cost is found by multiplying the cost per gear by the number of gears. So the formula is

$$C = Nx$$

Example 6 The energy E (in foot-pounds) required to lift an object y ft off the floor is found by multiplying the distance y by the weight w of the object (in pounds). Write the formula for E.

Solution. Since w is the weight, and y the height, we get

$$E = wy$$

Subscripts

Variables that represent similar quantities are often designated by letters with subscripts. Suppose, for example, that a circuit contains two variable resistors. Using the **subscripts** 1 and 2,

$$R_1 \quad \text{and} \quad R_2$$

can be used to represent the respective resistances. Although R_1 and R_2 are different variables, using a common letter suggests that they represent physical quantities of the same kind.

Example 7 The combined resistance of two resistors in series is the sum of the two resistances R_1 and R_2. (See Figure 4.2.) So if R denotes the combined resistance, then

$$R = R_1 + R_2$$

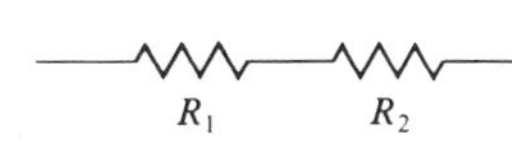

Figure 4.2

(Since the letter R does not have a subscript, it is a variable different from R_1 and R_2.)

Exercises / Section 4.1

In Exercises 1–8, identify the constants and variables in each formula.

1. $P = 2b + c$

2. $B = 4x + 3y$

3. $Q = 2s_1 + 3s_2$

4. $S = t_1 + 2t_2$

5. $L = \dfrac{a_1 + a_2}{2}$

6. $N = \dfrac{n(n + 1)}{2}$

7. $Z = 4F_1 + 6F_2$

8. $L = p_1 + \dfrac{1}{2}p_2$

9. Given that

$$m = \frac{y_2 - y_1}{x_2 - x_1}$$

find m for $y_1 = -4$, $y_2 = -1$, $x_1 = -2$, and $x_2 = 3$.

10. Given that $X = 2ab$, find X for $a = -4$ and $b = 9$.

11. Given that $P = \dfrac{2.0}{3.0a - 2.0b}$, determine P if $a = 5.0$ and $b = 6.0$.

12. Given that $S = \dfrac{20t_1 + 2.5}{11t_2 - 6.2}$, determine S if $t_1 = 0.50$ and $t_2 = 0.95$.

13. The force F (in pounds) required to move a body is equal to the mass m (in slugs) of the body times the acceleration a of the body. If the body accelerates at the rate of 4 ft/s^2, write the formula for F.

14. Write the formula for the area A of a rectangle 4 m long and x meters wide.

15. Write the formula for the length l of a strip around a square frame if each side has length s.

16. If a screw costs 12¢, how much do eight screws cost? If one screw costs n cents, write the formula for the price P of m screws.

17. If a woman earns \$10/h, how much will she earn in one 8-hour day? Using r for the hourly rate, write the formula for the daily amount D earned.

18. If the cost of carpeting an office is \$20/ft^2, write the formula for the cost C of carpeting an office measuring x by y (in feet).

19. The cost of finishing a piece of metal is \$0.75/cm^2. Write the formula for the cost C of finishing a metal plate l cm long and w cm wide.

20. The volume V (in cubic meters) of a rectangular box is found by multiplying the length l by the width w by the height h. Write V in terms of literal symbols. Find V if $l = 3.05$ m, $w = 2.47$ m, and $h = 1.78$ m.

21. The simple interest i earned on a principal P (in dollars) is the principal times the interest rate r (expressed as a decimal) times the time t (in years) that the money was invested. Write the formula for i. Find i if $P =$ \$500, $r = 8.5\%$, and $t = 6.5$ years.

22. The voltage V (in volts) across a resistor is equal to the current I times the resistance R. Write the formula for V and find V if $I = 1.90$ A and $R = 25.6\ \Omega$.

23. The combined capacitance C of two capacitors in parallel (Figure 4.3) is the sum of the two capacitances C_1 and C_2. Write C in terms of literal symbols and find C if $C_1 = 2.19\ \mu$F and $C_2 = 3.24\ \mu$F.

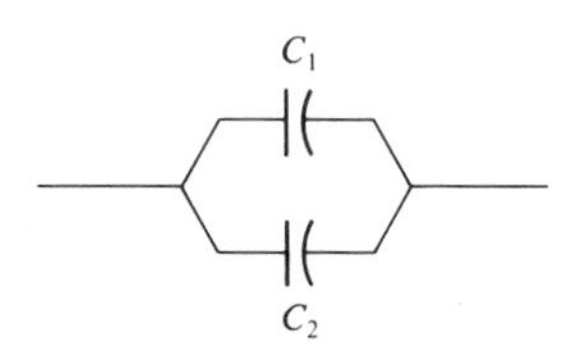

Figure 4.3

24. The resistance R (in ohms) of a resistor is equal to the voltage V across the resistor divided by the current I through the resistor. Write R in terms of V and I. Find R if $V = 63$ V and $I = 2.1$ A.

25. A transformer consists of two coils. The voltage V_1 in the first coil is equal to the voltage V_2 in the second coil times the quotient of the number of turns N_1 in the first coil and the number N_2 in the second. Write V_1 in terms of V_2, N_1, and N_2. If $V_2 = 20$ V, $N_1 = 75$ turns, and $N_2 = 100$ turns, find V_1.

4.2 Algebraic Expressions

We saw in the last section that algebraic statements contain literal symbols. To make our discussion easier, we now introduce some of the basic terms commonly used in algebra.

Power

Product
Factors

If two literal symbols a and b are multiplied, then ab is called the **product** and a and b the **factors.**

If a factor repeats, the product can be written using exponents. Let us recall the meaning of *power* from Chapter 1 (Section 1.11):

$$a \cdot a = a^2 \qquad \text{Read "}a\text{ squared"}$$

$$a \cdot a \cdot a = a^3 \qquad \text{Read "}a\text{ cubed"}$$

$$a \cdot a \cdot a \cdot a = a^4 \qquad \text{Read "}a\text{ to the fourth power"}$$

In general,

$$a \cdot a \cdot a \cdot \cdots \cdot a = a^n \qquad (n \text{ factors})$$

is read "a to the nth power" or simply "a to the nth."

> **Definition of power**
>
> $$a \cdot a \cdot a \cdot \cdots \cdot a = a^n \qquad (n \text{ factors})$$
>
> is read "a to the nth power"; a is called the **base** and n the **exponent.**

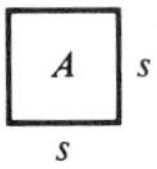

A s
s

Figure 4.4

If a number a occurs only once in a product, then a is raised to the first power, or a^1. However, since $a^1 = a$, the exponent 1 is not usually written.

Powers can be useful in writing formulas. For example, the area of a rectangle is found by multiplying the length by the width. For a square the length and width are equal (Figure 4.4). So $A = s \cdot s$, which is written $A = s^2$.

Powers may also occur in combination. Thus a^2b^3 represents $a \cdot a \cdot b \cdot b \cdot b$.

Coefficients

Coefficient

Sometimes we have a combination of numbers and literal symbols. If one of the factors is a number, that number is called the **numerical coefficient.** For example, 2 is the numerical coefficient of $2x$. The numerical coefficient is usually placed in front of the literal factor: We write $2x$, not $x2$.

Example 1

$3y$ has numerical coefficient 3

$4lw$ has numerical coefficient 4

$-7x^2y$ has numerical coefficient -7

$-10.4a^3b$ has numerical coefficient -10.4

Coefficients do not have to be numerical. Consider the product abc^2. Here a is the coefficient of bc^2, and ab is the coefficient of c^2. From now on, however, the term *coefficient* will usually refer to a numerical coefficient.

If no numerical coefficient is written, it is understood to be 1: Since $1 \cdot x = x$, we have $\boldsymbol{x = 1 \cdot x}$, so that x has coefficient 1.

$x = 1 \cdot x$

A combination of numbers and literal symbols connected by the four fundamental operations is called an **expression.** For example,

Expression

$$2x^2, \quad \frac{3a}{b^2}, \quad x + 4, \quad \text{and } A - 3.4\,P$$

are expressions. (Single numbers and literal symbols are also expressions. For example, y is an expression because $y = 1y$.)

Multinomials

An algebraic expression that does not involve addition or subtraction is called a **monomial.** For example,

Monomial

$$5, \quad x^2, \quad 3y^2, \quad \sqrt{5ab}, \quad \text{and } 3x/y$$

are monomials.

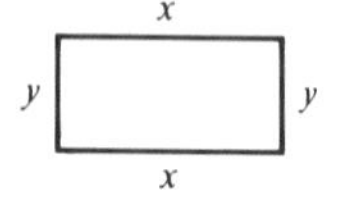

Figure 4.5

As we saw in our earlier discussion of formulas, sums of monomials often occur in algebra. For example, the perimeter P (total length of the sides) of a rectangle is given by

$$P = 2x + 2y$$

(See Figure 4.5.)

The sum of two or more monomials is called a **multinomial.** Each monomial in a multinomial, together with its positive or negative sign, is called a **term.** For example,

Multinomial

Term

$$2x - 3y^2$$

is a multinomial. The terms are $2x$ and $-3y^2$.

Some multinomials have been given special names to indicate the number of terms: A multinomial consisting of exactly two terms is called a **binomial** and one with exactly three terms, a **trinomial.** If there are more than three terms, the expression is simply called a **multinomial.**

Binomial

Trinomial

Example 2

$2x^3$ is a monomial. — One term

$3x^2 + 2y^3$ is a binomial. — Two terms

$xy - 3y^2 + 4z^3$ is a trinomial. — Three terms

Example 3

$-2.4z$ is a monomial. — One term

$5a - 3b^2$ is a binomial. — Two terms

$9 - 2c^2 - 4ab^3$ is a trinomial. — Three terms

$8x_1^2 - 3x_1x_2 + x_2^2 + 2$ is a multinomial.

By definition, the terms in a multinomial may contain quotients and roots.

Example 4 $2\sqrt{v_0}$ is a monomial.

$4.1 - 2.0\sqrt{V}$ is a binomial.

$2x - \frac{3x}{y} + \frac{5x^2}{y}$ is a trinomial.

We will see in Chapter 10 that quotients can be represented by using negative exponents, and roots by using fractional exponents. It follows that the terms in a multinomial can have negative or fractional exponents. If the terms have only positive integral exponents, then the expression is called a *polynomial*. For example, the expressions in Examples 2 and 3 are polynomials.

> A **polynomial** is a monomial or multinomial containing only terms with exponents that are positive integers.

We will discuss polynomials further in Chapter 5.

As already noted, multinomials often occur in formulas. Consider the next example.

Example 5 The tensile strength S (in pounds) of a piece of metal depends on the temperature T (in degrees Fahrenheit) and is given by the following formula:

$$S = 841 - 0.0842\sqrt{T}$$

Find the tensile strength when $T = 425°\text{F}$.

Solution. Substituting 425 for T, we get

$$S = 841 - 0.0842\sqrt{425}$$

The calculator sequence is

841 [−] 0.0842 [×] 425 [√] [=] → 839.26417

So the tensile strength is 839 lb, to three significant digits.

Exercises / Section 4.2

In Exercises 1–8, identify the numerical coefficients.

1. $3RC$
2. $4NP$
3. $-6xy^2$
4. $-8wz^2$
5. $-0.4LM$
6. $1.70pq$
7. $2.30t_1t_2$
8. $-7.63C_1C_2$

In Exercises 9–14, identify the numerical coefficient and the factors in each case.

9. $2ab$

10. $3xy$

11. $-4a^2b$

12. $-2m^2n$

13. $10pq^2$

14. $-11wz^2$

In Exercises 15–30, state in each case whether the expression is a monomial, binomial, or trinomial.

15. $5x^2y^3$

16. $4wxy$

17. $-7xyz$

18. $-20LMN$

19. $25x_1^2 - 16x_1x_2 + 8x_2^2$

20. $42T_1^2 + T_1T_2 - 2T_2^2$

21. $-7.4V_1^2 - 7.8V_1V_2 + 2.0V_2^2$

22. $0.12Z^2 - 0.04Z + 0.36$

23. $-3a^2 + b$

24. $7b^2 - c$

25. $2V_1^2 - 5V_2^2$

26. $-7.64m^2n$

27. $-10.6z^2q$

28. $\frac{1}{2}R_1^2 - 2R_1R_2 - 4R_2^2$

29. $\frac{1}{3}s^2 - s^3 + 3t^2$

30. $\frac{1}{4}N_1^2 + N_2$

31. Given that $S = 2b^2 - 5b$, find S if $b = 4$.

32. Given that $T = 2.0v_1^2 - 1.0v_2^2$, find T if $v_1 = 3.0$ and $v_2 = 2.0$.

33. Given that $P = \sqrt{x} - 2.0$, find P if $x = 9.0$.

34. $L = \sqrt{6.0a} - 4.0a$; find L if $a = 6.0$.

35. $Z = \frac{p}{q} + 4.6$; find Z if $p = 9.4$ and $q = 2.9$.

36. $Y = \frac{\sqrt{3x}}{y} + 2x$; find Y if $x = 3.2$ and $y = 6.5$.

37. The volume V of a cube is found by raising the length of a side s to the third power (Figure 4.6). Find the volume of a cubical tank with side 1.25 m in length.

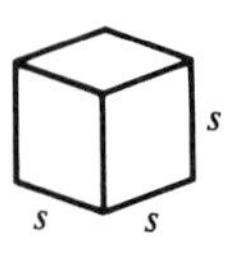

Figure 4.6

38. The distance s (in meters) from the ground of an object hurled upward is

$$s = -\frac{1}{2}gt^2 + v_0t$$

where $g = 9.8\ \text{m/s}^2$, v_0 the initial velocity, and t the time in seconds. Find s after 3.0 s if the initial velocity is 29 m/s.

39. The combined capacitance C (in microfarads) of two capacitors in series is

$$C = \frac{C_1C_2}{C_1 + C_2}$$

where C_1 and C_2 are the respective capacitances (Figure 4.7). Find C if $C_1 = 4.50\ \mu\text{F}$ and $C_2 = 3.64\ \mu\text{F}$.

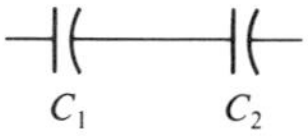

C_1 C_2

Figure 4.7

40. The combined resistance R (in ohms) of two resistors in parallel is found from the formula

$$\frac{1}{R} = \frac{1}{R_1} + \frac{1}{R_2}$$

where R_1 and R_2 are the respective resistances (Figure 4.8). Find R if $R_1 = 20.4\ \Omega$ and $R_2 = 30.9\ \Omega$.

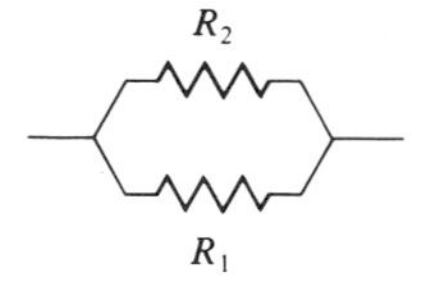

Figure 4.8

41. A box contains x gears costing \$10 each and y gears costing \$20 each. What does the binomial $10x + 20y$ represent?

42. The area A of a square of side s is $A = s^2$. Write the multinomial that represents the area of the shaded portion in Figure 4.9.

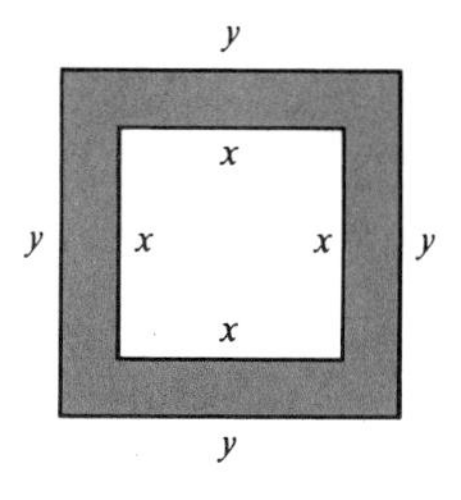

Figure 4.9

43. The cutting speed (CS) of a milling cutter (in feet per minute for a point on the surface of the cutter) is

$$CS = \pi D(RPM)$$

where D is the diameter of the cutter (in feet) and RPM (in revolutions per minute) the rate of rotation. Find the cutting speed of a piece of metal with an 8.0-inch diameter turning at 47 rev/min.

44. The horsepower (hp) transmitted by a belt is the product of the velocity v of the belt (in feet per minute), the belt width w (in inches), and the working stress s of the belt (in pounds per inch of belt width), divided by 33,000 foot-pounds per minute. Find the horsepower transmitted by a belt 6.0 in. wide moving at 110 ft/min, if the working stress is 35 lb/in.

45. A piece of wire x feet long is cut from a wire L feet long. Find an expression for the length of the other piece.

46. The period P (in seconds) of a pendulum is given by

$$P = \frac{2\pi}{\sqrt{g}}\sqrt{l}$$

where $g = 32\ \text{ft/s}^2$ and l the length of the pendulum in feet. Find P if $l = 4.8$ ft.

47. A firm pays each of 100 employees x dollars per month, each of 150 employees y dollars per month, and each of 50 employees z dollars per month. What does the polynomial $100x + 150y + 50z$ represent?

4.3 Addition and Subtraction of Multinomials

In this section we study addition and subtraction of multinomials. To this end we must first recall some of the basic laws from Chapter 1, but stated in algebraic form.

Commutative law

$$a + b = b + a \qquad ab = ba$$

Associative law

$$a + (b + c) = (a + b) + c \qquad a(bc) = (ab)c$$

Distributive law

$$a(b + c) = ab + ac$$
$$(b + c)a = ba + ca$$

To state the rules for addition and subtraction, we need the definition of *like terms,* stated next.

Like terms

Two or more terms are said to be **like terms** if the literal factors are the same.

For example,

$$2x \quad \text{and} \quad 7x$$

are like terms, since the literal factor is x in each case. On the other hand,

$$2x \quad \text{and} \quad 7y$$

are not like terms since the literal parts are different (x and y). Also, since x is not the same as x^2,

$$2x \quad \text{and} \quad 7x^2$$

are not like terms.

Example 1 Examples of like terms are

$$-2xy^2, \quad 3xy^2, \quad 5xy^2, \quad \text{and } 10xy^2$$

and

$$-4x^2y, \quad 6x^2y, \quad -3.2x^2y, \quad \text{and } 4.7x^2y$$

However,

$$-2x^2y \quad \text{and} \quad 4xy^2$$

are *not* like terms since the factors x and y have different exponents.

The definition of like terms is used in the following rule:

> Like terms are combined by adding or subtracting the numerical coefficients.

To see the reason for this rule, consider the sum

$$2x + 3x$$

By the distributive law in reverse,

$$2x + 3x = (2 + 3)x = 5x$$

In other words, $2x$ and $3x$ are added by adding the coefficients 2 and 3. However,

$$2x + 3y$$

cannot be combined since neither x nor y is common to both terms.

Example 2

$$-2x + 7x = 5x \qquad -2 + 7 = 5$$
$$-6y - 4y = -10y \qquad -6 - 4 = -10$$
$$-3ab + 8ab = 5ab \qquad -3 + 8 = 5$$

If a multinomial contains like terms, we can simplify the expression by combining like terms, as shown in the next example.

Example 3 Simplify the multinomial

$$-3T_a + 2T_b - 7T_a - 5T_b$$

(The subscripts a and b indicate that T_a and T_b are different variables.)

Solution. The like terms are $-3T_a$ and $-7T_a$, as well as $2T_b$ and $-5T_b$. Since

$$-3T_a - 7T_a = -10T_a$$

and

$$2T_b - 5T_b = -3T_b$$

we get

$$-3T_a + \mathbf{2T_b} - 7T_a - \mathbf{5T_b} = -10T_a - 3T_b$$

Multinomials are added by the rule given next:

Addition of multinomials

To add multinomials, add the coefficients of like terms.

Example 4 Perform the following addition:

$$(2x - 5y) + (-10x + 4y)$$

Solution. Since only like terms can be combined, let us write the multinomials in two rows to add corresponding coefficients:

$$\begin{array}{rl} 2x - 5y & \\ \underline{-10x + 4y} & \quad 2 + (-10) = -8 \\ -8x - 1y & \quad -5 + 4 = -1 \end{array}$$

Since $-1y = -1 \cdot y = -y$, the answer should be written $-8x - y$.

Example 4 illustrates the following rule:

$a = 1a$ and $-a = -1a$

Example 5 Perform the following addition:

$$(3\sqrt{s} - t) + (-4\sqrt{s} + 2t)$$

Solution. Since $-t = -1t$, the first binomial is written $3\sqrt{s} - 1t$:

$$\begin{array}{ll} 3\sqrt{s} - 1t & \quad 3 + (-4) = -1 \\ \underline{-4\sqrt{s} + 2t} & \quad -1 + 2 = 1 \\ -1\sqrt{s} + 1t = -\sqrt{s} + t & \end{array}$$

Common error Leaving out addition or subtraction signs between the terms of a multinomial.

Incorrect	*Correct*
Add: $2x + 3y$	Add: $2x + 3y$
$\underline{4x + 7y}$	$\underline{4x + 7y}$
$6x \quad 10y$ (+ missing)	$6x + 10y$

To subtract multinomials, we need to recall the following rule: To subtract signed numbers, change the sign of the subtrahend and proceed as in addition. The algebraic form of this rule is

$$a - b = a + (-b)$$

To subtract multinomials, we use the rule given next.

Subtraction of multinomials

To subtract one multinomial from another, change all the signs of the terms in the subtrahend (the multinomial to be subtracted) and proceed as in addition.

Example 6 Subtract $-4x + y - 3z$ from $6x + 2y - 6z$.

Solution. The subtrahend is $-4x + y - 3z$. Remember to change *all* the signs:

$$\begin{array}{r} 6x + 2y - 6z \\ \underline{4x - y + 3z} \\ 10x + y - 3z \end{array} \qquad \begin{array}{l} \\ -(-4x + y - 3z) = 4x - y + 3z \\ \text{Adding coefficients} \end{array}$$

Addition or subtraction of multinomials sometimes involves terms that cannot be combined, as shown in the next example.

Example 7 Subtract $4P_1 - 5\sqrt{P_3}$ from $2P_1 + 3P_2$.

Solution. We set up the subtraction so only like terms are in the same column:

$$\begin{array}{r} 2P_1 + 3P_2 \phantom{+ 5\sqrt{P_3}} \\ \underline{-4P_1 + 5\sqrt{P_3}} \\ -2P_1 + 3P_2 + 5\sqrt{P_3} \end{array} \qquad \begin{array}{l} \\ -(4P_1 - 5\sqrt{P_3}) = -4P_1 + 5\sqrt{P_3} \\ \text{Adding coefficients} \end{array}$$

Example 8 The voltage drops (in volts) across two resistors in series are given by $V_1 = 12.6\,R - 1.30$ and $V_2 = 15.3\,R + 2.76$, respectively. Find $V_1 + V_2$, which is the voltage drop across the combination.

Solution.
$$V_1 + V_2 = (12.6\,R - 1.30) + (15.3\,R + 2.76)$$
$$= 27.9\,R + 1.46$$

Exercises / Section 4.3

In Exercises 1–16, simplify each of the given expressions.

1. $3x + 2y - 6x + 7y$
2. $2a - 7b + 10a - 2b$
3. $7m - 2n - 12m - 3n$
4. $-8w - 3z + 5w - 6z$
5. $V_a - 2V_b - V_b + 9V_a - 6V_b$
6. $21R_1 + 25R_2 - 20R_2 - R_1 - 2R_1$
7. $26R + 30R^2 - 25R - R - 25R^2$
8. $16s - 10s + 2s^2 + 5s^2 - 6s$
9. $3mn - 2pq - mn + 2pq - 2mn$
10. $20I_1R_1 - 20I_2R_2 - 10I_1R_1 + 15I_2R_2$
11. $10mv - 16ab - 5mv + 8ab$
12. $5x^2y^2 + 3x^2y^2 - 7x^2z^2 - 8x^2z^2$
13. $-6\sqrt{z} + 8xz + 10\sqrt{z} - xz - 2xz$
14. $-9A^2 + 2\sqrt{AB} + 3\sqrt{AB} - A^2 + 10A^2$
15. $9cd - 2ef - 6cd - 3cd + 5ef$
16. $-10ar - 3bs - 4ar + 14ar + 3bs$

In Exercises 17–28, add the given multinomials.

17.
$$\begin{array}{rrr} x + & y & - 3z \\ 2x & & - 4z \\ \underline{-3x} & \underline{- 2y} & \underline{+ 6z} \end{array}$$

18.
$$\begin{array}{rrr} 2a & & - 2c \\ -6a & + 5b & - 3c \\ \underline{3a} & \underline{- 6b} & \underline{+ 5c} \end{array}$$

19. $3x - 2x^2,\ 5x - 3x^2,\ -4x + x^2$
20. $5a^2 - 2a + 4,\ 4a^2 - 2,\ 5a - 1$
21. $5R - 2V + 10\left(\frac{1}{C}\right),\ -6R + 5V - 7\left(\frac{1}{C}\right),\ R - \frac{1}{C}$
22. $10m^2 - 2\sqrt{mv},\ -8m^2 + 5\sqrt{mv},\ -7m^2 - 10\sqrt{mv}$
23. $V_b - V_c,\ 2V_a - 6V_c,\ -V_a + 5V_b + 5V_c$
24. $2S_a - S_c,\ -6S_b - 7S_c,\ 25S_a - 20S_c$
25. $3m_1 - 4m_2,\ -4m_1 + 3m_2,\ 5m_1 - 4m_2,\ 8m_1 - 3m_2$
26. $2R - 5\left(\frac{1}{W}\right),\ -2R - 2\left(\frac{1}{W}\right),\ 4R - 7\left(\frac{1}{W}\right),\ 3R + 5\left(\frac{1}{W}\right)$
27. $\frac{1}{2}x - \frac{1}{3}w + \frac{1}{4}z,\ \frac{1}{2}w - \frac{1}{2}z,\ x - \frac{1}{2}w - \frac{3}{4}z,\ -\frac{1}{2}x - \frac{1}{2}z$
28. $\frac{1}{3}R - \frac{1}{3}S + T,\ 3R - \frac{1}{3}S - \frac{1}{2}T,\ -\frac{1}{2}R - \frac{1}{4}T,\ \frac{2}{3}S - T$

In Exercises 29–36, subtract the second expression from the first.

29. $2t_1 - 6t_2 - 4t_3,\ 5t_1 - 5t_2 + t_3$
30. $16s_1 - 3s_1t_1 + 6t_2,\ -s_1 - 7s_1t_1 - t_2$
31. $3x^2y - 3xy^2 + 5x^2y^2,\ 15x^2y - 6xy^2 - 2x^2y^2$
32. $-3a^2b - 4ab^2 + 7ab,\ -5a^2b - 6ab^2 - 3ab$
33. $25mv - 16p,\ -5mv$
34. $-20op - 16pq,\ -10pq$
35. $10x^2 - 9m^2 - 3n^2,\ -5x^2 - 5m^2 - 10p^2$
36. $-3C_1 - 4C_2 - 4C_4,\ -8C_1 - 10C_2 + 5C_3 + 5C_4$

In Exercises 37–42, perform the indicated operations with a calculator.

37. $(3.06x - 7.04) + (-4.92x + 0.79)$

38. $(-1.20y - 10.4) + (-3.68y - 8.6)$

39. $(2.9x + 4.6y) - (-3.4x - 7.6y)$

40. $(4.9a - 2.7b) - (-3.4a + 1.6b)$

41. $(10.81V + 12.86) - (76.34V - 28.77) + (12.62V - 20.86)$

42. $(24.69R - 10.65C) + (-27.04R + 20.79C) - (14.40R + 17.26C)$

In the remaining exercises, use a calculator when convenient.

43. One solar panel is $(y - 2a)$ m long and another is $(3y + 4a)$ m long. Find an expression for the total length if the panels are placed end to end.

44. Simplify the following expression from a problem on kinetic energy:

$$3.46mv^2 - 7.68mv^2 + 9.81mv^2$$

45. Write an expression that represents $2R$ ohms less than the sum of $15R$ ohms and $30R$ ohms. Simplify the resulting expression.

46. Write an expression that represents $2.1I$ amperes more than the sum of $6.0I$ amperes and $4.7I$ amperes.

47. Write an expression for the perimeter of the rectangle in Figure 4.10.

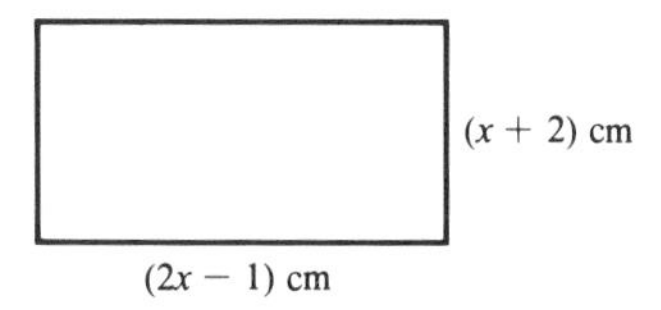

Figure 4.10

48. The cost (in dollars) of one type of home computer is $2x - y$ and the cost of another, more expensive type, is $5x + y$. Find an expression for the difference in the cost.

49. Two forces, $F_1 = 2f_1 + 6f_2$ and $F_2 = -3f_1 - 2f_2$, are acting on a beam. Write an expression for the net force $F_n = F_1 - F_2$.

50. The respective resistances of two variable resistors are $R_1 = 2.4t - 1.6$ and $R_2 = 6.7t - 3.4$, where t is in seconds. If the resistors are connected in series, find an expression for the combined resistance (the sum).

51. The resistance in each of two wires varies with temperature. If $R_1 = 0.00023T^2 + 0.0043T + 9.4$ and $R_2 = 0.00046T^2 + 0.0073T + 8.7$, T measured in degrees Celsius, find an expression for the total resistance if the wires are placed end to end. (See Exercise 50.)

52. The voltage across two or more resistors in series is equal to the sum of the voltages across the individual resistors. If the voltages are $V_1 = 7.2R + 1.8$, $V_2 = 10.1R + 2.4$, and $V_3 = 15.6R - 1.2$, find an expression for the voltage across the combination.

4.4 Symbols of Grouping

We have occasionally used parentheses to set off certain numbers or literal symbols. For example, the parentheses in the sum $-2 + (-4) = -6$ separate the negative number from the symbol for addition. In many cases parentheses indicate that certain operations are to be carried out first. In particular, since multiplication always precedes addition, parentheses are needed if a different order of operations is intended.

For example, the parentheses in the expression $2(3 + 5)$ indicate that the addition has to be carried out before the multiplication. Thus

$$2(3 + 5) = 2 \cdot 8 = 16$$

Without parentheses we have

$$2 \cdot 3 + 5 = 6 + 5 = 11$$

since multiplication is done first. Consequently,

$$2(3 + 5) \neq 2 \cdot 3 + 5$$

Symbols of grouping

Parentheses are called **symbols of grouping.** Other commonly used grouping symbols are brackets [ ] and braces { }, often occurring in combination. For example, the expression

$$a\{1 - [x + 4(x - y)]\}$$

has parentheses within brackets, all within a set of braces.

When simplifying certain algebraic expressions, symbols of grouping must often be removed. This is usually accomplished by means of the **distributive law.** To see how, let us simplify the expression $3[x + 2(x + 3)]$:

$$\begin{aligned} 3[x + 2(x + 3)] &= 3[x + 2x + 6] && \text{Distributive law} \\ &= 3[3x + 6] && x + 2x = 3x \\ &= 3(3x) + 3(6) && \text{Distributive law} \\ &= 9x + 18 \end{aligned}$$

This example, which is quite typical, leads to the following rule for removing grouping symbols.

> To **remove symbols of grouping,** work from the inside out.

Example 1 Simplify the expression

$$2[3(a - b) - b]$$

Solution.

$$\begin{aligned} 2[3(a - b) - b] &= 2[3(a) + 3(-b) - b] && \text{Distributive law} \\ &= 2[3a - 3b - b] && 3(-b) = -3b \\ &= 2[3a - 4b] && \text{Simplifying} \\ &= 2(3a) + 2(-4b) && \text{Distributive law} \\ &= 6a - 8b \end{aligned}$$

Of particular interest are grouping symbols preceded by a positive or negative sign. Consider, for example, the expression $a - (a + b)$. By the rule for subtraction, we change the sign of the subtrahend to get $a - a - b = -b$. A simple alternative is to write $-(a + b) = -1(a + b)$ and then use the distributive law to get $-1(a + b) = -a - b$. Similarly, $+(a + b) = +1(a + b) = a + b$. In summary,

$$-(a + b) = -1(a + b) \qquad +(a + b) = +1(a + b)$$

The general procedure for removing grouping symbols is given next.

Rule for removing symbols of grouping

1. To remove symbols of grouping preceded by a positive sign:
   a. Omit the symbols of grouping.
   b. Combine like terms.
2. To remove symbols of grouping preceded by a negative sign:
   a. Omit the symbols of grouping and change the signs of all the terms enclosed.
   b. Combine like terms.

Example 2 Remove symbols of grouping and simplify:

$$-[a - (b - 2a) + 4b]$$

Solution.

$$\begin{aligned} -[a - (+b - 2a) + 4b] &= -[a - b + 2a + 4b] && \text{Removing inner parentheses} \\ &= -[3a + 3b] && \text{Simplifying} \\ &= -3a - 3b && \text{Removing brackets (signs change)} \end{aligned}$$

Example 3 Simplify:

$$-\{s_1 - [1 + (s_1 + s_2 - 1)]\}$$

Solution.

$$\begin{aligned} &-\{s_1 - [1 + (s_1 + s_2 - 1)]\} \\ &= -\{s_1 - [1 + s_1 + s_2 - 1]\} && \text{Removing inner parentheses} \\ &= -\{s_1 - [s_1 + s_2]\} && \text{Simplifying} \\ &= -\{s_1 - [+s_1 + s_2]\} \\ &= -\{s_1 - s_1 - s_2\} && \text{Removing brackets} \\ &= -\{-s_2\} = s_2 && \text{Removing braces} \end{aligned}$$

Example 4 Simplify:

$$2\{R + [V - (2R - 3V)]\}$$

Solution.

$$\begin{aligned} &2\{R + [V - (2R - 3V)]\} \\ &= 2\{R + [V - 2R + 3V]\} && \text{Removing parentheses} \\ &= 2\{R + [4V - 2R]\} && \text{Simplifying} \\ &= 2\{R + 4V - 2R\} && \text{Removing brackets (no sign changes)} \\ &= 2\{-R + 4V\} && \text{Simplifying} \\ &= 2(-R) + 2(4V) && \text{Distributive law} \\ &= -2R + 8V \end{aligned}$$

Remark. Another common symbol of grouping is the bar used for fractions. This symbol of grouping will be discussed in Section 5.4.

Common errors

a. Changing the sign of only the first term when removing symbols of grouping preceded by a negative sign. For example,

$a - (b - c)$ should *not* be written as $a - b - c$

Instead,

$$a - (b - c) = a - (+b - c) = a - b + c$$

b. Multiplying only the first term when applying the distributive law. For example,

$2(x + y)$ is *not* equal to $2x + y$

Instead, we multiply *both* terms by 2:

$$2(x + y) = 2x + 2y$$

Example 5 If two forces F_1 and F_2 are acting on an object in opposite directions, then the combined force F_c is given by $F_c = F_1 - F_2$. If $F_1 = 3f_1 - 2f_2$ and $F_2 = -f_1 + 3f_2$, obtain an expression for F_c.

Solution. The combined force is

$$\begin{aligned} F_c &= F_1 - F_2 \\ &= (3f_1 - 2f_2) - (-f_1 + 3f_2) \\ &= 3f_1 - 2f_2 + f_1 - 3f_2 \\ &= 4f_1 - 5f_2 \end{aligned}$$

Exercises / Section 4.4

In Exercises 1–24, remove grouping symbols and simplify.

1. $-(x + y - 3x)$

2. $-(a + 2b - 4a)$

3. $-[a + (b - c)]$

4. $-[r - (s - t)]$

5. $-[-(R - C) - R]$

6. $-[-(a + b) + (2a + 2b)]$

7. $2[x - (a - x)]$

8. $3[-(a + b) + b]$

9. $-2[(R - V) - (2R + V)]$

10. $-4[-(2T_1 - 3T_2) + (-3T_1 - 2T_2)]$

11. $-[-(2x + y) - (x - y) + x]$

12. $-[(3a + 2b) - (-4a - b) + a]$

13. $-\{a - [b - (b - a) - a]\}$

14. $-\{2s + [-(s + 2t) - (s - 2t) - s]\}$

15. $-\{-[(x - 3z) - (-2x + 4z) - 3z] + x\}$

16. $-\{R_1 - [-(R_2 - R_1) + (2R_1 - 2R_2) - R_1] - R_2\}$

17. $(x - 3y + 2z) - (2x + 5y - 6z)$

18. $(-2x + 2y - 4z) - (7x - 3y + 10z)$

19. $(m_1 + 2m_2 - m_3) - (-m_1 - 4m_2 - 6m_3)$

20. $(-4V_1 - 4V_2 + 6V_3) - (3V_1 + 6V_2 - 12V_3)$

21. $(3x - 7w) - (4x + 4w) + (-8x + 11y)$

22. $-(3x - 5y) + (-4x - 8y) - (x - y)$

23. $-(C_1 - 2C_2) - (3C_1 - 4C_2) - (-4C_1)$

24. $-7L_1 - (-5L_1 + L_2) - (2L_1 + L_2) - L_1$

25. Remove the parentheses in the formula $C = \frac{5}{9}(F - 32)$ for converting degrees Fahrenheit to degrees Celsius.

26. The velocity v (in meters per second) of a body dropping from level y_1 to level y_2 is

$$v = \sqrt{19.6(y_2 - y_1)}$$

Write this formula without parentheses.

27. The following expression arose in a problem on amplifiers:

$$E_b - R - (E - E_1) - (E_1 - R)$$

Simplify this expression.

28. The voltage V across the components of a certain alternating current circuit is

$$V = (-2V_{ab} - 3V_{bc} + 4V_{cd}) - (-6V_{ab} + V_{bc} - 5V_{cd})$$

Simplify the expression for V.

29. Simplify the following expression from a problem in mechanics:

$$\left(\frac{1}{2}mv^2 - \frac{1}{3}\right) - \left(\frac{1}{4}mv^2 - \frac{1}{2}\right)$$

30. The following expression arose in an experiment in optics:

$$(0.021m_1 - 0.10m_2) + (0.038m_1 - 0.21m_2)$$

Simplify this expression.

4.5 Algebraic Expressions in BASIC (Optional)

In this section we discuss algebraic expressions written in BASIC. (For a brief discussion of the BASIC programming language, see Appendix A.)

In BASIC, a legal name for a variable can be any single capital letter or a single capital letter followed by a single digit. Examples of proper variable names are

A, B, C, X, Y, and Z

or

A1, B3, Y9, and Z0

In some versions of BASIC, these variable names can be longer than two characters.

The BASIC symbols for some of the algebraic operations are given in Table 4.1:

Table 4.1

Operation	Meaning	Example	Algebraic notation
+	Addition	A + B	$A + B$
−	Subtraction	A − B	$A - B$
*	Multiplication	A * B	AB or $A \cdot B$
/	Division	A/B	$\frac{A}{B}$ or A/B
↑	Power	A↑2	A^2
SQR(X)	Square root	SQR(X−Y)	$\sqrt{X - Y}$
ABS(X)	Absolute value	ABS(−2)	$\|-2\|$

In BASIC, operations are performed in the usual order:

1. Evaluation of quantities enclosed in parentheses.
2. Evaluation of powers.
3. Multiplication and division from left to right.
4. Addition and subtraction from left to right.

Example 1

Algebraic notation	*BASIC*
$\frac{x + 2}{5}$	(X + 2)/5
$2(a - b)^3$	2 * (A − B)↑3
$\frac{1}{R_1} + \frac{1}{R_2}$	1/R1 + 1/R2
$4\sqrt{b}$	4 * SQR(B)
$\sqrt{2 - t_1}$	SQR(2 − T1)

Exercises / Section 4.5

In Exercises 1–20, write each algebraic expression in BASIC.

1. $2x + 4y$
2. $7x - 3y$
3. $2b^2$
4. $5b^4$
5. $-4c^3 + 6d$
6. $3i^2 - 4k$
7. $\frac{2x - 4}{a}$
8. $\frac{4y - 7}{b}$
9. $2(x - 2y)^4$
10. $4(a - 4b)^3$
11. $\sqrt{b - 2}$
12. $\sqrt{1 - x}$
13. $\frac{(s - 4)^3}{\sqrt{t}}$
14. $\frac{(2 - 4z)^4}{\sqrt{a}}$
15. $\frac{2V_1 - 3V_2}{V_3}$
16. $\frac{2t_1 + t_2}{\sqrt{t_3}}$
17. $\frac{\sqrt{I_2 - I_1}}{2 - 3a}$
18. $\frac{\sqrt{C_2 - 3C_1}}{2 - 5C_4}$
19. $\frac{2}{\sqrt{v_1 - 2v_2}}$
20. $\frac{3}{(3w_2 - w_1)^4}$

In Exercises 21–30, write each BASIC expression in algebraic notation.

21. 3 * X↑2 − 4
22. 1 − 4 * Y↑3
23. (V2 − V1)/(1 − A)
24. (A1 − B1)/(A2 − B2)
25. (2 * X↑2 − Y)/(A − 3 * B)
26. (A − 2 * B)↑3/(X − Y)
27. SQR(T1 − T2)/(S1 + S2)
28. (2 + P)↑6/SQR(V)
29. (2 + SQR(A))/Y
30. (SQR(A − B) − 4)↑6

Review Exercises / Chapter 4

In Exercises 1–4, identify the constants and variables.

1. $L = 3C - 4D$
2. $S = 3t_1 - 6t_2$
3. $N = \frac{1}{4}(T_a - 2T_b)$
4. $A = \sqrt{4x^2 - b}$

In Exercises 5–8, identify the numerical coefficients.

5. $-4.7x^2y$
6. $6.46R^2C$
7. $-2.3T_1T_2$
8. $28pqr$

In Exercises 9–12, state in each case whether the expression is a monomial, binomial, or trinomial.

9. $-7x^2y$
10. $8xz^2 - 5xw^2$
11. $9a - 2b + 4c$
12. $10R_1 - 2R_1R_2 - 3R_2R_3$

In Exercises 13–16, simplify the given multinomials.

13. $6x^2 - 5y^2 - 3y^2 - 6z^2$
14. $-3P_a - 4P_b - 7P_a + 4P$
15. $-4a^2 - 5b^2 - 7a^2 - a^2 + 4b^2$
16. $3x^2y - 5xy^2 - 6xy - 3xy^2 + 4x^2y + 8xy$

In Exercises 17–22, add the given multinomials.

17. $-3a^2b - 4ab^2 + 6a^2b^2$, $6ab^2 + 2a^2b^2$, $4a^2b - 7a^2b^2$

18. $-6X_1 - 5X_3$, $10X_1 - 3X_2$, $-11X_2 - 12X_3$

19. $16R^2 - 17$, $-15R^2 - 14R$, $12R^2 + 20R + 5$

20. $2s^2 - 4t^2$, $-5r^2 + 5t^2$, $-8s^2 - 2r^2 + t^2$, $10r^2 - 2t^2$

21. $-7a - 6d + 4\sqrt{e}$, $8a - 5\sqrt{e} + f$, $-a + 2\sqrt{e}$

22. $-10xy + 2z$, $3w - 5z$, $8xy - 4z$, $-xy$

In Exercises 23–26, subtract the second expression from the first.

23. $25a^2 - c^2$, $20a^2 + 4d^2$

24. $5D_1 - 3D_2$, $-4D_2 - 4D_3$

25. $-18C_a - 14C_b$, $-16C_a - 18C$

26. $-10T_1 + 5T_2 - 4T_3$, $5T_1 + 6T_2 - 6T_3$

In Exercises 27–32, remove symbols of grouping and simplify.

27. $-[A - (B - 2A)]$

28. $2[a - (b - 2a)]$

29. $(5x - 2y - 3z) - (6x - 4y + 4z)$

30. $(2x^2 - 3y^2 - 4) - (-x^2 - 7y^2 + 1)$

31. $-\{R - [(2R - C) - (R + C)]\}$

32. $-\{p - [-(p + q) - (3p - q)] - q\}$

In the remaining exercises, use a calculator whenever convenient.

33. Given that $A = \frac{1}{2}ab$, find A if $a = 3.5$ and $b = 2.4$.

34. Given that $R = 2t^2 - 3t$, find R if $t = 2.6$.

35. $Z = \sqrt{R^2 + X^2}$; find Z if $R = 20.4$ and $X = 9.95$.

36. $L = 2.01x^2 - 4.09x$; find L if $x = 5.70$.

37. $(4.736x - 7.864) + (-2.841x + 4.079)$

38. $(10.68y + 41.71z) - (14.36y - 20.12z)$

39. The combined capacitance C (in microfarads) of two capacitors in parallel is the sum of the two capacitances C_1 and C_2. Find C if $C_1 = 0.0056t + 0.0012$ and $C_2 = 0.0089t + 0.0036$.

40. If one machine costs $20,000, how much do x machines cost?

41. A shipment contains x wrenches weighing 3 lb apiece and y wrenches weighing 5 lb apiece. What does the binomial $3x + 5y$ represent?

42. Recall that the volume V of a cube whose side has length s is $V = s^3$. Referring to Figure 4.11, what does the binomial $x^3 + y^3$ represent?

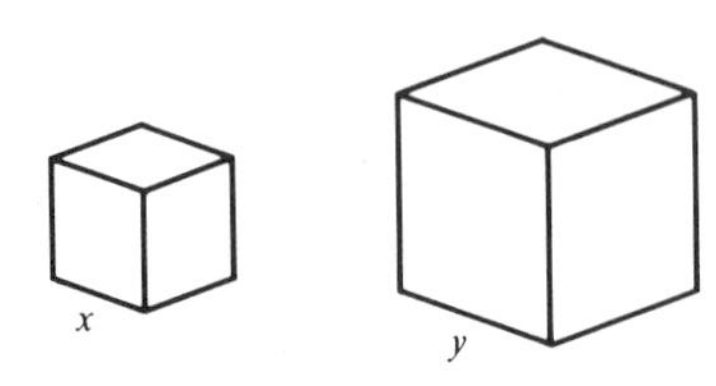

Figure 4.11

43. A certain impulsive force F is given by

$$F = -(m_A m_B - m_C m_D)$$

Simplify F by removing parentheses.

44. A bar is 12 ft long. If a piece x feet long is cut from the bar, find an expression for the remaining piece of length l.

45. One measure of the efficiency E of an engine is the heat output O over the heat input I, all subtracted from 1. Write the formula for E.

46. The cost of finishing a piece of material is $10.50/in.^{2}. Write the formula for the cost C of finishing a rectangular piece of material measuring a by b (in inches).

47. The resistance R (in ohms) in a wire is related to the temperature (in degrees Celsius) by the following formula:

$$R = 10.4 + 0.00360T + 0.000780T^2$$

Determine the resistance when $T = 98.4$°C.

48. Simplify the following expression from a problem in mechanics:

$$4m_1v_1 - 3m_2v_2 - 2m_1v_1 + 5m_2v_2$$

49. The voltage across one resistor is $V_1 = 1.23R - 1.73$, and the voltage across another resistor is $V_2 = 1.98R - 2.64$. Find the voltage across the combination if the resistors are connected in series ($V_1 + V_2$).

50. The output P (in watts) of a certain battery is given by $P = 12I + 25I^2$, where I is the current in amperes. Find P when $I = 1.5$ A.

51. The work done (in foot-pounds) in lifting an object from level y_1 to level y_2 is

$$W = mg(y_2 - y_1)$$

Write the expression for W without parentheses.

52. Simplify the following expression from a problem in mechanics:

$$10.3 - \left[\frac{1}{2}mv^2 - (2mv^2 - 12.3)\right]$$

CHAPTER 5

Multiplication and Division of Polynomials

In Chapter 4 we discussed algebraic expressions, addition and subtraction of multinomials, and simplification of algebraic expressions by removing symbols of grouping. In this chapter we complete our study of the four fundamental operations by discussing multiplication and division.

5.1 The Laws of Exponents

We discussed the meaning of *exponent* in Section 1.11 and again in Section 4.2. In this section we make a more extensive study of the properties of exponents. Our discussion consists of two parts:

1. Multiplication of two monomials
2. A discussion of other laws of exponents

First let us recall the definition of exponent from Section 4.2.

The nth power

$$a^n = a \cdot a \cdot a \cdot \cdots \cdot a \qquad (n \text{ factors})$$

a is called the **base** and n the **exponent**.

Multiplication of Monomials

The definition of exponent leads directly to the rule for multiplying powers. Consider, for example, the product x^2x^4. Note that

$$x^2 = x \cdot x \qquad \text{and} \qquad x^4 = x \cdot x \cdot x \cdot x$$

It follows that

$$x^2x^4 = (x \cdot x)(x \cdot x \cdot x \cdot x) = x^{2+4} = x^6 \qquad \text{Six factors}$$

In general, to multiply powers having the same base, we add the exponents, and write the base with this sum as an exponent. The general rule is given next.

To multiply powers having the same base:

1. Add the exponents.
2. Write the base with this sum as an exponent.

In symbols, $a^m a^n = a^{m+n}$.

Example 1 We add exponents in each case:

$$x^2x^5 = x^{2+5} = x^7 \qquad a^m a^n = a^{m+n}$$
$$y^3y^7 = y^{3+7} = y^{10} \qquad \text{Adding exponents}$$
$$xx^8 = x^1 \cdot x^8 = x^{1+8} = x^9 \qquad x = x^1$$

Remark. In Example 1 we used the fact that $x = x^1$; that is, if no exponent is written, it is understood to be 1. We already saw that if no coefficient is indicated, the coefficient is 1. The understood 1 may also occur in the denominator: $x = x/1$. Remember that

$$x = \frac{1x^1}{1}$$

Multiplication of two or more monomials can be understood by means of the commutative law. For example, to multiply $2x^3$ by $5x^4$, we may rearrange the factors as follows:

$$(2x^3)(5x^4) = 2 \cdot x^3 \cdot 5 \cdot x^4 = 2 \cdot 5 \cdot x^3 \cdot x^4 = 10x^7 \qquad x^3 \cdot 5 = 5 \cdot x^3$$

This observation leads to the following rule:

To multiply two monomials:

1. Multiply the coefficients.
2. Multiply the factors that have the same base.

Example 2 Multiply: $(-3a^4)(2a^7)$.

Solution.

$$\begin{aligned} (-3a^4)(2a^7) &= (-3 \cdot 2)(a^4a^7) && \text{Rearranging} \\ &= -6a^4a^7 && -3 \cdot 2 = -6 \\ &= -6a^4a^7 \\ &= -6a^{4+7} = -6a^{11} && \text{Adding exponents} \end{aligned}$$

Example 3

a. $(-2x^5)(4x^3) = (-2)(4)x^5x^3 = -8x^8$

b. $$\begin{aligned} (-2RV^2)(-6R^2V) &= (-2)(-6)RR^2V^2V \\ &= 12R^1R^2V^2V^1 && R = R^1,\ V = V^1 \\ &= 12R^3V^3 && \text{Adding exponents} \end{aligned}$$

c. $$\begin{aligned} (2a^2bc)(-3abc^2)(4a^2bc) &= (2)(-3)(4)(a^2a^1a^2)(b^1b^1b^1)(c^1c^2c^1) \\ &= -24a^5b^3c^4 \end{aligned}$$

Other Laws of Exponents

In our discussion of multiplication of monomials, we used the rule $a^ma^n = a^{m+n}$. In the remainder of this chapter we will also need the other laws of exponents:

Laws of exponents:

If m and n are positive integers, then

$$a^ma^n = a^{m+n} \tag{5.1}$$

$$\frac{a^m}{a^n} = a^{m-n} \quad (m > n,\ a \neq 0) \tag{5.2}$$

$$(a^m)^n = a^{mn} \tag{5.3}$$

$$(ab)^n = a^nb^n \tag{5.4}$$

$$\left(\frac{a}{b}\right)^n = \frac{a^n}{b^n} \quad (b \neq 0) \tag{5.5}$$

In addition to rule (5.1), we will discuss rules (5.3) and (5.4) in this section. Rules (5.2) and (5.5) are listed here only for completeness and will be discussed in detail in Section 5.3.

To understand rule (5.3),

$$(a^m)^n = a^{mn}$$

consider the expression $(x^3)^4$. By the definition of exponent

$$(x^3)^4 = x^3x^3x^3x^3 = x^{3+3+3+3} = x^{3\cdot 4} = x^{12}$$

In other words, when raising a power to a power, we multiply exponents.

Example 4

$(x^4)^3 = x^{4\cdot 3} = x^{12}$ Multiplying exponents

$(y^2)^5 = y^{2\cdot 5} = y^{10}$ Multiplying exponents

Rule (5.4), $(ab)^n = a^nb^n$, says that the power of a product is the product of the powers. To see why, consider the expression $(ab)^2$. By definition,

$$(ab)^2 = (ab)(ab)$$

Rearranging, we get $(ab)(ab) = (a \cdot a)(b \cdot b)$:

$$(ab)^2 = (ab)(ab) = (a \cdot a)(b \cdot b) = a^2b^2$$

Example 5

$$(xy)^3 = x^3y^3$$

since

$$(xy)^3 = (xy)(xy)(xy) = (x \cdot x \cdot x)(y \cdot y \cdot y) = x^3y^3$$

Similarly,

$$\begin{aligned}(x^2y)^3 &= (x^2y)(x^2y)(x^2y)\\ &= (x^2 \cdot x^2 \cdot x^2)(y \cdot y \cdot y)\\ &= (x^2)^3y^3\\ &= x^6y^3\end{aligned}$$

Common error Adding exponents when raising a power to a power. Thus

$(a^3)^5$ should not be written as a^8

The correct procedure is

$$(a^3)^5 = a^{3\cdot 5} = a^{15}$$

The rules discussed so far often occur in combination, as shown in the remaining examples.

Example 6 Simplify the following expressions:

a. $(-3x^2)^3$ **b.** $(-2a^2b^3)^4$

Solution.

a. $(-3x^2)^3 = (-3)^3(x^2)^3$ — $(ab)^n = a^n b^n$

$= -27(x^2)^3$ — $(-3)^3 = (-3)(-3)(-3) = -27$

$= -27(x^2)^3$

$= -27x^6$ — Multiplying exponents

b. $(-2a^2b^3)^4 = (-2)^4(a^2)^4(b^3)^4$ — $(ab)^n = a^n b^n$

$= 16(a^2)^4(b^3)^4$ — $(-2)^4 = 16$

$= 16a^8b^{12}$ — Multiplying exponents

Example 7 Perform the indicated operations and simplify:

$$(-3x^2y^3)^2(4x^4y^5)^3$$

Solution. The monomials have to be rewritten without the exponents outside the parentheses before the multiplication can be carried out:

$(-3x^2y^3)^2(4x^4y^5)^3 = [(-3)^2(x^2)^2(y^3)^2][4^3(x^4)^3(y^5)^3]$ — $(ab)^n = a^n b^n$

$= (9)(x^2)^2(y^3)^2 \cdot 64(x^4)^3(y^5)^3$ — $(-3)^2 = 9$, $4^3 = 64$

$= 9x^4y^6 \cdot 64x^{12}y^{15}$ — Multiplying exponents

$= 9 \cdot 64x^4x^{12}y^6y^{15}$ — Rearranging

$= 576x^{16}y^{21}$ — $9 \cdot 64 = 576$, adding exponents

Common error Not distinguishing between $(-ax)^2$ and $-(ax)^2$. Note that

$$(-ax)^2 = (-ax)(-ax) = a^2x^2$$

while

$$-(ax)^2 = -1(ax)^2 = -1(ax)(ax) = -1a^2x^2 = -a^2x^2$$

So $(-ax)^2 \neq -(ax)^2$.

Example 8 **a.** $(-2x)^2 = (-2)^2x^2 = 4x^2$

b. $(-2x)^3 = (-2)^3x^3 = -8x^3$

c. $-(-2x)^4 = -1(-2x)^4 = -1(-2)^4x^4 = -1(16)x^4 = -16x^4$

Example 9 a. $(-2a^2b)^4 = (-2)^4(a^2)^4b^4 = 16a^8b^4$
b. $-(2a^2b)^4 = -1(2a^2b)^4 = -1(2)^4(a^2)^4b^4 = -1 \cdot 16a^8b^4 = -16a^8b^4$
c. $(-2a^2b)^5 = (-2)^5(a^2)^5b^5 = -32a^{10}b^5$

Exercises / Section 5.1

In Exercises 1–36, perform the indicated operations and simplify.

1. x^2x^4 **2.** y^3y^4 **3.** a^5a^7
4. b^4b^5 **5.** $(2a^2)(-4a^3)$ **6.** $(-3b^3)(4b^2)$
7. $(-4s^3)(-3s^2)$ **8.** $(7t^4)(-4t^5)$ **9.** $(-10V^5)(-2V^2)$
10. $(8P^3)(-4P^6)$ **11.** $(-2xy^2)(4x^2y)$ **12.** $(3a^2b^3)(-2ab^2)$
13. $(4T_1T_2^3)(-5T_1^2T_2)$ **14.** $(-7V_a^2V_b^3)(-3V_a^4V_b)$ **15.** $(-ab)^2$
16. $(-2xy^2)^2$ **17.** $(-ab)^3$ **18.** $(-2xy^2)^3$
19. $(-mv^2)^3$ **20.** $(RI^2)^2$ **21.** $(-t_1^2t_2^3)^4$
22. $(-3C_1^2C_2^3)^3$ **23.** $(-2ab^2c)^2$ **24.** $(-4x^2yz^2)^3$
25. $(-5PQ^2)^2(-P^2Q)^3$ **26.** $(-M_1M_2^3)^3(-3M_1^2M_2)$ **27.** $(-7L_1^2L_2L_3^3)(-6L_1L_2^2L_3)$
28. $(-xy^2z)(-2x^2y)(5x^2y^2z)$ **29.** $(-ab)^3(2a^2b^3)^2$ **30.** $(-s^2t^3)(-3s^2t^2)(s^2t)^2$
31. $(-st^2)^3(-s^2t)^2(st)$ **32.** $(-V^2L)^2(-2VL^2)^2(-3VL^2)$ **33.** $(-2v_1v_2^3)^2(-v_1^2v_2)^3(-2v_1v_2^3)$
34. $(-2a^3b^4)^3(-a^3b^5)^2(a^2b^2)^3$ **35.** $(-4R^2C^3)^2(-R^3C^4)^2(-2RC)^3$ **36.** $(2M^2N)^4(-MN^2)^3(-3M^2N^2)^2$

In Exercises 37–44, use a calculator to perform the indicated operations. Round off the coefficients to the proper number of significant digits. (The numbers are approximate.)

37. $(5.764V_1^2V_2)(-4.731V_1^2V_2^3)$ **38.** $(-3.768p^3q^4)(-7.104p^4q^7)$ **39.** $(-0.076r^2s^3)(-0.021r^4s^4)$
40. $(0.192v_1^2v_2^3)(-0.864v_1^4v_2^5)$ **41.** $(-2.68x^2y^3)^2$ **42.** $(-1.063x^3y^2)^3$
43. $(-2.360R^2V^3)^5$ **44.** $(-0.74T_1^2T_2^4)^4$

5.2 Multiplication of Polynomials

In the last section all our multiplications involved monomials. In this section we discuss multiplication with **polynomials**. Recall that a polynomial is a monomial or multinomial containing only exponents that are positive integers.

Product of a Monomial and a Multinomial

Our first case is the product of a monomial and a multinomial. This case requires the distributive law $a(b + c) = ab + ac$. For example,

$$3x(x^2 - 4y) = 3x(x^2) + 3x(-4y)$$
$$= 3x^3 - 12xy$$

Consider another example.

Example 1 Perform the following multiplication:

$$-2xy^2(3xy^2 - 6x^2y^3)$$

Solution. We use the distributive law $a(b + c) = ab + ac$ with $a = -2xy^2$, $b = 3xy^2$, and $c = -6x^2y^3$. Then we get

$$\begin{aligned} -2xy^2(3xy^2 - 6x^2y^3) &= -2xy^2(3xy^2) - 2xy^2(-6x^2y^3) \\ &= -6x^2y^4 + 12x^3y^5 \end{aligned}$$

Example 2

$$\begin{aligned} (3V_1 - 2V_2)V_1 &= (3V_1)V_1 + (-2V_2)V_1 \\ &= 3V_1^2 - 2V_1V_2 \end{aligned}$$

The next example involves a sum of two products.

Example 3 Perform the indicated operations and simplify:

$$2(a - 5b) + 4(2a + b)$$

Solution. We perform the multiplication before the addition:

$$\begin{aligned} 2(a - 5b) + 4(2a + b) &= 2a - 10b + 8a + 4b && \text{Distributive law} \\ &= 10a - 6b && \text{Combining like terms} \end{aligned}$$

Product of Two Multinomials

The product of two polynomials containing more than one term can be found by repeated application of the distributive law. Consider the product

$$(x - 3y)(2x + y) \tag{5.6}$$

Referring to the distributive law $a(b + c) = ab + ac$, let $a = x - 3y$, $b = 2x$, and $c = y$. It follows that

$$\begin{aligned} a \cdot (b + c) &= a \cdot b + a \cdot c \\ (x - 3y) \cdot (2x + y) &= (x - 3y) \cdot (2x) + (x - 3y) \cdot (y) \end{aligned}$$

We now use the distributive law on the terms on the right side:

$$\begin{aligned} (x - 3y)(2x) + (x - 3y)(y) &= [(x)(2x) + (-3y)(2x)] \\ &\quad + [(x)(y) + (-3y)(y)] \\ &= (2x^2 - 6xy) + (xy - 3y^2) \\ &= 2x^2 - 5xy - 3y^2 \end{aligned} \tag{5.7}$$

A careful comparison of the terms in statements (5.6) and (5.7) tells us that to multiply two polynomials, we multiply every term in the first polynomial by every term in the second and add the resulting products.

This procedure can be made systematic by using a scheme that resembles multiplication in arithmetic. Let us return to the product $(x - 3y)(2x + y)$ and write the polynomials in two rows:

$$\begin{array}{r} x - 3y \\ \underline{2x + y} \end{array}$$

Working from left to right, starting with $2x$, multiply $2x$ by every term in the first row and place the resulting products in a new row below the line:

$$\begin{array}{r} x - 3y \\ \underline{2x + y} \\ 2x^2 - 6xy \end{array}$$

Next multiply $+y$ in the second row by every term in the first row and write the products in the fourth row:

$$\begin{array}{lll} x - 3y & & \text{First row} \\ \underline{2x + y} & & \text{Second row} \\ 2x^2 - 6xy & & \text{Third row} \\ xy - 3y^2 & & \text{Fourth row} \end{array}$$

Note especially that $-6xy$ and xy are like terms and are therefore placed in the same column to help with the addition step. After drawing another line and adding, we get

$$\begin{array}{rrrl} & x & -\ 3y & \\ & \underline{2x} & \underline{+\ y} & \\ 2x^2 & -\ 6xy & & \\ & \underline{xy} & \underline{-\ 3y^2} & \\ 2x^2 & -\ 5xy & -\ 3y^2 & \quad \text{since } xy = 1xy \end{array}$$

Consider another example.

Example 4 Perform the following multiplication:

$$(a^2 - 3ab + 2b^2)(a - 4b)$$

Solution. First we write the polynomials in two rows:

$$\begin{array}{r} a^2 - 3ab + 2b^2 \\ \underline{a - 4b} \end{array}$$

Again working from left to right, we multiply a (in the second row) by every term in the first row and write the resulting products in the row below the line:

$$\begin{array}{r}
a^2 - 3ab + 2b^2 \\
a - 4b \\
\hline
a^3 - 3a^2b + 2ab^2
\end{array}$$

Next, we multiply $-4b$ by every term in the first row and place the resulting products in the next row:

$$\begin{array}{rrrr}
a^2 & - 3ab & + 2b^2 & \\
 & a & - 4b & \\
\hline
a^3 & - 3a^2b & + 2ab^2 & \\
 & - 4a^2b & + 12ab^2 & - 8b^3 \\
\hline
a^3 & - 7a^2b & + 14ab^2 & - 8b^3
\end{array}$$

after drawing another line and adding.

A polynomial may have a missing power, indicating a coefficient of 0. For example, the polynomial $x^3 - 3x + 4$ can be written $x^3 + 0x^2 - 3x + 4$. Consider the next example.

Example 5 Perform the following multiplication:

$$(x^3 - 3x + 4)(3x - 2)$$

Solution. If we insert $0x^2$ for the missing second power, we get

$$\begin{array}{rrrrr}
 & x^3 & + 0x^2 & - 3x & + 4 \\
 & & & 3x & - 2 \\
\hline
3x^4 & + 0x^3 & - 9x^2 & + 12x & \\
 & - 2x^3 & + 0x^2 & + 6x & - 8 \\
\hline
3x^4 & - 2x^3 & - 9x^2 & + 18x & - 8
\end{array}$$

A good alternative is to leave a space for the missing power:

$$\begin{array}{rrrrr}
 & x^3 & & - 3x & + 4 \\
 & & & 3x & - 2 \\
\hline
3x^4 & & - 9x^2 & + 12x & \\
 & - 2x^3 & & + 6x & - 8 \\
\hline
3x^4 & - 2x^3 & - 9x^2 & + 18x & - 8
\end{array}$$

Longer polynomials are multiplied in the same manner. However, the intermediate products may require more than two rows.

Example 6

$$\begin{array}{rrrrr}
2x^2 - & 3xy + & 5y^2 & & \\
-x^2 + & 4xy - & 7y^2 & & \\
\hline
-2x^4 + & 3x^3y - & 5x^2y^2 & & \\
 & 8x^3y - & 12x^2y^2 + & 20xy^3 & \\
 & & - 14x^2y^2 + & 21xy^3 - & 35y^4 \\
\hline
-2x^4 + & 11x^3y - & 31x^2y^2 + & 41xy^3 - & 35y^4
\end{array}$$

Example 7 The voltage drop V (in volts) across a resistor is $V = IR$, where I is the current (in amperes) and R the resistance (in ohms). Suppose I and R vary with time t (in seconds) by the following relations: $I = 1.0t - 2.0$ and $R = 2.0t - 1.0$. Find an expression for V.

Solution. Since $V = IR$, we need to find the following product:

$$\begin{array}{lll}
1.0t & - 2.0 & \\
2.0t & - 1.0 & \\
\hline
2.0t^2 & - 4.0t & \\
 & - 1.0t & + 2.0 \\
\hline
2.0t^2 & - 5.0t & + 2.0
\end{array}$$

So $V = 2.0t^2 - 5.0t + 2.0$ (in volts).

Exercises / Section 5.2

In Exercises 1–54, perform the indicated operations.

1. $-x(x + y)$

2. $-y(x - y)$

3. $2x(4x - 3)$

4. $5x(x - 3x^2)$

5. $3(2P - Q)$

6. $R(-2R_1 - R_2)$

7. $-2x^2(-5x + 2x^2 - 7x^3)$

8. $-4x(-7x + 10x^2 - 2x^3)$

9. $(3ab - 4a^2b^2)(2ab^2)$

10. $(5c^2d^3 - 7cd^4)(4c^2d)$

11. $-7T_1T_2(-5T_1^2T_2 + 3T_1T_2^2)$

12. $-8C_1^2C_2(5C_1C_2^2 + 7C_1^2C_2^2)$

13. $2(x + 2) + 4(x - 1)$ (See Example 3.)

14. $3(x - 2) + 3(x + 2)$

15. $5(a - 4) + 2(2a - 1)$

16. $4(1 - 2b) + 2(2 + b)$

17. $3(a - 2b) - 2(4a - 3b)$

18. $4(R - 3C) - 5(R + 2C)$

19. $2(v - 4w) - 2(2v + w)$

20. $5(2s + 2t) - 3(s - 5t)$

21. $(x + 1)(x - 1)$

22. $(x + 2)(x - 2)$

23. $(2x + 3)(2x - 3)$

24. $(4x - 6)(4x - 6)$

25. $(2x - 4)(x + 3)$

26. $(x - 7)(3x + 4)$

27. $(x + 5y)(x - 4y)$

28. $(2x - 3y)(x + 2y)$

29. $(P - 2V)(P + 3V)$

30. $(2R - 6Q)(R - Q)$

31. $(S - T)^2$

32. $(2V - Q)^2$

33. $(T_a - 2T_b)^2$

34. $(2V_a - V_b)^2$

35. $(x^2 + x + 5)(x - 1)$

36. $(x^2 - 2x + 1)(x - 2)$

37. $(s^2 - 5s - 3)(2s + 5)$

38. $(r^2 - 4r + 1)(3r + 2)$

39. $(x^2 + xy + 2y^2)(x - y)$

40. $(x^2 - xy + y^2)(2x - y)$

41. $(s^2 - 3st + t^2)(2s - t)$

42. $(v^2 - vw - w^2)(v + 2w)$

43. $(3R^2 - 2RV - V^2)(2R + V)$

44. $(2m^2 + 2mn - n^2)(2m - 3n)$

45. $(2x^3 - 2x + 1)(x - 3)$

46. $(5x^3 - x - 1)(2x + 1)$

47. $(7x^3 + 3x^2 + 4)(x - 7)$

48. $(4x^3 + 2x^2 + 3)(2x - 4)$

49. $(x^2 + 4x + 1)(2x^2 - 6x - 2)$

50. $(2y^2 - 2y - 3)(y^2 + y + 1)$

51. $(3v^2 - v + 2)(2v^2 + v - 1)$

52. $(2L^2 - 2L + 1)(3L^2 + L - 4)$

53. $(x^2 + 2xy - 3y^2)(2x^2 + xy - 2y^2)$

54. $(2R^2 + RC + 3C^2)(R^2 - 2RC - 2C^2)$

In Exercises 55–60, use a calculator to perform the indicated multiplications. Round off the coefficients to the proper number of significant digits.

55. $2.03(1.64x + 7.83)$

56. $10.4(4.05a - 1.96)$

57. $-15.8(4.76 - 5.89V)$

58. $-8.4(7.1 - 4.5t)$

59. $(2.0x - 3.6)(1.5x + 2.7)$

60. $(7.30x + 2.40)(3.90x - 9.40)$

61. Find an expression for the total cost of $a + 2b$ calculators costing \$20 each and $2a + 3b$ calculators costing \$25 each.

62. A shipment contains $m + 3n$ ball bearings weighing 2 lb each and $2m + 4n$ ball bearings weighing 3 lb each. Find an expression for the total weight of the shipment.

63. The increase in energy of a certain body of mass m is

$$\frac{1}{2}m(v_2^2 - v_1^2)$$

Write this expression without parentheses.

64. The length l of a plate expands due to heating according to the formula $l = l_0(1 + aT)$. Write the expression for l without parentheses.

65. The deflection y (in inches) of a certain beam 24 ft in length is $y = kx(24 - x)$, where x is the distance from one end and k a constant. Write y as a polynomial.

66. The energy K of a diatomic molecule has the form $K = M(r_1 + r_2)^2$. Write the expression for K as a polynomial.

67. Find an expression for the area A of the concrete form in Figure 5.1.

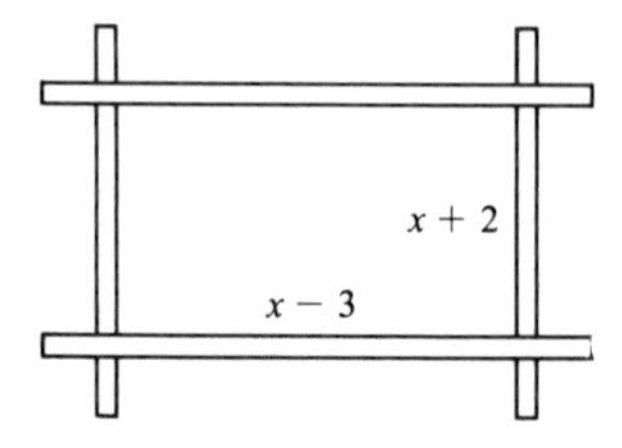

Figure 5.1

68. Find an expression for the area A of a square whose side has length $2x + 3y$.

69. The kinetic energy K of a body whose velocity v is much less than the velocity of light c is

$$K = (1 + \frac{1}{2}v^2)m_0c^2 - m_0c^2$$

Simplify the expression for K.

70. The resistance of a variable resistor is given by $R = 1.0t - 2.1$ and the current through the resistor by $i = 2.1t - 0.12$, where t is measured in seconds. Find an expression for the voltage across the resistor (voltage = current × resistance).

71. If an amount a of a substance is combined with an amount b of another substance, and if y is the amount that has already been converted after t seconds, then the rate R at which the reaction takes place is given by

$$R = k(a - y)(b - y)$$

where k is a constant. Express R as a polynomial.

5.3 Division of Monomials

In this section we study algebraic division of two monomials. For this operation we need rule (5.2):

$$\frac{a^m}{a^n} = a^{m-n} \qquad (m > n,\ a \neq 0)$$

This rule can be seen from the following example:

$$\frac{x^5}{x^2} = \frac{\overset{1}{\cancel{x}} \cdot \overset{1}{\cancel{x}} \cdot x \cdot x \cdot x}{\underset{1}{\cancel{x}} \cdot \underset{1}{\cancel{x}}} = x^{5-2} = x^3$$

In other words, if the exponent of the numerator is the greater, we simply subtract exponents to get the quotient.

Example 1

$$\frac{x^6}{x^4} = x^{6-4} = x^2 \qquad \text{Subtracting exponents}$$

$$\frac{x^{10}}{x^4} = x^{10-4} = x^6 \qquad \text{Subtracting exponents}$$

$$\frac{x^5}{x} = \frac{x^5}{x^1} = x^{5-1} = x^4 \qquad \text{Since } x^1 = x$$

When dividing two monomials, we divide the coefficients, as well as the literal factors.

Example 2

$$\frac{-4x^3y^6}{2x^2y^5} = \left(\frac{-4}{2}\right) x^{3-2}y^{6-5} = -2xy$$

$$\frac{-12a^7b^{12}}{-4a^4b^7} = \left(\frac{-12}{-4}\right) a^{7-4}b^{12-7} = 3a^3b^5$$

So far we have assumed that the exponents of the factors in the numerator are larger than the exponents of the factors in the denominator. If a factor in the numerator has a smaller exponent, we reduce the fraction to lowest terms. Consider, for example, the fraction

$$\frac{x^4}{x^7}$$

Since the exponent in the numerator is the smaller, we reduce the fraction by dividing numerator and denominator by x^4:

$$\frac{x^4}{x^7} = \frac{\frac{x^4}{x^4}}{\frac{x^7}{x^4}} = \frac{1}{x^{7-4}} = \frac{1}{x^3}$$

Note especially that

$$\frac{x^4}{x^4} = 1$$

since any number divided by itself is equal to 1.

In general,

$$\frac{a^n}{a^m} = \frac{1}{a^{m-n}} \qquad (m > n,\ a \neq 0) \tag{5.8}$$

Example 3

a. $\dfrac{x^3}{-x^{10}} = -\dfrac{\frac{x^3}{x^3}}{\frac{x^{10}}{x^3}} = -\dfrac{1}{x^7}$ or $\dfrac{x^3}{-x^{10}} = -\dfrac{x^3}{x^{10}} = -\dfrac{1}{x^{10-3}} = -\dfrac{1}{x^7}$

b. $\dfrac{-x^5}{x^{15}} = -\dfrac{\frac{x^5}{x^5}}{\frac{x^{15}}{x^5}} = -\dfrac{1}{x^{10}}$ or $\dfrac{-x^5}{x^{15}} = -\dfrac{x^5}{x^{15}} = -\dfrac{1}{x^{15-5}} = -\dfrac{1}{x^{10}}$

Example 3 illustrates the following rule:

$$\frac{-a}{b} = -\frac{a}{b} \quad \text{and} \quad \frac{a}{-b} = -\frac{a}{b}$$

(Negative signs should not be left in the numerator or denominator of a fraction.)

Example 4 Simplify the following fraction:

$$\frac{-8s^4t^6}{4s^6t^6}$$

Solution.

$$\frac{-8s^4t^6}{4s^6t^6} = -\frac{8s^4t^6}{4s^6t^6} \qquad \frac{-a}{b} = -\frac{a}{b}$$

$$= -\frac{8s^4\not{t^6}}{4s^6\not{t^6}} \qquad \text{Cancellation, } \frac{t^6}{t^6} = 1$$

$$= -\frac{2s^4}{s^6} \qquad \frac{8}{4} = 2$$

$$= -\frac{\frac{2s^4}{s^4}}{\frac{s^6}{s^4}} = -\frac{2}{s^2} \qquad \text{or } -\frac{2s^4}{s^6} = -\frac{2}{s^{6-4}} = -\frac{2}{s^2}$$

Some simplifications require rule (5.5):

$$\left(\frac{a}{b}\right)^n = \frac{a^n}{b^n}, \qquad b \neq 0$$

Example 5 Simplify:

$$\left(\frac{P^2}{V^3}\right)^2$$

Solution.

$$\left(\frac{P^2}{V^3}\right)^2 = \frac{(P^2)^2}{(V^3)^2} \qquad \left(\frac{a}{b}\right)^n = \frac{a^n}{b^n}$$

$$= \frac{P^4}{V^6} \qquad \text{Multiplying exponents}$$

The remaining examples illustrate certain combinations of operations.

Example 6 Reduce the following fraction:

$$\frac{(-2x^2y^2)^3}{(4xy^2)^2}$$

Solution.

$$\frac{(-2x^2y^2)^3}{(4xy^2)^2} = \frac{(-2)^3(x^2)^3(y^2)^3}{4^2x^2(y^2)^2} \qquad (ab)^n = a^nb^n$$

$$= \frac{-8(x^2)^3(y^2)^3}{16x^2(y^2)^2} \qquad (-2)^3 = -8;\ 4^2 = 16$$

$$= \frac{-8x^6y^6}{16x^2y^4} \qquad \text{Multiplying exponents}$$

$$= -\frac{8x^6y^6}{16x^2y^4} \qquad \frac{-a}{b} = -\frac{a}{b}$$

$$= -\frac{1}{2}x^4y^2 \qquad \text{Dividing coefficients and subtracting exponents}$$

Example 7 Perform the indicated operations and simplify:

$$\frac{(-3R^2V)^2(-6RV^2)}{(2R^3V^2)^4}$$

Solution.

$$\frac{(-3R^2V)^2(-6RV^2)}{(2R^3V^2)^4}$$

$$= \frac{(-3)^2(R^2)^2V^2(-6RV^2)}{2^4(R^3)^4(V^2)^4} \qquad (ab)^n = a^nb^n$$

$$= \frac{9(R^2)^2V^2(-6RV^2)}{16(R^3)^4(V^2)^4} \qquad (-3)^2 = 9;\ 2^4 = 16$$

$$= \frac{9R^4V^2(-6RV^2)}{16R^{12}V^8} \qquad \text{Multiplying exponents}$$

$$= \frac{-54R^5V^4}{16R^{12}V^8} \qquad \text{Multiplying coefficients and adding exponents}$$

$$= -\frac{27R^5V^4}{8R^{12}V^8} \qquad \frac{-a}{b} = -\frac{a}{b}\text{; dividing by 2}$$

$$= -\frac{27}{8R^7V^4} \qquad \text{Dividing numerator and denominator by } R^5V^4$$

Exercises / Section 5.3

Perform the indicated operations and simplify.

1. $\frac{x^7}{x^4}$
2. $\frac{a^6}{a^4}$
3. $\frac{2a^3}{a}$

4. $\dfrac{2V^6}{V^3}$

5. $\dfrac{8C^3D^4}{-4CD}$

6. $\dfrac{-12m^4n^5}{3mn^3}$

7. $\dfrac{-14a^6b}{-21a^2b^4}$

8. $\dfrac{-9p^4q^6}{15p^5q}$

9. $\dfrac{3a^3b}{ab}$

10. $\dfrac{5n^2m^2}{m^2}$

11. $\dfrac{-26T_1T_2^2}{13T_1^3T_2^4}$

12. $\dfrac{-18C_1^2C_2}{-12C_1^4C_2^5}$

13. $\dfrac{25x^2y^3z^6}{-15x^2yz^7}$

14. $\dfrac{-36a^3bc^2}{12abc}$

15. $\dfrac{-7C_1^2C_2^3C_3^4}{-21C_1C_2^4C_3^4}$

16. $\dfrac{26L_1^3L_2^4L_3}{39L_1^3L_2^3L_3^4}$

17. $\dfrac{(ab)^2}{a^2b}$

18. $\dfrac{(st)^3}{st^2}$

19. $\dfrac{-2xy^2}{(-2xy)^2}$

20. $\dfrac{-3a^2b}{(-3ab)^2}$

21. $\dfrac{(-2RC)^3}{(-2RC)^2}$

22. $\dfrac{(-4LM)^3}{(-4LM)^2}$

23. $\dfrac{(-xz)^4}{3x^2z}$

24. $\dfrac{(-m_1m_2)^3}{-2m_1m_2}$

25. $\dfrac{(-4R^2V)^3}{(-RV)^4}$

26. $\dfrac{(-3vw)^4}{(-vw)^3}$

27. $\dfrac{(-2x^2y^3z^4)^2}{(-xy^2z)^3}$

28. $\dfrac{(2a^3y^2n^3)^2}{(-3a^2y^2n^3)^3}$

29. $\dfrac{(-L_1^2L_2^2L_3^4)^2}{(-3L_1L_2^2L_3)^4}$

30. $\dfrac{(5P_1P_2^2P_3)^3}{(-5P_1P_2P_3^3)^4}$

31. $\left(\dfrac{x^2}{y}\right)^3$

32. $\left(\dfrac{a^3}{b^4}\right)^2$

33. $\left(\dfrac{2x^2}{y^3}\right)^3$

34. $\left(\dfrac{C}{4A^2}\right)^3$

35. $\left(\dfrac{-2V^2}{R^3}\right)^2$

36. $\left(\dfrac{N^4}{-3M^3}\right)^3$

5.4 Division of Polynomials

In the last section we considered division of monomials. In this section we discuss the remaining cases, the division of a polynomial with two or more terms by a monomial and the division of one multinomial by another.

Division of a Multinomial by a Monomial

As in the case of multiplication, division by a monomial depends on the distributive law $a(b + c) = ab + ac$. Just as division by 2 is equivalent to multiplication by $\frac{1}{2}$, so division by $x(x \neq 0)$ is the same as multiplication by $1/x$. For example,

$$\begin{aligned}
\frac{4x^3 - 2x^2y}{x} &= \frac{1}{x}(4x^3 - 2x^2y) \\
&= \frac{1}{x}(4x^3) + \frac{1}{x}(-2x^2y) \qquad \text{Distributive law} \\
&= \frac{4x^3}{x} - \frac{2x^2y}{x} \\
&= 4x^2 - 2xy
\end{aligned}$$

In other words, to divide a polynomial by a monomial, divide each term of the polynomial by the monomial.

Example 1

a. $$\frac{4a^3b^2 - 8a^4b}{2a^2b} = \frac{4a^3b^2}{2a^2b} + \frac{-8a^4b}{2a^2b} = 2ab - 4a^2$$

b. $$\frac{9R_1^2R_2 + 12R_1R_2^2 - 15R_1R_2}{3R_1R_2} = \frac{9R_1^2R_2}{3R_1R_2} + \frac{12R_1R_2^2}{3R_1R_2} + \frac{-15R_1R_2}{3R_1R_2}$$
$$= 3R_1 + 4R_2 - 5$$

Example 2

$$\frac{26R^2C^3 - 39RC^2 + 13RC}{13RC} = \frac{26R^2C^3}{13RC} + \frac{-39RC^2}{13RC} + \frac{13RC}{13RC}$$
$$= 2RC^2 - 3C + 1$$

Note especially that

$$\frac{13RC}{13RC} = 1$$

since **any number divided by itself is equal to 1.**

Common error Improper cancellation when the numerator is a sum or difference:

$$\frac{a + a^2b}{a} \neq 1 + a^2b$$

obtained by canceling the a's. The correct procedure is

$$\frac{a + a^2b}{a} = \frac{a}{a} + \frac{a^2b}{a} = 1 + ab$$

Division of Two Multinomials

Recall that a polynomial is a monomial or multinomial with exponents that are positive integers. If a polynomial involves only powers of the same variable, it can be written with *descending* powers of the variable. For example, the polynomial

$$4x^3 + 5x^2 - x + 2$$

has x^3 in the first term, x^2 in the second, x in the third, and no x in the fourth. The **degree** of a polynomial of a single variable is the highest power of that variable. For example, the polynomial $3x^4 + 2x^2 + 1$ is of fourth degree, while $x^2 + 3x + 1$ is of second degree. (If a polynomial contains more than one variable, the degree of a term is the sum of the powers. Thus $2x^3y^4$ is of degree 7.)

Degree

To divide one polynomial by another, we write the polynomials in descending powers of one of the variables and set up the problem so it resembles long division in arithmetic. For example, to perform the division $(7x + 2x^2 - 15) \div (x + 5)$, we write the polynomials in descending powers of x,

$$(2x^2 + 7x - 15) \div (x + 5)$$

and then set up the division problem as follows:

$$x + 5\overline{)2x^2 + 7x - 15}$$

The division can now be carried out as follows:

Step 1. Write both polynomials in descending powers of one of the variables.

$$+\ 5\overline{)\qquad + 7x - 15}$$

Step 2. Divide the first term $(2x^2)$ of the dividend by the first term of the divisor (x) to obtain the first term $(2x)$ of the quotient (placed on top).

$$\begin{array}{r} \mathbf{2x}\phantom{{}+ 7x - 15} \\ x + 5\overline{)\mathbf{2x^2} + 7x - 15} \end{array}$$

$$\frac{2x^2}{x} = 2x$$

Step 3. Multiply the quotient $(2x)$ by the divisor $(x + 5)$. Place the product in the row below.

$$\begin{array}{r} 2x\phantom{{}+ 7x - 15} \\ +\ 5\overline{)2x^2 + 7x - 15} \\ 2x^2 + 10x\phantom{{}- 15} \end{array}$$

$$2x(x + 5) = 2x^2 + 10x$$

Step 4. Draw a line and subtract the product obtained in Step 3. (Remember to change signs when subtracting.)

$$\begin{array}{r} 2x\phantom{{}+ 7x - 15} \\ +\ 5\overline{)2x^2 + 7x - 15} \\ \underline{2x^2 + 10x}\phantom{{}- 15} \\ -\ 3x\phantom{{}- 15} \end{array}$$

$$2x^2 - 2x^2 = 0$$
$$7x - 10x = -3x$$

Step 5. Bring down the next term of the dividend.

$$\begin{array}{r} 2x\phantom{{}+ 7x - 15} \\ +\ 5\overline{)2x^2 + 7x - \mathbf{15}} \\ \underline{2x^2 + 10x}\phantom{{}- 15} \\ -3x - \mathbf{15} \end{array}$$

Step 6. Divide the first term $(-3x)$ in the last row obtained in Step 5 by the first term of the divisor (x). This is the next term of the quotient (-3).

$$\begin{array}{r} 2x - 3\phantom{{}- 15} \\ x + 5\overline{)2x^2 + 7x - 15} \\ \underline{2x^2 + 10x}\phantom{{}- 15} \\ \mathbf{-3x} - 15 \end{array}$$

$$\frac{-3x}{x} = -3$$

Step 7. Multiply the quotient (-3) by the divisor ($x + 5$) and place the product in the next row.

$$\begin{array}{r} 2x \;-\; 3 \\ x + 5\overline{)2x^2 + \;\; 7x - 15} \\ \underline{2x^2 + 10x} \qquad \\ -3x - 15 \\ -3x - 15 \end{array}$$

$-3(x + 5) = -3x - 15$

Step 8. Repeat Steps 4–7 until the remainder is 0 or of a degree less than that of the divisor.

$$\begin{array}{r} 2x \;-\; 3 \\ x + 5\overline{)2x^2 + \;\; 7x - 15} \\ \underline{2x^2 + 10x} \qquad \\ -3x - 15 \\ \underline{-3x - 15} \\ \text{Subtracting} \qquad 0 \end{array}$$

Since the remainder is 0, we need not repeat Steps 5–7. The quotient is $2x - 3$.

Example 3 Perform the following division:

$$(2 - 3x + x^2) \div (x - 2)$$

Solution.

Step 1. Write the polynomials in descending powers of x.

$$x - 2\overline{)x^2 - 3x + 2}$$

Step 2. Divide the first term (x^2) of the dividend by the first term of the divisor (x), to obtain the first term of the quotient (x).

$$\begin{array}{r} x \qquad\quad \\ x - 2\overline{)x^2 - 3x + 2} \end{array}$$

$\frac{x^2}{x} = x$

Step 3. Multiply the quotient (x) by the divisor ($x - 2$). Place the product in the next row.

$$\begin{array}{r} x \qquad\quad \\ x - 2\overline{)x^2 - 3x + 2} \\ x^2 - 2x \qquad \end{array}$$

$x(x - 2) = x^2 - 2x$

Step 4. Draw a line and subtract:

$$x^2 - x^2 = 0$$

$$-3x - (-2x) = -3x + 2x$$

$$= -x$$

$$\begin{array}{r} x \qquad\quad \\ x - 2\overline{)x^2 - 3x + 2} \\ \underline{x^2 - 2x} \qquad \\ -x \qquad \end{array}$$

Step 5. Bring down the next term of the dividend.

$$\begin{array}{r} x \qquad\quad \\ x - 2\overline{)x^2 - 3x + 2} \\ \underline{x^2 - 2x} \qquad \\ -x + 2 \end{array}$$

Step 6. Divide the first term $(-x)$ in the last row obtained in Step 5 by the first term of the divisor (x). This is the next term of the quotient (-1).

$$\begin{array}{r} x - 1 \\ x - 2\overline{)x^2 - 3x + 2} \\ \underline{x^2 - 2x} \\ -x + 2 \end{array} \qquad \frac{-x}{x} = -1$$

Step 7. Multiply the quotient (-1) by the divisor $(x - 2)$ and place the product in the next row.

$$\begin{array}{r} x - 1 \\ x - 2\overline{)x^2 - 3x + 2} \\ \underline{x^2 - 2x} \\ -x + 2 \\ -x + 2 \end{array} \qquad -1(x - 2) = -x + 2$$

Step 8. Repeat Steps 4–7 until the remainder is 0 or of a degree less than that of the divisor.

$$\begin{array}{r} x - 1 \\ x - 2\overline{)x^2 - 3x + 2} \\ \underline{x^2 - 2x} \\ -x + 2 \\ \underline{-x + 2} \\ 0 \end{array}$$

Since the remainder is 0, the division is completed. So

$$(2 - 3x + x^2) \div (x - 2) = x - 1$$

The polynomials in the next example contain two variables.

Example 4 Divide:

$$(4xy^2 - 5x^2y + 2x^3 - y^3) \div (2x - y)$$

Solution. We write the polynomials in descending powers of x and proceed as before.

Step 1.

$$2x - y\overline{)2x^3 - 5x^2y + 4xy^2 - y^3}$$

Step 2. Note that the first term of the dividend $(2x^3)$ is divided by the first term of the divisor $(2x)$ to obtain the first term of the quotient.

$$\begin{array}{r} x^2 \qquad\qquad\qquad \\ 2x - y\overline{)2x^3 - 5x^2y + 4xy^2 - y^3} \end{array}$$

Steps 3–4. Now we multiply x^2 by $2x - y$, place the product in the next row, and subtract.

$$\begin{array}{r} x^2 \qquad\qquad\qquad \\ 2x - y\overline{)2x^3 - 5x^2y + 4xy^2 - y^3} \\ \underline{2x^3 - x^2y} \qquad\qquad \\ - 4x^2y \qquad\qquad \end{array}$$

Step 5. Bring down the next term.

$$
\begin{array}{r}
x^2 \\
2x - y\overline{)2x^3 - 5x^2y + 4xy^2 - y^3} \\
\underline{2x^3 - x^2y} \\
-4x^2y + 4xy^2
\end{array}
$$

Steps 6–7. Divide $-4x^2y$ by $2x$ to obtain $-2xy$. Then multiply: $-2xy \cdot (2x - y)$.

$$
\begin{array}{r}
x^2 - 2xy \\
2x - y\overline{)2x^3 - 5x^2y + 4xy^2 - y^3} \\
\underline{2x^3 - x^2y} \\
-4x^2y + 4xy^2 \\
-4x^2y + 2xy^2
\end{array}
$$

Step 8. Repeat Steps 4–7.

$$
\begin{array}{r}
x^2 - 2xy + y^2 \\
2x - y\overline{)2x^3 - 5x^2y + 4xy^2 - y^3} \\
\underline{2x^3 - x^2y} \\
-4x^2y + 4xy^2 \\
\underline{-4x^2y + 2xy^2} \\
2xy^2 - y^3 \\
\underline{2xy^2 - y^3} \\
0
\end{array}
$$

Subtract and bring down $-y^3$; $y^2 \cdot (2x - y)$

Since the remainder is 0, we conclude that

$$(4xy^2 - 5x^2y + 2x^3 - y^3) \div (2x - y) = x^2 - 2xy + y^2$$

We saw in Section 5.2 that a missing power indicates a coefficient of 0. Alternatively, we can leave a space, as shown in the next two examples.

Example 5 Divide:

$$(x^3 - 3xy^2 + 2y^3) \div (x + 2y)$$

Solution. We need to leave a space for the missing x^2 term:

$$
\begin{array}{r}
x^2 - 2xy + y^2 \\
x + 2y\overline{)x^3 - 3xy^2 + 2y^3} \\
\underline{x^3 + 2x^2y} \\
-2x^2y - 3xy^2 \\
\underline{-2x^2y - 4xy^2} \\
xy^2 + 2y^3 \\
\underline{xy^2 + 2y^3} \\
0
\end{array}
$$

Example 6 Divide:

$$(4R + 1 + 2R^4 - 5R^3) \div (2R^2 - 3R - 1)$$

Solution. We write the polynomials in descending powers of R and leave a space for the missing R^2 term:

$$\begin{array}{r} R^2 - R - 1 \\ 2R^2 - 3R - 1 \overline{)2R^4 - 5R^3 + 4R + 1} \\ \underline{2R^4 - 3R^3 - R^2} \\ -2R^3 + R^2 + 4R \\ \underline{-2R^3 + 3R^2 + R} \\ -2R^2 + 3R + 1 \\ \underline{-2R^2 + 3R + 1} \\ 0 \end{array}$$

If the remainder is not 0, we place the remainder over the divisor and add the resulting fraction to the quotient. We used the same procedure in Section 1.8 to change an improper fraction to a mixed number. For example, to convert $\frac{14}{3}$, we divide 14 by 3 to obtain 4 with remainder 2. So

$$\frac{14}{3} = 4 + \frac{2}{3} = 4\frac{2}{3}$$

Example 7 Divide:

$$(12V^3 - 7V^2 - 26V + 21) \div (4V - 5)$$

Solution.

$$\begin{array}{r} 3V^2 + 2V - 4 \\ 4V - 5 \overline{)12V^3 - 7V^2 - 26V + 21} \\ \underline{12V^3 - 15V^2} \\ 8V^2 - 26V \\ \underline{8V^2 - 10V} \\ -16V + 21 \\ \underline{-16V + 20} \\ 1 \end{array}$$

Since the remainder is 1, we write the result as

$$3V^2 + 2V - 4 + \frac{1}{4V - 5}$$

Example 8 The combined resistance of three resistors in parallel (Figure 5.2) can be found from the formula

$$\frac{1}{R} = \frac{R_2R_3 + R_1R_3 + R_1R_2}{R_1R_2R_3}$$

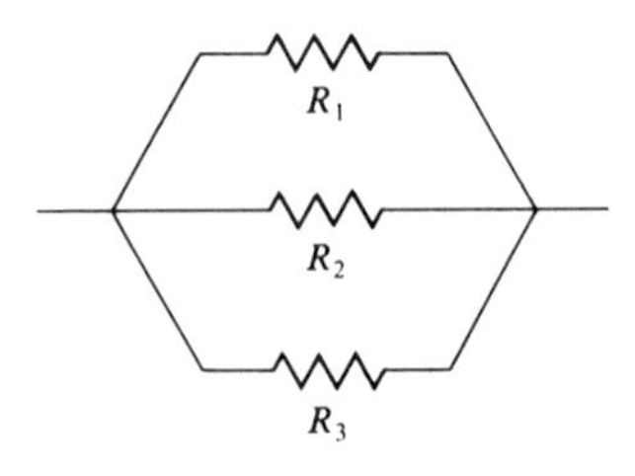

Figure 5.2

Simplify the expression for $1/R$ by division.

Solution.

$$\begin{aligned}\frac{1}{R} &= \frac{R_2R_3 + R_1R_3 + R_1R_2}{R_1R_2R_3}\\ &= \frac{R_2R_3}{R_1R_2R_3} + \frac{R_1R_3}{R_1R_2R_3} + \frac{R_1R_2}{R_1R_2R_3}\\ &= \frac{1}{R_1} + \frac{1}{R_2} + \frac{1}{R_3}\end{aligned}$$

Exercises / Section 5.4

In Exercises 1–68, perform the indicated divisions. (Use a calculator in Exercises 17–22.)

1. $\frac{x - y}{x}$
2. $\frac{x + y}{y}$
3. $\frac{2y^2 - 3y}{y}$
4. $\frac{4m^3 - 3m}{m}$
5. $\frac{6s^2 + 12s^3}{2s^2}$
6. $\frac{-14t^4 + 16t^5}{-2t^3}$
7. $\frac{10x^2y^3 - 15x^3y^2}{-5x^2y^2}$
8. $\frac{-12a^2b^2 + 20a^3b^4}{4a^2b^2}$
9. $\frac{10a^4x^5 - 2a^4x^6}{2a^4x^5}$
10. $\frac{4b^3y^6 + 8b^4y^7}{5b^3y^7}$
11. $\frac{-2ab + 2ab^2 - 4a^2b}{-2ab}$
12. $\frac{5s^2t^2 - 10s^3t^3 - 15st}{5st}$
13. $\frac{21C_1C_2 + 7C_1^2C_2 - 14C_1C_2^2}{7C_1C_2}$
14. $\frac{13V_1^2V_2^2 - 26V_1V_2 + 39V_1^3V_2^3}{13V_1V_2}$

15. $\dfrac{6m_0s^2 - 9m_0s + 12m_0^2s^2}{3m_0s}$

16. $\dfrac{9t^2z - 27t^2z^3 - 36t^3z^2}{9t^2z}$

17. $\dfrac{1.89x + 4.35}{1.23}$

18. $\dfrac{7.31a^2 - 6.43a}{2.36a}$

19. $\dfrac{21.4t - 8.46t^2}{10.6t}$

20. $\dfrac{0.75ab^2 + 1.8a^2b^2}{0.62ab}$

21. $\dfrac{0.7645V^2 - 0.1893V}{0.5406V}$

22. $\dfrac{1.014x^2y - 2.107xy}{10.37xy}$

23. $(x^2 - 5x + 6) \div (x - 2)$

24. $(x^2 - 6x + 8) \div (x - 4)$

25. $\dfrac{x + x^2 - 20}{x + 5}$

26. $\dfrac{x^2 - 3x - 18}{x + 3}$

27. $\dfrac{3y^2 - 7y + 4}{y - 1}$

28. $\dfrac{3 + 3m^2 - 6m}{m - 1}$

29. $(2s^2 + 5s - 3) \div (2s - 1)$

30. $(2t^2 - 7t - 4) \div (2t + 1)$

31. $(10z + 4z^2 - 6) \div (4z - 2)$

32. $(13w + 4w^2 - 12) \div (4w - 3)$

33. $(10r^2 + 16r - 8) \div (2r + 4)$

34. $(10y^2 + 18y + 8) \div (2y + 2)$

35. $(6n^2 - 19n + 15) \div (2n - 3)$

36. $(8a^2 + 14a - 15) \div (2a + 5)$

37. $(10x^2 - 13xy - 3y^2) \div (2x - 3y)$

38. $(6m^2 + 5mn - 4n^2) \div (3m + 4n)$

39. $(4s^2 + st - 3t^2) \div (4s - 3t)$

40. $(8a^2 - 2ab - 15b^2) \div (2a - 3b)$

41. $(6c^2 - 5cd - 4d^2) \div (2c + d)$

42. $(2v^2 + vw - 3w^2) \div (v - w)$

43. $(x^3 - x^2 - 13x - 3) \div (x + 3)$

44. $(x^3 - 5x^2 + 8x - 4) \div (x - 2)$

45. $(5y + 2y^3 - 7y^2 + 2) \div (y - 2)$

46. $(z^2 + 2z^3 - 2z + 8) \div (z + 2)$

47. $(3R^3 + 10R^2 - 4R + 16) \div (3R^2 - 2R + 4)$

48. $(3V^3 - 13V^2 + 14V - 6) \div (3V^2 - 4V + 2)$

49. $(4xz^2 + 3z^3 - 16x^2z + 8x^3) \div (4x^2 - 2xz - z^2)$

50. $(12x^2y - 3y^3 + 4x^3 + 7xy^2) \div (2x^2 + 3xy - y^2)$

51. $(9S^3 - 12S^2T + 10ST^2 - 4T^3) \div (3S^2 - 2ST + 2T^2)$

52. $(6m^3 + 7m^2n + 6mn^2 + 2n^3) \div (3m^2 + 2mn + 2n^2)$

53. $(R^2 - V^2) \div (R - V)$

54. $(T_1^2 - T_2^2) \div (T_1 + T_2)$

55. $(6a^3 - 4ab^2 - 2b^3) \div (2a - 2b)$

56. $(8x^3 - 8xy^2 - 3y^3) \div (2x + y)$

57. $(4s^3 - 11st^2 - 3t^3) \div (2s + 3t)$

58. $(9r^3 - 10rs^2 + 4s^3) \div (3r - 2s)$

59. $(x^3 + 2x^2y - 16y^3) \div (x - 2y)$

60. $(a^3 - 3a^2b + 20b^3) \div (a + 2b)$

61. $(x^3 - y^3) \div (x - y)$

62. $(x^3 + y^3) \div (x + y)$

63. $(x^4 - 2x^2y^2 + 2y^4) \div (x^2 - y^2)$

64. $(2x^4 - x^2y^2 - 2y^4) \div (2x^2 + y^2)$

65. $(2a^4 - 2a^2b^2 - 4b^4) \div (a^2 + 2b^2)$

66. $(4v^4 + 2v^2d^2 + 3d^4) \div (2v^2 - 2d^2)$

67. $(V^4 - W^4) \div (V^2 - W^2)$

68. $(4R^4 - C^4) \div (2R^2 + C^2)$

69. The following is a special case of Balmer's formula used in the study of the hydrogen spectrum:

$$u = 4k\left(\frac{n^2 - 4}{4n^2}\right)$$

Rewrite Balmer's formula by performing the indicated division.

70. If R is the combined resistance of two resistors R_1 and R_2 in parallel, then

$$\frac{1}{R} = \frac{R_2 + R_1}{R_2 R_1}$$

Rewrite the expression for $1/R$ by performing the indicated division.

71. The area of a rectangle is $A = x^3 + 4x^2 + 5x + 2$, and the length is $l = x + 2$. Find an expression for the width w (width $= A \div l$).

72. If 5 units cost \$20, how much does 1 unit cost? If $x - 3$ units cost $x^2 - x - 6$ dollars, how much does 1 unit cost?

73. If you can drive 90 mi on 3 gal of gasoline, how many miles per gallon do you get? If you can drive $x^2 + 2x - 15$ miles on $x + 5$ gallons, how many miles per gallon do you get?

Review Exercises / Chapter 5

In Exercises 1–46, perform the indicated operations.

1. $(-4s^2t)(-8st^3)$

2. $(5M^3T^2)(-3M^4T)$

3. $(-2C_1^2C_2^3)^3$

4. $(-2C_1^2C_2^3)^4$

5. $(3v^2w)^2(-2v^3w^2)^3$

6. $(-1.3a^3b)^2(2.4a^4b)^2$

7. $5a^2(7a^2c - 5ac^2 + 6a^3c^3)$

8. $-2p^2(-3pq^2 + 6p^2q - 5p^2q^2)$

9. $(2x - 3y)(4x + y)$

10. $(3a + b)(2a + 3b)$

11. $(L - 2v)(L + 2v)$

12. $(3m - 2n)(3m + 2n)$

13. $(3s^2 + 2st - t^2)(2s - t)$

14. $(2R^2 - 6RV + V^2)(3R - 2V)$

15. $(2C_1 + 3C_2)^2$

16. $(T_1 - 6T_2)^2$

17. $(1.30x - 2.40y)(5.70x - 4.10y)$

18. $(3.41x + 3.63y)(1.86x - 2.74y)$

19. $(2R^3 + R^2C - 3RC^2 + C^3)(2R + 2C)$

20. $(3M^3 - 2M^2N + MN^2 - 2N^3)(2M + N)$

21. $\dfrac{26x^4y^6}{-13x^2y}$

22. $\dfrac{-36a^8b^4}{12ab^2}$

23. $\dfrac{-18R^3N^4T^2}{-6RN^5T^6}$

24. $\dfrac{-21T_a^2T_b^4}{-14T_a^3T_b^3}$

25. $\dfrac{(-2L^2M^3)^3}{(-2L^3M^4)^2}$

26. $\dfrac{(-4a^3z^4)^2}{(-3a^2z^4)^3}$

27. $\dfrac{(-2abc)^3}{(-2abc)^2}$

28. $\dfrac{(-cde)^2}{(-cde)^3}$

29. $\left(\dfrac{-2x}{y^3}\right)^3$

30. $\left(\dfrac{-2x^3}{y^4}\right)^4$

31. $\dfrac{7.64c^2 - 8.36c}{1.76c}$

32. $\dfrac{10.46p^2 + 20.06pq}{5.344p}$

33. $\dfrac{4F^2 - 6F^3}{2F^2}$

34. $\dfrac{3D^2 + 6D^3}{-3D^2}$

35. $\dfrac{3a^2b - 3ab + 6ab^2}{-3ab}$

36. $\dfrac{8m^3v^2 + 4mv + 16m^2v^3}{-4mv}$

37. $(2x^2 + xy - 6y^2) \div (2x - 3y)$

38. $(2x^2 + 9xy - 5y^2) \div (2x - y)$

39. $(6x^2y + 8x^3 - 7xy^2 + 3y^3) \div (2x + 3y)$

40. $(3R^3 + 4R^2C - 3RC^2 + 2C^3) \div (R + 2C)$

41. $(6a^3 + 11a^2b - 2b^3) \div (2a + b)$

42. $(2a^3 + 3a^2b + 4b^3) \div (a + 2b)$

43. $(4m^3 - 10m^2n + 12mn^2 - 9n^3) \div (2m^2 - 2mn + 3n^2)$

44. $(6M^3 - 7M^2N + N^3) \div (3M^2 - 2MN - N^2)$

45. $(4x^2 - 2x - 1) \div (2x + 1)$

46. $(3R - 4R^2 + 2R^3 - 2) \div (R - 1)$

47. The energy W required to lift a body of mass m from a level y_1 to y_2 is

$$W = mg(y_2 - y_1)$$

Write the expression for W without parentheses.

48. The temperature T in a certain bar 10 cm long is $T = ax(100 - x^2)$, where x is the distance from the left end. Write T as a polynomial.

49. A string of length l is fixed at both ends. At some instant it is vibrating so that the velocity at each point is $v = 4x(l - x)$, where x is the distance from one end. Write v as a polynomial.

50. When measuring the specific heat c of a substance, the following expression arises: $(m_1 + m_2c)(t_2 - t_1)$. Perform the indicated multiplication.

51. Van der Waal's equation, which expresses a relationship between the temperature T of a gas, the volume V, and the pressure P is given by

$$T = \frac{1}{R}\left(P + \frac{a}{V^2}\right)(V - b)$$

Multiply the factors in the expression for T.

52. Given two inductances, $L_1 = 0.32t + 0.16$ and $L_2 = 0.24t - 0.10$ (t in seconds), find an expression for the average L of the inductances ($L = (L_1 + L_2)/2$).

53. If the current i (in amperes) in a circuit is $i = 2.40t - 1.60$, then the power P (in watts) delivered to a 10.0-Ω resistor is $P = 10.0i^2$. Find an expression for P.

54. The following expression arose in a problem in mechanics:

$$\frac{2mv^2 - 4m^2v^4}{2mv^2}$$

Simplify this expression by division.

55. Four resistors in parallel give rise to the expression

$$\frac{R_2R_3R_4 + R_1R_3R_4 + R_1R_2R_4 + R_1R_2R_3}{R_1R_2R_3R_4}$$

Simplify by performing the indicated division.

56. A box contains $x + 3$ machine parts of equal weight. If the total weight (in pounds) of the contents is $x^3 + 5x^2 + 7x + 3$, how much does each machine part weigh?

57. A small firm pays each of 10 employees $(2n + 3)$ dollars per week and each of 20 employees $(3n + 4)$ dollars per week. Find a simplified expression for the total weekly wages.

CHAPTER 6

Introduction to Geometry

In our earlier chapters we encountered geometric figures from time to time, particularly squares and rectangles. In this chapter we study other geometric figures, as well as some of the terms and definitions used in geometry.

6.1 Basic Concepts; Angles and Triangles

The basic geometric concepts are those of a *point,* a *line,* and a *plane*. These concepts are too basic to be defined in terms of other geometric concepts and are usually accepted without formal definition. Although not formally defined, these concepts do serve as models for certain physical objects. For example, we may think of a line as a stretched wire, a plane as a flat region such as a wall or a table top, and a point as a period or dot.

Line

A **line** (also called a **straight line**) extends indefinitely in two directions (Figure 6.1). A portion of a line has a definite length but no thickness.

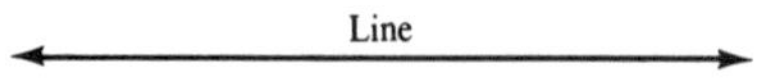

Line

Figure 6.1

Point

Plane

Two intersecting lines meet at a **point**. For example, the lines l_1 and l_2 in Figure 6.2 meet at point P. A point has position but no dimensions. The lines l_1 and l_2 in Figure 6.2 lie in the **plane** of the paper. A plane is flat and extends indefinitely in all directions.

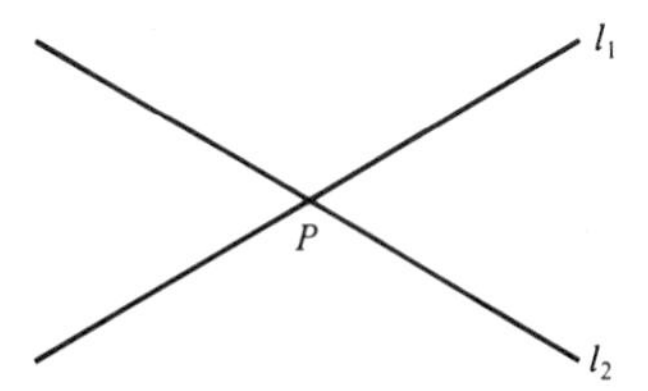

Figure 6.2

Line segment
Endpoints
Ray

Given two points A and B, there exists exactly one line passing through these points (Figure 6.3). This line, denoted by AB, is referred to as "line AB." The portion of the line between A and B is called a **line segment** (Figure 6.4). A line segment includes the **endpoints** A and B. If a line segment is extended in one direction, it becomes a **ray**. For example, the ray AB in Figure 6.5 has A for its endpoint and extends indefinitely to the right.

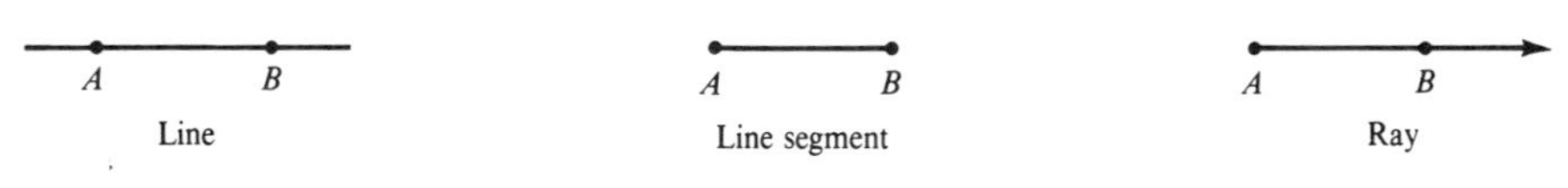

Figure 6.3 **Figure 6.4** **Figure 6.5**

Remark on notation. A common notation for a line through A and B is $\overleftrightarrow{AB}$; for a line segment, $\overline{AB}$; and for a ray with endpoint A, $\overrightarrow{AB}$. We will use the notation AB for all three.

Angles

Angle
Vertex
Sides

A figure consisting of two rays with a common endpoint is called an **angle**. (See Figure 6.6.) The point A is called the **vertex** of the angle and the two rays are called the **sides** of the angle. The angle in Figure 6.6 is denoted by the symbol $\angle A$. The angle in Figure 6.7 may also be denoted by $\angle ABC$, where B, the middle letter, is the vertex. A third way to denote an angle is to use a small letter or number inside the angle: $\angle x$ (Figure 6.7).

Figure 6.6 **Figure 6.7**

Angle measure

For a given angle, imagine that one side is rotated about the vertex until it coincides with the other side. The amount of rotation needed to bring the two sides together is called the **measure of the angle**. The amount of rotation is often suggested by a circular arrow or a circular arc (Figure 6.8).

Figure 6.8

A common unit of measure is the *degree*. (The symbol for degree is °; thus 5° means "5 degrees.") One complete rotation of a ray about its endpoint, called a **whole angle**, is defined to be an angle of 360°. (See Figure 6.9.) One-half of a complete rotation is called a **straight angle** and measures 180° (Figure 6.9). (Note that a straight angle forms a straight line.) One-fourth of a complete rotation is a **right angle** and measures 90° (Figure 6.9).

Whole angle
Straight angle
Right angle

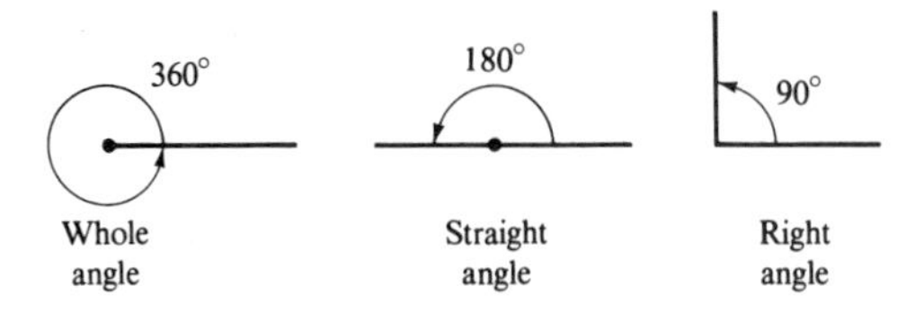

Figure 6.9

Since a complete rotation is 360°, it follows that 1° is $\frac{1}{360}$ of a complete rotation.

Degree measure

1° is equal to $\dfrac{1}{360}$ of a complete rotation.

If two lines meet so that they form a right angle, the lines are said to be **perpendicular**. The little square in Figure 6.10 is often used to denote perpendicularity. Point A is called the **foot** of the perpendicular. (Note that if two lines are perpendicular, they form four right angles.)

Perpendicular
Foot

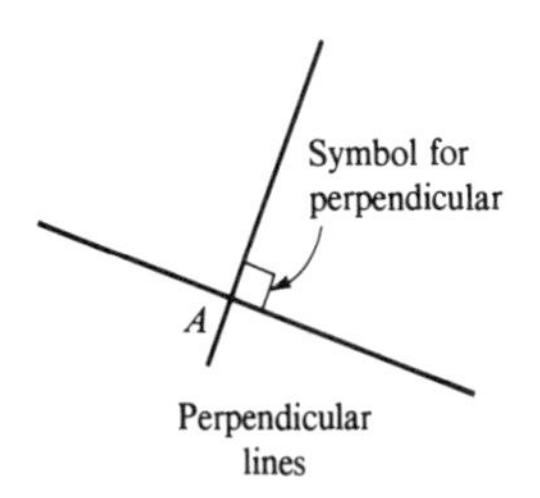

Figure 6.10

Example 1

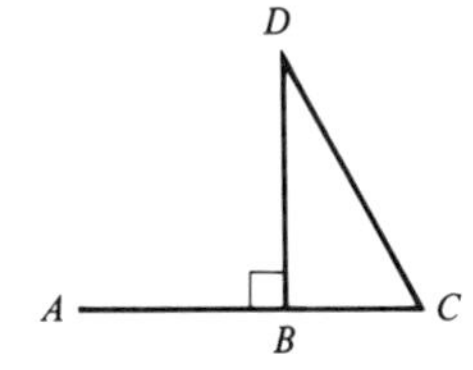

Figure 6.11

In Figure 6.11, $\angle ABD$ is a right angle, $\angle CBD$ is also a right angle, and $\angle ABC$ is a straight angle, since ABC forms a straight line.

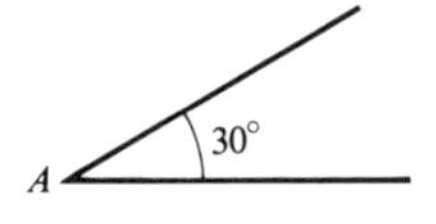

Figure 6.12

Remark on notation. A common notation for the measure of $\angle A$ is $m\angle A$. For example, the measure of the angle in Figure 6.12 may be expressed by $m\angle A = 30°$. We will not use this correct, but cumbersome, notation, using instead the symbol $\angle A$ for both the angle and the measure of the angle.

Angle measurement can be performed by using a **protractor**, like the one pictured in Figure 6.13.

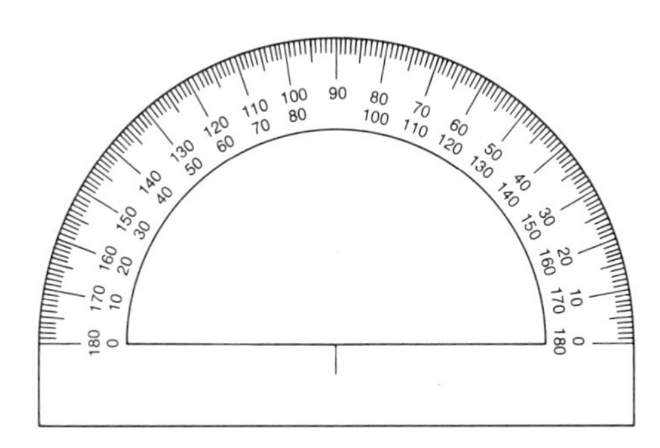

Figure 6.13

Degrees, Minutes, and Seconds

Minute, Second

We have seen that 1° measures $\frac{1}{360}$ of a complete rotation. A degree is divided into 60 **minutes** and a minute into 60 **seconds**. Minutes are denoted by the symbol ′ and seconds by the symbol ″. So a measure of 20 degrees, 40 minutes, 50 seconds is written 20°40′50″.

The availability of scientific calculators has led to extensive use of decimal parts of a degree. For example, 25°30′ can be written 25.5°, since 30′ is half a degree. Some calculators accept angles expressed in degrees and minutes, but many do not. As a result, we need to be able to change a given measure from one form to the other, based on the following relationship:

$$1° = 60'$$

Consider the next example.

Example 2 Express 0.35° in minutes.

Solution. Since $1° = 60'$, we have

$$0.35° \cdot \frac{60'}{1°} = (0.35)(60') = 21'$$

Although not shown, the degree symbols cancel.

Example 3 Express 48′ in degree measure.

Solution. Since $1° = 60'$, we get

$$48' \cdot \frac{1°}{60'} = \left(\frac{48}{60}\right)^{\circ} = 0.80°$$

If an angle measure contains a whole part and a fractional part, only the fractional part is converted.

Example 4 Change 15°25′ to decimal degrees.

Solution. $15°25' = 15° + 25' = 15° + 25'\left(\frac{1°}{60'}\right) = 15° + \left(\frac{25}{60}\right)^{\circ}$

The sequence is

15 [+] 25 [÷] 60 [=] → 15.416667

So $15°25' = 15.42°$.

Example 5 Change 38.54° to degrees and minutes.

Solution.

$$\begin{aligned} 38.54° &= 38° + 0.54° = 38° + 0.54°\left(\frac{60'}{1°}\right) \\ &= 38° + (0.54)(60') = 38° + 32.4' \\ &= 38° + 32' \\ &= 38°32' \end{aligned}$$

Rounded to the nearest minute

Triangles

Collinear

Triangle

Vertex

Sides

Points that lie on a straight line are called **collinear**. Points that do not lie on a straight line are called **noncollinear**. If three noncollinear points A, B, and C are joined by line segments, the resulting figure is a **triangle**. (See Figure 6.14.) The triangle is denoted by $\triangle ABC$. The points A, B, and C are called the **vertices** (plural of **vertex**) of the triangle. The line segments AB, AC, and BC are called the **sides** of the triangle. Note that the sides of a triangle form three interior angles with vertices at A, B, and C, respectively (Figure 6.15).

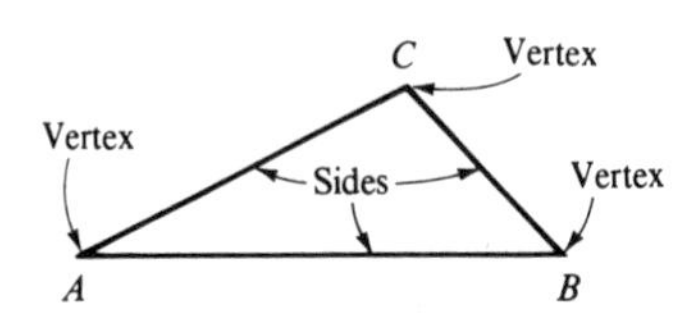

Figure 6.14

One of the fundamental properties of triangles is that the sum of the angles equals 180°. (You will see the reason for this property in Chapter 14, in our study of parallel lines.) So if two angles of a triangle are known, we can obtain the third angle by subtracting the sum of the two known angles from 180°.

In $\triangle ABC$

$$\angle A + \angle B + \angle C = 180°$$

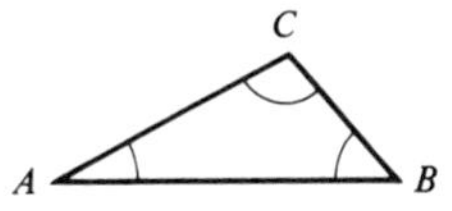

C

A B

Figure 6.15

Example 6 Determine the measure of A in the triangle shown in Figure 6.16.

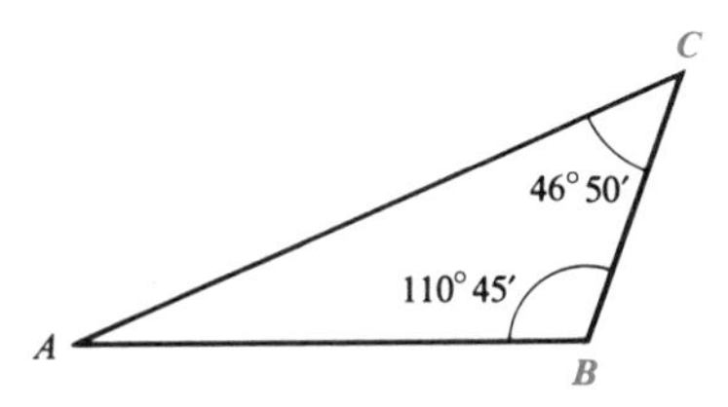

Figure 6.16

Solution. We first find the sum of the two known angles:

$$\begin{aligned} \angle B &= 110°45' \\ \angle C &= \ \ 46°50' \\ \hline \text{Total:} &\ \ 156°95' \end{aligned}$$

Since $60' = 1°$, we can rewrite the sum as follows:

$$\begin{aligned} 156°95' &= 156° + (60' + 35') \\ &= 156° + 1° + 35' \\ &= 157°35' \end{aligned}$$

This sum has to be subtracted from 180°. In order to do so, we write

$$180° = 179°60'$$

Then we get

$$\begin{aligned} &179°60' \\ &\underline{157°35'} \qquad \text{Subtracting} \\ &\ \ 22°25' = \angle A \end{aligned}$$

Equilateral triangle

A triangle is said to be an **equilateral triangle** if all three sides have the same length and, as a result, all angles have the same measure—namely, $180° \div 3 = 60°$ (Figure 6.17).

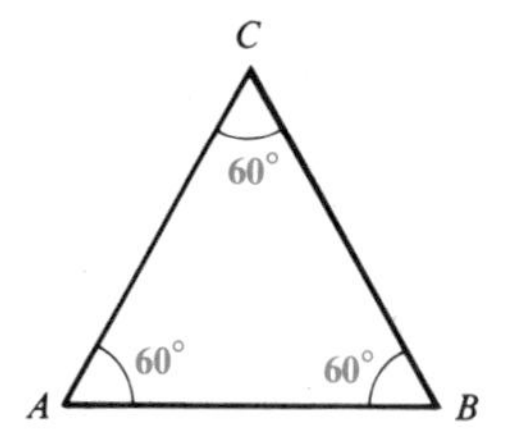

Figure 6.17

Isosceles triangle
Base angles

A triangle is said to be **isosceles** if two sides have the same length (Figure 6.18; the equal sides are labeled s). In an isosceles triangle the **base angles** (angles opposite the equal sides) are equal. (The base angles do not have to measure 60°.)

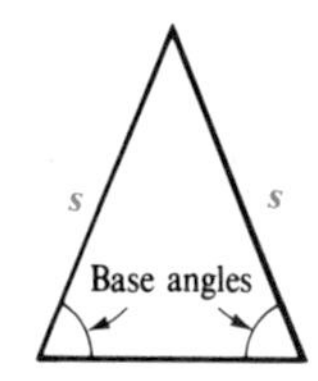

Figure 6.18

Scalene triangle
Right triangle
Hypotenuse
Legs

In a **scalene triangle** no two sides are equal. (See Figure 6.22.)

A **right triangle** has one right angle (Figure 6.19). The side opposite the right angle is called the **hypotenuse**. The remaining sides of a right triangle are called the **legs**.

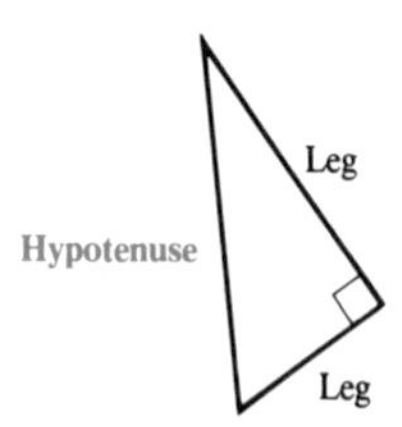

Figure 6.19

Example 7 The triangle in Figure 6.20 is an isosceles triangle since two of its sides are equal (6 cm).

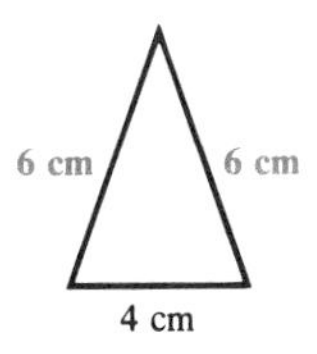

Figure 6.20

The triangle in Figure 6.21 is a right triangle.

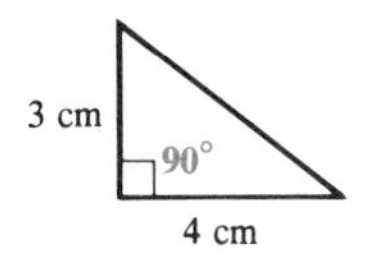

Figure 6.21

Example 8 The foundation wall in Figure 6.22 forms a scalene triangle. (The triangle is scalene because no two sides are equal.)

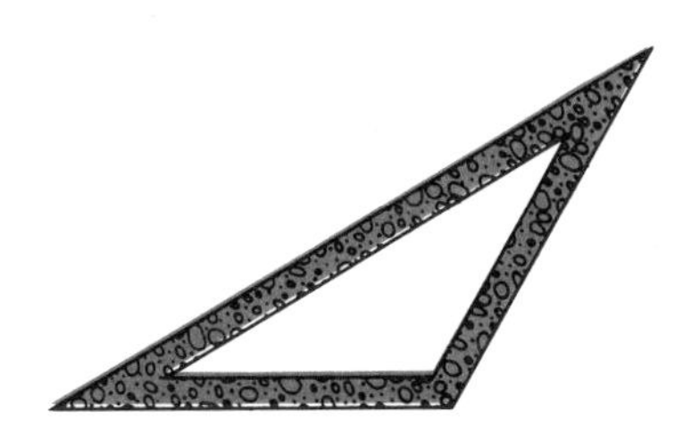

Figure 6.22

Exercises / Section 6.1

In Exercises 1–16, convert the given angle measures to degrees and minutes. (See Examples 2 and 5.)

1. 0.25° **2.** 0.78° **3.** 0.4° **4.** 0.9° **5.** 0.634°
6. 0.379° **7.** 2.18° **8.** 6.24° **9.** 12.68° **10.** 20.47°
11. 48.15° **12.** 40.7° **13.** 60.9° **14.** 70.85° **15.** 41.46°
16. 87.69°

In Exercises 17–32, convert the given angle measures to decimal degrees. (See Examples 3 and 4.)

17. 14′ **18.** 28′ **19.** 48′ **20.** 36′ **21.** 40′
22. 20′ **23.** 10°54′ **24.** 20°07′ **25.** 70°32′ **26.** 82°42′
27. 64°12′ **28.** 72°38′ **29.** 21°29′ **30.** 40°55′ **31.** 19°28′
32. 49°21′

In Exercises 33–44, two angles of a triangle are given. Find the third angle. (See Example 6.)

33. 20°, 30° **34.** 32°, 21° **35.** 18.2°, 39.7° **36.** 6.4°, 48.1° **37.** 12°20′, 18°30′
38. 26°45′, 15°11′ **39.** 34°56′, 16°40′ **40.** 22°44′, 19°36′ **41.** 16°17′, 25°50′ **42.** 25°16′, 36°49′
43. 15°35′, 41°43′ **44.** 21°16′, 29°17′

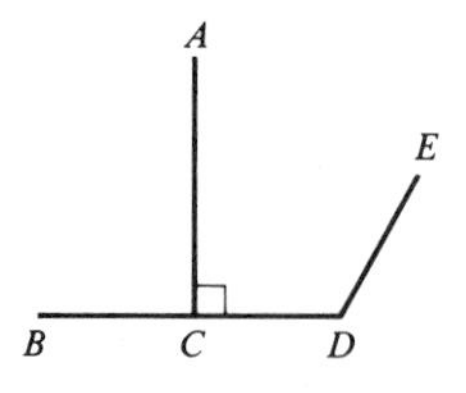

Figure 6.23

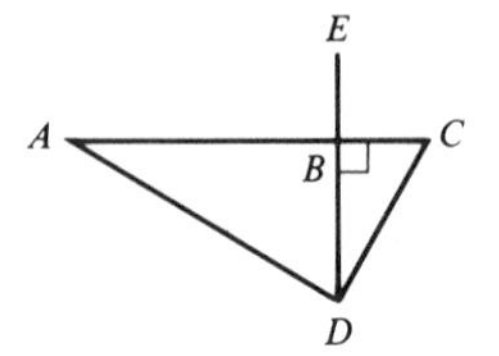

Figure 6.24

45. In Figure 6.23, identify two right angles. (See Example 1.)
46. In Figure 6.23, identify a straight angle. (See Example 1.)
47. In Figure 6.24, identify two straight angles. (See Example 1.)
48. In Figure 6.24, identify two right angles.
49. In Figure 6.24, identify a pair of perpendicular lines.
50. In Figure 6.23, identify a pair of perpendicular lines.

In Exercises 51–60 (Figures 6.25–6.34), determine whether the triangles are isosceles, right, or scalene.

51.

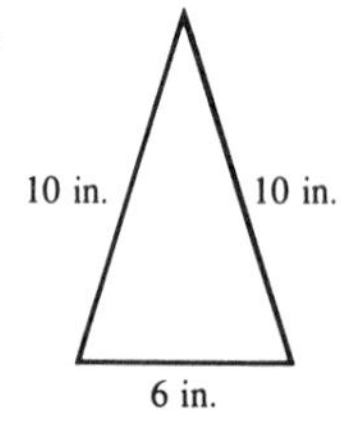

Figure 6.25

52.

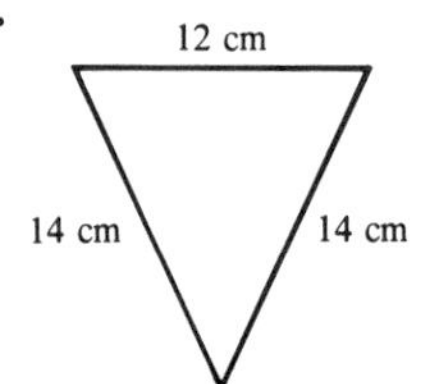

Figure 6.26

53.

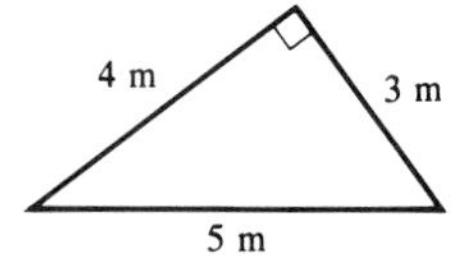

Figure 6.27

54.

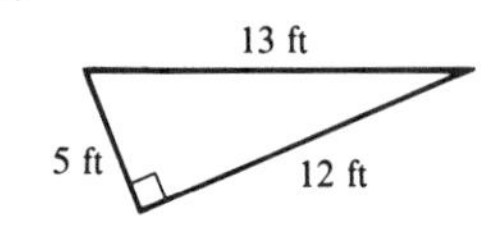

Figure 6.28

55.

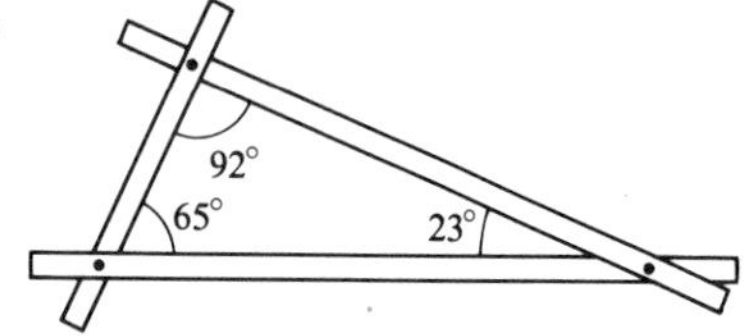

Figure 6.29

56.

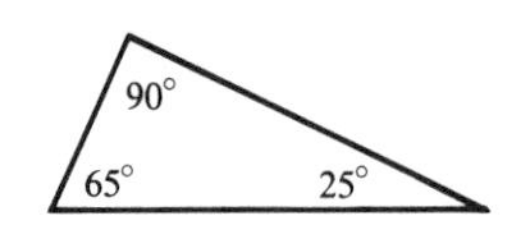

Figure 6.30

57.

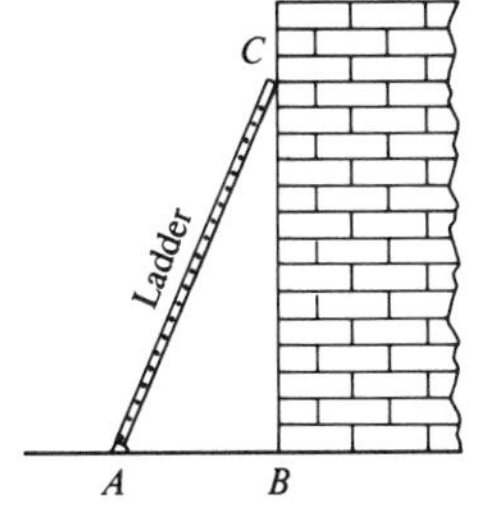

Figure 6.31

58.

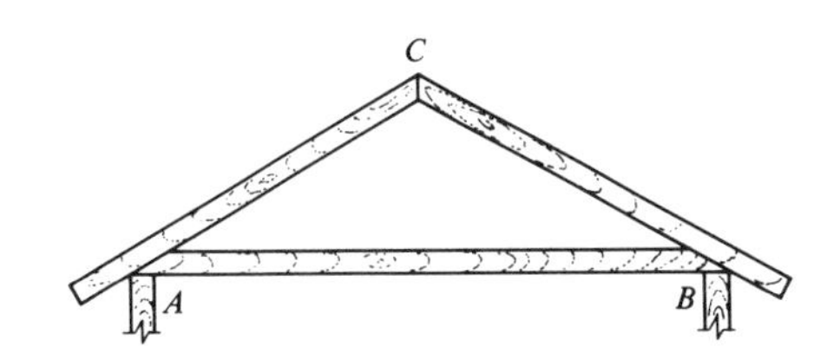

Figure 6.32

59.

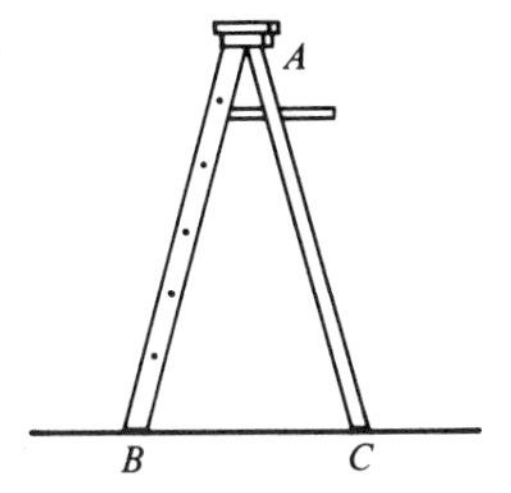

Figure 6.33

60.

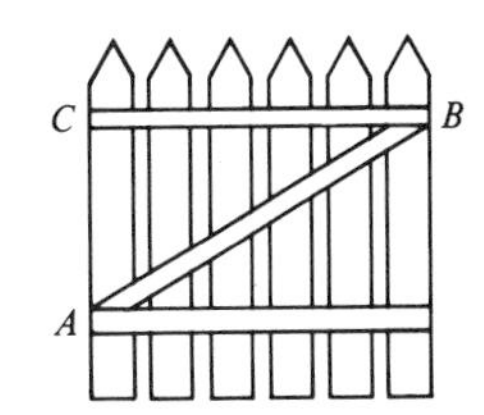

Figure 6.34

6.2 Quadrilaterals and Circles

We saw in the last section that a triangle is a three-sided figure. In the first part of this section we study certain figures with four sides.

Quadrilaterals

Quadrilateral

Parallel lines

A four-sided figure in a plane is called a **quadrilateral**. (See Figure 6.35.) We are particularly interested in quadrilaterals that have at least one pair of parallel sides. Two lines in a plane are said to be **parallel** if they do not intersect. For example, the lines l_1 and l_2 in Figure 6.36 are parallel; this is denoted by $l_1 \parallel l_2$. Two line segments are parallel if the line segments extended form parallel lines. In particular, two sides of a quadrilateral are parallel if the sides extended do not intersect.

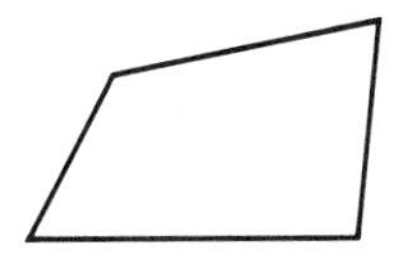

Figure 6.35

Figure 6.36

Parallelogram

Our first example of such a quadrilateral is a parallelogram. A **parallelogram** is a quadrilateral with two pairs of parallel sides. (See Figure 6.37.) Two parallel sides are also said to be **opposite** each other.

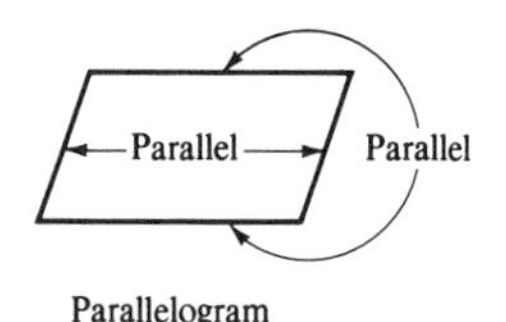

Parallelogram

Figure 6.37

There are several special parallelograms. A **rhombus** is a parallelogram with four equal sides (Figure 6.38). A **rectangle** is a parallelogram in which the intersecting sides are perpendicular (Figure 6.39). The larger side is called the **length** of the rectangle and the shorter side the **width**.

Finally, a **square** is a rectangle with four equal sides (Figure 6.40).

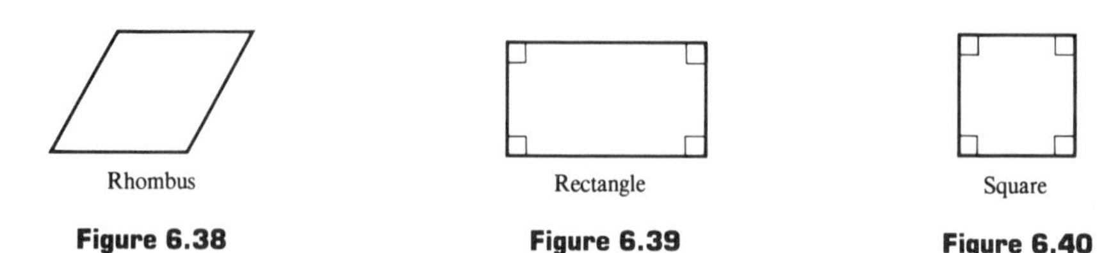

Rhombus — **Figure 6.38**

Rectangle — **Figure 6.39**

Square — **Figure 6.40**

Trapezoid

A **trapezoid** is a quadrilateral with exactly one pair of parallel sides. The parallel sides are called the **bases** of the trapezoid. (See Figure 6.41.)

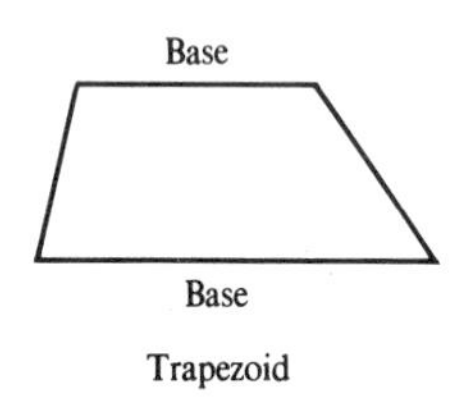

Trapezoid

Figure 6.41

Example 1 In the trapezoid $ABCD$ in Figure 6.42, $DC \parallel AB$. So DC and AB are the bases.

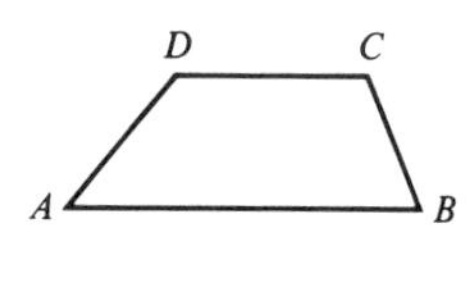

Figure 6.42

In the examples and exercises we use the following property of parallelograms:

> In a **parallelogram** the opposite sides are parallel and equal.

Example 2 In the parallelogram in Figure 6.43, $AB \parallel DC$ and $AD \parallel BC$. Note that the opposite pairs are equal.

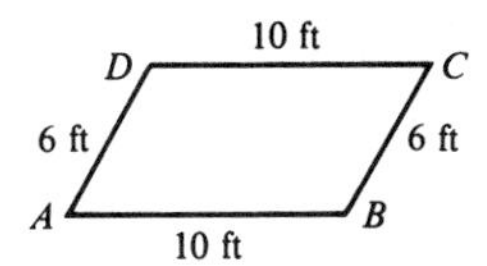

Figure 6.43

Example 3 Consider the concrete forming in Figure 6.44. Given that the inside has the shape of a rectangle, determine the lengths of AB and BC.

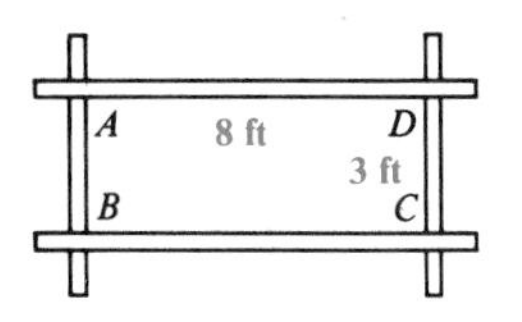

Figure 6.44

Solution. Since a rectangle is a parallelogram, the opposite sides are equal. We therefore have $AB = 3$ ft and $BC = 8$ ft.

Circles

Circle

Center

Radius

Diameter

The last figure we will consider in this section is the circle. A **circle** is a figure in a plane, all points of which are the same distance from a fixed point called the **center**. (See Figure 6.45.) The distance from the center is called the **radius**. For example, in Figure 6.46, O is the center and OA a radius, denoted by r. The distance BC in Figure 6.46 is called the **diameter** D. The diameter is therefore twice the radius, or $D = 2r$.

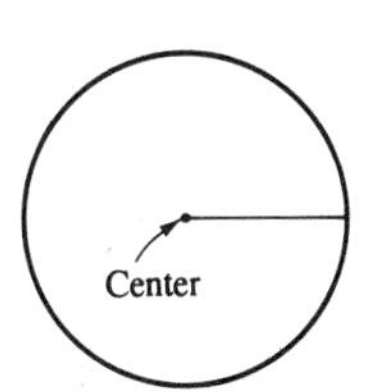

Figure 6.45

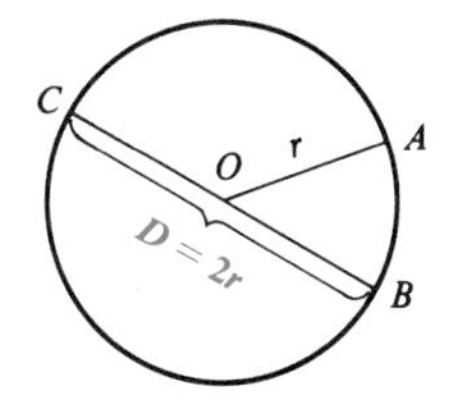

Figure 6.46

Example 4 Find the radius of the flywheel in Figure 6.47.

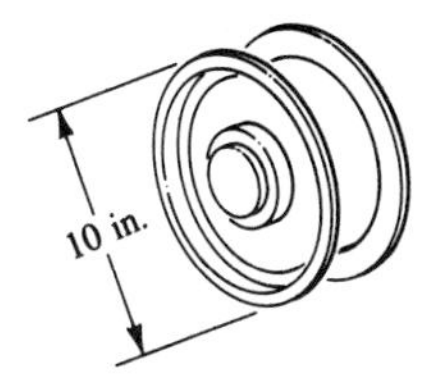

Figure 6.47

Solution. Since the diameter is twice the radius, $r = 5$ in.

Exercises / Section 6.2

1. Identify the parallel sides in Figure 6.48.

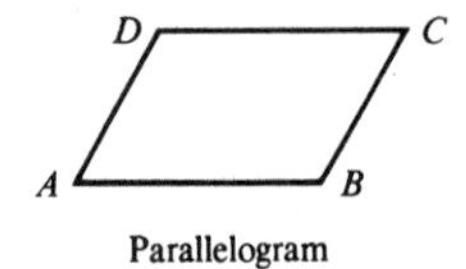

Figure 6.48

2. Identify the parallel sides in Figure 6.49.

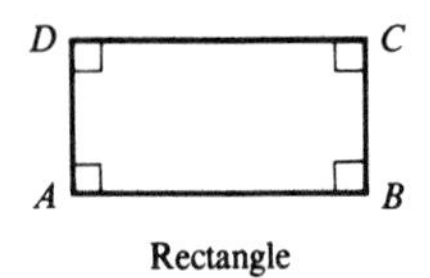

Figure 6.49

3. Determine the lengths of sides BC and CD in Figure 6.50.

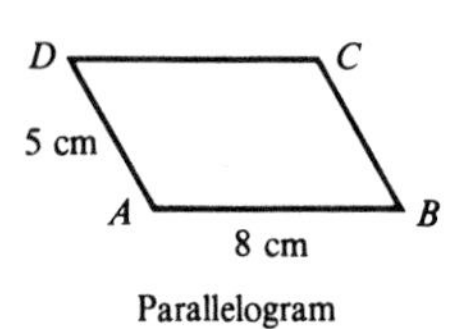

Figure 6.50

4. Determine the lengths of the remaining sides in Figure 6.51.

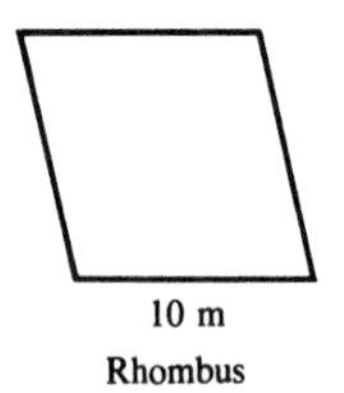

Figure 6.51

5. Determine the lengths of the remaining sides in Figure 6.52.

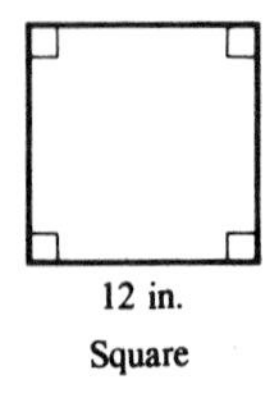

Figure 6.52

6. Identify the bases of the trapezoid in Figure 6.53. (*DC* is parallel to *AB*.)

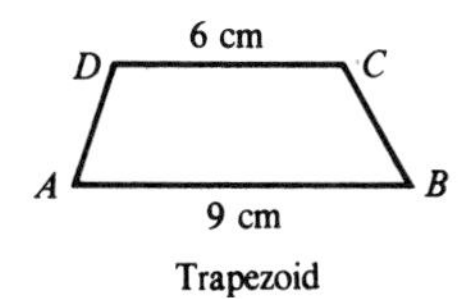

Figure 6.53

7. Determine the lengths of sides *AD* and *DC* in Figure 6.54.

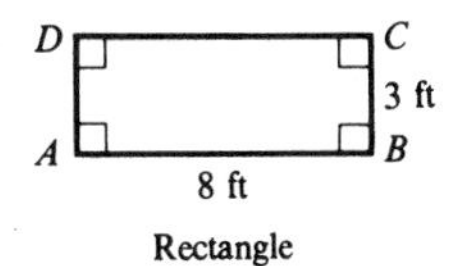

Figure 6.54

8. Determine the lengths of sides *AB* and *AD* in Figure 6.55.

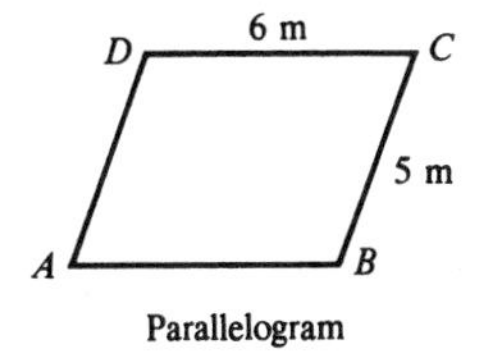

Figure 6.55

9. Identify the parallel sides in Figure 6.56.

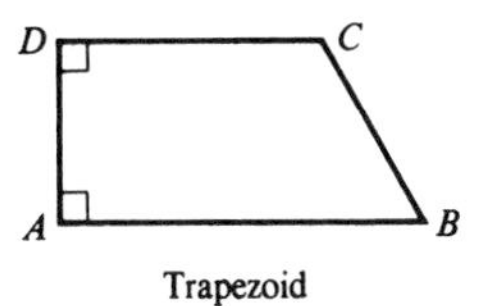

Figure 6.56

10. Identify the remaining right angles in Figure 6.57.

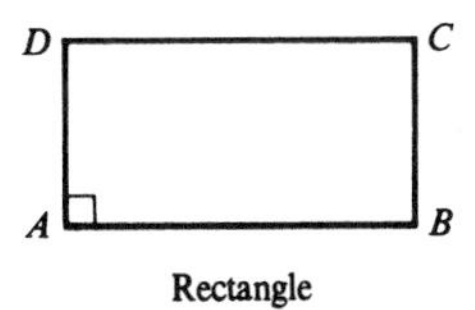

Figure 6.57

11. In Figure 6.58, identify
    a. one rectangle
    b. one parallelogram
    c. two trapezoids

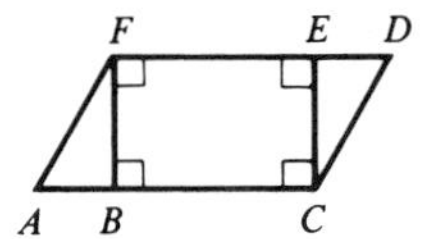

Figure 6.58

12. In Figure 6.59, identify
    a. one rectangle
    b. two trapezoids

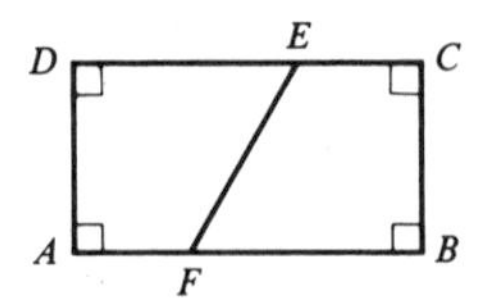

Figure 6.59

13. Find the diameter of the circle in Figure 6.60.

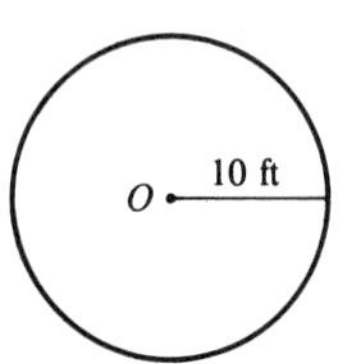

Figure 6.60

14. Find the radius of the circle in Figure 6.61.

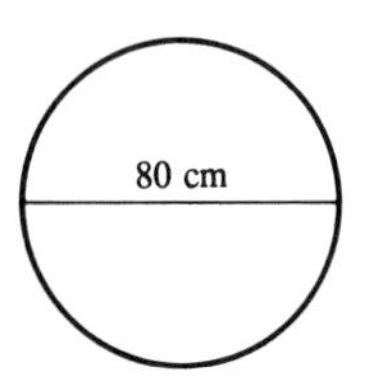

Figure 6.61

15. Find the radius of the circle in Figure 6.62.

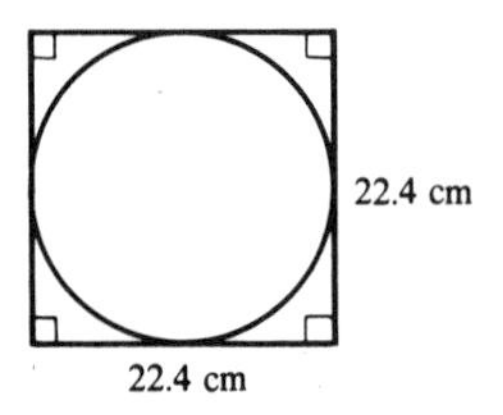

Figure 6.62

16. Find the width of the rectangle in Figure 6.63.

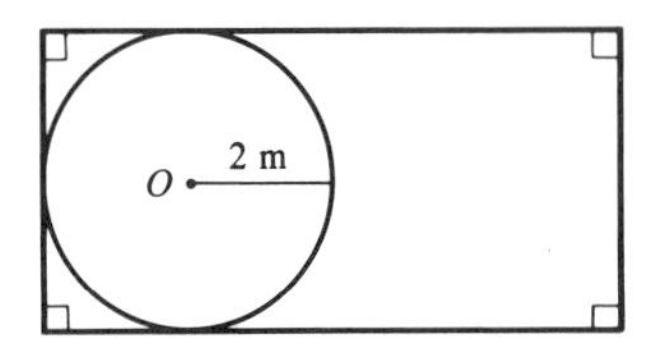

Figure 6.63

17. Find the radius of the wheel in Figure 6.64.

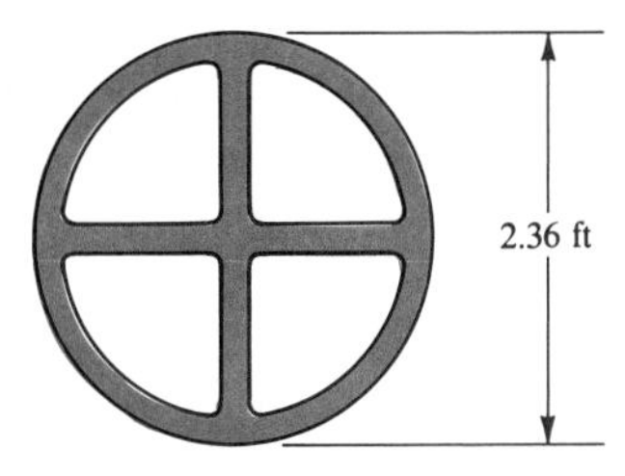

Figure 6.64

18. Find the height of the I-beam in Figure 6.65.

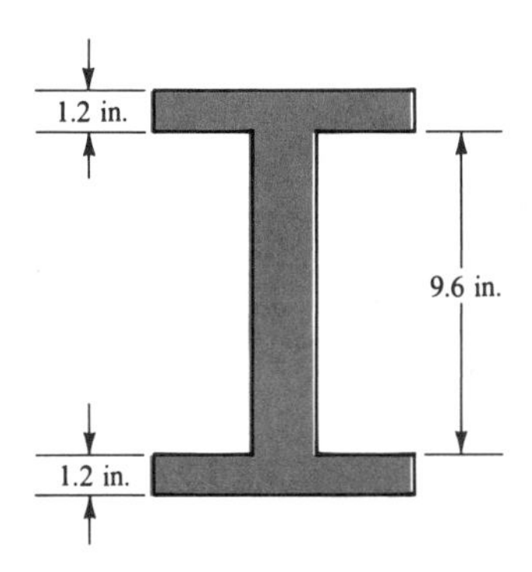

Figure 6.65

19. A sailboat travels east for 2 mi and then continues northeast for another mile. Next, it heads west for 2 mi and then southwest for another mile. What geometric figure is formed by this path?

20. A surveyor walks north for 2.5 km, turns right (at 90°), walks 1 km, turns right again, and walks for another 1.5 km. He then turns and walks in a straight line back to the starting point. What is the geometric figure formed by this path?

21. The moon is in a nearly circular orbit around the earth. If the diameter of the orbit is 478,000 mi, what is the distance from the center of the earth to the moon?

22. The radius of the earth is about 4000 mi. What is the earth's diameter?

6.3 Perimeter

So far we have concentrated mostly on geometric concepts and definitions. In the remainder of this chapter we will study some of the basic measures associated with geometric figures.

Perimeter

Our first basic measure is the perimeter, already introduced in Chapter 1. The **perimeter** of a geometric figure is the distance around it. For example, the perimeter P of the triangular fence in Figure 6.66 is

$$P = 10 \text{ m} + 15 \text{ m} + 18 \text{ m} = 43 \text{ m}$$

Consider another example.

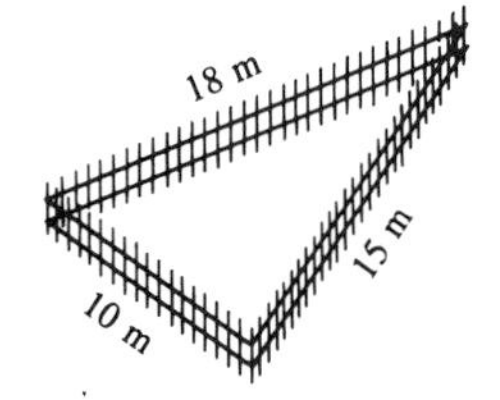

Figure 6.66

Example 1 Find the perimeter of the parallelogram in Figure 6.67.

Solution. Since the opposite sides of a parallelogram are equal, the perimeter is

$$P = 7.4 \text{ cm} + 7.4 \text{ cm} + 20.3 \text{ cm} + 20.3 \text{ cm}$$
$$= 55.4 \text{ cm}$$

or

$$P = 2(7.4 \text{ cm}) + 2(20.3 \text{ cm}) = 55.4 \text{ cm}$$

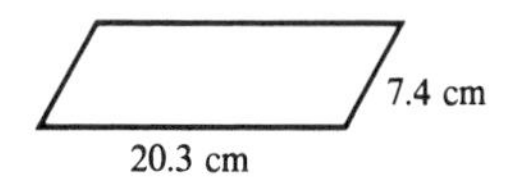

Figure 6.67

Circumference

The perimeter of a circle is called the **circumference.** The formula for the circumference is based on the fact that for any circle, the circumference divided by the diameter is equal to the constant value $\pi \approx 3.1416$. So if C is the circumference and D the diameter, then

$$\frac{C}{D} = \pi$$

An equivalent form is $C = \pi D$. Since the diameter of a circle is twice the radius, we also have $C = 2\pi r$ (Figure 6.68).

Circumference of a circle

$$C = 2\pi r \qquad \text{or} \qquad C = \pi D$$
$$\pi \approx 3.1416$$

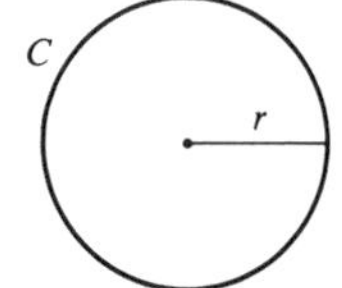

Figure 6.68

Example 2 Find the circumference of the circle in Figure 6.69.

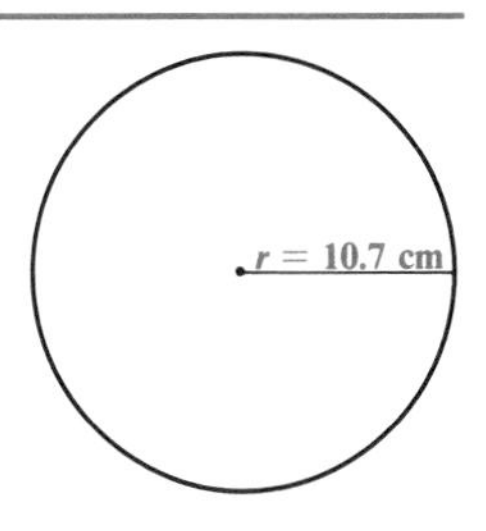

Figure 6.69

Solution. Using the formula $C = 2\pi r$, we get

$$\begin{aligned} C &= 2\pi(10.7 \text{ cm}) \\ &= 2(3.1416)(10.7 \text{ cm}) \\ &= 67.2 \text{ cm} \end{aligned}$$

Recall from Section 1.12 that scientific calculators have a special key for π. For the calculation in Example 2 the sequence is

2 [×] [π] [×] 10.7 [=] → 67.230083

To three significant digits, $C = 67.2$ cm.

Example 3 Find the circumference of the circular cross section of the pipe shown in Figure 6.70.

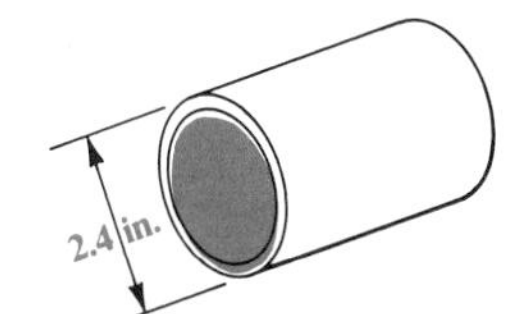

Figure 6.70

Solution. Since the diameter is given, we use the formula $C = \pi D$:

$$C = \pi(2.4 \text{ in.})$$

The sequence is

[π] [×] 2.4 [=] → 7.5398224

So the circumference is $C = 7.5$ in.

Exercises / Section 6.3

In Exercises 1–10 (Figures 6.71–6.80), find the perimeter of each of the given figures.

1.

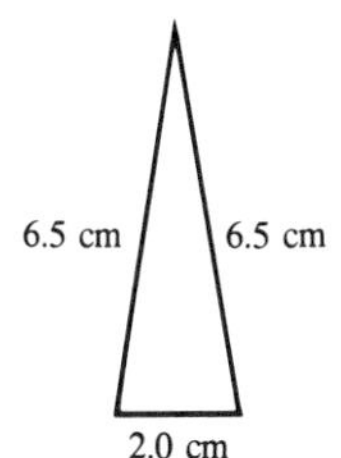

Figure 6.71

2.

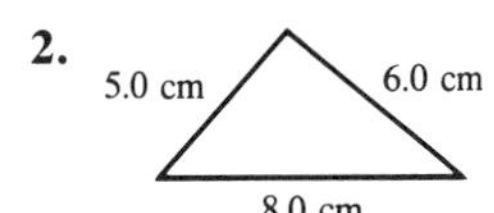

Figure 6.72

3.

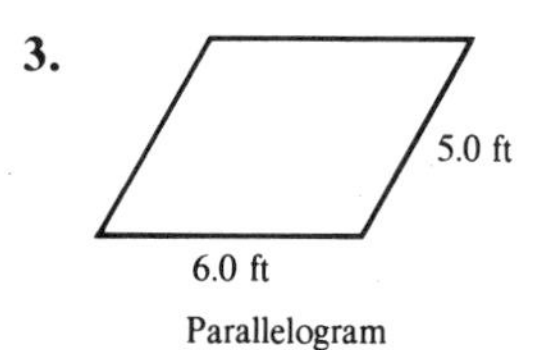

Figure 6.73

4.

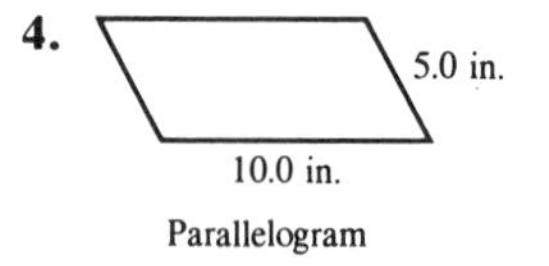

Figure 6.74

5.

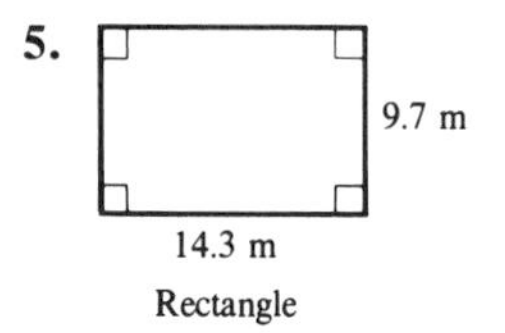

Figure 6.75

6.

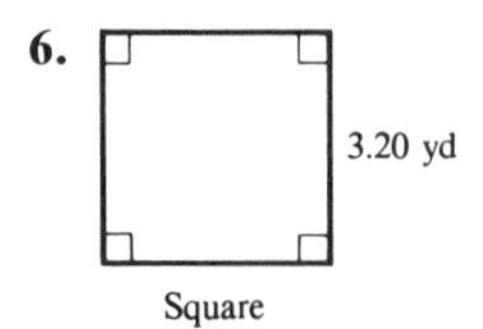

Figure 6.76

7.

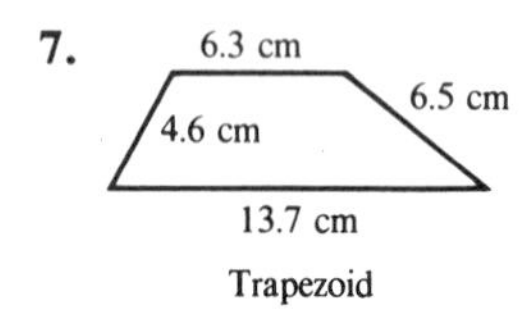

Figure 6.77

8.

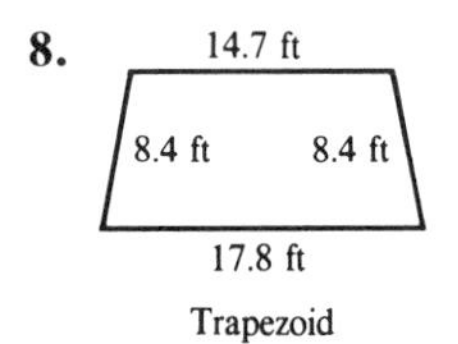

Figure 6.78

9.

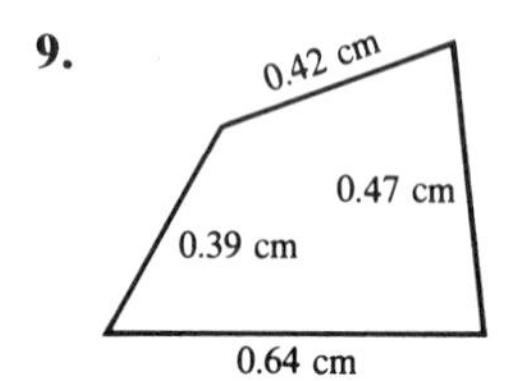

Figure 6.79

10.

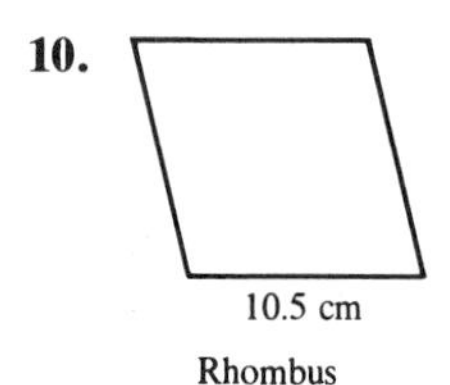

Figure 6.80

In Exercises 11–16, find the perimeter of each figure described.

11. Square: each side is 4.6 cm long

12. Rectangle: length = 20.46 cm, width = 8.4 cm

13. Trapezoid: bases are 8.92 in. and 15.3 in., respectively; remaining sides are 4.5 in. each

14. Rhombus: each side is 20.4 in. long

15. Rectangle: length = 2.4 ft, width = 1.6 ft

16. Square: each side is 10.8 m long

In Exercises 17–26, find the circumference of each circle. Illustrations are given for the first two (Figures 6.81 and 6.82).

17. $r = 2.1$ in.

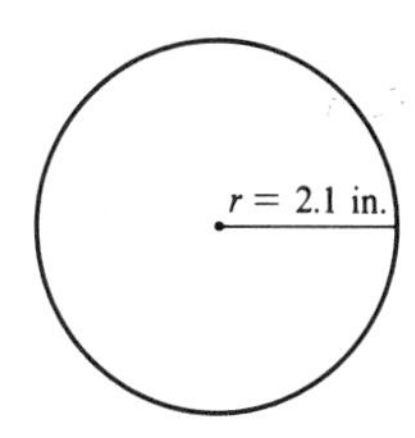

Figure 6.81

18. $D = 4.1$ cm

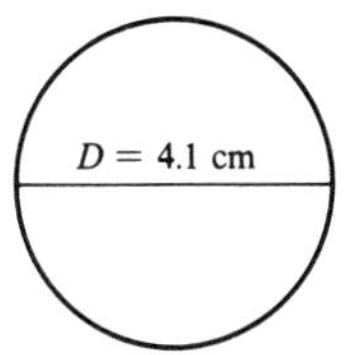

Figure 6.82

19. $D = 6.40$ ft

20. $r = 9.60$ ft

21. $r = 2.7$ cm

22. $D = 7.9$ cm

23. $D = 6.147$ in.

24. $r = 10.46$ in.

25. $r = 18.67$ m

26. $D = 59.64$ m

27. A machine part consists of a rectangle surmounted by a semicircle (Figure 6.83). Find the perimeter.

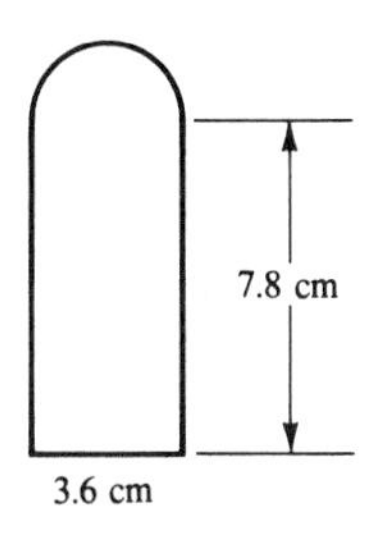

Figure 6.83

28. Find the circumference of each wheel in Figure 6.84.

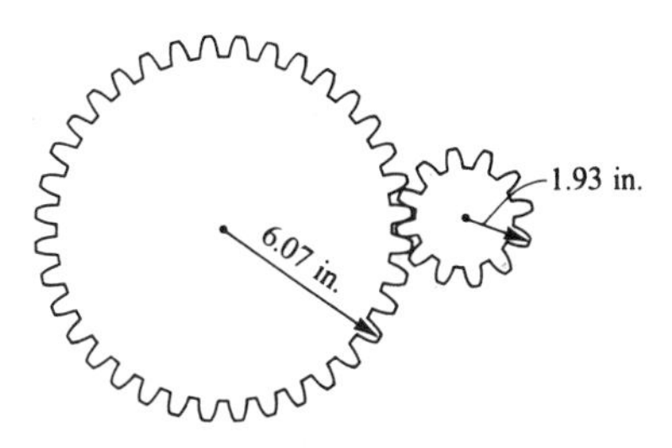

Figure 6.84

29. Find the inner and outer circumference of the washer shown in Figure 6.85.

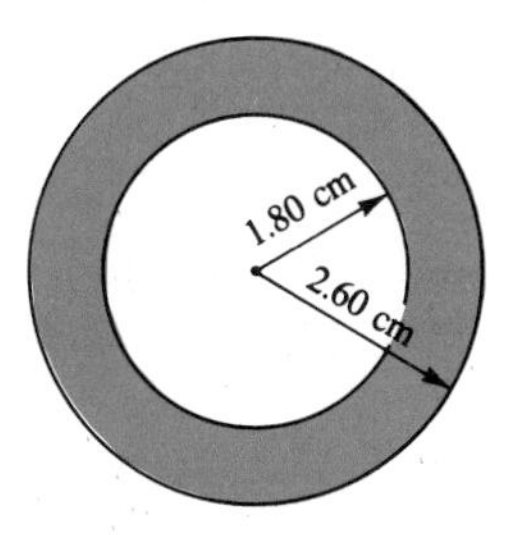

Figure 6.85

30. Find the circumference of the top of the tank shown in Figure 6.86.

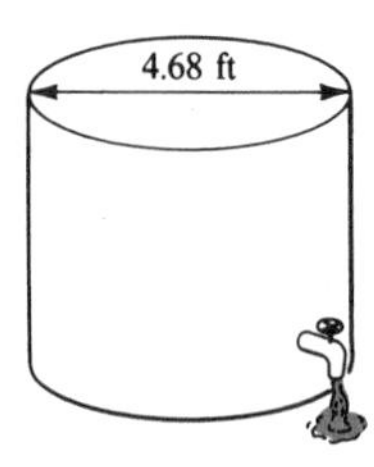

Figure 6.86

31. A rectangular computer chip measures 2.3 mm by 4.7 mm. Find the perimeter.
32. A rectangular area adjacent to a river is to be enclosed on the other three sides by a fence. The enclosed area measures 12 m by 18 m and the long side is along the river. How many feet of fence will be needed?
33. A communications satellite remains 22,300 mi above a point on the equator at all times. Given that the radius of the earth is about 3960 mi, find the length of the orbit of the satellite.
34. The circumference of a flywheel is 16.5 in. What is the radius? (*Hint:* $D = C/\pi$.)
35. The distance around the equator is 24,880 mi. Determine the diameter of the earth to three significant digits. (See Exercise 34.)
36. A piece of cardboard 18 in. long and 14 in. wide is made into a box by cutting 2-in. squares from each corner and bending up the sides. What is the perimeter of the bottom of the box?
37. A racetrack has the form of a rectangle with a semicircle at each end. If the rectangular part is 112 m long and 75.4 m wide, find the length of the track.

6.4 Area and Volume

In this section we study two other important measures associated with geometric figures, **area** and **volume.**

Area

The formulas for finding the areas of the figures discussed in this section are based on the definition of the area of a rectangle. Suppose the rectangle shown in Figure 6.87 is divided into little squares each measuring 1 in. on the side. The area of each square is therefore 1 in.^{2}. The total area consists of two rows of three squares each, for a total of $2 \times 3 = 6$ squares. So the total

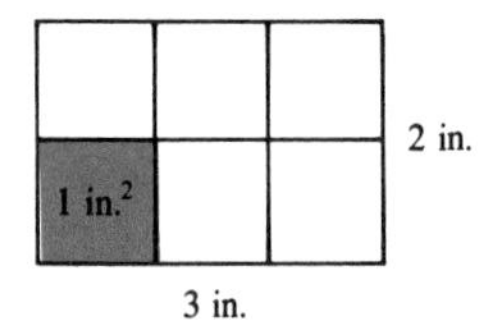

Figure 6.87

area is 6 in.^{2}. In other words, we multiply the length (3 in.) by the width (2 in.) to obtain an area of (3 in.) $\times$ (2 in.) = 6 in.^{2}. The general case is given next (Figure 6.88).

Area of a rectangle

$$A = lw$$

Figure 6.88

Note that the **length and width have to be measured in the same units of length.**

Since a square is merely a rectangle for which the length and width are the same, the area A of a square is $A = s^2$, where s is the length of a side (Figure 6.89).

Area of a square

$$A = s^2$$

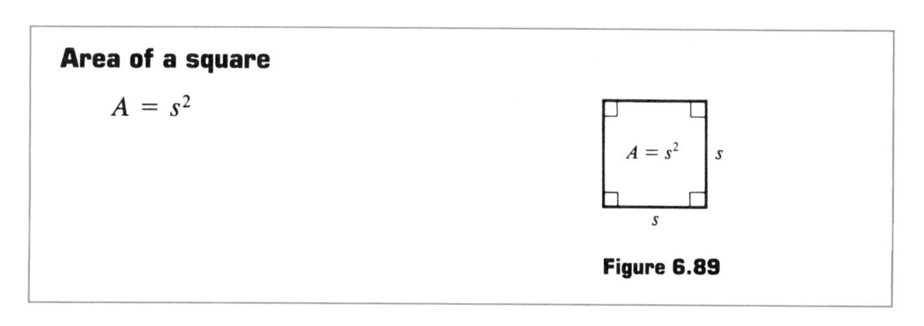

Figure 6.89

Example 1 A rectangle is 25 cm long and 5 cm wide. Determine the area.

Solution. From the formula $A = lw$, we get

$$A = (25 \text{ cm})(5 \text{ cm}) = 125 \text{ cm}^2$$

The definition of the area of a rectangle leads to the formula for the area of a parallelogram. In the parallelogram shown in Figure 6.90(a), the dotted

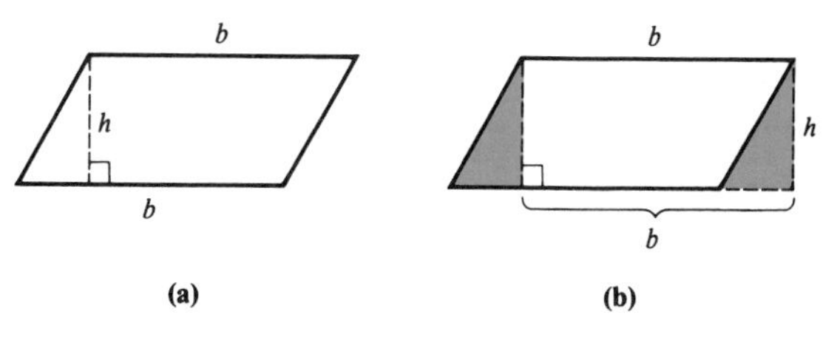

Figure 6.90

line h is perpendicular to the bottom side and determines a right triangle. If the triangle is moved to the right, as shown in Figure 6.90(b), we obtain a rectangle with the same area, namely $A = bh$. In Figure 6.90(a), h is called the **altitude** or **height** of the parallelogram and b the **base.** (See also Figure 6.91.)

Altitude, Height
Base

The area of a **parallelogram** is equal to the product of the base and the height:

$$A = bh$$

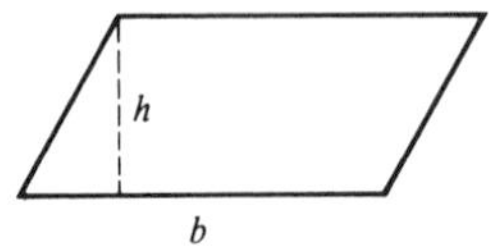

Figure 6.91

Example 2 Find the area of the parallelogram in Figure 6.92.

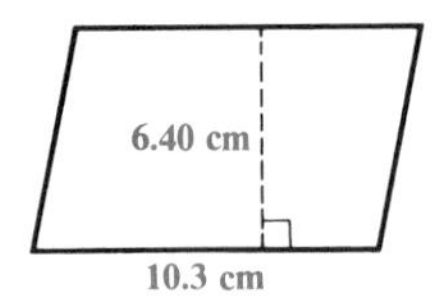

Figure 6.92

Solution. Since $h = 6.40$ cm and $b = 10.3$ cm, we obtain, from the formula $A = bh$,

$$A = (10.3 \text{ cm})(6.40 \text{ cm}) = 65.9 \text{ cm}^2$$

The area of a triangle is found from the area of a parallelogram, as follows: Divide the parallelogram into two equal triangles by drawing the **diagonal** shown in Figure 6.93. Each of the two triangles has an area equal to one-half the area of the parallelogram, or $A = \frac{1}{2}bh$. (See also Figure 6.94.) As in the case of the parallelogram, b is called the **base** and h the **altitude** or **height.** Note that the altitude is the perpendicular from a vertex to the side opposite or the side opposite extended.

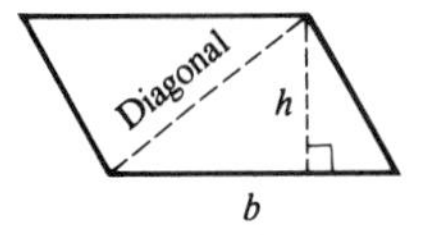

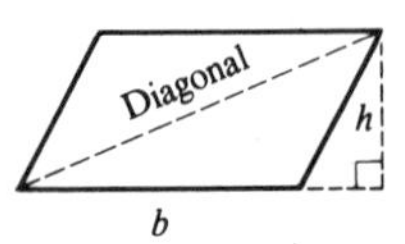

Figure 6.93

The area of a **triangle** is equal to one-half the product of the base and the altitude (see Figure 6.94):

$$A = \frac{1}{2}bh$$

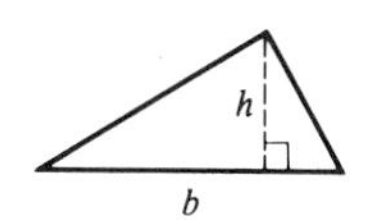

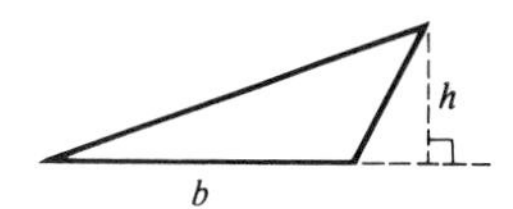

Figure 6.94

Example 3 The area of the triangle in Figure 6.95 is

$$A = \frac{1}{2}(3.6 \text{ ft})(2.5 \text{ ft}) = 4.5 \text{ ft}^2$$

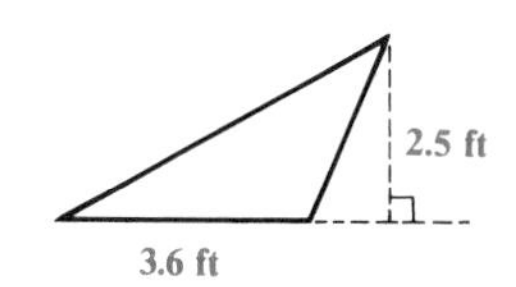

Figure 6.95

Of particular interest is the area of a right triangle. Since the legs of a right triangle are perpendicular, either leg can be used as the base. The other leg then becomes the altitude.

Example 4 Find the area of the right triangle in Figure 6.96.

Figure 6.96

Solution. If $b = 8.64$ in. and $h = 7.29$ in., we get

$$A = \frac{1}{2}(8.64 \text{ in.})(7.29 \text{ in.}) = 31.5 \text{ in.}^2$$

If $b = 7.29$ in. and $h = 8.64$ in., then

$$A = \frac{1}{2}(7.29 \text{ in.})(8.64 \text{ in.}) = 31.5 \text{ in.}^2$$

The area of a trapezoid can be obtained from the area of a triangle. Suppose we divide the trapezoid in Figure 6.97 into two triangles by the indicated diagonal. Since the bases of the trapezoid are parallel, h is perpendicular to b_1 and b_2. It follows that the area of triangle 1 is $\frac{1}{2}b_1h$ and the area of triangle 2 is $\frac{1}{2}b_2h$. The total area is $\frac{1}{2}b_1h + \frac{1}{2}b_2h$. By the distributive law,

$$\frac{1}{2}b_1h + \frac{1}{2}b_2h = \frac{1}{2}h(b_1 + b_2)$$

which is the area of the trapezoid (Figure 6.98).

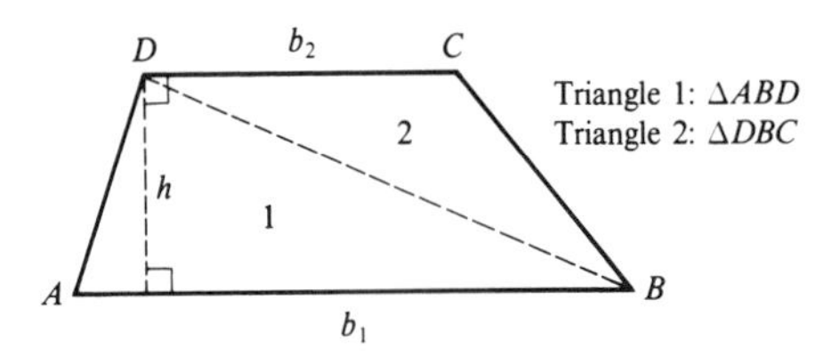

Figure 6.97

The area of a **trapezoid** is equal to one-half the product of the altitude and the sum of the bases:

$$A = \frac{1}{2} h(b_1 + b_2)$$

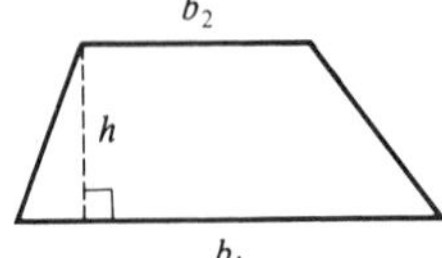

Figure 6.98

Example 5 Find the area of the trapezoid in Figure 6.99.

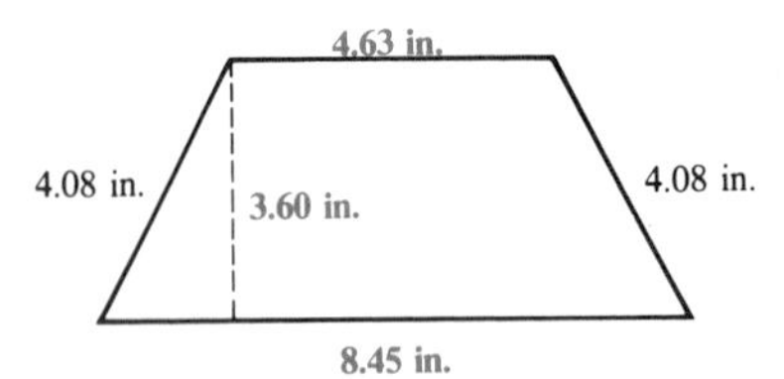

Figure 6.99

Solution. Note that $b_1 = 8.45$ in., $b_2 = 4.63$ in., and $h = 3.60$ in. The lengths of the remaining sides are not needed to find the area. We get

$$A = \frac{1}{2}(3.60)(8.45 + 4.63) = 23.5 \text{ in.}^2$$

The area A of a circle is given by $A = \pi r^2$. To see why, imagine that the circle has been divided into many small sections (Figure 6.100) and that the sections are arranged as shown in Figure 6.101. The region in Figure 6.101 is approximately rectangular. The length is equal to one-half the circumference: $\frac{1}{2}(2\pi r) = \pi r$. Since the width is r, the area is $(\pi r)(r) = \pi r^2$ (Figure 6.102).

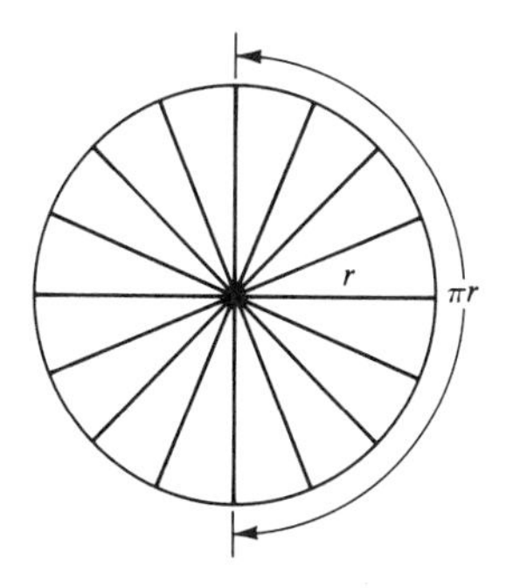

Figure 6.100

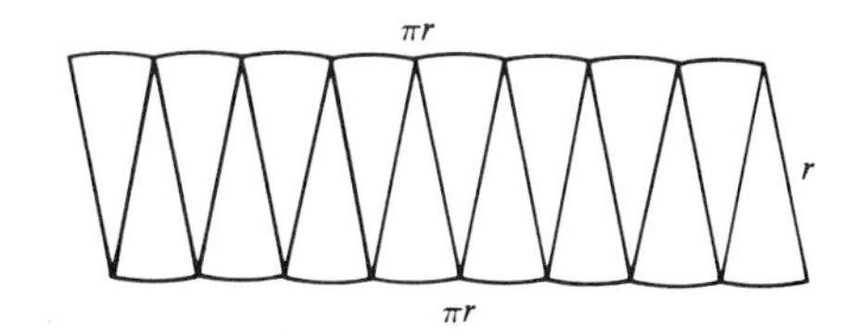

Figure 6.101

Area of a circle

$$A = \pi r^2$$

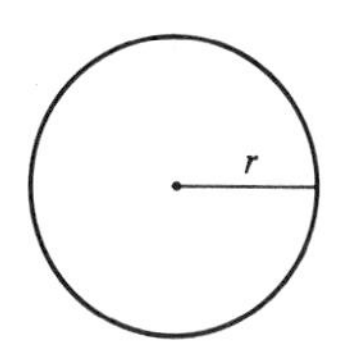

Figure 6.102

Example 6 Find the area of a circle with a diameter of 7.460 cm.

Solution. Since $D = 7.460$ cm, we have $r = \frac{1}{2}(7.460 \text{ cm}) = 3.730$ cm. The area is therefore

$$A = \pi(3.730 \text{ cm})^2$$

The sequence is

$\boxed{\pi}\ \boxed{\times}\ 3.730\ \boxed{x^2}\ \boxed{=} \rightarrow 43.708664$

It follows that $A = 43.71 \text{ cm}^2$ (four significant digits).

Volume of a Solid

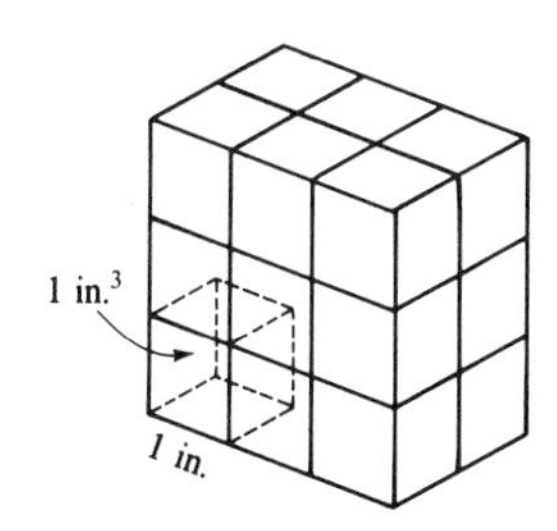

Figure 6.103

The volume of a solid is the number of cubic units contained in the solid. For example, in the solid shown in Figure 6.103, each cube has a volume of 1 in.^{3}. The total volume is equal to the number of cubes: $3 \times 2 \times 3 = 18$ in.^{3}.

Rectangular solid

The solid in Figure 6.103 is called a **rectangular solid.** Its volume is found by multiplying length, width, and height (Figure 6.104).

Volume of a rectangular solid

$$V = lwh$$

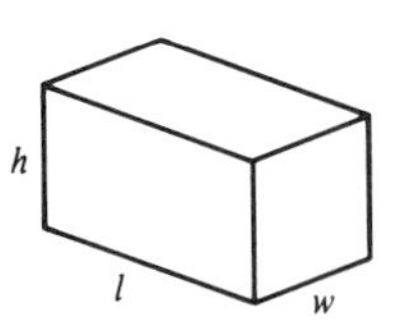

Figure 6.104

Example 7 A casting is in the shape of a rectangular solid 8.46 in. long, 6.02 in. wide, and 4.95 in. high. Given that the cost is 23¢ per cubic inch, determine the cost of the casting.

Solution. From $V = lwh$, the volume is

$$V = (8.46 \text{ in.})(6.02 \text{ in.})(4.95 \text{ in.}) = 252.1 \text{ in.}^3$$

Since the unit cost is \$0.23/in.³, the cost of the casting is

$$(252.1 \text{ in.}^3)\left(0.23 \frac{\text{dollars}}{\text{in.}^3}\right) = \$57.98$$

Cube

If the length, width, and height of a rectangular solid are equal, then the solid is a **cube.** So if s is the length of the side, then $V = s \cdot s \cdot s = s^3$ (Figure 6.105).

Volume of a cube

$$V = s^3$$

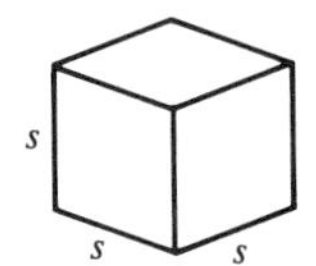

Figure 6.105

Example 8 Find the area of one side of the washer in Figure 6.106.

Solution. The area of the washer can be found by subtracting the area of the smaller circle from the area of the larger circle. Thus

$$A = A_{\text{larger}} - A_{\text{smaller}}$$
$$= \pi(3.4 \text{ cm})^2 - \pi(2.5 \text{ cm})^2$$

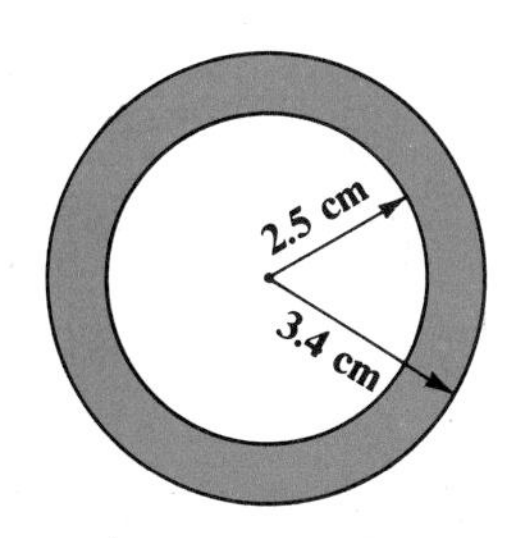

Figure 6.106

The sequence is

$\boxed{\pi}\ \boxed{\times}\ 3.4\ \boxed{x^2}\ \boxed{-}\ \boxed{\pi}\ \boxed{\times}\ 2.5\ \boxed{x^2}\ \boxed{=} \rightarrow 16.681857$

So $A = 17\ \text{cm}^2$.

Exercises / Section 6.4

In Exercises 1–16, find the areas of the indicated figures.

1. Square: $s = 10$ cm

2. Square: $s = 20$ cm

3. Rectangle: $l = 21$ ft, $w = 42$ ft

4. Rectangle: $l = 16$ cm, $w = 24$ cm

5. Parallelogram: $b = 7.22$ in., $h = 4.63$ in.

6. Parallelogram: $b = 5.05$ in., $h = 4.30$ in.

7. Triangle: $b = 4.703$ cm, $h = 7.006$ cm

8. Triangle: $b = 10.06$ cm, $h = 2.738$ cm

9. Right triangle: legs are 4.70 ft and 8.40 ft, respectively

10. Right triangle: legs are 10.3 m and 20.4 m, respectively

11. Trapezoid: $h = 25.6$ cm, $b_1 = 30.4$ cm, $b_2 = 40.6$ cm

12. Trapezoid: $h = 8.9$ in., $b_1 = 4.6$ in., $b_2 = 7.4$ in.

13. Circle: $r = 8.4$ ft

14. Circle: $r = 10.6$ ft

15. Circle: $D = 4.30$ m

16. Circle: $D = 26.4$ m

In Exercises 17–20, find the volume of each solid.

17. Rectangular solid: $l = 7.432$ cm, $w = 5.861$ cm, $h = 3.018$ cm

18. Rectangular solid: $l = 12.4$ cm, $w = 11.6$ cm, $h = 10.1$ cm

19. Cube: $s = 2.76$ in.

20. Cube: $s = 5.806$ ft

In Exercises 21–28 (Figures 6.107–6.114), find the area of each given figure.

21.

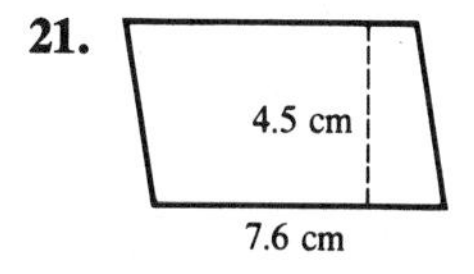

Figure 6.107

22.

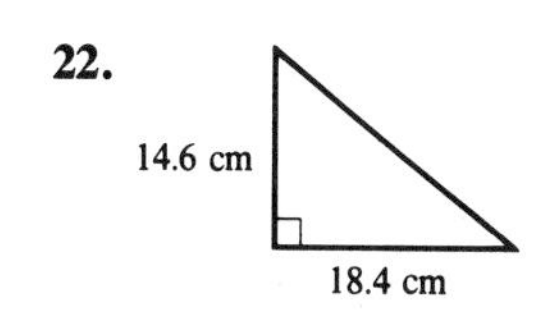

Figure 6.108

23.

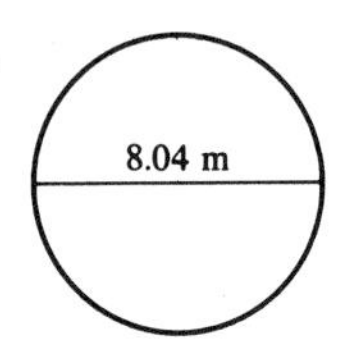

Figure 6.109

24.

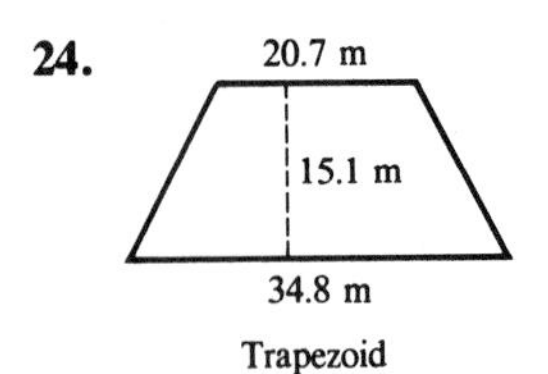

Figure 6.110

25.

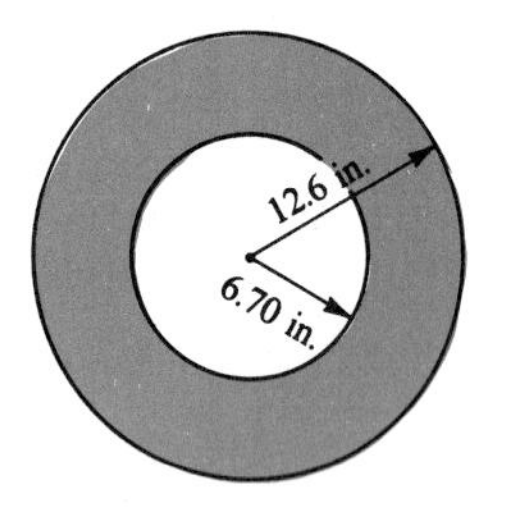

Figure 6.111

26.

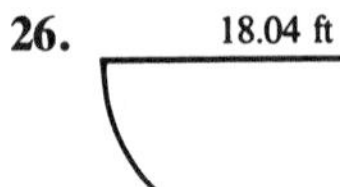

Figure 6.112

27.

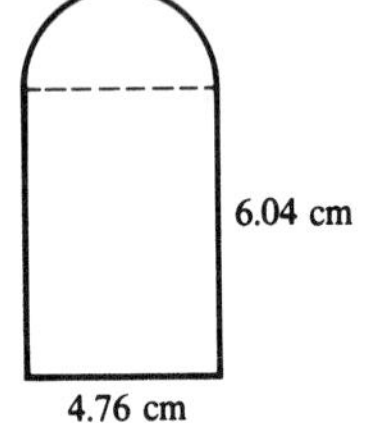

Figure 6.113

28.

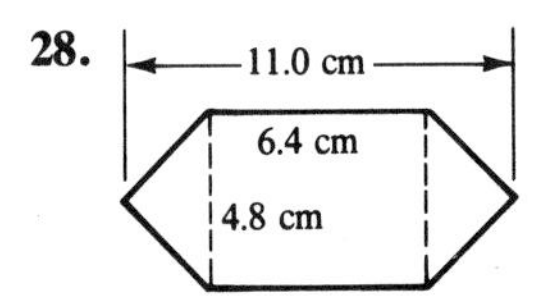

Figure 6.114

29. Find the area of the gate in Figure 6.115.

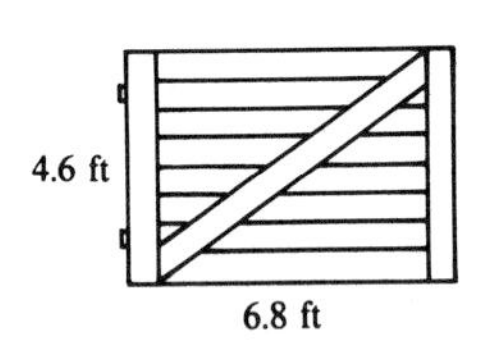

Figure 6.115

30. Find the area of the triangle determined by the horizontal beam and the rafters in Figure 6.116.

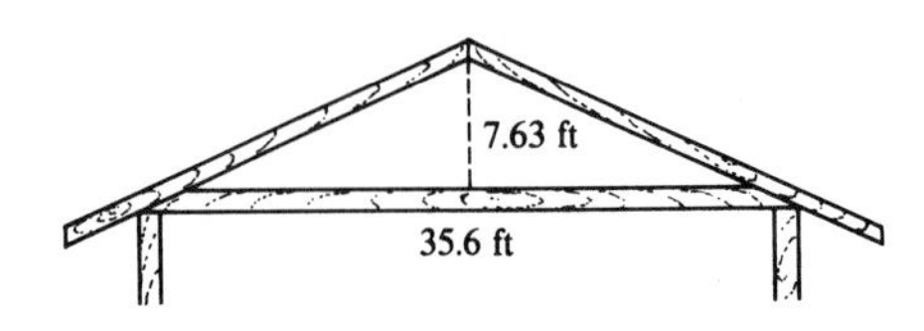

Figure 6.116

31. Find the area of the region enclosed by the fence in Figure 6.117.

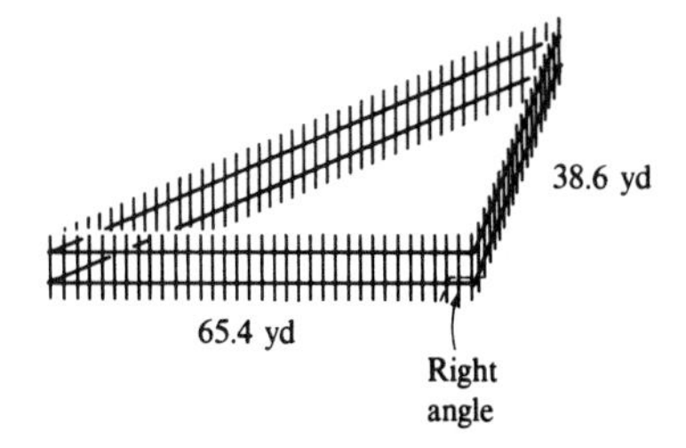

Figure 6.117

32. Find the area of the machine part pictured in Figure 6.118.

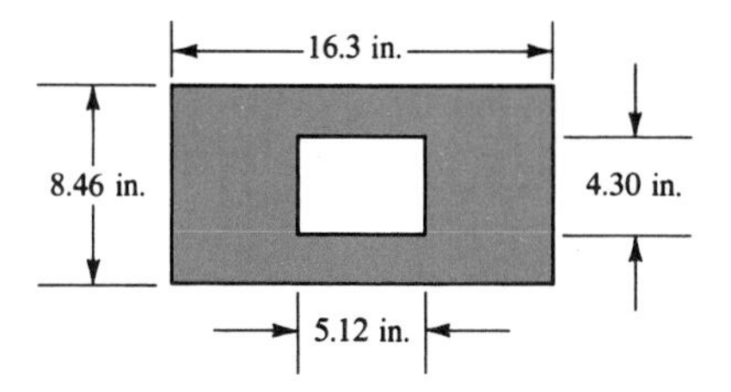

Figure 6.118

33. Determine the total area of the front of the garage in Figure 6.119.

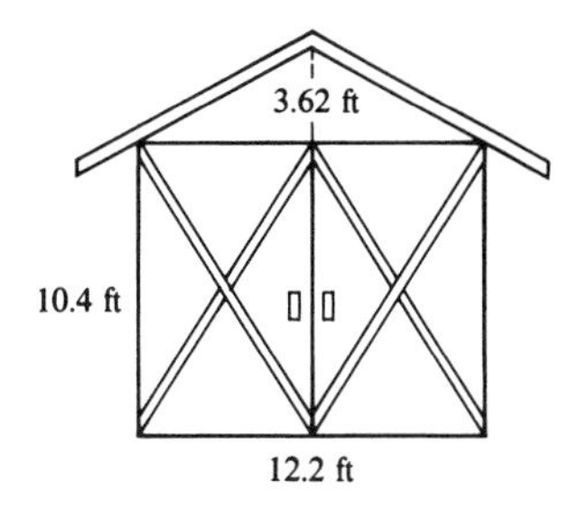

Figure 6.119

34. Find the area of the lot in Figure 6.120.

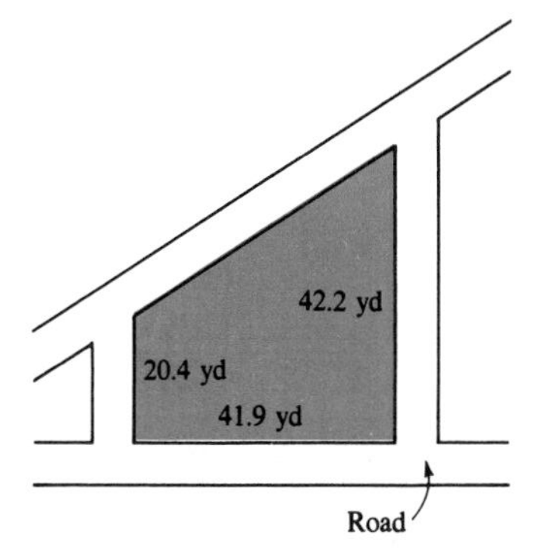

Figure 6.120

35. Determine the area of the cross section of the I-beam in Figure 6.121.

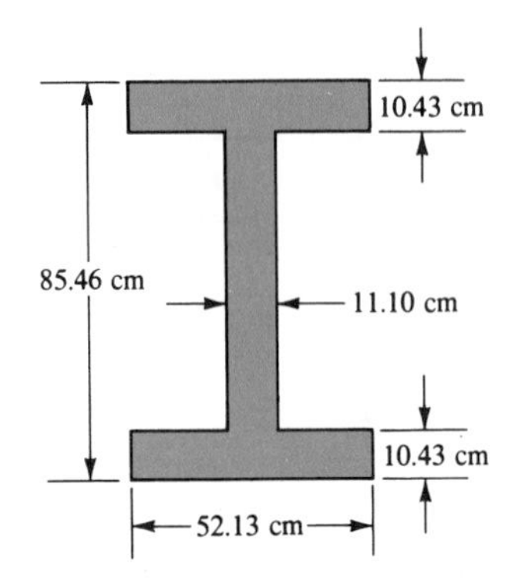

Figure 6.121

36. Find the total area of the wheels in Figure 6.122.

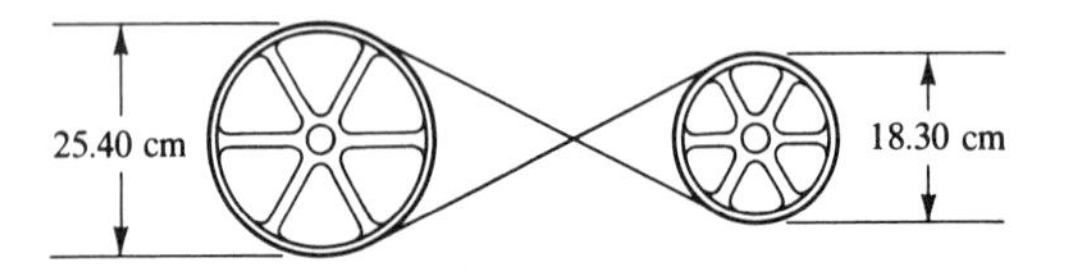

Figure 6.122

37. Find the area of the gasket in Figure 6.123 (exclusive of holes).

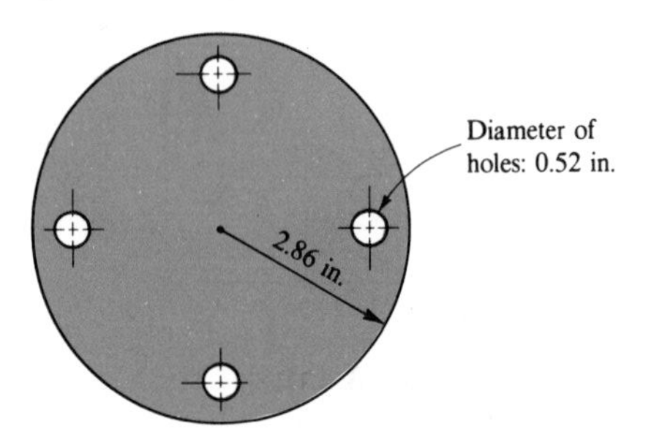

Figure 6.123

38. Find the area of the shaded portion in Figure 6.124.

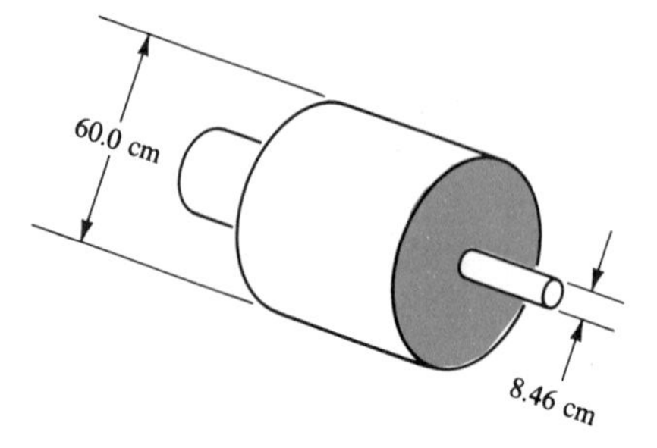

Figure 6.124

39. Find the volume of the solid in Figure 6.125.

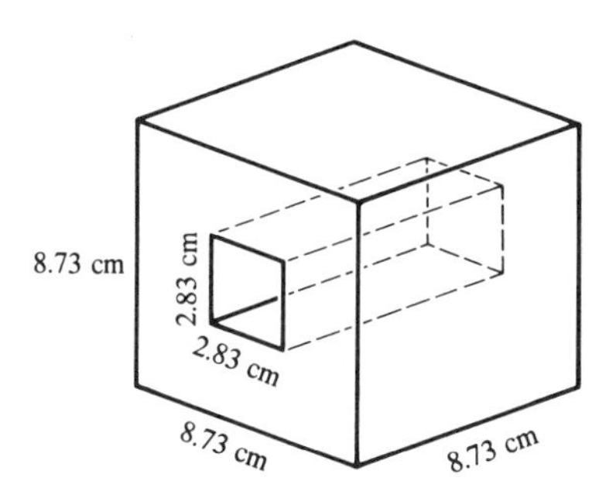

Figure 6.125

40. Find the volume of the magnet in Figure 6.126.

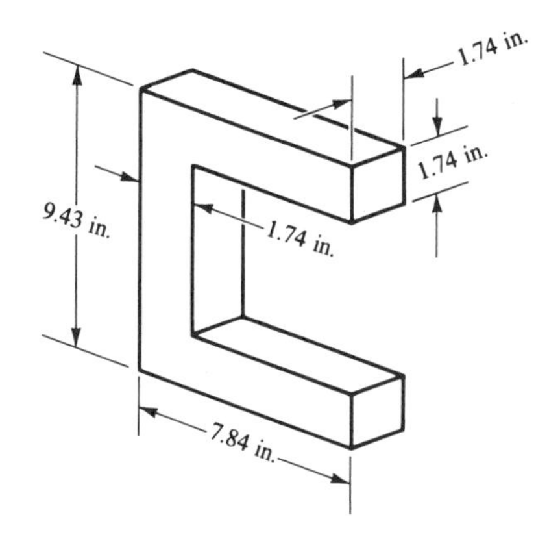

Figure 6.126

41. The cost of finishing a metal surface is \$5.00/cm^2. What is the cost of finishing one side of a metal plate in the shape of a right triangle whose sides measure 3.0 cm, 4.0 cm, and 5.0 cm, respectively?

42. A plate of uniform thickness weighs 4.6 oz/in.^{2}. Find the weight (in pounds) of a rectangular metal plate measuring 9.4 in. by 7.6 in. that has a circular hole (diameter 2.1 in.) in the center.

43. A picture frame measures 8.60 in. by 12.4 in. on the outside and 6.60 in. by 10.4 in. on the inside. What is the area of the frame?

44. Find the area of one side of a washer with outside radius 4.64 cm and inside radius 2.36 cm.

45. Water weighs 62.4 lb/ft^3. What is the weight of the water in a full rectangular tank 15.4 ft long, 12.6 ft wide, and 10.5 ft high?

46. If the tank in Exercise 45 is filled to a level of 6.00 ft, how much does the water in the tank weigh?

47. A flat metal ring has outside diameter 6.10 cm and inside diameter 2.50 cm. Find the area of one side of the ring.

48. A small copy machine is shipped to an office in a cubical box measuring 3.5 ft on the side. Find the volume of the box.

Review Exercises / Chapter 6

In Exercises 1–6, convert the given decimal degrees to degrees and minutes.

1. 0.35° **2.** 0.68° **3.** 2.04°

4. 7.36° **5.** 12.78° **6.** 10.49°

In Exercises 7–12, convert the given angle measures to decimal degrees.

7. 10′ **8.** 28′ **9.** 44′

10. 32′ **11.** 60°48′ **12.** 45°55′

In Exercises 13–16, two angles of a triangle are given. Find the third angle.

13. 60°45′, 70°18′ **14.** 20°30′, 70°40′ **15.** 16°50′, 17°20′

16. 100°10′, 20°40′

In Exercises 17–20, find the perimeters of the indicated figures.

17. Parallelogram: sides measure 10.4 ft and 8.40 ft

18. Rectangle: l = 4.7 cm, w = 8.5 cm

19. Square: s = 20.4 in.

20. Triangle: sides measure 40.3 cm, 18.4 cm, and 25.6 cm

In Exercises 21–26, find the areas of the indicated figures.

21. Rectangle: l = 6.4 cm, w = 4.8 cm

22. Parallelogram: b = 8.34 cm, h = 4.70 cm

23. Rhombus: b = 3.40 ft, h = 2.60 ft

24. Square: s = 25.4 in.

25. Circle: r = 3.76 cm

26. Circle: D = 10.4 in.

In Exercises 27–35 (Figures 6.127–6.135), find the perimeter or circumference and the area of each figure.

27.

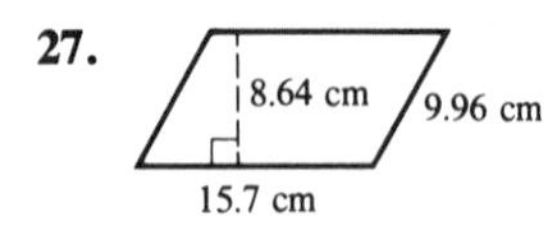

Parallelogram

Figure 6.127

28.

3.40 in.

3.90 in.

Rhombus

Figure 6.128

29.

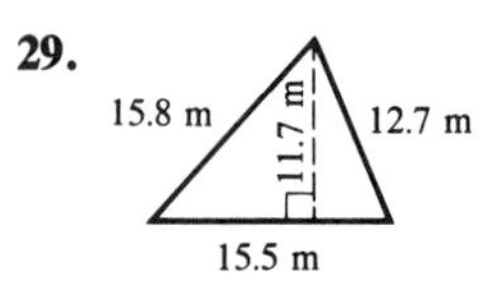

Figure 6.129

30.

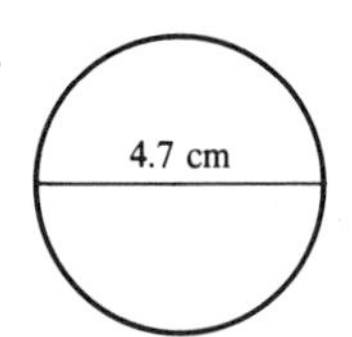

Figure 6.130

31.

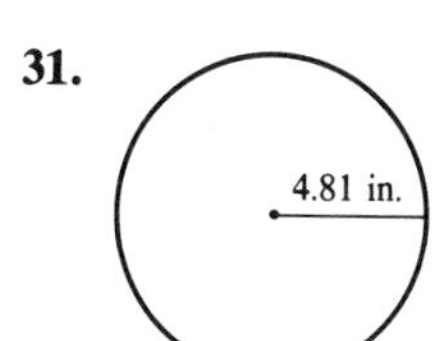

Figure 6.131

32.

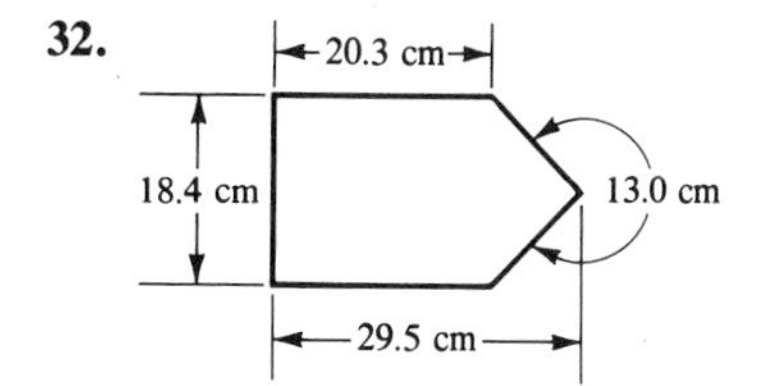

Figure 6.132

33.

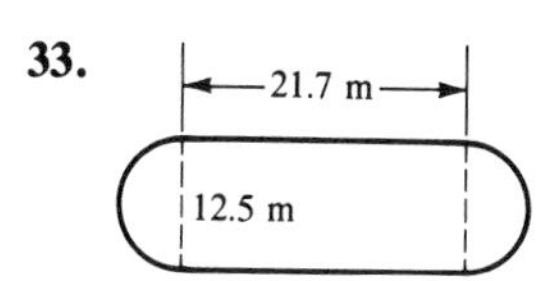

Figure 6.133

34.

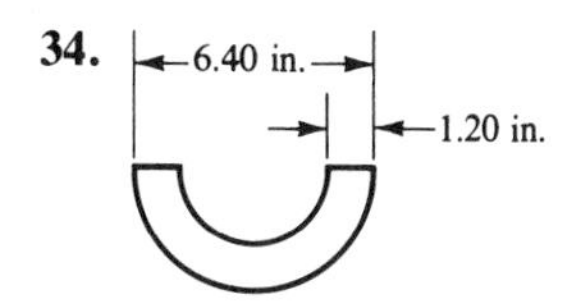

Figure 6.134

35.

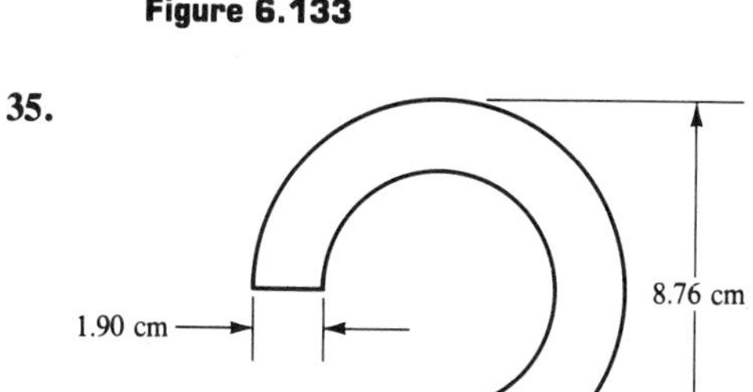

Figure 6.135

36. Find the area of the top of the bolt in Figure 6.136.

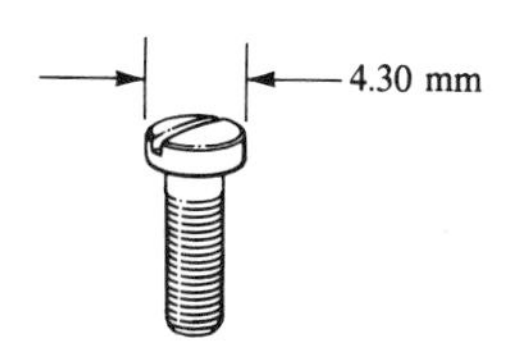

Figure 6.136

37. Find the area of the cross section of the wall of the pipe in Figure 6.137.

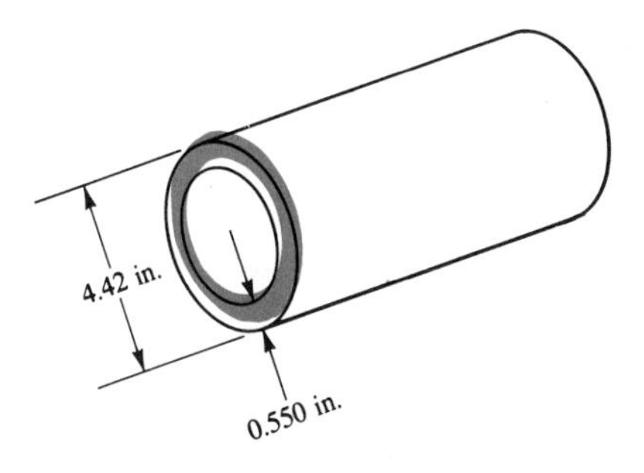

Figure 6.137

38. Find the circumference of the circular part of the motor in Figure 6.138.

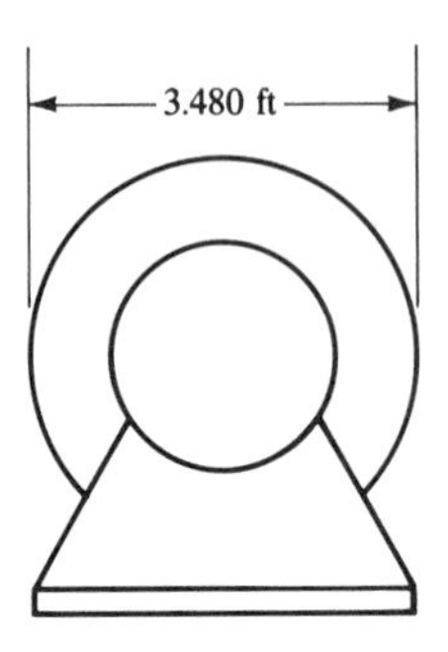

Figure 6.138

39. Find the area of a baseball diamond (Figure 6.139).

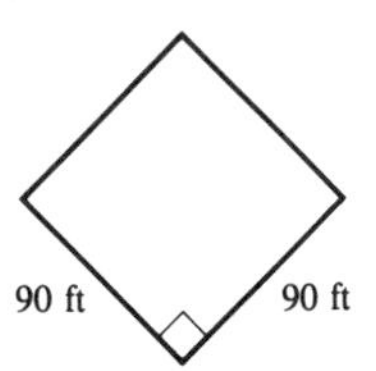

Figure 6.139

40. Find the volume of the die in Figure 6.140.

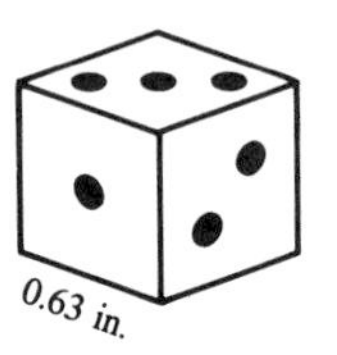

Figure 6.140

41. Find the volume of the block of ice shown in Figure 6.141.

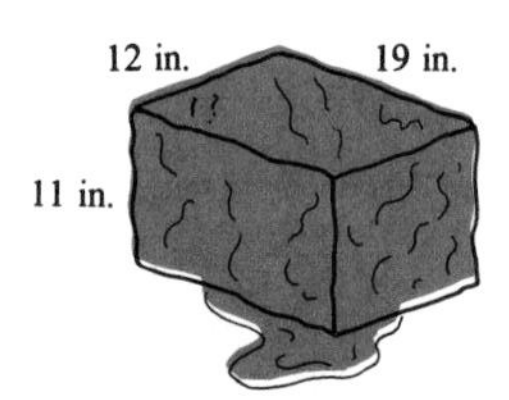

Figure 6.141

42. Find the circumference of the pulley in Figure 6.142.

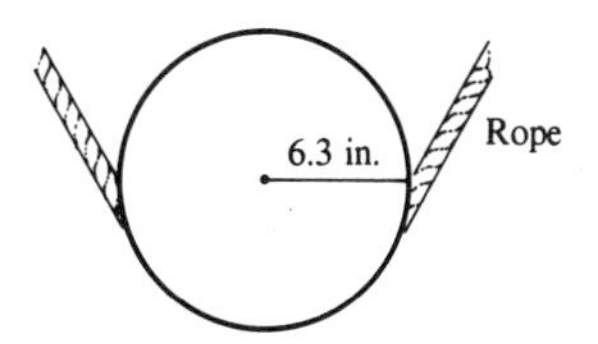

Figure 6.142

43. The density of gold is 0.697 lb/in.^{3}. Find the weight of a gold cube measuring 2.3 in. on the side.

44. A rectangular solid is made of copper, whose density is 8.9 g/cm^3. If the solid is 8.4 cm long, 6.3 cm wide, and 4.7 cm high, what is its weight in kilograms?

45. An office window is in the shape of a rectangle surmounted by a semicircle. If the rectangular part is 3.4 ft wide and 5.6 ft high, what is the perimeter of the window?

46. Determine the distance around the equator, given that the radius of the earth is 3960 mi.

47. Determine the diameter of the moon to three significant digits, given that the distance around its "equator" is 10,930 km.

48. A flat rubber ring has an outside diameter of 10.4 cm and an inside diameter of 8.60 cm. Find the area of one side.

49. Find the cost of the machine part pictured in Figure 6.143, given that the material costs 48¢/in.^{2}.

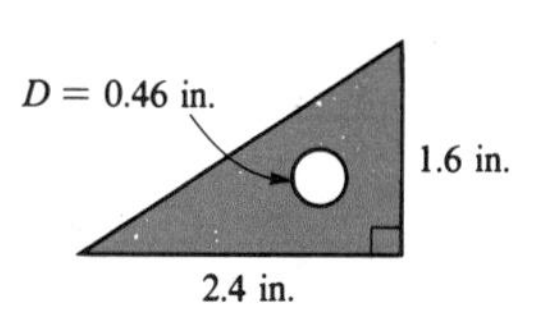

Figure 6.143

50. Find the length of the belt joining the two wheels in Figure 6.144, given that the diameter of each wheel is 1.50 ft.

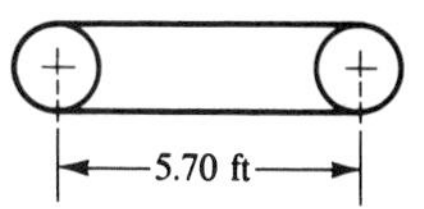

Figure 6.144

Cumulative Review Exercises / Chapters 4–6

1. A rectangular metal plate weighs w lb/cm^2. Find an expression for the weight of the plate, given that its length is x cm and its width $x - 2$ cm.
2. Which of the following expressions is a binomial and which is a trinomial?

   a. $3a^2 + 2ab - 4b^2$ b. $T_aT_b - 4T_a^2T_b^2$

In Exercises 3–16, perform the indicated operations.

3. $5a\sqrt{b} - 10a\sqrt{b}$
4. $(40.6R - 14.6) - (18.3R + 27.3)$
5. $-(C_1 + 2C_2) + (3C_1 - 6C_2) - (5C_1 - 10C_2)$
6. $-\{a - [-b - (a + b) - 2a]\}$
7. $(5L^3M^2)(-10L^2M^3)$
8. $(-2ab^2)^3(-4a^2b)^2$
9. $-4.62a(5.22a - 3.65a^2)$
10. $3(R - 6C) - 2(3R + 4C)$
11. $(2x^3 + 3x - 2)(x + 4)$
12. $(3x^2 - 2x + 4)(2x^2 + 3x - 1)$
13. $\dfrac{(3m^2n^4)^2}{9m^6n^2}$
14. $\dfrac{-10P^3V^2 + 15P^4V^4 + 5P^2V}{5P^2V}$
15. $\dfrac{6w^2 - 24z^2}{3w + 6z}$
16. $\dfrac{10x^3 - 21x^2 + 17x - 12}{2x - 3}$
17. Convert 42°38′ to decimal degrees.
18. Find the area of a parallelogram with base 4.83 cm and height 3.69 cm.
19. Find the area of a circle with diameter 12.60 in.
20. Find the area of a trapezoid with the following dimensions: $b_1 = 2.4$ in., $b_2 = 4.6$ in., and $h = 3.7$ in.
21. Determine the perimeter of the following figure:

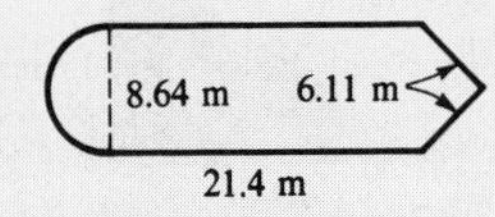

22. The cost (in dollars) of one type of computer is $5.0C + 4.2$, and the cost of another is $7.3C - 1.6$. Find an expression for the total cost.
23. Find an expression for the area of a triangular metal plate with base $2x - 4y$ (in feet) and height $4x + 3y$ (in feet).
24. If $2x + 3$ gears cost $2x^2 - x - 6$ dollars, find an expression for the cost of one gear.
25. An oil tank in the shape of a rectangular solid is 12.4 ft long, 6.74 ft wide, and 8 ft high, and it is filled with oil to a level of 5.24 ft. Given that oil weighs 50.0 lb/ft^3, find the weight of the oil in the tank.

CHAPTER 7

Introduction to Equations and Inequalities

This chapter is devoted to a study of basic equations and inequalities. This area of algebra is particularly useful in science and technology.

7.1 Simple Equations

An **equation** is a statement which says that two expressions are equal. Examples of equations are

1. $\frac{1}{2} = \frac{2}{4}$
2. $\frac{2}{5} = 0.4$
3. $x(x + 1) = x^2 + x$
4. $2x + 3 = 5$

The statement that defines an equation may be true for some values and not true for others. For example, equation 4,

$$2x + 3 = 5$$

is true for $x = 1$. (If $x = 1$, then $2x + 3 = 2(1) + 3 = 5$.) It is false for all other values of x. On the other hand, equation 3,

$$x(x + 1) = x^2 + x$$

is an example of the distributive law and is valid for all values of x.

The letter x in these equations is called a **variable**. Equation 3, as we have seen, is valid for all values of x and is called an *identity*. Equation 4 is valid only if $x = 1$ and is therefore called a *conditional equation*. In this

Equation

chapter we study only conditional equations. For that reason we use the word **equation** to mean conditional equation.

> The letter in an equation is called the **variable** or the **unknown**. The value of the unknown for which equality holds is called the **solution** or **root** and is said to **satisfy** the equation. Finding the solution of an equation is called **solving** the equation.

To see how an equation can be solved for the unknown, consider the equation

$$x - 2 = 3$$

We want to find the value of x for which equality holds. We know that

$$x - 2 \text{ (the left side)} = 3 \text{ (the right side)}$$

To eliminate the -2 on the left side, we add 2 to both sides. After adding 2, the left side is still equal to the right side. Therefore,

$$x - 2 + \mathbf{2} = 3 + \mathbf{2}$$

Performing the indicated operations, we get

$$x + 0 = 5$$

or

$$x = 5$$

So $x = 5$ is the solution. To see if the solution we found is correct, we substitute 5 for x in the given equation:

$$x - 2 = 3 \quad \text{Given equation}$$
$$\mathbf{5} - 2 = 3 \quad x = 5$$
$$3 = 3$$

The solution checks. Consider another example.

Example 1 Solve the equation

$$3x + 2 = 11$$

Solution. The first step in solving this equation is to subtract 2 from both sides:

$$3x + 2 = 11 \quad \text{Given equation}$$
$$3x + 2 - \mathbf{2} = 11 - \mathbf{2} \quad \text{Subtracting 2 from both sides}$$
$$3x = 9 \quad \text{Combining terms}$$

To obtain x on the left side, we divide both sides by 3:

$$\frac{3x}{3} = \frac{9}{3} \quad \text{Dividing both sides by 3}$$

$$x = 3 \quad \text{Simplifying}$$

So $x = 3$ is the solution. As a check, we substitute 3 in the given equation:

$$3x + 2 = 11 \quad \text{Given equation}$$

$$3(3) + 2 = 11 \ \checkmark \quad x = 3$$

These examples show that we can solve an equation by performing the same algebraic operation on both sides. It is helpful to compare an equation to a balance scale. If a scale is in balance and if we add 2 lb to each side, the scale remains balanced. If we subtract 3 lb from each side, the scale again remains balanced. The scale also remains balanced if we double the weights on each side or cut the weights on each side in half.

In the same way, to solve an equation, we add the same quantity to both sides, subtract the same quantity from both sides, or multiply or divide both sides by the same nonzero constant until only x remains on one side. Consider another example.

Example 2 Solve the equation

$$2x + 4 = 5x - 3$$

Solution. In this equation the unknown appears on both sides of the equation. Thus we first subtract $5x$ from both sides, so the unknown appears on one side only:

$$2x + 4 = 5x - 3 \quad \text{Given equation}$$

$$2x + 4 - 5x = 5x - 3 - 5x \quad \text{Subtracting } 5x$$

$$-3x + 4 = -3 \quad \text{Combining like terms}$$

$$-3x + 4 - 4 = -3 - 4 \quad \text{Subtracting 4}$$

$$-3x = -7 \quad \text{Simplifying}$$

$$\frac{-3x}{-3} = \frac{-7}{-3} \quad \text{Dividing by } -3$$

$$x = \frac{7}{3}$$

Check: $2x + 4 = 5x - 3$

Left side

$$2\left(\frac{7}{3}\right) + 4 = \frac{14}{3} + 4 = \frac{14}{3} + \frac{12}{3} = \frac{26}{3}$$

Right side

$$5\left(\frac{7}{3}\right) - 3 = \frac{35}{3} - 3 = \frac{35}{3} - \frac{9}{3} = \frac{26}{3} \quad ✓$$

The equations we have considered so far are called **first-degree equations** since the unknown x is raised to the first power. The first-degree equations of the type studied in this section can be solved by the following procedure:

1. Remove symbols of grouping.
2. Collect all terms containing the unknown on one side of the equation. Combine like terms.
3. Collect all constants on the other side. Combine the constants.
4. In the resulting equation $ax = b$, divide both sides by a to obtain the solution $x = b/a$.
5. Check the solution by substituting the value of x into the original equation.

Example 3 Solve the equation

$$2 - 6x = x - 7$$

Solution. This equation is similar to the equation in Example 2. If we proceed by subtracting x from both sides, we obtain $2 - 6x - x = x - 7 - x$. As an alternative, we can simplify the procedure a little by writing the x terms on the right side, thereby avoiding negative coefficients.

$2 - 6x = x - 7$ Given equation

Step 1. Since there are no symbols of grouping, this step does not apply.

Step 2. $2 - 6x + 6x = x - 7 + 6x$ Adding $6x$

$2 = 7x - 7$ Combining like terms

Step 3. $2 + 7 = 7x - 7 + 7$ Adding 7

$9 = 7x$ Simplifying

Step 4. $\dfrac{9}{7} = \dfrac{7x}{7}$ Dividing by 7

$\dfrac{9}{7} = x$

$x = \dfrac{9}{7}$ Switching sides

Step 5. *Check:* $2 - 6x = x - 7$

Left side	*Right side*
$2 - 6\left(\dfrac{9}{7}\right) = 2 - \dfrac{54}{7}$	$\dfrac{9}{7} - 7 = \dfrac{9}{7} - \dfrac{49}{7}$
$= \dfrac{14}{7} - \dfrac{54}{7}$	$= -\dfrac{40}{7}$ ✓
$= -\dfrac{40}{7}$	

The equation in the next example contains symbols of grouping.

Example 4 Solve the equation

$4(x + 6) = 2(2 - 3x)$

Solution. $4(x + 6) = 2(2 - 3x)$ Given equation

Step 1. $4x + 24 = 4 - 6x$ Removing parentheses

Step 2. $4x + 24 + 6x = 4 - 6x + 6x$ Adding $6x$

$10x + 24 = 4$ Combining like terms

Step 3. $10x + 24 - 24 = 4 - 24$ Subtracting 24

$10x = -20$

Step 4. $\dfrac{10x}{10} = \dfrac{-20}{10}$ Dividing by 10

$x = -2$

Step 5. *Check:* $4(x + 6) = 2(2 - 3x)$

Left side	*Right side*
$4(-2 + 6) = 4(4) = 16$	$2[2 - 3(-2)] = 2[2 + 6] = 2 \cdot 8 = 16$ ✓

If, after collecting terms, the coefficient of x is -1, we divide (or multiply) both sides by -1 to obtain the coefficient $+1$.

Example 5

$$4x = 5x - 2 \qquad \text{Given equation}$$
$$4x - 5x = 5x - 5x - 2 \qquad \text{Subtracting } 5x$$
$$-x = -2$$
$$\frac{-x}{-1} = \frac{-2}{-1} \qquad \text{Dividing by } -1$$
$$x = 2$$

Common error Forgetting to perform an operation on *both* sides of the equation. For example, to solve the equation $x - 2 = 4$, it is not correct to add 2 just to the left side to get $x - 2 + 2 = 4$. Instead, 2 has to be added to *both* sides:

$$x - 2 + 2 = 4 + 2$$
$$x = 6$$

Exercises / Section 7.1

In Exercises 1–32, solve each of the given equations. The coefficients are assumed to be exact, so the answers should be written in fractional, rather than in decimal form.

1. $x + 4 = 0$
2. $x - 7 = 3$
3. $x - 8 = 7$
4. $x + 9 = 3$
5. $2x = 4$
6. $3x = 12$
7. $7x = -21$
8. $6x = -24$
9. $5x = 2$
10. $7x = 9$
11. $2x - 3 = 7$
12. $2x + 5 = 13$
13. $5 - 3x = -10$
14. $3 + 4x = 15$
15. $2x - 5 = -2$
16. $3 - 2x = -6$
17. $7x + 6 = 14$
18. $4x - 7 = -4$
19. $2x - 4 = 7x + 6$
20. $3x + 6 = -2x + 21$
21. $3 - 2x = 4x - 9$
22. $1 - 7x = -x - 17$
23. $5 - 2x = -4x - 7$
24. $3 - x = -3x - 9$
25. $x + 4 = 2 - 4x$
26. $3x + 2 = -x - 7$
27. $2x - 4 = 3 - 2x$
28. $x - 10 = -3x - 8$
29. $2(x + 3) = 3(x + 2)$
30. $x - 4 = 2(5x - 2)$
31. $2(1 - 2x) = 6x - 2$
32. $5(x + 3) = 2(2x - 4)$

In Exercises 33–40, use a calculator to solve the given equations. Assume that the numbers are approximate.

33. $2.56x = -7.74$
34. $8.02x = -9.16$
35. $10.4x - 8.71 = 13.3$
36. $25.6x - 12.7 = 45.6$
37. $18.45 - 15.64x = 10.41x - 20.39$
38. $23.75x - 20.45 = 50.69 - 32.94x$
39. $2.1(2.4x - 3.0) = 4.6(8.7x + 5.5)$
40. $6.9(3.7x + 6.2) = 8.5(1.2x - 2.6)$

7.2 More Equations

In the last section we studied the solution of simple first-degree equations. In this section we study first-degree equations with (1) fractions and (2) literal coefficients. In addition, we use various letters to represent the unknown.

Equations with Fractions

A simple example of an equation with a fractional coefficient is

$$\frac{1}{3}x = 4$$

Using the method of the last section, we could divide both sides of the equation by $\frac{1}{3}$ to obtain

$$\frac{\frac{1}{3}x}{\frac{1}{3}} = \frac{4}{\frac{1}{3}}$$

$$x = \frac{4}{1} \cdot \frac{3}{1} = 12$$

It is much easier, however, to multiply both sides by 3, the reciprocal of $\frac{1}{3}$. Then we get

$$3 \cdot \frac{1}{3}x = 3 \cdot 4$$

$$x = 12 \qquad \text{Since } 3 \cdot \frac{1}{3} = 1$$

Consider another example.

Example 1

a. $\frac{3}{4}x = -3$

$$4\left(\frac{3}{4}x\right) = 4(-3) \qquad \text{Multiplying by 4}$$

$$3x = -12 \qquad \text{Since } 4 \cdot \frac{3}{4} = 3$$

$$x = -4 \qquad \text{Dividing by 3}$$

b. $-\frac{1}{6}x = -4$

$$(-6)\left(-\frac{1}{6}x\right) = (-6)(-4) \qquad \text{Multiplying by } -6$$

$$x = 24 \qquad (-6)\left(-\frac{1}{6}\right) = 1$$

If an equation contains more than one fraction, we can **clear fractions** by multiplying both sides of the equation by the lowest common denominator (LCD) of all the fractions. Consider the next example.

Example 2 Solve the following equation for y:

$$\frac{1}{2}y - 3 = \frac{1}{4}y - \frac{1}{2}$$

Solution. Note that the unknown is represented by the letter y, instead of the letter x used so far.

The lowest common denominator of $\frac{1}{2}$ and $\frac{1}{4}$ is 4. If we multiply both sides of the equation by 4, the fractions will be eliminated. *Remember to multiply all the terms.*

$$\frac{1}{2}y - 3 = \frac{1}{4}y - \frac{1}{2} \qquad \text{Given equation}$$

$$4\left(\frac{1}{2}y - 3\right) = 4\left(\frac{1}{4}y - \frac{1}{2}\right) \qquad \text{Multiplying by 4}$$

$$4\left(\frac{1}{2}y\right) + 4(-3) = 4\left(\frac{1}{4}y\right) + 4\left(-\frac{1}{2}\right) \qquad \text{Distributive law}$$

$$2y - 12 = y - 2 \qquad \text{Simplifying}$$

$$2y - y - 12 = y - y - 2 \qquad \text{Subtracting } y$$

$$y - 12 = -2 \qquad \text{Combining terms}$$

$$y - 12 + 12 = -2 + 12 \qquad \text{Adding 12}$$

$$y = 10$$

The solution can be checked in the usual way.

Procedure for solving first-degree equations

1. Remove symbols of grouping.
2. Clear fractions by multiplying both sides by the LCD of all the fractions.
3. Collect all terms containing the unknown on one side. Combine like terms.
4. Collect all constants on the opposite side. Combine the constants.
5. Multiply or divide both sides to make the coefficient of the unknown +1.
6. Check the solution by substituting the value of the unknown into the original equation.

A step-by-step procedure for solving first-degree equations is given on p. 216. Please note, however, that some equations do not require every step for the solution. For example, if an equation does not contain parentheses, Step 1 does not apply. Moreover, **the steps do not always have to be performed in the given order,** as we will see.

Example 3 Solve the following equation:

$$\frac{1}{3}\left(T - \frac{3}{2}\right) = 2\left(\frac{1}{3}T - \frac{3}{4}\right)$$

Solution.

$$\frac{1}{3}\left(T - \frac{3}{2}\right) = 2\left(\frac{1}{3}T - \frac{3}{4}\right) \quad \text{Given equation}$$

Step 1.

$$\frac{1}{3}T + \frac{1}{3}\left(-\frac{3}{2}\right) = 2\left(\frac{1}{3}T\right) + 2\left(-\frac{3}{4}\right) \quad \text{Distributive law}$$

$$\frac{1}{3}T - \frac{1}{2} = \frac{2}{3}T - \frac{3}{2} \quad \text{Simplifying}$$

Step 2.

$$6\left(\frac{1}{3}T - \frac{1}{2}\right) = 6\left(\frac{2}{3}T - \frac{3}{2}\right) \quad \text{LCD} = 6$$

$$6\left(\frac{1}{3}T\right) + 6\left(-\frac{1}{2}\right) = 6\left(\frac{2}{3}T\right) + 6\left(-\frac{3}{2}\right) \quad \text{Distributive law}$$

$$2T - 3 = 4T - 9 \quad \text{Simplifying}$$

Step 3.

$$2T - 3 - 4T = 4T - 9 - 4T \quad \text{Subtracting } 4T$$

$$-2T - 3 = -9$$

Step 4.

$$-2T - 3 + 3 = -9 + 3 \quad \text{Adding 3}$$

$$-2T = -6$$

Step 5.

$$\frac{-2T}{-2} = \frac{-6}{-2} \quad \text{Dividing by } -2$$

$$T = 3$$

Step 6. *Check:*

Left side

$$\frac{1}{3}\left(T - \frac{3}{2}\right) = \frac{1}{3}\left(3 - \frac{3}{2}\right) = \frac{1}{3}\left(\frac{3}{2}\right) = \frac{1}{2}$$

Right side

$$2\left(\frac{1}{3}T - \frac{3}{4}\right) = 2\left(\frac{1}{3}\cdot 3 - \frac{3}{4}\right) = 2\left(1 - \frac{3}{4}\right) = 2\left(\frac{1}{4}\right) = \frac{1}{2} \checkmark$$

Common error Not multiplying the entire side of an equation when clearing fractions. For example, the equation

$$\frac{1}{2}x + 2 = \frac{1}{4} - x$$

should *not* be written

$$2x + 2 = 1 - x$$

obtained by multiplying $\frac{1}{2}x$ and $\frac{1}{4}$ by 4. The correct procedure is

$$4\left(\frac{1}{2}x + 2\right) = 4\left(\frac{1}{4} - x\right)$$
$$2x + 8 = 1 - 4x$$

Equations with Literal Coefficients

If the equation $ax = b$ is to be solved for x, then the coefficient a is treated as a **constant.** The method of solution is the same as for equations with numerical coefficients. Dividing both sides by a, we get

$$\frac{ax}{a} = \frac{b}{a}$$

or

$$x = \frac{b}{a}$$

(It is understood that $a \neq 0$. In fact, it is assumed in this section that the values of the literal expressions are restricted so that division by zero is avoided.)

The procedure for solving first-degree equations is valid for equations containing letters. You may find the procedure more difficult to apply, at least initially. But it is important for you to practice this procedure so that the solution of formulas in the next section becomes easier.

Example 4 Solve for x: $ax + 2 = b$

Solution. Since the equation contains only one term with an unknown, we need only to subtract 2 from both sides and divide both sides by a:

$$ax + 2 = b \quad \text{Given equation}$$
$$ax + 2 - 2 = b - 2 \quad \text{Subtracting 2}$$
$$ax = b - 2$$
$$\frac{ax}{a} = \frac{b - 2}{a} \quad \text{Dividing by } a$$
$$x = \frac{b - 2}{a}$$

Note that the *entire* right side, $b - 2$, must be divided by a. This operation is indicated by using a bar.

Example 5 Solve for y:

$$2(2b - y) = 3(y + b)$$

Solution. $2(2b - y) = 3(y + b)$ Given equation

Step 1. $4b - 2y = 3y + 3b$ Distributive law

Step 2. Not necessary (no fractions to clear).

Step 3. $4b - 2y + 2y = 3y + 3b + 2y$ Adding $2y$

$$4b = 5y + 3b$$

Step 4. $4b - 3b = 5y + 3b - 3b$ Subtracting $3b$

$$b = 5y$$

Step 5. $\frac{b}{5} = \frac{5y}{5}$ Dividing by 5

$$\frac{b}{5} = y$$

$$y = \frac{b}{5}$$ Switching sides

Step 6. Although highly desirable, checking a solution containing literal terms would be difficult at this point; we will study algebraic fractions in Chapter 9.

Example 6 Solve the following equation for x:

$$\frac{1}{2}(ax + 2) = 3\left(\frac{1}{4}ax - 1\right)$$

Solution. $\frac{1}{2}(ax + 2) = 3\left(\frac{1}{4}ax - 1\right)$

Step 1. $\frac{1}{2}(ax) + \frac{1}{2} \cdot 2 = 3\left(\frac{1}{4}ax\right) + 3(-1)$ Distributive law

$$\frac{1}{2}ax + 1 = \frac{3}{4}ax - 3$$

Step 2. $4\left(\frac{1}{2}ax + 1\right) = 4\left(\frac{3}{4}ax - 3\right)$ LCD = 4

$$2ax + 4 = 3ax - 12$$

Step 3. $2ax + 4 - 2ax = 3ax - 12 - 2ax$ Subtracting $2ax$

$$4 = ax - 12$$

Step 4. $4 + 12 = ax - 12 + 12$ Adding 12

$$16 = ax$$

Step 5. $\frac{16}{a} = \frac{ax}{a}$ Dividing by a

$$\frac{16}{a} = x$$

$$x = \frac{16}{a}$$

It is not always necessary, or even desirable, to perform the steps in the order suggested in the step-by-step summary. The next example illustrates a different order of steps.

Example 7 Solve the equation

$$\frac{1}{2}(x + 1) = \frac{1}{4}(x + 2)$$

Solution. Rather than eliminating symbols of grouping, it may be better to clear fractions first (Step 2):

$$4 \cdot \frac{1}{2}(x + 1) = 4 \cdot \frac{1}{4}(x + 2) \qquad \text{Step 2}$$

$$2(x + 1) = 1 \cdot (x + 2)$$

$$2x + 2 = x + 2 \qquad \text{Step 1}$$

$$2x + 2 - 2 = x + 2 - 2 \qquad \text{Step 4}$$

$$2x = x$$

$$2x - x = x - x \qquad \text{Step 3}$$

$$x = 0$$

Exercises / Section 7.2

In Exercises 1–40, solve the given equations.

1. $x - 4 = 6$

2. $x + 6 = -2$

3. $4x = 6$

4. $8x = 10$

5. $\frac{1}{2}x = 3$

6. $\frac{1}{3}x = -1$

7. $\frac{2}{3}x = -6$

8. $\frac{3}{4}x = 3$

9. $4x + 3 = -x - 2$

10. $2x - 3 = 1 - 4x$

11. $2 - 3x = -x - 6$

12. $1 - 4x = x + 6$

13. $2(1 - 3y) = 5 - 3y$

14. $3(2y - 3) = 8y - 5$

15. $3 - 7z = 4(z - 2)$

16. $2 - 5z = 3(1 - 2z)$

17. $3(2w + 1) = 2(1 - 3w)$

18. $4(1 - w) = 3(2w + 1)$

19. $\frac{2}{3}m = \frac{1}{6}$

20. $-\frac{3}{8}n = \frac{1}{4}$

21. $-\frac{3}{5}x = 6$

22. $-\frac{3}{5}x = \frac{7}{10}$

23. $\frac{1}{2}x + \frac{1}{2} = \frac{1}{4}$

24. $\frac{1}{2} - \frac{1}{2}x = \frac{1}{6}$

25. $\frac{2}{3}y + 1 = \frac{1}{2}$

26. $\frac{1}{2}z + 1 = \frac{3}{4}$

27. $1 - \frac{1}{5}t = \frac{3}{5}$

28. $1 - \frac{2}{5}s = \frac{1}{5}$

29. $\frac{5}{6}T + \frac{1}{2} = \frac{1}{3}T$

30. $\frac{3}{8}u - \frac{1}{4} = \frac{1}{2}u$

31. $\frac{1}{3}m - \frac{1}{6} = \frac{1}{6}m + \frac{1}{12}$

32. $\frac{1}{2} - \frac{1}{6}n = -\frac{1}{3}$

33. $\frac{1}{3}x - \frac{1}{4} = \frac{1}{12} - \frac{1}{6}x$

34. $\frac{1}{2}x + \frac{1}{6} = \frac{1}{4}x - 2$

35. $\frac{1}{2}\left(3z - \frac{3}{2}\right) = -\frac{1}{2}z - 1$

36. $\frac{1}{3}\left(y - \frac{1}{2}\right) = \frac{1}{6}y - \frac{1}{2}$

37. $\frac{1}{6}(1 - w) = \frac{1}{3}w - \frac{1}{2}$

38. $\frac{1}{2}\left(\frac{1}{2}m - 1\right) = \frac{1}{3}m - 2$

39. $\frac{2}{3}\left(n - \frac{1}{2}\right) = \frac{3}{2}(2 - n)$

40. $\frac{5}{2}\left(v - \frac{2}{5}\right) = \frac{1}{3}\left(v - \frac{1}{2}\right)$

In Exercises 41–80, solve each equation for x.

41. $bx = 1$

42. $cx = 2$

43. $ax = -3$

44. $bx = -3$

45. $ax = b$

46. $bx = c$

47. $x + b = 2$

48. $x - c = 1$

49. $x - 2a = b$

50. $x + 4a = c$

51. $2x + b = 1$

52. $3x - c = 2$

53. $ax + 1 = 4$

54. $bx - 3 = 7$

55. $cx + 6 = d$

56. $dx - 7 = c$

57. $bx - a = m$

58. $1 - 2ax = s$

59. $2 - ax = -t$

60. $ax + b = 6$

61. $bx - 3 = c$

62. $ax - 4 = 2ax - 1$

63. $2ax - 6 = 3ax - 3$

64. $5ax - b = 2ax - 2b$

65. $6ax - 3c = 4ax + c$

66. $3ax - 8d = 7ax + d$

67. $2(ax + 1) = 3(ax - 3)$

68. $3(bx - 2) = 2(2bx - 1)$

69. $4(1 - cx) = 3(cx - 2)$

70. $4(ax + b) = -ax - b$

71. $2(2ax - b) = b - 3ax$

72. $3(b - 6ax) = ax - 3b$

73. $\frac{1}{2}(bx - 4) = \frac{1}{3}(bx - 3)$

74. $\frac{1}{2}(1 - 2bx) = \frac{1}{3}\left(bx - \frac{1}{2}\right)$

75. $\frac{1}{6}(1 - 3ax) = \frac{1}{3}\left(\frac{2}{3} - ax\right)$

76. $\frac{1}{2}\left(\frac{1}{4}ax - \frac{1}{2}\right) = \frac{1}{4}(ax - 2)$

77. $\frac{1}{5}\left(\frac{3}{2}cx - \frac{5}{2}\right) = \frac{1}{2}\left(-\frac{1}{5}cx - \frac{1}{5}\right)$

78. $\frac{1}{4}\left(1 - \frac{1}{2}ax\right) = \frac{1}{2}\left(\frac{1}{2}ax - \frac{1}{2}\right)$

79. $\frac{1}{10}(2 - ax) = \frac{1}{5}\left(ax - \frac{3}{2}\right)$

80. $\frac{1}{3}\left(1 - \frac{3}{2}ax\right) = \frac{1}{2}(ax - 3)$

7.3 Formulas

In the last section we learned to solve equations containing literal coefficients. In this section we study literal equations called *formulas,* which are solved for one letter in terms of the rest.

Formula

A **formula** is a relationship between variables, often expressing a geometric property or a physical law. For example, the voltage V across a resistor is equal to the current I times the resistance R, or

$$V = IR$$

In what sense can this formula be solved? Suppose we divide both sides of the equation by I. Then

$$\frac{V}{I} = \frac{IR}{I} \quad \text{Dividing by } I$$

$$\frac{V}{I} = \frac{\cancel{I}R}{\cancel{I}} \quad \text{Canceling}$$

$$R = \frac{V}{I} \quad \text{Switching sides}$$

This "solution" is actually a new formula. This formula says that the resistance R is equal to the voltage V across the resistor divided by the current I. This example shows that there may be good reasons for solving a formula. The term *formula rearrangement* is sometimes used instead of *solution* of a formula.

Example 1 The circumference C of a circle is equal to π times the diameter D, or

$$C = \pi D$$

Solve this formula for π.

Solution. Dividing both sides by D, we get

$$C = \pi D \quad \text{Given formula}$$

$$\frac{C}{D} = \frac{\pi D}{D} \quad \text{Dividing by } D$$

$$\frac{C}{D} = \pi$$

$$\pi = \frac{C}{D}$$

This formula is actually the definition of π (circumference divided by diameter).

The formulas discussed in this section can be solved by a procedure that is exactly the same as the procedure for solving first-degree equations:

1. Remove symbols of grouping.
2. Clear fractions, if necessary.
3. Collect on one side of the equation the terms containing the letter to be solved for; collect the remaining terms on the other side.
4. Divide both sides by the coefficient of the letter to be solved for.

Example 2 The area A of a trapezoid is given by $A = \frac{1}{2}h(b_1 + b_2)$. Solve this formula for b_1.

Solution. As usual, the order in which the steps are performed can be changed. In this example, clearing fractions is a good first step. However, for the purpose of illustration, let us stick to the order given in the summary.

$$A = \frac{1}{2}h(b_1 + b_2) \quad \text{Given formula}$$

Step 1. $$A = \frac{1}{2}hb_1 + \frac{1}{2}hb_2 \quad \text{Distributive law}$$

$$A = \frac{1}{2}hb_1 + \frac{1}{2}hb_2$$

Step 2. $$2A = 2 \cdot \frac{1}{2}hb_1 + 2 \cdot \frac{1}{2}hb_2 \quad \text{Multiplying by 2}$$

$$2A = hb_1 + hb_2$$

Step 3. $$2A - hb_2 = hb_1 + hb_2 - hb_2 \quad \text{Subtracting } hb_2$$

$$2A - hb_2 = hb_1$$

Step 4. $\dfrac{2A - hb_2}{h} = \dfrac{hb_1}{h}$ Dividing by h

$$b_1 = \frac{2A - hb_2}{h}$$

Example 3 Solve the following formula for D:

$$S = \frac{a(A + 3D)}{b}$$

Solution.

Step 1. $S = \dfrac{aA + 3aD}{b}$ Distributive law

Step 2. $bS = \dfrac{aA + 3aD}{\not{b}} \cdot \dfrac{\not{b}}{1}$ Multiplying by b

Step 3. $bS - aA = aA + 3aD - aA$ Subtracting aA

$bS - aA = 3aD$

Step 4. $\dfrac{bS - aA}{3a} = \dfrac{3aD}{3a}$ Dividing by $3a$

$$D = \frac{bS - aA}{3a}$$

As already noted, the order of the steps listed in the summary should not always be rigidly followed. This is shown in the next example.

Example 4 The number of teeth N in a gear is related to the pitch diameter D of the gear and the outside diameter D_0 by the formula

$$N = \frac{2D}{D_0 - D}$$

Solve this formula for D_0.

Solution. Since there are no parentheses in the formula, we clear fractions by multiplying both sides by the denominator of the right side, $D_0 - D$. This operation introduces $N(D_0 - D)$ on the left side, and we then go ahead and remove the grouping symbols.

$N = \dfrac{2D}{D_0 - D}$ Given formula

$N(D_0 - D) = 2D$ Multiplying by $D_0 - D$

$ND_0 - ND = 2D$ Distributive law

$$ND_0 - ND + ND = 2D + ND \quad \text{Adding } ND$$
$$ND_0 = 2D + ND$$
$$\frac{ND_0}{N} = \frac{2D + ND}{N} \quad \text{Dividing by } N$$
$$D_0 = \frac{2D + ND}{N}$$

Exercises / Section 7.3

In Exercises 1–30, solve each formula for the indicated letter.

1. $3a = b$, a

2. $5C = D$, C

3. $pq = s$, p

4. $st = 5$, s

5. $\frac{R_1}{R_2} = 3T$, R_2

6. $\frac{N_1}{N_2} = 2M$, N_2

7. $AD - Q = R$, D

8. $VW - B = C$, W

9. $\frac{1}{2} m_1 m_2 - 4 = m_3$, m_1

10. $\frac{1}{3} t_1 t_2 - t_3 = 2$, t_1

11. $\frac{1}{4} - \frac{1}{2} xy = \frac{1}{2} y$, x

12. $\frac{1}{3} - \frac{1}{2} xy = \frac{1}{6} y$, x

13. $2(a + b) = c$, b

14. $-3(c - d) = r$, d

15. $P(a - 2b) = 3$, a

16. $Q(r + st) = 4$, r

17. $Q = \frac{3a - 2b}{3}$, a

18. $N = \frac{2n - 3m}{4}$, n

19. $A = \frac{3v - 4}{a}$, v

20. $L = \frac{3s + b}{a}$, s

21. $P_0 = \frac{2(P_1 + 2P_2)}{a}$, P_2

22. $V_0 = \frac{3(2V_1 - V_2)}{b}$, V_1

23. $M = \frac{4(v - 3)}{s - t}$, s

24. $N = \frac{3r}{2s - t}$, s

25. $P_0 = \frac{2(P_1 - P_2)}{P_2}$, P_1

26. $L = \frac{c(b + 2)}{a - 2}$, a

27. $V = \frac{Pr}{a} + b$, P

28. $C = n + \frac{Mm}{n}$, M

29. $T_1 = T_2 - \frac{aT_0}{b}$, T_0

30. $R_1 = rR_2 - \frac{3R_0}{t}$, R_0

31. Boyle's law, relating the volume and pressure of a confined gas, is

$$P = \frac{k}{V}$$

Solve this formula for V.

32. In the study of the photoelectric effect, the formula

$$T = k(\nu - \nu_0) \qquad \nu = \text{nu}$$

arises. Solve this formula for ν.

33. An investment of P dollars accumulates to A dollars in t years according to the formula

$$A = P + Prt$$

where r is the simple interest rate. Solve this formula for t.

34. The volume V of a sphere of radius r is

$$V = \frac{4}{3}\pi r^3$$

Solve this formula for π.

35. The following formula arises in the study of atomic spectra:

$$E_d = \frac{(K + 1)H^2}{I}$$

Solve this formula for K.

36. The following formula comes from an agricultural study:

$$y = \frac{WHx}{x^2 + W^2}$$

Solve this formula for H.

37. The formula for converting degrees Celsius to degrees Fahrenheit is given by

$$F = \frac{9}{5}C + 32$$

Solve the formula for C. (The resulting formula is used to convert degrees Fahrenheit to degrees Celsius.)

38. The resistance R in a wire of length l and diameter D is

$$R = k\frac{l}{D^2}$$

where k is a constant. Solve this formula for l.

39. The output P of a battery is

$$P = VI - RI^2$$

Solve this formula for V.

40. The relationship between the tensile strength S (in pounds) of a piece of material and its temperature T (in degrees Fahrenheit) was found to be

$$S = 580 - 0.000078T$$

Solve for T.

41. The velocity v at any time t of a body hurled downward with initial velocity v_0 is

$$v = v_0 + gt$$

Solve for t.

42. The velocity of air in a bronchial tube is

$$v = kr^2(a - r)$$

Solve for a.

43. A tape of thickness T is wrapped around a core of radius r_1. If the outer radius is r_2, then the length L of the tape is

$$L = \frac{\pi}{T}(r_2^2 - r_1^2)$$

Solve for r_1^2.

44. The gravitational force between two masses m_1 and m_2 at a distance d is given by

$$F = \frac{Gm_1m_2}{d^2}$$

Solve for G.

45. The following formula arises in the study of an object moving in a vertical circle:

$$T = \frac{mv^2}{r} - mg$$

Solve for v^2.

46. The amount of heat Q conducted through a wall is found from the formula

$$Q = \frac{kA(t_1 - t_2)}{l}$$

Solve for t_1.

47. The power loss in a transmission line is calculated from the formula

$$C = k_1A + \frac{k_2}{A}$$

Solve for k_2.

48. The determination of the drag D of an airplane is found by using the formula

$$D = av^2 + \frac{b}{v^2}$$

Solve for b.

49. The volume V of a gas under constant pressure varies with the temperature T; the equation is

$$V = V_0[1 + b(T - T_0)]$$

Solve for T.

50. The energy required to compress a gas is given by the formula

$$W = \frac{p_1V_1 - p_2V_2}{b - 1}$$

Solve for V_1.

51. If n_1 items cost p_1 dollars each and n_2 items cost p_2 dollars each, then the average price P per item is

$$P = \frac{n_1 p_1 + n_2 p_2}{n_1 + n_2}$$

Solve this formula for p_1.

7.4 Applications of First-Degree Equations

The usefulness of algebra in technology has already been demonstrated in our discussion of formulas. Equations can also be used to solve problems that lead to algebraic equations. Such problems, often called "word problems," are solved by translating the statement in ordinary language into algebraic language.

Suppose we are told that one number is 2 more than another number and that the sum of the two numbers is 20. To find these numbers, we have to write the given statement as an equation. To do this, we let x represent the first number. Then, the second number, being two more than the first, is $x + 2$. Since the sum of the two numbers is 20, we write

$$\begin{aligned} x + (x + 2) &= 20 && \text{Algebraic statement} \\ 2x + 2 &= 20 && \text{Solving} \\ 2x &= 18 \\ x &= 9 \end{aligned}$$

It follows that $x + 2 = 11$, so that the two numbers are 9 and 11.

Translating a verbal statement into an equation is not easy. To help you in your work, consider the following guidelines.

Guidelines for solving word problems

1. Read the problem carefully to make sure you understand what is asked. Drawing a figure may help.
2. Identify the unknown quantity or quantities. Assign a letter to one of the unknown quantities and express the others, if any, in terms of this same letter.
3. Determine what quantities are equal. Use this information to write an equation.
4. Solve the equation.
5. Check the result in the original problem.

Example 1 One programmable calculator costs \$40 more than another, and the total cost of the two is \$210. Find the cost of each.

Solution.

Step 1. We are asked to find the cost of each calculator. The total cost is \$210; the difference in the cost is \$40.

Step 2. There are two unknown quantities: the cost of each calculator. Let's denote the cost of the cheaper calculator by x (in dollars). Since the other calculator costs \$40 more, the cost can be expressed as $x + 40$ (in dollars).

Step 3. What quantities are equal? According to the given statement, the cost of the first calculator plus the cost of the second is equal to \$210, the total cost. So

$$x + (x + 40) = 210 \text{ (in dollars)}$$

Step 4. We now solve the equation obtained in Step 3:

$$\begin{aligned} x + (x + 40) &= 210 \\ 2x + 40 &= 210 && x + x = 2x \\ 2x + 40 - 40 &= 210 - 40 && \text{Subtracting 40} \\ 2x &= 170 \\ x &= 85 && \text{Dividing by 2} \\ x + 40 &= 125 \end{aligned}$$

It follows that the cost of the first calculator is \$85 and the cost of the second is \$125.

Step 5. According to the guidelines, we check our solution in the original problem, not in the equation. This is important, since the equation we set up in Step 3 may already be wrong.

Given the cost of each calculator, \$125 and \$85, note that the difference is

$$\$125 - \$85 = \$40$$

and that the sum is

$$\$125 + \$85 = \$210$$

Our solution satisfies the conditions in the problem.

Example 2 The length of a rectangle is twice the width, and the perimeter of the rectangle is 48 cm. Find the dimensions.

Solution.

Figure 7.1

Step 1. Is the statement of the problem clear? The rectangle is twice as long as it is wide (Figure 7.1). The distance around the rectangle is 48 cm.

Step 2. We are looking for the length and width of the rectangle. Both are unknown quantities. Let's denote the width by x (in centimeters). We are told that the length is twice the width. So the length must be $2x$ (in centimeters). (See Figure 7.2).

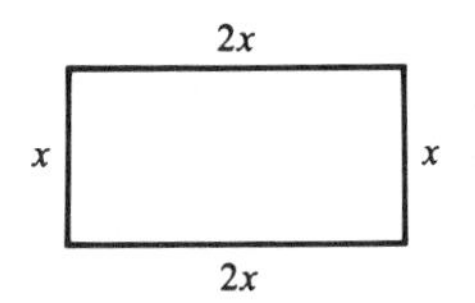

Figure 7.2

Step 3. What quantities are equal? We know that the perimeter (distance around the rectangle) is 48 cm. Using Figure 7.2, we see that

$$2x + x + 2x + x = 48 \text{ (in centimeters)}$$

This is the equation.

Step 4. We solve the equation obtained in Step 3:

$$2x + x + 2x + x = 48$$
$$6x = 48 \quad \text{Combining terms}$$
$$x = 8 \text{ cm} \quad \text{Dividing by 6}$$
$$2x = 16 \text{ cm}$$

So the dimensions are 8 cm × 16 cm.

Step 5. Since the length of the rectangle is 16 cm and the width 8 cm, the perimeter is

$$2(16 \text{ cm}) + 2(8 \text{ cm}) = 32 \text{ cm} + 16 \text{ cm} = 48 \text{ cm}$$

Also, 2(8 cm) = 16 cm. So the solution checks.

For our next example, we need to recall that the combined resistance R_T of two or more resistors in series is equal to the sum of the individual resistances. (See Figure 7.3.)

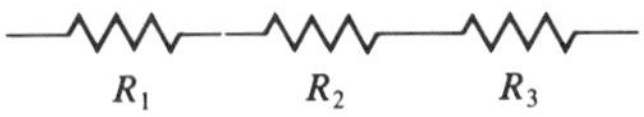

R_1 R_2 R_3

$$R_T = R_1 + R_2 + R_3$$

Figure 7.3

Example 3 Three resistors connected in series have a combined resistance of 11.4 Ω. The resistance of the second resistor is twice that of the first and the resistance of the third resistor is 3.0 Ω more than that of the first. Determine the three resistances.

Solution.

Step 1. Referring to Figure 7.3, we are looking for three resistances whose sum is 11.4 Ω (since the resistors are connected in series).

Step 2. The unknown quantities are the three resistances. If we denote the resistance of the first resistor by R (in ohms), then the resistance of the second, being twice that of the first, is $2R$ (in ohms). The resistance of the third is $R + 3.0$ (in ohms). These observations are summarized as: Let

$$R = \text{resistance of first resistor}$$

then

$$2R = \text{resistance of second}$$

and

$$R + 3.0 = \text{resistance of third}$$

Step 3. It follows from Step 1 and Figure 7.3 that the sum of the unknown quantities is 11.4 Ω. From Step 2 we get the equation

$$R + 2R + (R + 3.0) = 11.4 \qquad \text{Sum of three resistances}$$
$$R + 2R + R + 3.0 = 11.4$$
$$4R + 3.0 = 11.4 \qquad \text{Combining terms}$$
$$4R + 3.0 - 3.0 = 11.4 - 3.0 \qquad \text{Subtracting 3.0}$$
$$4R = 8.4$$
$$R = 2.1\ \Omega \qquad \text{Dividing by 4}$$
$$2R = 4.2\ \Omega$$
$$R + 3.0 = 5.1\ \Omega$$

Step 5. We check our answer in the original problem: From $R = 2.1\ \Omega$,

$$2(2.1\ \Omega) = 4.2\ \Omega \text{(resistance of second resistor)}$$
$$2.1\ \Omega + 3.0\ \Omega = 5.1\ \Omega \text{(resistance of third resistor)}$$

Finally,

$$2.1\ \Omega + 4.2\ \Omega + 5.1\ \Omega = 11.4\ \Omega$$

which is the required sum.

Another interesting technical application of equations involves mixing two chemical substances or combining two alloys to form a new alloy. Consider the next example.

Example 4 How many milliliters of a 6% alcohol solution (by volume) must be combined with 40 mL of a 12% alcohol solution to obtain a 10% solution?

Solution.
Step 1. The two alcohol solutions to be mixed contain 6% and 12% alcohol (by volume), respectively. The right combination will produce a solution containing 10% alcohol. The key to a "mixing problem" is to work with the actual quantities. In our problem *the amount of alcohol before mixing is the same as the amount after mixing* (even though the percent of alcohol varies).

Step 2. Let x = amount (in milliliters) of 6% solution.

Step 3. As noted in Step 1, the amount of alcohol stays the same. *Before mixing*, the amount of alcohol in the 12% solution is 12% of 40 mL, or 0.12(40) mL. Similarly, $0.06x$ is the amount of alcohol in the 6% solution. *After mixing*, we have a total of $(x + 40)$ mL, of which $0.10(x + 40)$ mL is alcohol. (See Figure 7.4.)

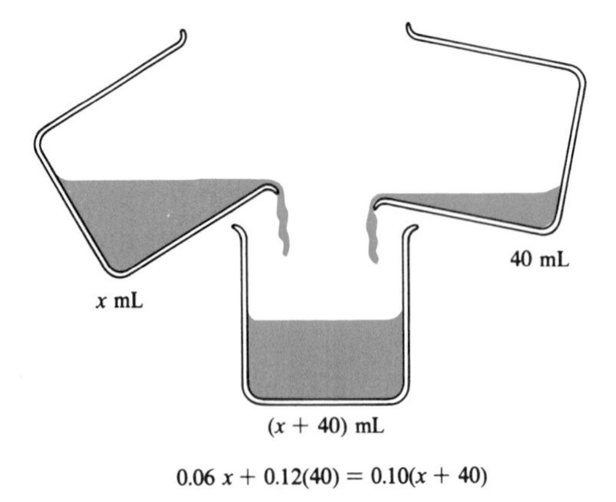

Figure 7.4

These observations lead to the following equation:

$$\text{amount before mixing} = \text{amount after mixing}$$
$$0.06x + 0.12(40) = 0.10(x + 40)$$

Step 4. To solve this equation, we multiply both sides by 100:

$$\begin{aligned}
100[0.06x + 0.12(40)] &= 100(0.10)(x + 40) \\
6x + 12(40) &= 10(x + 40) \\
6x + 480 &= 10x + 400 \\
6x + 480 - 6x &= 10x + 400 - 6x \\
480 &= 4x + 400 \\
480 - 400 &= 4x + 400 - 400 \\
80 &= 4x \\
20 &= x
\end{aligned}$$

So the desired amount is 20 mL.

Step 5. As always, we check in the original problem:

$$\begin{aligned}
6\% \text{ of } 20 \text{ mL} &= 0.06(20) = 1.2 \text{ mL} \\
12\% \text{ of } 40 \text{ mL} &= 0.12(40) = 4.8 \text{ mL} \\
10\% \text{ of } (20 + 40) \text{ mL} &= 0.10(60) = 6.0 \text{ mL}
\end{aligned}$$

It follows that

$$\text{amount before mixing} = \text{amount after mixing}$$
$$1.2 \text{ mL} + 4.8 \text{ mL} = 6.0 \text{ mL}$$

To use algebraic equations for solving certain problems in motion, we need to observe that

$$\text{distance traveled} = \text{rate} \times \text{time}$$

or

$$d = rt$$

For example, if you travel at the rate of 40 mi/h for 3 h, then the distance covered is

$$d = 40 \frac{\text{mi}}{\text{h}} \times 3 \text{ h} = 120 \text{ mi}$$

From $d = rt$, we also have

$$t = \frac{d}{r}$$

that is, the time required to travel a distance d at rate r is d divided by r.

Example 5 A car traveling at 40 mi/h leaves a certain intersection 45 min before a second car traveling at 52 mi/h. How long will it take for the second car to overtake the first?

Solution.

Step 1. Note that the cars start from the same intersection, but they leave at different times. As a result, the first car will have been on the road 45 min longer when the second car catches up with it.

Step 2. Let x be the time (in hours) required for the second car to overtake the first. Since 45 min = $\frac{3}{4}$ h, $x + \frac{3}{4}$ represents the number of hours the first car travels in the meantime. So

x = the time (in hours) required for the second car to overtake the first

and

$x + \frac{3}{4}$ = the length of time that the first car travels

Step 3. The first car travels at 40 mi/h for $(x + \frac{3}{4})$ hours, and the second car travels at 52 mi/h for x hours. Since the cars start from the same intersection, they travel the same distance. Since $d = rt$, we have

$$(\text{rate} \times \text{time})_{\text{first car}} = (\text{rate} \times \text{time})_{\text{second car}}$$

$$40\left(x + \frac{3}{4}\right) = 52x$$

Step 4.

$$40x + 40 \cdot \frac{3}{4} = 52x \qquad \text{Distributive law}$$

$$40x + 30 = 52x$$

$$30 = 12x \qquad \text{Subtracting } 40x$$

$$\frac{30}{12} = x \qquad \text{Dividing by 12}$$

$$x = 2.5 \text{ h}$$

Step 5. *Check:*

Distance covered by faster car: $(2.5\text{ h})\left(52\,\dfrac{\text{mi}}{\text{h}}\right) = 130\text{ mi}$

Distance covered by slower car: $(2.5\text{ h} + 0.75\text{ h})\left(40\,\dfrac{\text{mi}}{\text{h}}\right) = 130\text{ mi}$

Suppose an inlet valve can fill a tank in 6 h. Then $\frac{1}{6}$ of the tank is filled in 1 h. Similarly, if a tank can be drained in x hours, then the fractional part drained in 1 h is $1/x$. These observations are helpful in the problem illustrated next.

Example 6 A chemical tank can be filled in 14 min and drained in 6 min. If the tank is initially full and both drain and inlet valve are open, how long will it take to drain the tank?

Solution. Let x = time to drain the tank. Then $1/x$ is the fractional part drained after 1 min. Similarly, after 1 min, $\frac{1}{14}$ of the tank is filled and $\frac{1}{6}$ is drained. So the net amount drained after 1 min is

$$\frac{1}{6} - \frac{1}{14}$$

It follows that

$$\begin{aligned}\frac{1}{x} &= \frac{1}{6} - \frac{1}{14} \\ &= \frac{7}{42} - \frac{3}{42} = \frac{4}{42} \qquad \text{LCD} = 42 \\ &= \frac{2}{21}\end{aligned}$$

To solve for x, observe that whenever two nonzero numbers are equal, their reciprocals are equal; that is,

$$\text{if } a = b, \text{ then } \frac{1}{a} = \frac{1}{b}$$

In our problem, since

$$\frac{1}{x} = \frac{2}{21}$$

we have

$$\frac{x}{1} = \frac{21}{2}$$

or

$$x = \frac{21}{2} = 10\frac{1}{2}\text{ h}$$

Exercises / Section 7.4

Exercises 1–6 are number problems.

1. The sum of two numbers is 26. If one number is 2 more than the other, what are the numbers?
2. The sum of two numbers is 41. If one number is 3 less than the other, what are the numbers?
3. The sum of two numbers is 51. Given that one number is twice the other, find the two numbers.
4. The sum of two numbers is 52. Find the numbers, given that one is three times as large as the other.
5. If one number is 2 more than three times the other and if the sum of the numbers is 38, what are the two numbers?
6. The sum of two numbers is 43. One number is 2 less than four times the other. Find the numbers.

For Exercises 7–14, see Example 1.

7. Two calculators cost a total of $38. One calculator costs $8 less than the other. Determine the cost of each.
8. Two wrenches cost a total of $29. One wrench costs $11 more than the other. What is the cost of each?
9. One machine part costs twice as much as the other. The total cost is $7.17. Find the cost of each.
10. Together two gears cost $13.45. If one gear costs 1.5 times as much as the other, determine the cost of each.
11. The sum of two capacitances is 0.8188 F. Given that one capacitance is 0.3556 F more than the other, find each capacitance.
12. Two currents differ by 0.52 A and add up to 4.20 A. Find the two currents.
13. Two years from now a card sorter will be twice as old as it was 4 years ago. How old is the card sorter?
14. Two less than twice a given number is 10 less than three times the number. What is the number?

For Exercises 15–20, see Example 2.

15. The perimeter of a rectangular pattern is 54 cm. The length exceeds the width by 3 cm. Find the dimensions.
16. The width of a rectangle is 8 in. less than its length. The perimeter is 88 in. Find the dimensions.
17. The length of a rectangular computer chip exceeds the width by 0.7 mm. Find the dimensions, given that the perimeter is 7.0 mm.
18. The perimeter of a rectangular casting is 32.0 in. If the width is 1.2 in. less than the length, what are the dimensions?
19. Determine the dimensions of a rectangular metal plate, given that the perimeter is 34.8 cm and the width is one-half of the length.
20. The perimeter of a rectangular molding is 79.2 cm. If the length is twice the width, what are the dimensions?

For Exercises 21–24, see Example 3.

21. Two resistors are connected in series. The resistance of one is 14.2 Ω more than that of the other and the combined resistance is 76.8 Ω. Find the resistance of each.
22. The combined resistance of two resistors in series is 58.4 Ω. The resistance of one is three times that of the other. Find the resistance of each.
23. The combined resistance of three resistors in series is 41.3 Ω. The resistance of the second resistor is 4.5 Ω more than that of the first, and the third is 6.2 Ω more than that of the first. Find the resistance of each.

24. Three resistors are connected in series. The resistance of the second is twice that of the first, and the resistance of the third is 4.6 Ω more than that of the second. Find the resistance of each, given that the combined resistance is 114.6 Ω.

For Exercises 25–32, see Example 4.

25. How many pounds of an alloy containing 10% brass (by weight) have to be combined with 20 lb of an alloy containing 16% brass to obtain an alloy containing 12% brass?

26. How many liters of a brine solution containing 12% salt (by volume) have to be added to 16 L of a brine solution containing 18% salt to produce a 14% solution?

27. One brine solution contains 20% salt by volume and another 15%. How many gallons of each have to be mixed to obtain 20 gal containing 18% salt?

28. An alloy contains 9% brass (by weight) and another 17%. How many kilograms of each must be combined to form 50 kg of an alloy containing 13% brass?

29. How many milliliters of a 5% sulfuric acid solution (by volume) must be mixed with 19 mL of an 8% solution to produce a 6% solution?

30. How many milliliters of a 4% and an 8% hydrochloric acid solution (by volume) must be mixed to obtain 36 mL of a 5% solution?

31. A technician orders 50 mL of an 8% sulfuric acid solution but receives a 10% solution by mistake. How much must be drained off and replaced by distilled water to obtain the right concentration?

32. How much distilled water must be added to 80 L of a 12% solution to obtain a 10% solution?

For Exercises 33–42, see Example 5.

33. One car leaves a parking structure 30 min before another car. The first car averages 30 mi/h and the second 40 mi/h. How long will it take for the second car to overtake the first?

34. A car averaging 45 mi/h leaves a certain tollgate at 1:05 P.M. A second car, averaging 50 mi/h, leaves the same tollgate at 1:25 P.M. At what time will the second car catch up with the first?

35. Joan can row at the rate of 4 mi/h in still water. Rowing downstream for 2 h, she can go three times as far as rowing upstream for 2 h. How fast is the river flowing?

36. Traveling downstream for 2 h, a boat can go twice as far as traveling upstream for 2 h. Determine the rate of flow of the river, given that the speed of the boat is 15 km/h in still water.

37. Jim can row at the rate of 6 km/h in still water. Rowing downstream, he can travel twice as fast as rowing upstream. Find the rate at which the river is flowing.

38. A boat traveling at the rate of 5 mi/h in still water can travel four times as fast downstream as it can upstream. What is the rate of flow of the river?

39. At 3:00 P.M. two cars are 200 mi apart and traveling toward each other at 36 mi/h and 44 mi/h, respectively. At what time will they meet?

40. A river flows at the rate of 3 km/h. A boat traveling downstream for 2 h requires 6 h to return to the starting point. Find the speed of the boat.

41. A boat traveling downstream for 3 h requires 15 h to return to the starting point. Find the speed of the boat, given that the river flows at 4 km/h.

42. A river flows at the rate of 4 mi/h. A ship leaves a dock at 8:00 A.M. and heads downstream. At 11:00 A.M. she turns back and arrives at the dock at 8:00 P.M. What is the speed of the ship?

For Exercises 43–48, see Example 6.

43. A large chemical tank can be filled in 12 h and drained in 9 h. If the tank is initially full and both drain and inlet valve are open, how long will it take to drain the tank?

44. A fuel tank can be filled in 9 min and drained in 15 min. If the tank is initially empty and both drain and inlet valve are open, how long will it take to fill the tank?

45. One inlet can fill a tank in 3 h and another in 6 h. The tank is empty when both inlet valves are opened. How long will it take to fill the tank?

46. One drain can empty a tank in 8 min and another in 12 min. The tank is full when both drains are opened. How long will it take to empty the tank?

47. One inlet can fill a tank in 10 h, another in 6 h. If both inlets are open and the tank is initially $\frac{1}{3}$ full, how long will it take to fill the tank?

48. Working alone, one office worker can process a set of payroll records in 6 h; another can do the work in 9 h. If they work together, how long will it take them to do the job?

7.5 Basic Inequalities

Inequality

At the beginning of this chapter we defined an equation as an **equality** between two expressions. An **inequality** is a statement which says that one expression is greater than or less than another expression. Many situations in technology require the solution of inequalities.

As we saw in Section 3.1, the symbol $<$ means "less than" and the symbol $>$ means "greater than." For example, $3 < 5$ means "3 is less than 5" and $6 > -2$ means "6 is greater than -2." The direction of the inequality

Basic properties of inequalities

1. If $a < b$, then $a + c < b + c$, for any number c.
2. If $a < b$, then $a - c < b - c$, for any number c.
3. If $a < b$ and $c > 0$, then

$$ac < bc$$

and

$$\frac{a}{c} < \frac{b}{c}$$

4. If $a < b$ and $c < 0$, then

$$ac > bc$$

and

$$\frac{a}{c} > \frac{b}{c}$$

Similar statements hold for the opposite sense, $>$.

Sense of inequality

is called the **sense** of the inequality. For example, the inequalities $x < 2$ and $-3 < 0$ are said to have the *same* sense, while $5 > 2$ and $x < 7$ are said to have the *opposite* sense.

In this section we are mainly interested in inequalities involving a variable. The values of the variable that satisfy the inequality make up the

Solution

solution of the inequality. Consider, for example, the inequality

$$x - 2 > 0$$

Note that if $x = 2$, then $x - 2 = 0$. So x has to be greater than 2 in order for $x - 2$ to be greater than 0. It follows that the solution consists of all x such that $x > 2$.

The inequalities in this section can be solved in a manner similar to that we used for solving equations. Before illustrating the procedure, let us consider some of the basic properties of inequalities given on page 237.

Let us illustrate these properties by means of a few numerical examples.

Example 1 To illustrate Property 1, consider the inequality

$$-2 < 5$$

Then

$$-2 + 4 < 5 + 4 \quad \text{Adding 4 to both sides}$$

and

$$2 < 9 \quad \text{Combining numbers}$$

To illustrate Property 2, we get from $-2 < 5$,

$$-2 - 7 < 5 - 7 \quad \text{Subtracting 7 from both sides}$$

and

$$-9 < -2 \quad \text{Combining numbers}$$

As noted in the box, similar properties hold for the opposite sense. Thus from

$$8 > 2$$

we have

$$8 + 6 > 2 + 6$$
$$14 > 8$$

and

$$8 - 4 > 2 - 4$$
$$4 > -2$$

Example 2 To illustrate Property 3, consider the inequality

$$3 < 9$$

If we multiply both sides by 4, a *positive* number, we get

$$3 \cdot 4 < 9 \cdot 4 \qquad \text{Sense preserved}$$

or

$$12 < 36$$

On the other hand, if we multiply both sides of the inequality $3 < 9$ by -4, a *negative* number, we get

$$3(-4) > 9(-4) \qquad \text{Sense reversed}$$

or

$$-12 > -36$$

This illustrates Property 4. Note that the sense of the inequality is reversed when multiplying both sides by -4.

Summary. Adding the same number to both sides of an inequality (or subtracting the same number from both sides) preserves the sense of the inequality. Multiplying both sides by a positive number preserves the sense of the inequality, but multiplying both sides by a negative number reverses the sense.

The basic properties can be used to solve certain inequalities. The procedure is similar to that used for solving equations: We add the same number to both sides, subtract the same number from both sides, and so on, until only x remains on one side. Consider the next example.

Example 3 Solve the inequality

$$x - 3 < 1$$

Solution.

$$x - 3 < 1 \qquad \text{Given inequality}$$
$$x - 3 + 3 < 1 + 3 \qquad \text{Adding 3 to both sides}$$
$$x < 4$$

The solution therefore consists of all numbers x such that $x < 4$ (Figure 7.5).

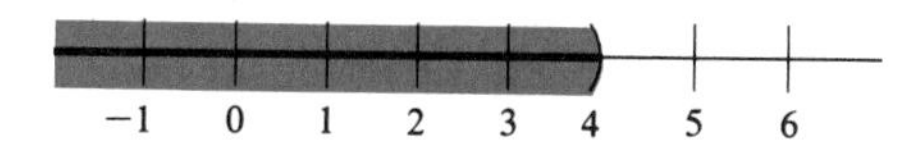

Figure 7.5

Example 4 Solve the inequality

$$15 - x > -3 - 7x$$

Solution.

$15 - x > -3 - 7x$	Given inequality
$15 - x + 7x > -3 - 7x + 7x$	Adding $7x$ to both sides
$15 + 6x > -3$	Combining terms
$15 - 15 + 6x > -3 - 15$	Subtracting 15 from both sides
$6x > -18$	
$\frac{6x}{6} > \frac{-18}{6}$	Dividing both sides by +6 (sense preserved)
$x > -3$	

(See Figure 7.6.)

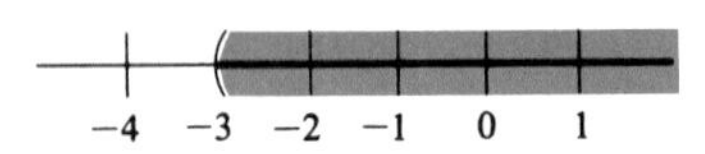

Figure 7.6

Note especially that dividing both sides by 6 preserves the sense of the inequality.

The next example illustrates division by a negative number.

Example 5 Solve the inequality

$$-5x - 3 < 4 - 3x$$

Solution.

$-5x - 3 < 4 - 3x$	Given inequality
$-5x - 3 + 3x < 4 - 3x + 3x$	Adding $3x$
$-2x - 3 < 4$	
$-2x - 3 + 3 < 4 + 3$	Adding 3
$-2x < 7$	
$\frac{-2x}{-2} > \frac{7}{-2}$	Dividing by −2 (sense reversed)
$x > -\frac{7}{2}$	

Note that dividing both sides by −2 reverses the sense of the inequality.

As in the case of equations, if the inequality contains symbols of grouping, these must first be removed.

Example 6 Solve the inequality

$$-2(3 - 4x) > 3(x - 1)$$

Solution.

$$-2(3 - 4x) > 3(x - 1) \quad \text{Given inequality}$$
$$-6 + 8x > 3x - 3 \quad \text{Distributive law}$$
$$-6 + 8x - 3x > 3x - 3x - 3 \quad \text{Subtracting } 3x$$
$$-6 + 5x > -3$$
$$-6 + 6 + 5x > -3 + 6 \quad \text{Adding 6}$$
$$5x > 3$$
$$\frac{5x}{5} > \frac{3}{5} \quad \text{Dividing by } +5$$
$$x > \frac{3}{5}$$

The concepts of equality and inequality can be combined: The symbol $\leq$ means "less than or equal to" and the symbol $\geq$ means "greater than or equal to." The technique for solving inequalities containing the symbols $\leq$ or $\geq$ is identical to the technique discussed in Examples 3–6.

Example 7

$$2x + 4 \leq 5x - 8 \quad \text{Given inequality}$$
$$-3x + 4 \leq -8 \quad \text{Subtracting } 5x$$
$$-3x \leq -12 \quad \text{Subtracting 4}$$
$$\frac{-3x}{-3} \geq \frac{-12}{-3} \quad \text{Sense reversed}$$
$$x \geq 4$$

Example 8 The current i (in amperes) in a certain circuit is given by $i = 2 - 3t$, where $t \geq 0$ is measured in seconds. Determine when the current is negative or zero.

Solution. The current is *negative or zero* whenever

$$i \leq 0$$

Thus

$$2 - 3t \leq 0$$
$$-3t \leq -2 \quad \text{Subtracting 2}$$

$$\frac{-3t}{-3} \geq \frac{-2}{-3} \qquad \text{Dividing by } -3$$

$$t \geq \frac{2}{3}$$

We conclude that $i \leq 0$ whenever $t \geq \frac{2}{3}$ s.

Exercises / Section 7.5

In Exercises 1–6, perform the indicated operations on *both sides* of the given inequality.

1. $-2 < -1$, add 2

2. $-2 > -6$, subtract 3

3. $-1 < 5$, multiply by 6

4. $-2 < 0$, multiply by -4

5. $2 < 10$, divide by -2

6. $-3 < 6$, divide by 3

In Exercises 7–26, solve the given inequalities.

7. $x - 2 < 6$

8. $x + 4 > 7$

9. $2x + 1 < -3$

10. $3x - 4 < 4$

11. $-2x > -4$

12. $-4x > 8$

13. $2x - 3 < 2 - 7x$

14. $2 - x < 3x - 4$

15. $7 - 2x > x + 3$

16. $-x - 3 < 3 - 4x$

17. $6x + 1 \leq 1 - 3x$

18. $2x + 4 \geq -x + 6$

19. $7x - 6 \geq x - 3$

20. $1 - x \leq 5 + 3x$

21. $2(x - 3) < 5(1 - x)$

22. $-2(1 - x) > 3(2 - 3x)$

23. $-3(x - 2) \geq -2(1 - 3x)$

24. $5(3 + x) \geq -2(1 - 4x)$

25. $\frac{1}{2} - \frac{1}{6}x \leq \frac{1}{3}x - \frac{1}{2}$ (*Hint:* Clear fractions.)

26. $\frac{1}{4}x - \frac{1}{12} \geq \frac{1}{2}x - 2$

27. The current i (in amperes) in a circuit is $i = 4t - 6$, where t is measured in seconds. Determine when the current satisfies the condition $i \geq 0$.

28. The resistance R (in ohms) in a certain wire is $R = 0.84t + 1.0$ (t in seconds). Determine when $R > 9.0\ \Omega$.

29. The resistance R in a wire is related to the temperature T (in degrees Celsius) by the formula

$$R = 0.34T + 1.7$$

Find the temperatures for which $R \geq 18\ \Omega$.

30. The formula for converting degrees Fahrenheit to degrees Celsius is $C = \frac{5}{9}(F - 32)$. Determine the Fahrenheit temperatures corresponding to $C \geq 5°\text{C}$.

7.6 Writing Equations and Inequalities in BASIC (Optional)

To write an equation or inequality in BASIC, we need the following symbols for equality and inequality:

Algebraic symbol	BASIC symbol
$=$	=
$<$	<
$\leq$	<=
$>$	>
$\geq$	>=

Strictly speaking, the symbol = in BASIC means "replace." For example, the program step

```
20  X = X + 1
```

means "replace the value of X in a specified storage location by X + 1."

Example 1

Algebraic notation	*BASIC*
$y = x^2 + 2$	Y = X↑2 + 2
$C = \frac{5}{9}(F - 32)$	C = (5/9)*(F − 32)
$x - 6 < 2$	X − 6 < 2
$y - 3 \geq 4$	Y − 3 >= 4

Exercises / Section 7.6

Write the following equations or inequalities in BASIC.

1. $y = x + 4x^2$

2. $y = 1 - 3x^2$

3. $P = \frac{k}{V}$

4. $A = P + Prt$

5. $E = \frac{(K + 1)H^2}{I}$

6. $y = \frac{WHx}{x^2 + W^2}$

7. $P = VI - RI^2$

8. $v = v_1 + gt$

9. $v = kr^2(a - r)$

10. $T = \frac{mv^2}{r} - mg$

11. $\frac{1}{R} = \frac{1}{R_1} + \frac{1}{R_2}$

12. $A = \frac{1}{2}(b_1 + b_2)h$

13. $Q = \frac{kA(t_1 - t_2)}{l}$

14. $H = \frac{KA(t_2 - t_1)}{L}$

15. $x - 2 \leq 2x + 3$

16. $x + 4 > 1 - x$

17. $z - 2 < 4 - z$

18. $y + 3 \geq 1 - y$

19. $1 - 3w < 3$

20. $w \leq 3$

21. $2 - 3w \geq 7$

Review Exercises / Chapter 7

In Exercises 1–14, solve the given equations.

1. $2 - 4x = -6$

2. $4x - 8 = 16$

3. $5 - 4x = x - 4$

4. $x - 7 = 3 - 6x$

5. $\frac{1}{3}x = -\frac{1}{2}$

6. $-\frac{1}{6}x = \frac{1}{12}$

7. $2(2 - y) = 3(2y + 1)$

8. $6(-z + 2) = 5(2z - 3)$

9. $-\frac{2}{5}w = \frac{1}{10}$

10. $\frac{3}{4}v = -\frac{3}{8}$

11. $\frac{1}{6} - \frac{1}{2}u = -\frac{1}{3}$

12. $\frac{1}{3}y + \frac{1}{4} = \frac{1}{12} - \frac{1}{6}y$

13. $\frac{1}{2}\left(\frac{1}{3}s - 1\right) = \frac{1}{3} - \frac{1}{6}s$

14. $\frac{2}{3}(1 - t) = \frac{4}{3}t - \frac{1}{3}$

In Exercises 15–24, solve each equation for x.

15. $x + 2b = 1$

16. $x - 3a = 2$

17. $cx = -d$

18. $bx - 2 = n$

19. $cx - b = 1$

20. $nx - b = c$

21. $2(2x - b) = 3(b - 2x)$

22. $4(a - 2x) = 3(a - 3x)$

23. $\frac{1}{2}(ax - 6) = \frac{1}{3}(3 - ax)$

24. $\frac{1}{2}\left(\frac{1}{2}ax - \frac{1}{3}\right) = \frac{3}{4}ax - \frac{1}{12}$

In Exercises 25–32, solve each formula for the indicated letter.

25. $\frac{T_1}{T_2} = 2N$, T_2

26. $st - 4 = w$, t

27. $\frac{1}{2}s_1s_2 - s_3 = 1$, s_1

28. $\frac{1}{2} + \frac{1}{4}AB = \frac{1}{2}B$, A

29. $P = \frac{2a - 4b}{c}$, a

30. $Q = \frac{3n + 4m}{d}$, n

31. $P_0 = \frac{a(P_1 + P_2)}{b}$, P_1

32. $L = \frac{s - t}{ab}$, t

In Exercises 33–42, solve the given inequalities.

33. $-\frac{1}{2}x < \frac{1}{4}$

34. $-\frac{1}{3}x > \frac{1}{12}$

35. $2x - 3 < -4x + 7$

36. $x - 2 > 1 - x$

37. $1 - 7x \geq 3x - 10$

38. $2 - 5x \leq 7 - 6x$

39. $\frac{1}{2}x - \frac{1}{3} \geq \frac{1}{4}x$

40. $\frac{1}{3}x < \frac{1}{2} - \frac{1}{6}x$

41. $2(4 - 2x) < 3(x - 2)$

42. $3(2 - 5x) > 4(x - 4)$

In Exercises 43–48, solve the given formulas for the indicated letter.

43. Modern physics (kinetic energy):

$$T = c^2(m - m_0),\ m_0$$

44. Radioactivity:

$$E_m = m_0c^2 + T_m,\ m_0$$

45. Physics (motion):

$$s = s_0 + v_0t + \frac{1}{2}gt^2,\ v_0$$

46. Heat transfer:

$$H = \frac{KA(t_2 - t_1)}{L},\ t_2$$

47. Optics:

$$\frac{q - f}{q} = \frac{f}{p},\ p$$

48. Nuclear reactions:

$$\frac{T'}{T} = \frac{m_2}{m_1 + m_2},\ m_1$$

49. The resistance R (in ohms) in a certain wire is related to the temperature T (in degrees Celsius) by

$$R = 2.5 + 0.0000082T$$

Solve for T.

50. The surface area A of a cylinder is

$$A = 2\pi rh + 2\pi r^2$$

Solve this formula for h.

51. One machine part costs $2\frac{1}{2}$ times as much as another. The total cost is \$18.41. Find the cost of each.

52. The combined resistance of two resistors in series is 60.95 Ω. The resistance of one is 30% more than that of the other. Find the two resistances.

53. How many pounds of an alloy containing 8% zinc (by weight) must be combined with 50 lb of an alloy containing 14% zinc to form an alloy containing 12% zinc?

54. A 6% sulfuric acid solution (by volume) and a 10% sulfuric acid solution are to be mixed to produce 35 mL of an 8% solution. How much of each solution is needed?

55. A car leaves a certain intersection at 12:10 P.M. and travels at an average speed of 40 mi/h. At 12:55 P.M. a second car leaves the same intersection and travels at 48 mi/h. At what time will the second car overtake the first?

56. A river flows at the rate of 4 km/h. A motorboat traveling downstream for 3 h requires 6 h to get back. Determine the speed of the boat.

57. A tank can be filled in 8 h and drained in 12 h. If the tank is initially one-fourth full and both drain and inlet valve are open, how long will it take to fill the tank?

58. At 1:00 P.M. a technician turns on the inlet valve for a chemical tank, which normally requires 12 min to fill. He discovers at 1:08 P.M. that he forgot to close the drain. Given that the drain can empty a full tank in 16 min, at what time will the tank be filled?

59. The deflection d (in centimeters) of a certain beam 30 ft long is

$$d = 0.024x$$

where x is the distance from one end. For what values of x is $d \geq 0.50$ ft?

CHAPTER 8

Special Products and Factoring

The main purpose of this chapter is to study an operation called *factoring*. Factoring is essentially multiplication in reverse and is particularly useful for performing operations with fractions. To this end we need to improve our ability to carry out certain basic multiplications quickly. These basic multiplications are also called *special products*. Special products occur often enough in mathematics and technology to be of value in their own right.

8.1 The Distributive Law and Common Factors

The Distributive Law

As noted in the introduction, our main goal in this chapter is to study an important operation called *factoring*. However, to perform this operation, we must first study *special products*.

Our first special product is the distributive law, which we have seen many times before:

Distributive law

$$a(b + c) = ab + ac$$

For example, $2(3 + 4) = 2 \cdot 3 + 2 \cdot 4 = 6 + 8 = 14$. Consider another example.

Example 1

a. $2(x - 3y) = 2(x - 3y) = 2(x) + 2(-3y) = 2x - 6y$

b. $x(xy + 2x^2y^3) = x(xy + 2x^2y^3) = x(xy) + x(2x^2y^3) = x^2y + 2x^3y^3$

The distributive law can be applied even if the polynomials contain several terms, as shown in the next example.

Example 2

$$\begin{aligned} -2a(4 - 2a + 5a^2) &= -2a(4 - 2a + 5a^2) \\ &= -2a(4) - 2a(-2a) - 2a(5a^2) \\ &= -8a + 4a^2 - 10a^3 \end{aligned}$$

Example 3

$$\begin{aligned} &-2V_aV_b(-V_aV_b + 2V_a^2V_b - 4V_aV_b^2) \\ &\quad = -2V_aV_b(-V_aV_b) - 2V_aV_b(2V_a^2V_b) - 2V_aV_b(-4V_aV_b^2) \\ &\quad = 2V_a^2V_b^2 - 4V_a^3V_b^2 + 8V_a^2V_b^3 \end{aligned}$$

Common Factors

Factoring plays an important role in algebra. We must be able to factor in order to simplify certain expressions, to reduce a fraction to lowest terms, and to perform algebraic operations with fractions.

Factoring is really multiplication in reverse. For example, since

$$2(x + y) = 2x + 2y$$

it follows that

$$2x + 2y = 2(x + y)$$

The factor 2 is said to be a **common factor**.

Common factors can be removed from a sum by using the distributive law in reverse.

Common factor

$$ab + ac = a(b + c)$$

For example, the terms in the binomial

$$ax + ay$$

have common factor a. Factoring out this common factor, we get

$$ax + ay = a(x + y)$$

Consider another example.

Example 4 Factor the expression

$$3x^2 - 9xy$$

Solution. Since $3x^2 = 3x(x)$ and $-9xy = 3x(-3y)$, we see that $3x$ is a common factor. Thus

$$3x^2 - 9xy = 3x(x) + 3x(-3y) = 3x(x - 3y) = 3x(x - 3y)$$

Removing a factor from a polynomial sometimes leaves a factor $+1$ or -1. For example,

$$x + x^2 = 1x + x^2 = x(1 + x)$$

Example 5 Factor the expression

$$ax^2 - 2ax^3 - ax$$

Solution. Each term has factor ax. To remove this factor from every term, we need to write the last term as $-1ax$. Thus

$$ax^2 - 2ax^3 - ax = ax(x) + ax(-2x^2) + ax(-1)$$
$$= ax(x - 2x^2 - 1) = ax(x - 2x^2 - 1)$$

Common error Leaving out the factor 1. For example, $x^2 + x$ should not be written $x(x + 0)$ since $x(x + 0) = x(x) + x(0) = x^2 + 0 = x^2$. Instead,

$$x^2 + x = x(x + 1)$$

as you can readily check by multiplying.

Example 6 Factor the expression

$$12T_1^2T_2 + 15T_1T_2^2 + 3T_1T_2$$

Solution. Since each numerical coefficient is divisible by 3, each term has common factor $3T_1T_2$. Note that removing $3T_1T_2$ leaves the factor 1 in the trinomial:

$$12T_1^2T_2 + 15T_1T_2^2 + 3T_1T_2 = 3T_1T_2(4T_1 + 5T_2 + 1)$$

Example 7 Factor the polynomial

$$8V^2C^3 - 12V^3C^2 - 16V^4C^4$$

Solution. Note that each term in the polynomial contains the factors 4, V^2, and C^2:

$$\begin{aligned}&8V^2C^3 - 12V^3C^2 - 16V^4C^4\\ &\quad = 4V^2C^2(2C) + 4V^2C^2(-3V) + 4V^2C^2(-4V^2C^2)\\ &\quad = 4V^2C^2(2C - 3V - 4V^2C^2) = 4V^2C^2(2C - 3V - 4V^2C^2)\end{aligned}$$

Prime factor

We say that a polynomial has been factored if it has been written as a product of prime factors. A factor is said to be **prime** or **irreducible** if it contains no other factors except itself and 1. For example, $2x + 5y$ is prime.

Common error Failing to factor out all the common factors. For example,

$$5x^2y^3 - 15x^4y^4 = x^2y^3(5 - 15x^2y)$$

This result is not really incorrect. However, to obtain *prime* factors, we must also remove the factor 5 from each term:

$$5x^2y^3 - 15x^4y^4 = 5x^2y^3(1 - 3x^2y)$$

Example 8 For a short time period, the current i (in amperes) in a circuit is given by

$$i = 2.0t - 4.0t^2$$

where t is measured in seconds. Factor the expression for i.

Solution. The common factor is $2.0t$. So

$$\begin{aligned}i &= 2.0t - 4.0t^2\\ &= 2.0t(1) + 2.0t(-2.0t)\\ &= 2.0t(1 - 2.0t)\end{aligned}$$

Exercises / Section 8.1

In Exercises 1–14, perform the indicated multiplications in one step.

1. $3(x - 4y)$
2. $7(2x - y)$
3. $a(4x + 3y)$
4. $b(x - 7y)$
5. $x^2(-3x - 1)$
6. $y^2(1 - 4y)$
7. $AB(AB - 1)$
8. $P^2Q(1 + 2PQ^2)$
9. $T_1(2T_1T_2 - 4T_1^2T_2)$
10. $V_1(4V_1^2V_2 - 1)$
11. $2R^2S(-2RS^2 - 4R^2S^2)$
12. $-4ST^2(2S^2T - ST)$
13. $-7s^2t^2(3s^3t - 4)$
14. $-10p^2q^2(3pq - p^2q)$

In Exercises 15–40, factor each expression.

15. $3x - 6y$
16. $4x - 12y$
17. $9a - 15b$
18. $15a + 25b$
19. $2x^2y + 4xy^2$
20. $8a^3b - 12ab^2$
21. $2ax^2 - 4ax^3$
22. $14s^2t^2 + 21s^3t$
23. $15p^3q^4 + 25p^3q$

24. $16A^4B^7 - 24AB^3$

25. $2x^4 - 4x^3 + 10x^2$

26. $9y^2 + 21y^3 - 18y^4$

27. $ay^4 - 2ay^3 - ay^2$

28. $2bz^7 - 6bz^4 - 8bz^2$

29. $A^2B^2 + AB$

30. $P^3Q^3 + PQ$

31. $T_a^2T_b^2 - T_aT_b^2$

32. $V_a^2V_b^2 - V_aV_b^3$

33. $3x + 6x^4 - 12x^5$

34. $4y - 8y^3 - 16y^4$

35. $21L^4V - 14L^2V^3 + 7L^2V$

36. $25A^3B^2 - 15A^2B^3 + 5AB^2$

37. $3T_1T_2^2 - 9T_1^2T_2^2 - 18T_1^3T_2^2$

38. $4mn + 16m^2n - 20m^3n$

39. $11vw - 11v^2w + 22vw^2$

40. $14xz + 14x^2z^2 - 28xz^3$

41. The increase in momentum of an object when the velocity increases from v_1 to v_2 is

$$M = mv_2 - mv_1$$

Factor the expression for M.

42. The pressure p on a dam at distance y from the bottom is

$$p = Dgh - Dgy$$

where h is the height of the dam. Factor the expression for p.

43. A bar of length l_0 expands due to heating to $l = l_0 + l_0aT$, where T is the increase in temperature and a the coefficient of thermal expansion. Factor the expression for l.

44. The distance s covered by a falling body is

$$s = v_0t + \frac{1}{2}gt^2$$

where v_0 is the initial velocity and t the time. Factor the expression for s.

45. The energy required to lift an object from level y_1 to level y_2 is

$$W = mgy_2 - mgy_1$$

Factor the expression for W.

46. The resistance R of a certain variable resistor varies with time according to the formula

$$R = 2t^3 + 4t^2 - 8t$$

Factor the expression for R.

47. The voltage E in the circuit in Figure 8.1 is given by $E = IR_1 + IR_2 + IR_3$. Factor the expression for E.

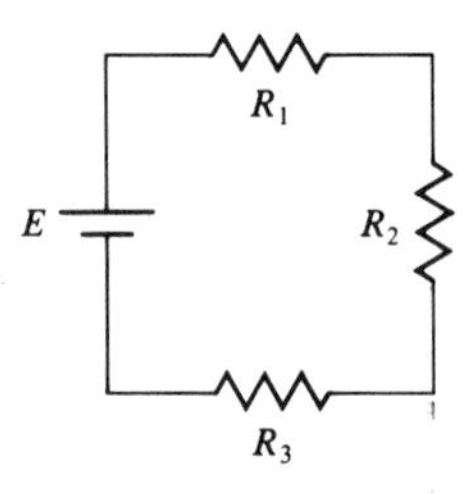

Figure 8.1

48. The formula for the output of a battery is $P = VI - RI^2$. Factor the expression for P.

49. Factor the following expression from a problem in cost analysis:

$$ax - 2ax^2 + 3ax^3$$

50. The force required to stretch a spring from length l_1 to length l_2 is $kl_2 - kl_1$. Factor this expression.

8.2 The Difference of Two Squares

The main purpose of this section is to discuss factoring of a *difference of two squares*. To do this type of factoring, we must first consider the corresponding multiplication.

The Product of a Sum and Difference of Two Terms

Given two algebraic quantities a and b, if the *sum* $a + b$ is multiplied by the *difference* $a - b$, we get:

$$\begin{array}{lll} & a - b & \\ & a + b & \\ \hline a^2 & - ab & \\ & + ab & - b^2 \\ \hline a^2 & & - b^2 \end{array}$$

In words: *The product of a sum and difference of two terms is equal to the difference of their squares.* In symbols:

$$(a - b)(a + b) = a^2 - b^2$$

Example 1 Perform the following multiplication:

$$(2x - 3)(2x + 3)$$

Solution. In the form $(a - b)(a + b) = a^2 - b^2$, we have $a = 2x$ and $b = 3$. So

$$(2x - 3)(2x + 3) = (2x)^2 - 3^2 = 4x^2 - 9$$

The next example is similar.

Example 2

a. $(1 - 4t)(1 + 4t) = 1^2 - (4t)^2 = 1 - 16t^2$

b. $(3R - 2V)(3R + 2V) = (3R)^2 - (2V)^2 = 9R^2 - 4V^2$

c. $(a^2 + 4b)(a^2 - 4b) = (a^2)^2 - (4b)^2 = a^4 - 16b^2$

Factoring the Difference of Two Squares

Since factoring is multiplication in reverse, factoring a difference of two squares leads to the product of the sum and difference of the terms. For example,

$$x^2 - 4 = x^2 - 2^2 = (x - 2)(x + 2)$$

as you can verify by multiplication.

> **Difference of two squares**
>
> $$a^2 - b^2 = (a - b)(a + b)$$

The best way to factor a difference of two squares is to write each term as a square, as shown in the next example.

Example 3 Factor the expression $x^2 - 16y^2$.

Solution. Note that x^2 is the square of x and $16y^2$ is the square of $4y$ [since $(4y)^2 = 4^2y^2 = 16y^2$]. We now get

$$x^2 - 16y^2 = x^2 - (4y)^2 = (x - 4y)(x + 4y)$$

or

$$x^2 - 16y^2 = (x + 4y)(x - 4y)$$

Example 4 **a.** $1 - 9y^2 = 1^2 - (3y)^2 = (1 - 3y)(1 + 3y)$
b. $25V^2 - 1 = (5V)^2 - 1^2 = (5V + 1)(5V - 1)$
c. $9a^2b^2 - 4 = (3ab)^2 - 2^2 = (3ab + 2)(3ab - 2)$

The next example involves a difference of two squares with fractional coefficients.

Example 5 Factor the expression

$$\frac{1}{4}a^2x^4 - \frac{1}{9}a^2x^2$$

Solution. Since $\frac{1}{4} = (\frac{1}{2})^2$ and $\frac{1}{9} = (\frac{1}{3})^2$, we may write

$$\frac{1}{4}a^2x^4 - \frac{1}{9}a^2x^2 = \left(\frac{1}{2}ax^2\right)^2 - \left(\frac{1}{3}ax\right)^2$$
$$= \left(\frac{1}{2}ax^2 - \frac{1}{3}ax\right)\left(\frac{1}{2}ax^2 + \frac{1}{3}ax\right)$$

Sometimes we need to factor out a common factor before a given expression can be written as a difference of two squares. For example,

$$2x^2 - 8 = 2(x^2 - 4) = 2(x - 2)(x + 2)$$

> In all factoring problems, first check for common factors.

Example 6 Factor $ax^4y^8 - a$.

Solution. As written, the expression is not a difference of two squares. However, note that each term has common factor a, which must first be removed:

$$\begin{aligned} ax^4y^8 - a &= a(x^4y^8 - 1) \\ &= a[(x^2y^4)^2 - 1^2] \\ &= a(x^2y^4 + 1)(x^2y^4 - 1) \end{aligned}$$

Sometimes the factored expressions can be factored even further, as shown in the next example.

Example 7 The energy E radiated by a blackbody is $E = kT^4 - kT_0^4$, where T is the temperature of the body, T_0 the temperature of the surrounding medium, and k a constant. Factor the expression for E.

Solution. To factor the expression for E, we must first remove the common factor k:

$$\begin{aligned} E &= kT^4 - kT_0^4 \\ &= k(T^4 - T_0^4) && \text{Common factor } k \\ &= k[(T^2)^2 - (T_0^2)^2] && \text{Difference of two squares} \\ &= k(T^2 - T_0^2)(T^2 + T_0^2) \\ &= k(T^2 - T_0^2)(T^2 + T_0^2) \\ &= k(T - T_0)(T + T_0)(T^2 + T_0^2) \end{aligned}$$

since $T^2 - T_0^2$ is also a difference of two squares.

Note. The sum of two squares

$$a^2 + b^2$$

is **not** factorable.

Exercises / Section 8.2

In Exercises 1–16, perform the indicated multiplications in one step.

1. $(x-3)(x+3)$
2. $(y-4)(y+4)$
3. $(2x+b)(2x-b)$
4. $(3y-c)(3y+c)$
5. $(5V-1)(5V+1)$
6. $(1+7P)(1-7P)$
7. $\left(\frac{1}{2}W+2\right)\left(\frac{1}{2}W-2\right)$
8. $\left(\frac{1}{4}m-10\right)\left(\frac{1}{4}m+10\right)$
9. $(1-ay)(1+ay)$
10. $(1+bx)(1-bx)$
11. $(2LV-4)(2LV+4)$
12. $(2-3RC)(2+3RC)$
13. $(aw+z)(aw-z)$
14. $(Dn-l)(Dn+l)$
15. $(aV_0+2b)(aV_0-2b)$
16. $(3P_1-4P_2)(3P_1+4P_2)$

In Exercises 17–44, factor each of the given expressions.

17. x^2-16
18. w^2-25
19. $9y^2-4$
20. $16x^2-9$
21. $4W^2-1$
22. $9V^2-1$
23. $4-9V_0^2$
24. $9-16C_1^2$
25. z^2-4b^2
26. m^2-9n^2
27. $\frac{1}{4}x^2-y^2$
28. $\frac{1}{16}p^2-q^2$
29. $\frac{1}{4}L^2-\frac{1}{9}M^2$
30. $\frac{1}{16}T_0^2-\frac{1}{4}T_1^2$
31. $\frac{x^2}{a^2}-b^2$
32. $\frac{w^2}{b^2}-4C^2$
33. $8V_0^2-2V_1^2$
34. $3C_1^2-27C_2^2$
35. az^2-4aw^2
36. $4cm^2-cn^2$
37. $V_aV_b^2-V_aV_c^2$
38. $\frac{2}{s^2}T_1^2-\frac{2}{s^2}T_2^2$
39. s^4-t^4
40. $3V^4-3Q^4$
41. $kV_1^4-kV_2^4$
42. $kP_1^4-kP_2^4$
43. $16aR^4-aC^4$
44. cV^4-cL^4

45. When the velocity of a rocket increases from v_1 to v_2, then the force due to air resistance increases by

$$kv_2^2-kv_1^2$$

Factor this expression.

46. Factor the following expression from a problem in fluid flow:

$$8av_1^2-2av_2^2$$

47. A lead pipe of length l has inner radius r_1 and outer radius r_2. Factor the expression for the volume of the wall of the pipe:

$$V=\pi r_2^2l-\pi r_1^2l$$

48. A motorboat of mass m increases its velocity from v_1 to v_2. The resulting increase in the resistance force N_R is given by $N_R=av_2^2-av_1^2$. Factor the expression for N_R.

49. The formula

$$\frac{1}{2}mv_2^2-\frac{1}{2}mv_1^2=\frac{1}{2}M\omega_2^2-\frac{1}{2}M\omega_1^2 \qquad \omega = \text{omega}$$

arises in the study of conservation of energy. Factor each side of the formula.

50. Factor the following expression, which arises in the study of the hydrogen atom:

$$\frac{k}{l^2}-\frac{k}{n^2}$$

51. The square of the velocity v of a body of mass m oscillating on a spring is

$$v^2 = \frac{k}{m}A^2 - \frac{k}{m}x^2$$

where x is the displacement of the mass and A the amplitude of the oscillation (Figure 8.2). Factor the expression for v^2.

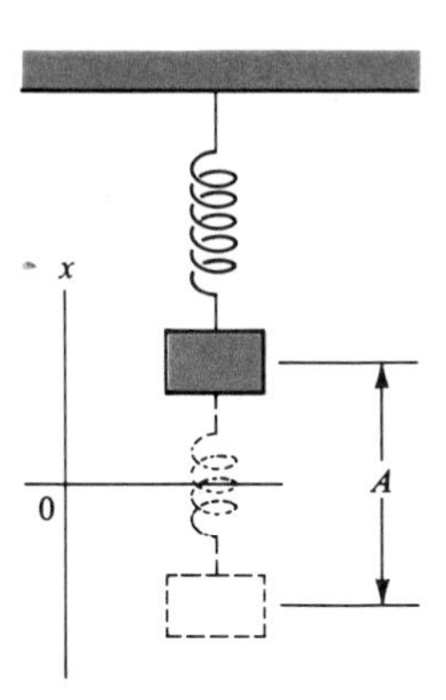

Figure 8.2

52. As the voltage across a resistor increases from V_1 to V_2, the power delivered to the resistor increases by

$$\frac{V_2^2}{R} - \frac{V_1^2}{R}$$

Factor this expression.

53. When moving two electrons closer together, the force required to overcome the repulsive action between the electrons is

$$\frac{k}{d_1^2} - \frac{k}{d_2^2}$$

Factor this expression.

54. For a body of temperature T_1, surrounded by a wall of temperature T_2, the net rate of loss R of energy per unit area by radiation is

$$R = e\sigma T_1^4 - e\sigma T_2^4 \qquad \sigma = \text{sigma}$$

Factor the expression for R.

8.3 The Product of Two Binomials

In Chapter 5 we first studied multiplication of two polynomials. In this section we learn to multiply two binomials directly in the following sense: Given two binomials, we obtain the complete product without writing any intermediate steps. Being able to multiply two binomials rapidly is of great help in factoring trinomials, which is the topic of the next section.

First let us recall the procedure for multiplication given in Chapter 5. To multiply $x - 2$ by $2x + 3$, we set up the multiplication as follows:

$$\begin{array}{r} x - 2 \\ \underline{2x + 3} \\ 2x^2 - 4x \\ \underline{ 3x - 6} \\ 2x^2 - x - 6 \end{array}$$

Note especially that both terms of $x - 2$ are multiplied by both terms of $2x + 3$. Written on the same line, the multiplication has the following form:

$$(x - 2)(2x + 3) = 2x^2 + [3x + (-4x)] - 6$$

Now observe that the right side consists of the product of the first terms (x and $2x$), plus the product of the last terms (-2 and 3), plus a middle term which is the sum of two products: the product of the outer terms (x and 3) and the product of the inner terms (-2 and $2x$). This procedure is easier to see from a diagram:

First, Last, Inner, Outer (brackets in diagram)

First Outer Inner Last

$$(x - 2)(2x + 3) = (x)(2x) + [3x + (-4x)] + (-2)(3)$$

$$= 2x^2 - x - 6$$

This scheme can be remembered by thinking of the word **FOIL**, where F stands for "first," O for "outer," I for "inner," and L for "last." In using the FOIL scheme, *the middle terms should be added mentally*.

Example 1 Multiply: $(3x - 4)(x - 1)$.

Solution. Using the FOIL scheme, we obtain the following diagram:

F, L, I, O (brackets in diagram)

Add mentally

F O I L

$$(3x - 4)(x - 1) = (3x)(x) + [(3x)(-1) + (-4)(x)] + (-4)(-1)$$

$$= 3x^2 + [-3x - 4x] + 4$$

$$= 3x^2 - 7x + 4$$

Example 2 Multiply: $(3V - 2R)(4V + 5R)$.

Solution. By the FOIL scheme we get

$$(3V - 2R)(4V + 5R) = \underbrace{(3V)(4V)}_{F} + \overbrace{[\underbrace{(3V)(5R)}_{O} + \underbrace{(-2R)(4V)}_{I}]}^{\text{Add mentally}} + \underbrace{(-2R)(5R)}_{L}$$

$$= 12V^2 + 7VR - 10R^2$$

Example 3 Multiply: $(5s + 4p)(s - 7p)$.

Solution. Using FOIL again, we get

$$(5s + 4p)(s - 7p) = \underbrace{(5s)(s)}_{F} + \overbrace{[\underbrace{(5s)(-7p)}_{O} + \underbrace{(4p)(s)}_{I}]}^{\text{Add mentally}} + \underbrace{(4p)(-7p)}_{L}$$

$$= 5s^2 - 31sp - 28p^2$$

The square of a binomial may be treated as a special case of the product of two binomials.

Example 4 Multiply: $(2x - 3y)^2$.

Solution. If we write the product as $(2x - 3y)(2x - 3y)$, we can use the FOIL scheme again:

$$\begin{aligned}(2x - 3y)(2x - 3y) &= (2x)(2x) + [(2x)(-3y) + (-3y)(2x)] \\ &\quad + (-3y)(-3y) \\ &= 4x^2 - 12xy + 9y^2\end{aligned}$$

The square of a binomial can also be found by using the forms given next:

$$(a + b)^2 = a^2 + 2ab + b^2$$
$$(a - b)^2 = a^2 - 2ab + b^2$$

In words, the square of the sum (or difference) of two terms is the square of the first term, plus (or minus) twice the product of the two terms, plus the square of the second term.

Example 4 **Alternate solution.** To multiply $(2x - 3y)^2$, we get from the form $(a - b)^2 = a^2 - 2ab + b^2$

$$(2x - 3y)^2 = (2x)^2 - 2[(2x)(3y)] + (3y)^2$$
$$= 4x^2 - 12xy + 9y^2$$

Example 5 Multiply: $(T_1 + 4T_2)^2$.

Solution. Using the form

$$(a + b)^2 = a^2 + 2ab + b^2$$

we get

$$(T_1 + 4T_2)^2 = T_1^2 + 2[(T_1)(4T_2)] + (4T_2)^2$$
$$= T_1^2 + 8T_1T_2 + 16T_2^2$$

Alternate solution. Using the FOIL scheme, we get

$$(T_1 + 4T_2)^2 = (T_1 + 4T_2)(T_1 + 4T_2) = T_1^2 + [(T_1)(4T_2) + (4T_2)(T_1)] + (4T_2)(4T_2)$$
$$= T_1^2 + 8T_1T_2 + 16T_2^2$$

Exercises / Section 8.3

Use the FOIL scheme to perform the indicated multiplications. Obtain each product without writing any intermediate steps.

1. $(x - 2)(x + 4)$
2. $(x - 1)(x + 3)$
3. $(2x - 1)(x - 3)$
4. $(3y - 2)(y + 4)$
5. $(2y - 7)(y + 1)$
6. $(x + 4)(2x + 1)$
7. $(2x - 4y)(3x + y)$
8. $(4x - 3y)(x + 2y)$
9. $(3x + y)(4x + 2y)$
10. $(6x - 2y)(x + 3y)$
11. $(4V + 2W)(3V - 2W)$
12. $(5S_a - 2S_b)(S_a - S_b)$
13. $(6R_1 - 2R_2)(5R_1 - R_2)$
14. $(2s + 2t)(3s + t)$
15. $(7w - 2z)(5w + 3z)$
16. $(6v - 2w)(3v + 5w)$
17. $(3P + 2Q)(5P - Q)$
18. $(4p - 3q)(2p - 3q)$
19. $(5r + s)(3r + 4s)$
20. $(6v_1 - 5v_2)(2v_1 - 7v_2)$
21. $(2x - 5)^2$
22. $(x - 3y)^2$
23. $(2V - 4)^2$
24. $(3P + 2Q)^2$

25. $(5R_1 + 2R_2)^2$

26. $(3V_1 + V_2)^2$

27. $(4v_1 - 3v_2)^2$

28. $(7 - V)^2$

29. $(1 - 6L)^2$

30. $(5M - 1)^2$

8.4 Factoring Trinomials

In the last section we used the FOIL scheme to multiply two binomials to obtain a trinomial. In this section we reverse the procedure: Starting with a trinomial, we use the FOIL scheme to obtain the binomial factors.

To see how we do this, consider the trinomial

$$x^2 + x - 6$$

If this expression is factorable, the factors will be first-degree binomials of the form

$$(x \quad)(x \quad)$$

To find the other terms in the binomial factors, let's keep in mind the FOIL scheme and set up the following diagram:

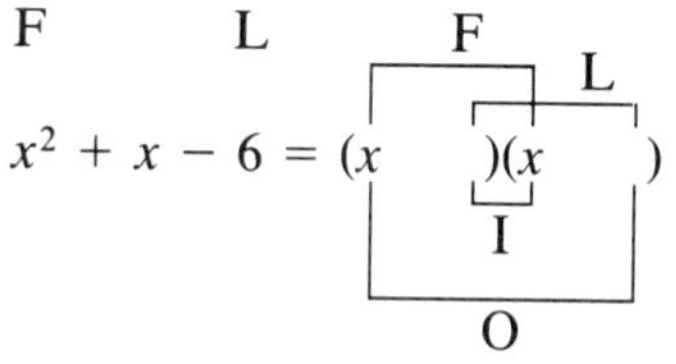

F L F L

$$x^2 + x - 6 = (x \quad)(x \quad)$$

I

O

Do you see the problem? Since the product of the L terms is -6, we seem to have several possibilities. For example, since $-6 \cdot 1 = -6$, we could try

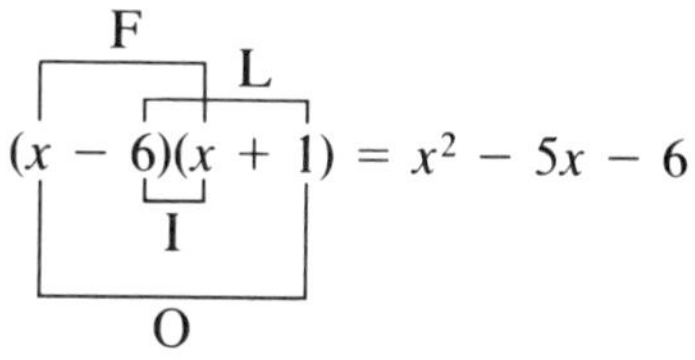

F L

$$(x - 6)(x + 1) = x^2 - 5x - 6$$

I

O

But, while the F terms and the L terms check, the middle term is way off ($-5x$ instead of x). Noting that $(3)(-2) = -6$, while $3 + (-2) = 1$, let's try

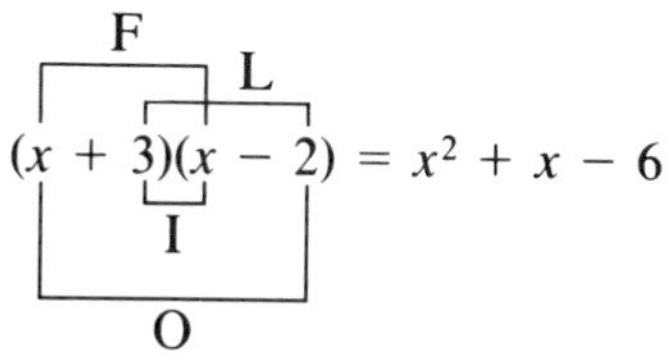

F L

$$(x + 3)(x - 2) = x^2 + x - 6$$

I

O

This result is correct.

Note especially that to factor $x^2 + x - 6$, we need two numbers whose *product* is -6 and whose *sum* is 1. These numbers are 3 and -2. Consider another example.

Example 1 Factor the trinomial $x^2 + 6xy + 8y^2$.

Solution. Keeping in mind the FOIL scheme, we need two numbers whose *product* is 8 and whose *sum* is 6. These numbers are 2 and 4.

$$x^2 + 6xy + 8y^2 = (x + 2y)(x + 4y)$$

(F, O, I, L brackets indicate the First, Outer, Inner, and Last products.)

Note that the middle term is $6xy$, as required.

The other combination fails:

$$(x + 8y)(x + y) = x^2 + 9xy + y^2$$

(The middle term does not check.)

The idea in the next two examples is similar.

Example 2 Factor $x^2 - 7xy + 12y^2$.

Solution. Since the last term has a positive coefficient, the L terms must agree in sign. This common sign is negative since the middle term has a negative coefficient. So we need two numbers whose *product* is 12 and whose *sum* is -7. These numbers are -3 and -4.

$$x^2 - 7xy + 12y^2 = (x - 3y)(x - 4y)$$

The middle term is $-7xy$, as required.

Example 3 Factor $x^2 + 3xy - 18y^2$.

Solution. This time we need two numbers whose *product* is -18 and whose *sum* is 3. These numbers are 6 and -3.

$$x^2 + 3xy - 18y^2 = (x - 3y)(x + 6y)$$

These examples suggest that there is a fairly simple method for factoring trinomials. Unfortunately, whenever the coefficient of the first term is a number different from 1, then the number of possible combinations may be greatly increased.

Example 4 Factor $4x^2 + 5x - 6$.

Solution. The first term can be written $4x^2 = (2x)(2x)$ or $4x^2 = (4x)(x)$. Suppose we try

$$(2x \quad)(2x \quad)$$

In view of the positive middle term, a good choice may be

$$(2x - 2)(2x + 3) \qquad -6 = -2 \cdot 3$$

(F, O, I, L)

However, this choice leads to $4x^2 + 2x - 6$. The combination -3 and 2 leads to a negative middle term, which makes matters worse.

Since $-6 = 6(-1)$, let us try 6 and -1:

$$(2x + 6)(2x - 1)$$

(F, O, I, L)

This choice results in $4x^2 + 10x - 6$, which is again incorrect. The combination -6 and 1 also fails.

Having exhausted all the possibilities, let us try

$$(4x \quad)(x \quad)$$

Using the same factors of -6, we might try

$$(4x - 3)(x + 2) = 4x^2 + 5x - 6$$

(F, O, I, L)

This result is correct.

Example 5 Factor $4a^2 - 20at + 25t^2$.

Solution. Since $4a^2 = (4a)(a)$, we could start by trying

$$(4a \quad)(a \quad)$$

as we did in Example 4. Since $25t^2 = (-5t)(-5t)$, let us try

$$(4a - 5t)(a - 5t) = 4a^2 - 25at + 25t^2$$

(F, O, I, L)

However, the middle term does not check. The combination -25 and -1 also fails.

For our next trial, let's start with

$$(2a \quad\quad)(2a \quad\quad)$$

We now have

$$(2a - 5t)(2a - 5t) = 4a^2 - 20at + 25t^2$$

(F, O, I, L)

This result is correct. Since the factors are identical, the answer should be written as a square of a binomial. Thus

$$4a^2 - 20at + 25t^2 = (2a - 5t)^2$$

It was noted in the last section that we should always be on the alert for common factors. Consider the next example.

Example 6 Factor $12s_1^2 - 2s_1s_2 - 4s_2^2$.

Solution. Since all the coefficients are even numbers, each term has common factor 2. So

$$12s_1^2 - 2s_1s_2 - 4s_2^2 = 2(6s_1^2 - s_1s_2 - 2s_2^2)$$

The first term of the trinomial, $6s_1^2$, leads to several possibilities. Let's try

$$2(3s_1 \quad\quad)(2s_1 \quad\quad)$$

The last term in the trinomial, $-2s_2^2$, can be written $(2s_2)(-s_2)$ or $(-2s_2)(s_2)$. For the first choice we get

$$2(3s_1 + 2s_2)(2s_1 - s_2) = 2(6s_1^2 + s_1s_2 - 2s_2^2)$$

which is incorrect. For the second choice we get

$$2(3s_1 - 2s_2)(2s_1 + s_2) = 2(6s_1^2 - s_1s_2 - 2s_2^2)$$

Since the middle term checks, we conclude that

$$12s_1^2 - 2s_1s_2 - 4s_2^2 = 2(3s_1 - 2s_2)(2s_1 + s_2)$$

Not every polynomial is factorable. Recall that a nonfactorable polynomial is called **prime** or **irreducible.** Consider, for example, the polynomial $x^2 + x + 3$. The only possible combinations are

$$(x + 3)(x + 1) = x^2 + 4x + 3$$

and

$$(x - 3)(x - 1) = x^2 - 4x + 3$$

Since neither combination works, the polynomial must be prime.

Example 7 The deflection d of a certain beam is given by

$$d = 2a^2x^2 - 8aLx + 6L^2$$

Factor the expression for d.

Solution. Since each term has common factor 2, we write $d = 2(a^2x^2 - 4aLx + 3L^2)$. By the FOIL scheme we now have

F, L, I, O

$$2(ax - 3L)(ax - L)$$

It follows that $d = 2(ax - 3L)(ax - L)$.

Exercises / Section 8.4

In Exercises 1–56, factor each of the given trinomials.

1. $x^2 + x - 2$

2. $x^2 - 3x + 2$

3. $y^2 - y - 6$

4. $z^2 - 5z + 6$

5. $L^2 + 3L - 4$

6. $P^2 - 2P - 8$

7. $x^2 - 3xy - 4y^2$

8. $x^2 - 7xy + 10y^2$

9. $x^2 + 7xy + 6y^2$

10. $x^2 - 7xy + 12y^2$

11. $x^2 - 10xy + 24y^2$

12. $x^2 + 5xy - 24y^2$

13. $p^2 + pq - 30q^2$

14. $s^2 - 3st - 54t^2$

15. $2x^2 - 11xy + 12y^2$

16. $3x^2 - 5xy - 2y^2$

17. $2V^2 + 11Vb - 6b^2$

18. $5b^2 - 11bn + 2n^2$

19. $4v_1^2 + 4v_1v_2 - 3v_2^2$
20. $4L_a^2 - 7L_aL_b - 2L_b^2$
21. $4s_1^2 - 5s_1s_2 + s_2^2$
22. $4t_1^2 - 8t_1t_2 + 3t_2^2$
23. $5T^2 + 9Tn - 2n^2$
24. $2x^2 + 13xz + 6z^2$
25. $6x^2 + xw - 2w^2$
26. $6w^2 + 2wb - 4b^2$
27. $6k^2 + 33ka - 18a^2$
28. $2a^2 - 12ab + 16b^2$
29. $3x^2 + 4x + 2$
30. $2y^2 + 3y + 4$
31. $x^2 + 6xy + 9y^2$
32. $a^2 - 4ab + 4b^2$
33. $4m^2 - 4mn + n^2$
34. $9y^2 - 12yz + 4z^2$
35. $2x^2 - 12xy + 16y^2$
36. $4x^2 + 14xy + 12y^2$
37. $12a^2 - 39ab + 9b^2$
38. $9R^2 - 21RS + 12S^2$
39. $4ax^2 + axy - 3ay^2$
40. $4bx^2 - 10bxy - 6by^2$
41. $24x^2 - 20xy + 4y^2$
42. $16x^2 - 20xy - 24y^2$
43. $10x^2 + 26xy - 12y^2$
44. $2x^2 - 14xy + 24y^2$
45. $2y^2 + 12yz - 32z^2$
46. $2n^2 + 4nz - 48z^2$
47. $3w^2 + 18wz + 24z^2$
48. $3P^2 - 24PQ + 36Q^2$
49. $2a^2 - 12ab + 18b^2$
50. $3s^2 - 12st + 12t^2$
51. $2L^2 + 3L + 5$
52. $V^2 + 2V + 2$
53. $4aV^2 - 12aV + 9a$
54. $b - 4bP + 4bP^2$
55. $9T_a^2 + 12T_aT_b + 4T_b^2$
56. $4V_1^2 + 20V_1V_2 + 25V_2^2$
57. The deflection d of a beam is given by

$$d = 2a^2x^2 - 7aLx + 6L^2$$

Factor the expression for d.

58. For a short time interval, the current i in a circuit is given by $i = 6t^2 - 16t + 8$, where t is measured in seconds. Factor the expression for i.
59. A stone is hurled upward from a height of 32 ft at 56 ft/s. Its distance s (in feet) above the ground is $s = 16t^2 - 56t - 32$, where t is measured in seconds. Factor the expression for s.
60. The resistance R in a certain variable resistor varies with time t according to the relation $R = 32.0t^2 - 8.00t - 60.0$. Factor the expression for R.
61. Two resistors have a combined resistance of 18 Ω when connected in series and 4 Ω when connected in parallel. To find the two resistances, it is necessary to solve the equation $R^2 - 18R + 72 = 0$. Factor the left side of this equation.
62. The weekly profit P of a company is $P = 2x^4 - 10x^3$ $(x \geq 0)$. Factor the expression for P.

Review Exercises / Chapter 8

In Exercises 1–20, perform the indicated operations.

1. $a(s - 2t)$
2. $b(2x - 3y)$
3. $2a^2b(-5ab^2 + 2ab - 3a^3b^2)$
4. $xy^3(3xy + 2x^2y^3 - x^3y^2)$

5. $V_aV_b(2V_a^2V_b - 4V_aV_b - 1)$

6. $T_1T_2(T_1^3T_2 - T_1T_2^2 + 1)$

7. $(y - 5)(y + 5)$

8. $(PQ - 1)(PQ + 1)$

9. $\left(\frac{1}{2}x - 1\right)\left(\frac{1}{2}x + 1\right)$

10. $\left(\frac{1}{4}a - \frac{1}{2}b\right)\left(\frac{1}{4}a + \frac{1}{2}b\right)$

11. $(4S_0 - T_0)(4S_0 + T_0)$

12. $(3t_1 - 2t_2)(3t_1 + 2t_2)$

13. $(x - 4)(x + 5)$

14. $(y + 6)(y - 3)$

15. $(2z - w)(z - 3w)$

16. $(3t + 2)(2t + 1)$

17. $(2R_1 - R_2)(4R_1 + R_2)$

18. $(3T_a - T_b)(T_a + 4T_b)$

19. $(4S - T)^2$

20. $(V - 4R)^2$

In Exercises 21–54, factor each of the given expressions.

21. $ab^2 - 2a$

22. $cd - d^2$

23. $3P^2Q^2 - 6PQ$

24. $4ST - 8S^2T^2$

25. $3x^2w^2 + xw$

26. $4n^2p^2 - np$

27. $35L^2M - 28LM^2 - 14L^2M^2$

28. $28S_a^3S_b^3 - 16S_a^2S_b^2 - 12S_aS_b$

29. $13RV + 26R^2V^3 - 39R^3V^2$

30. $14st^3 - 28s^3t - 42s^2t^2$

31. $P^2 - 16$

32. $9V^2 - 4$

33. $1 - 16V_0^2$

34. $1 - 25t_0^2$

35. $\frac{1}{4}s^2 - 1$

36. $\frac{1}{16}p^2 - 1$

37. $ax^2 - ay^2$

38. $2V_a^2 - 2V_b^2$

39. $3S^4 - 3T^4$

40. $cN_1^4 - cN_2^4$

41. $2y^2 - y - 1$

42. $w^2 + 3w - 10$

43. $4P^2 - 8P + 3$

44. $4L^2 - 5LP + P^2$

45. $2s^2 - 11st - 6t^2$

46. $5x^2 - 9xy - 2y^2$

47. $2a^2 - 6ab + 4b^2$

48. $4m^2 - 14mn + 12n^2$

49. $15y^2 + 39yw - 18w^2$

50. $3s_1^2 + 18s_1s_2 - 48s_2^2$

51. $18P^2 - 12PV + 2V^2$

52. $3 - 12L + 12L^2$

53. $36a^2 - 24ab + 4b^2$

54. $2a^2 + 3ab + 2b^2$

55. In the theory of relativity, the relation between mass and energy is

$$E = mc^2 - m_0c^2$$

Factor the expression for E.

56. The volume V of the bulb of a mercury thermometer is

$$V = V_0 + V_0bt$$

Factor the expression for V.

57. Factor the following expression for T_{max}, which occurs in the study of the photoelectric effect:

$$T_{max} = k\nu - k\nu_0$$

58. The difference E_d in the energy radiated by a filament at temperatures T_1 and T_2, respectively, is given by $E_d = kT_1^4 - kT_2^4$, where k is a constant. Factor the expression for E_d.

59. If the current through a resistor increases from i_1 to i_2, then the increase P_d in the power delivered to the resistor is

$$P_d = Ri_2^2 - Ri_1^2$$

Factor the expression for P_d.

60. For a short time interval, the resistance R of a variable resistor is given by $R = 18.0t^2 - 21.0t - 60.0$, where t is measured in seconds. Factor the expression for R.

61. If m denotes the month in the year, the profit P on a certain commodity is given by $P = 3m^2 - 21m - 24$. Factor the expression for P.

62. The equation for the transverse shearing stress of a rectangular beam has the form

$$T = K\left(1 - \frac{4a^2}{h^2}\right)$$

Factor the right side.

CHAPTER 9

Algebraic Fractions and Fractional Equations

In Chapter 1 we studied various operations with arithmetic fractions. In this chapter we turn to the corresponding operations with algebraic fractions. Algebraic fractions play a role in algebra similar to the role played by arithmetic fractions in everyday life.

9.1 Equivalent Fractions

Equivalent fractions

Recall from Chapter 1 that one fraction is **equivalent** to another fraction if multiplying (or dividing) the numerator and denominator of one fraction by the same nonzero constant results in the other fraction. For example,

$$\frac{4}{6} = \frac{2}{3} \quad \text{since} \quad \frac{4}{6} = \frac{2 \cdot 2}{2 \cdot 3}$$

For algebraic fractions we have, similarly,

$$\frac{4x}{6y} = \frac{2x}{3y} \quad \text{since} \quad \frac{4x}{6y} = \frac{2 \cdot 2x}{2 \cdot 3y}$$

and

$$\frac{ax}{ay} = \frac{x}{y} \quad \text{since} \quad \frac{ax}{ay} = \frac{a \cdot x}{a \cdot y}$$

These examples illustrate the following fundamental principle:

> The value of a fraction stays the same if the numerator and denominator are multiplied or divided by the same nonzero quantity.

This principle is used to reduce a given fraction to lowest terms.

> A fraction is said to be **reduced to lowest terms** if its numerator and denominator have no common factor except 1.

For example, the fractions

$$\frac{2}{3} \quad \text{and} \quad \frac{5}{9}$$

are reduced to lowest terms since the numerator and denominator have no common factor except 1.

However, the numerator and denominator of the fraction

$$\frac{9x^2}{15x}$$

have common factor $3x$:

$$\frac{9x^2}{15x} = \frac{3x(3x)}{5(3x)} = \frac{3x}{5}$$

which is obtained by dividing the numerator and denominator by $3x$. The resulting fraction cannot be reduced further.

Similarly,

$$\frac{14xy}{21xy^2} = \frac{2(7xy)}{3y(7xy)} = \frac{2}{3y}$$

which is obtained by dividing the numerator and denominator by $7xy$.

Cancellation

The simplest way to carry out this reduction is to cross out the common factor. This procedure is called **cancellation.** In the last example, we cancel as follows:

$$\frac{14xy}{21xy^2} = \frac{2\overset{1}{\cancel{(7xy)}}}{3y\underset{1}{\cancel{(7xy)}}} = \frac{2}{3y}$$

(From now on we assume that in every fraction the values of the variables are so chosen that the denominator is not equal to zero.)

Example 1 Reduce the following fractions:

a. $\dfrac{25a^3b}{15a^2b}$ **b.** $\dfrac{26s^2t^3}{13st^4}$

Solution.

a. The common factors of the numerator and denominator are 5, a^2, and b. So

$$\frac{25a^3b}{15a^2b} = \frac{5a(5a^2b)}{3(5a^2b)} = \frac{5a\overset{1}{\cancel{(5a^2b)}}}{3\underset{1}{\cancel{(5a^2b)}}} = \frac{5a}{3}$$

b. The common factor is $13st^3$:

$$\frac{26s^2t^3}{13st^4} = \frac{2s(13st^3)}{t(13st^3)} = \frac{2s\overset{1}{\cancel{(13st^3)}}}{t\underset{1}{\cancel{(13st^3)}}} = \frac{2s}{t}$$

The cancellation procedure can also be used to reduce fractions containing polynomials. To do so, however, **both numerator and denominator must be factored first.**

Example 2 Reduce the following fraction to lowest terms:

$$\frac{x^2 + 2x - 8}{x^2 + 3x - 10}$$

Solution. Using the FOIL scheme, we first factor numerator and denominator and then cancel:

$$\frac{x^2 + 2x - 8}{x^2 + 3x - 10} = \frac{(x - 2)(x + 4)}{(x - 2)(x + 5)} = \frac{\overset{1}{\cancel{(x - 2)}}(x + 4)}{\underset{1}{\cancel{(x - 2)}}(x + 5)} = \frac{x + 4}{x + 5}$$

The 1 resulting from the cancellation is often omitted:

$$\frac{\cancel{(x - 2)}(x + 4)}{\cancel{(x - 2)}(x + 5)} = \frac{x + 4}{x + 5}$$

Sometimes a fraction contains factors that differ only in sign. For example,

$$\frac{2x - 3}{3 - 2x}$$

is readily reduced if we observe that $3 - 2x = -1(2x - 3)$. Using this we get

$$\frac{2x - 3}{3 - 2x} = \frac{\overset{1}{\cancel{2x - 3}}}{-1\underset{1}{\cancel{(2x - 3)}}} = \frac{1}{-1} = -1$$

(Since the only factor in the numerator cancels, we need to place a 1 near the canceled factors.)

To avoid difficulties with signs, let us note the following properties:

$$a - b = -1(b - a)$$

$$\frac{a}{-b} = -\frac{a}{b} \qquad \frac{-a}{b} = -\frac{a}{b}$$

Example 3 Reduce:

$$\frac{L^2 - W^2}{2W - 2L}$$

Solution.

$$\frac{L^2 - W^2}{2W - 2L} = \frac{(L - W)(L + W)}{2W - 2L} \quad \text{Difference of two squares}$$

$$= \frac{(L - W)(L + W)}{2(W - L)} \quad \text{Common factor 2}$$

$$= \frac{(L - W)(L + W)}{2(-1)(L - W)} \quad a - b = -1(b - a)$$

$$= \frac{\cancel{(L - W)}(L + W)}{2(-1)\cancel{(L - W)}} \quad \text{Cancellation}$$

$$= \frac{L + W}{-2} \quad 2(-1) = -2$$

$$= -\frac{L + W}{2} \quad \frac{a}{-b} = -\frac{a}{b}$$

(The negative sign should not be left in the denominator.)

Common error Canceling terms even though numerator or denominator have not been written as a product. For example,

$$\frac{x + 2}{x + 3} \text{ is not equal to } \frac{2}{3} \text{ or to } \frac{1 + 2}{1 + 3}$$

In other words, the x cannot be canceled since x is a term, *not a factor*. The fraction is already in simplest form.

Similarly,

$$\frac{a + b}{a} \neq 1 + b$$

The a does not cancel since the numerator is not factored. Although the fraction is already in simplest form, an acceptable alternative is

$$\frac{a+b}{a} = \frac{a}{a} + \frac{b}{a} = 1 + \frac{b}{a}$$

The danger of a wrong cancellation is further illustrated in the next example.

Example 4 Reduce the fraction

$$\frac{x+2}{(x+3)x+2}$$

Solution. Here it is particularly tempting to cancel $x + 2$. However, note that $(x + 3)x + 2$ is not the same as $(x + 3)(x + 2)$. Rather, $(x + 3)x + 2$ is the *sum* of $(x + 3)x$ and 2. So we need to rewrite the denominator as a product first and then cancel.

$$\begin{aligned}
\frac{x+2}{(x+3)x+2} &= \frac{x+2}{x^2+3x+2} && (x+3)x = x^2 + 3x \\
&= \frac{x+2}{(x+2)(x+1)} && \text{Factoring} \\
&= \frac{\overset{1}{\cancel{x+2}}}{\underset{1}{\cancel{(x+2)}}(x+1)} && \text{Cancellation} \\
&= \frac{1}{x+1}
\end{aligned}$$

Note that a 1 has to be placed near the canceled factors since the only factor in the numerator cancels.

Example 5 The voltage drop (in volts) across a certain resistor is $V = 6t^2 + 7t + 2$, where t is measured in seconds. If the resistance R (in ohms) is given by $R = 2t + 1$ (t in seconds), find a simplified expression for the current $i = V/R$.

Solution. From the formula $i = V/R$, we get

$$\begin{aligned}
i &= \frac{6t^2 + 7t + 2}{2t + 1} \\
&= \frac{(2t+1)(3t+2)}{(2t+1)} \\
&= \frac{\overset{1}{\cancel{(2t+1)}}(3t+2)}{\underset{1}{\cancel{(2t+1)}}} \\
&= 3t + 2 \quad \text{(in amperes)}
\end{aligned}$$

Exercises / Section 9.1

In Exercises 1–38, reduce each fraction to lowest terms.

1. $\dfrac{4x^2}{6x^4}$

2. $\dfrac{8y^7}{12y^3}$

3. $\dfrac{16w}{14w^2}$

4. $\dfrac{21z^4}{14z^4}$

5. $\dfrac{-7x^2y^6}{28x^2y^3}$

6. $\dfrac{16w^7y^3}{-24w^4y^3}$

7. $\dfrac{2a^4LW^5}{a^2L^2W^6}$

8. $\dfrac{aN^3P^3}{3a^6N^3P^4}$

9. $\dfrac{x^2 - 2x}{x^3 - 2x^2}$

10. $\dfrac{y^2 + y}{y^4 + y^3}$

11. $\dfrac{s^5 - 2s^4t}{s^3 - 2s^2t}$

12. $\dfrac{6p - 6q}{12p - 12q}$

13. $\dfrac{2x - 2y}{4y - 4x}$

14. $\dfrac{8x - 16y}{8y - 4x}$

15. $\dfrac{x^2 - 2x - 8}{xy + 2y}$

16. $\dfrac{y^2 + 2y - 8}{2y + 8}$

17. $\dfrac{x^2 - y^2}{x + y}$

18. $\dfrac{a - b}{a^2 - b^2}$

19. $\dfrac{V^2 - W^2}{W - V}$

20. $\dfrac{s^2 - t^2}{2t - 2s}$

21. $\dfrac{4T^2 - T_0^2}{T_0 - 2T}$

22. $\dfrac{3t - 1}{1 - 9t^2}$

23. $\dfrac{v_1^2 + 2v_1v_2 + v_2^2}{v_1 + v_2}$

24. $\dfrac{s_1 - s_2}{s_1^2 - 2s_1s_2 + s_2^2}$

25. $\dfrac{m^2 - mn - 2n^2}{m^2 + 2mn + n^2}$

26. $\dfrac{a^2 - 2ac + c^2}{a^2 + 2ac - 3c^2}$

27. $\dfrac{f^2 - g^2}{g^2 + gf - 2f^2}$

28. $\dfrac{x^2 + 2xy - 3y^2}{-2x^2 + 3xy - y^2}$

29. $\dfrac{2T_1^2 - 7T_1T_2 + 3T_2^2}{3T_1^2 - 5T_1T_2 - 12T_2^2}$

30. $\dfrac{3v^2 + 4vv_0 - 4v_0^2}{2v^2 + 7vv_0 + 6v_0^2}$

31. $\dfrac{6s_1^2 + 5s_1s_2 - s_2^2}{12s_1^2 + 4s_1s_2 - s_2^2}$

32. $\dfrac{4s^2 + 4st - 3t^2}{8s^2 - 2st - t^2}$

33. $\dfrac{x + 4}{(4x + 17)x + 4}$

34. $\dfrac{V + 2}{(2V + 5)V + 2}$

35. $\dfrac{a - 3}{(2a - 5)a - 3}$

36. $\dfrac{m - 2}{(3m - 5)m - 2}$

37. $\dfrac{x + 2}{(3x + 7)x + 2}$

38. $\dfrac{P + 5}{(3P + 16)P + 5}$

39. The following fraction arises in the study of the ballistic pendulum (which is used for measuring the velocity of a bullet):

$$\frac{\frac{1}{2}mv^2 + \frac{1}{2}Mv^2}{\frac{1}{2}mv^2}$$

Reduce this fraction.

40. Simplify the following expression from a problem in mechanics:

$$\frac{Mv_0^2 + mv_0^2}{M^2 + 2Mm + m^2}$$

41. The resistance R (in ohms) of a certain variable resistor is $R = 1 + 3t$ (t in seconds) and the voltage drop (in volts) across the resistor is $V = 2 + 11t + 15t^2$. Find a simplified expression for the current i (in amperes), which is given by

$$i = \frac{V}{R} = \frac{2 + 11t + 15t^2}{1 + 3t}$$

42. The area of a rectangle is $2x^2 + 6x - 20$ and the width is $2x - 4$. Reduce the expression for the length l, which is given by

$$l = \frac{2x^2 + 6x - 20}{2x - 4}$$

43. The area of a rectangle is $2a^2 + 9ab + 10b^2$, and the length is $a + 2b$. Find a simplified expression for the width w, where

$$w = \frac{2a^2 + 9ab + 10b^2}{a + 2b}$$

9.2 Multiplication and Division of Fractions

In this section we discuss multiplication and division of fractions. The rules for these operations carry over from arithmetic. (See Section 1.6.)

Multiplication

$$\frac{a}{b} \cdot \frac{c}{d} = \frac{ac}{bd}$$

Division

$$\frac{a}{b} \div \frac{c}{d} = \frac{a}{b} \cdot \frac{d}{c} = \frac{ad}{bc}$$

Multiplication

The following example illustrates the rule for multiplication:

$$\frac{x^2}{2} \cdot \frac{4}{x} = \frac{4x^2}{2x} = \frac{(2x)\overset{1}{\cancel{(2x)}}}{\underset{1}{\cancel{2x}}} = 2x$$

However, just as in the case of arithmetic fractions, it is better to cancel before multiplying. For example, if we multiply

$$\frac{2}{x+y} \quad \text{and} \quad \frac{x+y}{x}$$

by the rule, we get

$$\frac{2}{x+y} \cdot \frac{x+y}{x} = \frac{2x+2y}{x^2+xy} = \frac{2\cancel{(x+y)}}{x\cancel{(x+y)}} = \frac{2}{x}$$

Now, rather than multiplying and then factoring again, it is better to cancel the factors in the original problem:

$$\frac{2}{\underset{1}{\cancel{x+y}}} \cdot \frac{\overset{1}{\cancel{x+y}}}{x} = \frac{2 \cdot 1}{1 \cdot x} = \frac{2}{x}$$

To multiply two or more fractions, factor the expressions in the numerator and denominator of each given fraction and cancel. Then multiply the resulting fractions by the rule

$$\frac{a}{b} \cdot \frac{c}{d} = \frac{ac}{bd}$$

Consider the following example.

Example 1 Perform the following multiplication:

$$\frac{x^2 - xy}{a} \cdot \frac{2}{xy - y^2}$$

Solution.

$$\frac{x^2 - xy}{a} \cdot \frac{2}{xy - y^2} = \frac{x(x-y)}{a} \cdot \frac{2}{y(x-y)} \qquad \text{Common factors}$$

$$= \frac{x\cancel{(x-y)}}{a} \cdot \frac{2}{y\cancel{(x-y)}} \qquad \text{Cancellation}$$

$$= \frac{x \cdot 2}{a \cdot y} = \frac{2x}{ay} \qquad \text{Multiplication}$$

Example 2 Perform the following multiplication:

$$\frac{P_2^2 - P_1^2}{P_1P_2} \cdot \frac{P_1^2P_2^2}{P_1^2 + 2P_1P_2 - 3P_2^2}$$

Solution.

$$\frac{P_2^2 - P_1^2}{P_1P_2} \cdot \frac{P_1^2P_2^2}{P_1^2 + 2P_1P_2 - 3P_2^2}$$

$$= \frac{(P_2 - P_1)(P_2 + P_1)}{P_1P_2} \cdot \frac{P_1^2P_2^2}{(P_1 - P_2)(P_1 + 3P_2)}$$ Difference of two squares; Factoring trinomial

$$= \frac{-1(P_1 - P_2)(P_2 + P_1)}{P_1P_2} \cdot \frac{P_1^2P_2^2}{(P_1 - P_2)(P_1 + 3P_2)}$$ $P_2 - P_1 = -1(P_1 - P_2)$

$$= \frac{-1\cancel{(P_1 - P_2)}(P_2 + P_1)}{\cancel{P_1}\cancel{P_2}} \cdot \frac{P_1^{\cancel{2}}P_2^{\cancel{2}}}{\cancel{(P_1 - P_2)}(P_1 + 3P_2)}$$ Cancellation

$$= -\frac{P_1P_2(P_1 + P_2)}{P_1 + 3P_2}$$ $P_2 + P_1 = P_1 + P_2$

Common error Multiplying fractions first instead of canceling first. (See Example 3.)

Example 3 Perform the following multiplication:

$$\frac{x - 1}{x - 2} \cdot \frac{x^2 + x - 6}{x^2 + 4x - 5}$$

Solution. As noted, a common error is to multiply the given fractions right away to get

$$\frac{(x - 1)(x^2 + x - 6)}{(x - 2)(x^2 + 4x - 5)} = \frac{x^3 - 7x + 6}{x^3 + 2x^2 - 13x + 10}$$

The polynomials in the resulting fraction, however, cannot be factored by any method we have considered so far. Avoid this problem by first factoring the polynomials in the given fractions and then canceling:

$$\frac{x - 1}{x - 2} \cdot \frac{x^2 + x - 6}{x^2 + 4x - 5} = \frac{x - 1}{x - 2} \cdot \frac{(x - 2)(x + 3)}{(x - 1)(x + 5)}$$ Factoring

$$= \frac{\cancel{x - 1}}{\cancel{x - 2}} \cdot \frac{\cancel{(x - 2)}(x + 3)}{\cancel{(x - 1)}(x + 5)}$$ Canceling

$$= \frac{x + 3}{x + 5}$$

Division

By the rule for division,

$$\frac{a}{b} \div \frac{c}{d} = \frac{a}{b} \cdot \frac{d}{c}$$

The division procedure requires one more step than multiplication: *To divide two fractions, invert the divisor and then proceed as in multiplication.* Put another way, multiply by the reciprocal of the divisor.

Example 4

$$\frac{R^2V}{S} \div \frac{R^2V^2}{S^2} = \frac{R^2V}{S} \cdot \frac{S^2}{R^2V^2} \qquad \text{Inverting divisor}$$

$$= \frac{\cancel{R^2}\cancel{V}}{\cancel{S}} \cdot \frac{\cancel{S^2}}{\cancel{R^2}V^{\cancel{2}}} \qquad \text{Cancellation}$$

$$= \frac{S}{V} \qquad \text{Multiplication}$$

The next example illustrates division of fractions containing polynomials.

Example 5 Perform the following division:

$$\frac{2a^2 + 3ab - 5b^2}{m^2 - n^2} \div \frac{a^2 - 3ab + 2b^2}{3m + 3n}$$

Solution. We invert the divisor and then proceed as in multiplication:

$$\frac{2a^2 + 3ab - 5b^2}{m^2 - n^2} \div \frac{a^2 - 3ab + 2b^2}{3m + 3n}$$

$$= \frac{2a^2 + 3ab - 5b^2}{m^2 - n^2} \cdot \frac{3m + 3n}{a^2 - 3ab + 2b^2} \qquad \text{Inverting the divisor}$$

$$= \frac{(a - b)(2a + 5b)}{m^2 - n^2} \cdot \frac{3m + 3n}{(a - b)(a - 2b)} \qquad \text{Factoring trinomials}$$

$$= \frac{(a - b)(2a + 5b)}{(m - n)(m + n)} \cdot \frac{3(m + n)}{(a - b)(a - 2b)} \qquad \text{Common factor 3; Difference of two squares}$$

$$= \frac{\cancel{(a - b)}(2a + 5b)}{(m - n)\cancel{(m + n)}} \cdot \frac{3\cancel{(m + n)}}{\cancel{(a - b)}(a - 2b)} \qquad \text{Cancellation}$$

$$= \frac{3(2a + 5b)}{(m - n)(a - 2b)}$$

Exercises / Section 9.2

In Exercises 1–26, perform the indicated multiplications and divisions and simplify.

1. $\frac{x^2}{a^2} \cdot \frac{a^4}{x^3}$

2. $\frac{y^3}{b^2} \cdot \frac{2b}{y}$

3. $\frac{4m^2n}{3} \cdot \frac{6}{4m^3n}$

4. $\frac{3R^4C^2}{5} \cdot \frac{15}{R^6C^2}$

5. $\frac{7x^2z}{4ab} \div \frac{14x^2z^3}{8a^2b}$

6. $\frac{axy}{4pq} \div \frac{ax^3y}{8pq^2}$

7. $\frac{x+y}{x} \cdot \frac{x^2}{2x+2y}$

8. $\frac{ax+ay}{b} \cdot \frac{2b}{x+y}$

9. $\frac{2w+2z}{ab^3} \div \frac{3w+3z}{ab^2}$

10. $\frac{2v}{aM+aN} \div \frac{3v}{bM+bN}$

11. $\frac{a-3b}{x^2-w^2} \cdot \frac{x-w}{2a-6b}$

12. $\frac{v^2-4w^2}{4ab} \cdot \frac{2a^2b^2}{v+2w}$

13. $\frac{P^2+2PQ+Q^2}{x^2-4} \div \frac{P^2-Q^2}{x+2}$

14. $\frac{b^2-9}{x^2-2ax+a^2} \div \frac{b+3}{x^2-a^2}$

15. $\frac{a^2+2ab-3b^2}{c^2-16} \cdot \frac{c+4}{a^2+ab-2b^2}$

16. $\frac{2s^2+st-10t^2}{p-2q} \cdot \frac{p^2-4q^2}{s^2-4t^2}$

17. $\frac{3v^2+5vs-2s^2}{2p+1} \div \frac{6v^2+vs-s^2}{4p^2-1}$

18. $\frac{a^2L^2-1}{2p^2+pt-t^2} \div \frac{aL-1}{2p^2+9pt-5t^2}$

19. $\frac{1-4n^2}{6R^2-RC-2C^2} \cdot \frac{9R^2-12RC+4C^2}{2n-1}$

20. $\frac{9V^2-4V_0^2}{6l^2-5lz+z^2} \cdot \frac{4l^2+4lz-3z^2}{2V_0-3V}$

21. $\frac{2x^2-5xy-3y^2}{3C-Q} \div \frac{x^2-xy-6y^2}{Q-3C}$

22. $\frac{s_1-3s_2}{4m^2-5mn+n^2} \div \frac{3s_2-s_1}{4n^2-15nm-4m^2}$

23. $\frac{V_1^2-3V_1V_2+2V_2^2}{1-s^2} \cdot \frac{1+s}{2V_1-2V_2} \cdot \frac{3}{V_1-2V_2}$

24. $\frac{1+w}{9r-3s} \cdot \frac{3r^2-4rs+s^2}{ac^3} \cdot \frac{a^2c^3}{w^2-1}$

25. $\frac{3-N}{4+p} \cdot \frac{2N+6}{3p^2-5p-28} \cdot \frac{p^2-16}{N^2-9}$

26. $\frac{ab-2b^2}{x-3y} \cdot \frac{R^2C}{4b^2-a^2} \cdot \frac{2x^2-xy-15y^2}{bRC^2}$

27. The combined resistance of two parallel resistors is the reciprocal of

$$\frac{R_1+R_2}{R_1R_2}$$

Find an expression for the combined resistance.

28. The focal length of a lens is equal to the reciprocal of

$$\frac{p+q}{pq}$$

where p is the distance from the object to the lens and q the distance from the image to the lens. Find an expression for the focal length.

29. A certain configuration of objects has a mass of

$$\frac{2m_1 + 2m_2}{m_3}$$

and undergoes an acceleration of

$$\frac{am_1m_2m_3}{m_1^2 - m_2^2}$$

The force F exerted by this system is equal to the product of mass and acceleration. Find a simplified expression for F.

30. The energy E required to lift an object of weight

$$\frac{d}{2d + 8} \quad \text{a distance of} \quad \frac{d^2 - 16}{d^2}$$

is equal to the product of weight and distance. Find a simplified expression for E.

31. The mass of an object is ab^3/c^2 and its volume is ab/c. Find a simplified expression for its density (mass ÷ volume).

32. The pressure on a horizontally submerged plate is equal to the force against the plate divided by the area. If the force against the plate is

$$\frac{F^2}{F^2 - 1}$$

and the area of the plate is

$$\frac{F}{F + 1}$$

find a simplified expression for the pressure.

33. The resistance R (in ohms) of a variable resistor is

$$R = \frac{4t^2 + 4t + 1}{t + 3}$$

and the voltage drop V (in volts) across the resistor is $V = 2t + 1$ (t measured in seconds). Find an expression for the current $i = V/R$ and simplify.

34. The average power P (in watts) is defined to be the energy E required to perform a certain task divided by the average time T required. If

$$E = \frac{9t^2 + 12t + 4}{t + 1} \quad \text{(in joules)}$$

and

$$T = \frac{3t + 2}{2t + 2} \quad \text{(in seconds)}$$

find an expression for P and simplify.

35. The discovery of deuterium depended on the fact that the motion of the nucleus of an atom reduces the mass of an electron spinning about the nucleus. If m is the mass of an electron, then the reduced mass of the electron is given by

$$m' = \frac{mM}{M + m}$$

where M is the mass of the nucleus. Find a simplified expression for m'/m.

9.3 Addition and Subtraction of Fractions

In this section we study addition and subtraction of fractions. First, recall from Chapter 1, Section 1.7, that we add two or more fractions with the same denominator by adding the numerators of all the fractions and placing this sum over the denominator that is common to them. For example,

$$\frac{2}{7} + \frac{3}{7} = \frac{2 + 3}{7} = \frac{5}{7}$$

Subtraction is performed similarly:

$$\frac{9}{11} - \frac{2}{11} = \frac{9 - 2}{11} = \frac{7}{11}$$

If the denominators of the fractions are not the same, then we change each fraction to an equivalent fraction so the denominators are all the same. Only then can we proceed with the addition or subtraction. For example, to subtract $\frac{1}{6}$ from $\frac{1}{3}$, we change $\frac{1}{3}$ to $\frac{2}{6}$ and get

$$\frac{1}{3} - \frac{1}{6} = \frac{2}{6} - \frac{1}{6} = \frac{2 - 1}{6} = \frac{1}{6}$$

The number 6 is called the **lowest common denominator (LCD).**

The Lowest Common Denominator

As noted, we combine two or more fractions by changing each fraction to an equivalent fraction with the same denominator as the others—that is, with a common denominator. However, we know from our previous work with fractions that the most convenient common denominator to use is the lowest common denominator.

> **Lowest common denominator**
>
> The lowest common denominator (LCD) of two or more fractions is an expression that is divisible by every denominator and that has no more factors than needed to satisfy this condition.

For example, the common denominator of

$$\frac{a}{x} \quad \text{and} \quad \frac{b}{x^2}$$

is x^2, since x^2 is divisible by both x and x^2. In a simple case like this, the LCD can be determined by observation. For more complicated cases we need a systematic procedure:

> **To construct the lowest common denominator** of two or more algebraic fractions, we factor each of the denominators. Then the LCD is the product of the factors of the denominators, each factor having an exponent equal to the largest of the exponents of any of the factors.

For example, to find the lowest common denominator of

$$\frac{1}{12}, \quad \frac{7}{90}, \quad \text{and } \frac{17}{60}$$

we factor each of the denominators:

$$12 = 3 \times 4 = 3 \cdot 2^2$$
$$90 = 10 \times 9 = 2 \cdot 5 \cdot 3^2$$
$$60 = 15 \times 4 = 3 \cdot 5 \cdot 2^2$$

Note that the largest exponent on the factor 2 is 2 and on the 3 is also 2. So 2^2 and 3^2 are included in the LCD. Since the largest exponent on 5 is 1, the factor 5 is also included in the LCD. It follows that

$$\text{LCD} = 2^2 \cdot 3^2 \cdot 5 = 180$$

The procedure for algebraic fractions is similar.

Example 1 Find the LCD of the fractions

$$\frac{5}{6a^2b} \quad \text{and} \quad \frac{4}{9ab^3}$$

Solution. As in the case of arithmetic fractions, we first factor the denominators:

$$6a^2b = 2 \cdot 3 \cdot a^2 \cdot b$$
$$9ab^3 = 3^2 \cdot a \cdot b^3$$

To construct the LCD, observe that the factors are 2, 3, a, and b. The largest exponent on the factor 2 is 1 and the largest exponent on the factor 3 is 2. So 2 and 3^2 are included in the LCD. Also, the largest exponent on a is 2 and the largest exponent on b is 3. So a^2 and b^3 are included in the LCD. It follows that

$$\text{LCD} = 2 \cdot 3^2 \cdot a^2 \cdot b^3 = 18a^2b^3$$

If the fractions contain polynomial denominators, these have to be factored first. Otherwise the procedure for finding the LCD is similar.

Example 2 Find the LCD of the following fractions:

$$\frac{3x}{x^2 + 6x + 8}, \quad \frac{5}{x^2 + x - 12}, \quad \text{and} \quad \frac{1}{x^2 - 6x + 9}$$

Solution. Once again, we factor each of the denominators:

$$x^2 + 6x + 8 = (x + 2)(x + 4)$$
$$x^2 + x - 12 = (x - 3)(x + 4)$$
$$x^2 - 6x + 9 = (x - 3)^2$$

The largest exponent on the factor $x - 3$ is 2 and the largest exponent on each of the factors $x + 2$ and $x + 4$ is 1. So

$$\text{LCD} = (x - 3)^2(x + 2)(x + 4)$$

Addition and Subtraction of Fractions

To add (or subtract) two or more fractions, we find the LCD of the fractions and then change each fraction (if necessary) to an equivalent fraction having the LCD for its denominator. Next, we add (or subtract) the numerators of the fractions, placing the result over the LCD. Finally, we reduce the resulting fraction to lowest terms.

Example 3 Combine

$$\frac{5}{x - y} - \frac{3}{x + y}$$

Solution. Since the two fractions contain distinct prime denominators, each with exponent 1, we have

$$\text{LCD} = (x - y)(x + y)$$

To obtain the necessary equivalent fractions, we multiply numerator and denominator of the first fraction by $x + y$:

$$\frac{5}{x - y} = \frac{5(x + y)}{(x - y)(x + y)}$$

The second fraction is adjusted similarly:

$$-\frac{3}{x + y} = -\frac{3(x - y)}{(x + y)(x - y)}$$

The resulting equivalent fractions are now combined as follows:

$$\frac{5}{x-y} - \frac{3}{x+y}$$

$$= \frac{5(x+y)}{(x-y)(x+y)} - \frac{3(x-y)}{(x+y)(x-y)} \qquad \text{LCD} = (x-y)(x+y)$$

$$= \frac{5(x+y) - 3(x-y)}{(x-y)(x+y)} \qquad \text{Combining numerators}$$

To simplify this result, we multiply out the expressions in the numerator and combine like terms. The denominator, however, is left in its present form.

$$\frac{5(x+y) - 3(x-y)}{(x-y)(x+y)} = \frac{5x + 5y - 3x + 3y}{(x-y)(x+y)}$$

$$= \frac{2x + 8y}{(x-y)(x+y)}$$

Example 4 Combine

$$\frac{R}{R-C} - \frac{R^2}{R^2 - C^2}$$

Solution. If we factor the denominator of the second fraction, we get

$$\frac{R}{R-C} - \frac{R^2}{(R-C)(R+C)}$$

The LCD is the product of the factors $R - C$ and $R + C$, each with exponent 1:

$$\text{LCD} = (R-C)(R+C)$$

Note that the denominator of the second fraction is the LCD. So we need to change only the first fraction:

$$\frac{R}{R-C} = \frac{R(R+C)}{(R-C)(R+C)}$$

Combining the resulting fractions, we get

$$\frac{R}{R-C} - \frac{R^2}{R^2 - C^2}$$

$$= \frac{R(R+C)}{(R-C)(R+C)} - \frac{R^2}{(R-C)(R+C)} \qquad \text{LCD} = (R-C)(R+C)$$

$$= \frac{R(R+C) - R^2}{(R-C)(R+C)} \qquad \text{Combining numerators}$$

$$= \frac{R^2 + RC - R^2}{(R-C)(R+C)} \qquad \text{Distributive law}$$

$$= \frac{RC}{(R-C)(R+C)} \qquad \text{Simplifying numerator}$$

Note that the denominator is again left in factored form.

A good alternative to this procedure is to supply each fraction with any factor that is missing.

Example 4 **Alternate solution.** To combine the fractions

$$\frac{R}{R - C} - \frac{R^2}{R^2 - C^2}$$

we factor the denominator of the second fraction, as we did before:

$$\frac{R}{R - C} - \frac{R^2}{(R - C)(R + C)}$$

Now observe that both denominators contain the factor $R - C$. The denominator of the second fraction contains the factor $R + C$, but the denominator of the first fraction does not. Supplying this factor, we get

$$\frac{R(R + C)}{(R - C)(R + C)} - \frac{R^2}{(R - C)(R + C)} = \frac{R(R + C) - R^2}{(R - C)(R + C)}$$
$$= \frac{RC}{(R - C)(R + C)}$$

The next example illustrates both the first method and the alternative.

Example 5 Combine the following fractions:

$$\frac{1}{x - y} + \frac{1 - x}{x^2 - xy}$$

Solution. *First method:* The denominator of the second fraction has common factor x and can be written $x(x - y)$. Since $x - y$ is in the denominator of the first fraction,

$$\text{LCD} = x(x - y)$$

So the first fraction becomes

$$\frac{1}{x - y} = \frac{1 \cdot x}{(x - y) \cdot x} = \frac{x}{x(x - y)}.$$

We now get

$$\frac{1}{x - y} + \frac{1 - x}{x^2 - xy} = \frac{x}{x(x - y)} + \frac{1 - x}{x(x - y)}$$
$$= \frac{x + (1 - x)}{x(x - y)} = \frac{1}{x(x - y)}$$

Alternate method: From the factored form

$$\frac{1}{x-y} + \frac{1-x}{x(x-y)}$$

we see that both denominators contain the factor $x - y$. In addition, the denominator of the second fraction contains the factor x, but the denominator of the first fraction does not. Supplying this factor, we get

$$\frac{1}{x-y} + \frac{1-x}{x(x-y)} = \frac{1 \cdot x}{(x-y) \cdot x} + \frac{1-x}{x(x-y)}$$
$$= \frac{x+1-x}{x(x-y)} = \frac{1}{x(x-y)}$$

Example 6 Combine the following fractions:

$$\frac{2}{v+1} - \frac{2v}{v^2-1} + \frac{1}{v-1}$$

Solution. The denominator of the middle fraction is a difference of two squares and can be written as

$$v^2 - 1 = (v-1)(v+1)$$

It follows that

$$\text{LCD} = (v-1)(v+1)$$

If we now supply the missing factors in the first and third fraction, we get

$$\frac{2}{v+1} - \frac{2v}{(v-1)(v+1)} + \frac{1}{v-1}$$
$$= \frac{2(v-1)}{(v+1)(v-1)} - \frac{2v}{(v-1)(v+1)}$$
$$+ \frac{1 \cdot (v+1)}{(v-1)(v+1)} \qquad \text{LCD} = (v-1)(v+1)$$
$$= \frac{2(v-1) - 2v + (v+1)}{(v-1)(v+1)} \qquad \text{Combining numerators}$$
$$= \frac{2v - 2 - 2v + v + 1}{(v-1)(v+1)} \qquad \text{Distributive law}$$
$$= \frac{v-1}{(v-1)(v+1)} \qquad \text{Simplifying numerator}$$
$$= \frac{\overset{1}{\cancel{v-1}}}{\underset{1}{\cancel{(v-1)}}(v+1)} \qquad \text{Cancellation}$$
$$= \frac{1}{v+1}$$

Common errors

1. Forgetting to change all the signs when subtracting numerators. For example,

$$\frac{x}{y} - \frac{2-a}{y} \quad \text{is not equal to} \quad \frac{x-2-a}{y}$$

We need to subtract the *entire* numerator of the second fraction to obtain

$$\frac{x}{y} - \frac{2-a}{y} = \frac{x-(2-a)}{y} = \frac{x-2+a}{y}$$

2. Incorrectly writing

$$\frac{1}{x} + \frac{1}{y} \quad \text{as} \quad \frac{1}{x+y}$$

The correct procedure is

$$\frac{1y}{xy} + \frac{1x}{yx} = \frac{x+y}{xy} \qquad \text{LCD} = xy$$

3. Incorrectly writing

$$\frac{1}{x+y} \quad \text{as} \quad \frac{1}{x} + \frac{1}{y}$$

The fraction

$$\frac{1}{x+y}$$

is already in simplest form and cannot be split into two fractions.

Example 7 The energy E of an electron in an atom is

$$E = \frac{e^2}{4\pi\varepsilon_0 r} - \frac{e^2}{8\pi\varepsilon_0 r} \qquad \varepsilon = \text{epsilon}$$

where e is the charge of an electron, r is the radius of the orbit, and ε_0 is a constant. Simplify the expression for E.

Solution.

$$\begin{aligned} E &= \frac{e^2}{4\pi\varepsilon_0 r} - \frac{e^2}{8\pi\varepsilon_0 r} \\ &= \frac{2 \cdot e^2}{2 \cdot 4\pi\varepsilon_0 r} - \frac{e^2}{8\pi\varepsilon_0 r} \\ &= \frac{2e^2 - e^2}{8\pi\varepsilon_0 r} = \frac{e^2}{8\pi\varepsilon_0 r} \end{aligned}$$

Exercises / Section 9.3

In Exercises 1–16, find the lowest common denominator of the given fractions. *Do not combine the fractions.*

1. $\frac{3}{2x^2y}, \frac{a}{4xy^2}, \frac{c}{8xy}$

2. $\frac{x}{6a^3b}, \frac{y}{9a^2b^2}, \frac{z}{9a^2b^3}$

3. $\frac{L}{6R^2C^3}, \frac{2L}{15R^3C^2}, \frac{3L}{3R^2C^2}$

4. $\frac{a}{10m^4n}, \frac{b}{15m^3n^2}, \frac{c}{5mn}$

5. $\frac{a}{xy - y^2}, \frac{b}{x - y}$

6. $\frac{3}{2L - 2V}, \frac{4}{L - V}$

7. $\frac{3}{4P^2 - Q^2}, \frac{b}{2P - Q}$

8. $\frac{s}{s + 3t}, \frac{t}{s^2 - 9t^2}$

9. $\frac{1}{x^2 + x - 2}, \frac{2}{x - 1}$

10. $\frac{c}{y + 3}, \frac{d}{y^2 + y - 6}$

11. $\frac{2t}{2r^2 - rs - s^2}, \frac{t}{2r - 2s}$

12. $\frac{Q}{2P - 4W}, \frac{R}{2P^2 - 3PW - 2W^2}$

13. $\frac{2}{x - y}, \frac{x}{x + y}, \frac{4x}{2x^2 + 5xy + 3y^2}$

14. $\frac{3}{a - 2b}, \frac{6}{a + 2b}, \frac{8}{2a^2 + 7ab + 6b^2}$

15. $\frac{x}{x^2 + xy - 2y^2}, \frac{y}{2x^2 + 7xy + 6y^2}, \frac{xy}{2x^2 + xy - 3y^2}$

16. $\frac{1}{a^2 + 5ab + 6b^2}, \frac{5}{2a^2 + 5ab + 2b^2}, \frac{7}{2a^2 + 7ab + 3b^2}$

In Exercises 17–50, combine the given fractions and simplify. (Leave the denominator of the answer in factored form.)

17. $\frac{1}{2x} + \frac{3}{2x}$

18. $\frac{4}{5a} - \frac{1}{5a}$

19. $\frac{1}{2} - \frac{1}{x}$

20. $\frac{1}{a} + \frac{1}{3}$

21. $\frac{1}{a} + \frac{1}{a^2}$

22. $\frac{3}{b} - \frac{4}{b^2}$

23. $\frac{1}{2V} - \frac{1}{4V} + \frac{1}{V}$

24. $\frac{1}{6C} - \frac{1}{2C} + \frac{1}{C}$

25. $\frac{1}{2x} + \frac{1}{2x^2}$

26. $\frac{1}{4y} + \frac{1}{2y^2}$

27. $\frac{3}{2xy} - \frac{1}{4xy^2}$

28. $\frac{1}{6x^2y} - \frac{5}{12xy^3}$

29. $\frac{4}{9R^3C^2} + \frac{5}{12R^2C^4}$

30. $\frac{7}{10L^2V^4} - \frac{4}{15L^3V^3}$

31. $\frac{1}{a^3b^2c} + \frac{2}{a^2bc^3} - \frac{3}{ab^2c^2}$

32. $\frac{4}{r^4s^2t^3} - \frac{3}{r^3s^2t^4} + \frac{1}{r^2s^3t^2}$

33. $\frac{1}{x + 4} - \frac{1}{x - 2}$

34. $\frac{1}{y - 3} - \frac{1}{y + 5}$

35. $\frac{2}{L - 4} + \frac{4}{L - 7}$

36. $\frac{3}{C + 6} - \frac{1}{C - 5}$

37. $\frac{b}{a^2 + ab} + \frac{1}{a + b}$

38. $\frac{1}{p - q} - \frac{q}{p^2 - pq}$

39. $\frac{z}{w^2 - wz} - \frac{1}{w - z}$

40. $\frac{2}{L^2 - 2L} + \frac{L}{L - 2}$

41. $\dfrac{s}{s^2 + 3st + 2t^2} + \dfrac{1}{s + t}$

42. $\dfrac{4s_2}{s_1^2 + 2s_1s_2 - 3s_2^2} + \dfrac{1}{s_1 + 3s_2}$

43. $\dfrac{v}{v - 2} - \dfrac{2v^2}{v^2 - 4} - \dfrac{2}{v + 2}$

44. $\dfrac{T_0}{T_0 - 2} - \dfrac{2}{T_0 + 2} - \dfrac{T_0^2}{T_0^2 - 4}$

45. $\dfrac{T_1}{T_1 - T_2} - \dfrac{T_2}{T_1 + T_2} - \dfrac{2T_2^2}{T_1^2 - T_2^2}$

46. $\dfrac{3xy}{x^2 - y^2} + \dfrac{x}{y - x} + \dfrac{x - y}{x + y}$

47. $\dfrac{4}{x^2 - 7x + 12} + \dfrac{1}{3 - x}$

48. $\dfrac{3}{2 - x} + \dfrac{1}{2x^2 - 5x + 2}$

49. $\dfrac{3}{2x + 2y} + \dfrac{2}{x^2 - 2xy - 3y^2}$

50. $\dfrac{4}{2x^2 - xy - 6y^2} - \dfrac{2}{3x - 6y}$

51. Find the perimeter of a triangular plate with sides $\frac{5}{24}$ in., $\frac{5}{16}$ in., and $\frac{3}{8}$ in.

52. The deflection D of a certain beam is given by

$$D = K\left(\frac{x^3}{16} - \frac{x^4}{20}\right)$$

Write D as a single fraction.

53. The efficiency E of a Carnot engine is

$$E = 1 - \frac{T_2}{T_1}$$

where T_1 is the intake temperature and T_2 the exhaust temperature. Write E as a single fraction.

54. The displacement of a vibrating string varies with

$$\frac{t}{T} - \frac{x}{\lambda} \qquad \lambda = \text{lambda}$$

where λ is the wavelength, T the period, x the distance from one end, and t the time. Combine the two fractions.

55. Under certain conditions, the energy E of an atomic particle has the form

$$E = K\left(\frac{a^2}{L^2} + \frac{b^2}{W^2}\right)$$

Write E as a single fraction.

56. The time t required to make a round-trip d miles each way by a boat traveling at velocity V is

$$t = \frac{d}{V - v} - \frac{d}{V + v}$$

where v is the velocity of the river. Write t as a single fraction.

57. The combined resistance R of the resistors in Figure 9.1 is

$$R = R_1 + \frac{R_2R_3}{R_2 + R_3}$$

Write R as a single fraction.

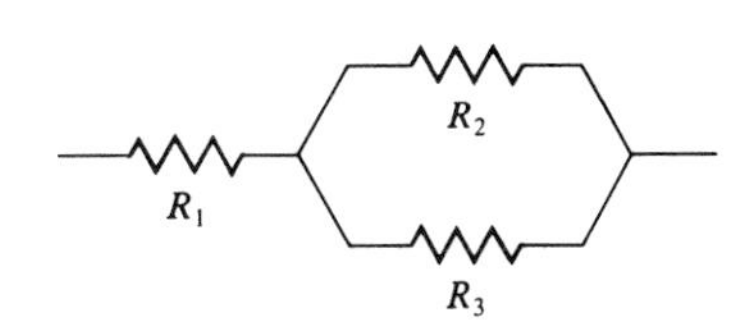

Figure 9.1

58. A hydrogen atom dropping from the nth energy level to the $(n - p)$th energy level emits a photon whose frequency is

$$k\left[\frac{1}{(n-p)^2} - \frac{1}{n^2}\right]$$

where k is a constant. Simplify this expression.

59. The distance between two heat sources with respective intensities a and b is L. The total intensity I of heat at a point between the sources is

$$I = \frac{a}{x^2} + \frac{b}{(L-x)^2}$$

where x is the distance from one of the sources. Express I as a single fraction.

9.4 Equations with Fractions

In Section 7.2 we studied the solution of equations with fractional coefficients. In this section we solve fractional equations in which the unknown appears in the denominator.

First let us recall the procedure for solving equations with fractional coefficients. Consider, for example, the equation

$$\frac{1}{2}x + 1 = \frac{1}{6}x - 2$$

The simplest way to solve this equation is to **clear fractions**. That is, we multiply both sides of the equation by 6, the LCD of all the fractions. Then we get

$$\begin{aligned} 6\left(\frac{1}{2}x + 1\right) &= 6\left(\frac{1}{6}x - 2\right) \\ 3x + 6 &= x - 12 \\ 3x - x &= -12 - 6 \\ 2x &= -18 \\ x &= -9 \end{aligned}$$

The procedure for clearing algebraic fractions in an equation is similar.

> **Clearing fractions:** To solve an equation containing fractions, *clear the fractions* by multiplying both sides of the equation by the LCD of all the fractions in the equation.

Example 1 Solve the following equation:

$$\frac{1}{x-2} - \frac{2}{x+3} = \frac{1}{x^2 + x - 6}$$

Solution. First we need to factor the denominator of the right side of the equation:

$$\frac{1}{x-2} - \frac{2}{x+3} = \frac{1}{(x+3)(x-2)}$$

We now see that the LCD $= (x+3)(x-2)$. Clearing fractions, we get

$$\frac{(x+3)(x-2)}{1}\left(\frac{1}{x-2} - \frac{2}{x+3}\right) = \frac{(x+3)(x-2)}{1} \cdot \frac{1}{(x+3)(x-2)}$$

$$\frac{(x+3)\cancel{(x-2)}}{1} \cdot \frac{1}{\cancel{x-2}} - \frac{\cancel{(x+3)}(x-2)}{1} \cdot \frac{2}{\cancel{x+3}} = \frac{\cancel{(x+3)}\cancel{(x-2)}}{1} \cdot \frac{1}{\cancel{(x+3)}\cancel{(x-2)}}$$

$$(x+3) - 2(x-2) = 1$$

$$x + 3 - 2x + 4 = 1 \qquad \text{Distributive law}$$

$$-x + 7 = 1$$

$$-x + 7 - 7 = 1 - 7 \qquad \text{Subtracting 7}$$

$$-x = -6$$

$$(-1)(-x) = (-1)(-6) \qquad \text{Multiplying by } -1$$

$$x = 6$$

Check:

Left side

$$\frac{1}{6-2} - \frac{2}{6+3} = \frac{1}{4} - \frac{2}{9} = \frac{9}{36} - \frac{8}{36} = \frac{1}{36}$$

Right side

$$\frac{1}{6^2 + 6 - 6} = \frac{1}{6^2} = \frac{1}{36} \quad \checkmark$$

Some fractional equations do not have any solution, as shown in the next example.

Example 2 Solve the equation

$$\frac{x+4}{x+3} + 2 = \frac{1}{x+3}$$

Solution. Since the LCD $= x + 3$, we get

$$\frac{x+3}{1}\left(\frac{x+4}{x+3} + 2\right) = \frac{x+3}{1} \cdot \frac{1}{x+3}$$

$$\frac{\cancel{x+3}}{1} \cdot \frac{x+4}{\cancel{x+3}} + (x+3)(2) = \frac{\cancel{x+3}}{1} \cdot \frac{1}{\cancel{x+3}}$$

$$x + 4 + (x+3)(2) = 1$$

$$x + 4 + 2x + 6 = 1$$

$$3x + 10 = 1$$
$$3x + 10 - 10 = 1 - 10$$
$$3x = -9$$
$$\frac{3x}{3} = \frac{-9}{3}$$
$$x = -3$$

Check: If we substitute $x = -3$ in the given equation, we obtain a 0 for each denominator. Since division by 0 is not allowed, we conclude that the equation has no solution. The only (apparent) root is called an **extraneous root.**

Extraneous root

Remark on extraneous roots. Whenever we multiply both sides of an equation by an expression containing the unknown, we may get an extraneous root. It is easy to see why: The equation

$$x - 1 = 0$$

has only one root, $x = 1$. If both sides are multiplied by $x - 2$, we get

$$(x - 1)(x - 2) = 0$$

which has two roots, $x = 1$ and $x = 2$. The "new" root, $x = 2$, does not, of course, satisfy the given equation $x - 1 = 0$. While we normally avoid multiplying both sides of an equation by an expression containing the unknown, such multiplications are necessary with fractional equations (to clear fractions), and extraneous roots may result.

Formulas

Many formulas containing fractions can be solved for one letter in terms of the other letters by first clearing fractions. In the process, it may be necessary to factor out the letter we are solving for. For example, if a certain equation simplifies to

$$ab + ac = d$$

and if we want to solve for a, then we must factor the left side to obtain

$$a(b + c) = d$$

We then divide both sides of the equation by the coefficient $b + c$ to obtain

$$a = \frac{d}{b + c}$$

Consider the next example.

Example 3 Solve the following formula for V_1:

$$\frac{1}{V_0} - \frac{1}{V_1 + V_2} = k$$

Solution. We clear fractions by multiplying both sides by the LCD = $V_0(V_1 + V_2)$:

$$\frac{V_0(V_1 + V_2)}{1}\left(\frac{1}{V_0} - \frac{1}{V_1 + V_2}\right) = kV_0(V_1 + V_2) \qquad \text{Clearing fractions}$$

$$\frac{\cancel{V_0}(V_1 + V_2)}{1}\frac{1}{\cancel{V_0}} - \frac{V_0\cancel{(V_1 + V_2)}}{\cancel{V_1 + V_2}} = kV_0(V_1 + V_2) \qquad \text{Cancellation}$$

$$(V_1 + V_2) - V_0 = kV_0(V_1 + V_2)$$

$$V_1 + V_2 - V_0 = kV_0V_1 + kV_0V_2$$

$$V_1 + V_2 - V_0 - kV_0V_1 = kV_0V_1 + kV_0V_2 - kV_0V_1 \qquad \text{Subtracting } kV_0V_1$$

$$V_1 + V_2 - V_0 - kV_0V_1 = kV_0V_2$$

$$V_1 + V_2 - V_0 - kV_0V_1 - V_2 + V_0 = kV_0V_2 - V_2 + V_0$$

$$V_1 - kV_0V_1 = kV_0V_2 - V_2 + V_0$$

$$V_1(1 - kV_0) = kV_0V_2 - V_2 + V_0 \qquad \text{Common factor } V_1$$

$$\frac{V_1(1 - kV_0)}{1 - kV_0} = \frac{kV_0V_2 - V_2 + V_0}{1 - kV_0} \qquad \text{Dividing by } 1 - kV_0$$

$$V_1 = \frac{kV_0V_2 - V_2 + V_0}{1 - kV_0}$$

Less complicated formulas can often be solved by using the fact that whenever two nonzero numbers are equal, then their reciprocals are equal, or

If $a = b$, then $\frac{1}{a} = \frac{1}{b}$.

This fact is used in the next example.

Example 4 Solve the following formula for N:

$$\frac{1}{N} + \frac{1}{P + Q} = \frac{1}{P - Q}$$

Solution. We could start by clearing fractions, as usual. However, note that the letter N occurs only in the first term. So we can solve the equation by first solving for $1/N$ and then taking the reciprocals of both sides:

$$\frac{1}{N} + \frac{1}{P+Q} - \frac{1}{P+Q} = \frac{1}{P-Q} - \frac{1}{P+Q}$$

$$\frac{1}{N} = \frac{1}{P-Q} - \frac{1}{P+Q}$$

$$\frac{1}{N} = \frac{P+Q}{(P-Q)(P+Q)} - \frac{P-Q}{(P+Q)(P-Q)}$$

$$\frac{1}{N} = \frac{P+Q-P+Q}{(P-Q)(P+Q)}$$

$$\frac{1}{N} = \frac{2Q}{P^2-Q^2}$$

$$N = \frac{P^2-Q^2}{2Q} \quad \text{Taking reciprocals}$$

Example 5 After the engine of a motorboat is shut off, its velocity v after t seconds can be obtained from the formula

$$\frac{1}{v} = \frac{1}{v_0} + kt$$

where v_0 is the velocity when the engine is shut off and k a constant. Solve this formula for v.

Solution. The LCD for the fractions in the formula is vv_0. Clearing fractions, we get

$$vv_0\left(\frac{1}{v}\right) = vv_0\left(\frac{1}{v_0} + kt\right) \quad \text{Clearing fractions}$$

$$\frac{\cancel{v}v_0}{\cancel{v}} = \frac{v\cancel{v_0}}{\cancel{v_0}} + vv_0kt \quad \text{Distributive law}$$

$$v_0 = v + vv_0kt$$

$$v_0 = v(1 + v_0kt) \quad \text{Common factor } v$$

$$v(1 + v_0kt) = v_0 \quad \text{Switching sides}$$

$$\frac{v(1 + v_0kt)}{1 + v_0kt} = \frac{v_0}{1 + v_0kt} \quad \text{Dividing by } 1 + v_0kt$$

$$\frac{v\cancel{(1 + v_0kt)}}{\cancel{1 + v_0kt}} = \frac{v_0}{1 + v_0kt} \quad \text{Cancellation}$$

$$v = \frac{v_0}{1 + v_0kt}$$

Exercises / Section 9.4

In Exercises 1–24, solve the given equations.

1. $\frac{1}{x} + \frac{1}{2} = 1$

2. $\frac{1}{x} - \frac{1}{2} = \frac{1}{2}$

3. $3 + \frac{8}{x} = \frac{1}{3}$

4. $\frac{2}{3} - \frac{3}{x} = 1$

5. $\frac{1}{x+1} = \frac{1}{2x}$

6. $\frac{2}{x-2} = \frac{1}{x}$

7. $\frac{4}{3x} = \frac{1}{x+3}$

8. $\frac{3}{5x} = \frac{1}{x-2}$

9. $\frac{x}{x-2} = \frac{x+2}{x+4}$

10. $\frac{x-1}{x+3} = \frac{x}{x+2}$

11. $\frac{2x}{x+3} = \frac{2x+4}{x+2}$

12. $\frac{x+1}{x-2} = \frac{x+3}{x-2}$

13. $\frac{2x}{x-2} + \frac{1}{x+3} = 2$

14. $\frac{1}{x+1} - \frac{2x}{x-2} = -2$

15. $\frac{2}{x+2} - \frac{3x}{x-1} + 3 = 0$

16. $\frac{x}{x-1} - \frac{2}{x-3} = 1$

17. $\frac{x+1}{x} + \frac{x-2}{x+1} = 2$

18. $\frac{x}{x-2} + \frac{x-1}{x} = 2$

19. $\frac{x}{x-2} - \frac{x+1}{x^2-4} = 1$

20. $\frac{2}{x-3} + \frac{x+1}{x^2-9} = \frac{1}{x+3}$

21. $\frac{2}{x-4} + \frac{2}{x+4} = \frac{x+1}{x^2-16}$

22. $\frac{2}{x+2} + \frac{x}{x-3} = \frac{x^2}{x^2-x-6}$

23. $\frac{4}{x-3} - \frac{x+30}{x^2+x-12} = \frac{5}{x+4}$

24. $\frac{3}{x+4} - \frac{1}{x+2} = \frac{3-x}{x^2+6x+8}$

In Exercises 25–38, solve the given formulas for the indicated letter.

25. $\frac{1}{a} + \frac{1}{b} = 2$, a

26. $\frac{1}{P} - \frac{1}{V} = 1$, P

27. $\frac{1}{L} + \frac{1}{2R} = \frac{1}{C}$, R

28. $\frac{1}{S} - \frac{2}{5T} = 1$, S

29. $\frac{1}{N} + 1 = \frac{R}{R-Q}$, N

30. $1 - \frac{1}{y} = \frac{a}{a-b}$, y

31. $\frac{2}{z} + \frac{1}{a+b} = \frac{1}{a-b}$, z

32. $\frac{1}{C_1} = \frac{1}{C_2} + \frac{1}{C_3} + \frac{1}{C_4}$, C_2

33. $v_2 = \frac{v_1}{v_1+v_2}$, v_1

34. $\frac{1}{R_0} = \frac{R_1+1}{R_1}$, R_1

35. $s = \frac{m_1-m_2}{m_1+m_2}$, m_1

36. $\frac{1}{V} = \frac{p_2+p_1}{p_2-p_1}$, p_2

37. $\dfrac{1}{a-b} + 1 = c, \quad a$

38. $2r - \dfrac{1}{c-d} = 1, \quad c$

39. Charles's law, which arises in the study of the expansion of a gas, is

$$\frac{P_1V_1}{T_1} = \frac{P_2V_2}{T_2}$$

Solve this formula for T_1.

40. The average acceleration a of a body whose velocity is v_1 at time t_1 and whose velocity increases to v_2 at time t_2 is given by

$$a = \frac{v_2 - v_1}{t_2 - t_1}$$

Solve this formula for t_2.

41. The energy E required to expand a gas is computed from the formula

$$E = \frac{p_1V_1 - p_2V_2}{a - 1}$$

Solve this formula for p_1.

42. The following formula arises in the loran system of navigation:

$$r = a - \frac{b}{y}$$

Solve this formula for y.

43. If two thin lenses having respective focal lengths f_1 and f_2 are placed in contact, then the focal length f of the combination is related to f_1 and f_2 by the formula

$$\frac{1}{f} = \frac{1}{f_1} + \frac{1}{f_2}$$

Solve for f.

44. The combined resistance R of two resistors in parallel can be found from the formula

$$\frac{1}{R} = \frac{1}{R_1} + \frac{1}{R_2}$$

Solve for R.

45. The combined capacitance C of the capacitors in Figure 9.2 can be found from the formula

$$\frac{1}{C} = \frac{1}{C_1} + \frac{1}{C_2 + C_3}$$

Solve for C.

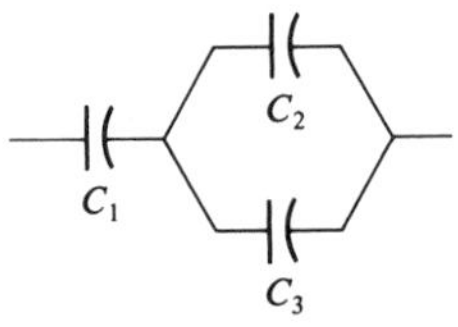

Figure 9.2

46. A special form of Bernoulli's principle from fluid dynamics is

$$\frac{p}{Dg} + \frac{v^2}{2g} + y = k$$

Solve for D.

47. If I_M and I_m represent the maximum and minimum quantities of light striking a photocell, then the *relative polarization* of the incident light is

$$P = \frac{I_M - I_m}{I_M + I_m}$$

Solve for I_M.

48. The focal length f of a double convex lens is related to the respective radii r_1 and r_2 of the spherical surfaces of the lenses by the formula

$$\frac{1}{f} = (1 - n)\left(\frac{1}{r_1} + \frac{1}{r_2}\right)$$

where n is the index of refraction. Solve for r_1.

49. The pitch diameter D of a gear is given by

$$D = \frac{D_0 N}{N + 2}$$

where D_0 is the outside diameter of the gear and N the number of teeth. Solve for N.

Review Exercises / Chapter 9

In Exercises 1–8, reduce each given fraction to lowest terms.

1. $\dfrac{-8V^8W^6}{24V^{10}W^4}$

2. $\dfrac{b^3C_1^2C_2^3}{b^4C_1^5C_2}$

3. $\dfrac{P_0^2 - P_0^3}{P_0 - P_0^2}$

4. $\dfrac{v_1^2 - v_2^2}{v_1 + v_2}$

5. $\dfrac{4m_1 - m_2}{16m_1^2 - m_2^2}$

6. $\dfrac{2s^2 - st - t^2}{2s - 2t}$

7. $\dfrac{p^2 - q^2}{3p^2 - 8pq + 5q^2}$

8. $\dfrac{2w^2 - 3wy - 2y^2}{3w^2 - 7wy + 2y^2}$

In Exercises 9–34, perform the indicated operations and simplify.

9. $\dfrac{y^2}{b^4} \cdot \dfrac{b^2}{y^6}$

10. $\dfrac{3R^6C^2}{5} \cdot \dfrac{20}{R^6C^4}$

11. $\dfrac{4a^2b^2}{3vw^2} \div \dfrac{6ab^6}{v^4w}$

12. $\dfrac{15m^3n^2}{8P^2Q^3} \div \dfrac{9mn}{16P^4Q}$

13. $\dfrac{a^2}{bp - bq} \cdot \dfrac{2p - 2q}{a^4}$

14. $\dfrac{4V^2 - L^2}{M^3} \cdot \dfrac{3M}{2V + L}$

15. $\dfrac{2s - 6t}{4a^3} \div \dfrac{s^2 - 9t^2}{8a^2}$

16. $\dfrac{ac^4}{3m + 12n} \div \dfrac{3ac^3}{m^2 - 16n^2}$

17. $\dfrac{b^2V^2 - 1}{2c^2 + 7cd - 4d^2} \cdot \dfrac{2c^2 + cd - d^2}{1 - bV}$

18. $\dfrac{1 - d^2v^2}{2s_1^2 + 3s_1s_2 - 2s_2^2} \cdot \dfrac{2s_1^2 - 3s_1s_2 + s_2^2}{dv - 1}$

19. $\dfrac{3a^2 - 5ab + 2b^2}{x^2 - 4y^2} \div \dfrac{6a^2 - ab - 2b^2}{x + 2y}$

20. $\dfrac{x^2 - 4x - 5}{2s + 2t} \cdot \dfrac{x^2 - 3x + 2}{4x + 4} \cdot \dfrac{s + t}{x^2 - 7x + 10}$

21. $\dfrac{1}{2V} - \dfrac{3}{4V}$

22. $\dfrac{2}{x^2 y} - \dfrac{1}{xy^2}$

23. $\dfrac{5}{6abc} + \dfrac{1}{9a^2bc^2}$

24. $\dfrac{2}{R_1^2 R_2} + \dfrac{4}{R_1 R_2^3}$

25. $\dfrac{1}{x - 4} - \dfrac{2}{x + 3}$

26. $\dfrac{1}{V - 6} + \dfrac{1}{V - 5}$

27. $\dfrac{3}{2s_1 - 2s_2} - \dfrac{1}{s_1 - s_2}$

28. $\dfrac{2}{3V_1 + 3V_2} + \dfrac{2}{V_1 + V_2}$

29. $\dfrac{s}{s - 3} - \dfrac{18}{s^2 - 9} - \dfrac{s}{s + 3}$

30. $\dfrac{32}{v^2 - 16} - \dfrac{v}{v - 4} + \dfrac{v}{v + 4}$

31. $\dfrac{2V_a + 2V_b}{V_a^2 + 2V_aV_b - 8V_b^2} - \dfrac{1}{V_a + 4V_b}$

32. $\dfrac{3}{2s - 2t} - \dfrac{t}{s^2 + st - 2t^2}$

33. $\dfrac{2}{y^2 + yz - 6z^2} + \dfrac{3}{y^2 + 4yz + 3z^2}$

34. $\dfrac{4}{R^2 - 2RS - 3S^2} + \dfrac{1}{2R^2 - 5RS - 3S^2}$

In Exercises 35–42, solve each equation.

35. $\dfrac{1}{2x} - \dfrac{1}{3x} = \dfrac{1}{12}$

36. $\dfrac{x}{x - 1} + \dfrac{1}{x} = 1$

37. $\dfrac{x}{x + 3} + \dfrac{2}{x + 1} = 1$

38. $\dfrac{x}{x - 3} + \dfrac{1}{x + 1} = 1$

39. $\dfrac{3}{x - 1} - \dfrac{x + 2}{x^2 - 1} = \dfrac{1}{x + 1}$

40. $\dfrac{x + 1}{x^2 - 4} + \dfrac{1}{x - 2} = \dfrac{1}{x + 2}$

41. $\dfrac{1}{x - 3} - \dfrac{1}{x^2 - 5x + 6} = \dfrac{2}{x - 2}$

42. $\dfrac{1}{x - 2} - \dfrac{2}{x + 3} = \dfrac{4}{x^2 + x - 6}$

In Exercises 43–46, solve each formula for the indicated letter.

43. $\dfrac{1}{V} - \dfrac{3}{P} = \dfrac{1}{4}$, P

44. $\dfrac{P}{N} - \dfrac{1}{S - T} = \dfrac{1}{S + T}$, N

45. $P_0 = \dfrac{P_1}{P_1 - P_2}$, P_1

46. $\dfrac{1}{L} = \dfrac{l_2 - l_1}{l_2 + l_1}$, l_2

47. Simplify the following fraction from a problem in mechanics:

$$\frac{\frac{1}{2} m_1 v^2(m_1 + m_2) - \frac{1}{2} m_1^2 v^2}{\frac{1}{2} m_1 v^2(m_1 + m_2)}$$

48. A company produces

$$\frac{aN^2}{4a + 4b}$$

items at a cost of $a/(2a + 2b)$ dollars. Find a simplified expression for the cost of each item.

49. The mass of a system moving at velocity $ab/(m_1 + m_2)$ is

$$\frac{m_1^2 - m_2^2}{a^2b^2}$$

Find a simplified expression for the momentum, defined to be mass times velocity.

50. The deflection D of a beam is

$$D = k\left(\frac{x^3}{12} - \frac{x^2}{16}\right)$$

Write D as a single fraction.

51. The combined capacitance of the capacitors in Figure 9.3 is given by

$$C_1 + \frac{C_2C_3}{C_2 + C_3}$$

Write this expression as a single fraction.

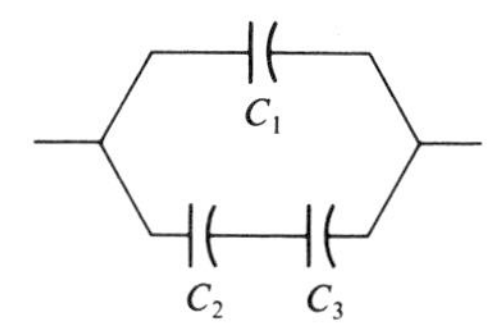

Figure 9.3

52. The following expression arises in the study of the quantum theory of light:

$$\frac{m_0c}{h}\left(\frac{1}{\lambda} - \frac{1}{\lambda'}\right) \qquad \lambda = \text{lambda}$$

Write this expression as a single fraction.

53. According to Hooke's law, the force F required to stretch a spring x units is given by $F = kx$; the constant k is called the *spring constant*. If two springs with respective spring constants k_1 and k_2 are connected as shown in Figure 9.4, then the spring constant k of the combination is related to k_1 and k_2 by the formula

$$\frac{1}{k} = \frac{1}{k_1} + \frac{1}{k_2}$$

Solve this formula for k.

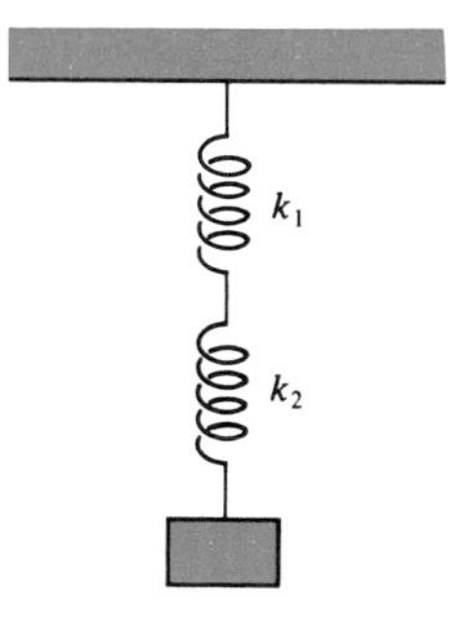

Figure 9.4

54. The combined capacitance C of two capacitors in series can be found from the formula

$$\frac{1}{C} = \frac{1}{C_1} + \frac{1}{C_2}$$

where C_1 and C_2 are the individual capacitances. Solve for C.

55. The focal length f of a lens is related to the object distance p and the image distance q by the formula

$$\frac{1}{f} = \frac{1}{p} + \frac{1}{q}$$

Solve for q.

56. The frequency of the sound of an ambulance siren when the ambulance is approaching is given by

$$F_1 = \frac{132{,}400}{331 - v}$$

and, when the ambulance is moving away, it is given by

$$F_2 = \frac{132{,}400}{331 + v}$$

(this is the Doppler effect). Find a simplified expression for the difference $F_1 - F_2$ in the frequency.

57. The relationship between the pressure P and the volume V of a gas is

$$\left(P + \frac{a}{V^2}\right)(V - b) = K$$

where a, b, and K are constants. Solve this formula for P.

Cumulative Review Exercises / Chapters 7–9

1. Solve for y: $\frac{1}{2}(3y - 4) = \frac{1}{6} - \frac{1}{3}y$

2. Solve for x: $ax - \frac{1}{2} = 3\left(ax - \frac{b}{2}\right)$

3. Solve the following inequality:

$$-2(x + 5) \geq -3(1 - 4x)$$

In Exercises 4–7, factor each of the given expressions.

4. $17S_a^4S_b^2 - 51S_a^2S_b^3 + 17S_aS_b$

5. $3 - 3T_0^2$

6. $2a^2 - 18a + 28$

7. $6t_1^2 + 28t_1t_2 - 48t_2^2$

8. Reduce the following fraction:

$$\frac{2V^2 - 2P^2}{3P^2 + 5PV - 8V^2}$$

In Exercises 9–12, perform the indicated operations.

9. $\frac{4a^2 + 2ab}{a - b} \cdot \frac{b^2 - a^2}{2a + b}$

10. $\frac{3x^2 - 7x + 4}{4x^2 - 8x - 5} \div \frac{2x^2 - 5x + 3}{2x^2 + 7x - 30}$

11. $\frac{w}{w - 1} + \frac{1}{w}$

12. $\frac{A + 1}{A - 3} - \frac{6A + 6}{A^2 - 9} + \frac{A}{A + 3}$

13. Solve for m_0: $E = (m - m_0)c^2$

14. Solve for b_1: $A = \frac{1}{2}h(b_1 + b_2)$

15. Solve for P: $\frac{3}{P - 3} - \frac{P + 2}{P^2 + P - 12} = \frac{4}{P + 4}$

16. Solve for V_1: $\frac{2}{V_1 - V_2} - \frac{1}{V_1 + V_2} = \frac{aV_1}{V_1^2 - V_2^2}$

17. The voltage V in a certain one-loop circuit is

$$V = IR_1 + IR_2 + IR_3$$

Factor the expression for V.

18. A metal disk of radius r_2 has a balancing hole of radius r_1. The area A of one side is

$$A = \pi r_2^2 - \pi r_1^2$$

Factor the expression for A.

19. The following expression arises in the theory of blackbody radiation: $aT_0^4 - aT_1^4$. Factor this expression.

20. Just as electrical components can be connected in parallel, thin tubes that branch out and come together again are said to be connected in parallel. Each tube offers a resistance to the flow of the fluid. If r_1 and r_2

denote the respective resistances of two tubes in parallel, then the reciprocal of the combined resistance r is

$$\frac{1}{r} = \frac{1}{r_1} + \frac{1}{r_2}$$

Solve this equation for r_1.

21. In a certain circuit the voltage V varies with time t by the formula

$$V = \frac{t^2 + 4t + 4}{t + 4}$$

and the resistance R varies according to

$$R = \frac{t + 2}{2t + 8}$$

Find a simplified expression for the current $i = V/R$.

22. A piece of wire 113.4 cm long is bent into the shape of a rectangle 2.5 times as long as it is wide. Find the dimensions of the rectangle.

23. One brine solution contains 10% salt by volume and another 15%. How many liters of each must be mixed to produce 100 L containing 12% salt?

CHAPTER 10

Exponents and Radicals

10.1 Zero and Negative Integral Exponents

We first encountered exponents and some of the fundamental operations with exponents in Section 5.1. In this section we continue with a study of both zero and negative exponents. Before we do so, however, we must first recall the basic laws of exponents from Section 5.1.

Laws of exponents

$$a^m a^n = a^{m+n} \tag{10.1}$$

$$\frac{a^m}{a^n} = a^{m-n}, \qquad a \neq 0 \tag{10.2}$$

$$(a^m)^n = a^{mn} \tag{10.3}$$

$$(ab)^n = a^n b^n \tag{10.4}$$

$$\left(\frac{a}{b}\right)^n = \frac{a^n}{b^n}, \qquad b \neq 0 \tag{10.5}$$

The next example illustrates the basic laws.

Example 1

(10.1): $a^3 a^4 a^5 = a^{3+4+5} = a^{12}$ Adding exponents

(10.2): $\dfrac{a^8}{a^5} = a^{8-5} = a^3$ Subtracting exponents

(10.3): $(a^3)^6 = a^{3\cdot 6} = a^{18}$ Multiplying exponents

(10.4): $(2x)^3 = 2^3x^3 = 8x^3$ $\quad (ab)^n = a^n b^n$

(10.5): $\left(\frac{y}{3}\right)^3 = \frac{y^3}{3^3} = \frac{y^3}{27}$ $\quad \left(\frac{a}{b}\right)^n = \frac{a^n}{b^n}$

Reminder: $a = a^1$, so that $aa^4 = a^{1+4} = a^5$.

The laws of exponents are often used in combination, as shown in the next example.

Example 2 Simplify

$$\frac{(2xy^3)^2}{(3x^4y^2)^2}$$

Solution.

$$\frac{(2xy^3)^2}{(3x^4y^2)^2} = \frac{2^2x^2(y^3)^2}{3^2(x^4)^2(y^2)^2} \qquad (ab)^n = a^n b^n$$

$$= \frac{4x^2(y^3)^2}{9(x^4)^2(y^2)^2} \qquad 2^2 = 4,\ 3^2 = 9$$

$$= \frac{4x^2y^6}{9x^8y^4} \qquad \text{Multiplying exponents}$$

$$= \frac{4y^2\cancel{(x^2y^4)}}{9x^6\cancel{(x^2y^4)}} \qquad \text{Cancellation}$$

$$= \frac{4y^2}{9x^6}$$

As already noted, our main goal in this section is to study operations with zero and negative exponents. The problem is that our definition of power requires that an exponent be positive. As a result, an expression such as a^0 has no meaning unless the definition of exponent is extended.

To make zero and negative exponents useful in mathematics and technology, the laws of exponents must hold for these exponents as well. In particular, if the rule for multiplication is to hold for a^0, we must have

$$a^0a^2 = a^{0+2} = a^2$$

But since $1 \cdot a^2 = a^2$, it follows that $a^0 = 1$. Similarly, by the law for division,

$$\frac{a^m}{a^m} = a^{m-m} = a^0$$

But since any (nonzero) number divided by itself is equal to 1, it follows again that $a^0 = 1$.

The definition of zero exponent now leads to the definition of negative exponent. Consider, for example, the expression a^{-2}. If a^{-2} is multiplied by

a^2, then we get

$$a^{-2}a^2 = a^{-2+2} = a^0 = 1$$

But if $a^{-2}a^2 = 1$, then a^{-2} must be the reciprocal of a^2, that is,

$$a^{-2} = \frac{1}{a^2}$$

As a check, note that

$$a^{-2}a^2 = \frac{1}{a^2} \cdot a^2 = 1$$

These observations suggest the definition $a^{-n} = 1/a^n$.

Zero and negative exponents

$$a^0 = 1, \qquad a \neq 0 \tag{10.6}$$

$$a^{-n} = \frac{1}{a^n}, \qquad a \neq 0 \tag{10.7}$$

Since zero and negative exponents were defined with the basic laws of exponents in mind, these laws hold for zero and negative exponents. Consider the next example.

Example 3

(10.1): $x^{-2}x^6x^0 = x^{-2+6+0} = x^4$ — Adding exponents

(10.2): $\dfrac{x^{-3}}{x^{-7}} = x^{-3-(-7)} = x^{-3+7} = x^4$ — Subtracting exponents

(10.3): $(x^{-4})^3 = x^{-4\cdot 3} = x^{-12}$ — Multiplying exponents

(10.4): $(x^{-3}y^5)^{-2} = (x^{-3})^{-2}(y^5)^{-2}$ — $(ab)^n = a^nb^n$

$= x^6y^{-10}$ — Multiplying exponents

(10.5): $\left(\dfrac{x^{-6}}{y^4}\right)^3 = \dfrac{(x^{-6})^3}{(y^4)^3}$ — $\left(\dfrac{a}{b}\right)^n = \dfrac{a^n}{b^n}$

$= \dfrac{x^{-18}}{y^{12}}$ — Multiplying exponents

In more complicated expressions, a factor with a negative exponent may also occur in the denominator. To simplify such expressions, it is helpful to observe that

$$\frac{1}{a^{-3}} = \frac{1}{\frac{1}{a^3}} = \frac{1}{1} \cdot \frac{a^3}{1} = a^3$$

When a **factor** is moved from the numerator of a fraction to the denominator or from the denominator to the numerator, the sign of the exponent is changed:

$$\frac{a^n}{1} = \frac{1}{a^{-n}} \quad \text{and} \quad \frac{a^{-n}}{1} = \frac{1}{a^n} \tag{10.8}$$

Consider the next example:

Example 4

a. $\dfrac{1}{a^{-5}b^4} = a^5b^{-4}$ $\quad \dfrac{1}{a^{-5}} = a^5, \quad \dfrac{1}{b^4} = \dfrac{b^{-4}}{1} = b^{-4}$

b. $\dfrac{1}{3x^4y} = 3^{-1}x^{-4}y^{-1}$ By rule (10.8)

c. $\dfrac{2^{-1}}{v_1^2v_2^{-4}} = \dfrac{v_2^4}{2v_1^2}$

d. $\dfrac{3^{-1}V^{-2}}{C^{-3}L^5} = \dfrac{C^3}{3L^5V^2}$

Although not done in this example, final results are usually written without negative exponents.

The next example illustrates the evaluation of numerical expressions containing negative exponents.

Example 5 Evaluate:

a. $(2^{-1})^{-3}$ b. $5^0 \cdot 3^{-2}$ c. $\dfrac{9^{-2}}{3^{-6}}$

Solution.

a. $(2^{-1})^{-3} = 2^{(-1)(-3)}$ Multiplying exponents

$= 2^3 = 8$

b. $5^0 \cdot 3^{-2} = 1 \cdot 3^{-2}$ Since $5^0 = 1$

$= \dfrac{1}{3^2} = \dfrac{1}{9}$ By rule (10.8)

c. Since $9 = 3^2$, we have

$$9^{-2} = (3^2)^{-2} = 3^{-4}$$

So

$$\frac{9^{-2}}{3^{-6}} = \frac{(3^2)^{-2}}{3^{-6}} = \frac{3^{-4}}{3^{-6}} \qquad \text{Multiplying exponents}$$

$$= 3^{-4-(-6)} \qquad \text{Subtracting exponents}$$

$$= 3^{-4+6}$$

$$= 3^2 = 9$$

This example suggests that evaluation of expressions can be done most easily by writing the constants in prime factored form.

While negative exponents are highly useful in performing algebraic operations, final results are usually expressed without them.

Example 6 Simplify and write without negative exponents:

$$\frac{2^{-4}R^{-2}C^{-4}}{3^{-1}R^4C^{-3}}$$

Solution.

$$\frac{2^{-4}R^{-2}C^{-4}}{3^{-1}R^4C^{-3}} = \frac{3R^{-2}C^{-4}}{2^4R^4C^{-3}} \qquad \text{By rule (10.8)}$$

$$= \frac{3R^{-2}C^{-4}}{16R^4C^{-3}} \qquad 2^4 = 16$$

$$= \frac{3C^{-4}}{16R^4R^2C^{-3}} \qquad R^{-2} = \frac{1}{R^2}$$

$$= \frac{3C^{-4}}{16R^6C^{-3}} \qquad R^4R^2 = R^6$$

$$= \frac{3C^3}{16R^6C^4} \qquad \text{By rule (10.8)}$$

$$= \frac{3}{16R^6C} \qquad \text{Dividing numerator and denominator by } C^3$$

The given expression can also be simplified by subtracting exponents. From

$$\frac{2^{-4}R^{-2}C^{-4}}{3^{-1}R^4C^{-3}} = \frac{3R^{-2}C^{-4}}{2^4R^4C^{-3}}$$

we get

$$\frac{3R^{-2-4}C^{-4-(-3)}}{2^4} = \frac{3R^{-6}C^{-1}}{16} \qquad \text{Subtracting exponents; } 2^4 = 16$$

$$= \frac{3}{16R^6C} \qquad \text{By rule (10.8)}$$

Example 7 Simplify the expression

$$\frac{(3s^2t^{-2})^{-2}}{(2s^{-6}t^4)^{-3}}$$

Solution.

$$\frac{(3s^2t^{-2})^{-2}}{(2s^{-6}t^4)^{-3}} = \frac{3^{-2}(s^2)^{-2}(t^{-2})^{-2}}{2^{-3}(s^{-6})^{-3}(t^4)^{-3}} \qquad (ab)^n = a^n b^n$$

$$= \frac{3^{-2}(s^2)^{-2}(t^{-2})^{-2}}{2^{-3}(s^{-6})^{-3}(t^4)^{-3}}$$

$$= \frac{2^3(s^2)^{-2}(t^{-2})^{-2}}{3^2(s^{-6})^{-3}(t^4)^{-3}} \qquad 3^{-2} = \frac{1}{3^2}, \frac{1}{2^{-3}} = 2^3$$

$$= \frac{2^3s^{-4}t^4}{3^2s^{18}t^{-12}} \qquad \text{Multiplying exponents}$$

$$= \frac{8s^{-4}t^4}{9s^{18}t^{-12}} \qquad 2^3 = 8, 3^2 = 9$$

$$= \frac{8t^4t^{12}}{9s^{18}s^4} \qquad \text{By rule (10.8)}$$

$$= \frac{8t^{16}}{9s^{22}} \qquad \text{Adding exponents}$$

Starting with the expression

$$\frac{8s^{-4}t^4}{9s^{18}t^{-12}}$$

we can also simplify by subtracting exponents:

$$\frac{8s^{-4}t^4}{9s^{18}t^{-12}} = \frac{8s^{-4-18}t^{4-(-12)}}{9} \qquad \text{Subtracting exponents}$$

$$= \frac{8s^{-22}t^{16}}{9}$$

$$= \frac{8t^{16}}{9s^{22}} \qquad s^{-22} = \frac{1}{s^{22}}$$

So far all expressions with exponents have been factors. If a fraction contains a sum (or difference) of terms with negative exponents, then the terms in the numerator and denominator must be combined first. Consider the next example.

Example 8 Simplify

$$\frac{1 - x^{-1}}{1 + x^{-2}}$$

Solution. Since neither the numerator nor the denominator is a product, the negative exponents cannot be eliminated by using rule (10.8). Instead, we write x^{-1} as $1/x$ and x^{-2} as $1/x^2$ and combine the terms in the numerator and denominator separately:

$$\frac{1 - x^{-1}}{1 + x^{-2}} = \frac{1 - \dfrac{1}{x}}{1 + \dfrac{1}{x^2}}$$

$$= \frac{\dfrac{1 \cdot x}{x} - \dfrac{1}{x}}{\dfrac{1 \cdot x^2}{x^2} + \dfrac{1}{x^2}} \qquad \begin{matrix}\text{LCD} = x \\ \text{LCD} = x^2\end{matrix}$$

$$= \frac{\dfrac{x - 1}{x}}{\dfrac{x^2 + 1}{x^2}} \qquad \text{Combining fractions}$$

$$= \frac{x - 1}{x} \cdot \frac{x^2}{x^2 + 1} \qquad \text{Inverting divisor}$$

$$= \frac{x - 1}{\cancel{x}} \cdot \frac{x^{\cancel{2}}}{x^2 + 1}$$

$$= \frac{x(x - 1)}{x^2 + 1}$$

Common error **a.** Incorrectly writing

$$\frac{1}{a^{-1} + b^{-1}} \quad \text{as} \quad a + b$$

Rule (10.8) does not apply since a^{-1} and b^{-1} are *not factors*. Instead

$$\frac{1}{a^{-1} + b^{-1}} = \frac{1}{\dfrac{1}{a} + \dfrac{1}{b}} = \frac{1}{\dfrac{b}{ab} + \dfrac{a}{ab}} \qquad \text{LCD} = ab$$

$$= \frac{1}{\dfrac{b + a}{ab}} = \frac{ab}{b + a} = \frac{ab}{a + b}$$

b. Writing

$$a^{-1} \quad \text{as} \quad -\frac{1}{a}$$

Instead,

$$a^{-1} = \frac{1}{a}$$

(A negative exponent does not make the quantity negative.)

Example 9 The velocity in meters per second of a body of mass 1 kg falling from rest and subject to a retarding force due to air resistance is given by

$$v = \frac{9.8}{k}(1 - e^{-kt}) \qquad (e = 2.71 \text{ approximately})$$

where k is a constant that depends on the size and shape of the object and t the time in seconds. Write v without negative exponents and simplify.

Solution.

$$\begin{aligned} v &= \frac{9.8}{k}(1 - e^{-kt}) && \text{Given formula} \\ &= \frac{9.8}{k}\left(1 - \frac{1}{e^{kt}}\right) && a^{-1} = \frac{1}{a} \\ &= \frac{9.8}{k}\left(\frac{e^{kt}}{e^{kt}} - \frac{1}{e^{kt}}\right) && \text{LCD} = e^{kt} \\ &= \frac{9.8}{k}\left(\frac{e^{kt} - 1}{e^{kt}}\right) && \text{Combining fractions} \\ &= \frac{9.8(e^{kt} - 1)}{ke^{kt}} && \text{Multiplying} \end{aligned}$$

Exercises / Section 10.1

In Exercises 1–16, evaluate the given expressions.

1. 3^{-2}
2. 4^{-2}
3. 7^{-1}
4. 6^{-2}
5. $3^0 \cdot 2^{-4}$
6. $4^0 \cdot 5^{-3}$
7. $(2^{-1})^{-2}$
8. $(3^{-1})^2$
9. $(6^{-2})^{-1}$
10. $(3^{-2})^2$
11. $\frac{5^{-4}}{5^{-6}}$
12. $\frac{2^{-6}}{2^{-4}}$
13. $\frac{3^0 \cdot 3^{-7}}{3^{-5}}$
14. $\frac{6^0 \cdot 4^{-1}}{4^{-3}}$
15. $\frac{(2^{-2})^{-1}}{3^{-1}}$
16. $\frac{3^{-2}}{(3^{-2})^2}$

In Exercises 17–68, simplify the given expressions and write the results without zero or negative exponents.

17. $\frac{3x^6}{x^4}$
18. $\frac{2a}{a^4}$
19. $\frac{21V^3}{14V^4}$
20. $\frac{-18P^4}{-15P^3}$
21. $(-2x)^4$
22. $(-3b^2)^3$
23. $-(a^2b)^2$
24. $-(2cd)^2$
25. $\frac{x^2}{x^{-2}}$
26. $\frac{y^{-1}}{y^{-6}}$
27. $\frac{a^{-3}b^2}{b^{-3}}$
28. $\frac{x^0y^{-3}}{x^2y}$

29. $\dfrac{a^0m^{-3}n^{-2}}{m^2n^{-1}}$

30. $a^{-1}b^{-2}c^0$

31. $\dfrac{1}{2}d^{-2}e^{-3}f^0$

32. $(vw^{-2})^{-3}$

33. $(V_0^2V^{-4})^{-2}$

34. $(2P_aP_b^2)^{-3}$

35. $(-L^{-2}M^{-2})^{-4}$

36. $(-p^{-2}q^{-1})^2$

37. $(a^0b^{-2}c^3)^{-2}$

38. $(p^0P^{-2}R^{-1})^{-2}$

39. $(2x^0y^{-2}z^{-3})^2$

40. $(3b^0m^{-2}n^{-3})^2$

41. $\dfrac{2^{-1}x^{-2}y}{3x^3y^{-4}}$

42. $\dfrac{4w^2z^{-3}}{3^{-2}w^{-2}z^2}$

43. $\dfrac{2^{-1}s_1^{-1}s_2}{4^{-1}s_1s_2^{-3}}$

44. $\dfrac{3^{-1}t_1t_2^{-2}}{3t_1^{-1}t_2^{-1}}$

45. $\dfrac{3A^{-1}B^{-2}}{3^{-1}A^{-2}B^3}$

46. $\dfrac{2^{-1}rs^{-1}}{4^{-1}r^{-2}s^{-2}}$

47. $\dfrac{24x^4y^{-2}z^{-6}}{48x^2y^3z^{-8}}$

48. $\dfrac{24a^{-1}n^4z^4}{32a^{-3}n^2z^{-2}}$

49. $\dfrac{4^{-1}v_1^3v_2^4v_3^6}{8^{-2}v_1^4v_2^{-4}v_3^{-1}}$

50. $\dfrac{3^{-2}x_1^{-6}x_2^5x_3^{-4}}{27^{-1}x_1x_2^{-3}x_3^{-3}}$

51. $\dfrac{(d^{-1}e)^2}{(-de^{-1})^3}$

52. $\dfrac{(-2v^{-3}w)^2}{(7vw^2)^{-1}}$

53. $\dfrac{1}{(-3st)^{-2}}$

54. $\dfrac{a}{(-2a^{-3}b)^{-3}}$

55. $\dfrac{(2T^{-1}V)^{-1}}{(3T^{-2}V)^2}$

56. $\dfrac{(4d^{-1}m^2)^{-2}}{(3dm^{-4})^{-3}}$

57. $\dfrac{1}{1 - n^{-1}}$

58. $\dfrac{1 - a^{-2}}{3}$

59. $\dfrac{b^{-4} - 2}{b}$

60. $\dfrac{c - c^{-1}}{c^2}$

61. $\dfrac{a^{-1} + b^{-1}}{a^{-1}}$

62. $\dfrac{1 - x^{-3}}{x^{-3}}$

63. $\dfrac{1 - P^{-2}}{1 + P^{-2}}$

64. $\dfrac{2 + V^{-1}}{2 + V^{-2}}$

65. $\dfrac{2R^{-1} + 2}{2R^{-1} + 3}$

66. $\dfrac{R^{-1} + R^{-2}}{R^{-1} + R^{-3}}$

67. $\dfrac{a^{-2} + a^2}{a^2}$

68. $\dfrac{b^{-4} - b}{b^2}$

69. The radioactive element carbon-14, which is used to determine the age of a fossil, decays according to the formula

$$P = P_0(2^{-t/5580})$$

Write this formula without the negative exponent.

70. The *present value* of an investment is the amount of money P that needs to be invested to yield a prescribed amount S at the end of n years and is given by

$$P = S(1 + i)^{-n}$$

where i is the interest rate compounded annually. Write this formula without the negative exponent.

71. According to Boyle's law, the relationship between the pressure and volume of a gas is

$$P = kV^{-1}$$

Solve this formula for V.

72. The diameter of a certain molecule is 4×10^{-8} cm. Write this number as a decimal.

73. The combined resistance R of two resistors in parallel can be expressed as

$$R = (R_1^{-1} + R_2^{-1})^{-1}$$

Simplify this expression.

74. A decelerating body moves along a line according to the formula $d = 1/(1 + 2t)$, where d is the distance (in meters) from the origin and t the time (in seconds). If v is the velocity, it can be shown using calculus that $|v| = 2(1 + 2t)^{-2}$. Find $|v|$ when $t = 4$ s.

75. The relationship between the object distance p, the image distance q, and the focal length f of a lens can be expressed as

$$f^{-1} = p^{-1} + q^{-1}$$

If $p = 6$ cm and $q = 12$ cm, find f.

76. The Lyman series, which occurs in the study of the hydrogen atom, is

$$\lambda^{-1} = R(1 - n^{-2}) \qquad \lambda = \text{lambda}$$

Solve for λ and simplify.

10.2 Scientific Notation

Many scientific and technical applications involve very large and very small numbers. For example, the mass of a lithium nucleus is

$$0.00000000000000000000001164 \text{ g}$$

Such numbers can be written in a more convenient way by using scientific notation.

Scientific notation

A number in scientific notation has the form

$$n \times 10^k, \qquad 1 \leq n < 10$$

where k is a positive or negative integer.

In other words, a number in scientific notation is expressed as a product of a number between 1 and 10 and a power of 10. Let's consider some of these powers:

$$10 = 10^1 \qquad \frac{1}{10} = 10^{-1}$$

$$100 = 10^2 \qquad \frac{1}{100} = \frac{1}{10^2} = 10^{-2}$$

$$1000 = 10^3 \qquad \frac{1}{1000} = \frac{1}{10^3} = 10^{-3}$$

$$10{,}000 = 10^4 \qquad \frac{1}{10{,}000} = \frac{1}{10^4} = 10^{-4}$$

To see how a typical large number such as 50,000 can be written in scientific notation, observe that

$$50{,}000 = 5 \times 10{,}000 = 5 \times 10^4$$

Similarly,

$$51{,}000 = 5.1 \times 10{,}000 = 5.1 \times 10^4 \qquad \text{Proper form}$$

Note that 5.1 is a number between 1 and 10, so that 5.1×10^4 has the proper form. On the other hand, 51×10^3, while equal to 51,000, is not in the proper form since 51 is greater than 10.

Now consider a small number such as 0.0019. By first writing this number as a common fraction, we get

$$0.0019 = \frac{19}{10{,}000} = \frac{19}{10^4} = \frac{\frac{19}{10}}{\frac{10^4}{10}}$$

$$= \frac{1.9}{10^3} = 1.9 \times 10^{-3} \qquad \text{Proper form}$$

The procedure for writing a number in scientific notation is given next.

> To write a number in **scientific notation**
>
> $$n \times 10^k, \qquad 1 \le n < 10$$
>
> move the decimal point to the right of the first nonzero digit. The number of places that the decimal point is moved is equal to the exponent k:
>
> 1. If the decimal point is moved to the left, k is positive.
> 2. If the decimal point is moved to the right, k is negative.

Example 1 Write 2,345,000 in scientific notation.

Solution. According to the rule, the decimal point is placed to the right of 2 (to obtain 2.345). To place the decimal point in this position, we must move the original decimal point 6 places to the *left*. So $k = +6$. The result is

$$2345000 = 2.345 \times 10^6$$

(6 places)

As a check, note that

$$2.345 \times 10^6 = (2.345)(1{,}000{,}000) = 2{,}345{,}000$$

Example 2 Write 0.000000497 in scientific notation.

Solution. The decimal point is placed to the right of the first nonzero digit, to obtain 4.97. To place the decimal point in this position, we have to move the original decimal point 7 places to the *right*. So $k = -7$. The result is

$$0.000000497 = 4.97 \times 10^{-7}$$

7 places

Example 3 **a.** $98{,}440{,}000 = 9.844 \times 10^7$ (decimal point moved 7 places to the left)
b. $0.000057 = 5.7 \times 10^{-5}$ (decimal point moved 5 places to the right)

To change a number from scientific to regular decimal notation, we reverse this procedure.

Example 4 To change

$$6.4 \times 10^6$$

to decimal notation, we move the decimal 6 places to the right. To do so, we must add zeros:

$$6.4 \times 10^6 = 6{,}400{,}000$$

As a check,

$$6.4 \times 10^6 = (6.4)(1{,}000{,}000) = 6{,}400{,}000$$

Example 5 **a.** $4.5 \times 10^4 = 45{,}000$ (decimal point moved 4 places)
b. $2.6 \times 10^{-3} = 0.0026$ (decimal point moved 3 places)

In our discussion of significant digits (Section 2.1), we noted that final zeros are significant only if they lie to the right of the decimal point. Thus 3.100 has four significant digits. The number 3100 has only two significant digits, unless otherwise specified. For example, if a measurement is given as 3100 ft, accurate to the nearest foot, then 3100 has four significant digits. Scientific notation takes care of questionable final zeros in a simple and natural way.

> When a number is written in scientific notation, $n \times 10^k$, n indicates the number of significant digits.

So the measurement 3100 ft is written 3.1×10^3 ft to indicate two significant digits, and 3100 ft is written 3.100×10^3 ft to indicate four significant digits.

Example 6 The number 3,140,000 is written:

a. 3.14×10^6 to indicate three significant digits
b. 3.140×10^6 to indicate four significant digits
c. 3.1400×10^6 to indicate five significant digits

Multiplication and division can be carried out directly with numbers written in scientific notation. However, final answers may have to be changed to obtain the proper form. Consider the next example.

Example 7 Multiply and write the result in scientific notation:

$$(3.89 \times 10^{12}) \cdot (9.26 \times 10^{-5})$$

Solution. Rearranging the factors, we get

$$\begin{aligned}(3.89 \times 10^{12}) \cdot (9.26 \times 10^{-5}) &= (3.89 \cdot 9.26)(10^{12} \cdot 10^{-5}) \\ &= 36.0214 \times 10^7 \\ &= 36.0 \times 10^7 \quad \text{Three significant digits}\end{aligned}$$

Since the coefficient of 10^7 has to be a number between 1 and 10, we need to write 36.0 as 3.60×10. We now get

$$\begin{aligned}36.0 \times 10^7 = (3.60 \times 10) \times 10^7 &= 3.60 \times (10 \times 10^7) \\ &= 3.60 \times 10^8\end{aligned}$$

Example 8 Perform the following division:

$$(4.360 \times 10^{-6}) \div (8.327 \times 10^{20})$$

Solution. Writing the division problem in fractional form, we get

$$\begin{aligned}\frac{4.360 \times 10^{-6}}{8.327 \times 10^{20}} &= \frac{4.360}{8.327} \times \frac{10^{-6}}{10^{20}} \\ &= \frac{4.360}{8.327} \times 10^{-6-20} \\ &= 0.5236 \times 10^{-26} \quad \text{Four significant digits}\end{aligned}$$

To obtain the proper form, we write 0.5236 as 5.236/10 to get

$$\begin{aligned}0.5236 \times 10^{-26} &= \frac{5.236}{10} \times 10^{-26} \\ &= (5.236 \times 10^{-1}) \times 10^{-26} \\ &= 5.236 \times 10^{-27}\end{aligned}$$

Addition and subtraction of numbers in scientific notation can be performed by the rule for adding and subtracting polynomials.

Example 9 Combine the following numbers:

$$(3.96 \times 10^{-6}) + (7.25 \times 10^{-6}) - (4.25 \times 10^{-6})$$

Solution.

Numbers	*Algebraic comparison*
3.96×10^{-6}	$3.96x$
$+7.25 \times 10^{-6}$	$+\ 7.25x$
-4.25×10^{-6}	$-\ 4.25x$
6.96×10^{-6}	$6.96x$

Note that to add or subtract numbers written in scientific notation, the powers of 10 have to be the same. (Only like terms can be added or subtracted.)

Calculator Operations

Scientific calculators are programmed to perform operations with numbers in scientific notation. In fact, very large and very small numbers must first be expressed in scientific notation before they can even be entered. Further, very large and very small numbers resulting from a calculation are automatically displayed in this form.

To enter a number expressed in scientific notation, use [EE], [EXP], [EEX], or a similar key to enter the exponent. For example, to enter 3.862×10^{-16}, use the following sequence:

3.862 [EE] 16 [+/−] → 3.862 −16

Note the space between 3.862 and −16 in the display.

Example 10 Use a calculator to perform the following multiplication:

$$(6.35 \times 10^{8}) \cdot (7.27 \times 10^{15}) \cdot (8.12 \times 10^{-12})$$

Solution. The sequence is

6.35 [EE] 8 [×] 7.27 [EE] 15 [×] 8.12 [EE] 12 [+/−]
[=] → 3.7485574 13

So the product to three significant digits is 3.75×10^{13}.

Example 11 Use a calculator to perform the following operations:

$$\frac{(0.000000082) \cdot (0.0000000013)}{0.00000000046}$$

Solution. For most calculators these numbers contain too many digits to be entered in decimal form and must therefore be expressed in scientific notation:

$$0.000000082 = 8.2 \times 10^{-8}$$
$$0.0000000013 = 1.3 \times 10^{-9}$$
$$0.000000000046 = 4.6 \times 10^{-11}$$

The sequence is

8.2 [EE] 8 [+/−] [×] 1.3 [EE] 9 [+/−] [÷] 4.6 [EE] 11 [+/−]
[=] → 2.3173913 −6

The answer is 2.3×10^{-6}.

Example 12 Elementary particles called μ-mesons (μ = mu) are created in the upper levels of the earth's atmosphere by high-speed cosmic rays from outer space. A typical μ-meson has a speed of 2.994×10^8 m/s and a lifetime of 2×10^{-6} s (when it decays into an electron). How far does it travel before decaying?

Solution. To find the distance, we use the relationship $d = rt$, so that

$$d = \left(2.994 \times 10^8 \frac{\text{m}}{\text{s}}\right) \cdot (2 \times 10^{-6} \text{ s})$$

The sequence is

2.994 [EE] 8 [×] 2 [EE] 6 [+/−] [=] → 598.8

We conclude that a typical μ-meson travels 600 m before decaying.

Exercises / Section 10.2

In Exercises 1–10, express the given numbers in scientific notation.

1. 320,000
2. 968,300,000
3. 0.0000570
4. 0.0000063
5. 976,120,000
6. 340,000,000
7. 0.00000100
8. 0.0000002000
9. 3,000,000
10. 1,000,000,000

In Exercises 11–20, express the given numbers in decimal notation.

11. 3.2×10^4
12. 4.76×10^6
13. 3.19×10^{-7}
14. 5.38×10^{-8}
15. 1.76×10^8
16. 2.34×10^4
17. 6.37×10^{-4}
18. 1.27×10^{-6}
19. 5.02×10^{-5}
20. 3.00×10^{-7}

In Exercises 21–38, perform the indicated operations without using a calculator. Express each result in scientific notation.

21. $(2.30 \times 10^4) \cdot (9.16 \times 10^5)$

22. $(7.00 \times 10^7) \cdot (8.10 \times 10^4)$

23. $(4.60 \times 10^{20}) \cdot (7.10 \times 10^{-10})$

24. $(6.30 \times 10^{-15}) \cdot (7.60 \times 10^5)$

25. $(3.40 \times 10^{-8}) \cdot (4.70 \times 10^{-7})$

26. $(8.920 \times 10^{-10}) \cdot (2.110 \times 10^{-8})$

27. $\dfrac{2.200 \times 10^{16}}{4.400 \times 10^5}$

28. $\dfrac{1.200 \times 10^4}{3.000 \times 10^{-5}}$

29. $\dfrac{3.2 \times 10^{-12}}{4.5 \times 10^{-7}}$

30. $\dfrac{1.1 \times 10^{-15}}{2.8 \times 10^{-9}}$

31. $(4.7 \times 10^{-8}) + (1.2 \times 10^{-8})$

32. $(3.8 \times 10^{12}) + (2.7 \times 10^{12})$

33. $(9.3 \times 10^6) + (5.7 \times 10^6)$

34. $(6.8 \times 10^{-4}) - (6.4 \times 10^{-4})$

35. $(4.30 \times 10^{-5}) + (6.40 \times 10^{-5}) - (2.60 \times 10^{-5})$

36. $(1.36 \times 10^7) - (3.84 \times 10^7) - (9.43 \times 10^7)$

37. $(2.41 \times 10^8) + (5.60 \times 10^8) + (7.00 \times 10^8)$

38. $(1.10 \times 10^{-6}) - (2.83 \times 10^{-6}) + (8.40 \times 10^{-6})$

In Exercises 39–44, perform the indicated operations with a calculator.

39. $(2.16 \times 10^{-18}) \cdot (3.14 \times 10^{-9}) \cdot (9.08 \times 10^6)$

40. $(7.73 \times 10^{-15}) \cdot (6.34 \times 10^{11}) \cdot (4.03 \times 10^{-10})$

41. $\dfrac{2.647 \times 10^{-4}}{(3.426 \times 10^{-7})(4.006 \times 10^{16})}$

42. $\dfrac{(2.020 \times 10^{-4}) \cdot (3.564 \times 10^7)}{2.993 \times 10^{-6}}$

43. $\dfrac{(9.43 \times 10^{-4}) \cdot (4.34 \times 10^{-10})}{(2.03 \times 10^{-3}) \cdot (5.43 \times 10^{12})}$

44. $\dfrac{(5.07 \times 10^4) \cdot (6.30 \times 10^9)}{(2.67 \times 10^8) \cdot (8.49 \times 10^{-14})}$

In Exercises 45–54, use a calculator to perform the indicated operations. Write the answers in scientific notation.

45. $(9{,}286{,}000{,}000) \cdot (8{,}400{,}000{,}000)$

46. $(0.0000000763) \cdot (0.00000000473)$

47. $(0.0000000020) \cdot (0.000000038)$

48. $(9{,}320{,}000{,}000) \cdot (10{,}972{,}000{,}000)$

49. $(0.000000000120) \cdot (0.0000000076)$

50. $(0.0000000072) \cdot (0.0000000284)$

51. $\dfrac{5{,}360{,}000{,}000{,}000}{0.00000068}$

52. $\dfrac{0.000000983}{2{,}400{,}000{,}000}$

53. $\dfrac{(0.000000910) \cdot (0.000000084)}{9{,}200{,}000{,}000{,}000}$

54. $\dfrac{(74{,}000{,}000{,}000) \cdot (0.00000046)}{0.000000000000364}$

In each statement of Exercises 55–66, change the given number from decimal notation to scientific notation or from scientific notation to decimal notation.

55. A communications satellite is 22,300 mi above the surface of the earth.

56. One light year is equal to a distance of 5,872,000,000,000 mi.

57. A certain computer has 2.64×10^5 bytes of memory.

58. A fast computer can do an addition in 5.5×10^{-10} s.

59. The faintest audible sound has an intensity of 1×10^{-12} W/m^2.

60. The mass of the earth is 5.98×10^{24} kg.

61. The speed of light is 3×10^{10} cm/s.

62. The oil shale reserves in the United States are estimated to be 2×10^9 tons.

63. The diameter of an oxygen molecule is 3×10^{-8} cm.

64. The diameter of the sun is 864,000 mi.

65. The gravitational constant G is $G = 0.00000000006670 \dfrac{\text{N} \cdot \text{m}^2}{\text{kg}^2}$.

66. The capacity of a certain capacitor is 0.0000048 F.

In Exercises 67–70, use the relationships

$$d = rt \qquad \text{and} \qquad t = \frac{d}{r}$$

(See Example 12.)

67. Given that the speed of light is 1.86×10^5 mi/s and the distance from the earth to the sun is 9.3×10^7 mi, find the time in minutes for the sun's light to reach the earth.

68. The distance from the earth to the moon is 2.39×10^5 mi. Given that a radio signal travels at 1.86×10^5 mi/s, how long does it take for a radio signal from the moon to reach the earth?

69. A lightyear is the distance that light travels in one year. Given that the speed of light is 3×10^5 km/s, what is 1 lightyear in kilometers?

70. The star closest to our solar system is Proxima Centauri, 2.5×10^{13} mi away. How many years would it take a rocket traveling at 30,000 mi/h to reach the nearest star?

71. If a computer performs a single operation (such as addition) in 7.6×10^{-9} s, how many operations can it perform in an hour?

72. The frequency of a certain TV signal is 5.90×10^7 cycles per second (hertz). What is the frequency in cycles per hour?

73. The mass of the earth is 5.98×10^{24} kg, and the volume of the earth is 1.08×10^{21} m³. Determine the average density of the earth in kilograms per cubic meter.

74. The force F of gravitational attraction between two masses m_1 and m_2 is given by

$$F = G \frac{m_1 m_2}{d^2}$$

where d is the distance between the masses and G the gravitational constant. (Refer to Exercise 65.) Find the gravitational force (in newtons) between two 10.0-kg masses that are 2.00 cm apart.

10.3 Radicals

We first encountered square roots in Section 1.12 on calculator operations. In this section we discuss roots of higher order.

Square root

First recall that the **square root** of a number a, denoted by $\sqrt{a}$, is a number that, when squared, is equal to a. For example, $\sqrt{4} = 2$, since $2^2 = 4$. Similarly, $\sqrt{9} = 3$ and $\sqrt{25} = 5$.

*n*th root

Radical, Radicand

Index

Principal square root

The ***n*th root** of a number, denoted by $\sqrt[n]{a}$, is another number b such that $b^n = a$. For example, $\sqrt[3]{8} = 2$ (since $2^3 = 8$) and $\sqrt[4]{16} = 2$ (since $2^4 = 16$). The expression $\sqrt[n]{a}$ is called a **radical;** a is called the **radicand,** and n is called the **index** of the radical. If $n = 2$, we omit the index and write $\sqrt{a}$.

The **principal square root** is the positive square root. The need for this definition arises from the fact that $\sqrt{16}$ could be given as either 4 or -4, since $4^2 = 16$ and $(-4)^2 = 16$. However, whenever the symbol $\sqrt{\ }$ is used, it is understood that only the positive square root is intended. So the number 4 is the principal square root of 16. Similarly, $\sqrt{25} = 5$, not -5, and $\sqrt{49} = 7$, not -7.

Example 1

$\sqrt{36} = 6$ Since $6^2 = 36$

$\sqrt{625} = 25$ Since $25^2 = 625$

$\sqrt{\frac{1}{4}} = \frac{1}{2}$ Since $\left(\frac{1}{2}\right)^2 = \frac{1}{4}$

$\sqrt{\frac{4}{9}} = \frac{2}{3}$ Since $\left(\frac{2}{3}\right)^2 = \frac{4}{9}$

If the index n is even, then the principal nth root is again the positive nth root. So $\sqrt[4]{16} = 2$, not -2.

If n is an odd index, then the nth root can be a negative number. For example, while $\sqrt[3]{8} = 2$, we have $\sqrt[3]{-8} = -2$, since $(-2)^3 = -8$. In general, if ***n* is odd,** then

1. $\sqrt[n]{a}$ is positive if a is positive
2. $\sqrt[n]{a}$ is negative if a is negative

As already noted, if n is even (and a positive), then $\sqrt[n]{a}$ is positive. The case where n is even and a negative will be discussed in Section 10.5.

Example 2

$\sqrt[3]{27} = 3$ Since $3^3 = 27$

$\sqrt[3]{-27} = -3$ Since $(-3)^3 = -27$

Example 3

$\sqrt[5]{32} = 2$ Since $2^5 = 32$

$\sqrt[5]{-32} = -2$ Since $(-2)^5 = -32$

$\sqrt[3]{-\frac{1}{8}} = -\frac{1}{2}$ Since $\left(-\frac{1}{2}\right)^3 = -\frac{1}{8}$

Perfect root

If the nth root of a number can be found exactly, it is called a **perfect *n*th root.** For example, $\sqrt[5]{32} = 2$ is a perfect fifth root. A perfect root is a **rational**

number. A number is rational if it can be written as a fraction m/n, where m and n are integers. For example,

$$\sqrt{\frac{4}{9}} = \frac{2}{3} \quad \text{and} \quad \sqrt{4} = 2 = \frac{2}{1}$$

are rational numbers.

Irrational number

If the root of a number cannot be found exactly, then the number is **irrational.** For example, the number $\sqrt{2}$ is not a perfect square root and is therefore irrational.

Example 4

a. The number $\sqrt{16}$ is rational since $\sqrt{16} = 4$ exactly.
b. $\sqrt{0.04} = 0.2$ exactly, so $\sqrt{0.04}$ is rational.
c. $\sqrt{10} \approx 3.16$. (The symbol $\approx$ means "approximately equal to.") Since $\sqrt{10}$ is not a perfect square root, it is an irrational number. (The number 3.16 is only an approximation.)

Recall from Section 1.12 that the square root can be found by pressing the square root key. So to find $\sqrt{10}$, use the sequence

10 [√‾] → 3.1622777

Exercises / Section 10.3

In Exercises 1–20, find the indicated roots without a calculator.

1. $\sqrt{1}$ **2.** $\sqrt[3]{-1}$ **3.** $\sqrt{81}$
4. $\sqrt{121}$ **5.** $\sqrt[3]{-125}$ **6.** $\sqrt{64}$
7. $\sqrt{49}$ **8.** $\sqrt[3]{-64}$ **9.** $\sqrt[3]{-\frac{1}{27}}$
10. $\sqrt[3]{64}$ **11.** $\sqrt[6]{64}$ **12.** $\sqrt[3]{125}$
13. $\sqrt[4]{81}$ **14.** $\sqrt{\frac{1}{16}}$ **15.** $\sqrt{\frac{9}{64}}$
16. $\sqrt{\frac{25}{49}}$ **17.** $\sqrt{0.09}$ **18.** $\sqrt{0.16}$
19. $\sqrt{0.25}$ **20.** $\sqrt{0.81}$

In Exercises 21–26, use a calculator to find the indicated square roots accurate to three decimal places.

21. $\sqrt{5}$ **22.** $\sqrt{7}$ **23.** $\sqrt{11}$
24. $\sqrt{15}$ **25.** $\sqrt{19}$ **26.** $\sqrt{29}$

In Exercises 27–44, state whether the given numbers are rational (r) or irrational (i).

27. $\sqrt{36}$ **28.** $\sqrt{49}$ **29.** $\sqrt{37}$

30. $\sqrt{51}$
31. $\sqrt[3]{\frac{1}{27}}$
32. $\sqrt[3]{\frac{1}{64}}$
33. $\sqrt[3]{\frac{1}{29}}$
34. $\sqrt[3]{\frac{1}{65}}$
35. $\sqrt{0.36}$
36. $\sqrt{0.64}$
37. $\sqrt{3.6}$
38. $\sqrt{6.4}$
39. $\sqrt[5]{\frac{1}{32}}$
40. $\sqrt[5]{-\frac{1}{32}}$
41. $\sqrt[3]{25}$
42. $\sqrt[4]{49}$
43. $\sqrt[3]{32}$
44. $\sqrt[4]{32}$

10.4 Fractional Exponents

In Section 10.1 we discussed zero exponents and exponents that are positive or negative integers. In this section we continue with **fractional exponents.**

Consider the fractional exponent $4^{1/2}$. What could such an exponent mean? If the laws of exponents are to hold for fractional powers, then we could argue that

$$(4^{1/2})^2 = 4$$

which is obtained by multiplying $\frac{1}{2}$ and 2. Now compare the following results:

$$(4^{1/2})^2 = 4 \qquad \text{and} \qquad 2^2 = 4$$

This comparison suggests that $4^{1/2} = 2$. Similarly,

$$(9^{1/2})^2 = 9 \qquad \text{and} \qquad 3^2 = 9$$

suggesting that $9^{1/2} = 3$. Since $\mathbf{\sqrt{4} = 2}$ and $\mathbf{\sqrt{9} = 3}$, it follows that $\mathbf{4^{1/2} = \sqrt{4}}$ and $\mathbf{9^{1/2} = \sqrt{9}}$. In general,

$$a^{1/2} = \sqrt{a}$$

This definition can be further generalized to

$$a^{1/n} = \sqrt[n]{a}$$

By the rule for multiplying exponents, we also have

$$a^{m/n} = (a^m)^{1/n} = (a^{1/n})^m$$

It follows that

$$a^{m/n} = \sqrt[n]{a^m} = (\sqrt[n]{a})^m$$

These results are summarized next.

Fractional exponents

$$a^{1/n} = \sqrt[n]{a} \tag{10.9}$$

$$a^{m/n} = \sqrt[n]{a^m} = (\sqrt[n]{a})^m \tag{10.10}$$

Since fractional exponents are really radicals in disguise, numerical expressions containing fractional exponents are evaluated by changing the expression to radical form. It is usually easier to find the root first, and then the power. For example,

$$8^{2/3} = (\sqrt[3]{8})^2 = 2^2 = 4$$

Example 1 Evaluate:

a. $27^{2/3}$ **b.** $8^{-1/3}$ **c.** $16^{-3/2}$

Solution.

a. Changing to radical form, we get

$$27^{2/3} = (\sqrt[3]{27})^2 = 3^2 = 9$$

b. By the definition of negative exponent, we get

$$8^{-1/3} = \frac{1}{8^{1/3}} = \frac{1}{\sqrt[3]{8}} = \frac{1}{2}$$

c. $16^{-3/2} = \dfrac{1}{16^{3/2}} = \dfrac{1}{(\sqrt{16})^3} = \dfrac{1}{4^3} = \dfrac{1}{64}$

Since the definition of fractional exponent was formulated with the laws of exponents in mind, the laws of exponents in Section 10.1 hold for fractional powers.

Example 2 Simplify the following expressions:

a. $xx^{1/4}$ **b.** $\dfrac{a^{2/3}}{a^{1/2}}$ **c.** $(V^{-1/2})^{2/3}$

Solution.

a. Since

$$1 + \frac{1}{4} = \frac{4}{4} + \frac{1}{4} = \frac{5}{4}$$

we have

$$xx^{1/4} = x^1x^{1/4} = x^{1+1/4} = x^{5/4} \quad \text{Adding exponents}$$

b. $\dfrac{a^{2/3}}{a^{1/2}} = a^{2/3-1/2} = a^{4/6-3/6} = a^{1/6}$ Subtracting exponents

c. $(V^{-1/2})^{2/3} = V^{(-1/2)(2/3)} = V^{-1/3}$ Multiplying exponents

---

As in the case of integral exponents, final results are ordinarily written without zero or negative exponents.

Example 3 Simplify

$$\frac{16^{-1/4}R^{1/2}C^{-1/4}}{2^{-1}R^{-1/4}C^{-1/3}}$$

Solution. Let us first write the expression without negative exponents:

$$\frac{16^{-1/4}R^{1/2}C^{-1/4}}{2^{-1}R^{-1/4}C^{-1/3}} = \frac{2R^{1/2}C^{-1/4}}{16^{1/4}R^{-1/4}C^{-1/3}}$$

$$= \frac{2R^{1/2}C^{-1/4}}{2R^{-1/4}C^{-1/3}} \qquad 16^{1/4} = \sqrt[4]{16} = 2$$

$$= \frac{R^{1/2}R^{1/4}C^{-1/4}}{C^{-1/3}} \qquad \frac{1}{R^{-1/4}} = R^{1/4}$$

$$= \frac{R^{1/2}R^{1/4}C^{-1/4}}{C^{-1/3}}$$

$$= \frac{R^{1/2}R^{1/4}C^{1/3}}{C^{1/4}} \qquad C^{-1/4} = \frac{1}{C^{1/4}}, \frac{1}{C^{-1/3}} = C^{1/3}$$

$$= R^{1/2+1/4}C^{1/3-1/4} \qquad \text{Adding and subtracting exponents}$$

Since

$$\frac{1}{2} + \frac{1}{4} = \frac{2}{4} + \frac{1}{4} = \frac{3}{4}$$

and

$$\frac{1}{3} - \frac{1}{4} = \frac{4}{12} - \frac{3}{12} = \frac{1}{12}$$

we get

$$R^{3/4}C^{1/12}$$

Another way to simplify

$$\frac{R^{1/2}C^{-1/4}}{R^{-1/4}C^{-1/3}}$$

is to subtract exponents directly to obtain

$$R^{1/2-(-1/4)}C^{-1/4-(-1/3)} = R^{(1/2+1/4)}C^{(-1/4+1/3)} = R^{3/4}C^{1/12}$$

Example 4 Simplify and write without zero or negative exponents:

$$\frac{3^0T_a^{-5/6}T_b^{-1/9}T_c^0}{3^2T_a^{1/3}T_b^{-1/15}}$$

Solution. Since $3^0 = 1$ and $T_c^0 = 1$, we get

$$\frac{T_a^{-5/6}T_b^{-1/9}}{9T_a^{1/3}T_b^{-1/15}} \qquad 3^0 = 1,\ T_c^0 = 1 \quad 3^2 = 9$$

Subtracting respective exponents, we get

$$-\frac{5}{6} - \frac{1}{3} = -\frac{5}{6} - \frac{2}{6} = -\frac{7}{6} \qquad T_a \text{ exponents}$$

and

$$-\frac{1}{9} - \left(-\frac{1}{15}\right) = -\frac{1}{9} + \frac{1}{15} = -\frac{5}{45} + \frac{3}{45} = -\frac{2}{45} \qquad T_b \text{ exponents}$$

We now have

$$\begin{aligned}\frac{T_a^{-5/6}T_b^{-1/9}}{9T_a^{1/3}T_b^{-1/15}} &= \frac{1}{9}\,T_a^{-5/6-1/3}T_b^{-1/9-(-1/15)}\\ &= \frac{1}{9}\,T_a^{-7/6}T_b^{-2/45}\\ &= \frac{1}{9T_a^{7/6}T_b^{2/45}}\end{aligned}$$

The next example requires the rule $(ab)^n = a^n b^n$.

Example 5 Simplify and write without zero or negative exponents

$$(s^{-1}t^{-1/3}u^{-1/2}v)^{-3}$$

Solution. The expression inside the parentheses can be written

$$\frac{v}{st^{1/3}u^{1/2}}$$

However, because of the large number of negative exponents, it is better to multiply exponents first. Then we get

$$\begin{aligned}(s^{-1}t^{-1/3}u^{-1/2}v)^{-3} &= (s^{-1})^{-3}(t^{-1/3})^{-3}(u^{-1/2})^{-3}v^{-3} && (ab)^n = a^n b^n\\ &= s^3tu^{3/2}v^{-3} && \text{Multiplying exponents}\\ &= \frac{s^3tu^{3/2}}{v^3}\end{aligned}$$

Calculator Operations

To evaluate a power of a number with a calculator, use the key [y^x] or [x^y]. For example, to evaluate 2^5, use the sequence

2 [y^x] 5 [=] → 32

Expressions with fractional exponents are evaluated similarly. For example, to evaluate $27^{-1/3}$, a correct sequence is

$$27 \boxed{y^x} \boxed{(} 1 \boxed{\div} 3 \boxed{+/-} \boxed{)} \boxed{=} \rightarrow 0.3333333$$

Example 6 Evaluate

$$(3^{2/3} + 4)^{-3/7}$$

Solution. The sequence is

$$3 \boxed{y^x} \boxed{(} 2 \boxed{\div} 3 \boxed{)} \boxed{+} 4 \boxed{=} \boxed{y^x} \boxed{(} 3 \boxed{\div} 7 \boxed{+/-} \boxed{)} \boxed{=} \rightarrow 0.4613587$$

Example 7 The *half-life* of a radioactive element is the time required for half of a given amount of the element to decay. If N_0 is the given amount and H the half-life, the amount left varies with time according to the formula

$$N = N_0(2)^{-t/H}$$

Given that the half-life of radium is 1590 years, how much of 1250 g of radium will be left after 150 years?

Solution. We are given that $N_0 = 1250$ g and $H = 1590$ years. So by the formula we get

$$N = 1250(2)^{-150/1590}$$

The sequence is

$$2 \boxed{y^x} \boxed{(} 150 \boxed{+/-} \boxed{\div} 1590 \boxed{)} \boxed{\times} 1250 \boxed{=} \rightarrow 1170.8761$$

We conclude that after 150 years, 1170 g of radium will be left.

Exercises / Section 10.4

In Exercises 1–8, change each expression to radical form.

1. $V^{1/2}$
2. $P^{1/3}$
3. $z^{1/4}$
4. $b^{1/5}$
5. $s^{2/3}$
6. $t^{3/4}$
7. $m^{5/3}$
8. $n^{2/5}$

In Exercises 9–32, evaluate each expression without using a calculator.

9. $9^{1/2}$
10. $25^{1/2}$
11. $9^{-1/2}$
12. $25^{-1/2}$
13. $27^{1/3}$
14. $27^{-1/3}$
15. $\left(\frac{4}{9}\right)^{-1/2}$
16. $\left(\frac{1}{64}\right)^{-2/3}$
17. $(-8)^{1/3}$
18. $(-64)^{2/3}$
19. $(-32)^{3/5}$
20. $(-32)^{-2/5}$
21. $(4^{1/2})(8^{-1/3})$
22. $(100^{1/2})(9^{-1/2})$
23. $\dfrac{8^{1/3} \cdot 9^{1/2}}{4^{-1/2}}$

24. $\frac{8^{2/3}}{4^{-3/2}}$ **25.** $\frac{16^{-3/4}}{8^{-4/3}}$ **26.** $\frac{32^{-1/5}}{16^{-1/4}}$

27. $\frac{25^{-3/2}}{81^{-1/2}}$ **28.** $\frac{25^{3/2}}{125^{-1/3}}$ **29.** $\frac{(-125)^{1/3}}{27^{2/3}}$

30. $\frac{(-64)^{2/3}}{(-8)^{1/3}}$ **31.** $\frac{(-8)^{2/3}}{4^{3/2}}$ **32.** $\frac{16^{3/4}}{(-27)^{2/3}}$

In Exercises 33–60, simplify the given expressions and write the results without zero or negative exponents.

33. $x^{1/6}x^{5/6}$ **34.** $y^{1/4}y^{3/4}$ **35.** $\frac{a}{a^{1/4}}$

36. $\frac{b^{2/3}}{b}$ **37.** $a^0b^{-1/6}b^{1/3}$ **38.** $c^0d^{-1/6}d^{1/9}$

39. $(-8)^{1/3}V^{-1/2}V^{1/3}$ **40.** $(-27)^{1/3}P^{-1/4}P^{1/8}$ **41.** $\frac{5^0p^{-1/12}}{6^0p^{1/6}}$

42. $\frac{7^{-1}L^{-1/10}}{3^0L^{-1/5}}$ **43.** $\frac{8^{-1/3}s_1^2s_2^{-1/6}}{4^{-1/3}s_1^{1/2}s_2^{-2}}$ **44.** $\frac{2^{1/2}t_1^{-2}t_2^{1/6}}{2^{1/4}t_1^{-3/2}t_2}$

45. $\frac{4^0s^{-2}t^{-4}}{4^{1/2}s^{-1/3}t^{-2/3}}$ **46.** $\frac{3^{1/2}R^{-7}C^{-4/9}}{9^{1/2}R^{-3}C^{2/3}}$ **47.** $\frac{4^{1/2}R_a^3R_b^{-1/4}}{4R_a^{5/2}R_b^{5/8}}$

48. $\frac{16^{-1/2}v_1^{1/10}v_2^{-2/9}}{5^0v_1^{-3/10}v_2^{-4}}$ **49.** $(x^2y^{-1/2})^2$ **50.** $(a^{2/3}b^{-1/3})^3$

51. $(t^{-2/3}w^{-1/3})^{-6}$ **52.** $(16s^{-2}t^{-3/4})^{-2}$ **53.** $(3p^{-3}q^{1/2})^{-1}$

54. $(4^{-1}L^{-1/3}T^{-2/3})^{-3}$ **55.** $(2R^{-2}S^{1/2})^{-1/2}$ **56.** $(16p^{-4}v^8)^{-1/2}$

57. $(16m^{-8}n^4)^{-1/4}$ **58.** $(V^{-4/3}W^{8/3})^{3/4}$ **59.** $(M^{-3/2}N^6)^{2/3}$

60. $(L^{-4/3}V^{-4})^{3/2}$

In Exercises 61–74, use a calculator to evaluate the given expressions.

61. $(13)^4$ **62.** $(11)^6$ **63.** $(2.76)^{1/3}$

64. $(8.407)^{1/4}$ **65.** $(72.4)^{-1/2}$ **66.** $(89.6)^{-1/3}$

67. $[1.00 + (2.46)^3]^{4/3}$ **68.** $[(1.14)^3 + 3.04]^{5/3}$ **69.** $(4^{1/3} - 3^{1/2})^{1/3}$

70. $(7^{2/7} + 5^{3/4})^{1/3}$ **71.** $(2^{1.40} + 4)^{1.72}$ **72.** $(3^{3.60} + 7)^{-2.46}$

73. $(3^{2.60} + 4)^{-1.73}$ **74.** $(5 - 2^{-1.7})^{2.5}$

75. In an adiabatic expansion of gas (no gain or loss of heat) the temperatures and volumes are related by the formula

$$\frac{T_1}{T_2} = \left(\frac{V_1}{V_2}\right)^{1-a}$$

Show that this formula can be written

$$\frac{T_1}{T_2} = \left(\frac{V_2}{V_1}\right)^{a-1}$$

76. In some searchlights the shape of a cross section of the mirror can be described by the equation

$$\sqrt{x} + \sqrt{y} = \sqrt{a}$$

Write the equation with fractional exponents.

77. The shape of certain gears can be described by using the equation

$$x^{2/3} + y^{2/3} = a^{2/3}$$

Write this equation with radicals instead of fractional exponents.

78. The equation $E = T/(x^2 + a^2)^{3/2}$ gives the electric-field intensity on the axis of a uniformly charged ring, where T is the total charge on the ring and a the radius of the ring. Write E in the form of a radical.

79. The pressure in a tank is

$$P = 3.60\ V^{-1}$$

Find P (in atmospheres) when $V = 4.8$ L.

80. The tensile strength S (in pounds) of a certain wire varies with temperature according to the formula

$$S = 843 - 0.0600T^{3/2}$$

Find S when $T = 90.0$°F.

81. The weight w (in pounds) of steam flowing through a hole each second is given by Grashof's formula

$$w = 0.0165AP^{0.95}$$

where A (in square inches) is the cross-sectional area of the hole and P (in pounds per square inch) the pressure. Find w given that $A = 8.0$ in.^{2} and $P = 85$ lb/in.^{2}.

82. The charge q (in coulombs) in a certain capacitor varies with time t (in seconds) according to the equation

$$q = (1.0 + 2.1t^{4/3})^{1/2}$$

Find q at $t = 4.0$ s.

83. At a certain instant, the current i (in amperes) in a circuit is given by

$$i = (10.0)^{-1}[1 - (2.72)^{-1.09}]$$

Find the value of i.

84. The temperature T (in degrees Celsius) of an object placed in an oven varies with time according to the formula

$$T = 95 - 82\left(\frac{4}{3}\right)^{-t}$$

Find the temperature when $t = 2.6$ min.

85. The half-life of the radioactive element polonium is 140 days. If there are 95.0 g of the substance initially, how much will be left after 250 days? (See Example 7.)

10.5 Simplifying Radicals

In the last section we discovered that fractional exponents are really radicals in disguise. Because of this, fractional exponents are a convenient way to simplify certain radical expressions.

However, for some simplifications, the radical form is more convenient. In this section we consider two such simplifications: (1) removing certain factors from the radical and (2) rationalizing the denominator. (Complex numbers are also discussed briefly.)

To perform these operations, we need the following basic laws of radicals:

Laws of radicals

$$\sqrt[n]{ab} = \sqrt[n]{a}\,\sqrt[n]{b} \tag{10.11}$$

$$\sqrt[n]{a^n} = (\sqrt[n]{a})^n = a \tag{10.12}$$

$$\sqrt[n]{\frac{a}{b}} = \frac{\sqrt[n]{a}}{\sqrt[n]{b}}, \qquad b \neq 0 \tag{10.13}$$

To justify rule (10.11), let us write the radicals as fractional exponents. Since $(ab)^n = a^n b^n$, we get

$$\sqrt[n]{ab} = (ab)^{1/n} = a^{1/n}b^{1/n} = \sqrt[n]{a}\,\sqrt[n]{b}$$

Similarly, for rule (10.13),

$$\sqrt[n]{\frac{a}{b}} = \left(\frac{a}{b}\right)^{1/n} = \frac{a^{1/n}}{b^{1/n}} = \frac{\sqrt[n]{a}}{\sqrt[n]{b}}, \qquad b \neq 0$$

To check rule (10.12), note that

$$\sqrt[n]{a^n} = (a^n)^{1/n} = a^{n(1/n)} = a^1 = a$$

and

$$(\sqrt[n]{a})^n = (a^{1/n})^n = a^{(1/n)n} = a^1 = a$$

Removing Perfect *n*th-Power Factors

To see how these rules can be used to simplify certain radicals, consider $\sqrt{50}$. Since $50 = 25 \cdot 2$ and $\sqrt{25} = 5$, it follows that

$$\sqrt{50} = \sqrt{25 \cdot 2} = \sqrt{25}\,\sqrt{2} = 5\sqrt{2} \qquad \text{By rule (10.11)}$$

Since the number under the radical sign is smaller, $5\sqrt{2}$ is considered to be simpler than $\sqrt{50}$. Such a simplification is always possible if the radicand can be written as a product, one factor of which is a perfect root. For example,

$$\sqrt[3]{x^4} = \sqrt[3]{x^3x} = \sqrt[3]{x^3}\,\sqrt[3]{x} = x\sqrt[3]{x}$$

Example 1 Simplify $\sqrt{32A^5}$.

Solution. Note that $32 = 16 \cdot 2$, while $\sqrt{16} = 4$. So $\sqrt{32} = \sqrt{16 \cdot 2} = \sqrt{16}\,\sqrt{2} = 4\sqrt{2}$. Similarly, $A^5 = A^4A$ and $\sqrt{A^4} = (A^4)^{1/2} = A^{4(1/2)} = A^2$. It follows that

$$\begin{aligned}\sqrt{32A^5} &= \sqrt{16 \cdot 2A^4A} = \sqrt{(16A^4)(2A)} \\ &= \sqrt{16A^4}\,\sqrt{2A} = 4A^2\sqrt{2A} = 4A^2\sqrt{2A}\end{aligned}$$

In general, x^n is a perfect root if the exponent is divisible by the index. For example, $\sqrt[3]{x^6} = (x^6)^{1/3} = x^{6(1/3)} = x^2$ (6 is divisible by 3), and $\sqrt[4]{a^{16}} = (a^{16})^{1/4} = a^{16(1/4)} = a^4$ (16 is divisible by 4). Consider another example.

Example 2 Simplify $\sqrt[3]{16s^5t^{10}}$.

Solution. Since we can remove a power only if the exponent is divisible by the index, we write the given radical as follows:

$$\begin{aligned}
\sqrt[3]{16s^5t^{10}} &= \sqrt[3]{8 \cdot 2s^3s^2t^9t} \\
&= \sqrt[3]{(8s^3t^9)(2s^2t)} \\
&= \sqrt[3]{8s^3t^9}\,\sqrt[3]{2s^2t} && \sqrt[3]{ab} = \sqrt[3]{a}\,\sqrt[3]{b} \\
&= (8s^3t^9)^{1/3}\,\sqrt[3]{2s^2t} && \sqrt[3]{a} = a^{1/3} \\
&= 8^{1/3}(s^3)^{1/3}(t^9)^{1/3}\,\sqrt[3]{2s^2t} && (ab)^n = a^nb^n \\
&= 2st^3\sqrt[3]{2s^2t} && \text{Multiplying exponents, } 8^{1/3} = 2
\end{aligned}$$

Example 3 Simplify $\sqrt[4]{V_a^9V_b^{18}}$.

Solution. First observe that $\sqrt[4]{V_a^8} = (V_a^8)^{1/4} = V_a^2$ and that $\sqrt[4]{V_b^{16}} = (V_b^{16})^{1/4} = V_b^4$. We now have

$$\begin{aligned}
\sqrt[4]{V_a^9V_b^{18}} &= \sqrt[4]{V_a^8V_aV_b^{16}V_b^2} \\
&= \sqrt[4]{(V_a^8V_b^{16})(V_aV_b^2)} \\
&= \sqrt[4]{V_a^8V_b^{16}}\,\sqrt[4]{V_aV_b^2} && \sqrt[4]{ab} = \sqrt[4]{a}\,\sqrt[4]{b} \\
&= (V_a^8V_b^{16})^{1/4}\,\sqrt[4]{V_aV_b^2} && \sqrt[4]{a} = a^{1/4} \\
&= (V_a^8)^{1/4}(V_b^{16})^{1/4}\,\sqrt[4]{V_aV_b^2} && (ab)^n = a^nb^n \\
&= V_a^2V_b^4\,\sqrt[4]{V_aV_b^2} && \text{Multiplying exponents}
\end{aligned}$$

Complex Numbers

So far we have assumed that the square root of a number can be found only if the number is positive. There are reasons for this. Suppose we tried to find $\sqrt{-4}$. We can see that 2 does not work, since $2^2 = +4$. However, -2 does not work either, since $(-2)^2 = +4$, not -4. Yet square roots of negative numbers are quite useful in the study of electricity and other areas. To realize this advantage, we have to introduce a new kind of number that is *not a real number*. This number, denoted by the letter j, is defined to be $j = \sqrt{-1}$. The number j is called the **basic imaginary unit**. The basic imaginary unit enables us to extend the *real numbers* (the numbers we have considered so far) to a larger system that includes square roots of negative numbers.

Returning to $\sqrt{-4}$, note that

$$\sqrt{-4} = \sqrt{4(-1)} = \sqrt{4}\,\sqrt{-1} = 2\sqrt{-1} = 2j$$

The number $\sqrt{-4} = 2j$ is called an *imaginary number.*

Basic imaginary unit

$$j = \sqrt{-1} \qquad \text{or} \qquad j^2 = -1$$

$$\sqrt{-a} = \sqrt{a}\,j, \qquad a > 0$$

Example 4 Write the following imaginary numbers in terms of j:

a. $\sqrt{-9}$ **b.** $\sqrt{-25}$ **c.** $\sqrt{-8}$ **d.** $\sqrt{-27}$

Solution.

a. $\sqrt{-9} = \sqrt{9(-1)} = \sqrt{9}\,\sqrt{-1} = 3\sqrt{-1} = 3j$

b. $\sqrt{-25} = \sqrt{25(-1)} = \sqrt{25}\,\sqrt{-1} = 5\sqrt{-1} = 5j$

c. $\sqrt{-8} = \sqrt{8(-1)} = \sqrt{4 \cdot 2 \cdot (-1)} = \sqrt{4 \cdot 2}\,\sqrt{-1} = 2\sqrt{2}j$

d. $\sqrt{-27} = \sqrt{9 \cdot 3 \cdot (-1)} = \sqrt{9 \cdot 3}\,\sqrt{-1} = 3\sqrt{3}j$

Complex number

If a and b are real numbers, then $a + bj$ is called a **complex number**. As just noted, the numbers we have encountered prior to this section are real numbers. Complex numbers are not real numbers; rather, they must be viewed as an extension of the real numbers. As a result, imaginary numbers are not imaginary in the usual sense of the word. The word *imaginary* is merely the traditional name for the square root of a negative number, and it reflects the resistance that once existed toward this concept.

Examples of complex numbers are $2 - 5j$, $-3 + \sqrt{2}j$, and $\sqrt{5} - 7j$. We will see more complex numbers in the next chapter.

Rationalizing Denominators

Another kind of simplification of radicals can be performed by using rule (10.12). Consider the fraction $1/\sqrt{2}$. We can eliminate the radical in the denominator by multiplying the numerator and denominator by $\sqrt{2}$:

$$\frac{1}{\sqrt{2}} = \frac{1}{\sqrt{2}}\,\frac{\sqrt{2}}{\sqrt{2}} = \frac{\sqrt{2}}{(\sqrt{2})^2} = \frac{\sqrt{2}}{2}$$

Rationalizing the denominator

since $(\sqrt{2})^2 = 2$ by rule (10.12). Eliminating the radical in the denominator is called **rationalizing the denominator**. A rationalized form is regarded as simpler than a form that is not rationalized.

Example 5 Simplify by rationalizing the denominator:

$$\frac{\sqrt{x}}{\sqrt{2ab}}$$

Solution. If we multiply the numerator and denominator by $\sqrt{2ab}$, we get

$$\frac{\sqrt{x}}{\sqrt{2ab}} = \frac{\sqrt{x}}{\sqrt{2ab}}\frac{\sqrt{2ab}}{\sqrt{2ab}} \qquad \text{Rationalizing the denominator}$$

$$= \frac{\sqrt{x}\sqrt{2ab}}{(\sqrt{2ab})^2} \qquad \text{Multiplying fractions}$$

$$= \frac{\sqrt{x}\sqrt{2ab}}{2ab} \qquad (\sqrt{2ab})^2 = 2ab$$

$$= \frac{\sqrt{2abx}}{2ab} \qquad \text{Rule (10.11): } \sqrt{a}\,\sqrt{b} = \sqrt{ab}$$

Simplest form

A radical expression is said to be in **simplest form** if:

1. The radicand does not contain any perfect nth power factors
2. No radical appears in the denominator

Example 6 Simplify the following expression:

$$\frac{\sqrt{3R^3V}}{\sqrt{5RV}}$$

Solution. We first rationalize the denominator by multiplying numerator and denominator by $\sqrt{5RV}$:

$$\frac{\sqrt{3R^3V}}{\sqrt{5RV}} = \frac{\sqrt{3R^3V}}{\sqrt{5RV}}\frac{\sqrt{5RV}}{\sqrt{5RV}} \qquad \text{Rationalizing the denominator}$$

$$= \frac{\sqrt{3R^3V}\,\sqrt{5RV}}{(\sqrt{5RV})^2} \qquad \text{Multiplying fractions}$$

$$= \frac{\sqrt{(3R^3V)(5RV)}}{5RV} \qquad \sqrt{a}\,\sqrt{b} = \sqrt{ab},\ (\sqrt{a})^2 = a$$

$$= \frac{\sqrt{15R^4V^2}}{5RV}$$

$$= \frac{R^2V\sqrt{15}}{5RV} \qquad \sqrt{R^4} = R^2,\ \sqrt{V^2} = V$$

$$= \frac{R\sqrt{15}}{5} \qquad \text{Reducing the fraction}$$

A good alternative is to use rule (10.13),

$$\frac{\sqrt[n]{a}}{\sqrt[n]{b}} = \sqrt[n]{\frac{a}{b}}$$

to write

$$\frac{\sqrt{3R^3V}}{\sqrt{5RV}} = \sqrt{\frac{3R^3V}{5RV}} \qquad \frac{\sqrt{a}}{\sqrt{b}} = \sqrt{\frac{a}{b}}$$

$$= \sqrt{\frac{3R^2}{5}} \qquad \text{Reducing the fraction}$$

$$= \sqrt{\frac{3R^2 \cdot 5}{5 \cdot 5}}$$

$$= \frac{\sqrt{15R^2}}{\sqrt{5^2}} \qquad \sqrt{\frac{a}{b}} = \frac{\sqrt{a}}{\sqrt{b}}$$

$$= \frac{\sqrt{15R^2}}{5} \qquad \sqrt{5^2} = 5$$

$$= \frac{R\sqrt{15}}{5} \qquad \sqrt{R^2} = R$$

(We will see more of the alternative procedure in the next section.)

Example 6 shows that the square root of a fraction can be rationalized by using the following principle:

$$\sqrt{\frac{a}{b}} = \sqrt{\frac{ab}{bb}} = \frac{\sqrt{ab}}{\sqrt{b^2}} = \frac{\sqrt{ab}}{b}$$

Example 7

$$\sqrt{\frac{3}{7}} = \sqrt{\frac{3 \cdot 7}{7 \cdot 7}} = \frac{\sqrt{21}}{\sqrt{7^2}} = \frac{\sqrt{21}}{7}$$

$$\sqrt{\frac{3R}{C}} = \sqrt{\frac{3RC}{C \cdot C}} = \frac{\sqrt{3RC}}{\sqrt{C^2}} = \frac{\sqrt{3RC}}{C}$$

$$\sqrt{\frac{2P_1}{5P_2}} = \sqrt{\frac{(2P_1)(5P_2)}{(5P_2)(5P_2)}} = \frac{\sqrt{10P_1P_2}}{\sqrt{(5P_2)^2}} = \frac{\sqrt{10P_1P_2}}{5P_2}$$

Exercises / Section 10.5

In Exercises 1–40, simplify the given radicals.

1. $\sqrt{8}$ **2.** $\sqrt{18}$ **3.** $\sqrt{48}$ **4.** $\sqrt{75}$ **5.** $\sqrt{125}$

6. $\sqrt{98}$ **7.** $\sqrt{12}$ **8.** $\sqrt{24}$ **9.** $\sqrt{72}$ **10.** $\sqrt{40}$

11. $\sqrt{28}$ **12.** $\sqrt{27}$ **13.** $\sqrt[3]{54}$ **14.** $\sqrt[3]{16}$ **15.** $\sqrt[3]{32}$

16. $\sqrt[4]{32}$ **17.** $\sqrt[5]{64}$ **18.** $\sqrt[3]{24}$ **19.** $\sqrt[4]{48}$ **20.** $\sqrt[6]{128}$

21. $\sqrt{x^5}$ **22.** $\sqrt{y^7}$ **23.** $\sqrt{8V^3}$ **24.** $\sqrt{12P^3}$ **25.** $\sqrt{s_1^3 s_2}$

26. $\sqrt{2t_1^3 t_2}$ **27.** $\sqrt{16m^2n^3}$ **28.** $\sqrt{25p^5q^2}$ **29.** $\sqrt{50v_1^2v_2^3v_3}$ **30.** $\sqrt{49w_1w_2^2w_3^3}$

31. $\sqrt[3]{8s^4t^3}$ **32.** $\sqrt[3]{27p^7q^3}$ **33.** $\sqrt[4]{32T_a^5T_b^4}$ **34.** $\sqrt[4]{16V_a^6V_b^5}$ **35.** $\sqrt[5]{32L_1^{10}L_2^{16}}$

36. $\sqrt[5]{T_1^5T_2^7T_3^{10}}$ **37.** $\sqrt[3]{27x^9y^{12}z^{17}}$ **38.** $\sqrt[3]{8v^{12}w^{10}z^5}$ **39.** $\sqrt[4]{32a^8b^{10}c^{14}}$ **40.** $\sqrt[4]{t_1^5t_2^6t_3^{12}t_4^{20}}$

In Exercises 41–52, write each given expression in the form bj. (See Example 4.)

41. $\sqrt{-16}$ **42.** $\sqrt{-36}$ **43.** $\sqrt{-64}$ **44.** $\sqrt{-81}$ **45.** $\sqrt{-32}$

46. $\sqrt{-24}$ **47.** $\sqrt{-72}$ **48.** $\sqrt{-75}$ **49.** $\sqrt{-12}$ **50.** $\sqrt{-28}$

51. $\sqrt{-20}$ **52.** $\sqrt{-18}$

In Exercises 53–80, simplify the given expressions by rationalizing denominators.

53. $\dfrac{1}{\sqrt{6}}$ **54.** $\dfrac{1}{\sqrt{10}}$ **55.** $\dfrac{a}{\sqrt{b}}$ **56.** $\dfrac{b}{\sqrt{p}}$ **57.** $\dfrac{2}{\sqrt{2}}$

58. $\dfrac{5}{\sqrt{5}}$ **59.** $\dfrac{4}{s\sqrt{t}}$ **60.** $\dfrac{7}{2\sqrt{T}}$ **61.** $\dfrac{1}{\sqrt{T_0T_1}}$ **62.** $\dfrac{3}{\sqrt{2V_0V_1}}$

63. $\dfrac{L}{\sqrt{3RC}}$ **64.** $\dfrac{R_1}{\sqrt{R_1R_2}}$ **65.** $\dfrac{a}{\sqrt{2ab}}$ **66.** $\dfrac{MN}{\sqrt{3N}}$ **67.** $\dfrac{\sqrt{C_1C_2}}{\sqrt{C_1}}$

68. $\dfrac{\sqrt{RC}}{\sqrt{R}}$ **69.** $\dfrac{\sqrt{8v}}{\sqrt{2z}}$ **70.** $\dfrac{\sqrt{3s^2}}{\sqrt{st}}$ **71.** $\sqrt{\dfrac{p}{q}}$ **72.** $\sqrt{\dfrac{5V}{3}}$

73. $\sqrt{\dfrac{7P}{5}}$ **74.** $\sqrt{\dfrac{n^2}{st}}$ **75.** $\sqrt{\dfrac{4p^4}{s_1s_2^2}}$ **76.** $\sqrt{\dfrac{8s^4}{t^2v^2w}}$ **77.** $\sqrt{\dfrac{3t_1t_2^2}{2t_1t_2}}$

78. $\sqrt{\dfrac{12ST}{P}}$ **79.** $\sqrt{\dfrac{2V_1}{3V_2}}$ **80.** $\sqrt{\dfrac{7st}{2t^3}}$

81. The frequency f of a series resonance circuit is given by

$$f = \frac{1}{2\pi\sqrt{LC}}$$

Express this formula without a radical in the denominator.

82. The number N of vibrations per unit time of a mass hanging on a spring (Figure 10.1) is

$$N = \frac{1}{2\pi}\sqrt{\frac{k}{m}}$$

Write the formula for N without a radical in the denominator.

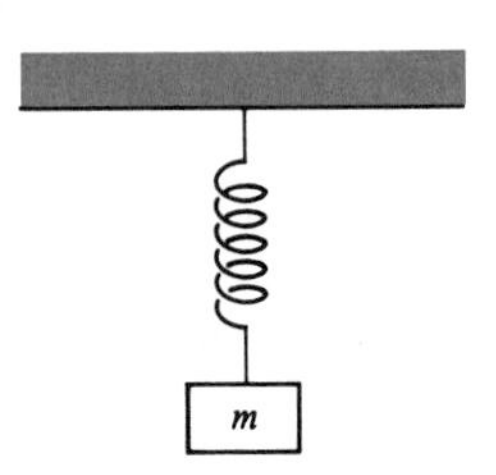

Figure 10.1

83. Rationalize the following expression from a problem in mechanics:

$$\sqrt{\frac{v_0}{mv^2}}$$

84. The radius of a circle is

$$r = \sqrt{\frac{A}{\pi}}$$

Rationalize the expression for r.

85. A manufacturing process is said to be *out of control* if it produces more than

$$3\sqrt{\frac{p(1-p)}{n}}$$

defectives, where p is the proportion of defectives produced under normal operating conditions. Simplify this expression by rationalizing the denominator.

86. The impedance Z in a circuit containing a resistor and capacitor is

$$Z = \sqrt{R^2 + \left(\frac{1}{2\pi fC}\right)^2}$$

Simplify this formula.

87. If the mass m in Figure 10.2 is released from rest and falls through a distance y, then the velocity v is

$$v = \sqrt{\frac{2gmy}{m + \frac{I}{r^2}}}$$

where I is the moment of inertia of the wheel and g the acceleration due to gravity. Simplify the expression for v.

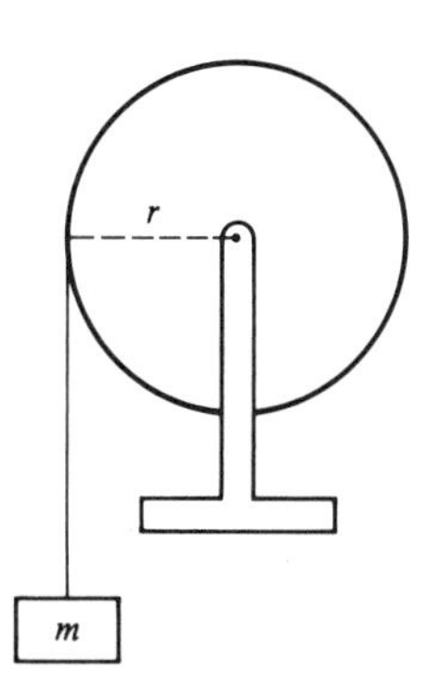

Figure 10.2

88. The maximum current I_m in an alternating current circuit is

$$I_m = \frac{V_m}{\sqrt{R^2 + (X_L - X_C)^2}}$$

Find I_m if $V_m = 120$ V, $R = 250\ \Omega$, $X_L = 110\ \Omega$, and $X_C = 190\ \Omega$.

89. Let c denote the velocity of light in kilometers per second. A body of mass 1 kg moves at half the velocity of light relative to a stationary observer. According to the special theory of relativity, the mass of the object relative to this observer is

$$m = \frac{1}{\sqrt{1 - \frac{(c/2)^2}{c^2}}} \quad \text{(in kilograms)}$$

Simplify the expression for m.

10.6 Operations with Radicals

In this section we make a brief study of the four fundamental operations with radicals.

Addition and Subtraction of Radicals

Addition and subtraction of radicals are similar to the corresponding operations with polynomials. For example, just as

$$2x + 3x = 5x$$

we have

$$2\sqrt{x} + 3\sqrt{x} = 5\sqrt{x}$$

In some cases the radicals have to be simplified before the terms can be combined.

Example 1 Add: $\sqrt{2} + \sqrt{8}$.

Solution. Since $\sqrt{8} = \sqrt{4 \cdot 2} = \sqrt{4}\,\sqrt{2} = 2\sqrt{2}$, we get

$$\sqrt{2} + \sqrt{8} = \sqrt{2} + \sqrt{4 \cdot 2} = \sqrt{2} + 2\sqrt{2} = 3\sqrt{2}$$

To see the last step more clearly, consider the following additions:

Numbers	*Algebraic comparison*
$\sqrt{2} + 2\sqrt{2} = 3\sqrt{2}$	$x + 2x = 3x$

The idea in the next example is similar.

Example 2

$$\begin{aligned} &\sqrt{3} + \sqrt{50} + \sqrt{12} - \sqrt{32} \\ &= \sqrt{3} + \sqrt{25 \cdot 2} + \sqrt{4 \cdot 3} - \sqrt{16 \cdot 2} \\ &= \sqrt{3} + 5\sqrt{2} + 2\sqrt{3} - 4\sqrt{2} \\ &= (\sqrt{3} + 2\sqrt{3}) + (5\sqrt{2} - 4\sqrt{2}) \\ &= 3\sqrt{3} + \sqrt{2} \end{aligned}$$

$x + 2x = 3x,\ 5x - 4x = x$

Addition and subtraction can also be performed with terms that have literal factors.

Example 3 Combine terms:

$$2\sqrt[3]{8a^4} + \sqrt{25b} + 3\sqrt[3]{a} - \sqrt{16b}$$

Solution. We simplify the radicals and combine like terms:

$$\begin{aligned}
&2\sqrt[3]{2^3a^3a} + \sqrt{5^2b} + 3\sqrt[3]{a} - \sqrt{4^2b}\\
&\quad = 2\sqrt[3]{2^3a^3}\,\sqrt[3]{a} + \sqrt{5^2}\,\sqrt{b} + 3\sqrt[3]{a} - \sqrt{4^2}\,\sqrt{b}\\
&\quad = 2(2a)\sqrt[3]{a} + 5\sqrt{b} + 3\sqrt[3]{a} - 4\sqrt{b}\\
&\quad = (4a\sqrt[3]{a} + 3\sqrt[3]{a}) + (5\sqrt{b} - 4\sqrt{b}) && \text{Grouping like terms}\\
&\quad = (4a + 3)\sqrt[3]{a} + (5\sqrt{b} - 4\sqrt{b}) && \text{Factoring } \sqrt[3]{a}\\
&\quad = (4a + 3)\sqrt[3]{a} + \sqrt{b} && \text{Combining terms}
\end{aligned}$$

Multiplication and Division of Radicals

Radicals with the same index can be multiplied by rule (10.11), as we already saw in the previous section:

$$\sqrt[n]{a}\,\sqrt[n]{b} = \sqrt[n]{ab}$$

For example,

$$\sqrt[3]{x^2}\,\sqrt[3]{x^2} = \sqrt[3]{x^2 \cdot x^2} = \sqrt[3]{x^4} = \sqrt[3]{x^3x} = \sqrt[3]{x^3}\,\sqrt[3]{x} = x\sqrt[3]{x}$$

Example 4 Multiply

$$\sqrt[4]{3L^3M^3}\sqrt[4]{2LM^3}$$

Solution. We combine the two radicals by the rule

$$\sqrt[n]{a}\,\sqrt[n]{b} = \sqrt[n]{ab}$$

and simplify the resulting radical:

$$\begin{aligned}
\sqrt[4]{3L^3M^3}\,\sqrt[4]{2LM^3} &= \sqrt[4]{(3L^3M^3)(2LM^3)}\\
&= \sqrt[4]{6L^4M^6} = \sqrt[4]{L^4M^4(6M^2)}\\
&= \sqrt[4]{L^4M^4}\,\sqrt[4]{6M^2} = LM\sqrt[4]{6M^2}
\end{aligned}$$

Binomials containing radicals can be multiplied by the FOIL scheme, as shown next.

Example 5 Multiply: $(2 - \sqrt{3}) \cdot (1 - \sqrt{6})$

Solution. Using the FOIL scheme, we get

$$(2-\sqrt{3})(1-\sqrt{6}) = \overset{F}{(2)(1)} + \overset{O}{(2)(-\sqrt{6})} + \overset{I}{(-\sqrt{3})(1)} + \overset{L}{(-\sqrt{3})(-\sqrt{6})}$$

$$= 2 - 2\sqrt{6} - \sqrt{3} + \sqrt{3 \cdot 6}$$
$$= 2 - 2\sqrt{6} - \sqrt{3} + \sqrt{18}$$
$$= 2 - 2\sqrt{6} - \sqrt{3} + 3\sqrt{2}$$

Example 6 Perform the following division:

$$\frac{\sqrt{5s_1t_1}}{\sqrt{10s_1}}$$

Solution. To divide radicals, we use rule (10.13)

Division of radicals

$$\frac{\sqrt[n]{a}}{\sqrt[n]{b}} = \sqrt[n]{\frac{a}{b}}$$

as we already saw in the last section:

$$\frac{\sqrt{5s_1t_1}}{\sqrt{10s_1}} = \sqrt{\frac{5s_1t_1}{10s_1}} = \sqrt{\frac{t_1}{2}} = \sqrt{\frac{t_1 \cdot 2}{2 \cdot 2}} = \frac{\sqrt{2t_1}}{2}$$

We can see from Example 6 that the division of two radicals with the same index is essentially a problem in rationalizing the denominator.

To describe the procedure for rationalizing a fraction with a binomial in the denominator, it is helpful to introduce the word **conjugate**.

Conjugate

The conjugate of $a + b$ is $a - b$ and the conjugate of $a - b$ is $a + b$.

This definition can be used to state the rule for rationalizing binomial denominators containing square roots.

To **rationalize** a fraction in which the denominator contains a sum or difference of two terms, at least one of which is a square root, multiply the numerator and denominator by the conjugate of the denominator.

This technique is illustrated in the remaining examples.

Example 7 Simplify the fraction

$$\frac{1}{\sqrt{2}-1}$$

Solution. By the conjugate rule, we multiply the numerator and denominator of the given fraction by $\sqrt{2}+1$, the conjugate of $\sqrt{2}-1$:

$$\frac{1}{\sqrt{2}-1} = \frac{1}{\sqrt{2}-1} \cdot \frac{\sqrt{2}+1}{\sqrt{2}+1} \qquad \text{Rationalizing the denominator}$$

$$= \frac{\sqrt{2}+1}{(\sqrt{2}-1)(\sqrt{2}+1)} \qquad \text{Multiplying the fractions}$$

$$= \frac{\sqrt{2}+1}{(\sqrt{2})^2-(1)^2} \qquad (a-b)(a+b) = a^2-b^2$$

$$= \frac{\sqrt{2}+1}{2-1} = \frac{\sqrt{2}+1}{1} = \sqrt{2}+1$$

Example 8 Divide $\sqrt{a}$ by $\sqrt{a}+\sqrt{b}$.

Solution. Writing the division problem as a fraction and rationalizing the denominator, we get

$$\frac{\sqrt{a}}{\sqrt{a}+\sqrt{b}} = \frac{\sqrt{a}}{\sqrt{a}+\sqrt{b}} \cdot \frac{\sqrt{a}-\sqrt{b}}{\sqrt{a}-\sqrt{b}} \qquad \text{Rationalizing the denominator}$$

$$= \frac{\sqrt{a}(\sqrt{a}-\sqrt{b})}{(\sqrt{a}+\sqrt{b})(\sqrt{a}-\sqrt{b})} \qquad \text{Multiplying}$$

$$= \frac{\sqrt{a}(\sqrt{a}-\sqrt{b})}{(\sqrt{a})^2-(\sqrt{b})^2} \qquad (x+y)(x-y) = x^2-y^2$$

$$= \frac{(\sqrt{a})^2-\sqrt{a}\sqrt{b}}{a-b}$$

$$= \frac{a-\sqrt{ab}}{a-b} \qquad (\sqrt{a})^2 = a,\ \sqrt{a}\sqrt{b} = \sqrt{ab}$$

Example 9 The time required to drain the contents of a vessel from level L_1 to level L_2 through an opening in the bottom is given by

$$t = \frac{k(L_1-L_2)}{\sqrt{2g}(\sqrt{L_1}+\sqrt{L_2})}$$

where g is the acceleration due to gravity and k a constant. Simplify the expression for t.

Solution. Since the conjugate of $\sqrt{L_1} + \sqrt{L_2}$ is $\sqrt{L_1} - \sqrt{L_2}$, we get

$$t = \frac{k(L_1 - L_2)}{\sqrt{2g}(\sqrt{L_1} + \sqrt{L_2})} \cdot \frac{\sqrt{L_1} - \sqrt{L_2}}{\sqrt{L_1} - \sqrt{L_2}}$$

$$= \frac{k(L_1 - L_2)(\sqrt{L_1} - \sqrt{L_2})}{\sqrt{2g}[(\sqrt{L_1})^2 - (\sqrt{L_2})^2]}$$

$$= \frac{k\cancel{(L_1 - L_2)}(\sqrt{L_1} - \sqrt{L_2})}{\sqrt{2g}\cancel{(L_1 - L_2)}}$$

$$= \frac{k(\sqrt{L_1} - \sqrt{L_2})}{\sqrt{2g}} \frac{\sqrt{2g}}{\sqrt{2g}}$$

$$= \frac{k\sqrt{2g}(\sqrt{L_1} - \sqrt{L_2})}{2g}$$

Exercises / Section 10.6

In Exercises 1–20, combine the given radicals.

1. $\sqrt{5} - 2\sqrt{5}$
2. $\sqrt{7} - 3\sqrt{7}$
3. $\sqrt{2} + \sqrt{50}$
4. $\sqrt{2} + \sqrt{32}$
5. $2\sqrt{18} - \sqrt{32} + \sqrt{50}$
6. $\sqrt{12} + \sqrt{27} - 4\sqrt{3}$
7. $\sqrt{27} + \sqrt{72} - 2\sqrt{75} + \sqrt{8}$
8. $\sqrt[3]{2} + 3\sqrt{75} - \sqrt[3]{16} - \sqrt{48}$
9. $\sqrt{a} + \sqrt{16a}$
10. $\sqrt{4b} - \sqrt{9b}$
11. $\sqrt{25V} - \sqrt{16V} - \sqrt{36V}$
12. $2a\sqrt{P} - \sqrt{a^2P}$
13. $\sqrt{75} + \sqrt{16V} - \sqrt{48} + \sqrt{9V}$
14. $\sqrt[3]{16} + \sqrt{25R} + \sqrt[3]{54} - \sqrt{36R}$
15. $\sqrt{a^2b} + \sqrt{4a^2b} + \sqrt{16a^2b}$
16. $\sqrt{P^2V} - 2\sqrt{9P^2V} + \sqrt{16P^2V}$
17. $\sqrt{R^2C} + 4\sqrt{16N} - 2\sqrt{R^2C} - \sqrt{25N}$
18. $\sqrt{a^3b} - \sqrt{a^5b} - 2\sqrt{ab}$
19. $\sqrt{50} + \sqrt[3]{a} - \sqrt{32} + 2\sqrt[3]{ab^3}$
20. $\sqrt{s^3t} + \sqrt{s^5t} - 4\sqrt{st}$

In Exercises 21–42, perform the indicated multiplications and simplify.

21. $\sqrt{2a}\sqrt{2b}$
22. $\sqrt{3c}\sqrt{3d}$
23. $\sqrt{2y}\sqrt{6yz}$
24. $\sqrt{2V}\sqrt{10VR}$
25. $\sqrt[3]{2t^2}\sqrt[3]{4t^2}$
26. $\sqrt[3]{3V_1^2V_2}\sqrt[3]{9V_2^2}$
27. $\sqrt[3]{9L^2S^2}\sqrt[3]{12L^2S}$
28. $\sqrt[4]{4R^3C^2}\sqrt[4]{8R^3C}$
29. $\sqrt[4]{8S_a^2S_b^3S_c}\sqrt[4]{2S_a^3S_bS_c^2}$
30. $\sqrt[3]{54V_a^2V_b^2}\sqrt[3]{V_aV_b^2}$
31. $\sqrt{3}(1 + \sqrt{3})$
32. $\sqrt{2}(1 + \sqrt{3})$
33. $(1 - \sqrt{3})(1 + \sqrt{3})$
34. $(1 + \sqrt{2})(1 - \sqrt{5})$
35. $(\sqrt{a} - b)(\sqrt{a} + b)$
36. $(\sqrt{b} - 1)(\sqrt{b} + 1)$
37. $(\sqrt{2} + 1)(\sqrt{2} - 3)$
38. $(1 + \sqrt{5})(2 - 3\sqrt{5})$
39. $(1 - 2\sqrt{6})(2 + \sqrt{6})$
40. $(\sqrt{2} + \sqrt{3})(\sqrt{2} - \sqrt{3})$
41. $(\sqrt{6} + \sqrt{2})(2\sqrt{6} - 3\sqrt{2})$
42. $(3\sqrt{7} - 2)(1 + 2\sqrt{3})$

In Exercises 43–52, perform the indicated divisions and simplify.

43. $\dfrac{\sqrt{a}}{\sqrt{5}}$
44. $\dfrac{\sqrt{x}}{\sqrt{ab}}$
45. $\dfrac{\sqrt{3}}{\sqrt{t}}$

46. $\frac{\sqrt{5}}{\sqrt{7}}$

47. $\frac{2\sqrt{ab}}{\sqrt{b}}$

48. $\frac{\sqrt{x}}{\sqrt{6}}$

49. $\frac{\sqrt{2a}}{\sqrt{7}}$

50. $\frac{\sqrt{3R}}{\sqrt{C}}$

51. $\frac{\sqrt{5C_1}}{\sqrt{C_2}}$

52. $\frac{\sqrt{aS_a}}{\sqrt{S_a}}$

In Exercises 53–66, perform the indicated divisions and simplify.

53. $\sqrt{3} \div (1 - \sqrt{5})$

54. $\sqrt{6} \div (\sqrt{3} + 2)$

55. $\sqrt{7} \div (1 - \sqrt{2})$

56. $\sqrt{2} \div (2 - \sqrt{3})$

57. $\frac{3}{\sqrt{5} + \sqrt{3}}$

58. $\frac{4}{\sqrt{3} - \sqrt{2}}$

59. $\frac{\sqrt{3} + 1}{\sqrt{5} - 1}$

60. $\frac{1 - \sqrt{2}}{1 + \sqrt{3}}$

61. $\frac{a}{\sqrt{a} + b}$

62. $\frac{b}{\sqrt{c} - d}$

63. $\frac{\sqrt{R}}{1 - \sqrt{C}}$

64. $\frac{\sqrt{T_0}}{\sqrt{T_1} - 1}$

65. $\frac{\sqrt{3} - \sqrt{t}}{\sqrt{3} + \sqrt{t}}$

66. $\frac{2 - \sqrt{V}}{2 + \sqrt{V}}$

67. Simplify the following expression from a problem in the study of motion:

$$\frac{\sqrt{v}}{\sqrt{m} + \sqrt{v}}$$

68. Simplify the following expression from a problem in fluid dynamics:

$$\frac{2}{1 - \frac{\rho}{\sqrt{v}}} \qquad \rho = \text{rho}$$

69. The coefficient of reflection R is the amplitude of the reflected wave divided by the amplitude of the incident wave. If two ropes whose masses per unit length are μ_1 and μ_2, respectively, lie along the x-axis and are joined at the origin, then

$$R = \frac{\sqrt{\mu_1} - \sqrt{\mu_2}}{\sqrt{\mu_1} + \sqrt{\mu_2}} \qquad \mu = \text{mu}$$

Simplify the expression for R.

70. Simplify the following expression, which arose in a problem in the study of the expansion of a gas:

$$\frac{\sqrt{V_2}}{\sqrt{V_2} - \sqrt{V_1}}$$

Review Exercises / Chapter 10

In Exercises 1 and 2, change to radical form.

1. $a^{1/4}$

2. $T^{2/3}$

In Exercises 3–12, evaluate the given expressions.

3. 6^{-2} **4.** $(-2)^{-4}$ **5.** $4^0 \cdot 6^{-1}$ **6.** $\frac{3^0 \cdot 5^{-3}}{5^{-4}}$

7. $\frac{(3^{-2})^{-1}}{4^{-1}}$ **8.** $16^{-3/4}$ **9.** $27^{-2/3}$ **10.** $\frac{5^0 \cdot 4^{-1/2}}{5^{-1}}$

11. $\frac{4^{3/2}}{8^{-1/3}}$ **12.** $\frac{(-32)^{-1/5}}{25^{-3/2}}$

In Exercises 13–24, simplify the given expressions and write the results without zero or negative exponents.

13. $(-2cd^2)^3$ **14.** $\frac{c^0x^4}{x^{-6}y^{-1}}$ **15.** $\frac{6^{-1}A^{-1}B^{-1}}{2^{-2}A^{-3}B^4}$ **16.** $\frac{(x^{-2}z^{-1})^{-3}}{3^{-1}x^{-2}}$

17. $\frac{(T_1^{-3}T_2^2)^{-2}}{T_1^{-3}T_2}$ **18.** $\frac{V_aV_b^{-8}}{(V_aV_b^2)^{-2}}$ **19.** $\frac{R^{-1}+3}{R^{-2}+2}$ **20.** $\frac{S^2 - S^{-2}}{S^2}$

21. $\frac{2^{-1/2}s_1^{-2}s_2^{1/3}}{2^{1/4}s_1^{1/2}s_2^{-1/6}}$ **22.** $\frac{4^0p^{-2/3}q^{-1/8}}{4^{-1/2}p^{-1/9}q^{-1/4}}$ **23.** $(V^{-3/4}R^{-1/4})^{-4}$ **24.** $(8M^{-1/3}N^{-6})^{-2/3}$

In Exercises 25–28, express the given numbers in scientific notation.

25. 43,000,000 **26.** 983,000,000 **27.** 0.00000012 **28.** 0.0000000000930

In Exercises 29–32, express the given numbers in decimal notation.

29. 3.60×10^{-8} **30.** 1.400×10^{-7} **31.** 2.3×10^6 **32.** 6.34×10^{10}

In Exercises 33–44, use a calculator to perform the indicated operations.

33. $(5.60 \times 10^{-14}) \cdot (7.8 \times 10^{-8})$

34. $(7.63 \times 10^5) \cdot (9.30 \times 10^{-20})$

35. $\frac{1.84 \times 10^{-7}}{2.83 \times 10^{-15}}$

36. $\frac{3.9 \times 10^{-10}}{4.66 \times 10^6}$

37. $2.40 \times 10^{-9} + 9.63 \times 10^{-9}$

38. $4.76 \times 10^{-6} - 8.43 \times 10^{-6}$

39. $(0.000000000084) \cdot (0.0000000092)$

40. $\frac{9{,}200{,}000}{0.00000842}$

41. $\frac{0.000008464}{0.000000000000742}$

42. $(1.236)^{3.407}$

43. $(2.00 + \sqrt[3]{1.40})^{1.76}$

44. $[(3.64)^{2.76} - (2.84)^{-1.24}]^{0.0574}$

In Exercises 45–48, find the indicated roots without a calculator.

45. $\sqrt[3]{-64}$

46. $\sqrt[4]{\frac{1}{16}}$

47. $\sqrt{0.36}$

48. $\sqrt[3]{0.027}$

In Exercises 49–52, state whether the given numbers are rational (r) or irrational (i).

49. $\sqrt{81}$

50. $\sqrt{22}$

51. $\sqrt{\frac{1}{121}}$

52. $\sqrt{\frac{1}{10}}$

In Exercises 53–60, simplify the given expressions.

53. $\sqrt{32a^3b^6}$

54. $\sqrt[4]{16P_1^6P_2^9}$

55. $\sqrt[5]{64m^6n^{12}}$

56. $\dfrac{1}{\sqrt{20}}$

57. $\dfrac{P}{\sqrt{2Q}}$

58. $\dfrac{LV}{\sqrt{50LV}}$

59. $\sqrt{\dfrac{3s_1s_2}{9s_2}}$

60. $\sqrt{\dfrac{10A^2B}{16B}}$

In Exercises 61 and 62, write the given expressions in the form *bj*.

61. $\sqrt{-48}$

62. $\sqrt{-108}$

In Exercises 63–76, perform the indicated operations.

63. $6\sqrt{2} - \sqrt{50}$

64. $\sqrt{72} + \sqrt{C^2V} - \sqrt{32} + \sqrt{36C^2V}$

65. $\sqrt{3a}\sqrt{6b}$

66. $\sqrt[3]{6P^2V}\sqrt[3]{9P^2V^2}$

67. $\sqrt[4]{8M^3N^2}\sqrt[4]{8M^2N^2}$

68. $(A - \sqrt{B})(A + \sqrt{B})$

69. $(1 - 2\sqrt{5})(2 + 3\sqrt{5})$

70. $(\sqrt{3} - \sqrt{5})(2\sqrt{3} + 3\sqrt{5})$

71. $\dfrac{\sqrt{P}}{\sqrt{3P}}$

72. $\dfrac{\sqrt{5S}}{\sqrt{S}}$

73. $\dfrac{\sqrt{b}}{a + \sqrt{b}}$

74. $\dfrac{\sqrt{5} - \sqrt{2}}{\sqrt{5} + \sqrt{2}}$

75. $\dfrac{1 - \sqrt{5}}{2 + \sqrt{7}}$

76. $\dfrac{\sqrt{T_1}}{\sqrt{T_1} - \sqrt{T_2}}$

In Exercises 77–81, change the given numbers from scientific to decimal notation or from decimal to scientific notation.

77. The speed of light is 186,000 mi/s.

78. The average distance from the earth to the sun is 93,000,000 mi.

79. The capacitance of a certain capacitor is 3.6×10^{-6} F.

80. A typical steel wire will break if pulled with a force of 6×10^4 lb/in.^{2}.

81. The resistance of a copper wire is 0.000063 Ω/in.

82. In determining the period of a pendulum, the expression $(1 - n^{-1})^{-1}$ has to be simplified. Carry out this simplification.

83. The Pascher series, which arises in the study of the hydrogen atom, is

$$\lambda^{-1} = R(3^{-2} - n^{-2}) \qquad \lambda = \text{lambda}$$

Solve this formula for λ and simplify the resulting expression.

84. The nearest galaxy, the Andromeda Nebula, is 1.17×10^{19} mi away. Given that the velocity of light is 186,000 mi/s, about how many years does it take for the light from this galaxy to reach us?

85. The percent efficiency of an engine is given by

$$\text{Eff}(\%) = 100\left(1 - \frac{1}{(V_2/V_1)^{\gamma-1}}\right) \qquad \gamma = \text{gamma}$$

Show that this formula can be written

$$\text{Eff}(\%) = 100\left(\frac{V_1^{1-\gamma} - V_2^{1-\gamma}}{V_1^{1-\gamma}}\right)$$

86. The current i in a circuit varies with time t (in seconds) according to

$$i = (15.5)^{-1}[1 - (2.72)^{(-15.5/0.00348)t}]$$

Find i when $t = 1.50$ ms.

87. The period of a pendulum is

$$T = 2\pi\sqrt{\frac{L}{g}}$$

Write this formula without a radical in the denominator.

88. According to the special theory of relativity, the mass m of a body moving at velocity v relative to a stationary observer is

$$m = \frac{m_0}{\sqrt{1 - \dfrac{v^2}{c^2}}}$$

Simplify this formula.

CHAPTER 11

Quadratic Equations

The equations we have encountered so far have been of first degree. In this chapter we study equations of second degree. Such equations arise in many areas of technology.

11.1 Solution by Factoring and Pure Quadratic Equations

An equation is called a **quadratic equation** if the unknown is raised to the second power. For example, the equation $2x^2 + 3x - 1 = 0$ is a quadratic equation because of the term $2x^2$.

Many problems in science and technology lead to quadratic equations. For example, if an object is hurled into the air at 18 ft/s from a height of 5 ft, then the distance s (in feet) above the ground is given by $s = -16t^2 + 18t + 5$, where t is measured in seconds. Other applications will be studied in Section 11.3.

The definition of quadratic equation is given next.

Quadratic equation

A quadratic equation has the form

$$ax^2 + bx + c = 0, \qquad a \neq 0 \tag{11.1}$$

where a, b, and c are constants.

The form $ax^2 + bx + c = 0$ is called the **standard form** of the quadratic equation. For example, if we subtract 6 from both sides of the quadratic equation

$$2x^2 - 3x = 6$$

we obtain the standard form

$$2x^2 - 3x - 6 = 0$$

For the time being we will confine ourselves to equations in which the left side of $ax^2 + bx + c = 0$ is factorable. To solve such an equation, we factor the left side and set each factor equal to zero. To justify this procedure, we use the following property of the real number system:

$$ab = 0 \quad \text{if and only if} \quad a = 0 \quad \text{or} \quad b = 0$$

For example, the equation $x^2 - x - 6 = 0$ can be written

$$(x - 3)(x + 2) = 0$$

Now observe that if $x = 3$, then $x - 3 = 0$. As a result,

$$(x - 3)(x + 2) = (3 - 3)(3 + 2) = 0 \cdot 5 = 0$$

It follows that $x = 3$ satisfies the equation. Similarly, if $x = -2$, then $x + 2 = 0$, and

$$(x - 3)(x + 2) = (-2 - 3)(-2 + 2) = -5 \cdot 0 = 0$$

So $x = -2$ also satisfies the equation. We see that the equation has two roots, $x = 3$ and $x = -2$.

The procedure for solving quadratic equations by factoring is summarized next.

Method of solution by factoring

1. Write the equation in standard form by collecting all the terms on the left side.
2. Factor the expression on the left side.
3. Set each factor equal to zero.
4. Solve the resulting two first-degree equations.
5. Check the roots in the original equation.

Example 1 Solve the following equation by the method of factoring:

$$x^2 + 2x = 24$$

Solution.

Step 1. To obtain the standard form, we subtract 24 from both sides:

$x^2 + 2x = 24$	Given equation
$x^2 + 2x - 24 = 24 - 24$	Subtracting 24 from both sides
$x^2 + 2x - 24 = 0$	Standard form

Step 2. $(x - 4)(x + 6) = 0$ Factoring left side

Step 3. $x - 4 = 0$ $\quad x + 6 = 0$ Setting each factor equal to zero

Step 4. $x = 4$ $\quad x = -6$ Solving each first-degree equation

Step 5. The roots should be checked in the original equation, not in the later versions, since an error could have occurred in any of the steps.

$$x = 4: \quad 4^2 + 2 \cdot 4 = 16 + 8 = 24$$
$$x = -6: \quad (-6)^2 + 2(-6) = 36 - 12 = 24 \checkmark$$

While the definition of quadratic equation was given in terms of the variable x, in technical problems we often use different letters, as we have seen many times before.

Example 2 Solve the equation

$$2t^2 = 24 - 13t$$

Solution.

Step 1. Changing the equation to standard form, we get

$$2t^2 = 24 - 13t$$
$$2t^2 - 24 + 13t = 24 - 13t - 24 + 13t$$
$$2t^2 + 13t - 24 = 0$$

Step 2. $(2t - 3)(t + 8) = 0$ Factoring left side

Step 3. $2t - 3 = 0$ $\quad t + 8 = 0$ Setting each factor equal to zero

Step 4. $2t - 3 + 3 = 0 + 3$ $\quad t + 8 - 8 = 0 - 8$ Solving each first-degree equation

$$2t = 3 \qquad t = -8$$
$$t = \frac{3}{2} \qquad t = -8$$

Step 5. *Check:*

Left side / *Right side*

$$t = \frac{3}{2}: \quad 2\left(\frac{3}{2}\right)^2 = 2\left(\frac{9}{4}\right) = \frac{9}{2} \qquad 24 - 13\left(\frac{3}{2}\right) = \frac{24 \cdot 2}{2} - \frac{13 \cdot 3}{2} = \frac{48 - 39}{2} = \frac{9}{2} \checkmark$$

$$t = -8: \quad 2(-8)^2 = 2(64) = 128 \qquad 24 - 13(-8) = 24 + 104 = 128 \checkmark$$

In the next example, the left side leads to two identical factors.

Example 3 Solve the equation

$$9R^2 - 12R + 4 = 0$$

Solution.

Step 1. Since the given equation is already in standard form, no rearrangement is necessary.

Step 2. If we factor the left side, we obtain two identical factors:

$$(3R - 2)^2 = 0$$

or

$$(3R - 2)(3R - 2) = 0$$

Step 3. $\quad 3R - 2 = 0 \qquad 3R - 2 = 0$

Step 4. $\quad 3R = 2 \qquad 3R = 2$

$$R = \frac{2}{3} \qquad R = \frac{2}{3}$$

Double root

The solution $R = \frac{2}{3}, \frac{2}{3}$ is called a **double root.**

Step 5. We substitute $R = \frac{2}{3}$ into the given equation:

$$9\left(\frac{2}{3}\right)^2 - 12\left(\frac{2}{3}\right) + 4 = 9\left(\frac{4}{9}\right) - 12\left(\frac{2}{3}\right) + 4 = 4 - 8 + 4 = 0 \quad ✓$$

If $c = 0$ in the standard form $ax^2 + bx + c = 0$, the left side has common factor x, as shown in the next example.

Example 4

$2x^2 - x = 0$		Given equation
$x(2x - 1) = 0$		Common factor x
$x = 0$	$2x - 1 = 0$	Solving the resulting first-degree equations
$x = 0$	$x = \frac{1}{2}$	

So the roots are $x = 0$ and $x = \frac{1}{2}$.

Common error Canceling x when solving $ax^2 + bx = 0$. For example, if both sides of the equation $x^2 - 2x = 0$ are divided by x, we get

$$\frac{x^2}{x} - \frac{2x}{x} = \frac{0}{x}$$

$$x - 2 = 0 \quad \text{or} \quad x = 2$$

and the root $x = 0$ is lost. The correct procedure is to *factor* the x and set each factor equal to 0. Thus

$$x^2 - 2x = 0$$
$$x(x - 2) = 0$$
$$x = 0 \qquad x - 2 = 0$$
$$x = 2$$

So the roots are $x = 0$ and $x = 2$.

Pure Quadratic Equations

If $b = 0$ in the standard form $ax^2 + bx + c = 0$, we get

$$ax^2 + c = 0$$

Pure quadratic equation

which is called a **pure quadratic equation.**

Pure quadratic equations can sometimes be solved by the method of factoring. For example, the equation $x^2 - 4 = 0$ can be written

$$(x - 2)(x + 2) = 0$$

so that $x = 2, -2$. However, the method of factoring does not work for the equation in the next example.

Example 5 Solve the pure quadratic equation

$$x^2 - 5 = 0$$

Solution. This equation cannot be factored in the usual sense. However, we can still obtain the solution if we first solve the equation for x^2 and then take the square root of both sides:

$x^2 - 5 = 0$	Given equation
$x^2 - 5 + 5 = 0 + 5$	Adding 5 to both sides
$x^2 = 5$	Simplifying
$\sqrt{x^2} = \pm\sqrt{5}$	Taking the square root of each side
$x = \pm\sqrt{5}$	

Plus or minus

Remark. The symbol $\pm$ means "plus or minus"; thus, the solution could also be written $x = \sqrt{5}, -\sqrt{5}$. Without the minus sign we get only $x = \sqrt{5}$, which is a positive number by the definition of principal square root.

The procedure for solving pure quadratic equations is summarized next.

Solving pure quadratic equations

1. Solve the equation for x^2.
2. Take the square root of both sides of the equation.

Example 6 Solve the pure quadratic equation

$$3x^2 - 1 = 0$$

Solution. $3x^2 - 1 = 0$ Given equation

Step 1. $3x^2 - 1 + 1 = 0 + 1$ Adding 1 to both sides

$$3x^2 = 1$$

$$3x^2 = 1$$

$$x^2 = \frac{1}{3}$$ Dividing by 3

Step 2. $\sqrt{x^2} = \pm\sqrt{\frac{1}{3}}$ Taking square roots

$$x = \pm\frac{\sqrt{1}}{\sqrt{3}} = \pm\frac{1}{\sqrt{3}} \qquad \sqrt{\frac{a}{b}} = \frac{\sqrt{a}}{\sqrt{b}}$$

$$x = \pm\frac{1}{\sqrt{3}} \cdot \frac{\sqrt{3}}{\sqrt{3}}$$ Rationalizing the denominator

$$x = \pm\frac{\sqrt{3}}{3}$$

To check our solution, we can substitute both roots at the same time:

$$3\left(\pm\frac{\sqrt{3}}{3}\right)^2 - 1 = 3\,\frac{(\sqrt{3})^2}{3^2} - 1 \qquad (\pm a)^2 = a^2$$

$$= 3 \cdot \frac{3}{9} - 1 = \frac{9}{9} - 1 = 1 - 1 = 0 \ \checkmark$$

Pure quadratic equations may lead to imaginary roots. In this case, the roots have to be written in the form aj. For example,

$$\sqrt{-9} = \sqrt{9(-1)} = \sqrt{9}\,\sqrt{-1} = 3j$$

(See Section 10.5.)

Example 7 Solve the pure quadratic equation

$$4y^2 + 5 = 0$$

Solution. $4y^2 + 5 = 0$

Step 1. $4y^2 + 5 - 5 = 0 - 5$ Subtracting 5 from both sides

$$4y^2 = -5$$

$$4y^2 = -5$$

$$y^2 = -\frac{5}{4}$$ Dividing by 4

Step 2.

$$\sqrt{y^2} = \pm\sqrt{-\frac{5}{4}} \qquad \text{Taking square roots}$$

$$y = \pm\sqrt{\frac{5}{4}(-1)}$$

$$= \pm\sqrt{\frac{5}{4}}\sqrt{-1} \qquad \sqrt{ab} = \sqrt{a}\sqrt{b}$$

$$y = \pm\frac{\sqrt{5}}{\sqrt{4}}j \qquad \sqrt{\frac{a}{b}} = \frac{\sqrt{a}}{\sqrt{b}},\ \sqrt{-1} = j$$

$$y = \pm\frac{\sqrt{5}}{2}j$$

Check:

$$4\left(\pm\frac{\sqrt{5}}{2}j\right)^2 + 5 = 4\left(\frac{\sqrt{5}}{2}\right)^2(j)^2 + 5 \qquad (\pm a)^2 = a^2$$

$$= 4\left(\frac{\sqrt{5}}{2}\right)^2(-1) + 5 \qquad j^2 = -1$$

$$= 4\,\frac{(\sqrt{5})^2}{2^2}(-1) + 5 \qquad \left(\frac{a}{b}\right)^2 = \frac{a^2}{b^2}$$

$$= 4\,\frac{5}{4}(-1) + 5 = -5 + 5 = 0 \checkmark$$

Common error Forgetting the negative root when solving $x^2 - a = 0$. The roots are $x = \pm\sqrt{a}$, not $x = \sqrt{a}$ alone.

Exercises / Section 11.1

In Exercises 1–20, solve the given pure quadratic equations.

1. $x^2 - 9 = 0$ **2.** $x^2 - 16 = 0$ **3.** $x^2 - 49 = 0$

4. $x^2 - 121 = 0$ **5.** $x^2 - 5 = 0$ **6.** $x^2 - 2 = 0$

7. $9x^2 - 1 = 0$ **8.** $4x^2 - 1 = 0$ **9.** $16x^2 - 3 = 0$

10. $25x^2 - 7 = 0$ **11.** $3x^2 - 2 = 0$ **12.** $6x^2 - 5 = 0$

13. $y^2 + 4 = 0$ **14.** $z^2 + 1 = 0$ **15.** $m^2 + 7 = 0$

16. $n^2 + 5 = 0$ **17.** $4a^2 + 7 = 0$ **18.** $3b^2 + 5 = 0$

19. $7R^2 + 1 = 0$ **20.** $5C^2 + 2 = 0$

In Exercises 21–56, solve the given quadratic equations by the method of factoring.

21. $x^2 - x = 0$ **22.** $x^2 + 2x = 0$ **23.** $3x^2 + x = 0$

24. $3x^2 + 2x = 0$ **25.** $4x^2 = 7x$ **26.** $5x^2 = 2x$

27. $x^2 - 3x + 2 = 0$ **28.** $x^2 + 2x - 3 = 0$ **29.** $x^2 + 3x - 4 = 0$

30. $x^2 - 7x + 10 = 0$ **31.** $x^2 + 7x + 12 = 0$ **32.** $x^2 + 10x + 16 = 0$

33. $2x^2 - 3x + 1 = 0$

34. $2x^2 - x - 1 = 0$

35. $3x^2 + x - 2 = 0$

36. $3x^2 - 4x + 1 = 0$

37. $2y^2 - 7y + 6 = 0$

38. $2z^2 - 3z - 5 = 0$

39. $3s^2 - 7s = 6$

40. $3t^2 + 4t = 4$

41. $3v^2 + 8v + 4 = 0$

42. $3w^2 + 14w + 8 = 0$

43. $4T^2 + 7T = 2$

44. $4V^2 = 5V + 6$

45. $5R^2 = 13R + 6$

46. $6S^2 = 2 - S$

47. $C^2 + 6C + 9 = 0$

48. $r^2 - 8r + 16 = 0$

49. $4p^2 - 4p + 1 = 0$

50. $4q^2 + 12q + 9 = 0$

51. $9m^2 = 6m - 1$

52. $4n^2 + 20n + 25 = 0$

53. $4D^2 = 4D + 3$

54. $4D^2 = 4D + 15$

55. $6v^2 = -19v - 10$

56. $6v^2 = 35 - 11v$

11.2 The Quadratic Formula

The method of solving quadratic equations by factoring, which was discussed in the last section, has one serious drawback: The left side of the equation has to be factorable. Nonfactorable equations can be solved by using a formula called the **quadratic formula,** discussed in this section.

Let's start with the standard form

$$ax^2 + bx + c = 0, \qquad a \neq 0$$

To solve this equation for x in terms of the letters a, b, and c, we first transform the equation into a pure quadratic equation:

$$ax^2 + bx + c - c = 0 - c \qquad \text{Subtracting } c$$

$$ax^2 + bx = -c \qquad \text{Simplifying}$$

$$\frac{ax^2}{a} + \frac{bx}{a} = \frac{-c}{a} \qquad \text{Dividing by } a$$

$$x^2 + \frac{bx}{a} = -\frac{c}{a}$$

Next, we add the quantity

$$\frac{b^2}{4a^2}$$

to both sides, thereby making the left side a perfect square:

$$x^2 + \frac{bx}{a} + \frac{b^2}{4a^2} = -\frac{c}{a} + \frac{b^2}{4a^2}$$

$$\left(x + \frac{b}{2a}\right)^2 = -\frac{c}{a} + \frac{b^2}{4a^2} \qquad \text{Factoring left side}$$

$$= -\frac{4ac}{4a^2} + \frac{b^2}{4a^2} \qquad \text{LCD} = 4a^2$$

$$= \frac{b^2 - 4ac}{4a^2} \qquad \text{Adding fractions}$$

$$\sqrt{\left(x + \frac{b}{2a}\right)^2} = \pm\sqrt{\frac{b^2 - 4ac}{4a^2}}$$ Taking the square root of both sides

$$x + \frac{b}{2a} = \pm\frac{\sqrt{b^2 - 4ac}}{\sqrt{4a^2}}$$ $\sqrt{\frac{x}{y}} = \frac{\sqrt{x}}{\sqrt{y}}$

$$= \pm\frac{\sqrt{b^2 - 4ac}}{2a}$$ $\sqrt{4a^2} = 2a$

$$x + \frac{b}{2a} - \frac{b}{2a} = -\frac{b}{2a} \pm \frac{\sqrt{b^2 - 4ac}}{2a}$$ Subtracting $\frac{b}{2a}$

$$x = \frac{-b \pm \sqrt{b^2 - 4ac}}{2a}$$ Combining fractions

The last equation is called the **quadratic formula** and can be used to solve any quadratic equation.

Quadratic formula

The roots of the quadratic equation

$$ax^2 + bx + c = 0, \qquad a \neq 0$$

are given by

$$x = \frac{-b \pm \sqrt{b^2 - 4ac}}{2a}$$

While the quadratic formula gives the solution of a quadratic equation in one step, some simplification is usually necessary. Consider the next example.

Example 1 Solve the equation

$$x^2 - 2x - 1 = 0$$

Solution. Let us first identify the constants a, b, and c:

$$\overset{a}{1}x^2 - \overset{b}{2}x - \overset{c}{1} = 0 \qquad ax^2 + bx + c = 0$$

We see that $a = 1$, $b = -2$, and $c = -1$. So by the quadratic formula

$$x = \frac{-b \pm \sqrt{b^2 - 4ac}}{2a} = \frac{-(-2) \pm \sqrt{(-2)^2 - 4(1)(-1)}}{2(1)}$$

$$= \frac{2 \pm \sqrt{4 + 4}}{2} = \frac{2 \pm \sqrt{8}}{2}$$

Since $\sqrt{8} = \sqrt{4 \cdot 2} = \sqrt{4}\sqrt{2} = 2\sqrt{2}$, we have

$$x = \frac{2 \pm 2\sqrt{2}}{2} = \frac{2(1 \pm \sqrt{2})}{2} \qquad \text{Factoring 2}$$

$$= \frac{\not{2}(1 \pm \sqrt{2})}{\not{2}} = 1 \pm \sqrt{2} \qquad \text{Cancellation}$$

The roots are therefore given by $x = 1 + \sqrt{2}$ and $x = 1 - \sqrt{2}$, usually written $x = 1 \pm \sqrt{2}$.

Check:

$\underline{x = 1 - \sqrt{2}:}$

$$x^2 - 2x - 1 = (1 - \sqrt{2})^2 - 2(1 - \sqrt{2}) - 1$$

$$= 1 - 2\sqrt{2} + (\sqrt{2})^2 - 2(1 - \sqrt{2}) - 1 \qquad (a - b)^2 = a^2 - 2ab + b^2$$

$$= 1 - 2\sqrt{2} + 2 - 2 + 2\sqrt{2} - 1 = 0 \ \checkmark$$

$\underline{x = 1 + \sqrt{2}:}$

$$x^2 - 2x - 1 = (1 + \sqrt{2})^2 - 2(1 + \sqrt{2}) - 1$$

$$= 1 + 2\sqrt{2} + (\sqrt{2})^2 - 2(1 + \sqrt{2}) - 1 \qquad (a + b)^2 = a^2 + 2a + b^2$$

$$= 1 + 2\sqrt{2} + 2 - 2 - 2\sqrt{2} - 1 = 0 \ \checkmark$$

The given equation may first have to be written in standard form before the quadratic formula can be applied.

Example 2 Solve the equation

$$2R^2 = 3 - 5R$$

Solution. Adding $-3 + 5R$ to both sides, we get

$$2R^2 - 3 + 5R = 3 - 5R - 3 + 5R$$

$$2R^2 + 5R - 3 = 0$$

Note that $a = 2$, $b = 5$, and $c = -3$. So by the quadratic formula,

$$R = \frac{-(5) \pm \sqrt{5^2 - 4(2)(-3)}}{2(2)}$$

$$= \frac{-5 \pm \sqrt{25 + 24}}{4} = \frac{-5 \pm \sqrt{49}}{4} = \frac{-5 \pm 7}{4}$$

The roots are therefore given by

$$R = \frac{-5 + 7}{4} = \frac{2}{4} = \frac{1}{2}$$

and

$$R = \frac{-5 - 7}{4} = \frac{-12}{4} = -3$$

In this example the solution could also have been found by the method of factoring. From the standard form $2R^2 + 5R - 3 = 0$, we get

$$(2R - 1)(R + 3) = 0$$

$$R = \frac{1}{2}, -3$$

Some quadratic equations lead to **complex roots,** as shown in the next example.

Example 3 Solve the equation

$$5t^2 + 4t + 2 = 0$$

Solution. Since $a = 5$, $b = 4$, and $c = 2$, we get

$$t = \frac{-4 \pm \sqrt{4^2 - 4(5)(2)}}{2(5)}$$

$$= \frac{-4 \pm \sqrt{16 - 40}}{10} = \frac{-4 \pm \sqrt{-24}}{10}$$

Since $-24 = (4)(6)(-1)$, we get

$$\sqrt{-24} = \sqrt{4 \cdot 6 \cdot (-1)} = \sqrt{4}\sqrt{6}\sqrt{-1} = 2\sqrt{6}j$$

since $j = \sqrt{-1}$. The solution is now written

$$t = \frac{-4 \pm 2\sqrt{6}j}{10} = \frac{2(-2 \pm \sqrt{6}j)}{10}$$ Factoring 2

$$= \frac{-2 \pm \sqrt{6}j}{5}$$ Dividing the numerator and denominator by 2

Complex roots $$= -\frac{2}{5} \pm \frac{\sqrt{6}}{5}j$$ Dividing by 5

Let us check the root $t = -\frac{2}{5} + \frac{\sqrt{6}}{5}j$. To make our task easier, we first compute t^2:

$$t^2 = \left(-\frac{2}{5} + \frac{\sqrt{6}}{5}j\right)^2$$
$$= \left(-\frac{2}{5}\right)^2 + 2\left(-\frac{2}{5}\right)\left(\frac{\sqrt{6}}{5}j\right) + \left(\frac{\sqrt{6}}{5}j\right)^2$$
$$= \frac{4}{25} - \frac{4\sqrt{6}}{25}j + \frac{6}{25}j^2$$
$$= \frac{4}{25} - \frac{4\sqrt{6}}{25}j - \frac{6}{25} \qquad \text{Since } j^2 = -1$$
$$= -\frac{2}{25} - \frac{4\sqrt{6}}{25}j$$

We now have

$$5t^2 + 4t + 2 = 5\left(-\frac{2}{25} - \frac{4\sqrt{6}}{25}j\right) + 4\left(-\frac{2}{5} + \frac{\sqrt{6}}{5}j\right) + 2$$
$$= -\frac{2}{5} - \frac{4\sqrt{6}}{5}j - \frac{8}{5} + \frac{4\sqrt{6}}{5}j + 2$$
$$= \left(-\frac{2}{5} - \frac{8}{5} + 2\right) + \left(-\frac{4\sqrt{6}}{5}j + \frac{4\sqrt{6}}{5}j\right)$$
$$= \left(-\frac{10}{5} + 2\right) + 0 = -2 + 2 = 0 \quad \checkmark$$

The other root is checked similarly.

The next example illustrates the solution of an equation with a double root.

Example 4 Solve the equation

$$4D^2 - 4D + 1 = 0$$

Solution. From $a = 4$, $b = -4$, and $c = 1$, we get

$$D = \frac{-(-4) \pm \sqrt{(-4)^2 - 4(4)(1)}}{2(4)}$$
$$= \frac{4 \pm \sqrt{16 - 16}}{8}$$
$$= \frac{4 \pm 0}{8}$$
$$D = \frac{1}{2}, \frac{1}{2}$$

(Whenever the radical is zero, we get a double root.)

Common error Placing $2a$ under the radical but not under the $-b$. For example, when solving the equation $x^2 + x - 1 = 0$ it is not correct to write

$$x = -1 \pm \frac{\sqrt{1 + 4}}{2} \qquad \text{Incorrect}$$

The correct solution is

$$x = \frac{-1 \pm \sqrt{1 + 4}}{2} \qquad \text{Correct}$$

$$= \frac{-1 \pm \sqrt{5}}{2}$$

Discriminant

The expression $b^2 - 4ac$ under the radical sign in the quadratic formula is called the **discriminant.** The discriminant determines whether the roots are real or complex. We have seen that a given equation has two distinct real roots if $b^2 - 4ac > 0$ (Examples 1 and 2). The equation has complex roots if $b^2 - 4ac < 0$ (Example 3). Finally, the equation has a double root if $b^2 - 4ac = 0$ (Example 4).

The discriminant can also be used to determine if an equation is factorable: Put the equation in standard form and evaluate the discriminant $b^2 - 4ac$. If the discriminant is a perfect square, then the equation is factorable. However, solving the equation itself is not a great deal more work than finding the discriminant. So if you cannot readily factor the left side, it is better to use the quadratic formula.

Calculator Operation

The numerical values of the roots

$$x = \frac{-b \pm \sqrt{b^2 - 4ac}}{2a}$$

can be found easily with a calculator: Begin by evaluating the radical $\sqrt{b^2 - 4ac}$, and store the resulting value. Next, add $-b$ and divide the sum by $2a$.

To obtain the other root, transfer the contents of the memory to the display, change the sign, and proceed as before. Consider the next example.

Example 5 Use a calculator to solve the equation

$$5.89x^2 + 2.46x - 4.31 = 0$$

Solution. The roots are

$$x = \frac{-2.46 \pm \sqrt{(2.46)^2 - 4(5.89)(-4.31)}}{2(5.89)}$$

We first find the value of the radical and store this value in memory. Then we subtract 2.46 and divide the result by 2(5.89). The sequence is

2.46 [x²] [−] 4 [×] 5.89 [×] 4.31 [+/−]
[=] [√] [STO] [−] 2.46 [=] [÷] 2 [÷] 5.89 [=] → 0.6717156

Now we transfer the value in memory to the display, change the sign, and proceed as before. The sequence is

[MR] [+/−] [−] 2.46 [=] [÷] 2 [÷] 5.89 [=] → −1.0893727

Using three significant digits, the roots are $x = 0.672$ and $x = -1.09$.

Fractional Equations

Fractional equations sometimes lead to quadratic equations, as shown in this last example.

Example 6 Solve the equation

$$\frac{1}{R} + \frac{1}{R-6} = \frac{1}{4}$$

Solution. Recall that the best way to solve a fractional equation is to clear fractions. We multiply both sides of the equation by the LCD $= 4R(R - 6)$:

$$\frac{\mathbf{4R(R-6)}}{\mathbf{1}}\left(\frac{1}{R} + \frac{1}{R-6}\right) = \frac{\mathbf{4R(R-6)}}{\mathbf{1}}\frac{1}{4}$$

$$\frac{4\cancel{R}(R-6)}{1}\frac{1}{\cancel{R}} + \frac{4R\cancel{(R-6)}}{1}\frac{1}{\cancel{R-6}} = \frac{\cancel{4}R(R-6)}{1}\frac{1}{\cancel{4}} \qquad \text{Distributive law; cancellation}$$

$$4(R-6) + 4R = R(R-6)$$

$$4R - 24 + 4R = R^2 - 6R \qquad \text{Removing parentheses}$$

$$8R - 24 = R^2 - 6R \qquad 4R + 4R = 8R$$

$$0 = R^2 - 6R - 8R + 24$$

$$R^2 - 14R + 24 = 0 \qquad \text{Standard form}$$

$$(R-2)(R-12) = 0$$

$$R = 2, 12$$

Exercises / Section 11.2

In Exercises 1–46, solve the given equations by using the quadratic formula.

1. $x^2 - 7x + 12 = 0$ **2.** $x^2 + 2x - 15 = 0$ **3.** $2x^2 + 5x - 3 = 0$

4. $2x^2 - 7x - 15 = 0$ **5.** $x^2 - x - 1 = 0$ **6.** $x^2 - x - 3 = 0$

7. $x^2 = 1 - 2x$
8. $x^2 + 2x - 4 = 0$
9. $x^2 - 2x = 4$
10. $x^2 - 4x = 6$
11. $x^2 - 6x + 9 = 0$
12. $x^2 + 4x + 4 = 0$
13. $9x^2 + 6x + 1 = 0$
14. $4x^2 + 4x + 1 = 0$
15. $4x^2 = 20x - 25$
16. $9x^2 = 24x - 16$
17. $2x^2 - 5 = 0$
18. $3x^2 - 4 = 0$
19. $x^2 - 2x = 0$
20. $x^2 - 5x = 0$
21. $3t^2 = 5t$
22. $7v^2 = 9v$
23. $2v^2 - 2v - 1 = 0$
24. $2t^2 + 4t + 1 = 0$
25. $2p^2 = 4p + 1$
26. $2q^2 = 2q + 3$
27. $4s^2 - 2s - 1 = 0$
28. $4t^2 - 4t - 1 = 0$
29. $3R^2 = 4 - 2R$
30. $3C^2 + 4C - 2 = 0$
31. $3T^2 + T - 1 = 0$
32. $5V^2 = 2 - V$
33. $7m^2 - 4m = 2$
34. $7n^2 + 4n = 1$
35. $x^2 - 2x + 2 = 0$
36. $x^2 - 4x + 5 = 0$
37. $x^2 - 2x + 6 = 0$
38. $x^2 + 4x + 13 = 0$
39. $D^2 + D + 2 = 0$
40. $L^2 - L + 3 = 0$
41. $M^2 = 2M - 4$
42. $2q^2 = q - 2$
43. $3r^2 = -2r - 2$
44. $4R^2 = -2R - 3$
45. $5C^2 - 6C + 3 = 0$
46. $6L^2 - 3L + 2 = 0$

In Exercises 47–54, use a calculator to solve the given equations.

47. $1.02x^2 - 2.04x - 3.16 = 0$
48. $3.60x^2 + 1.96x - 2.07 = 0$
49. $3.96x^2 + 1.85x - 2.23 = 0$
50. $6.47x^2 - 4.34x - 1.09 = 0$
51. $10.6x^2 - 8.63x - 15.4 = 0$
52. $21x^2 - 9.7x - 18 = 0$
53. $12t^2 + 84t + 15 = 0$
54. $25T^2 + 75T + 12 = 0$

In Exercises 55–62, solve the given fractional equations.

55. $2x - \frac{2}{x} + 3 = 0$
56. $3x - \frac{2}{x} + 5 = 0$
57. $\frac{1}{s} - \frac{1}{s+1} = \frac{1}{12}$
58. $\frac{1}{T-1} - \frac{1}{T} = \frac{1}{6}$
59. $\frac{1}{R} + \frac{1}{R+6} = \frac{1}{4}$
60. $\frac{1}{y} - \frac{1}{y+1} = \frac{1}{20}$
61. $\frac{1}{P} + \frac{1}{P-4} = \frac{3}{8}$
62. $\frac{1}{V} + \frac{1}{V+3} = \frac{1}{2}$

11.3 Applications of Quadratic Equations

As noted earlier, many problems in technology lead to quadratic equations. The strategy for setting up quadratic equations from verbal problems is the same as for first-degree equations. Look back to Section 7.4 at this time to recall the guidelines.

For our first example, let us consider a situation in which the equation is given. (See also Exercises 1–8.)

Example 1 A rock is hurled upward from a point 4 m above the ground at the rate of 19 m/s. The distance s (in meters) above the ground varies with time t (in seconds) according to the equation $s = -5t^2 + 19t + 4$. (The instant when the rock is hurled upward corresponds to $t = 0$ s.) When will the rock strike the ground?

Solution. Since s is the distance above the ground, the rock is at ground level when $s = 0$. So the problem is to find the value of t for which

$$-5t^2 + 19t + 4 = 0$$

Solving this equation, we get

$$5t^2 - 19t - 4 = 0 \qquad \text{Multiplying both sides by } -1$$

$$(5t + 1)(t - 4) = 0 \qquad \text{Factoring}$$

$$5t + 1 = 0 \qquad t - 4 = 0 \qquad \text{Solving each first-degree equation}$$

$$t = -\frac{1}{5} \qquad t = 4$$

Since $t = 0$ corresponds to the time when the motion begins, the root $t = -\frac{1}{5}$ has no meaning here. It follows from the root $t = 4$ that the rock hits the ground after 4 seconds.

In most problems we must first set up the equation from the given information, as shown in the next example.

Example 2 Find two positive numbers whose difference is 11 and whose product is 80.

Solution. Let

$$x = \text{smaller number}$$

Then

$$x + 11 = \text{larger number}$$

Since the product is 80, we obtain the equation

$$x(x + 11) = 80$$

$$x^2 + 11x = 80$$

$$x^2 + 11x - 80 = 0 \qquad \text{Subtracting 80 from both sides}$$

$$(x + 16)(x - 5) = 0 \qquad \text{Factoring}$$

$$x = 5, -16$$

Since the numbers we are looking for are positive, we accept only the root $x = 5$. From $x + 11$, the second number is 16.

As a check, note that $16 \cdot 5 = 80$ and $16 - 5 = 11$.

In Examples 1 and 2, only one of the two roots obtained was meaningful. In the next example both roots are acceptable, so the problem has two distinct solutions.

Example 3 A strip of metal 32 cm wide is to be made into a trough by bending up equal sides. (See Figure 11.1.) If the cross-sectional area is to be 120 cm², find the width and depth of the trough.

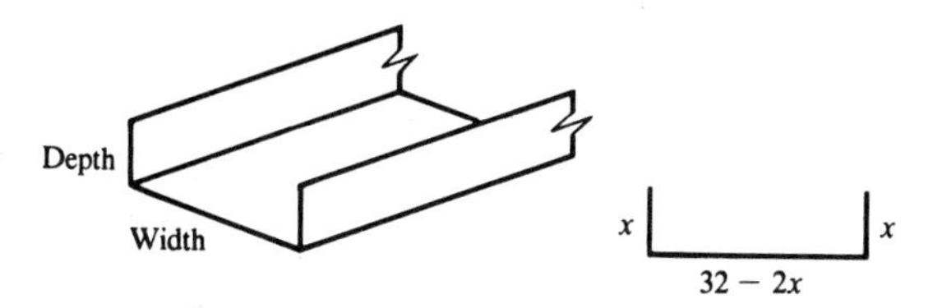

Figure 11.1

Solution. Let

$x =$ depth of trough

Since there are two equal vertical sides having a total length of $2x$ cm, the bottom is $(32 - 2x)$ cm in length. (See Figure 11.1.) Using the fact that the cross-sectional area is 120 cm², we obtain the equation

$$\begin{aligned} x(32 - 2x) &= 120 && A = l \cdot w \\ 32x - 2x^2 &= 120 && \text{Distributive law} \\ -2x^2 + 32x &= 120 \\ -2x^2 + 32x - 120 &= 0 && \text{Subtracting 120} \\ x^2 - 16x + 60 &= 0 && \text{Dividing by } -2 \\ (x - 10)(x - 6) &= 0 && \text{Factoring} \\ x &= 6, 10 \end{aligned}$$

Both roots are acceptable. If $x = 6$ cm, then $32 - 2x = 32 - 2(6) = 20$ cm. If $x = 10$ cm, then $32 - 2x = 32 - 2(10) = 12$ cm. So the dimensions are either

20 cm × 6 cm   or   12 cm × 10 cm

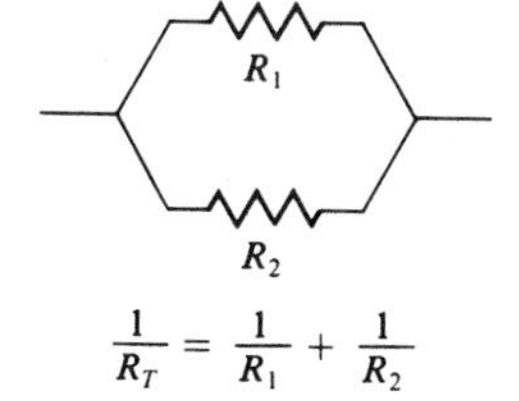

Figure 11.2

If two resistors with respective resistances R_1 and R_2 are connected in parallel, then the combined resistance R_T can be found from the formula

$$\frac{1}{R_T} = \frac{1}{R_1} + \frac{1}{R_2}$$

(See Figure 11.2.) These facts are needed in the next example.

Example 4 Two resistors connected in parallel have a combined resistance of 15.0 Ω. Given that the resistance of one resistor is 10.8 Ω more than that of the other, find the resistance of each.

Solution. Let

R = resistance of first resistor

Then

$R + 10.8$ = resistance of second resistor

Since the combined resistance is 15.0 Ω, we get from the above formula

$$\frac{1}{R} + \frac{1}{R + 10.8} = \frac{1}{15.0}$$

This is an example of a fractional equation that reduces to a quadratic equation after clearing fractions. Multiplying both sides by the LCD = $15.0R(R + 10.8)$, we get

$$\frac{15.0R(R + 10.8)}{1}\left(\frac{1}{R} + \frac{1}{R + 10.8}\right) = \frac{15.0R(R + 10.8)}{1}\,\frac{1}{15.0}$$

$$15.0(R + 10.8) + 15.0R = R(R + 10.8)$$

$$15.0R + 162 + 15.0R = R^2 + 10.8R \qquad \text{Distributive law}$$

$$30.0R + 162 = R^2 + 10.8R \qquad 15.0R + 15.0R = 30.0R$$

$$0 = R^2 + 10.8R - 30.0R - 162$$

$$R^2 - 19.2R - 162 = 0 \qquad \text{Standard form}$$

$$R = \frac{19.2 \pm \sqrt{(-19.2)^2 - 4(1)(-162)}}{2}$$

$$R = 25.5, -6.34$$

Since the negative root has no meaning here, we conclude that $R = 25.5$ Ω. The resistance of the second resistor is therefore $R + 10.8 = 25.5 + 10.8 = 36.3$ Ω.

As a check, note that

$$\frac{1}{25.5} + \frac{1}{36.3} = \frac{1}{15.0}$$

Example 5 Two card-sorting machines used simultaneously can sort a set of cards in 12 min. If one machine at a time is used, then one machine takes 7 min longer than the other to do the job. How long does each one take?

Solution. Let

x = time required for faster machine to do the job

Then

$x + 7$ = time required for slower machine

Now recall from Section 7.4 that we can obtain an equation by finding an expression for the fractional part that can be sorted in one time unit, in this case 1 min. So

$$\frac{1}{x} + \frac{1}{x+7} = \frac{1}{12}$$

$$\frac{12x(x+7)}{1}\left(\frac{1}{x} + \frac{1}{x+7}\right) = \frac{12x(x+7)}{1}\frac{1}{12} \qquad \text{LCD} = 12x(x+7)$$

$$12(x+7) + 12x = x(x+7)$$

$$12x + 84 + 12x = x^2 + 7x$$

$$24x + 84 = x^2 + 7x$$

$$0 = x^2 + 7x - 24x - 84$$

$$x^2 - 17x - 84 = 0$$

$$(x - 21)(x + 4) = 0$$

$$x = 21, -4$$

Disregarding the negative root, we get $x = 21$ min for the faster machine and $x + 7 = 21 + 7 = 28$ min for the slower machine.

Check:

$$\frac{1}{21} + \frac{1}{28} = \frac{1}{3 \cdot 7} + \frac{1}{4 \cdot 7} = \frac{4}{3 \cdot 7 \cdot 4} + \frac{3}{4 \cdot 7 \cdot 3}$$

$$= \frac{4}{84} + \frac{3}{84} = \frac{7}{84} = \frac{1}{12}$$

For the next example, we need to recall the relationships distance = rate × time and time = distance ÷ rate. In symbols,

$$d = rt \qquad \text{and} \qquad t = \frac{d}{r}$$

Example 6 A heavy machine is delivered by truck to a factory 200 mi away. The empty truck makes the return trip 10 mi/h faster and gets back in 1 h less time. Find the rate each way.

Solution. Let

x = rate going

So

$x + 10$ = rate returning

Since the distance is 200 mi for both trips, it follows from the relationship

$$t = \frac{d}{r}$$

that

$$\frac{200}{x}$$

is the time required to reach the factory and

$$\frac{200}{x + 10}$$

is the time required to return. Since the difference in the times is 1 h, we get

$$\frac{200}{x} - \frac{200}{x + 10} = 1$$

$$x(x + 10)\left(\frac{200}{x} - \frac{200}{x + 10}\right) = x(x + 10)1 \qquad \text{LCD} = x(x + 10)$$

$$200(x + 10) - 200x = x^2 + 10x$$

$$200x + 2000 - 200x = x^2 + 10x$$

$$2000 = x^2 + 10x$$

$$x^2 + 10x - 2000 = 0$$

$$(x + 50)(x - 40) = 0$$

$$x = 40, -50$$

Again disregarding the negative root, we obtain $x = 40$ for the slower rate and $x + 10 = 50$ for the faster rate. So the truck travels at 40 mi/h to the factory and at 50 mi/h on the way back.

Exercises / Section 11.3

For Exercises 1–8, see Example 1.

1. A rock is hurled upward from the ground at the rate of 48 ft/s. Its distance d (in feet) above the ground is given by $d = -16t^2 + 48t$, where t is the time in seconds. When will the rock strike the ground?
2. A ball is hurled upward at the rate of 18 m/s from a point 8 m above the ground. Its distance s (in meters) above the ground varies with time according to the equation $s = -5t^2 + 18t + 8$. Determine when the ball will strike the ground. (The motion starts at $t = 0$ s.)
3. The power P (in watts) delivered to a certain element is $P = 100i - 50i^2$. Find the values of the current i (in amperes) for which $P = 0$.
4. The current i (in amperes) varies with time (in seconds) according to $i = 8.4t^2 - 5.6t$. Determine when the current is zero.
5. The deflection d (in meters) of a certain beam is $d = 0.10x^2 - 0.30x$, where x $(0 \leq x \leq 4 \text{ m})$ is the distance from one end. Determine where the deflection is zero.
6. The deflection d of a certain beam is given by $d = 4ax^2 - 3alx$, where l is the length of the beam and x the distance from one end. Determine where the deflection is zero.

7. The power P (in watts) delivered to the resistor R_L in Figure 11.3 is given by

$$P = VI - I^2R$$

If $V = 95.2$ V and $R = 80.0\ \Omega$, find the current I (in amperes) for which $P = 25.3$ W.

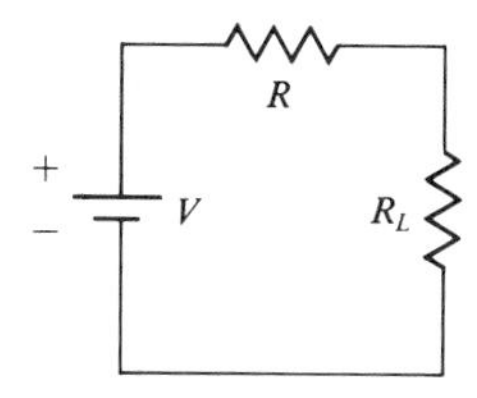

Figure 11.3

8. Repeat Exercise 7, given that $V = 90.0$ V, $R = 90.0\ \Omega$, and $P = 22.5$ W.

For Exercises 9–16, see Example 2.

9. Find two positive numbers whose difference is 6 and whose product is 91.
10. Find two positive numbers whose difference is 8 and whose product is 84.
11. Find, to three significant digits, two numbers whose sum is 15.3 and whose product is 56.9.
12. Find, to three significant digits, two positive numbers whose difference is 1.72 and whose product is 6.93.
13. The manager of a machine shop has an order for a metal plate meeting the following specifications: The length exceeds the width by 5.0 in. and the area is 84 in.^{2} Find the dimensions.
14. A plate in the shape of a parallelogram has an area of 108 cm^2. Find the base and height, given that the base exceeds the height by 3 cm.
15. The cost of carpeting an office at \$10/ft^2 was \$1500. Given that the length of the office exceeds the width by 5 ft, find the dimensions of the office.
16. The perimeter of a rectangular plate is 28 cm and its area is 45 cm^2. Find the dimensions.

For Exercises 17–22, see Example 3.

17. A metal strip 31 cm wide is to be bent into the form of a trough. Find the width and depth if the cross-sectional area of the trough is to be 120 cm^2. (There are two possible solutions.)
18. Repeat Exercise 17, given that the strip is 32 cm wide and the cross-sectional area 128 cm^2.
19. Two rectangular metal plates are welded together. The sum of the perimeter and the seam is 24 in. and the area is 24 in.^{2}. Find the dimensions. (*Hint:* See Figure 11.4.)

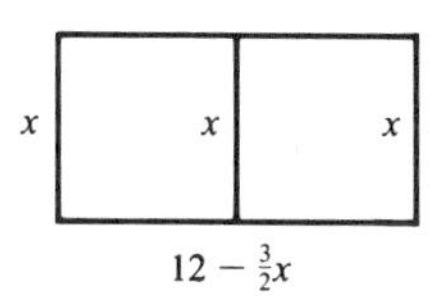

Figure 11.4

20. A rectangular metal casting 0.75 in. thick is to be made from 28.2 in. of forming. (See Figure 11.5.) Given that 31.6 in.^{3} are poured into the form, find the dimensions of the casting.

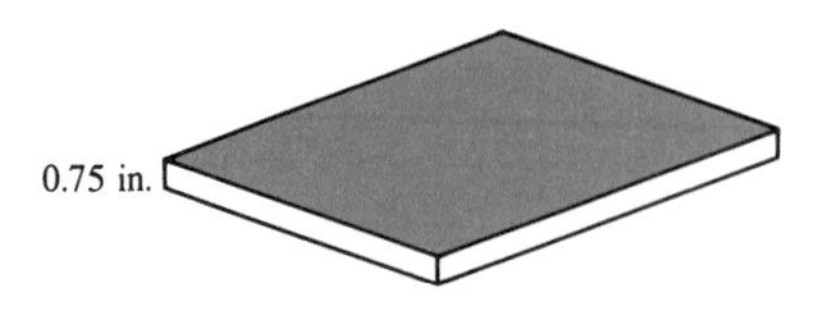

Figure 11.5

21. When a square metal plate is heated, its side increases by 0.30 in. and its area becomes 1.21 times as large as the original area. Find the original length of a side.

22. To cover the floor of a new storage area, 100 square tiles of a certain size are needed. If square tiles 2 in. longer on each side are used, only 64 tiles are needed. What is the size of the smaller tile?

For Exercises 23–26, see Example 4.

23. Two resistors connected in parallel have a combined resistance of 4 Ω. If the resistance of one resistor is 6 Ω more than that of the other, what is the resistance of each?

24. Two resistors connected in parallel have a combined resistance of 7.13 Ω. Find the resistance of each, given that the resistance of one resistor is 4.40 Ω more than that of the other.

25. If two capacitors are connected in series, then the combined capacitance C_T is related to C_1 and C_2 by the formula

$$\frac{1}{C_T} = \frac{1}{C_1} + \frac{1}{C_2}$$

(See Figure 11.6.) If the combined capacitance is 11.9 μF and the capacitance of one is 8.20 μF more than that of the other, find the capacitance of each.

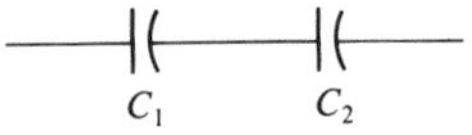

C_1 C_2

$$\frac{1}{C_T} = \frac{1}{C_1} + \frac{1}{C_2}$$

Figure 11.6

26. The thin lens equation relates the object distance p to the image distance q. If f is the focal length of the lens, then

$$\frac{1}{f} = \frac{1}{p} + \frac{1}{q}$$

If $f = 4.0$ cm and q is 6.0 cm less than p, find p and q.

For Exercises 27–30, see Example 5.

27. Two card sorters used simultaneously can sort a set of cards in 20 min. If only one machine is used, one machine requires 9 min more than the other. How long does each one take?

28. If both inlets of a chemical tank are open, the tank can be filled in 6 h. One of the inlets alone requires 5 h longer than the other to fill the tank. How long does each one take?

29. It takes two hours longer to fill a tank than to drain it. If the tank is initially full and both drain and inlet valve are open, the tank is drained in 12 h. How long does it take to fill the tank when the drain is closed?

30. Working together, two men can unload a boxcar in 4 h. Working alone, one man requires 6 h more than the other. How long does it take for each man to do the job alone?

For Exercises 31–39, see Example 6.

31. To stay physically fit, a woman hikes 10 km every Saturday. On one occasion she is in a hurry and walks 1 km/h faster than usual, completing the hike in $\frac{1}{2}$ h less time. Find her usual rate.

32. A boat sails to an island 90 km away at a constant rate. If the rate were increased by 2 km/h, the boat would arrive $\frac{1}{2}$ h sooner. Find the rate.

33. A technician drives to a conference early in the day. Due to heavy morning traffic, her average speed for the first 90 mi is 15 mi/h less than for the second 90 mi and requires 1 h more time. Find the two average speeds.

34. One car travels 30 mi/h faster than another car and covers 100 mi in 3 h less time. Find the rate of each.

35. A delivery truck travels 120 mi to a factory. The empty truck travels 10 mi/h faster on the return trip and arrives in 1 h less time. Find the rate each way.

36. A car travels 25 mi/h faster than a bicycle and arrives at a camping ground 120 mi away in 5 h less time. Find the rate of each.

37. A commuter travels a total of 110 mi per week in rush hour traffic. During other times of the day he could travel 8.00 mi/h faster and cover the same distance in 1.60 h less time. Find his usual rate.

38. Moving downstream, a boat travels 1.58 km/h faster than it does when moving upstream and covers a distance of 25.0 mi in 48 min less time. Find the rate each way.

39. A woman rows 15.6 mi downstream. Returning upstream, her rate is 2.00 mi/h less and takes 1.07 h longer. Find the rate each way.

Review Exercises / Chapter 11

In Exercises 1–6, solve the given pure quadratic equations.

1. $x^2 - 49 = 0$
2. $x^2 + 49 = 0$
3. $4x^2 - 9 = 0$
4. $4x^2 + 9 = 0$
5. $9R^2 - 2 = 0$
6. $3C^2 + 5 = 0$

In Exercises 7–16, solve the given quadratic equations by the method of factoring.

7. $2x^2 - x = 0$
8. $x^2 + 5x = 0$
9. $x^2 - 2x - 3 = 0$
10. $x^2 - 7x + 12 = 0$
11. $2V^2 = 1 - V$
12. $3t^2 + 5t + 2 = 0$
13. $2t^2 - 5t + 2 = 0$
14. $4v^2 = 7v + 2$
15. $4R^2 + 4R + 1 = 0$
16. $4C^2 = 12C - 9$

In Exercises 17–34, solve the given equations by the method of factoring, if possible. Otherwise use the quadratic formula.

17. $x^2 - 7x = 0$
18. $V^2 - 4V + 4 = 0$
19. $S^2 - 2S - 1 = 0$
20. $T^2 + 2T + 2 = 0$
21. $z^2 = 2z + 15$
22. $2y^2 + 7y - 15 = 0$
23. $m^2 + 4m = 6$
24. $t^2 + 2t = -4$
25. $6t^2 + t - 15 = 0$
26. $3v^2 - 2v + 2 = 0$
27. $2P^2 = 3P + 2$
28. $3V^2 + 5V + 2 = 0$

29. $4N^2 + 2N - 3 = 0$

30. $5M^2 - 3M - 2 = 0$

31. $2w^2 - 3w - 1 = 0$

32. $3v^2 = 1 - v$

33. $5y^2 = 1 - 3y$

34. $3z^2 + 7z - 2 = 0$

In Exercises 35 and 36, use a calculator to solve the given equations.

35. $2.04x^2 - 7.30x + 1.74 = 0$

36. $24x^2 + 23x - 1.8 = 0$

In Exercises 37 and 38, solve the given fractional equations.

37. $\frac{1}{x} - \frac{1}{x + 16} = \frac{1}{12}$

38. $\frac{1}{x + 8} + \frac{1}{x} = \frac{1}{3}$

39. The deflection d of a certain beam is given by

$$d = 2L^2 - 3Lx - 2x^2, \qquad L > 0$$

Find the value of x for which the deflection is zero.

40. The power P (in watts) dissipated in a certain circuit is $P = 93.4i - 35.2i^2$. Find the value of the current i (in amperes) for which $P = 10.4$ W.

41. A rock is hurled upward from the ground at $t = 0$ s at the rate of 64 ft/s. Its distance s (in feet) above the ground is $s = -16t^2 + 64t$, where t is measured in seconds. When will the rock strike the ground?

42. A triangular metal plate meets the following specifications: Its base exceeds the height by 2.00 cm and its area is 40.0 cm^2. Find the base and height.

43. Two capacitors connected in series have a combined capacitance of 8.0 μF. The capacitance of one capacitor is 12 μF more than that of the other. Find the capacitance of each. (Recall that $1/C_T = 1/C_1 + 1/C_2$.)

44. It takes 4 h longer to fill a chemical tank than to drain it. If the tank is initially full and both drain and inlet valves are open, then the tank can be drained in 15 h. How long does it take to drain the full tank if the inlet valve is closed?

45. A car travels 120 mi at a uniform rate. If the speed were increased by 10 mi/h, the same distance could be covered in 36 min less time. Find the rate.

46. A car gets 5 mi/gal less in the city than on the highway. Driving a total of 300 mi in city traffic requires 2 gal more gas than driving the same distance on the highway. Determine the gas mileage in the city.

47. A technician buys a number of shares of stock for \$600. If she had paid \$2 less per share, she could have bought 10 more shares. How many shares did she buy?

CHAPTER 12

Functions, Graphs, and Functional Variation

12.1 Functions

Many problems in technology deal with variable quantities. An important relationship between variable quantities is that of a **function,** discussed in this section.

Definition of Function

Suppose a resistor of 50 Ω is in series with a variable resistor R. Then the combined resistance R_T is given by $R_T = 50 + R$ (in ohms). This formula gives the effective resistance of the combination for *any* value of R. Now observe that if $R = 10\ \Omega$, then $R_T = (50 + 10)\ \Omega = 60\ \Omega$. Similarly, if $R = 80\ \Omega$, then $R_T = (50 + 80)\ \Omega = 130\ \Omega$. In other words, every value of R yields a *unique* value of R_T. The relationship $R_T = 50 + R$ is called a *function*.

> **Definition of function**
>
> If two variables x and y are so related that for every value of the variable x there corresponds one, and only one, value of the variable y, then we call y a **function** of x. The variable x is called the **independent** variable and y is called the **dependent** variable.

The variables x and y are representative and can stand for any physical quantity. As the function $R_T = 50 + R$ has shown, it is sometimes convenient to use different letters altogether.

Example 1 The equation $y = x^2 + 1$ is a function since for every value of x we get one, and only one, value of y. The independent variable is x and the dependent variable is y.

Example 2 The area A of a square of side s is

$$A = s^2$$

This formula defines a function since for every $s > 0$ there exists a unique value of A. The independent variable is s and the dependent variable is A. (A also exists for $s \leq 0$, but using such values makes no sense in this problem.)

The main purpose of the function concept is to define various operations. More precisely, a function describes an operation on the independent variable that yields a unique value of the dependent variable, regardless of the letters used. Thus $R_T = 50 + R$ and $y = 50 + x$ represent the same function. Even though the letters are different, the operations performed on the independent variable are the same: The value of the independent variable plus 50 is equal to the value of the dependent variable.

Notation for Functions

To make our discussion of functions easier, we need to introduce an appropriate notation, given next.

Notation for a function: If y is a function of x, we write $y = f(x)$, which is read "y equals f of x."

(The notation $y = f(x)$ is a special notation for a function and does not mean f multiplied by x.)

Example 3 By the functional notation $y = f(x)$, the function

$$y = 3 - 2x^2$$

can be written

$$f(x) = 3 - 2x^2$$

In other words, $y = 3 - 2x^2$ and $f(x) = 3 - 2x^2$ are two different ways of representing the same function.

Example 4 The equation $s = 16t^2 - 20t$ can be expressed as $s = f(t)$, where $f(t) = 16t^2 - 20t$. So $s = 16t^2 - 20t$ and $f(t) = 16t^2 - 20t$ are two different ways of writing the same function.

One of the main advantages of functional notation is that function values corresponding to particular values of the independent variable can be specified in a simple and natural way. Consider, for example, the function $y = x^2$. If $x = 3$, then $y = 3^2 = 9$. Using the notation $f(x) = x^2$, we may write

$$f(3) = 3^2 = 9$$

Consider another example.

Example 5 Given the function $f(x) = \sqrt{2 + x}$, find

$$f(0), \quad f(1), \quad \text{and } f(2)$$

Solution. In each case, we assign the specified value to the independent variable x. To find $f(0)$, we let $x = 0$ in the expression $\sqrt{2 + x}$ to obtain

$$f(x) = \sqrt{2 + x}$$

$$f(0) = \sqrt{2 + 0} = \sqrt{2} \qquad x = 0$$

Similarly,

$$f(1) = \sqrt{2 + 1} = \sqrt{3} \qquad x = 1$$

and

$$f(2) = \sqrt{2 + 2} = \sqrt{4} = 2 \qquad x = 2$$

As we have seen, letters other than x and y can be used for the variables. For example, $y = 1 + \sqrt{x}$ and $w = 1 + \sqrt{z}$ represent the same function. If more than one function occurs in a particular discussion, we may have to use another letter in place of f. For example, $y = g(x)$, $y = F(x)$, and $y = h(x)$ all represent functions, but so do $y = f(z)$, $w = g(z)$, and $r = F(t)$.

Example 6 If $G(v) = 2v^2 + 3v$, find

$$G(1), \quad G(-2), \quad \text{and } G(0)$$

Solution. Replacing v by the specified values, we get

$$G(v) = 2v^2 + 3v$$

$$G(1) = 2(1)^2 + 3(1) = 2 + 3 = 5$$

$$G(-2) = 2(-2)^2 + 3(-2) = 8 - 6 = 2$$

$$G(0) = 2(0)^2 + 3(0) = 0$$

Occasionally the independent variable is replaced by an algebraic expression, as shown in the next example.

Example 7 If $H(z) = 1 - z^2$, find

$$H(a) \quad \text{and} \quad H(a+1)$$

Solution.

$$H(a) = 1 - a^2$$

$$H(a+1) = 1 - (a+1)^2 = 1 - (a^2 + 2a + 1)$$
$$= 1 - a^2 - 2a - 1 = -a^2 - 2a$$

Domain

By the definition of function, if $y = f(x)$ is a function, then every value of the variable x yields a unique value of y. However, for mathematical or physical reasons some values of x may not be usable. For example, if $y = 1/x$, then x cannot be zero (to avoid division by zero). The permissible values of the independent variable are referred to as the **domain** of the function. The resulting function values are called the **range** of the function.

Domain
Range

Example 8 Find the domain of the function

$$P = \frac{k}{V} \qquad (k \text{ constant})$$

relating the pressure P to the volume V of a confined gas.

Solution. Since we cannot have a negative volume, V must be positive or zero for physical reasons. However, to avoid division by zero, the value $V = 0$ must also be excluded. So the domain of the function consists of all values of V such that $V > 0$.

We noted in Chapter 11 that square roots of negative numbers are used in certain applications. However, in our discussion of functions in this chapter, we will assume that all functions are real-valued. As a result, the values of the independent variable must be restricted in such a way that the resulting values of the dependent variable are real. Consider the next example.

Example 9 Find the domain of the function $f(x) = \sqrt{4 - x}$.

Solution. To avoid imaginary numbers, the values of x must be restricted so that the radicand (the expression under the radical sign) is positive or zero:

$$4 - x \geq 0$$
$$4 - x + x \geq 0 + x \qquad \text{Adding } x \text{ to both sides}$$
$$4 \geq x$$
$$x \leq 4$$

We conclude that the domain of the function $f(x) = \sqrt{4 - x}$ is the set of all values of x such that $x \leq 4$.

Remark. So far we have emphasized the definition of function. However, not every equation represents a function. For example, the equation

$$y^2 - 4x^2 = 0$$

can be written

$$y^2 = 4x^2$$

or

$$y = \pm\sqrt{4x^2} = \pm 2x$$

So if $x = 1$, then $y = \pm 2(1) = \pm 2$. We see, then, that the value $x = 1$ yields two different y-values, $+2$ and -2. It follows that $y^2 - 4x^2 = 0$ is not a function.

Example 10 The force F due to air resistance of a certain falling body is 0.25 times the square of the velocity v. Express the force as a function of the velocity.

Solution. The square of the velocity is v^2. So from the given information,

$$F = 0.25v^2$$

Exercises / Section 12.1

In Exercises 1–8, identify the independent and dependent variables. (See Examples 1 and 2.)

1. $y = x^3 - 3$

2. $w = 1 - 3z^4$

3. $E = 3R$

4. $s = 16t^2$

5. $A = \pi r^2$

6. $V = 4\pi h$

7. $C = A + \dfrac{2}{A}$

8. $C = \dfrac{5}{9}(F - 32)$

In Exercises 9–16, write the given function by replacing the independent variable by the form $f(x)$. (See Examples 3 and 4.)

9. $y = 3x - 2$

10. $y = x^2 - 3x + 7$

11. $z = \sqrt{7v + 1}$

12. $z = \dfrac{1}{w + 3}$

13. $s = -7t^2 + 3$

14. $A = 3r^2 + 4$

15. $w = \dfrac{1}{\sqrt[3]{s + 2}}$

16. $L = \sqrt{1 - 3n}$

In Exercises 17–30, find the indicated function values. (See Examples 5–7.)

17. $f(x) = 2x;\ \ f(0), f(1), f(-1)$

18. $f(x) = x^2;\ \ f(-1), f(-2), f(0)$

19. $g(x) = \dfrac{1}{x + 1};\ \ g(1), g(2), g(5)$

20. $h(x) = \dfrac{1}{x - 2};\ \ h(0), h(3), h(4)$

21. $F(t) = \sqrt{t + 3};\ \ F(-3), F(0), F(1)$

22. $F(s) = \sqrt{1 - 2s};\ \ F(-3), F(-2), F(0)$

23. $q(t) = \dfrac{1}{\sqrt{t + 2}};\ \ q(-1), q(2), q(7)$

24. $i(t) = 1 - 3t^2;\ \ i(0), i(1), i(2)$

25. $Z(v) = v^2 - 2;\ \ Z(0), Z(2), Z(5)$

26. $R(w) = 1 + 2w^2;\ \ R(0), R(2), R(6)$

27. $f(x) = 1 + x;\ \ f(a), f(b), f(a - 1)$

28. $g(x) = x^2 - 1;\ \ g(c), g(c - 1)$

29. $F(u) = \sqrt{2u^2 - 1};\ \ F(a), F(a - 1)$

30. $G(v) = \dfrac{1}{v - 1};\ \ G(b), G(b^2 + 1)$

In Exercises 31–40, state the domain of each function. (See Examples 8 and 9.)

31. $f(x) = \dfrac{1}{x + 1}$

32. $h(x) = \dfrac{1}{x - 2}$

33. $S(t) = \dfrac{1}{(t - 1)(t - 2)}$

34. $g(s) = \dfrac{1}{(s - 3)(s - 4)}$

35. $F(x) = \sqrt{1 - x}$

36. $h(z) = \sqrt{z + 2}$

37. $g(t) = \sqrt{t + 3}$

38. $G(v) = \sqrt{v - 2}$

39. $P(r) = \sqrt{r - 3}$

40. $H(y) = \sqrt{4 - y}$

41. Express the area A of a square as a function of the side s.

42. Express the volume V of a cube as a function of the edge s.

43. Express the area A of a circle as a function of the diameter D.

44. The base of a rectangle is 4 units long. Express the area A as a function of the height h.

45. The cost of finishing a metal plate is \$5/in.^{2}. Express the cost C as a function of the area A.

46. The demand D of a certain product is 4.5 times the reciprocal of the price P. Express D as a function of P.

47. The energy E radiated by a filament is equal to the fourth power of the temperature T multiplied by a constant k. Express E as a function of T.

48. A computer consultant charges a base fee of \$60 plus \$30 per hour. Express the cost C of a consultation as a function of time t.

12.2 The Rectangular Coordinate System

Origin

We saw in Section 3.1 that the real numbers can be placed on a line. We start with an arbitrary point, called the **origin,** and place the positive integers at equal intervals on the right side of the origin and the negative integers on the left side. (See Figure 12.1.) Other real numbers are placed at appropriate points between the integers. For example, the number $\frac{3}{2}$ is located midway between 1 and 2 and the number $-\frac{5}{2}$ midway between -3 and -2.

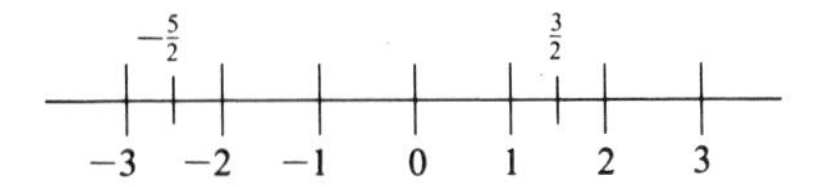

Figure 12.1

Rectangular coordinate system
x-axis, *y*-axis
Coordinate axes
Quadrants

To plot a point in the plane, we construct a reference system called the **rectangular coordinate system.** This system consists of two number lines that intersect at right angles at their respective origins. (See Figure 12.2.) The horizontal line is called the ***x*-axis** and the vertical line, the ***y*-axis.** Together these lines are called the **coordinate axes.** Note that on the x-axis, the positive numbers are located on the right side of the origin and the negative numbers on the left. On the y-axis, the positive numbers are located above the origin and the negative numbers below the origin. The coordinate axes divide the plane into four parts, called **quadrants,** numbered I, II, III, and IV. (See Figure 12.2.)

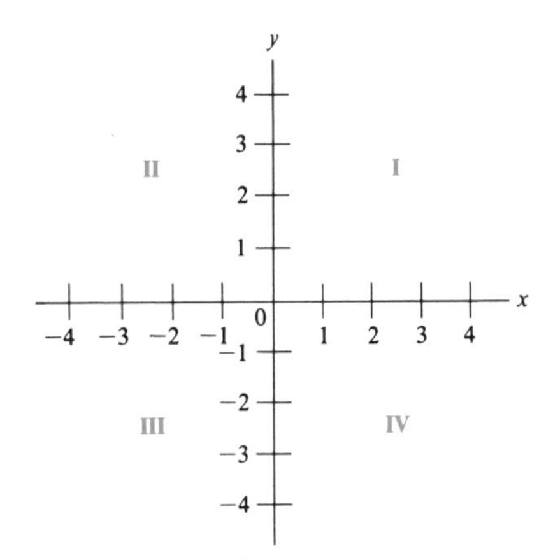

Figure 12.2

x-coordinate
y-coordinate

Points in the plane can be located by describing the perpendicular distances and directions from the coordinate axes. The distance from the y-axis, with proper sign, is called the ***x*-coordinate,** or **abscissa.** The distance from the x-axis, with proper sign, is called the ***y*-coordinate,** or **ordinate.** To describe a point, we place the coordinates in parentheses, with the x-coordinate first: (x, y). For example, the point P in Figure 12.3 is denoted by $(2, 4)$ and the point Q by $(-2, 3)$.

Plotting points

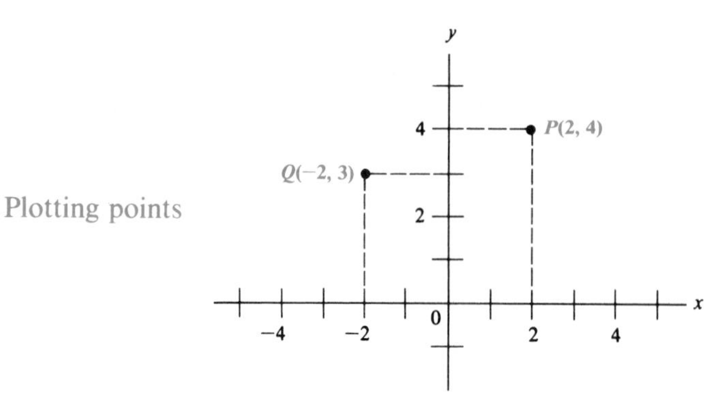

Figure 12.3

Example 1 Plot the following points:

$$A(5, 2), \quad B(-3, 1), \quad C(-2, -6), \quad \text{and } D(4, -4)$$

Solution. For the point $A(5, 2)$, the x-coordinate is 5 and the y-coordinate, 2. Note that the point is located in quadrant I. (See Figure 12.4.)

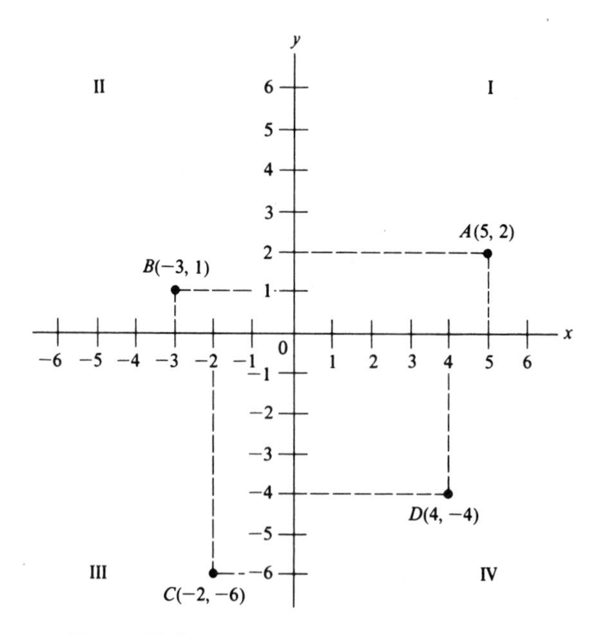

Figure 12.4

For the point $B(-3, 1)$, the x-coordinate is -3 and the y-coordinate is 1. The point is located in the second quadrant (II). The points C and D are plotted similarly.

The rectangular coordinate system is sometimes used to describe geometric figures, as shown in the next example.

Example 2 The following points form the vertices of a parallelogram:

$$A(-4, -4), \quad B(-3, -2), \quad C(4, -2), \quad \text{and } D(3, -4)$$

Draw the resulting figure.

Solution. The vertices and the resulting parallelogram are shown in Figure 12.5.

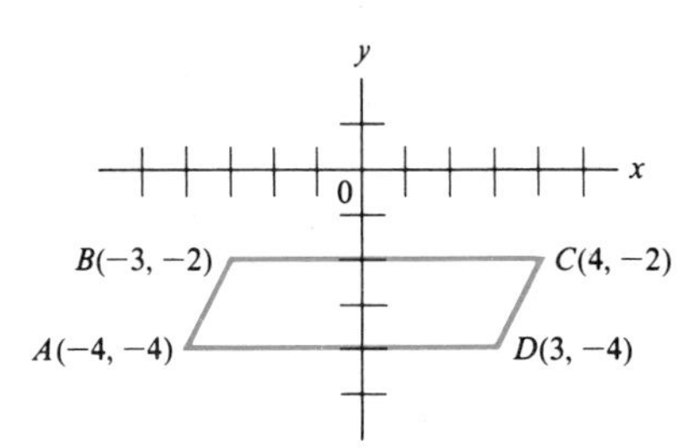

Figure 12.5

Exercises / Section 12.2

In Exercises 1–6, plot the given set of points.

1. $(3, 4), (-1, 7), (4, -6), (-2, 3)$

2. $(-1, 6), (2, 1), (-7, -1), (3, -6)$

3. $(-2, -2), (7, 6), (-6, 4), (2, -1)$

4. $(4, 4), (-3, -3), (5, -4), (-6, 8)$

5. $\left(\frac{1}{2}, 0\right), (-1.2, -3.1), \left(\frac{4}{5}, \frac{10}{3}\right), (-2.4, 5.6)$

6. $\left(\frac{4}{3}, 0\right), (4.4, -5.2), \left(\frac{16}{5}, \frac{9}{4}\right), (-4.8, -3.2)$

7. Plot the points $(5, 0), (0, -6), (0, 0)$.

8. Plot the points $(-4, 0), (0, 3)$.

9. Determine the points in the plane for which (a) the x-coordinates are zero; (b) the y-coordinates are zero.

10. Determine the quadrants in which (a) y/x is positive; (b) y/x is negative.

11. Determine the points for which $y = x$ and $y = -x$.

12. Find the set of all points satisfying the following conditions: (a) $x = 2$; (b) $y = 3$.

In Exercises 13–18, draw the indicated figures. (See Example 2.)

13. The triangle $A(2, 4), \quad B(-3, 6), \quad C(-2, -1)$

14. The right triangle $A(-3, 5), \quad B(2, 5), \quad C(-3, 2)$

15. The right triangle $A(-5, 5), \quad B(5, 9), \quad C(9, -1)$

16. The parallelogram $A(-5, 2), \quad B(-2, 8), \quad C(8, 4), \quad D(5, -2)$

17. The rectangle $A(-1, -3), \quad B(-3, 3), \quad C(6, 6), \quad D(8, 0)$

18. The square $A(-1, 5), \quad B(3, 1), \quad C(7, 5), \quad D(3, 9)$

12.3 The Graph of a Function

In this section we combine the idea of a function with that of a coordinate system.

If a function is given by an equation of the form $y = f(x)$, we assign various values to the independent variable x and compute the corresponding y-values. We then plot the resulting points and connect them by a smooth curve, thereby obtaining the **graph** of the function. The graph of a function gives a revealing picture of the behavior of the function.

> The **graph** of a function $y = f(x)$ consists of all points (x, y) whose coordinates satisfy the equation $y = f(x)$.

Example 1 Graph the function $y = x + 1$.

Solution. To graph this function, we need to plot enough points to obtain a smooth curve. To this end, we construct a table of values by assigning different values to x and finding the corresponding y-values. For example, if $x = -2$, then

$$y = x + 1 = -2 + 1 = -1$$

So the point $(-2, -1)$ lies on the curve. (See Figure 12.6.) Similarly if $x = -1$, then

$$y = x + 1 = -1 + 1 = 0$$

So the point $(-1, 0)$ also lies on the curve. These and some additional pairs of values are listed in the following table:

x:	-2	-1	0	1	2
y:	-1	0	1	2	3

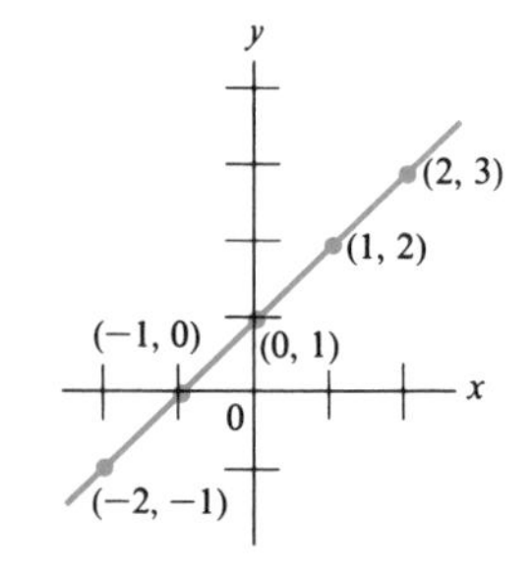

Figure 12.6

If we plot the points indicated in the table and connect the points by a smooth curve, we obtain the graph shown in Figure 12.6.

Linear function

The function in Example 1 is known as a **linear function,** since the graph is a straight line. The general form of a linear function is

$$f(x) = ax + b$$

where a and b are constants.

For many graphs, particularly for graphs of linear functions, it is useful to obtain those points at which the graph crosses the coordinate axes. These points are called **intercepts.**

Definition of intercept: An **intercept** is a point at which the graph crosses a coordinate axis. A point $(a, 0)$ is called an ***x*-intercept,** and a point $(0, b)$ is called a ***y*-intercept.**

Example 2 Find the intercepts of the following linear function and sketch the graph:

$$y = 2x - 3$$

Solution. To find the x-intercept, we let $y = 0$ and solve for x:

$$y = 2x - 3$$

$$0 = 2x - 3$$

$$0 + 3 = 2x - 3 + 3 \quad \text{Adding 3 to both sides}$$

$$3 = 2x$$

$$\frac{3}{2} = x \quad \text{Dividing by 2}$$

It follows that the x-intercept is $(\frac{3}{2}, 0)$.

To find the y-intercept, we let $x = 0$ and solve for y:

$$y = 2x - 3$$

$$y = 2(0) - 3 = -3$$

So the y-intercept is $(0, -3)$.

Since two distinct points determine a line, the two intercepts are sufficient to draw the graph. It is good practice, however, to plot one additional point as a check. Letting $x = 1$, we get

$$y = 2(1) - 3 = -1$$

so the point $(1, -1)$ lies on the line.

The two intercepts and the check point are listed in the following table:

x:	$\frac{3}{2}$	0	1
y:	0	−3	−1

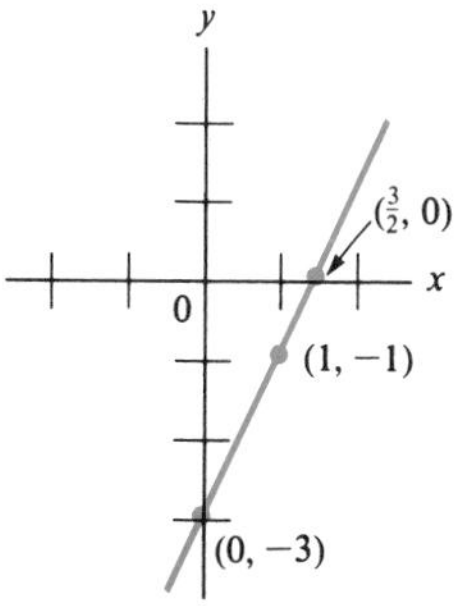

y

x

0

$(\frac{3}{2}, 0)$

$(1, -1)$

$(0, -3)$

Figure 12.7

The graph of the resulting line is shown in Figure 12.7.

The function in the next example is of second degree and is therefore referred to as a *quadratic function.*

Example 3 Find the intercepts and sketch the graph of

$$y = x^2 - 4x + 3$$

Solution. Intercepts: If $x = 0$, then $y = 3$. So the y-intercept is $(0, 3)$. If $y = 0$, then

$$\begin{aligned} x^2 - 4x + 3 &= 0 \\ (x - 1)(x - 3) &= 0 \\ x &= 1, 3 \end{aligned}$$

The x-intercepts are, therefore, $(1, 0)$ and $(3, 0)$, and the y-intercept is $(0, 3)$.

In addition to the intercepts, we plot a few other points from the following table:

x:	-1	2	4	5
y:	8	-1	3	8

The graph, shown in Figure 12.8, is called a **parabola.**

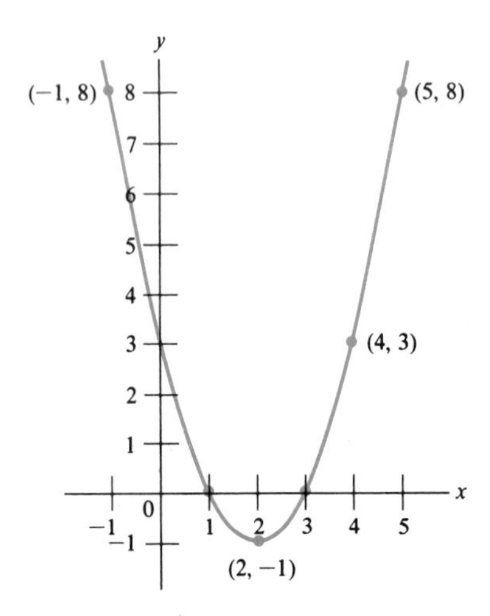

Figure 12.8

Quadratic function

The function in Example 3 is known as a **quadratic function,** whose general form is

$$f(x) = ax^2 + bx + c$$

where a, b, and c are constants.

For many functions it is helpful to determine the domain before attempting the sketch. Consider the next example.

Example 4 Determine the intercepts and domain of the function $y = \sqrt{4 - x}$. Sketch the graph.

Solution. Intercepts: If $x = 0$, then $y = \sqrt{4 - 0} = \sqrt{4} = 2$. If $y = 0$, then

$$0 = \sqrt{4 - x}$$
$$0^2 = (\sqrt{4 - x})^2 \qquad \text{Squaring both sides}$$
$$0 = 4 - x \qquad (\sqrt{a})^2 = a$$
$$x = 4$$

So the x-intercept is $(4, 0)$ and the y-intercept is $(0, 2)$.

Domain: To avoid imaginary values, x cannot be greater than 4. For example, if $x = 4$, then $y = \sqrt{4 - 4} = 0$, but if $x = 5$, then $y = \sqrt{4 - 5} = \sqrt{-1} = j$, a pure imaginary number. It follows that the domain of the function is $x \leq 4$. As a result, no values greater than 4 appear in the following table:

x:	-5	-3	-1	0	1	2	3	4
y:	3	$\sqrt{7}$	$\sqrt{5}$	2	$\sqrt{3}$	$\sqrt{2}$	1	0

The graph is shown in Figure 12.9.

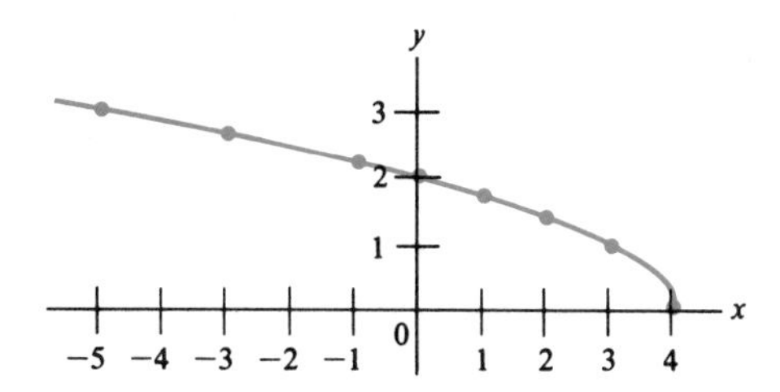

Figure 12.9

Example 5 Graph the function

$$y = 1 + \frac{2}{x}$$

Solution. Note that the value $x = 0$ leads to division by 0. Since the function is defined for all other values of x, the domain consists of all x such that $x \neq 0$.

Since $x \neq 0$, there are no y-intercepts. To find the x-intercepts, let $y = 0$ and solve for x:

$$0 = 1 + \frac{2}{x}$$
$$0 \cdot x = \left(1 + \frac{2}{x}\right) x \qquad \text{Multiplying by } x$$
$$0 = x + 2$$
$$-2 = x$$
$$x = -2$$

So the x-intercept is $(-2, 0)$. Some other points are plotted from the following table:

x:	-4	-3	-2	-1	$-\frac{1}{2}$	$\frac{1}{2}$	1	2	3	5
y:	$\frac{1}{2}$	$\frac{1}{3}$	0	-1	-3	5	3	2	$\frac{5}{3}$	$\frac{7}{5}$

The graph, shown in Figure 12.10, is called a **hyperbola.**

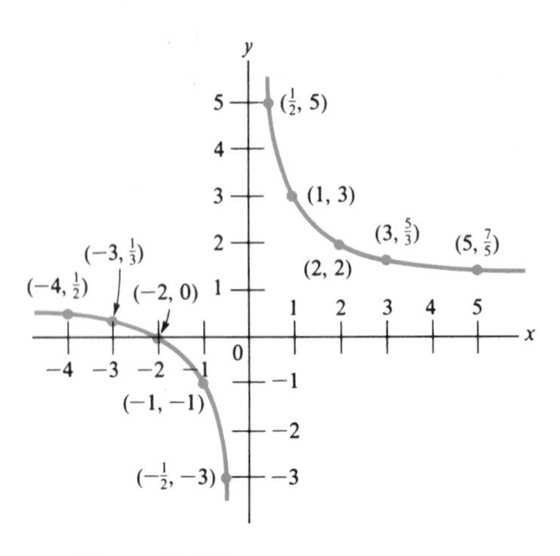

Figure 12.10

Example 6 Graph the function $y = 2^x$.

Solution. If $x = 0$, then $y = 2^0 = 1$. So the y-intercept is $(0, 1)$. Since $2^x > 0$ for all x, there are no x-intercepts.

For positive x-values, y increases rapidly as x increases. (See the table of values.) Note that negative x-values lead to negative exponents. For example, if $x = -2$, then

$$y = 2^{-2} = \frac{1}{2^2} = \frac{1}{4}$$

Similarly, if $x = -4$, then

$$y = 2^{-4} = \frac{1}{2^4} = \frac{1}{16}$$

So the graph gets ever closer to the x-axis as we move to the left.

The graph, shown in Figure 12.11, is plotted from the following table of values:

x:	-4	-3	-2	-1	0	1	2	3
y:	$\frac{1}{16}$	$\frac{1}{8}$	$\frac{1}{4}$	$\frac{1}{2}$	1	2	4	8

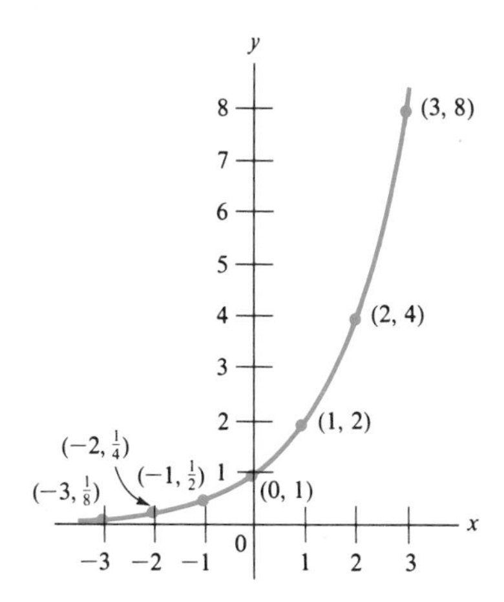

Figure 12.11

Exponential function

The function $y = 2^x$ in Example 6 is an example of an **exponential function,** whose general form is

$$y = a^x, \qquad a > 0, a \neq 1$$

Example 7

In the circuit in Figure 12.12, the voltage drop across the 20-Ω resistor is $20I$ volts, and the voltage drop across the variable resistor is RI volts (by Ohm's law). By Kirchhoff's voltage law, the voltage drop across the combination, $20I + RI$, is equal to the impressed voltage 10 V, or $20I + RI = 10$. Solving for I, we get

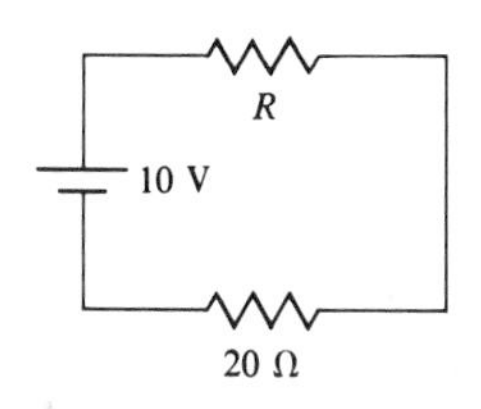

Figure 12.12

$$20I + RI = 10$$

$$(20 + R)I = 10 \qquad \text{Factoring } I$$

$$I = \frac{10}{20 + R} \qquad \text{Dividing by } 20 + R$$

The power across the variable resistor is given by

$$P = I^2R = \left(\frac{10}{20 + R}\right)^2 R = \frac{100R}{(20 + R)^2}$$

Sketch the function

$$P = \frac{100R}{(20 + R)^2}$$

and estimate the setting on R so that it takes maximum power.

Solution. We establish the following table of values:

R:	0	5	10	15	17	20	23	25	30	80
P:	0	0.8	1.1	1.22	1.24	1.25	1.24	1.23	1.20	0.8

The graph is shown in Figure 12.13. According to the figure, the power is at a maximum when $R = 20\ \Omega$.

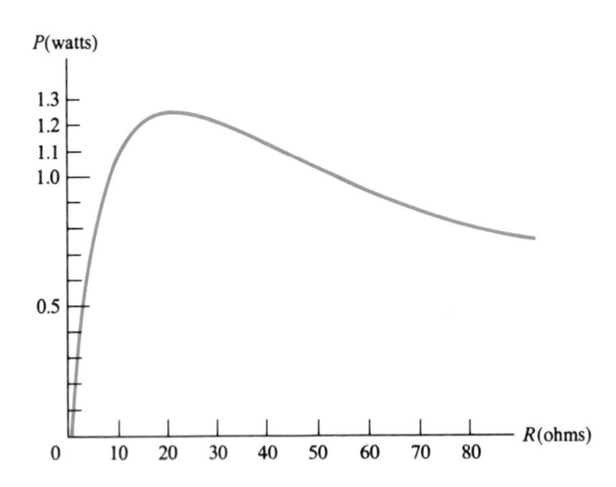

Figure 12.13

Exercises / Section 12.3

In Exercises 1–28, sketch the graph of each given function.

1. $y = x - 1$

2. $y = 2x + 4$

3. $y = 2x + 3$

4. $y = 3x + 4$

5. $y = 1 - 2x$

6. $y = 2 - 3x$

7. $y = 2x^2$

8. $y = 2x^2 + 1$

9. $y = x^2 + x - 2$

10. $y = x^2 + x - 6$

11. $y = 2x^2 - 11x + 12$

12. $y = 2x^2 - 5x + 2$

13. $y = \sqrt{x}$

14. $y = \sqrt{x + 1}$

15. $y = \sqrt{1 - x}$

16. $y = \sqrt{2 - x}$

17. $y = 2 + \dfrac{1}{x}$

18. $y = 2 + \dfrac{2}{x}$

19. $y = 1 - \dfrac{2}{x}$

20. $y = 2 - \dfrac{1}{x}$

21. $y = 3^x$

22. $y = 3^{-x}$

23. $y = 2^{-x}$

24. $y = 4^x$

25. $y = \left(\dfrac{1}{3}\right)^x$

26. $y = \left(\dfrac{1}{2}\right)^x$

27. $y = 1 - 3^{-x}$

28. $y = 2 - 2^{-x}$

29. The force F required to stretch a spring x units is $F = kx$. Given that $k = 2$, draw the graph of the function, assuming F to be in pounds and x in inches.

30. Deliveries of a certain item are made to a store on the first day of the month. The supply S throughout the month is given by

$$S = 50 - 2t, \qquad t \geq 0$$

Sketch the graph. What is the significance of the t-intercept? The S-intercept?

31. The area of a rectangle is 1 square unit. If P denotes the perimeter, show that

$$P = 2x + \frac{2}{x}, \qquad x > 0$$

(See Figure 12.14). Sketch this function.

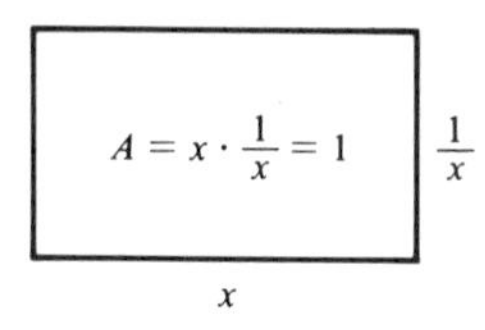

Figure 12.14

32. The charge on a certain capacitor is

$$q(t) = \frac{1}{t - 1}$$

Sketch the graph of $q(t)$ for $t > 1$.

33. If a body is dropped from rest, its distance s (in feet) from the starting point as a function of time t (in seconds) is $s = 16t^2$, $t \geq 0$. Sketch the graph.

34. A projectile is shot directly upward with a velocity of 30 m/s. Its distance s (in meters) above the ground is given by $s = 30t - 5t^2$ ($t \geq 0$). Sketch the curve and estimate the highest point reached by the projectile. (See Example 7.)

35. The period T (in seconds) of a certain pendulum as a function of its length L (in meters) is given by $T = 2\sqrt{L}$. Sketch the graph.

36. The current i in a 4-Ω resistor as a function of the power P delivered to the resistor is $i = \frac{1}{2}\sqrt{P}$. Sketch the graph.

37. The power delivered to a certain resistor is

$$P = \frac{16R}{(R + 2)^2}$$

Sketch the graph and estimate the maximum power. (See Example 7.)

38. The velocity v (in feet per second) of a falling body subject to a retarding force due to air resistance is

$$v = 10(1 - 2^{-0.2t}), \qquad t \geq 0$$

where t is the time in seconds. Sketch the graph.

39. A certain body cools according to the formula

$$T = 10(3^{-0.062t}), \qquad t \geq 0$$

where T is the temperature in degrees Celsius and t the time in minutes. Sketch the curve.

12.4 Graphical Solution of Equations

Earlier we studied the solution of equations of first and second degree. In this section we will see how approximate solutions of equations can be determined graphically.

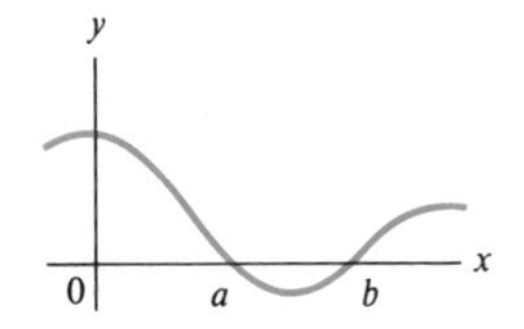

Figure 12.15

Consider the graph of the function $y = f(x)$. Recall from the last section that the x-intercepts are the points where the graph crosses the x-axis. (See Figure 12.15.) These intercepts are found by letting y be equal to zero and solving for x. In other words, the x-intercepts are the roots of the equation $f(x) = 0$.

Many equations of the form $f(x) = 0$ are difficult or even impossible to solve by algebraic means. When this is the case, we may still be able to obtain an approximate solution by graphing the equation $y = f(x)$ and estimating the intercepts from the graph. The next example illustrates the technique.

Example 1 Solve the equation $x^2 + 2x - 7 = 0$ graphically.

Solution. We first graph the equation $y = x^2 + 2x - 7$ from the following table of values:

x:	-5	-4	-3	-2	-1	0	1	2	3
y:	8	1	-4	-7	-8	-7	-4	1	8

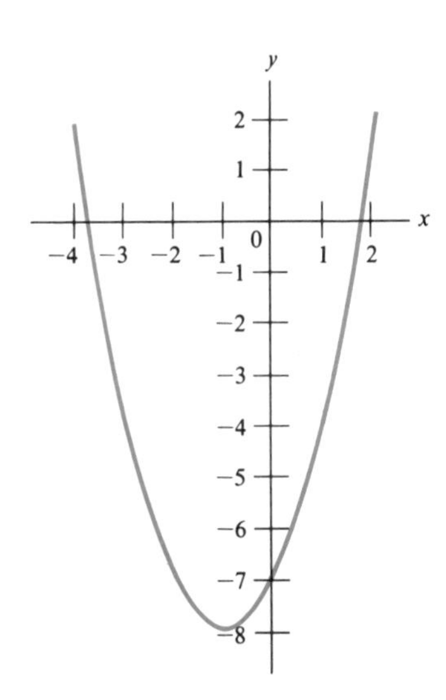

Figure 12.16

According to the graph shown in Figure 12.16, the x-intercepts are $(-3.8, 0)$ and $(1.8, 0)$ to the nearest tenth of a unit. The approximate roots are therefore $x = -3.8$ and $x = 1.8$.

The equation in Example 1 could have been solved by using the quadratic formula. The equation in the next example, however, cannot be solved by any method studied so far.

Example 2 A spherical ball of radius r and specific gravity c will sink in water to a depth h ($0 < h \leq 2r$) according to the equation $h^3 - 3rh^2 + 4r^3c = 0$. Graphically determine the depth to which a ball of radius 1 ft and specific gravity 0.75 will sink.

Solution. If $r = 1$ and $c = 0.75$, the equation $h^3 - 3rh^2 + 4r^3c = 0$ becomes

$$h^3 - 3(1)h^2 + 4(1^3)(0.75) = 0$$
$$h^3 - 3h^2 + 3 = 0$$

Now let $y = h^3 - 3h^2 + 3$ and construct the following table of values:

h:	0	0.5	1	1.2	1.5	2
y:	3	2.4	1	0.4	−0.4	−1

Since the diameter of the ball is 2 ft, we need to plot only values of h between 0 and 2. The graph is shown in Figure 12.17.

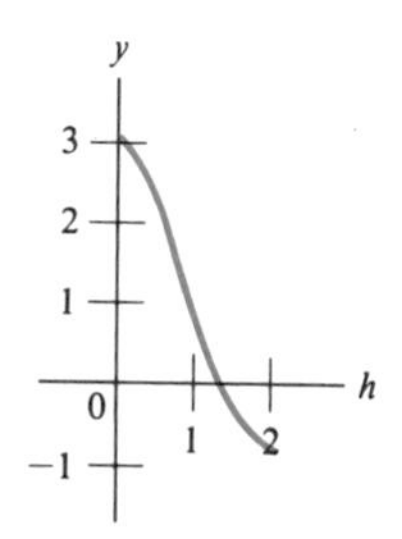

Figure 12.17

According to the graph, $y = 0$ when $h = 1.3$ (to the nearest tenth). It follows that the ball will sink to a depth of 1.3 feet.

Exercises / Section 12.4

In Exercises 1–10, solve the given equations graphically. Estimate each root to the nearest tenth.

1. $x^2 - 2x - 2 = 0$
2. $x^2 - 4x - 1 = 0$
3. $x^3 - 3x - 4 = 0$
4. $x^3 + 3x - 6 = 0$
5. $x^3 + 2x + 6 = 0$
6. $2x^3 + 3x - 6 = 0$
7. $3 - 2^x = 0$
8. $3^x - 2 = 0$
9. $x - 2^{-x} = 0$
10. $2x - 3^{-x} = 0$

In Exercises 11–14, determine all roots graphically to the nearest tenth.

11. A ball of radius 3 in. and specific gravity 0.51 will sink in water to a depth h ($0 < h \leq 6$ in.) according to the equation $h^3 - 9h^2 + 55 = 0$. (See Example 2.) Determine the depth to which the ball will sink.

12. An object is taken out of a freezer kept at $-10°F$ and placed in a room. The temperature T as a function of time t (in minutes) is given by $T = 10 - 20(2^{-0.11t})$. When will the temperature be 0°F?

13. Starting at $t = 0$, the current i (in amperes) in a certain circuit as a function of time t (in seconds) is given by $i = 3 - 4(2^{-0.2t})$. Determine when the current is zero.

14. A box is constructed from a square piece of cardboard 6 in. on the side by cutting equal squares from each corner and turning up the sides. (See Figure 12.18.) Determine the side of the square that must be cut out if the volume of the box is to be 15 in.^{3}. (There are two solutions.)

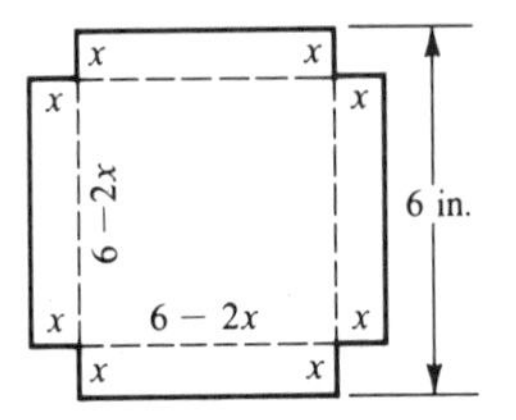

Figure 12.18

12.5 Functions in BASIC (Optional)

We saw in Section 4.5 that SQR(X) is the square root function in BASIC. Later we will see other examples of functions that are stored in the computer (library functions). This section is concerned with functions that you define yourself.

To define a function, use the DEF (define) statement, which has the following form:

line number DEF FNA(X)

The function name, FNA, must consist of three letters, the first two of which are FN and the third any letter from A to Z. For the variable, any legal variable name can be used in most systems.

Example 1 Write the function

$$f(x) = 2x^3 + 4x - 5$$

in BASIC.

Solution. Using the name FNA, we get

DEF FNA(X) = 2 * X ↑ 3 + 4 * X − 5

The same function can be defined by the following statement:

DEF FNB(Y) = 2 * Y ↑ 3 + 4 * Y − 5

The computer will evaluate function values in the usual way, as shown in the next example.

Example 2 Consider the programming sequence

```
5∅ DEF FNA(X) = 2 + 3 * X ↑ 2
6∅ PRINT 4 * SQR(FNA (2))
```

Line 50 defines the function

$$f(x) = 2 + 3x^2$$

Line 60 causes the value 2 to be substituted for x in line 50. The number printed is therefore

$$4\sqrt{f(2)} = 4\sqrt{2 + 3(2)^2} = 14.96662955$$

A convenient way to generate a table of values is by means of the FOR-NEXT statement. (See Appendix A.)

12.6 Ratio and Proportion

The purpose of this section is to discuss two related concepts: **ratio** and **proportion**.

Ratio

Suppose a room is 10 ft long and 8 ft wide. Then the *ratio* of the width of the room to the length is

$$\frac{\text{width}}{\text{length}} = \frac{8 \text{ ft}}{10 \text{ ft}} = \frac{4}{5}$$

This ratio says that the room is only 4/5 as wide as it is long. As we will see, ratios occur often in technology.

> A **ratio** is a quotient of two quantities.

As the room example shows, if the ratio involves measurements of the same kind, then the units divide out and we obtain a dimensionless number. Consider another example.

Example 1 The *specific gravity* of a substance is the ratio of the density of the substance to the density of water (62.4 lb/ft^3). Given that the density of copper is 555 lb/ft^3, find its specific gravity.

Solution. Since this ratio involves measurements of the same kind, we obtain the following dimensionless number:

$$\text{specific gravity of copper} = \frac{555\ \dfrac{\text{lb}}{\text{ft}^3}}{62.4\ \dfrac{\text{lb}}{\text{ft}^3}} = \frac{555\ \dfrac{\cancel{\text{lb}}}{\cancel{\text{ft}^3}}}{62.4\ \dfrac{\cancel{\text{lb}}}{\cancel{\text{ft}^3}}} = \frac{555}{62.4} = 8.89$$

Ratios often involve different units of measure. For example, suppose you are walking at the rate of 4 mi/h. Then the ratio of distance to time is

$$\frac{4 \text{ miles}}{1 \text{ hour}} = 4\ \frac{\text{mi}}{\text{h}}$$

This ratio is actually the rate at which you are walking.

Similarly, if you can drive 90 mi on 3 gal of gasoline, then your gas mileage is

$$\frac{90 \text{ mi}}{3 \text{ gal}} = 30\ \frac{\text{mi}}{\text{gal}}$$

Consider another example.

Example 2 The force against a horizontal submerged plate with an area of 25.3 ft^2 is 278 lb. Find the **pressure** p, defined to be

$$p = \frac{\text{force}}{\text{area}}$$

Solution. From the given formula,

$$p = \frac{278 \text{ lb}}{25.3 \text{ ft}^2} = 11.0\ \frac{\text{lb}}{\text{ft}^2}$$

Other examples of standard ratios can be found in the exercises.

Proportion

Our other concept in this section is that of **proportion**, defined next.

> A **proportion** is an equality between two ratios.

In other words, a proportion has the form

$$\frac{a}{b} = \frac{c}{d}$$

For example, $\frac{2}{3}$ is equal to $\frac{4}{6}$. This equality can be expressed as a proportion:

$$\frac{2}{3} = \frac{4}{6}$$

The reason proportions are of interest to us is that in many problems one ratio is known but only part of another ratio is known. By setting up a proportion, the remaining part can be determined, as shown in the next example.

Example 3 The ratio of an unknown number to 5 is the same as the ratio of 9 to 15. Find the number.

Solution. Let us denote the unknown number by x. Since the ratio of x to 5 is the same as the ratio of 9 to 15, we get

$$\frac{x}{5} = \frac{9}{15}$$

We now solve the resulting equation for x:

$$\frac{x}{5} = \frac{9}{15}$$

$$\frac{x}{5} \cdot 5 = \frac{9}{15} \cdot 5 \quad \text{Multiplying by 5}$$

$$\frac{x}{\cancel{5}} \cdot \cancel{5} = \frac{9}{\underset{3}{\cancel{15}}} \cdot \cancel{5} \quad \text{Cancellation}$$

$$x = 3$$

Remark. A convenient way to solve for the unknown in a proportion is by *cross-multiplication*. For example, in the equation in Example 3, we cross-multiply as follows:

$$\frac{x}{5} \gtrless \frac{9}{15}$$

$$x \cdot 15 = 5 \cdot 9 \quad \text{Cross-multiplying}$$

$$15x = 45$$

$$x = 3 \quad \text{Dividing by 15}$$

Proportions are sometimes used for simple conversions. A particularly useful case involves the relationship

$$60\ \frac{\text{mi}}{\text{h}} = 88\ \frac{\text{ft}}{\text{s}}$$

Remark. When setting up a proportion, make sure that the same units of measurement are used for the two numerators and the same units of measurement are used for the two denominators.

Example 4 Convert 45.6 mi/h to feet per second.

Solution. Let x (in feet per second) be the unknown rate. Then

$$\frac{x}{45.6\ \frac{\text{mi}}{\text{h}}} = \frac{88\ \frac{\text{ft}}{\text{s}}}{60\ \frac{\text{mi}}{\text{h}}} \qquad \begin{array}{l} x \text{ in feet per second} \\ \text{Same units} \end{array}$$

Multiplying both sides by 45.6 mi/h, we get

$$\begin{aligned} x &= \frac{88\ \frac{\text{ft}}{\text{s}}}{60\ \frac{\text{mi}}{\text{h}}} \times \frac{45.6\ \frac{\text{mi}}{\text{h}}}{1} \\ &= \frac{(88)(45.6)}{60}\ \frac{\text{ft}}{\text{s}} = 66.9\ \frac{\text{ft}}{\text{s}} \end{aligned}$$

Cross-multiplication yields

$$\frac{x}{45.6\ \frac{\text{mi}}{\text{h}}} = \frac{88\ \frac{\text{ft}}{\text{s}}}{60\ \frac{\text{mi}}{\text{h}}}$$

$$x \cdot 60\ \frac{\text{mi}}{\text{h}} = 88\ \frac{\text{ft}}{\text{s}} \cdot 45.6\ \frac{\text{mi}}{\text{h}} \qquad \text{Cross-multiplying}$$

$$x = \frac{(88)(45.6)}{60}\ \frac{\text{ft}}{\text{s}} = 66.9\ \frac{\text{ft}}{\text{s}}$$

Other technical applications of proportions are given in the remaining examples and exercises.

Example 5 The ratio of the number of teeth of two gears is 4 to 7. If the larger gear has 35 teeth, how many teeth does the smaller gear have?

Solution. Let

x = number of teeth in smaller gear

Then

$$\frac{x}{35} = \frac{4}{7}$$

$$7x = 4 \cdot 35 \qquad \text{Cross-multiplying}$$

$$x = \frac{4 \cdot 35}{7} = 20 \text{ teeth}$$

Example 6 The *mechanical advantage* (MA) of a pipe wrench is the ratio of the length of the handle of the wrench to the radius of the pipe (Figure 12.19). If the length of the handle is 12 in. and the radius of the pipe is $\frac{1}{2}$ in., then

$$\text{MA} = \frac{12 \text{ in.}}{\frac{1}{2} \text{ in.}} = 24$$

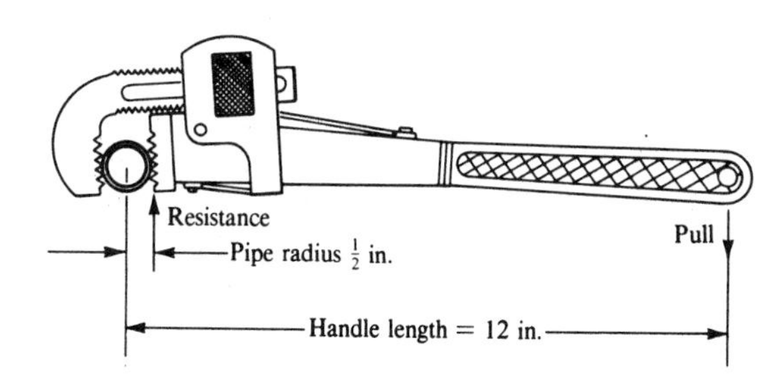

Figure 12.19

Alternatively,

$$\text{MA} = \frac{\text{force required to loosen pipe}}{\text{force on handle}}$$

If the pipe offers a resistance of 672 lb, what pull on the handle will loosen the pipe?

Solution. Let

x = pull required on handle

Then

$$\frac{672 \text{ lb}}{x} = \frac{24}{1} \qquad \text{MA} = 24$$

$$24x = 672 \text{ lb} \qquad \text{Cross-multiplying}$$

$$x = \frac{672 \text{ lb}}{24} = 28 \text{ lb}$$

Exercises / Section 12.6

Use a calculator in this exercise set.

1. The *Mach number* is the ratio of the velocity of an object to the velocity of sound, 330 m/s. What is the Mach number of a rocket traveling at 2500 m/s?
2. The *density* of a substance is the ratio of its mass to its volume. Find the density of gold, given that 2.60 cm^3 of gold has a mass of 50.18 g.
3. The *pressure* is defined as the ratio of force to area. The force against a horizontally submerged plate with an area of 5.34 ft^2 is 4360 lb. Find the pressure on the plate.
4. The *compression ratio* of a car engine is the ratio of the cylinder volume to the compressed volume. If the cylinder volume is 45.0 $in.^3$ when the piston is at the bottom of its stroke and 6.85 $in.^3$ when it is at the top of its stroke, find the compression ratio.
5. The *specific gravity* of a substance is the ratio of the density of the substance to the density of water. Determine the specific gravity of kerosene, which weighs 51 lb/ft^3, given that water weighs 62.4 lb/ft^3.
6. The *pitch* of a roof is the ratio of the rise of the rafter to the span of the roof (see Figure 12.20). Find the pitch of a roof with a rise of 6.5 ft and a span of 26 ft.

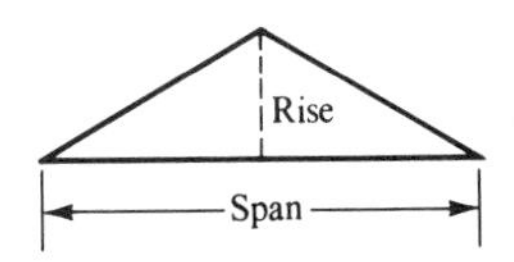

Figure 12.20

7. Convert 55 mi/h to feet per second. (See Example 4.)
8. Convert 21.7 mi/h to feet per second.
9. Convert 25.2 ft/s to miles per hour.
10. Convert 50.3 ft/s to miles per hour.
11. Given 1 in. = 2.54 cm (exactly), convert 3.16 in. to centimeters.
12. Given that 1 in. = 2.54 cm (exactly), convert 13.9 cm to inches.
13. The resistance (in ohms) of a wire is proportional to its length. If a wire 12 ft long has a resistance of 0.10 Ω, determine the resistance of a wire 20 ft long if it is made of the same material.
14. A machine can sort 25 cards in 12 s. How many cards can it sort in 1 min?
15. If a car uses 1.4 qt of oil in 520 mi, how much oil will it use in 860 mi?
16. The cost of a casting is proportional to its weight. If a 16-lb casting costs $1.20, how much does a 28-lb casting cost?
17. On a blueprint, 3 in. represents 12 ft. If the length of a shop measures 1 ft on the blueprint, what is the actual length of the shop?
18. It takes 20 lb of a certain base to neutralize 25 lb of sulfuric acid. How many pounds of this base are needed to neutralize 15 lb of this acid?
19. A tree casts a shadow 10 ft long. At the same time, a 12-in. ruler casts a shadow 4 in. long. How tall is the tree?

20. A flagpole casts a shadow 8.0 ft long. At the same time, a yardstick casts a shadow 6.4 in. long. How tall is the flagpole?

21. If you peddle your bicycle at the rate of 2.4 revolutions per second (rev/s) to attain a speed of 12 mi/h, how fast must you peddle your bicycle to attain a speed of 16 mi/h?

22. If a 5-day vacation costs $750, how much does a 12-day vacation cost?

23. Consider the wrench in Example 6. If the pipe offers a resistance of 1120 lb, what pull on the handle will loosen the pipe?

24. If the handle of the wrench in Example 6 is pulled with a force of 100 lb, what is the resulting force on the pipe?

25. Suppose the price of a bag of peanuts is proportional to the weight. If a 16-oz bag costs $2.50, how much will a 9-oz bag cost?

26. If your car uses 2.4 gal of gas driving 89 mi, how much gas will you need to drive 110 mi?

27. A Wheatstone bridge, shown in Figure 12.21, is used for measuring resistance. In the figure, R_1 and R_2 have a known constant resistance, R is an adjustable resistor, and X is the unknown resistance. R is adjusted so that the current from a to b is zero, as measured by galvanometer G. The relationship between the resistances is given by the proportion

$$\frac{R_1}{R_2} = \frac{R}{X}$$

Given that $R_1 = 20.0\ \Omega$ and $R_2 = 50.0\ \Omega$, and if R is found to be $34.6\ \Omega$, determine the resistance X.

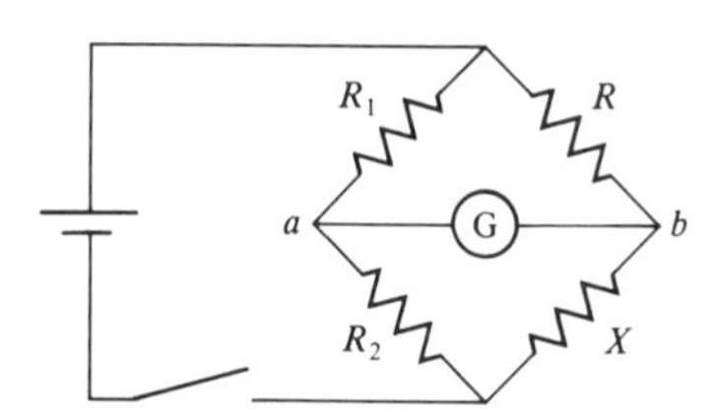

Figure 12.21

28. A transformer consists of two coils. The current in one coil induces a current in the other coil. (See Figure 12.22.) The following proportion gives the relationship between the voltages in each coil and the number of windings:

$$\frac{V_1}{V_2} = \frac{N_1}{N_2}$$

For the first coil, $V_1 = 80$ V and $N_1 = 500$. If $N_2 = 1200$ (second coil), determine the voltage V_2.

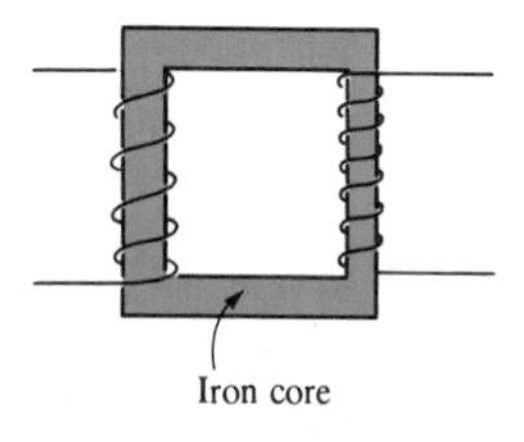

Figure 12.22

12.7 Variation

We have seen in our study of functions that two variable quantities can be related in different ways. In this section we will study a special kind of relationship between variables called **variation**.

To see what is meant by variation, suppose two variables x and y have a fixed ratio k regardless of the values of x and y. This relationship can be written $y/x = k$ or

$$y = kx$$

This functional relationship is called a *direct variation*.

Direct variation

If two variables x and y are related so that $y = kx$, then y is said to vary directly as x or to be directly proportional to x. The constant k is called the **constant of proportionality**.

The first step in a problem on variation is to determine the constant k from the given information. Once k is found, we get an equation that can be used to determine any additional pairs of values.

Example 1 Suppose y varies directly as x. Given that $y = 8$ when $x = 2$, find y when $x = 6$.

Solution. The first step is to determine the proportionality constant k. Since

$$y = kx \qquad \text{Direct variation}$$

we let $x = 2$ and $y = 8$ and solve for k:

$$8 = k \cdot 2 \qquad \text{or} \qquad k = 4$$

The equation $y = kx$ now becomes

$$y = 4x$$

Finally, if $x = 6$, then

$$y = 4 \cdot 6 = 24$$

Variations frequently occur in technical applications as shown in the remaining examples.

Example 2 Hooke's law states that the force required to stretch a spring is directly proportional to the elongation of the spring. Suppose a force of 3.0 lb stretches a spring 5.2 in. What is the force required to stretch the spring 7.8 in.?

Solution. Denote the force by F and the extension by x. Then

$$F = kx$$

Now let $F = 3.0$ lb and $x = 5.2$ in. to obtain the constant k:

$$3.0 = k(5.2) \qquad \text{or} \qquad k = 0.58$$

The equation becomes

$$F = 0.58x$$

To determine the required force, we let $x = 7.8$ in. Then

$$F = 0.58(7.8) = 4.5 \text{ lb}$$

Remark. The constant of proportionality k has units. In our example,

$$k = \frac{3.0 \text{ lb}}{5.2 \text{ in.}} = 0.58 \frac{\text{lb}}{\text{in.}}$$

While our definition of direct variation was stated in the form $y = kx$, the variable on the right side can take different forms. For example, the equation $y = k\sqrt{x}$ is read

"y is directly proportional to $\sqrt{x}$"

or

"y varies directly as $\sqrt{x}$"

Similarly, $y = kx^3$ is read

"y is directly proportional to x^3"

or

"y varies directly as x^3"

Example 3 The period T of a pendulum is directly proportional to the square root of its length L. This statement is written

$$T = k\sqrt{L}$$

The variation "y is directly proportional to $1/x$" has been given a special name, *inverse variation.*

Inverse variation

If two variables x and y are related so that $y = k/x$, then we say that y varies inversely as x or that y is inversely proportional to x.

Example 4 Boyle's law states that the volume V of a confined gas varies inversely as the pressure P at constant temperature. If $V = 95.0$ in.^{3} when $P = 45.0$ lb/in.^{2}, what is the volume when the pressure is increased to 75.2 lb/in.^{2}?

Solution. By the definition of inverse variation,

$$V = k\frac{1}{P} \quad \text{or} \quad V = \frac{k}{P}$$

From the given information, $V = 95.0$ and $P = 45.0$. So

$$95.0 = \frac{k}{45.0}$$

or

$$k = (95.0)(45.0) = 4275$$

The equation is therefore

$$V = \frac{4275}{P}$$

Letting $P = 75.2$, we get

$$V = \frac{4275}{75.2} = 56.8 \text{ in.}^3$$

The variations we have discussed may also occur in combination, called *joint variation*. For example, the relation $z = kxy$ can be expressed as "z varies jointly as x and y." However, we can also say "z is directly proportional to the product of x and y" or "z varies directly as the product of x and y."

Similarly, the relationship

$$z = \frac{kxy}{w}$$

is read "z varies directly as the product of x and y and inversely as w" or "z varies jointly as x and y and inversely as w."

Example 5 The force F of gravitational attraction between two masses m_1 and m_2 is directly proportional to the product of m_1 and m_2 and inversely proportional to the square of the distance d between them. This statement is written

$$F = G\frac{m_1 m_2}{d^2}$$

The constant of proportionality G is the gravitational constant. (See Section 10.2.)

Example 6 The resistance R in a wire varies directly as the length L and inversely as the square of the diameter D. If a 90.0-ft long wire with diameter 0.0250 in. has a resistance of 10.0 Ω, find the resistance in a wire 145 ft long with diameter 0.0160 in., if it is made of the same material.

Solution. From the given relationship we have

$$R = k\frac{L}{D^2}$$

To evaluate k, we let $R = 10.0\ \Omega$, $L = 90.0$ ft, and $D = 0.0250$ in. Then

$$10.0 = k\frac{90.0}{(0.0250)^2}$$

or

$$k = \frac{(10.0)(0.0250)^2}{90.0} = 6.944 \times 10^{-5}$$

Thus

$$R = 6.944 \times 10^{-5}\frac{L}{D^2}$$

To find the resistance in the 145-ft wire, we let $L = 145$ ft and $D = 0.0160$ in.:

$$R = (6.944 \times 10^{-5})\frac{145}{(0.0160)^2} = 39.3\ \Omega$$

Exercises / Section 12.7

In Exercises 1–14, express each statement in the form of an equation.

1. z is directly proportional to y.

2. y varies directly as x^2.

3. w varies directly as z^2.

4. N is directly proportional to $\sqrt{M}$.

5. P is directly proportional to S^3.

6. L is inversely proportional to N.

7. s is inversely proportional to $\sqrt{t}$.
8. p varies inversely as r^2.
9. A varies directly as b and inversely as c.
10. z varies directly as w and inversely as x.
11. Q is directly proportional to the product of p and q and inversely proportional to s.
12. D is directly proportional to the product of a and b and inversely proportional to c^3.
13. E varies directly as the product of v and w and inversely as $\sqrt{u}$.
14. R varies directly as the product of s and t^2 and inversely as v.

In the remaining exercises, use a calculator when convenient. In Exercises 15–22, find the equations from the given information.

15. z varies directly as x; $z = 6$ when $x = 2$.
16. w varies inversely as y; $w = 2$ when $y = 3$.
17. P is inversely proportional to V; $P = 2$ when $V = \frac{1}{2}$.
18. B is inversely proportional to d; $B = \frac{1}{3}$ when $d = 5$.
19. n is directly proportional to the product of s and t; if $s = 2$ and $t = 3$, then $n = 10$.
20. L varies directly as the product of R and C; if $R = 10$ and $C = \frac{1}{2}$, then $L = 5$.
21. F varies directly as the product of a and b and inversely as c; if $a = 2$, $b = 4$, and $c = \frac{1}{2}$, then $F = 8$.
22. G is directly proportional to the product of x_1 and x_2 and inversely proportional to $\sqrt{x_3}$; if $x_1 = 1$, $x_2 = 2$, and $x_3 = 4$, then $G = 6$.
23. The force exerted on a spring is directly proportional to the elongation of the spring. If a force of 4.5 lb is required to stretch the spring 1.5 in., what is the force required to stretch it 2.5 in.?
24. If a force of 10 N is required to stretch a spring 4 cm, what is the force required to stretch it 6 cm? (Refer to Exercise 23.)
25. The *strength* S of a rectangular beam varies directly as the product of the width w and the square of the depth d (Figure 12.23). Find an expression for S.

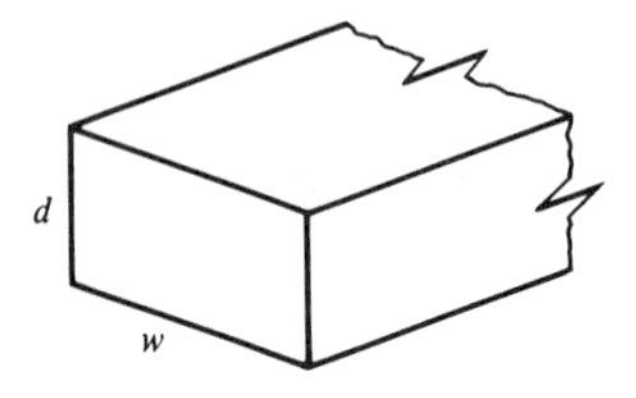

Figure 12.23

26. The *stiffness* T of a beam with a rectangular cross section is directly proportional to the product of the width w and the cube of the depth d (Figure 12.23). Write an expression for T.
27. If a body is dropped from rest, the distance d from the starting point is directly proportional to the square of the time t. If an object falls 36 ft in 1.5 s, how far does it fall in 4 s?

28. The period T of a pendulum is directly proportional to the square root of its length L. If $T = 6.03$ s when $L = 9.00$ m, find the relationship.

29. For two gears in mesh, the speed of each is inversely proportional to the number of teeth. If a gear having 30 teeth and turning at 100 rev/min is in mesh with a gear having 50 teeth, what is the speed of the second gear?

30. Within a small range, the demand for a product varies inversely as the price. If a store sells 90 units of a product per week at \$1.50 apiece, how many units can it expect to sell if the price is reduced to \$1.35?

31. Boyle's law states that the pressure of gas varies inversely as the volume (at constant temperature). If the pressure is 10.0 lb/ft^2 when the volume is 7.50 ft^3, what is the pressure when the volume is increased to 12.3 ft^3?

32. The kinetic energy of a moving particle is directly proportional to the square of the velocity. If the kinetic energy of a particle moving at 10.0 m/s is 20.0 J, what is the kinetic energy if the velocity is increased to 18.0 m/s?

33. The repulsive force F between two charged particles having like charges, q_1 and q_2, respectively, varies directly as the product of q_1 and q_2 and inversely as the square of the distance d between them.
   a. What is the effect on the force F if q_1 and q_2 are each doubled and d remains the same?
   b. What is the effect on F if d is doubled and q_1 and q_2 remain the same?

34. The intensity of illumination at a given point is directly proportional to the intensity of the light source and inversely proportional to the square of the distance from the light source. Suppose a 60-W bulb is placed 6.2 ft from an object. What size light bulb will provide the same illumination on the object at a distance of 8 ft?

35. The resistance R in a wire varies directly as the length L of the wire and inversely as the square of the diameter D. If a wire 80.5 ft long with diameter 0.0808 in. has a resistance of 9.25 Ω, find the resistance of a wire 65.0 ft long if it has a diameter of 0.1019 in. and is made of the same material.

36. If a wire 112 ft long with diameter 0.0634 in. has a resistance of 12.5 Ω, find the resistance of a wire 212 ft long if it has a diameter of 0.0808 in. and is made of the same material.

Review Exercises / Chapter 12

1. Identify the independent and dependent variables of the function $s = -16t^2 + 24t + 5$.

2. Given that $f(v) = \dfrac{1}{2v - 1}$, find $f(0)$, $f(2)$, $f(3)$.

3. Given that $G(w) = \sqrt{1 + 2w}$, find $G\left(-\frac{1}{2}\right)$, $G(0)$, $G(2)$, $G(4)$.

4. State the domain of the function $P(t) = \sqrt{2t + 4}$.

5. The base of a triangle is 5 units long. Write the area A of the triangle as a function of its height h.

6. Determine the points in the plane for which (a) the x-coordinates are zero; (b) the y-coordinates are negative.

In Exercises 7–11, sketch the graphs of the given functions.

7. $y = 2x - 4$ 8. $y = 2x^2 - 2$ 9. $y = \sqrt{4 - x}$ 10. $y = 2 - \dfrac{3}{x}$ 11. $y = 2^{-x}$

12. Solve the following equation graphically to the nearest tenth: $2x - 2^{-x} = 0$.

In Exercises 13–16, express each statement in the form of an equation.

13. y varies directly as x and inversely as z.

14. P is directly proportional to a and inversely proportional to $\sqrt{b}$.

15. Z varies directly as the product of s and t.

16. V is directly proportional to the product of R and C^2.

17. S varies directly as p and inversely as q. If $p = 3$ and $q = 6$, then $S = 10$. Write an equation expressing this relationship.

18. T is directly proportional to the product of v_1 and v_2 and inversely proportional to v_3. Write this relationship as an equation.

19. The temperature in degrees Fahrenheit (F) as a function of the temperature in degrees Celsius (C) is a linear function:

$$F = \frac{9}{5}C + 32$$

Sketch the graph.

20. The half-life of a radioactive substance is the time required for half of a given quantity to decay. If a radioactive substance has a half-life of 1 year, then 10 kg of the substance decays according to the formula

$$N = 10\left(\frac{1}{2}\right)^t, \qquad t > 0$$

Sketch the graph of this function.

21. A certain body is taken out of a freezer kept at $-15°$F. The temperature T of the body as a function of t (in minutes) is given by

$$T = 5 - 20(3^{-0.2t})$$

Determine when $T = 0$ graphically (to the nearest tenth).

22. The *efficiency* of an engine is defined as the ratio of output to input, commonly expressed as a percent. If the input of an engine is 16,000 W and the output 10,560 W, what is the efficiency?

23. The *percentage error* is the ratio of the error in a measurement to the correct value, multiplied by 100. If the side of a square is measured to be 12.0 in. with an error in measurement of +0.1 in., determine the resulting percentage error in the area.

24. A certain alloy is made of tin and lead in the ratio of 5 to 3. How many pounds of each are contained in 48 lb of the alloy?

25. A machine produces 16 parts in 25 min. How long will it take to produce 88 parts?

26. The capacitance C of two parallel plates varies inversely as the distance d between the plates. Write this relationship in the form of an equation.

27. The voltage V across a resistor is directly proportional to the current I. If $V = 20.8$ V when $I = 2.0$ A, find V when $I = 2.8$ A.

28. According to Newton's second law, the acceleration of a body varies directly as the force applied. Suppose an acceleration of 5.75 m/s^2 results from a force of 30.0 N. What is the force required to yield an acceleration of 10.0 m/s^2?

29. The attractive force F between two charged particles having unlike charges, q_1 and q_2, respectively, varies directly as the product of the charges and inversely as the square of the distance d between them. Find an expression for F.

30. According to Newton's law of cooling, the rate R of decrease of the temperature of a body is directly proportional to the difference between the temperature T of the body and the temperature T_0 of the surrounding medium. Write this statement in the form of an equation.

Cumulative Review Exercises / Chapters 10–12

1. Evaluate: **a.** $(-32)^{-1/5}$ **b.** $\dfrac{4^0 \cdot 9^{-1/2}}{3^{-1}}$

In Exercises 2–5, simplify the given expressions and write the results without zero or negative exponents.

2. $\dfrac{(a^{-2}b^{-3})^{-2}}{3^{-1}}$

3. $\dfrac{T_a^{-2}T_b^3}{(T_aT_b^{-1})^{-3}}$

4. $\dfrac{(8p^{-3/2}q^{1/2})^{-2/3}}{(3^{-1}p^{-1/2})^2}$

5. $\dfrac{1 - C^{-1}}{C^{-1} + C^{-2}}$

6. Use a calculator to evaluate

$$\frac{0.0000000007894}{639{,}000{,}000}$$

7. Simplify: $\sqrt[4]{32P_1^6P_2^{10}}$

8. Simplify: $\dfrac{V}{\sqrt{8Q}}$

In Exercises 9–14, perform the indicated operations with the radicals and simplify.

9. $\dfrac{\sqrt{4RC}}{\sqrt{8R}}$

10. $\sqrt{6mn}\,\sqrt{3m}$

11. $\sqrt[3]{9M^2N}\,\sqrt[3]{6MN}$

12. $\sqrt{4x} - \sqrt{16x} - \sqrt{25x}$

13. $(s - \sqrt{t})(3s + 2\sqrt{t})$

14. $\dfrac{\sqrt{5} + \sqrt{3}}{\sqrt{5} - \sqrt{3}}$

15. Solve the pure quadratic equation $4x^2 + 5 = 0$.

In Exercises 16–18, solve the given quadratic equations by factoring, if possible. Otherwise use the quadratic formula.

16. $2s^2 + s - 6 = 0$

17. $v^2 + 2v = 2$

18. $3t^2 - 2t + 2 = 0$

19. Determine the domain of the function $f(t) = \sqrt{2 - 4t}$; find $f(0)$ and $f(-4)$.

20. Graph the quadratic function $f(x) = x^2 - 2x - 3$.

21. Write the following statement as an equation: V is directly proportional to the product of v and w and inversely proportional to $\sqrt{t}$.

22. When a block of mass m slides down a frictionless inclined plane, its velocity v at the bottom of the plane is $v = (2gh)^{1/2}$, where h is the vertical height. Express this formula in radical form.

23. The frequency f of a vibrating string of length L is given by

$$f = \frac{1}{2L}\sqrt{\frac{T}{\rho}} \qquad \rho = \text{rho}$$

where T is the tension in the string and ρ is a constant. Simplify the formula for f by rationalizing the denominator.

24. A dam has a V-shaped notch with a 90° angle. If the water reaches a level of h meters above the notch, then the rate of flow r (in cubic meters per second) is $r = 1.37h^{5/2}$. Write r in the form of a radical.

25. Referring to Exercise 24, determine the rate of flow over the notch if the water reaches a level of 3.68 m.

26. The distance s from the origin of a particle moving along the x-axis as a function of time is $s = \sqrt{t}$. It is shown in calculus that the velocity v is found from the expression

$$\frac{\sqrt{t+h} - \sqrt{t}}{h}$$

by rationalizing the numerator and letting $h = 0$. Show that $v = 1/(2\sqrt{t})$.

27. It takes 4.76 h longer to empty a chemical tank than to fill it. If the tank is initially empty and both drain and inlet valve are open, then the tank can be filled in 20.0 h. How long does it take to fill the empty tank if the drain is closed?

28. The number π is defined to be the ratio of the circumference of a circle to the diameter. If a circle has circumference 10.92 cm and radius 1.738 cm, determine the value of π to four significant digits.

29. A radioactive substance decays according to the formula $N = 5.0(2^{-0.35t})$, where N (in grams) is the amount left after t minutes. Sketch the graph of this formula.

30. The resistance R in a wire varies directly as the length L and inversely as the square of the diameter D. If a wire 125 ft long with diameter 0.0808 in. has a resistance of 10.4 Ω, find the resistance in a wire 85.0 ft long if it has a diameter of only 0.0236 in. (The wires are made of the same material.)

CHAPTER 13

Simultaneous Linear Equations

In Chapter 7 we studied the solution of equations of first degree. In this chapter we study the solution of systems of two linear equations with two unknowns. Such systems are useful in many areas of technology.

13.1 General Discussion and Graphical Method of Solution

Recall from Section 12.3 that a function of the form $f(x) = ax + b$ is called a *linear function* since the graph of the equation $y = ax + b$ is a straight line. More generally, it can be shown that any equation of the form

$$ax + by = c$$

Linear equation

represents a straight line. This equation is called a **linear equation,** or a **linear equation in two variables.**

Consider, for example, the linear equation $2x - y = 2$. The resulting straight line, shown in Figure 13.1, is plotted from the following table of values:

x:	0	1	2
y:	-2	0	2

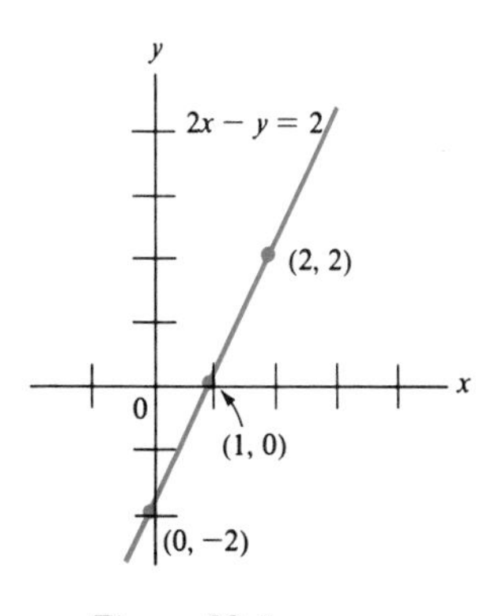

Figure 13.1

Note especially that the coordinates of the points $(0, -2)$, $(1, 0)$, and $(2, 2)$ given in the table satisfy the equation. In fact, the coordinates of *every* point on the line satisfy the equation.

Now consider another line, $x + y = 4$, shown in Figure 13.2. As before, the coordinates of every point on this line satisfy the equation $x + y = 4$.

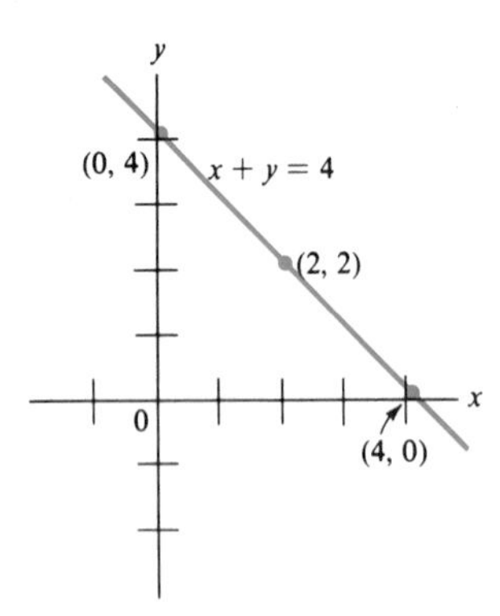

Figure 13.2

Both lines are shown in Figure 13.3. Judging from Figure 13.3, the lines intersect at $(2, 2)$. So the coordinates of the point $(2, 2)$ must satisfy *both* equations. In other words, $x = 2$ and $y = 2$ satisfy the pair of equations

$$2x - y = 2$$
$$x + y = 4$$

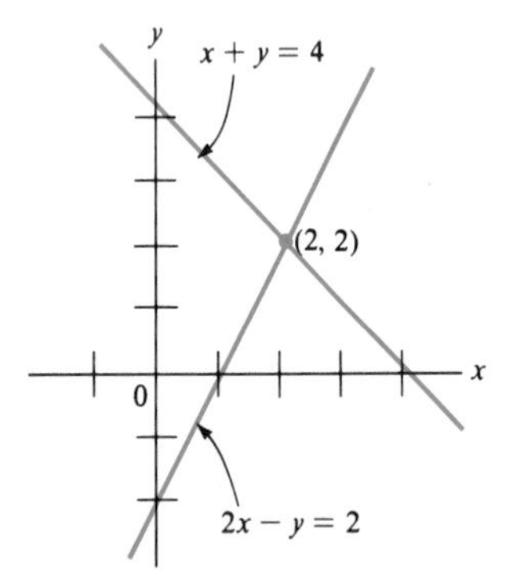

Figure 13.3

This pair of equations is called a system of two **simultaneous linear equations.** The solution $x = 2$, $y = 2$ is called a **common solution** of the system.

It follows from our discussion that a common solution (x, y) is the point of intersection of the two lines. If the lines are distinct, then this point is unique, so that the solution is also unique.

Simultaneous linear equations

A system of two simultaneous linear equations has the form

$$a_1x + b_1y = c_1$$
$$a_2x + b_2y = c_2$$

A pair of values (x, y) that satisfies both equations is called a **common solution** of the system.

Graphical solution: To find the common solution of a system of two linear equations, draw the two lines and determine from the graphs the coordinates of the point of intersection.

Recall from Section 12.3 that the simplest way to draw the graph of a line is to find the intercepts: To find the x-intercept, we let $y = 0$ and solve for x. To find the y-intercept, we let $x = 0$ and solve for y. Since two distinct points determine a straight line, the intercepts are the only points needed to draw the graph. However, it is good practice to plot a third point as a check. Let us return to the system

$$2x - y = 2$$
$$x + y = 4$$

For the first equation, if $y = 0$, then $2x = 2$ and $x = 1$. So $(1, 0)$ is the x-intercept. If $x = 0$, then $-y = 2$ and $y = -2$. So $(0, -2)$ is the y-intercept.

The intercepts and check points for this system are shown in the following tables:

$2x - y = 2$ and $x + y = 4$:

x	y		x	y
0	−2	y-intercept	0	4
1	0	x-intercept	4	0
2	2	check point	3	1

Let's consider another example.

Example 1 Determine the common solution of the system

$$2x + y = 5$$
$$x + 3y = 5$$

by drawing the graph of each line and estimating the point of intersection.

Solution. To graph these lines, we find the intercepts in each case and a check point. In the first equation, if $x = 0$, then $y = 5$. If $y = 0$, then $2x = 5$ and $x = \frac{5}{2}$, and so on. The intercepts and check points are given in the following tables:

$2x + y = 5$

x	y	
0	5	y-intercept
$\frac{5}{2}$	0	x-intercept
1	3	check point

$x + 3y = 5$

x	y
0	$\frac{5}{3}$
5	0
−1	2

The lines are shown in Figure 13.4.

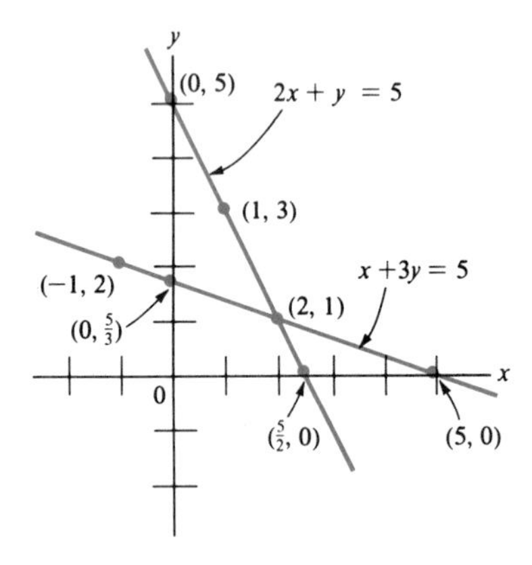

Figure 13.4

The lines appear to intersect at the point (2, 1). So $x = 2$ and $y = 1$ is the solution of the system.

As a check, let us substitute $x = 2$ and $y = 1$ in the given system:

$$2(2) + 1 = 5 \checkmark$$

$$2 + 3(1) = 5 \checkmark$$

Since (2, 1) satisfies both equations, it is the point of intersection and therefore the common solution of the system.

Example 2 Determine the solution of the system

$$2x - y = -4$$

$$x - 3y = 2$$

graphically to the nearest tenth of a unit.

Solution. The intercepts and check points are given in the following tables:

$2x - y = -4$

x	y	
0	4	y-intercept
-2	0	x-intercept
1	6	check point

$x - 3y = 2$

x	y	
0	$-\frac{2}{3}$	y-intercept
2	0	x-intercept
-1	-1	check point

Now draw the lines (Figure 13.5) and estimate the point of intersection. To the nearest tenth of a unit, the coordinates appear to be $x = -2.8$ and $y = -1.6$.

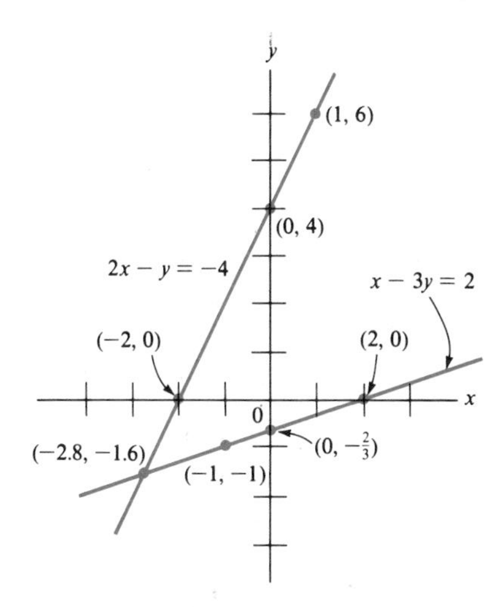

Figure 13.5

Recall from Chapter 6 that two lines in a plane are either parallel or intersecting. If two lines are parallel, they do not intersect, and the corresponding system of equations does not have a solution. Such a system is called **inconsistent**. If a unique solution exists, the system is called **consistent**.

Example 3 Graph the lines

$$2x - 3y = 6$$
$$4x - 6y = -12$$

Solution. $2x - 3y = 6$

x	y	
0	-2	y-intercept
3	0	x-intercept
6	2	check point

$4x - 6y = -12$

x	y	
0	2	y-intercept
-3	0	x-intercept
2	$\frac{10}{3}$	check point

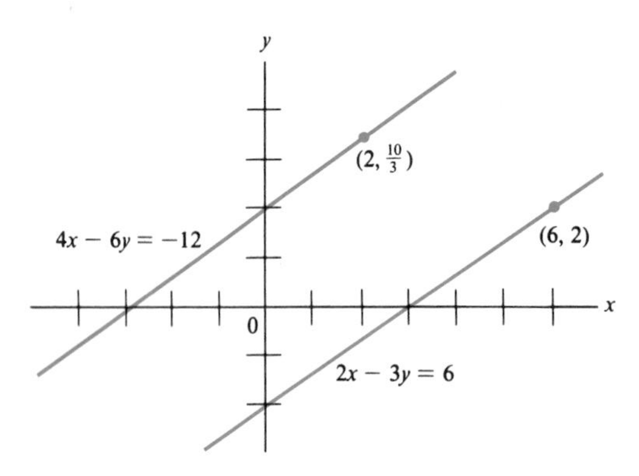

Figure 13.6

The lines, shown in Figure 13.6, appear to be parallel. (They are.) Since the lines do not intersect, there is no common solution. The given system is therefore inconsistent.

The systems in Examples 1 and 2 are consistent. The system in Example 3 is inconsistent. One more possibility exists: The system may have infinitely many solutions. Such a system is called **dependent.**

Example 4 Graph the lines

$$x - 2y = 4$$
$$3x - 6y = 12$$

Solution. $x - 2y = 4$

x	y	
0	-2	y-intercept
4	0	x-intercept
1	$-\frac{3}{2}$	check point

$3x - 6y = 12$

x	y	
0	-2	y-intercept
4	0	x-intercept
1	$-\frac{3}{2}$	check point

Note that the intercepts and check points are the same for both equations. So the two equations represent exactly the same line (Figure 13.7). It follows that the coordinates of any point on this line satisfy both equations, so the system has infinitely many solutions. The system is therefore dependent.

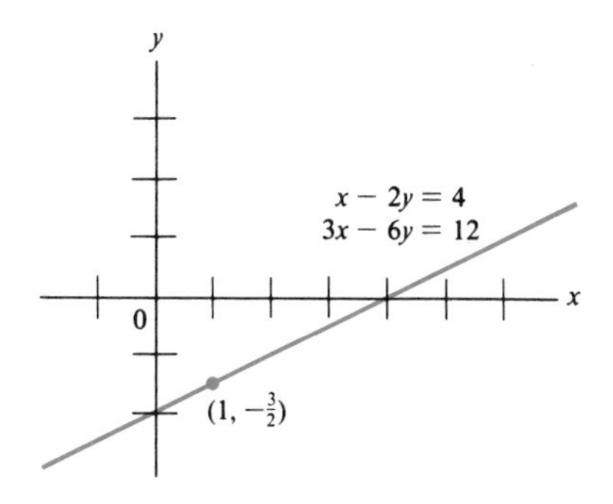

Figure 13.7

Exercises / Section 13.1

In Exercises 1–16, solve the given systems of equations graphically. Estimate the solutions to the nearest tenth of a unit.

1. $2x + y = 4$
   $x - y = 2$
2. $x + 3y = 6$
   $x - 2y = -4$
3. $x - y = 2$
   $2x - y = 3$
4. $x - y = 3$
   $x + y = 5$
5. $4x + y = 4$
   $3x - y = 6$
6. $x - 2y = 8$
   $3x - y = 6$
7. $x - 6y = -12$
   $2x - 3y = 6$
8. $x - 4y = 12$
   $-x - 6y = 12$
9. $2x - 3y = 12$
   $4x - y = -4$
10. $2x - y = 4$
    $x + y = 7$
11. $-x - 3y = 4$
    $2x + 2y = 5$
12. $x + 2y = 13$
    $x - 3y = 2$
13. $2x + 3y = 2$
    $3x + 2y = 1$
14. $3x + 3y = 19$
    $-x + 2y = 1$
15. $5x - 2y = 9$
    $4x - 3y = 4$
16. $x - 3y = 10$
    $3x - y = 3$

In Exercises 17–24, determine graphically which systems are inconsistent and which are dependent.

17. $2x + y = 4$
    $4x + 2y = 5$
18. $-6x + 2y = 9$
    $3x - y = 6$
19. $1.2x - 1.8y = 3.6$
    $3.6x - 5.4y = 16.2$
20. $-7.2x + 20.4y = 28.8$
    $3.6x - 10.2y = 28.8$
21. $x - 2y = 7$
    $-2x + 4y = -14$
22. $-2x + 3y = 9$
    $6x - 9y = -27$
23. $10x + 25y = 50$
    $2x + 5y = 10$
24. $12x - 8y = -24$
    $-3x + 2y = 6$

13.2 Solution by Addition or Subtraction

In the last section we studied the graphical solution of two simultaneous linear equations. In this section we will see how such systems can be solved by an algebraic procedure called **elimination by addition or subtraction.**

To see how this method works, consider the system

$$\begin{aligned} 2x + 3y &= 1 \\ x + 3y &= 2 \end{aligned}$$

Observe that the coefficients of y are both $+3$. So if the second equation is subtracted from the first, the variable y is eliminated:

$$\begin{aligned} 2x + 3y &= 1 \\ \underline{x + 3y} &\underline{= 2} \\ x \qquad &= -1 \qquad 2x - x = x, \quad 3y - 3y = 0, \quad 1 - 2 = -1 \end{aligned}$$

We conclude that $x = -1$. To obtain y, we substitute $x = -1$ into either of the given equations. Using the second equation, $x + 3y = 2$, we get

$$\begin{aligned} x + 3y &= 2 \\ -1 + 3y &= 2 \\ 3y &= 3 \\ y &= 1 \end{aligned}$$

The common solution is therefore $x = -1$, $y = 1$.

This example shows that whenever the coefficients of one of the variables are numerically equal, this variable can be eliminated by adding or subtracting the two equations. The resulting equation has only one unknown and can be readily solved. If the coefficients of the unknowns are not numerically equal, we use the procedure given next.

Method of addition or subtraction

Step 1. Multiply both sides of the equations (if necessary) by constants so chosen that the coefficients of one of the variables are numerically equal.
Step 2. If the coefficients have opposite signs, add the corresponding members of the equations. If the coefficients have like signs, subtract the corresponding members of the equations.
Step 3. Solve the resulting equation in one unknown for the unknown.
Step 4. Substitute the value found in Step 3 into either of the original equations and solve for the second unknown.
Step 5. Check the solution in the original system.

Example 1 Solve the following system by the method of addition or subtraction.

$$3x - 2y = 5$$
$$5x + 2y = 3$$

Solution.

Step 1. Not necessary in this system since the coefficients of y are numerically equal.

Step 2.
$$\begin{aligned} 3x - 2y &= 5 \\ \underline{5x + 2y} &\underline{= 3} \\ 8x \quad\quad &= 8 \end{aligned} \qquad \text{Adding}$$

(Since the y-coefficients have opposite signs, we add the equations to eliminate y.)

Step 3. From $8x = 8$, we get $x = 1$.

Step 4. To find the corresponding y-value, we substitute $x = 1$ into either of the original equations and solve for x. Using the first equation, we get

$$\begin{aligned} 3x - 2y &= 5 \\ 3(1) - 2y &= 5 && \text{Substituting } x = 1 \\ 3 - 2y &= 5 \\ -2y &= 5 - 3 && \text{Subtracting 3 from both sides} \\ -2y &= 2 \\ y &= -1 && \text{Dividing both sides by } -2 \end{aligned}$$

Step 5. As a check, let us substitute $x = 1$ and $y = -1$ in the given system:

$$3(1) - 2(-1) = 3 + 2 = 5 \checkmark$$
$$5(1) + 2(-1) = 5 - 2 = 3 \checkmark$$

The solution checks. Note that the system is **consistent.**

In the next example we need to multiply both sides of one of the equations by a constant (Step 1).

Example 2 Solve the system

$$2v + 2w = 5$$
$$v + 3w = 1$$

Solution. We can eliminate either variable. To eliminate w we can multiply the first equation by 3 and the second by 2 (to get equal w-coefficients) and then subtract the equations. Suppose we eliminate v by multiplying both sides of the second equation by 2.

$$\begin{array}{lrl} \text{Step 1.} & 2v + 2w = 5 & \\ & \underline{2v + 6w = 2} & 2(v + 3w) = 2(1) \\ \text{Step 2.} & -4w = 3 & \text{Subtracting} \end{array}$$

[Recall that in subtraction we change the sign of the subtrahend and add. Thus $2w - (+6w) = 2w + (-6w) = -4w$.]

Step 3. Solving the equation $-4w = 3$, we get

$$w = -\frac{3}{4}$$

Step 4. To find the corresponding v-value, we substitute $w = -\frac{3}{4}$ into either of the given equations and solve for v. Using the first equation, we get

$$\begin{aligned} 2v + 2w &= 5 & & \\ 2v + 2\left(-\frac{3}{4}\right) &= 5 & & \text{Substituting } w = -\frac{3}{4} \\ 2v - \frac{3}{2} &= 5 & & \\ 2\left(2v - \frac{3}{2}\right) &= 2(5) & & \text{Clearing fractions} \\ 4v - 3 &= 10 & & \\ 4v &= 10 + 3 = 13 & & \text{Adding 3 to both sides} \\ v &= \frac{13}{4} & & \text{Dividing by 4} \end{aligned}$$

The solution is therefore

$$\left(\frac{13}{4}, -\frac{3}{4}\right)$$

Step 5. As a check, we substitute these values into the original system:

$$2\left(\frac{13}{4}\right) + 2\left(-\frac{3}{4}\right) = \frac{13}{2} - \frac{3}{2} = \frac{10}{2} = 5 \quad \checkmark$$

$$\frac{13}{4} + 3\left(-\frac{3}{4}\right) = \frac{13}{4} - \frac{9}{4} = \frac{4}{4} = 1 \quad \checkmark$$

In some cases both equations have to be multiplied by a constant before one of the variables can be eliminated.

Example 3 Solve the system

$$3F_1 + 2F_2 = 1$$
$$4F_1 - 3F_2 = 2$$

Solution. Neither addition nor subtraction will eliminate either variable. To eliminate F_1, we multiply the first equation by 4 and the second by 3. To eliminate F_2, we multiply the first equation by 3 and the second by 2. Suppose we eliminate F_2.

Step 1. $9F_1 + 6F_2 = 3$ $\quad 3(3F_1 + 2F_2) = 3 \cdot 1$

$8F_1 - 6F_2 = 4$ $\quad 2(4F_1 - 3F_2) = 2 \cdot 2$

Step 2. $17F_1 = 7$ $\quad$ Adding

Step 3. $F_1 = \frac{7}{17}$ $\quad$ Dividing by 17

Step 4. From $3F_1 + 2F_2 = 1$, we get

$$3\left(\frac{7}{17}\right) + 2F_2 = 1 \qquad F_1 = \frac{7}{17}$$

$$\frac{21}{17} + 2F_2 = 1$$

$$2F_2 = 1 - \frac{21}{17} \qquad \text{Subtracting } \frac{21}{17}$$

$$2F_2 = \frac{17}{17} - \frac{21}{17} = -\frac{4}{17} \qquad \text{LCD} = 17$$

$$F_2 = -\frac{4}{17} \cdot \frac{1}{2} \qquad \text{Dividing by 2}$$

$$F_2 = -\frac{2}{17}$$

The solution is therefore given by

$$\left(\frac{7}{17}, -\frac{2}{17}\right)$$

Step 5. *Check:*

$$3\left(\frac{7}{17}\right) + 2\left(-\frac{2}{17}\right) = \frac{21}{17} - \frac{4}{17} = \frac{17}{17} = 1 \checkmark$$

$$4\left(\frac{7}{17}\right) - 3\left(-\frac{2}{17}\right) = \frac{28}{17} + \frac{6}{17} = \frac{34}{17} = 2 \checkmark$$

The next two examples illustrate inconsistent and dependent systems, respectively.

Example 4 Solve the system

$$2x - 3y = 6$$
$$-4x + 6y = 12$$

Solution. We multiply both sides of the first equation by 2:

$$\begin{aligned} 4x - 6y &= 12 \\ -4x + 6y &= 12 \\ \hline 0 &= 24 \quad \text{Adding} \end{aligned}$$

Since 0 is not equal to 24, it follows that the system has no solution. It is therefore inconsistent. The lines, shown in Figure 13.8, are parallel.

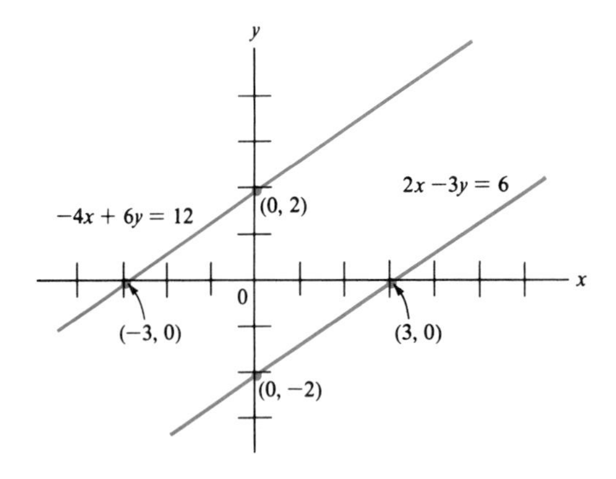

Figure 13.8

Example 5 Compare the system

$$2x - 3y = 6$$
$$-4x + 6y = -12$$

to the system in Example 4.

Solution.

$$\begin{aligned} 4x - 6y &= 12 \quad \text{Multiplying by 2} \\ -4x + 6y &= -12 \\ \hline 0 &= 0 \quad \text{Adding} \end{aligned}$$

This time no contradiction results. In fact, we have merely shown that the two equations represent exactly the same line. (See Figure 13.9.) As a result, the coordinates of any point on this line satisfy the given system, and we get infinitely many solutions. The system is therefore dependent.

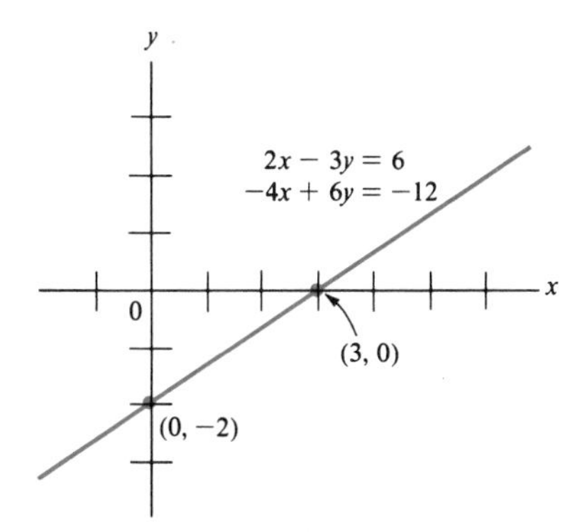

Figure 13.9

We can see from Examples 4 and 5 that a system is inconsistent or dependent if the respective coefficients of the variables are multiples of each other. This means that the ratios of the respective coefficients are equal. If this common ratio is also equal to the ratio of the constants, then the system is dependent. If the common ratio is different from the ratio of the constants, then the system is inconsistent.

Summary

$$a_1x + b_1y = c_1$$
$$a_2x + b_2y = c_2$$

System of equations

I. The system is **consistent** if

$$\frac{a_1}{a_2} \neq \frac{b_1}{b_2}$$

II. The system is **inconsistent** if

$$\frac{a_1}{a_2} = \frac{b_1}{b_2} \neq \frac{c_1}{c_2}$$

III. The system is **dependent** if

$$\frac{a_1}{a_2} = \frac{b_1}{b_2} = \frac{c_1}{c_2}$$

Exercises / Section 13.2

In Exercises 1–30, solve each system of equations by the method of addition or subtraction.

1. $2x - y = 6$
$x + y = 3$

2. $2x - 3y = 7$
$2x - 5y = 12$

3. $x + 2y = 1$
$3x - y = -4$

4. $2x - 3y = 5$
$x + 4y = 8$

5. $3x - 2y = 1$
$4x - 3y = 4$

6. $2x + 7y = 0$
$3x - 2y = 25$

7. $3x - 2y = 21$
$4x - 5y = 42$

8. $4x - 6y = 11$
$3x - 2y = 7$

9. $2x + 2y = 1$
$5x - 5y = 1$

10. $3x + 5y = 10$
$5x + 3y = 10$

11. $3x - 4y = 4$
$-6x + 8y = 2$

12. $2x - 5y = 4$
$-6x + 15y = -12$

13. $5x + 6y = 2$
$20x + 24y = 8$

14. $-9x - 12y = 1$
$3x + 4y = 1$

15. $3v_1 + 4v_2 = 1$
$2v_1 + 3v_2 = 4$

16. $-2V - 6W = 4$
$5V + 10W = 5$

17. $3s - 2t = 4$
$-7s + 5t = 1$

18. $-5R + 6C = 7$
$4R - 5C = 8$

19. $3A - 4B = 20$
$5A - 6B = 8$

20. $6R_1 + 3R_2 = -1$
$5R_1 + 3R_2 = 2$

21. $4s_1 - 7s_2 = 10$
$5s_1 - 8s_2 = 20$

22. $2W_1 + W_2 = 5$
$W_1 + 3W_2 = 5$

23. $F_1 + 2F_2 = 5$
$2F_1 + F_2 = 6$

24. $2F_1 + 3F_2 = 12$
$2F_1 + 4F_2 = 15$

25. $3C_1 + 4C_2 = 20$
$4C_1 + 2C_2 = 15$

26. $6a + 5b = -18$
$3a - b = 5$

27. $m + 4n = -1$
$5m + 12n = -7$

28. $6u + 5v = 2$
$12u - 5v = 1$

29. $2T_1 + 4T_2 = 26$
$4T_1 + 3T_2 = 39$

30. $2S_1 + 7S_2 = -3$
$-4S_1 + S_2 = 5$

In Exercises 31–34, use a calculator to solve each system of equations. (The numbers are approximate.)

31. $2.1x + 3.2y = 1.7$
$4.6x - 7.4y = 2.5$

32. $1.4x - 7.6y = 9.3$
$2.6x - 4.8y = 5.7$

33. $-4.88x + 10.4y = 12.6$
$8.72x - 12.1y = 19.5$

34. $15.3x - 8.76y = 25.3$
$-17.4x + 14.4y = 21.6$

13.3 Solution by Substitution

In the last section we studied the method of addition or subtraction. For some systems of equations another method of solution is sometimes more convenient. This is the method of **elimination by substitution,** discussed in this section.

Consider, for example, the system

$$x - 2y = 3$$
$$2x + y = 6$$

Note that the first equation is easily solved for x by adding $2y$ to both sides. Thus

$$x = 2y + 3$$

If we substitute this expression for x in the second equation, then x is eliminated:

$$2x + y = 6$$
$$2(2y + 3) + y = 6$$

The resulting equation is now solved for y:

$$2(2y + 3) + y = 6$$

$$4y + 6 + y = 6 \qquad \text{Removing parentheses}$$

$$5y + 6 = 6$$

$$5y = 6 - 6 \qquad \text{Subtracting 6}$$

$$5y = 0$$

$$y = 0 \qquad \text{Dividing by 5}$$

From $x = 2y + 3$, we now get $x = 2 \cdot 0 + 3 = 3$. So the solution is $(3, 0)$. The method of substitution is summarized next.

Method of substitution

Step 1. Solve one of the equations for one of the variables in terms of the other.
Step 2. Substitute the expression obtained in the other equation.
Step 3. Solve the resulting equation in one unknown for the unknown.
Step 4. Substitute the value of the unknown obtained in Step 3 in either of the original equations and solve for the other unknown.
Step 5. Check the solution in the original system.

Example 1 Solve the following system of linear equations by the method of substitution:

$$2x - y = 2$$

$$6x + 2y = 1$$

Solution. The first step is to solve one of the equations for one of the variables. The easiest way is to solve the first equation for y.

Step 1.

$$2x - y = 2 \qquad \text{First equation}$$

$$-y = 2 - 2x \qquad \text{Subtracting } 2x$$

$$y = -2 + 2x \qquad \text{Multiplying by } -1$$

Step 2.

$$6x + 2y = 1 \qquad \text{Second equation}$$

$$6x + 2(-2 + 2x) = 1 \qquad \text{Substituting } -2 + 2x \text{ for } y$$

Step 3. Solving for x:

$$6x + 2(-2 + 2x) = 1$$

$$6x - 4 + 4x = 1 \qquad \text{Removing parentheses}$$

$$10x - 4 = 1$$

$$10x = 1 + 4 \qquad \text{Adding 4}$$

$$10x = 5$$

$$x = \frac{5}{10} = \frac{1}{2} \qquad \text{Dividing by 10}$$

Step 4. From the first equation, written in the form $y = -2 + 2x$ (Step 1), we get

$$y = -2 + 2\left(\frac{1}{2}\right) \qquad \text{Substituting } x = \frac{1}{2}$$

$$= -2 + 1 = \mathbf{-1}$$

The solution is therefore given by

$$\left(\frac{1}{2}, -1\right)$$

Step 5. *Check:*

$$2\left(\frac{1}{2}\right) - (-1) = 1 + 1 = 2 \checkmark$$

$$6\left(\frac{1}{2}\right) + 2(-1) = 3 - 2 = 1 \checkmark$$

Example 2 Solve the following system by the method of substitution:

$$3w_1 - 4w_2 = 6$$
$$w_1 - 3w_2 = 4$$

Solution. In this problem it is best to solve the second equation for w_1.

Step 1.

$$w_1 - 3w_2 = 4 \qquad \text{Second equation}$$
$$w_1 = 3w_2 + 4 \qquad \text{Adding } 3w_2$$

Step 2.

$$3w_1 - 4w_2 = 6 \qquad \text{First equation}$$
$$3(3w_2 + 4) - 4w_2 = 6 \qquad \text{Substituting } 3w_2 + 4 \text{ for } w_1$$

Step 3. Solving for w_2:

$$9w_2 + 12 - 4w_2 = 6 \qquad \text{Removing parentheses}$$
$$5w_2 + 12 = 6$$
$$5w_2 = 6 - 12 \qquad \text{Subtracting 12}$$
$$5w_2 = -6$$
$$w_2 = -\frac{6}{5} \qquad \text{Dividing by 5}$$

Step 4. From the second equation, written in the form $w_1 = 3w_2 + 4$ (Step 1), we get

$$w_1 = 3\left(-\frac{6}{5}\right) + 4 \qquad \text{Substituting } w_2 = -\frac{6}{5}$$

$$= -\frac{18}{5} + \frac{20}{5} \qquad \text{LCD} = 5$$

$$= \frac{2}{5}$$

The solution is therefore given by

$$\left(\frac{2}{5}, -\frac{6}{5}\right)$$

Step 5. *Check:*

$$3\left(\frac{2}{5}\right) - 4\left(-\frac{6}{5}\right) = \frac{6}{5} + \frac{24}{5} = \frac{30}{5} = 6 \checkmark$$

$$\frac{2}{5} - 3\left(-\frac{6}{5}\right) = \frac{2}{5} + \frac{18}{5} = \frac{20}{5} = 4 \checkmark$$

We can see from these examples that the method of substitution is most convenient if one of the variables has coefficient 1 or -1. If none of the variables have coefficient 1 or -1, then solving for one of the variables often leads to an expression involving fractions. In such a case, the method of substitution is less convenient and it may be easier to solve the system by the method of addition or subtraction.

Exercises / Section 13.3

In Exercises 1–24, solve each system of equations by the method of substitution.

1. $x + 3y = 1$, $2x + 5y = 1$

2. $9x + 2y = 3$, $x + 2y = -5$

3. $3x + 2y = 5$, $2x + y = 2$

4. $2x + 3y = 11$, $3x + y = -1$

5. $4x + y = 1$, $6x + 5y = 12$

6. $2x - 3y = 9$, $x - 3y = 5$

7. $x - 3y = 4$, $2x - y = 3$

8. $x + 2y = 12$, $x - 3y = 2$

9. $3x + 4y = 21$, $-x + 2y = 3$

10. $x + 3y = 1$, $3x - y = -5$

11. $2x + y = 1$, $x + 3y = 8$

12. $x + 2y = 13$, $3x - y = -31$

13. $8v_1 - 10v_2 = -13$, $v_1 + 2v_2 = 0$

14. $-w_1 - 2w_2 = 3$, $2w_1 + w_2 = 4$

15. $2F_1 - 3F_2 = 4$, $6F_1 - F_2 = 4$

16. $-F_1 + 3F_2 = 3$, $2F_1 + 3F_2 = -3$

17. $2v_1 - v_2 = 1$, $4v_1 + 3v_2 = -18$

18. $3R + 8C = 3$, $-R + 4C = 3$

19. $4L - C = 1$, $8L + 3C = 12$

20. $5y + z = 5$, $10y - 5z = -11$

21. $w - 4z = 7$, $3w - 8z = 18$

22. $5s + v = 5$, $10s + 3v = 14$

23. $4V_1 - 3V_2 = 1$, $5V_1 - 4V_2 = 1$

24. $2R_1 - 3R_2 = 1$, $3R_1 - 2R_2 = 2$

In Exercises 25–44 solve the given systems of equations by either method.

25. $3x - 4y = 1$, $5x + 6y = 8$

26. $2x - y = 9$, $x + 2y = 7$

27. $-4x + 2y = 7$, $x - 3y = 2$

28. $5x - 4y = 5$
$2x - y = 1$

29. $3F_1 + 2F_2 = 5$
$F_1 - F_2 = 2$

30. $2v_1 - 4v_2 = 2$
$3v_1 - 5v_2 = 4$

31. $4p + 3q = 9$
$5p + 6q = 12$

32. $2m + 6n = 17$
$-4m + 7n = 23$

33. $3s_1 - 2s_2 = 3$
$6s_1 - 4s_2 = 7$

34. $2x - 4y = -1$
$-4x + 8y = 3$

35. $2x - 3y = 5$
$4x - 6y = 10$

36. $x - 4z = 2$
$-3x + 12z = -6$

37. $R - 3L = 5$
$2R - L = 5$

38. $R - 3C = 4$
$2R - C = 5$

39. $3S - 4T = 1$
$-2S + 6T = 1$

40. $3w + 2z = 3$
$4w + 7z = 5$

41. $y + 3z = 14$
$2y - z = 0$

42. $2F_1 - F_2 = 0$
$3F_1 + 2F_2 = 7$

43. $2T_1 + 2T_2 = 9$
$4T_1 + 3T_2 = 14$

44. $2C_1 + 3C_2 = 8$
$4C_1 + 3C_2 = 7$

13.4 Applications of Systems of Linear Equations

Systems of linear equations have many applications in technology. In addition, many problems leading to single linear equations can be solved more easily by using systems of equations. Consider the first example.

Example 1 The sum of two numbers is 25.1. One number is 2.7 more than 3 times the other number. Find the two numbers.

Solution. Since we are looking for two numbers, let us denote the first number by x and the second (other) number by y.

We are told that the sum is 25.1. So $x + y = 25.1$ is one of the equations. Furthermore, since 3 times the other number now becomes $3y$, the statement "2.7 more than 3 times the other number" can be expressed as $3y + 2.7$. It follows that $x = 2.7 + 3y$ is the second equation.

The resulting system of equations is therefore given by

$$x + y = 25.1$$
$$x = 2.7 + 3y$$

This system is readily solved by substitution:

$$x = 2.7 + 3y \quad \text{Second equation}$$
$$x + y = 25.1 \quad \text{First equation}$$
$$2.7 + 3y + y = 25.1 \quad \text{Substituting } 2.7 + 3y \text{ for } x$$
$$2.7 + 4y = 25.1$$
$$4y = 25.1 - 2.7 \quad \text{Subtracting 2.7}$$
$$4y = 22.4$$
$$y = \frac{22.4}{4} = 5.6 \quad \text{Dividing by 4}$$

Substituting $y = 5.6$ into the second equation ($x = 2.7 + 3y$), we get

$$x = 2.7 + 3(5.6) = 19.5$$

So the numbers are 19.5 and 5.6.

To check the solution, we return to the given problem: The sum of the two numbers is $19.5 + 5.6 = 25.1$. Three times the second number plus 2.7 is $3(5.6) + 2.7 = 19.5$, which is the first number.

In the next example we use the fact that the combined resistance of two or more resistors in series is equal to the sum of the individual resistances. (See Figure 13.10.)

Combined resistance $R_T = R_1 + R_2$

Figure 13.10

Example 2 Two resistors are connected in series. The resistance of the second is 12.6 Ω less than that of the first. The combined resistance is 138.2 Ω. Find the resistance of each.

Solution. Let

R_1 = resistance of first resistor

and

R_2 = resistance of second resistor

From Figure 13.10 and the given information,

$$\begin{aligned} R_1 + R_2 &= 138.2 \\ R_1 - R_2 &= 12.6 \end{aligned} \qquad \text{Resulting system}$$

$$\begin{aligned} 2R_1 &= 150.8 && \text{Adding} \\ R_1 &= 75.4\ \Omega && \text{Dividing by 2} \\ R_2 &= 62.8\ \Omega && R_2 = 138.2 - R_1 \end{aligned}$$

Example 3 The manager of a shop has an order for a rectangular metal plate meeting the following specifications: The length is to be 1.1 cm less than 3 times the width, and the perimeter is to be 28.2 cm. Find the required dimensions.

Solution. Let

x = width of rectangle

and

$$y = \text{length of rectangle}$$

(See Figure 13.11.) Note that the perimeter is

$$x + x + y + y = 2x + 2y = 28.2$$

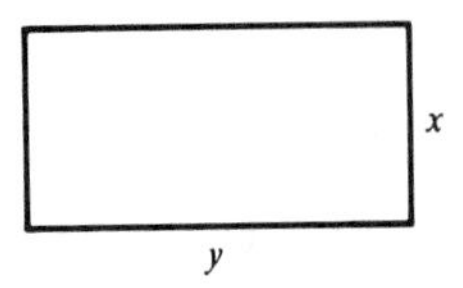

Figure 13.11

The condition "the length is to be 1.1 cm less than 3 times the width" can be written

$$y = 3x - 1.1$$

We now have the following system:

$$2x + 2y = 28.2 \qquad \text{Perimeter} = 28.2 \text{ cm}$$
$$y = 3x - 1.1$$

Let us solve this system by substitution:

$$\begin{aligned} y &= 3x - 1.1 && \text{Second equation} \\ 2x + 2y &= 28.2 && \text{First equation} \\ 2x + 2(3x - 1.1) &= 28.2 && \text{Substitution} \\ 2x + 6x - 2.2 &= 28.2 && \text{Removing parentheses} \\ 8x - 2.2 &= 28.2 && \\ 8x &= 28.2 + 2.2 && \text{Adding 2.2} \\ 8x &= 30.4 && \\ x &= 3.8 \text{ cm} && \text{Dividing by 8} \end{aligned}$$

Substituting $x = 3.8$ into the second equation ($y = 3x - 1.1$), we get

$$y = 3x - 1.1 = 3(3.8) - 1.1 = 10.3 \text{ cm}$$

We conclude that the length is 10.3 cm and the width 3.8 cm.

Mixture problems can also be solved more easily by using systems of equations.

Example 4 One alloy contains 20% brass (by weight) and another 25%. How many pounds of each must be combined to form 60 lb of an alloy containing 22% brass?

Solution. Let

$x =$ number of pounds of 20% alloy

and

$y =$ number of pounds of 25% alloy

Since the total weight is 60 lb, one of the equations is $x + y = 60$.

To obtain the other equation, we need to recall that it is best to work with the actual quantities, in this case the amount of brass. Thus $0.20x$ is the amount (in pounds) of brass in the 20% alloy and $0.25y$ is the amount in the 25% alloy. We now have the following system of equations:

$$0.20x + 0.25y = (0.22)(60)$$
$$x + y = 60$$

$$
\begin{aligned}
20x + 25y &= (22)(60) && \text{Multiplying by 100} \\
20x + 20y &= 1200 && \text{Multiplying by 20} \\
5y &= 120 && \text{Subtracting} \\
y &= 24 \text{ lb} \\
x &= 36 \text{ lb} && x = 60 - 24
\end{aligned}
$$

An important application of systems of linear equations is the analysis of electrical circuits by means of Kirchhoff's laws. Since a discussion of these laws is too lengthy to consider here, we will work with given systems of equations in the example and exercises.

Example 5 Consider the circuit in Figure 13.12. If the directions of the currents I_1 and I_2 are as shown in the diagram, then I_1 and I_2 satisfy the following conditions:

$$3I_1 + 4(I_1 - I_2) = -2$$
$$4(-I_1 + I_2) + 3I_2 = 2$$

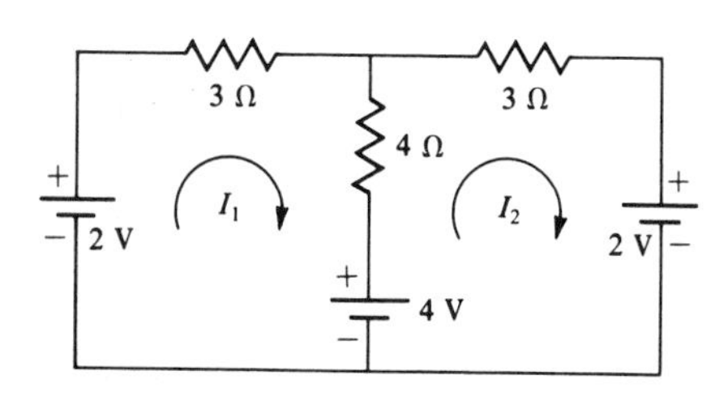

Figure 13.12

Find the currents.

Solution.

$$\begin{array}{rcll} 3I_1 + 4(I_1 - I_2) &=& -2 & \text{Given system} \\ 4(-I_1 + I_2) + 3I_2 &=& 2 & \\ \hline 3I_1 + 4I_1 - 4I_2 &=& -2 & \text{Removing parentheses} \\ -4I_1 + 4I_2 + 3I_2 &=& 2 & \\ \hline 7I_1 - 4I_2 &=& -2 & \text{Combining terms} \\ -4I_1 + 7I_2 &=& 2 & \\ \hline 28I_1 - 16I_2 &=& -8 & 4(7I_1 - 4I_2) = 4(-2) \\ -28I_1 + 49I_2 &=& 14 & 7(-4I_1 + 7I_2) = 7(2) \\ \hline 33I_2 &=& 6 & \text{Adding} \end{array}$$

$$I_2 = \frac{6}{33} = \frac{2}{11}\text{ A}$$

Substituting $I_2 = \frac{2}{11}$ in the equation $4(-I_1 + I_2) + 3I_2 = 2$, we get

$$4\left(-I_1 + \frac{2}{11}\right) + 3\left(\frac{2}{11}\right) = 2$$

$$-4I_1 + \frac{8}{11} + \frac{6}{11} = 2$$

$$-4I_1 + \frac{14}{11} = 2$$

$$-4I_1 = 2 - \frac{14}{11} = \frac{22}{11} - \frac{14}{11} = \frac{8}{11}$$

$$I_1 = -\frac{2}{11}\text{ A}$$

So $I_1 = (-\frac{2}{11})$A and $I_2 = (\frac{2}{11})$A. (The minus sign in I_1 indicates that the direction of the current I_1 is opposite to the direction originally assigned (Figure 13.12). The current I_2, however, is in the assigned direction.)

Many problems in technology involve **moments.** Consider the **lever** in Figure 13.13, which is supported at a point called the **fulcrum.** Neglecting the weight of the lever, a weight w at distance d from the fulcrum (Figure 13.13) has a **moment** given by weight times distance, or wd. The distance d is called the **moment arm.**

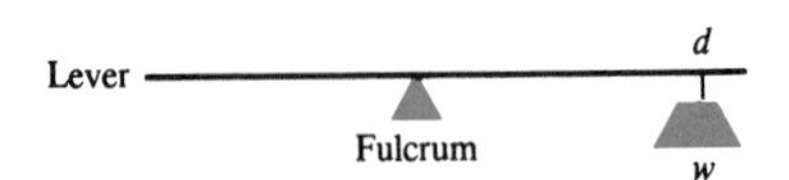

Figure 13.13

If two weights w_1 and w_2 are placed on opposite sides of the fulcrum at respective distances d_1 and d_2, then

$$w_1d_1 = w_2d_2$$

whenever the weights are balanced on the lever.

Moments can be added. For example, the total moment on the left side of the lever in Figure 13.14 is $w_1d_1 + w_2d_2$. For the weights on this lever to balance, the moment on the left must equal the moment on the right, or

$$w_1d_1 + w_2d_2 = w_3d_3$$

Figure 13.14

This formula can be extended to any number of weights.

Example 6 A weight of 1 lb and a lever are to be used to determine two other weights by placing the 1-lb weight on one side of the fulcrum and the unknown weights on the other side. Letting $w_3 = 1$ lb in Figure 13.14, it is observed that the lever balances when $d_3 = 24$ in., $d_1 = 4$ in., and $d_2 = 3$ in. Another balance is obtained when $d_3 = 26$ in., $d_1 = 2$ in., and $d_2 = 5$ in. Determine the weights w_1 and w_2.

Solution. From the relationship

$$w_1d_1 + w_2d_2 = w_3d_3$$

and the given measurements, we get the system

$$4w_1 + 3w_2 = 24 \cdot 1$$
$$2w_1 + 5w_2 = 26 \cdot 1$$

$$4w_1 + 3w_2 = 24$$
$$4w_1 + 10w_2 = 52 \qquad 2(2w_1 + 5w_2) = 2(26)$$
$$-7w_2 = -28 \qquad \text{Subtracting}$$
$$w_2 = 4 \text{ lb}$$
$$w_1 = 3 \text{ lb} \qquad 2w_1 + 5(4) = 26$$

Exercises / Section 13.4

1. The velocities v_1 and v_2 (in centimeters per second) of two colliding bodies were found to satisfy the relationships

$$3v_1 + 3v_2 = 15$$
$$4v_1 + 2v_2 = 30$$

Find the velocities.

2. Measurements of the tension (in pounds) of two supporting cables produced the following equations:

$$0.22T_1 + 0.46T_2 = 59$$
$$0.36T_1 + 0.58T_2 = 81$$

Find T_1 and T_2.

3. The relationship between the tensile strength S (in pounds) of a certain metal rod and the temperature T (in degrees Celsius) has the form $S = a - bT$. Experimenters found that if $T = 50.0°C$, then $S = 565.9$ lb; if $T = 100°C$, then $S = 565.8$ lb. Find the relationship.

4. The relationship between the length of a certain bar (measured in centimeters) and its temperature (in degrees Celsius) is known to be $L = aT + b$. Tests have shown that if $T = 15°C$, then $L = 50.0$ cm; if $T = 60°C$, then $L = 50.8$ cm. Find the relationship.

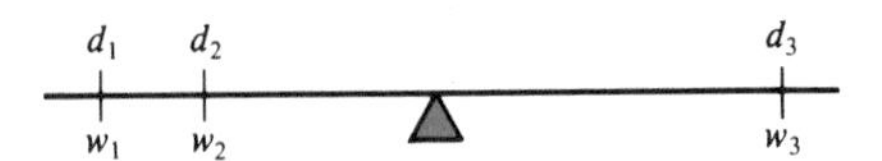

Figure 13.15

In Exercises 5 and 6, refer to Figure 13.15 to find w_1 and w_2 in each case. (See Example 6.)

5. $w_3 = 2.0$ N; a balance is obtained if $d_1 = 2.0$ m, $d_2 = 2.0$ m, and $d_3 = 3.5$ m, and if $d_1 = 2.0$ m, $d_2 = 1.0$ m, and $d_3 = 2.0$ m.

6. $w_3 = 2.0$ lb; a balance is obtained if $d_1 = 2.0$ in., $d_2 = 3.0$ in., and $d_3 = 4.0$ in., and if $d_1 = 4.0$ in., $d_2 = 1.0$ in., and $d_3 = 3.0$ in.

For Exercises 7–16, see Example 1.

7. The sum of two numbers is 32 and their difference is 4. Find the numbers.

8. The sum of two numbers is 55 and their difference is 11. Find the numbers.

9. One number is twice the other and their sum is 10.8. What are the numbers?

10. One number is 3 times the other and their sum is 44.8. What are the numbers?

11. A pipe 29 ft long is to be cut into two pieces so that one piece is 3 ft longer than the other piece. How must the pipe be cut?

12. A wire is 65 cm long. How must the wire be cut so that one piece is 13 cm longer than the other?

13. How must a 29-m cable be cut so that one part is 1 m longer than 3 times the other part?

14. Two gears have a total of 52 teeth. If the number of teeth of one is 4 more than twice the number of the other, find the number of teeth of each.

15. Two machines have a total of 62 moving parts. If one machine has 2 more than 3 times as many moving parts as the other, how many moving parts does each machine have?

16. Two machines have a total of 60 moving parts. Three times the number of moving parts of one machine is 5 more than twice the number of moving parts of the other machine. Find the number of moving parts of each machine.

For Exercises 17–20, see Example 2.

17. Two resistors connected in series have a combined resistance of 180 Ω. The resistance of one resistor is 20 Ω less than that of the other. Find the two resistances.

18. The combined resistance of two resistors in series is 110 Ω. Find the two resistances, given that the resistance of one is 10 Ω less than that of the other.

19. The sum of two currents is 3.9 A. It was found that twice the first current is 1.8 A more than 3 times the second. Find the two currents.

20. The sum of the voltages across two resistors is 55.1 V. It was found that 3 times the first voltage is 9.7 V less than 4 times the second. Find the two voltages.

For Exercises 21–26, see Example 3.

21. A rectangular metal plate has a perimeter of 18.4 cm. The width is 4.4 cm less than the length. Find the dimensions.

22. A machinist has an order for a rectangular metal plate with the following specifications: The length is 1.0 cm less than 3 times the width and the perimeter is 24.4 cm. Find the dimensions of the plate.

23. Twice the length of a machine shop is equal to 3 times the width. The perimeter is 100 ft. Find the dimensions.

24. The bottom of a chemical tank has the shape of a rectangle whose length is 2.5 times the width. The perimeter is 98 ft. Find the dimensions.

25. A rectangular computer chip is 3 times as long as it is wide. If the perimeter is 9.6 mm, find the dimensions.

26. Five times the width of a rectangular solar panel is 1.0 ft more than 3 times the length. Given that the perimeter is 26.0 ft, find the dimensions.

For Exercises 27–30, see Example 4.

27. A technician needs 100 mL of a 16% nitric acid solution (by volume). She has a 20% and a 10% solution (by volume) in stock. How many milliliters of each must she mix to obtain the required solution?

28. How many liters of a 5% solution (by volume) must be added to a 10% solution to obtain 20 L of an 8% solution?

29. One alloy contains 6% brass (by weight) and another 12% brass (by weight). How many pounds of each must be combined to form 50 lb of an alloy containing 10% brass?

30. How many pounds of an alloy containing 8% tin (by weight) must be combined with an alloy containing 14% tin (by weight) to form 36.8 lb of an alloy containing 10% tin?

In Exercises 31 and 32, find the currents in each of the circuits (Figures 13.16 and 13.17) by solving the system of equations given in each case. (See Example 5.)

31. $I_1 + 2I_1 - 2I_2 = 4$

$2I_2 - 2I_1 + 3I_2 = 2$

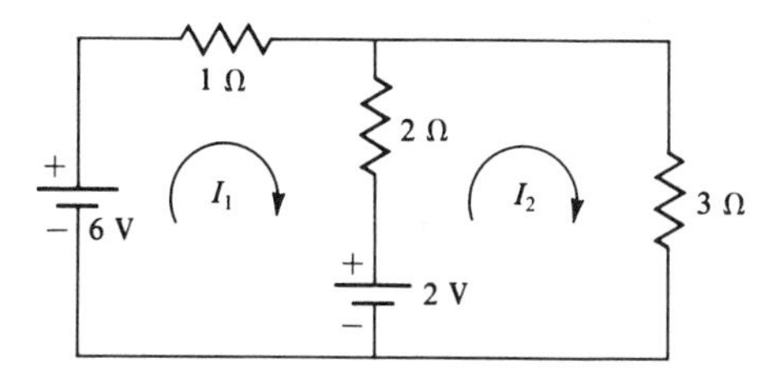

Figure 13.16

32. $2I_1 + 3I_1 + 3I_2 = 2$
$-3I_1 - 3I_2 - I_2 = -4$

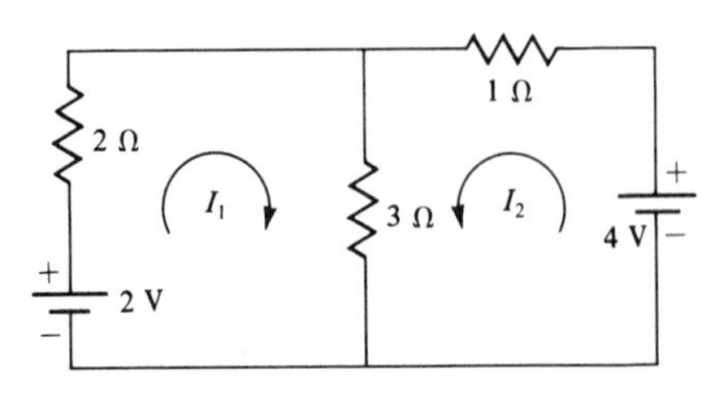

Figure 13.17

33. Tickets for an industrial exhibit cost \$5.00 for regular admission and \$4.00 for senior citizens. On one day 215 tickets were sold for a total intake of \$1050. How many tickets of each type were sold?

34. Holly has \$3.30 in nickels and dimes. If she has 9 more nickels than dimes, determine the number of each.

35. A certain amount of money was invested at 10% and the rest at 9%. If the first investment was \$1000 more than the second, and if the investment income was \$1050, how much was invested at each rate?

36. The price p (in dollars) of a commodity depends on the demand q (number of units sold). The equilibrium price of a certain item can be obtained from the system

$$2p + 3q = 243$$
$$10p + 5q = 415$$

Determine the price and the corresponding quantity sold.

Review Exercises / Chapter 13

In Exercises 1 and 2, solve the given systems of equations graphically. Estimate the solutions to the nearest tenth of a unit.

1. $x - 3y = 6$
$2x + y = 4$

2. $4x - 4y = -19$
$x - 5y = 5$

In Exercises 3–8, solve each system by the method of addition or subtraction.

3. $2x + 5y = 4$
$2x + 7y = 8$

4. $5x - 2y = 7$
$2x + 3y = 18$

5. $4x + 3y = 7$
$8x + 5y = 11$

6. $2x + 9y = 1$
$2x - 3y = 5$

7. $3v_1 - 8v_2 = 1$
$6v_1 - 4v_2 = -1$

8. $4F_1 - 3F_2 = 5$
$8F_1 - 3F_2 = 4$

In Exercises 9–14, solve each system of equations by the method of substitution.

9. $x + 3y = 3$
$-2x + 5y = 16$

10. $2x - 3y = 13$
$3x - y = 9$

11. $9x + y = 1$
$-3x + 2y = 9$

12. $3x - y = 4$
$6x - 5y = 14$

13. $2R - 4C = 5$
$-R + 8C = 2$

14. $4P - Q = 2$
$8P + 3Q = -11$

In Exercises 15–32, solve the given systems of equations by either method.

15. $-5x + 2y = 10$
$5x + 3y = 5$

16. $-6x + 2y = 9$
$6x + 5y = 26$

17. $-x + 2y = 11$
$2x + 5y = 14$

18. $3x - 2y = 5$
$6x - y = 4$

19. $5w - 3z = 3$
$15w + 6z = 4$

20. $4m + 3n = 4$
$12m - 6n = 7$

21. $-6C + 16V = 5$
$12C - 8V = 5$

22. $3L + 2C = 12$
$4L + 3C = 17$

23. $5p + 3q = -1$
$7p + 4q = -1$

24. $2s + 3t = 11$
$5s + 7t = 25$

25. $2r - s = 4$
$10r + 2s = -1$

26. $2a - 5b = 11$
$a - 4b = 10$

27. $2A_1 - 7A_2 = 5$
$-6A_1 + 21A_2 = -15$

28. $1.5D + 3E = 4$
$3D + 6E = 5$

29. $7R_1 - 3R_2 = 29$
$-5R_1 - 11R_2 = 45$

30. $2v_1 + 5v_2 = 33$
$-3v_1 + 4v_2 = 8$

31. $-3s_1 - s_2 = 5$
$6s_1 - 3s_2 = 10$

32. $4t_1 - 6t_2 = -1$
$11t_1 - 12t_2 = 1$

33. The resistance in a wire as a function of the temperature has the form $R = aT + b$, where R is measured in ohms and T in degrees Celsius. The following relationships were found experimentally:

$$1.52 = 10.0a + b$$
$$1.64 = 20.0a + b$$

Find the function.

34. The sum of two numbers is 32.2. Given that one number is 3.2 more than 4 times the other, find the numbers.

35. Two resistors connected in series have a combined resistance of 130 Ω. If the resistance of one is 30 Ω more than that of the other, find the two resistances.

36. An order for a set of rectangular plates must meet the following specifications: The length is to be 1.0 cm more than 3 times the width, and the perimeter is to be 29.2 cm. Find the dimensions.

37. A rectangular field is divided into two smaller fields by a dividing fence parallel to the shorter sides. If the field is 25 m longer than it is wide, and if 425 m of fence is available, what are the dimensions?

38. How many pounds of an alloy containing 6% copper (by weight) must be combined with an alloy containing 12% copper (by weight) to form 54 lb of an alloy containing 10% copper?

39. An office building has 20 offices. The smaller offices rent for $300 per month, and the larger offices for $420 per month. Given that the rental income is $7440 per month, how many of each type of office are there?

40. One consultant to a firm charges $200 per day, and another consultant charges $250 per day. After 13 days the total charged by the two consultants comes to $2950. Assuming that only one of the consultants was called in on any one day, how many days did each one work?

CHAPTER 14

Additional Topics from Geometry

In this chapter we continue our study of geometry begun in Chapter 6. Our approach will continue to be informal, concentrating mainly on definitions, formulas, and applications.

14.1 Angles and Parallel Lines

In this section we will introduce a few more concepts concerning angles. These concepts will be useful in our discussion of parallel lines.

Angles

Right angle

Acute angle

Obtuse angle

Recall from Chapter 6 that a **right angle** is an angle that measures 90°. Two related concepts are the following: An angle between 0° and 90° is called an **acute angle,** and an angle between 90° and 180° is called an **obtuse angle.** (See Figure 14.1.)

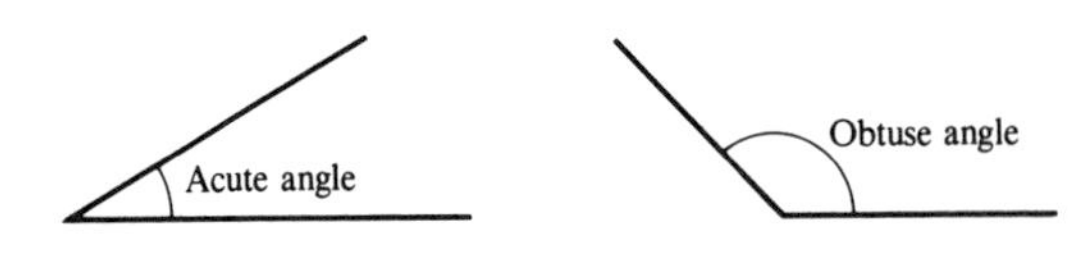

Figure 14.1

Example 1 An angle of 82° is acute since $0° < 82° < 90°$. An angle of 120° is obtuse since $90° < 120° < 180°$.

Two angles may be given special names depending on their sizes or relative positions.

Adjacent angles

1. Two angles are **adjacent angles** if they have a common vertex and one common side.

Vertical angles

2. A pair of angles formed by two intersecting lines and lying on opposite sides of the point of intersection are called **vertical angles.** (Two vertical angles are always equal.)

Complementary angles

3. Two angles are called **complementary angles** if their sum is 90°.

Supplementary angles

4. Two angles are called **supplementary angles** if their sum is 180°.

Example 2 In Figure 14.2, $\angle BAC$ and $\angle CAD$ are adjacent angles. Note that the angles have a common vertex A and common side AC.

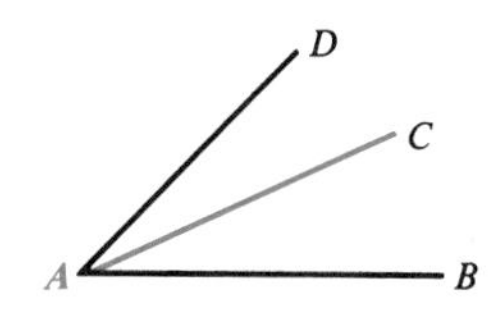

Figure 14.2

Example 3 In Figure 14.3, $\angle AOB$ and $\angle COD$ are vertical angles. Note that the angles are formed by two intersecting lines; $\angle AOB$ and $\angle COD$ are vertical angles because they lie on *opposite* sides of O. (Being vertical angles, $\angle AOB = \angle COD$.)

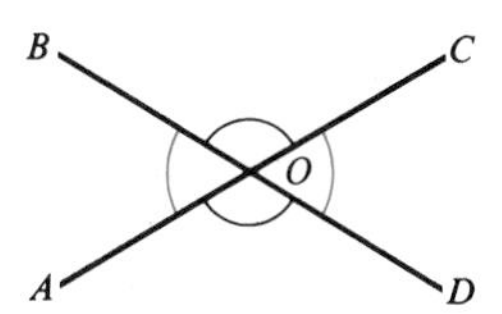

Figure 14.3

For the same reason, $\angle BOC$ and $\angle AOD$ are vertical angles and $\angle BOC = \angle AOD$.

Example 4 If $\angle A = 40°$ and $\angle B = 50°$, then $\angle A$ and $\angle B$ are complementary. (The sum of the angles is $40° + 50° = 90°$.)

If $\angle C = 110°$ and $\angle D = 70°$, then $\angle C$ and $\angle D$ are supplementary. (The sum of the angles is $110° + 70° = 180°$.)

Of particular interest are complementary angles and supplementary angles that are also adjacent. These cases are discussed in the next two examples.

Example 5 Consider the right angle ABC in Figure 14.4. Since $\angle 1 + \angle 2 = \angle ABC = 90°$, $\angle 1$ and $\angle 2$ are complementary.

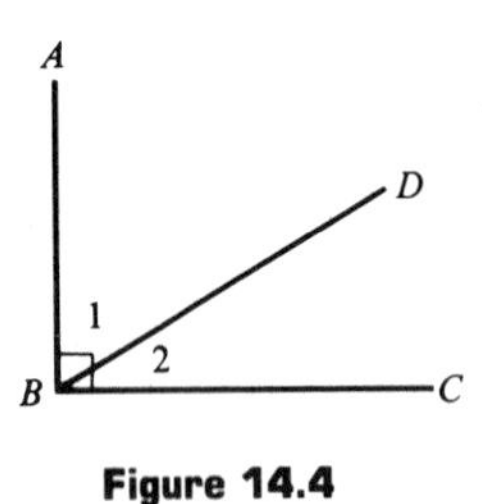

Figure 14.4

Example 6 Consider the straight angle ABC in Figure 14.5. Since $\angle 1 + \angle 2 = \angle ABC = 180°$, $\angle 1$ and $\angle 2$ are supplementary.

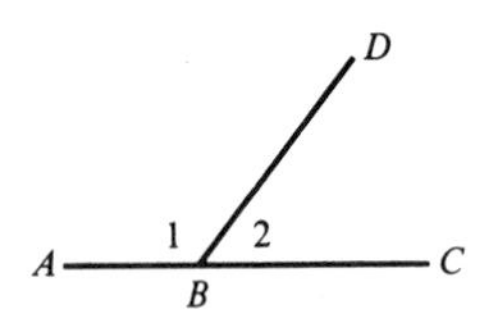

Figure 14.5

Complement

Supplement

If $\angle A$ and $\angle B$ are complementary, then $\angle A$ is the **complement** of $\angle B$ (and $\angle B$ is the complement of $\angle A$.) If $\angle A$ and $\angle B$ are supplementary, then $\angle A$ is the **supplement** of $\angle B$ (and $\angle B$ is the supplement of $\angle A$).

Example 7 Given that $\angle A = 50°40'$, find the complement and the supplement.

Solution. Since the sum of two complementary angles is 90°, the complement of $\angle A$ is

$$90° - 50°40' = 89°60' - 50°40' \qquad 90° = 89°60'$$

$$\begin{array}{r} 89°60' \\ -50°40' \\ \hline 39°20' \end{array} \qquad \text{Subtracting}$$

Since the sum of two supplementary angles is 180°, the supplement of $\angle A$ is

$$180° - 50°40' = 179°60' - 50°40' = 129°20' \qquad 180° = 179°60'$$

Parallel Lines

Parallel lines

Transversal

Vertical and adjacent angles are useful in the study of parallel lines. First let us recall from Chapter 6 that two lines in a plane are **parallel** if they do not intersect. A line crossing a pair of parallel or nonparallel lines is called a **transversal.** For example, in Figure 14.6, the lines are *AB* and *CD*, and the line *EF* is a transversal.

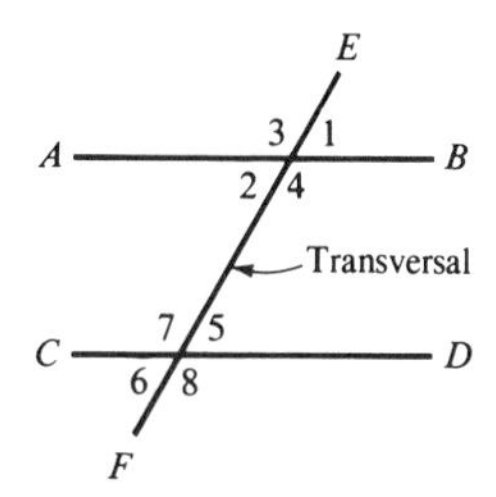

Figure 14.6

Alternate interior angles

Alternate exterior angles

Corresponding angles

When two lines are cut by a transversal, eight angles are formed. In Figure 14.6, the four angles between the lines *AB* and *CD* (that is, angles 2, 4, 5, and 7) are called **interior angles.** The four angles outside (angles 1, 3, 6, and 8) are called **exterior angles.** Two nonadjacent interior angles on opposite sides of the transversal are called **alternate interior angles.** Thus $\angle 2$ and $\angle 5$, as well as $\angle 4$ and $\angle 7$, are alternate interior angles. Similarly, $\angle 1$ and $\angle 6$ are **alternate exterior angles,** as are $\angle 3$ and $\angle 8$. Certain angles that occupy the same position relative to the lines are called **corresponding angles.** The respective corresponding angles are: $\angle 1$ and $\angle 5$, $\angle 3$ and $\angle 7$, $\angle 4$ and $\angle 8$, and $\angle 2$ and $\angle 6$.

The basic property of these pairs of angles is stated next.

> Whenever two parallel lines are cut by a transversal, then any pair of alternate interior, alternate exterior, or corresponding angles are equal.

This statement can also be reversed: If two lines are cut by a transversal, and if a pair of alternate interior, alternate exterior, or corresponding angles are equal, then the lines are parallel.

Example 8 Suppose $AB \parallel CD$ in Figure 14.7. Using the 60° angle in the figure, determine angles 1, 2, 3, 4, 5, and 6.

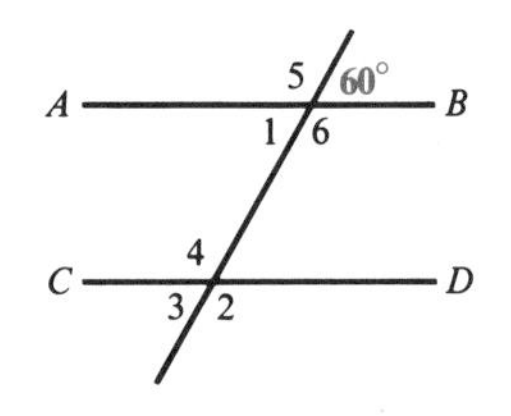

Figure 14.7

Solution. Since the given 60° angle and $\angle 1$ are vertical angles, $\angle 1 = 60°$.

Next, we note that $\angle 1$ and $\angle 5$ are adjacent angles whose sum is a straight angle. (See Example 6.) It follows that $\angle 5 = 180° - \angle 1 = 180° - 60° = 120°$.

Since $\angle 1$ and $\angle 3$ are corresponding angles, $\angle 3 = 60°$. Also, since $\angle 6$ and $\angle 5$ are vertical angles, $\angle 6 = 120°$. Since alternate interior angles are equal, $\angle 4 = \angle 6$, so that $\angle 4 = 120°$. Finally, $\angle 2 = 120°$, since $\angle 2$ and $\angle 4$ are vertical angles (and $\angle 2$ and $\angle 6$ corresponding angles).

Example 9 In Figure 14.8, the given alternate interior angles are equal. It follows that $L_1 \parallel L_2$.

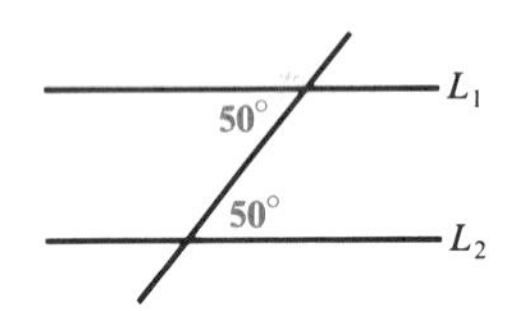

Figure 14.8

Exercises / Section 14.1

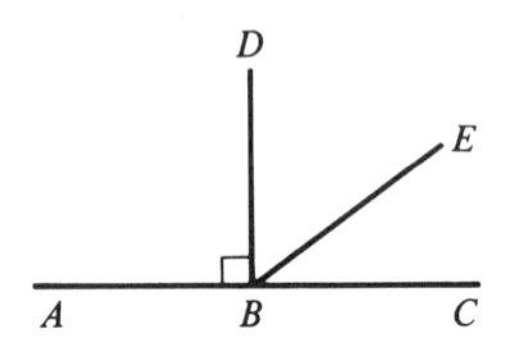

Figure 14.9

Exercises 1–10 refer to Figure 14.9. Identify the indicated angles in each case.

1. An acute angle with side BC
2. An acute angle with side BD
3. Two pairs of adjacent angles
4. Two right angles
5. The obtuse angle
6. One acute angle
7. One pair of supplementary angles
8. The pair of complementary angles

9. The complement of $\angle CBE$

10. The supplement of $\angle CBE$

11. If $\angle A = 119°34'$, what is the measure of the supplement of $\angle A$?

12. If $\angle A = 30°15'$, what is the measure of the complement of $\angle A$?

13. $\angle A$ and $\angle B$ are complementary angles whose difference is 10°. Find $\angle A$ and $\angle B$. (*Hint:* Let $x = \angle A$.)

14. $\angle A$ and $\angle B$ are supplementary and $\angle A$ is twice $\angle B$. Find $\angle A$ and $\angle B$.

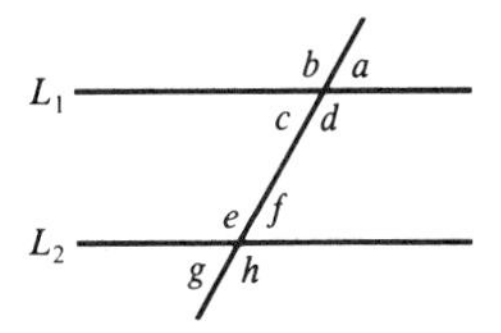

Figure 14.10

Exercises 15–22 refer to Figure 14.10. Identify the indicated angles as alternate interior, corresponding, and so on.

15. $\angle a$ and $\angle c$

16. $\angle g$ and $\angle f$

17. $\angle c$ and $\angle f$

18. $\angle d$ and $\angle e$

19. $\angle c$ and $\angle g$

20. $\angle a$ and $\angle f$

21. $\angle b$ and $\angle h$

22. $\angle a$ and $\angle g$

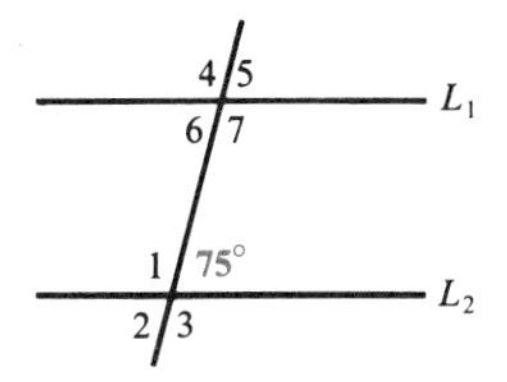

Figure 14.11

In Exercises 23–29, find the degree measures of the angles indicated. (Refer to Figure 14.11.)

23. $\angle 2$

24. $\angle 1$

25. $\angle 7$

26. $\angle 4$

27. $\angle 5$

28. $\angle 6$

29. $\angle 3$

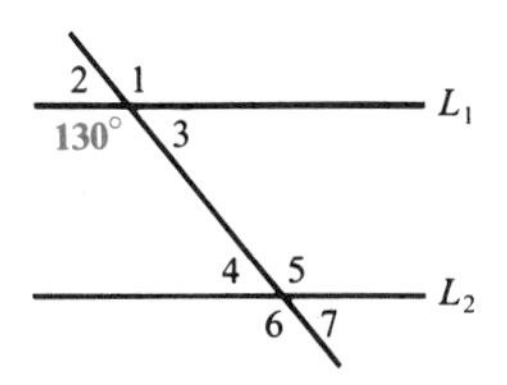

Figure 14.12

In Exercises 30–36, find the degree measures of the angles indicated. (Refer to Figure 14.12.)

30. $\angle 1$ **31.** $\angle 3$ **32.** $\angle 4$ **33.** $\angle 7$ **34.** $\angle 6$

35. $\angle 5$ **36.** $\angle 2$

37. In Figure 14.13, why is L_1 parallel to L_2?

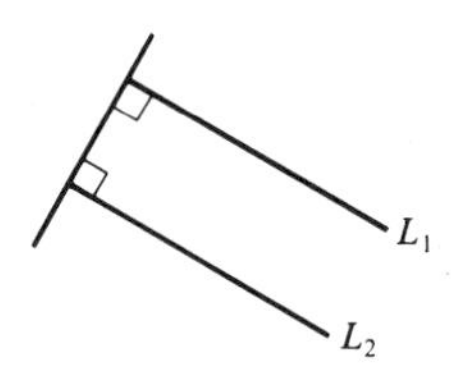

Figure 14.13

38. In Figure 14.14, why is L_1 parallel to L_2?

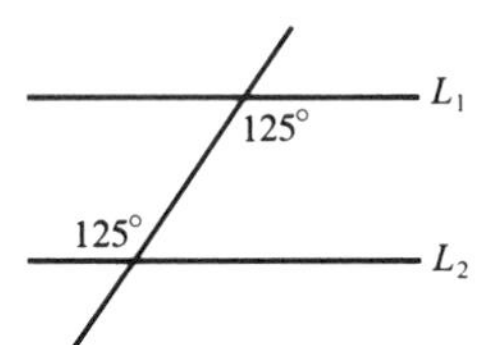

Figure 14.14

39. Explain why Figure 14.15 is a parallelogram.

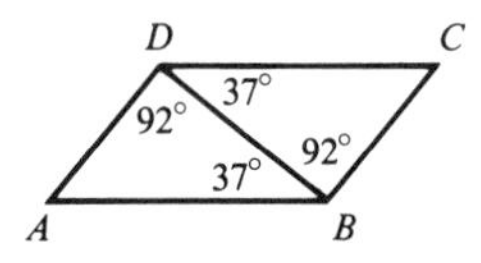

Figure 14.15

14.2 Properties of Polygons and Circles

In this section we discuss the properties of various plane figures.

Polygons

Polygon
Side
Vertex

A **polygon** is formed by three or more line segments that enclose a portion of a plane. Each line segment is called a **side** of the polygon and each point where the sides meet is called a **vertex.** Polygons are named according to the number of sides. Some of the types are shown in Figure 14.16.

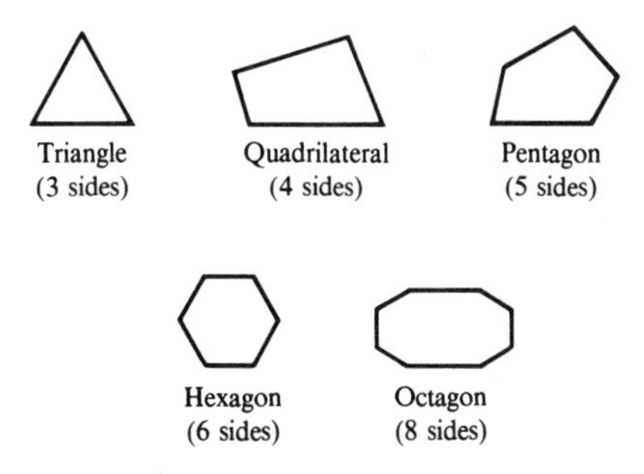

Figure 14.16

Convex

Two special types of polygons are convex and regular polygons. A polygon is said to be **convex** if any two points inside the polygon can be connected by a line segment that lies entirely inside. (See Figure 14.17.) A polygon that is not convex is called *concave*. Note that a polygon is convex if all interior angles are less than 180°.

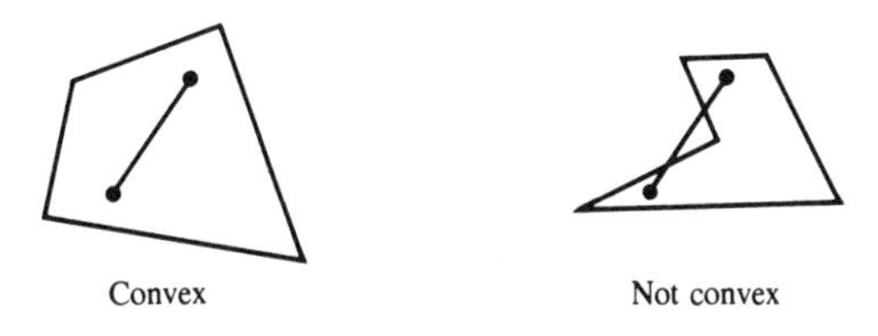

Figure 14.17

Regular polygon

A polygon is called **regular** if all sides have the same length and all interior angles are equal. Figure 14.18 is an example of a regular pentagon.

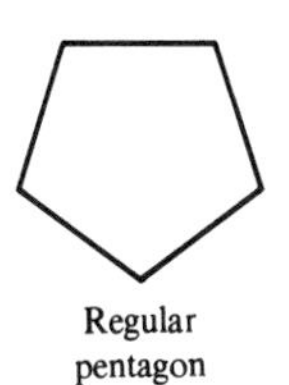

Figure 14.18

Triangles

An important type of polygon is a triangle, already introduced in Chapter 6. We discussed right, equilateral, and isosceles triangles. We will now study some additional properties of triangles.

A fundamental property of triangles is the following:

The sum of the interior angles of a triangle is 180°.

To see why, consider the triangle in Figure 14.19. Assume that $DE \parallel AB$ and consider the angles on top. Since $\angle 4$, $\angle 3$, and $\angle 5$ together form a straight angle, we have

$$\angle 4 + \angle 3 + \angle 5 = 180°$$

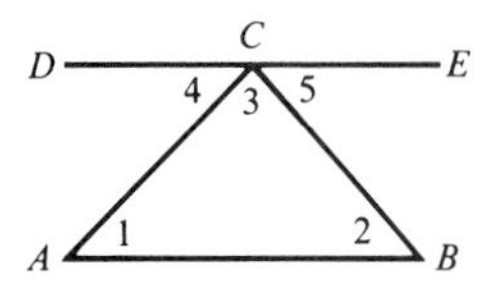

Figure 14.19

However, $\angle 1$ and $\angle 4$ are alternate interior angles, as are $\angle 2$ and $\angle 5$. So $\angle \mathbf{1} = \angle \mathbf{4}$ and $\angle \mathbf{2} = \angle \mathbf{5}$ Substituting in

$$\angle \mathbf{4} + \angle 3 + \angle \mathbf{5} = 180°$$

it follows that

$$\angle \mathbf{1} + \angle 3 + \angle \mathbf{2} = 180° \qquad \angle 4 = \angle 1,\ \angle 5 = \angle 2$$

Example 1

Find $\angle A$ in the triangle shown in Figure 14.20.

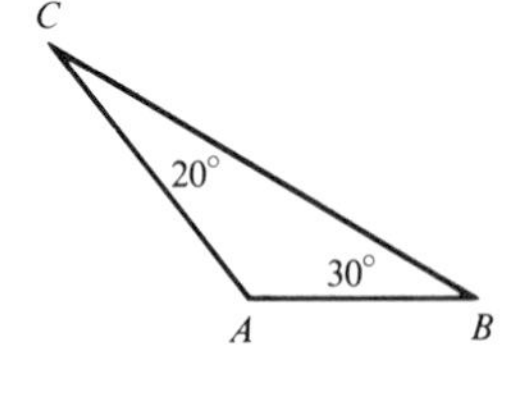

Figure 14.20

Solution. Since the sum of the angles of a triangle is 180°, we have

$$\angle A + \angle B + \angle C = 180°$$

From Figure 14.20, $\angle \mathbf{B} = \mathbf{30°}$ and $\angle \mathbf{C} = \mathbf{20°}$. So

$$\angle A + \mathbf{30°} + \mathbf{20°} = 180° \qquad \text{Substitution}$$
$$\angle A + 50° = 180°$$
$$\angle A + 50° - \mathbf{50°} = 180° - \mathbf{50°} \qquad \text{Subtracting } 50°$$
$$\angle A = 130°$$

Altitude, Height

Base

A few other terms associated with triangles are the following: The **altitude** or **height** of a triangle (denoted by h) is the line segment from a vertex perpendicular to the opposite side or its extension. The opposite side is called the **base.** Figure 14.21 illustrates the three cases.

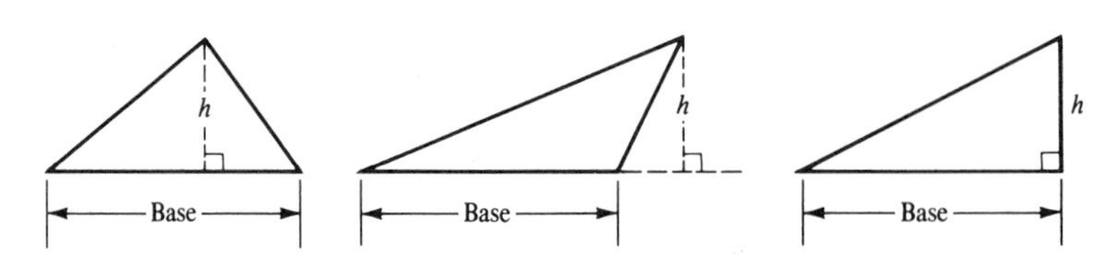

Figure 14.21

Note that every triangle has three altitudes. (See Figure 14.22; the altitudes are denoted by h_1, h_2, and h_3.)

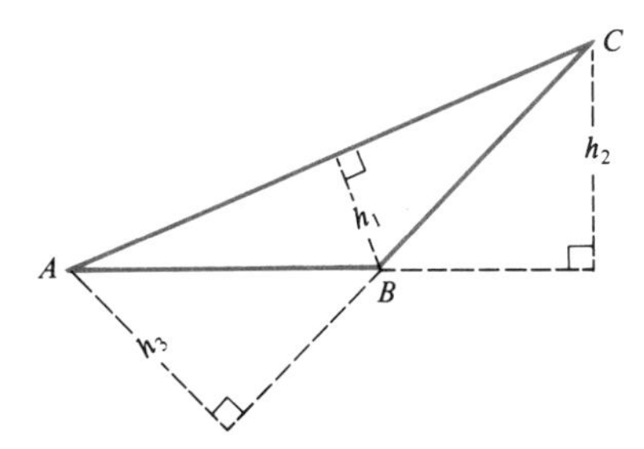

Figure 14.22

Median

A **median** of a triangle is a line segment from a vertex to the midpoint of the opposite side. Every triangle has three medians, which meet at a point called the *centroid* or *center of mass*. (See Figure 14.23.)

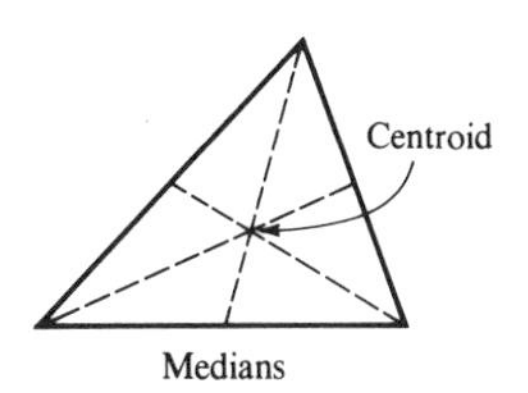

Figure 14.23

Bisector

A **bisector** of an angle divides the angle into two equal parts. The three bisectors of the angles of a triangle meet at a point. (See Figure 14.24.)

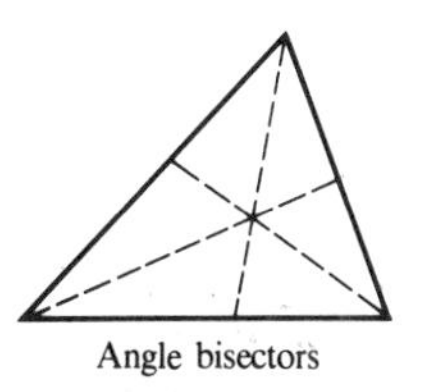

Figure 14.24

Quadrilaterals

In Chapter 6 we discussed five types of quadrilaterals: the square, the rectangle, the parallelogram, the rhombus, and the trapezoid. We will now consider an important property of quadrilaterals.

Diagonal

A **diagonal** of a quadrilateral is a line segment joining two nonadjacent vertices (Figure 14.25). A diagonal divides a quadrilateral into two triangles.

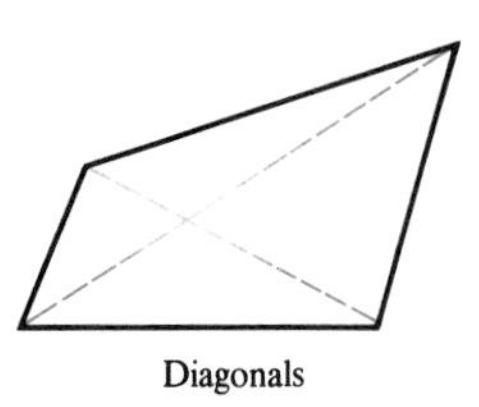

Figure 14.25

Since the sum of the angles of a triangle is 180°, it follows that:

> The sum of the interior angles of a quadrilateral is 360°.

Example 2 Find $\angle A$ in the quadrilateral shown in Figure 14.26.

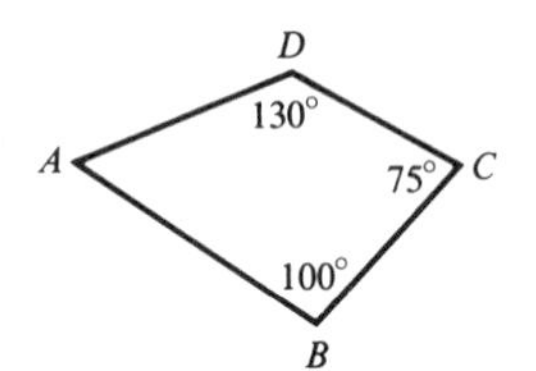

Figure 14.26

Solution. As noted above, the sum of the angles of a quadrilateral is 360°, or

$$\angle A + 100° + 75° + 130° = 360°$$
$$\angle A + 305° = 360°$$
$$\angle A = 360° - 305° = 55°$$

Circles

In Chapter 6 we defined the radius and diameter of a circle and then found the area and circumference of a given circle. In this section we introduce several new terms.

Concentric circles

Two circles may have the same center, as shown in Figure 14.27. Such circles are called **concentric.**

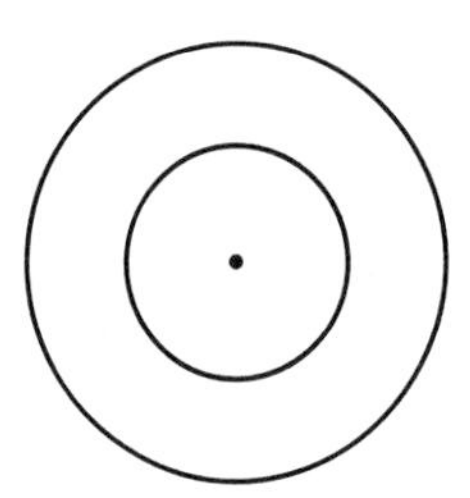

Figure 14.27

Tangent line

A line (in the same plane as a circle) that touches a circle at one point is called a **tangent line.** The common point is called the *point of tangency* (Figure 14.28). A noteworthy property of tangent lines is stated next.

A tangent line is perpendicular to the radius drawn to the point of tangency.

(See Figure 14.28.)

Secant line

A line passing through two points of a circle is called a **secant line** (Figure 14.28).

Chord

Finally, a line segment joining two points on a circle is called a **chord** (Figure 14.28).

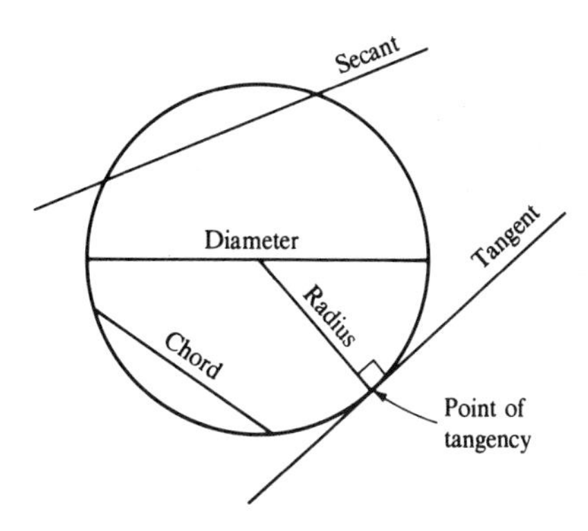

Figure 14.28

Central angle

Certain angles associated with circles have special names: A **central angle** is an angle formed by two radii (Figure 14.29). The vertex of a central angle is at the center.

Inscribed angle

An **inscribed angle** of a circle is an angle whose vertex is on the circle (Figure 14.29).

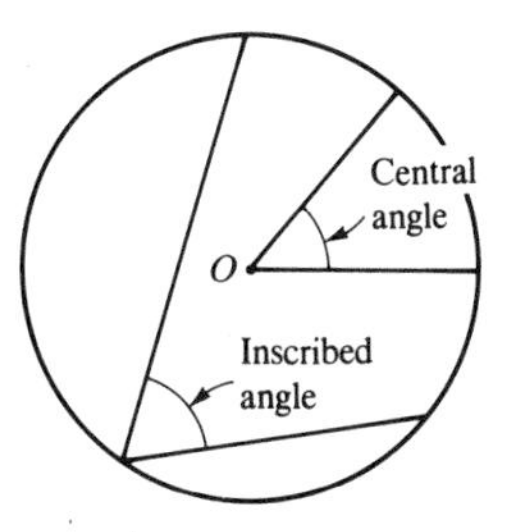

Figure 14.29

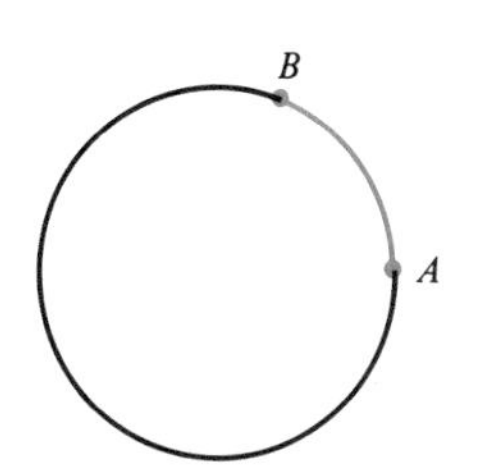

Figure 14.30

An **arc** of a circle is that (curved) portion of the circle between two points. The symbol used for arc is ⌢. Thus $\widehat{AB}$ is the arc in Figure 14.30.

Arcs are often measured in degrees: The number of degrees is equal to the degree measure of its central angle.

Example 3 Find the degree measure of the arcs of the circle in Figure 14.31.

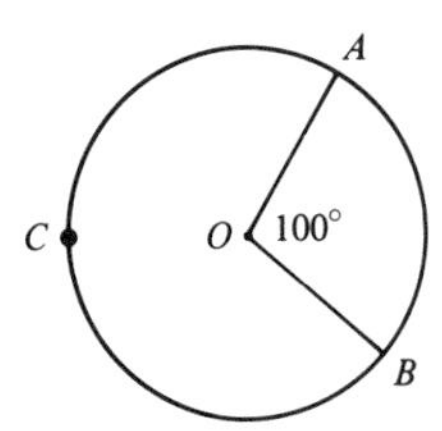

Figure 14.31

Solution. Since the measure of an arc is the degree measure of its central angle, we have

$$\overset{\frown}{AB} = 100°$$

For the larger arc, denoted by $\overset{\frown}{ACB}$, we get

$$\overset{\frown}{ACB} = 360° - 100° = 260°$$

Intercepted arc

The arc corresponding to an inscribed angle is called the **intercepted arc.** An inscribed angle has the property stated next.

> The degree measure of an inscribed angle is equal to one-half the degree measure of the intercepted arc.

Example 4 Find the degree measure of $\angle ABC$ in Figure 14.32.

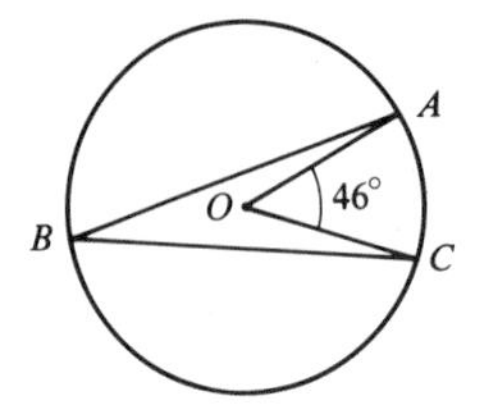

Figure 14.32

Solution. Since the central angle is 46°, we get

$$\overset{\frown}{AC} = 46°$$

By the property of inscribed angles stated above,

$$\angle ABC = \frac{1}{2}(46°) = 23°$$

Semicircle

An angle inscribed in a semicircle has a special property. A **semicircle** (or half-circle) is that part of a circle on one side of a diameter. **An angle inscribed in a semicircle is a right angle.** To see why, consider the inscribed angle A in Figure 14.33. Since the intercepted arc is 180°, we get

$$\angle A = \frac{1}{2}(180°) = 90°$$

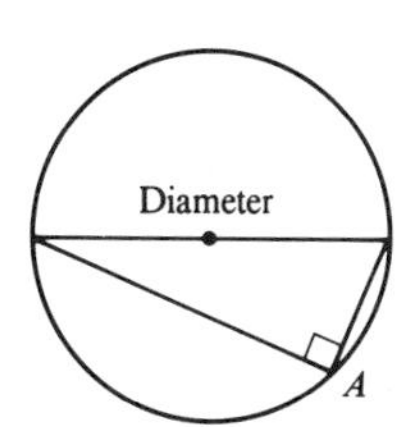

Figure 14.33

Exercises / Section 14.2

1. Identify the given figures (Figures 14.34–14.36).

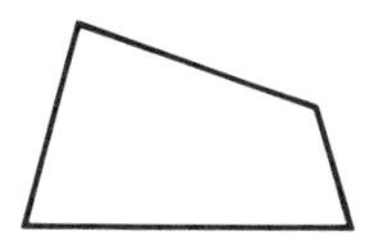

Figure 14.34

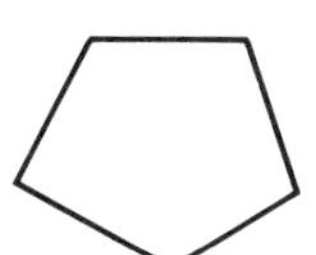

Figure 14.35

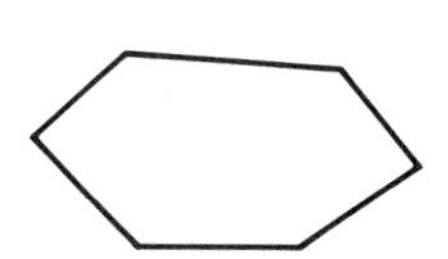

Figure 14.36

In Exercises 2–4, find $\angle A$ (Figures 14.37–14.39).

2.

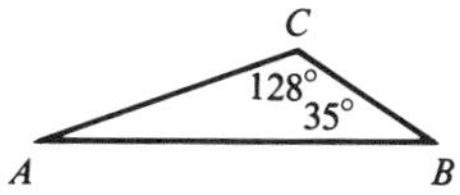

Figure 14.37

3.

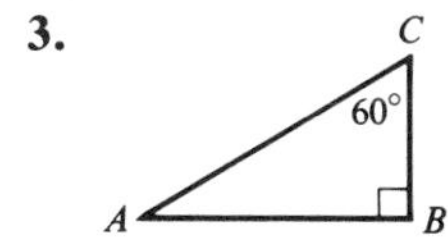

Figure 14.38

4.

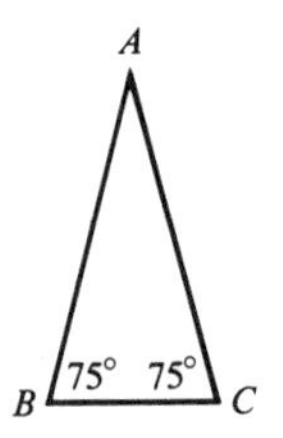

Figure 14.39

In Exercises 5–7, find $\angle A$ (Figures 14.40–14.42).

5.

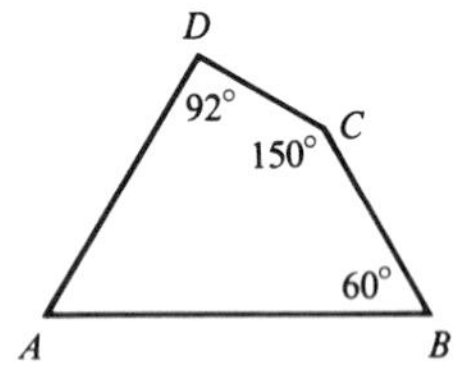

Figure 14.40

6.

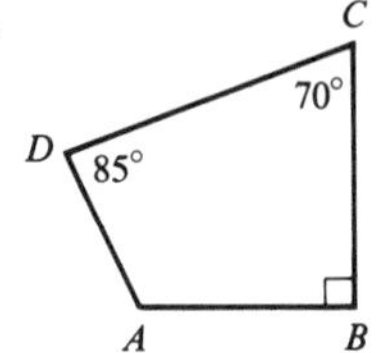

Figure 14.41

7.

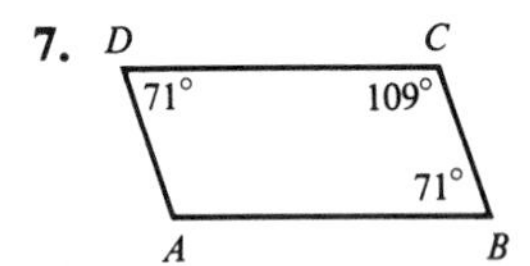

Figure 14.42

8. What is the sum of the angles of a pentagon?
9. The base angles of an isosceles triangle each measure 50°. What is the measure of the remaining angle?
10. A line is said to *bisect* a line segment if it passes through the midpoint of the line segment. If the diagonals of a quadrilateral bisect each other, what is the quadrilateral?
11. Find $\angle A$ (Figure 14.43).
12. Find $\angle CBD$ (Figure 14.44).

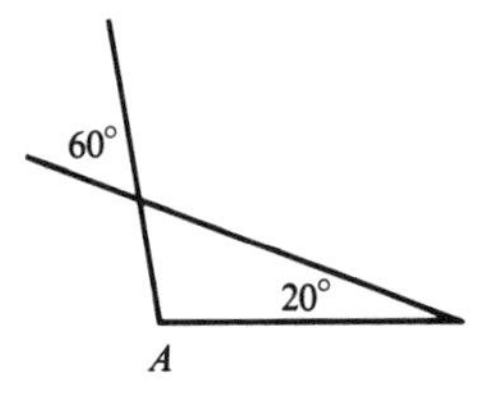

Figure 14.43

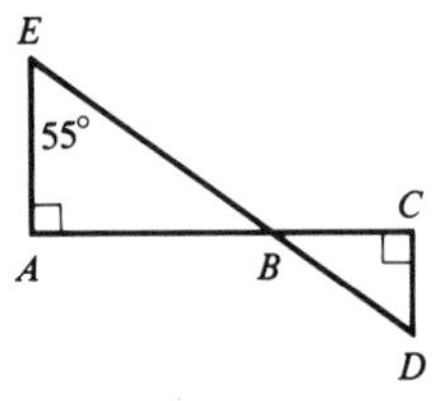

Figure 14.44

13. Find all the angles of the parallelogram $ABCD$ (Figure 14.45).

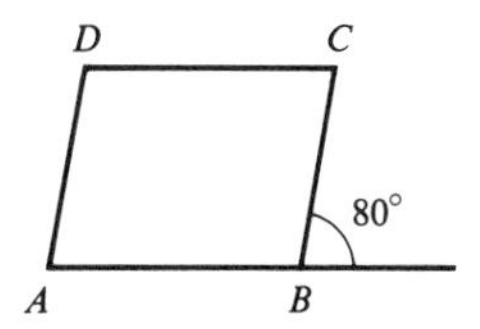

Figure 14.45

14. In Figure 14.46, $CE \parallel AB$; find $\angle A$ and $\angle B$.

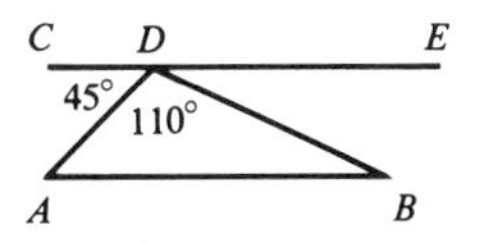

Figure 14.46

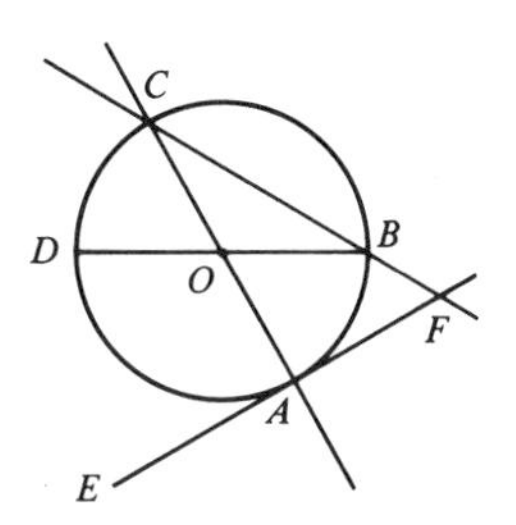

Figure 14.47

In Exercises 15–20, refer to Figure 14.47. Identify the following:

15. A tangent line

16. Two secant lines

17. A diameter

18. Two central angles

19. Two inscribed angles

20. Two chords

In Exercises 21–24, refer to Figure 14.47. Given that $\angle BOC = 120°$, find the degree measure of each of the following:

21. $\widehat{BC}$

22. $\widehat{BA}$

23. $\angle ACB$

24. $\angle CBD$

In Exercises 25–28, find $\angle A$ (Figures 14.48–14.51).

25.

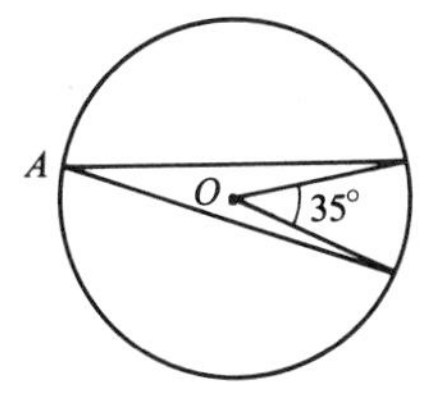

Figure 14.48

26.

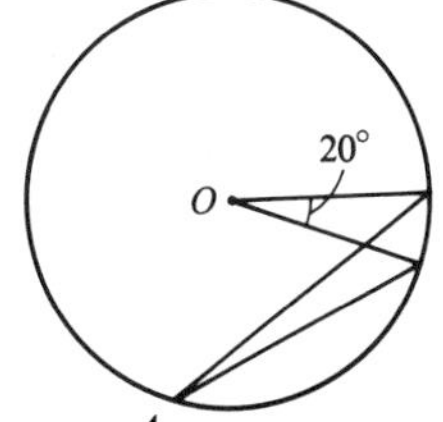

Figure 14.49

27.

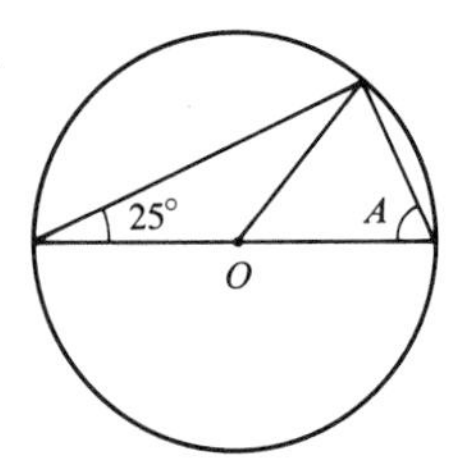

Figure 14.50

28. Given: $BC = DC$

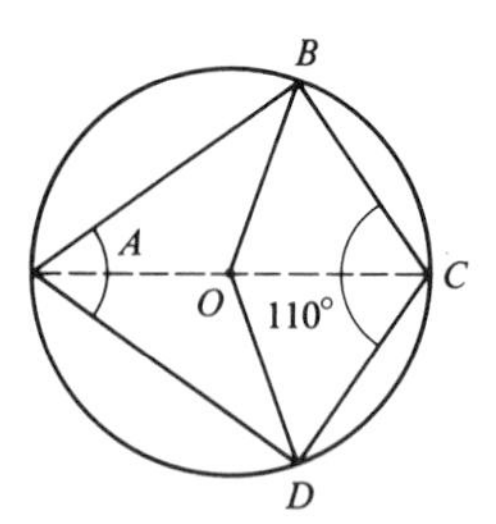

Figure 14.51

14.3 The Pythagorean Theorem

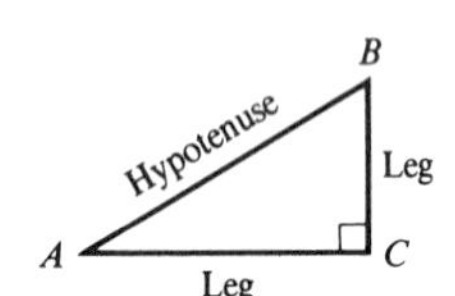

Figure 14.52

A right triangle has a property that is so useful in technical applications that we will devote an entire section to it. This property, called the *Pythagorean theorem,* is stated next. (Refer to Figure 14.52.)

> **Pythagorean theorem**
>
> The square of the length of the hypotenuse of a right triangle is equal to the sum of the squares of the lengths of the legs.

(Recall that the hypotenuse of a right triangle is the side opposite the right angle and that the other two sides are called the *legs*.)

Using the letters in Figure 14.53, the Pythagorean theorem can also be stated in symbols.

> **Pythagorean theorem:** $c^2 = a^2 + b^2$
>
> 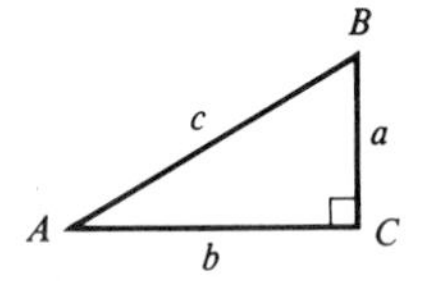
>
>
> Figure 14.53

To see why $c^2 = a^2 + b^2$ for any right triangle, let us draw triangle ABC four times in such a way that the resulting figure is a square of side $a + b$, as shown in Figure 14.54. First note that

$\angle 1 + \angle 2 = 90°$ Since $\angle A + \angle B = 90°$

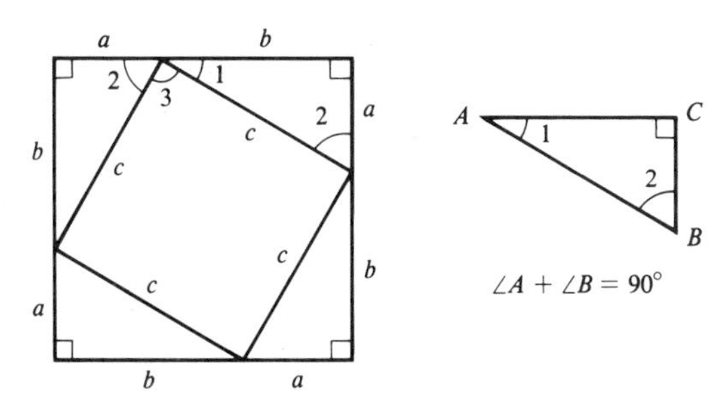

Figure 14.54

Also,

$$\angle 1 + \angle 3 + \angle 2 = 180^\circ \qquad \text{Straight angle}$$
$$\angle 1 + \angle 2 + \angle 3 = 180^\circ \qquad \text{Rearranging}$$
$$90^\circ + \angle 3 = 180^\circ \qquad \angle 1 + \angle 2 = 90^\circ$$
$$\angle 3 = 90^\circ \qquad \text{Subtracting } 90^\circ \text{ from both sides}$$

It follows that the quadrilateral inside the square is also a square. Its area is c^2. Furthermore, the area of each triangle is $\frac{1}{2}ab$, so that the total area of the large square is

$$c^2 + 4\left(\frac{1}{2}ab\right) \qquad \text{Four triangles}$$

However, since the large square has side $a + b$, this area is also given by $(a + b)^2$. Having obtained the area of the large square in two ways, we may now write

$$(a + b)^2 = c^2 + 4\left(\frac{1}{2}ab\right)$$
$$(a + b)^2 = c^2 + 2ab \qquad 4 \cdot \frac{1}{2} = 2$$
$$a^2 + 2ab + b^2 = c^2 + 2ab \qquad \text{Squaring the trinomial}$$
$$a^2 + b^2 = c^2 \qquad \text{Subtracting } 2ab \text{ from both sides}$$

The last equation is the Pythagorean theorem.

Whenever we know two sides of a right triangle, we can use the Pythagorean theorem to find the third side, as shown in the first two examples.

Example 1 Find the length of the hypotenuse of the triangle in Figure 14.55.

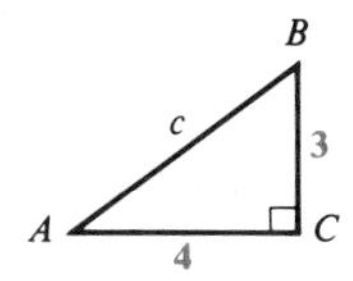

Figure 14.55

Solution. The unknown side is the hypotenuse c. Letting $a = 3$ and $b = 4$, we obtain from the Pythagorean theorem

$$
\begin{aligned}
c^2 &= a^2 + b^2 \\
c^2 &= 3^2 + 4^2 && a = 3,\ b = 4 \\
c^2 &= 9 + 16 = 25 \\
c &= \sqrt{25} = 5 && \text{Taking the square root of each side}
\end{aligned}
$$

(We take the positive square root because the length of a side of a triangle is usually considered positive.)

Example 2 Find x in the triangle in Figure 14.56.

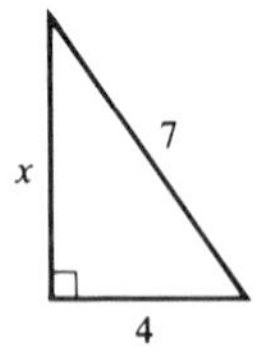

Figure 14.56

Solution. In this example the hypotenuse is given to be 7, but one of the sides is unknown. Denoting the unknown side by x, we get from the Pythagorean theorem,

$$
\begin{aligned}
x^2 + 4^2 &= 7^2 \\
x^2 + 16 &= 49 \\
x^2 &= 49 - 16 = 33 \\
x &= \sqrt{33}
\end{aligned}
$$

The Pythagorean theorem is useful in everyday life. Consider the next example.

Example 3 A baseball diamond is a 90-ft square. Find the distance from home plate to second base.

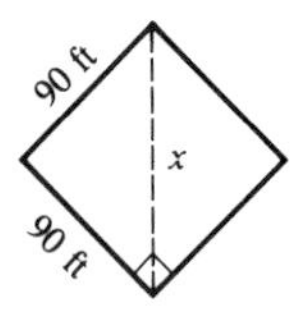

Figure 14.57

Solution. The baseball diamond is shown in Figure 14.57. The distance from second base to home plate is denoted by x. Since x is the hypotenuse, we get from the Pythagorean theorem,

$$
\begin{aligned}
x^2 &= 90^2 + 90^2 \\
x^2 &= 2 \cdot 90^2 && a^2 + a^2 = 2a^2 \\
x &= \sqrt{2 \cdot 90^2} && \text{Taking square roots} \\
x &= \sqrt{2}\sqrt{90^2} = \sqrt{2}(90) = 90\sqrt{2} \text{ ft}
\end{aligned}
$$

So $x = 127$ ft to the nearest foot.

The Pythagorean theorem is useful in many technical problems, as the next two examples show.

Example 4 A machine part in the shape of a right triangle is to have a hypotenuse of 9.70 cm and a base of 8.60 cm, as shown in Figure 14.58. Find the height h.

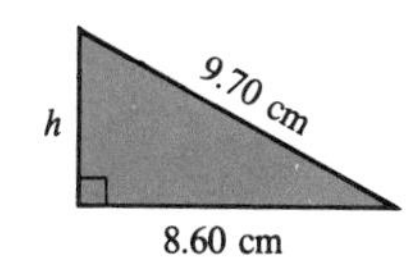

Figure 14.58

Solution. Since 9.70 cm is the length of the hypotenuse, we use the Pythagorean theorem to set up the following equation:

$$h^2 + 8.60^2 = 9.70^2$$
$$h^2 = 9.70^2 - 8.60^2$$
$$h = \sqrt{9.70^2 - 8.60^2}$$

The sequence is

9.70 [x²] [−] 8.60 [x²] [=] [√] → 4.4866469

So $h = 4.49$ cm.

Example 5 Suppose an alternating-current circuit (AC) contains a resistor and an inductor (coil). The resistance R of the resistor and the effective resistance X_L of the coil, called the *inductive reactance*, produces a total effective resistance Z, called the *impedance*. The relationship is $Z^2 = R^2 + X_L^2$. Since the relationship has the form of the Pythagorean theorem, a diagram is helpful in visualizing the relationship. (See Figure 14.59.) If $R = 25.0\ \Omega$ and $X_L = 19.3\ \Omega$, find Z.

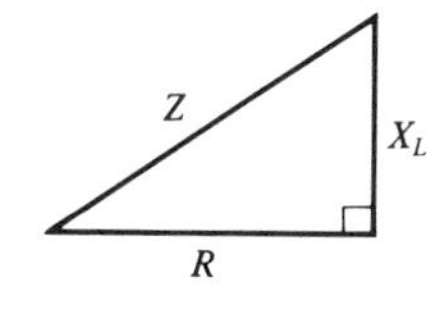

Figure 14.59

Solution. First we place the given values on the triangle in Figure 14.60. From the Pythagorean theorem,

$$Z^2 = (25.0)^2 + (19.3)^2$$

and

$$Z = \sqrt{(25.0)^2 + (19.3)^2}$$

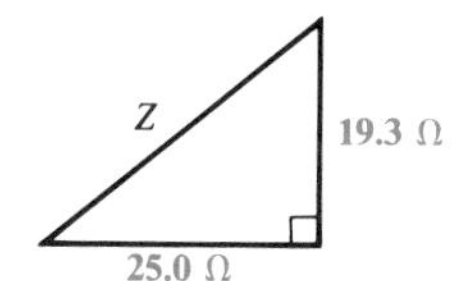

Figure 14.60

The sequence is

25.0 [x²] [+] 19.3 [x²] [=] [√] → 31.583065

So $Z = 31.6\ \Omega$.

Exercises / Section 14.3

(Use a calculator in this exercise set.)

In Exercises 1–20, refer to Figure 14.61. Use the Pythagorean theorem to determine the unknown side of the triangle in each case.

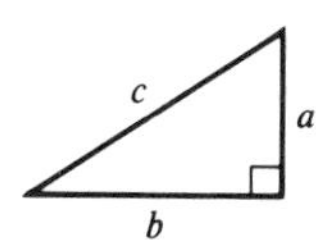

Figure 14.61

	a	b	c
1.	4	b	5
2.	5	12	c
3.	a	12	13
4.	9	b	15
5.	8	6	c
6.	a	10	26
7.	3.14	7.46	c
8.	1.73	9.46	c
9.	a	25.2	58.1
10.	a	73.9	99.3
11.	0.341	b	1.25
12.	0.542	b	0.906
13.	8.79	5.42	c
14.	10.1	12.0	c
15.	a	104	151
16.	a	406	979
17.	225	b	565
18.	1.05	b	2.00
19.	a	0.310	0.410
20.	0.430	b	0.870

21. A metal plate has the shape of a rectangle (Figure 14.62). Find the length of the diagonal shown.

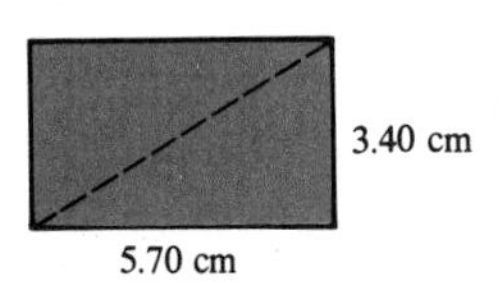

Figure 14.62

22. A rectangular computer chip measures 2.4 mm by 1.6 mm. Find the length of a diagonal.

23. An AC circuit contains a resistor and an inductor. If $R = 30.0\ \Omega$ and $X_L = 16.0\ \Omega$, find the impedance Z. (See Example 5.)

24. An AC circuit contains a resistor and a capacitor. If $R = 25.3\ \Omega$ and the *capacitive reactance* X_C is $30.1\ \Omega$, find Z ($Z^2 = R^2 + X_C^2$).

25. A flagpole 21.0 ft high casts a shadow 10.5 ft long. Find the distance from the top of the flagpole to the end of the shadow.

26. The tower in Figure 14.63 is 45 ft high and is supported by a guy wire anchored to the ground, as shown. Find the length of the wire.

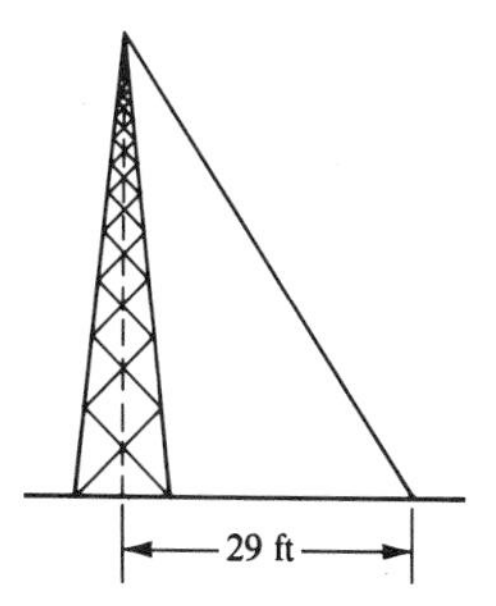

Figure 14.63

27. Find the length s of the support shown in Figure 14.64.

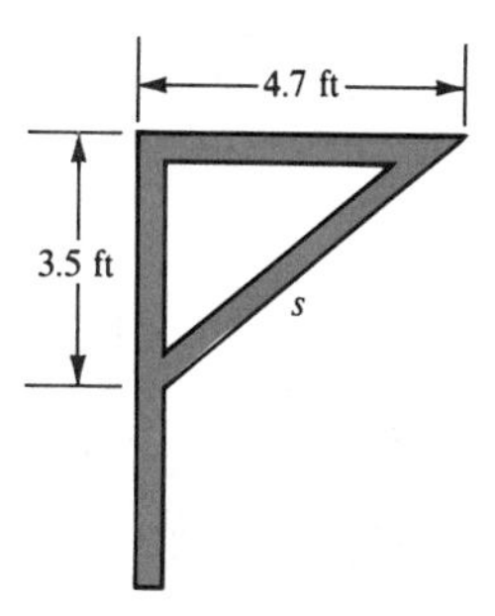

Figure 14.64

28. Find the distance d between the center of the gasket and the small hole shown in Figure 14.65.

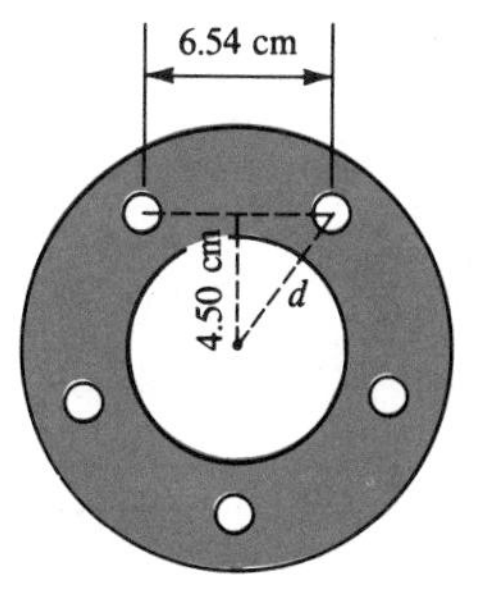

Figure 14.65

29. A straight road rises 25.0 m for every 285 m measured horizontally. What is the length of this stretch of road?

30. A boat travels 30.0 km due west, turns right, and continues for another 25.5 km. Find the direct-line distance from the starting point.

31. A boat travels 26.75 mi due east. How far south must it travel to be 40.60 mi from the starting point?

32. A piece of cable 15.36 ft long is hanging from the top of a pole 10.02 ft high. The cable is stretched taut and anchored to a point on the ground. Determine the distance from this point to the base of the pole.

33. Figure 14.66 shows a roof truss with a 21-ft span and a 12-ft rafter. Find the height h of the truss.

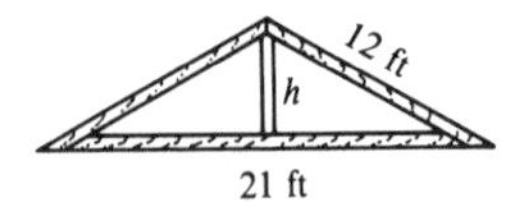

Figure 14.66

34. A hole is to be drilled in the metal plate shown in Figure 14.67. Find the distance from the base to the center of the hole.

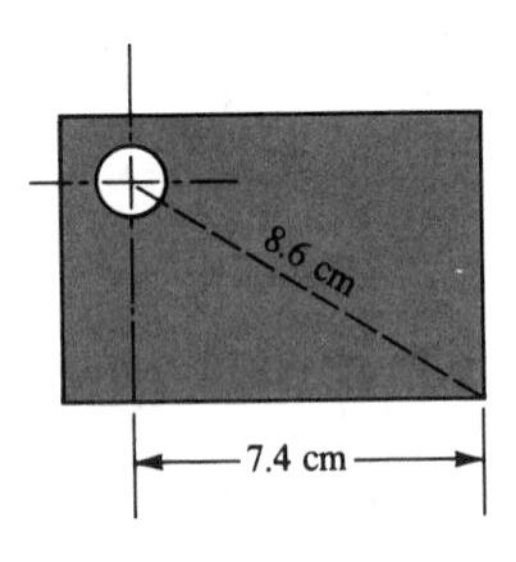

Figure 14.67

35. Find the distance between the centers of the pulleys in Figure 14.68.

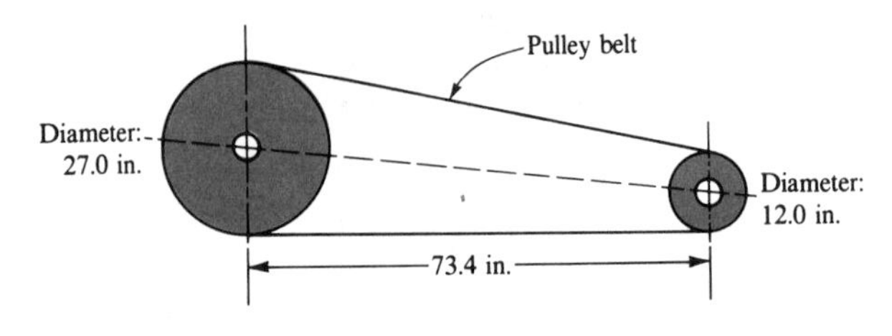

Figure 14.68

36. Determine the radius of the circular portion of the machine part in Figure 14.69.

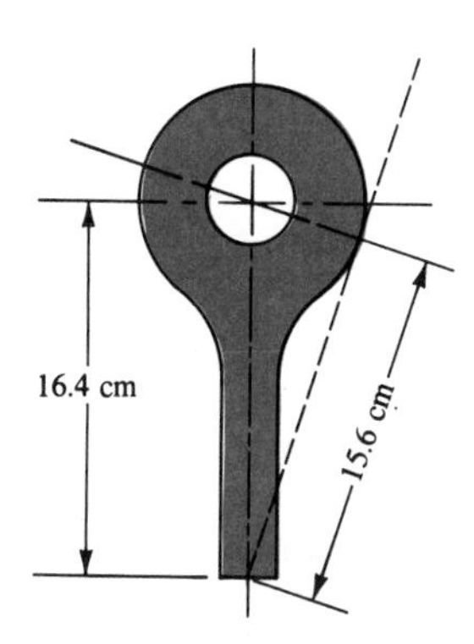

Figure 14.69

14.4 Congruent and Similar Triangles

So far we have studied various properties of single triangles. In this section we will study properties of pairs of triangles having the same shape. In some cases the triangles also have the same size, but in other cases they do not.

To compare two triangles, we need a convenient correspondence between the respective parts. Consider, for example, the triangles in Figure 14.70. Since $\angle A$ and $\angle A'$ (read "A prime") occupy the same relative positions, they are said to be corresponding angles. Similarly, $\angle B$ corresponds to $\angle B'$ and $\angle C$ corresponds to $\angle C'$.

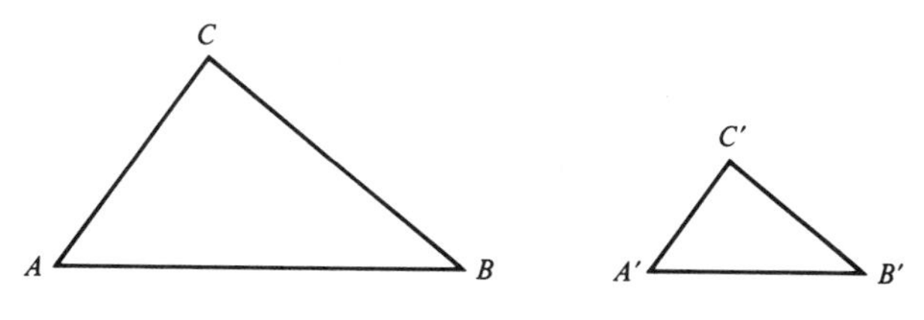

Figure 14.70

A similar correspondence exists between the sides. Thus AB corresponds to $A'B'$, AC corresponds to $A'C'$, and BC corresponds to $B'C'$.

While corresponding parts can be defined for any two triangles, in practice the correspondence is of interest only if the triangles are **similar** or **congruent**.

Similar triangles

Two triangles are **similar** if the corresponding angles are equal or if the corresponding sides are proportional.

Informally, two triangles are similar if they have the same shape. If the corresponding sides of two similar triangles are equal, then the triangles are **congruent**.

Congruent triangles

Two triangles are **congruent** if the corresponding angles and corresponding sides are equal.

Informally, two triangles are congruent if they have the same shape and size.

Notation. a. ΔABC is similar to $\Delta A'B'C'$ is written

$$\Delta ABC \sim \Delta A'B'C'$$

b. ΔABC is congruent to $\Delta A'B'C'$ is written

$$\Delta ABC \cong \Delta A'B'C'$$

To show that two triangles are similar, we have to show that one of the following conditions is met:

Properties of similar triangles

1. Corresponding angles are equal.
2. Corresponding sides are proportional, or

$$\frac{AB}{A'B'} = \frac{AC}{A'C'} = \frac{BC}{B'C'} \qquad \text{(Figure 14.71)}$$

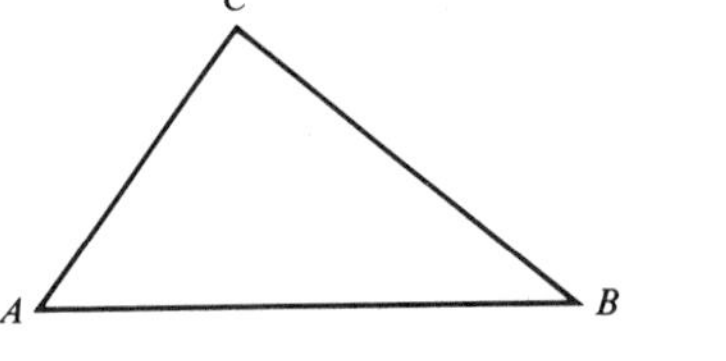

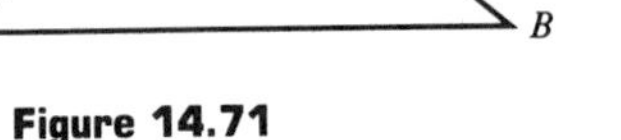

Figure 14.71

Example 1 Show that the triangles in Figure 14.72 are similar.

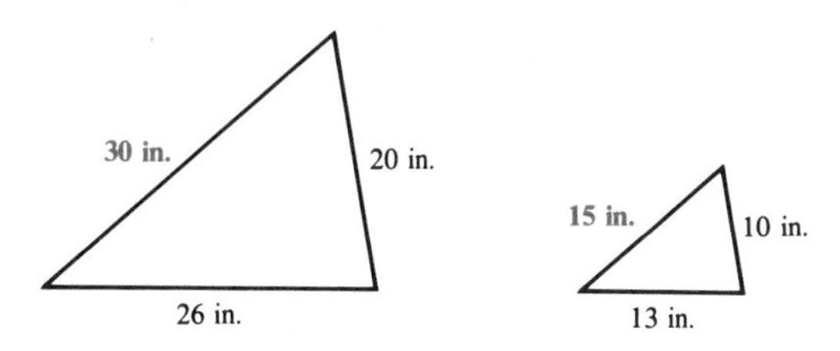

Figure 14.72

Solution. To show that two triangles are similar, it is sufficient to show that the corresponding sides are proportional. Note that the 30-inch side of the large triangle corresponds to the 15-inch side of the small triangle, the 26-inch side to the 13-inch side, and the 20-inch side to the 10-inch side. Forming the ratios of the corresponding sides, we get

$$\frac{\text{side of large triangle}}{\text{corresponding side of small triangle}} = \frac{\textbf{30 in.}}{\textbf{15 in.}} = \frac{26 \text{ in.}}{13 \text{ in.}}$$
$$= \frac{20 \text{ in.}}{10 \text{ in.}} = \frac{2}{1} = 2$$

Since the ratios are equal, the corresponding sides are proportional. The triangles are therefore similar by Property 2.

Example 2 In Figure 14.73, $AC \parallel ED$. Show that $\Delta ABC \sim \Delta EBD$.

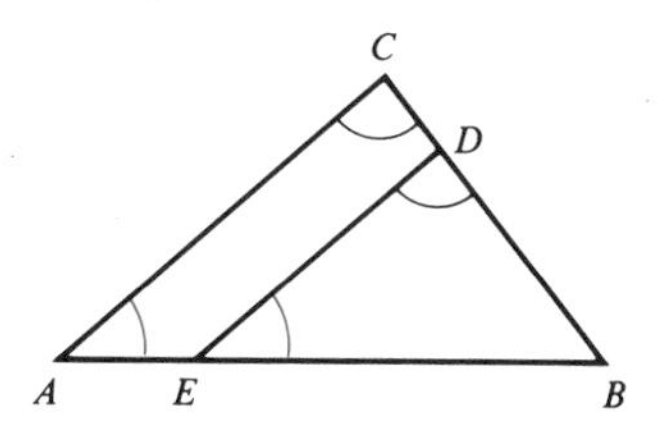

Figure 14.73

Solution. We are given that $AC \parallel ED$. These parallel lines are cut by transversals AB and CB. It follows that

$$\angle \boldsymbol{BAC} = \angle \boldsymbol{BED} \qquad \text{and} \qquad \angle ACB = \angle EDB$$

(If two parallel lines are cut by a transversal, the corresponding angles are equal.) Also, $\angle B$ is common to the two triangles, and this angle is equal to itself.

We have shown that the corresponding angles of the two triangles are equal. So

$$\Delta ABC \sim \Delta EBD$$

by Property 1.

Example 3

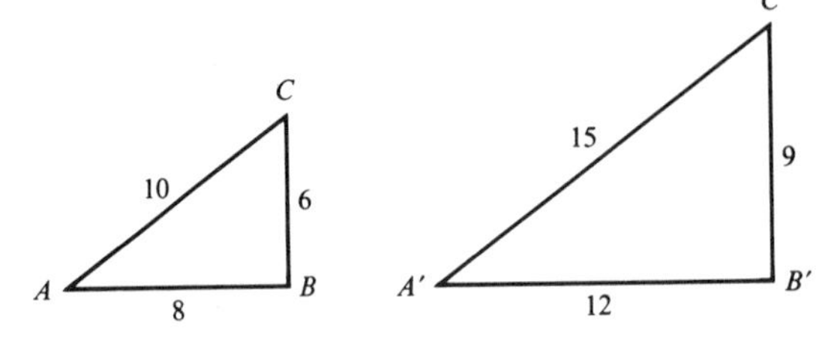

Figure 14.74

In Figure 14.74,

$$\frac{AB}{A'B'} = \frac{AC}{A'C'} = \frac{BC}{B'C'} = \frac{2}{3}$$

So

$$\Delta ABC \sim \Delta A'B'C'$$

by Property 2.

In the next example we show that two given triangles are congruent.

Example 4 Given the parallelogram in Figure 14.75, show that

$$\Delta ABC \cong \Delta CDA$$

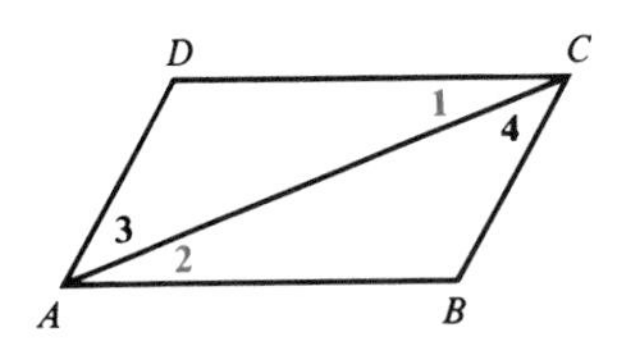

Figure 14.75

Solution. Since $ABCD$ is a parallelogram, $AB \parallel CD$ and $AD \parallel CB$. Since AC is a transversal, it follows that

$\angle 1 = \angle 2$ If two parallel lines are cut by a transversal, the alternate interior angles are equal.

and

$\angle 3 = \angle 4$

Since the sum of the angles of a triangle is 180°, we also have $\angle B = \angle D$. We have shown that all corresponding angles are equal, so that

$$\Delta ABC \sim \Delta CDA$$

Next, by Property 2 for similar triangles,

$$\frac{AD}{CB} = \frac{CD}{AB} = \frac{AC}{AC}$$

But since

$$\frac{AC}{AC} = 1$$

it follows that all corresponding sides are equal. So by the definition of congruence, we conclude that

$$\Delta ABC \cong \Delta CDA$$

In many problems, including technical applications, we use the properties of similar triangles to find the unknown parts of a triangle from a given similar triangle.

Example 5 Given that $\Delta ABC \sim \Delta A'B'C'$ (Figure 14.76), find $A'B'$ and $B'C'$.

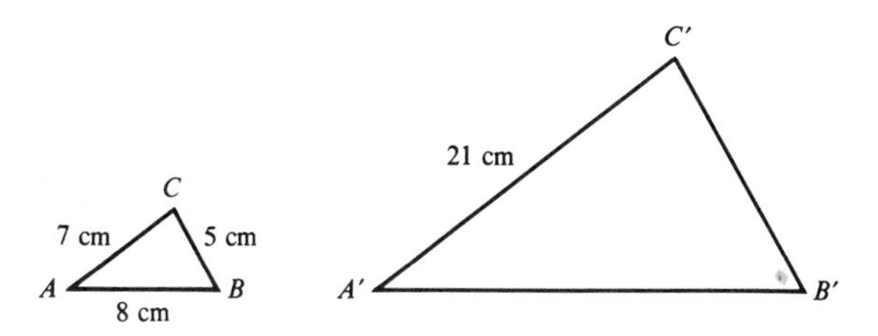

Figure 14.76

Solution. Since the triangles are similar, the corresponding sides are proportional:

$$\frac{A'B'}{AB} = \frac{A'C'}{AC} = \frac{B'C'}{BC}$$

By substituting the known quantities, we can solve for the unknown quantities. From

$$\frac{A'B'}{AB} = \frac{A'C'}{AC}$$

we get

$$\frac{A'B'}{8} = \frac{21}{7}$$

or

$$\frac{A'B'}{8} = 3 \qquad \text{and} \qquad A'B' = 24 \text{ cm}$$

From

$$\frac{B'C'}{BC} = \frac{A'C'}{AC}$$

we get

$$\frac{B'C'}{5} = \frac{21}{7}$$

or

$$\frac{B'C'}{5} = 3 \qquad \text{and} \qquad B'C' = 15 \text{ cm}$$

Example 6 A building casts a shadow 35 ft long. At the same time, a vertical yardstick casts a shadow 21 in. long. Find the height of the building.

Solution. Figure 14.77 illustrates the situation.

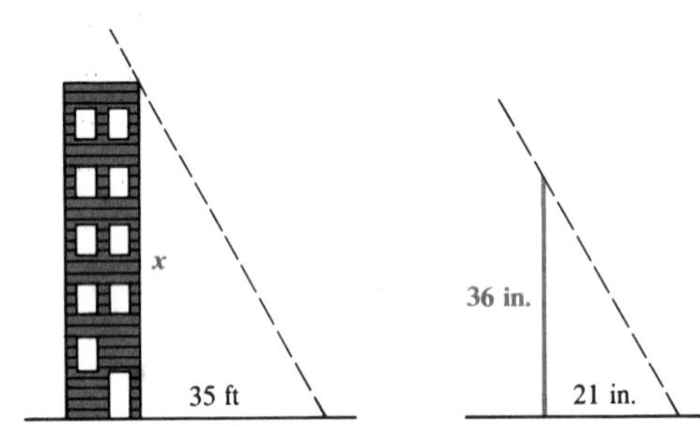

Figure 14.77

Since the rays of the sun are parallel, the top angles in each triangle are equal. Both triangles are right triangles, so that the remaining angles are also equal. It follows that the triangles are similar.

Denote the height of the building by x. Since 1 yd = 36 in., we now have

$$\frac{x}{36 \text{ in.}} = \frac{35 \text{ ft}}{21 \text{ in.}}$$

$$x(21 \text{ in.}) = (35 \text{ ft})(36 \text{ in.}) \qquad \text{Cross-multiplying}$$

$$x = \frac{(35 \text{ ft})(36 \cancel{\text{in.}})}{21 \cancel{\text{in.}}} \qquad \text{Dividing by 21 in.}$$

$$x = 60 \text{ ft}$$

A practical use of similar triangles (and other similar figures) in technology is that of a **scale drawing**. In a scale drawing all distances are proportional to the actual distances represented. For example, both drawings in Figure 14.77 are scale drawings.

Example 7 A blueprint of a machine has a scale of

$$\frac{3}{4} \text{ in.} = 4 \text{ ft}$$

The machine is $4\frac{1}{2}$ in. long on the blueprint. What is the actual length of the machine?

Solution. Let

$$x = \text{length of machine}$$

Then

$$\frac{x}{4\frac{1}{2} \text{ in.}} = \frac{4 \text{ ft}}{\frac{3}{4} \text{ in.}}$$

$$x\left(\frac{3}{4} \text{ in.}\right) = (4 \text{ ft})\left(4\frac{1}{2} \text{ in.}\right) \qquad \text{Cross-multiplying}$$

$$x = \frac{(4 \text{ ft})\left(4\frac{1}{2} \cancel{\text{in.}}\right)}{\frac{3}{4} \cancel{\text{in.}}} \qquad \text{Dividing by } \frac{3}{4} \text{ in.}$$

$$x = \left(4 \cdot \frac{9}{2} \cdot \frac{4}{3}\right) \text{ ft}$$

$$x = 24 \text{ ft}$$

Exercises / Section 14.4

In Exercises 1–4, the given triangles (Figures 14.78–14.81) are similar. Identify the corresponding parts.

1.

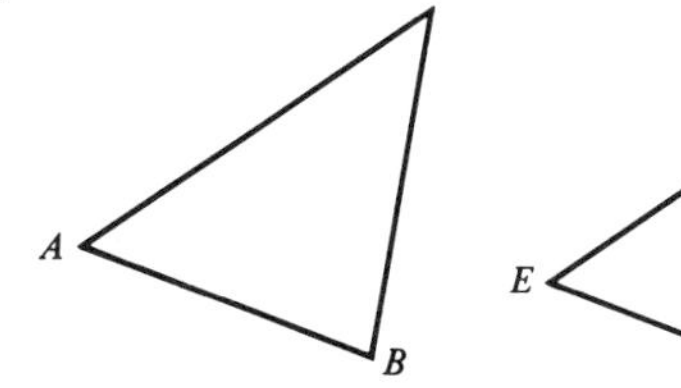

Figure 14.78

2.

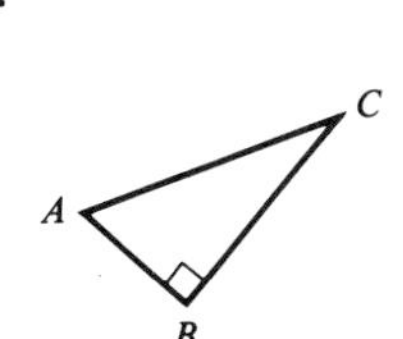

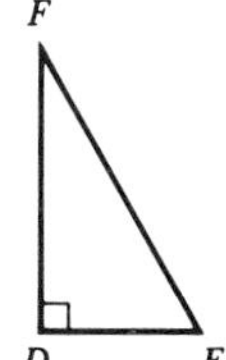

Figure 14.79

3.

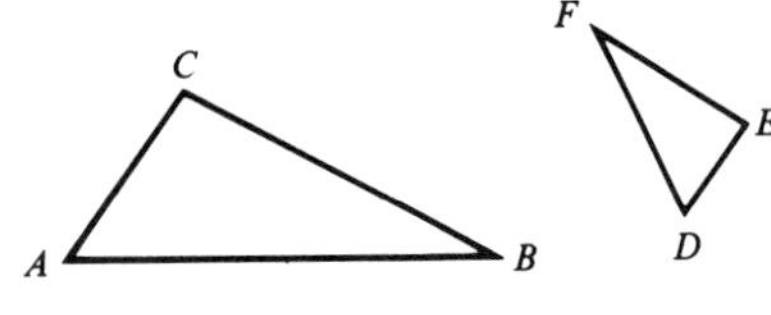

Figure 14.80

4.

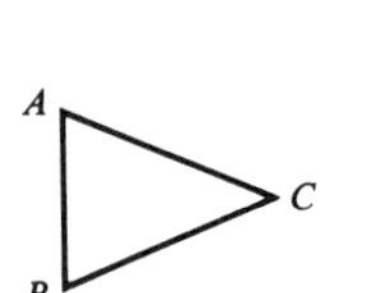
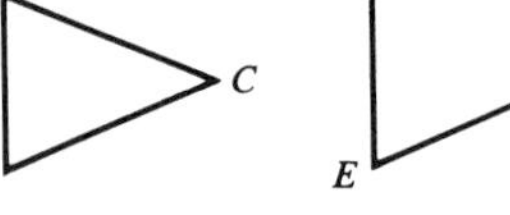

Figure 14.81

5. In Figure 14.82, $AC \parallel DE$. Show that $\Delta ABC \sim \Delta DBE$. (See Example 2.)

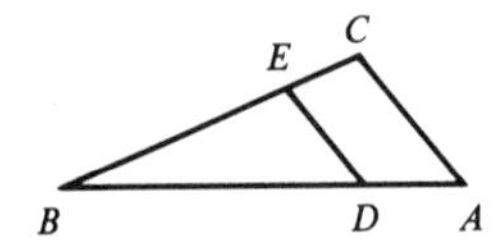

Figure 14.82

6. Show that the triangles in Figure 14.83 are similar.

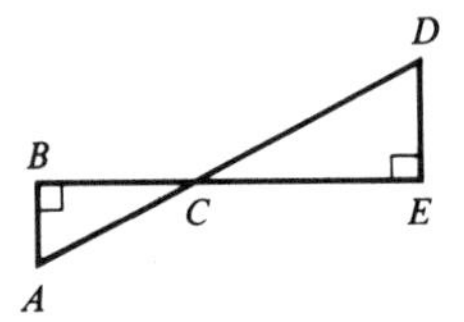

Figure 14.83

7. In Figure 14.84, $EC = CB$. Show that

$$\Delta ABC \cong \Delta DEC$$

(See Example 4.)

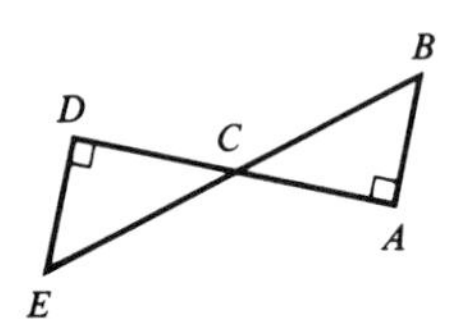

Figure 14.84

8. In Figure 14.85, ΔABC is isosceles. Show that

$$\Delta ADC \cong \Delta BDC$$

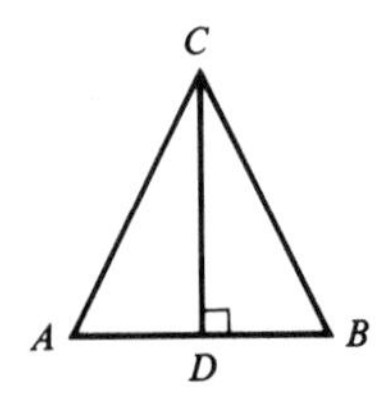

Figure 14.85

In Exercises 9–12, the given triangles (Figures 14.86–14.89) are similar. Find x in each case.

9.

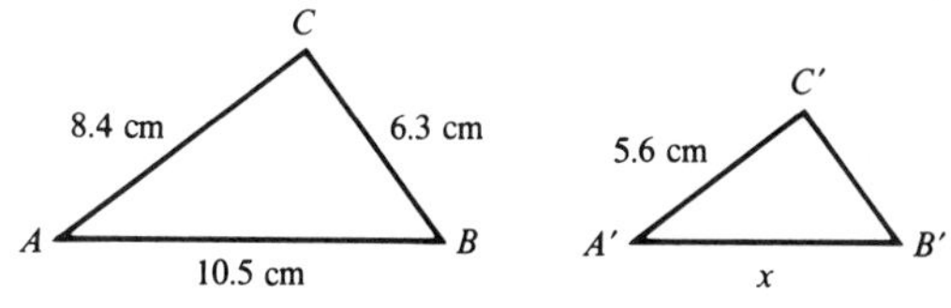

Figure 14.86

10.

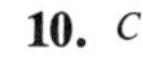

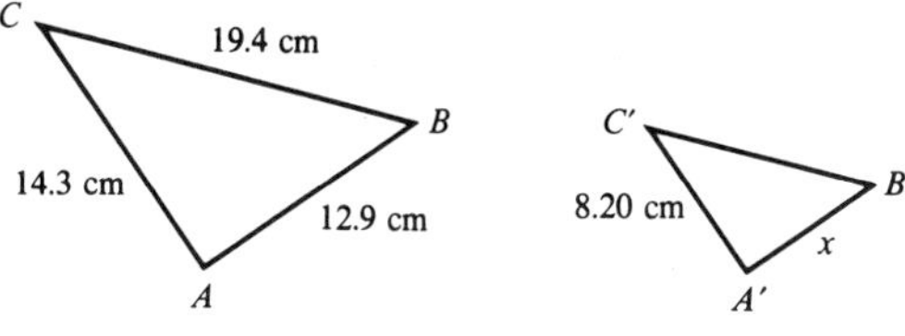

Figure 14.87

11.

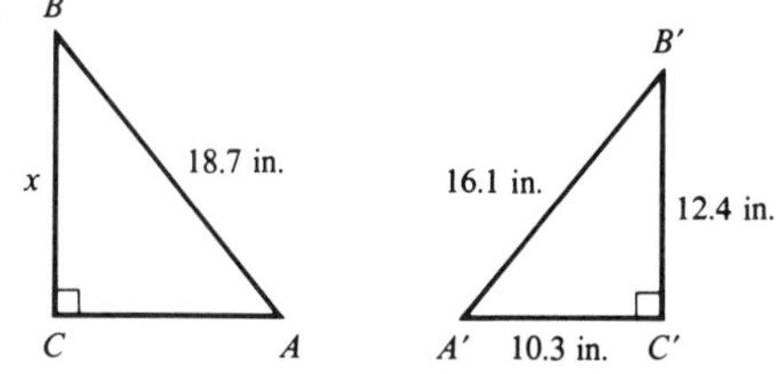

Figure 14.88

12.

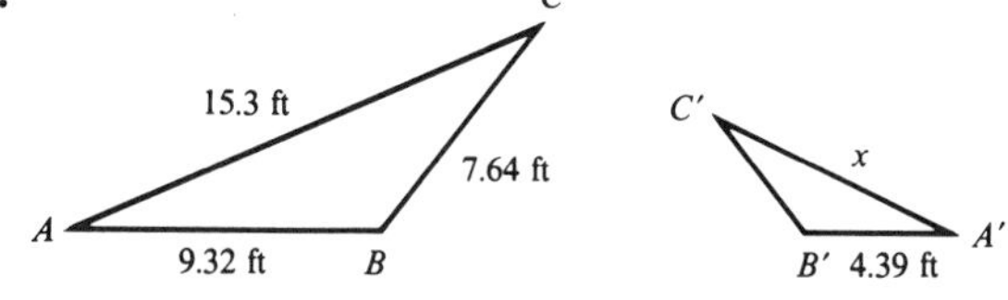

Figure 14.89

In Exercises 13–16 (Figures 14.90–14.93) find x in each case.

13.

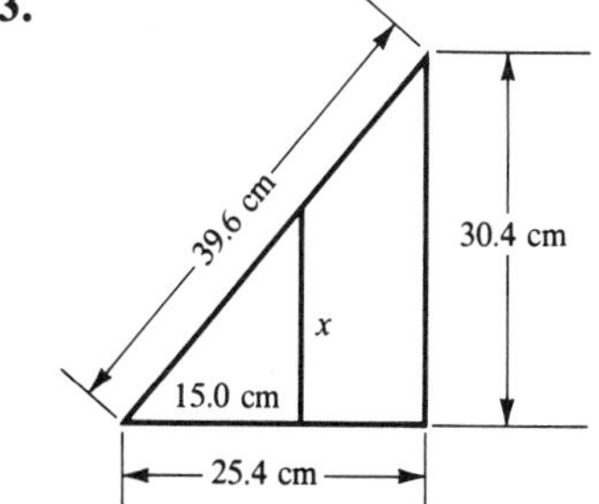

Figure 14.90

14.

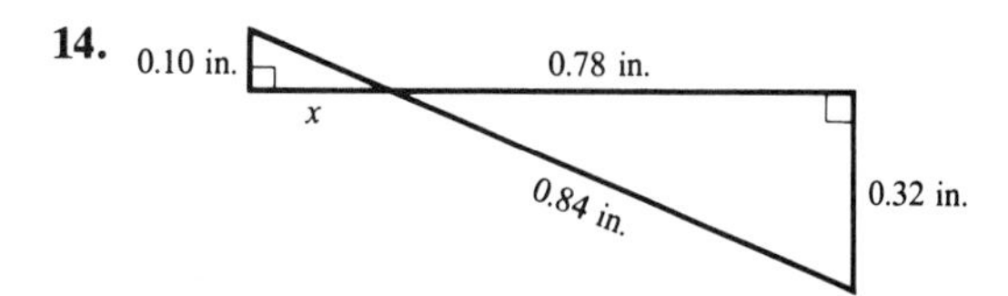

Figure 14.91

15.

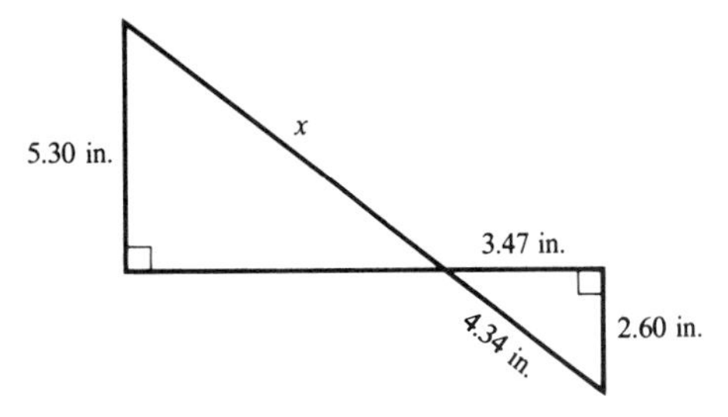

Figure 14.92

16.

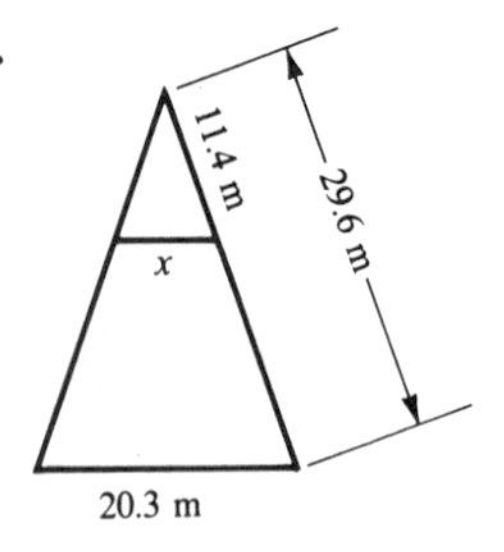

Figure 14.93

17. A tree casts a shadow 21 ft long. At the same time a girl 5.0 ft tall casts a shadow 4.0 ft long. How tall is the tree? (See Example 6.)

18. A pole casts a shadow 16 ft long. At the same time a 12-in. ruler casts a shadow 8.0 in. long. How tall is the pole?

19. In Figure 14.94, the top of the tree reflects in the mirror so that $\angle 1 = \angle 2$. Find the height of the tree.

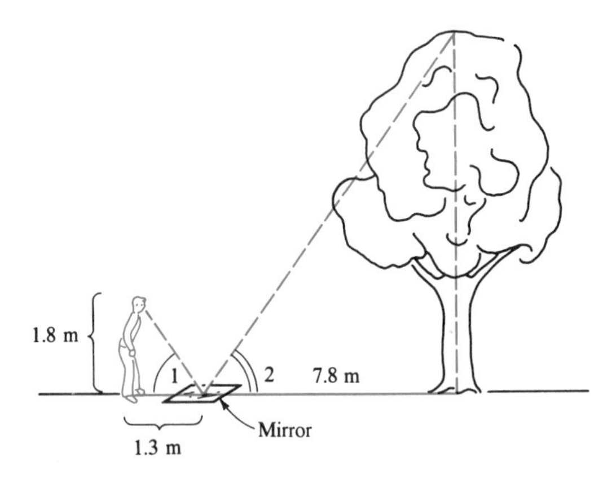

Figure 14.94

20. The image of a rectangular microprocessor chip appears under a microscope to have a diagonal of 4.5 cm and a length of 3.2 cm. If the length of the actual chip is 0.67 mm, find the length of the diagonal.

21. The scale on a blueprint is $1\frac{1}{4}$ in. = 3 ft. Determine the diameter of a gear measuring $\frac{1}{8}$ in. on the blueprint. (See Example 7.)

22. On the blueprint of an office building, the width of a certain office measures $2\frac{1}{4}$ in. If the scale is $1\frac{1}{2}$ in. = 10 ft, what is the width of the office?

23. Find the width of the river in Figure 14.95 from the given information.

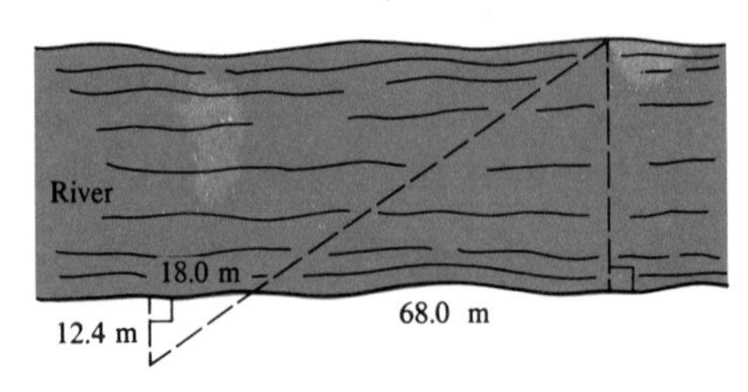

Figure 14.95

24. Find the distance across the lake in Figure 14.96.

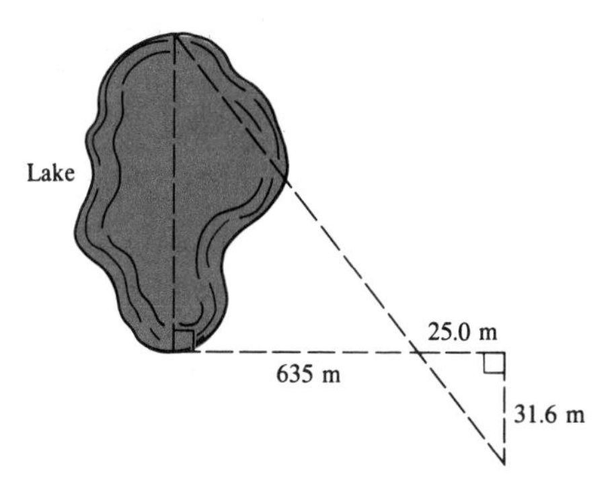

Figure 14.96

25. A trough has a cross section in the shape of an isosceles triangle measuring 51.0 cm across the top and 42.3 cm deep in the center. If the trough is 1.15 m long, what is the surface area of the water, if the water is 37.0 cm deep in the center?

26. An 8.0-m ladder is leaning against a wall. The third rung, which is 82 cm from the bottom of the ladder, is 75 cm above the ground. How far does the ladder reach up on the wall?

14.5 Solid Geometric Figures

Prisms

Most of the figures we have encountered so far have been plane figures. In this section we turn to solid figures. Our main goal in the study of solids is the determination of surface areas and volumes.

Polyhedron
Prism, Base

A **polyhedron** is a solid figure bounded by planes. A **prism** is a polyhedron whose **bases** (top and bottom) are parallel congruent polygons and whose sides are parallelograms. (See Figure 14.97.)

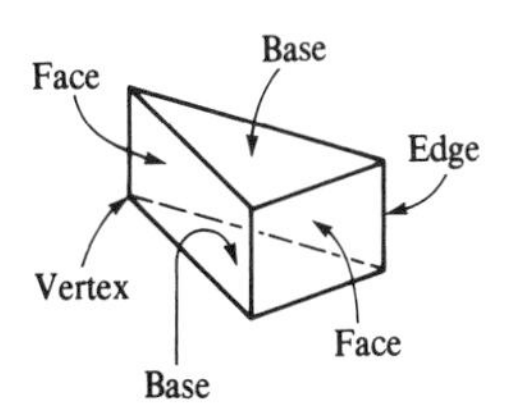

Figure 14.97

Face
Edge
Vertex
Rectangular solid

Each plane surface (including the bases) is called a **face**. The intersection of two faces is called an **edge**. The intersection of edges is called a **vertex** (Figure 14.97).

An important, special type of prism is a **rectangular solid**, all of whose faces are rectangles (Figure 14.98). The volume of a rectangular solid is found by multiplying length, width, and height. Since the area B of the base is $B = lw$, the volume of a rectangular solid can also be written $V = Bh$. This formula remains valid if the base is any polygon.

Volume of rectangular solid: $V = lwh$, where l is the length, w the width, and h the height.

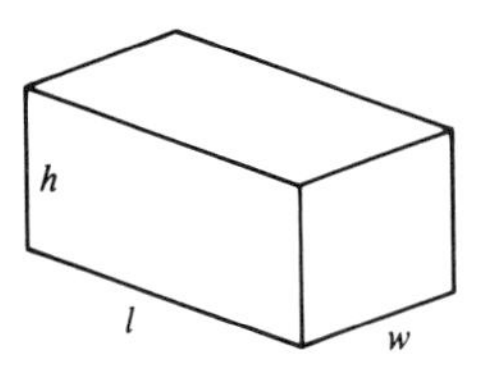

Figure 14.98

Volume of prism: $V = Bh$

Example 1 For the prism in Figure 14.99, $B = 36$ in.^{2} and $h = 10$ in. Find the volume.

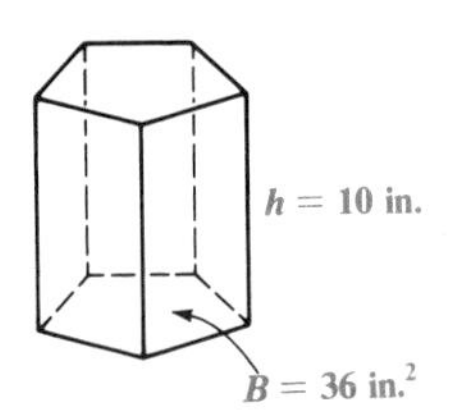

Figure 14.99

Solution. By the formula $V = Bh$, we have

$$\begin{aligned} V &= (36 \text{ in.}^2)(10 \text{ in.}) \\ &= 360 \text{ in.}^3 \end{aligned}$$

Example 2 Find the volume of the prism in Figure 14.100. (The base is a right triangle.)

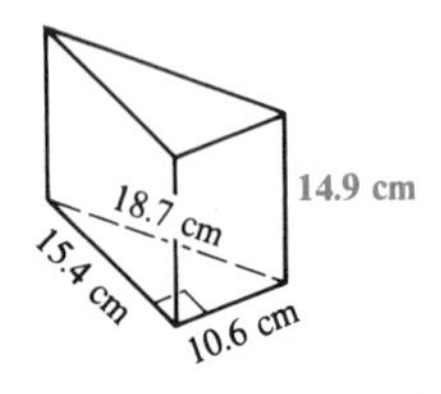

Figure 14.100

Solution. The base B of the prism is a triangle whose area is

$$B = \frac{1}{2}bh = \frac{1}{2}(10.6 \text{ cm})(15.4 \text{ cm})$$
$$= 81.62 \text{ cm}^2$$

By the formula $V = Bh$, we now get

$$V = (81.62 \text{ cm}^2)(14.9 \text{ cm}) \qquad h = 14.9 \text{ cm}$$
$$= 1220 \text{ cm}^3$$

using three significant digits.

Surface area
Lateral surface area

The **surface area** of a prism is the sum of the areas of the faces (including the bases). The **lateral surface area** is the total area of the sides (total surface area minus area of bases).

Example 3 Find the lateral surface area of the prism in Figure 14.100.

Solution. The lateral surface area is the total area of the sides.

(10.6 cm)(14.9 cm)	Right face
+ (15.4 cm)(14.9 cm)	Left face
+ (18.7 cm)(14.9 cm)	Back face
666 cm^2	Total

Cylinders

A **right circular cylinder** is formed by revolving a rectangle about one of its sides. (See Figure 14.101.)

As a result, the two bases of a right circular cylinder are circles, while the lateral surface is a curved surface perpendicular to the base.

As with a prism, the volume of a cylinder is found by multiplying the area of the base by the height h. Since the base is a circle, its area is πr^2. So the volume of the cylinder is $V = \pi r^2 h$.

Volume of cylinder: $V = \pi r^2 h$

where r is the radius of the base and h is the altitude

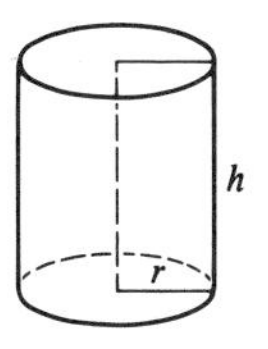

Figure 14.101

Example 4 The chemical tank in Figure 14.102 has a radius of 2.0 m. If the tank is filled to a level of 2.8 m, find the volume of the liquid.

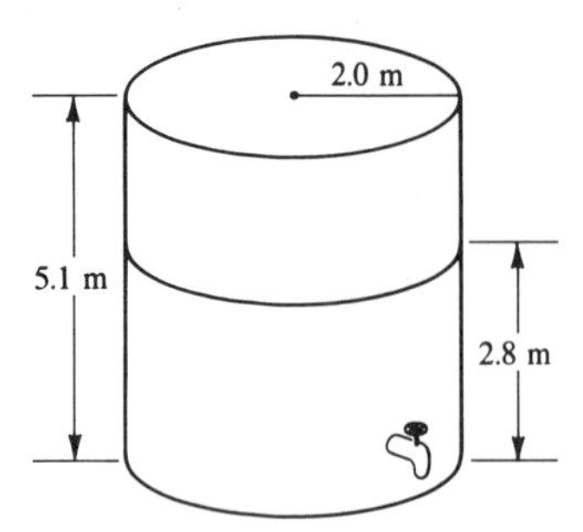

Figure 14.102

Solution. Since $r = 2.0$ m and $h = 2.8$ m, we get by the formula $V = \pi r^2 h$

$$V = \pi(2.0 \text{ m})^2(2.8 \text{ m})$$

The sequence is

$$\boxed{\pi}\ \boxed{\times}\ 2.0\ \boxed{x^2}\ \boxed{\times}\ 2.8\ \boxed{=} \rightarrow 35.185838$$

So the volume of the liquid is 35 m^3.

The lateral surface area of a cylinder is found by cutting the side vertically and making it flat, as shown in Figure 14.103. The width is thereby h and the length is equal to the circumference of the circle, $2\pi r$. Multiplying base and height of the rectangle, we get $2\pi rh$.

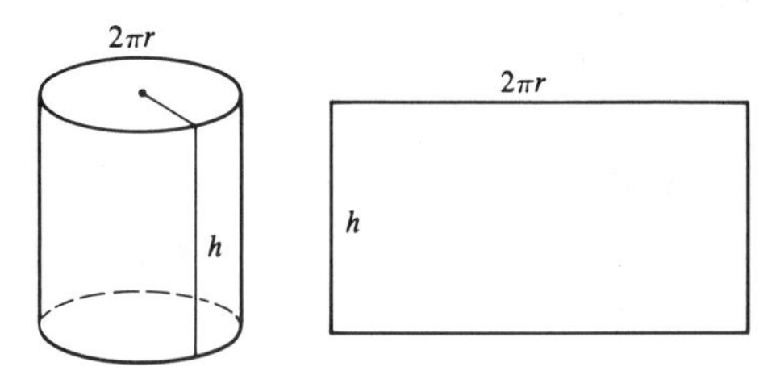

Figure 14.103

Lateral surface area of cylinder: $S = 2\pi rh$

Example 5 Find the lateral surface of the tank in Figure 14.102.

Solution. We are given that $r = 2.0$ m and $h = 5.1$ m. By the formula $S = 2\pi rh$, we have

$$S = 2\pi(2.0 \text{ m})(5.1 \text{ m}) = 64 \text{ m}^2$$

Pyramids and Cones

So far all our solids have had vertical sides. We will now study two types of solids with slanted sides and pointed tops: pyramids and cones.

Pyramid
Vertex
Altitude, Height

A **pyramid** is a solid whose base is a polygon and whose *lateral faces* are triangles that meet at a common point called the **vertex**. The **altitude** or **height** is the perpendicular distance from the vertex to the base. (See Figure 14.104.)

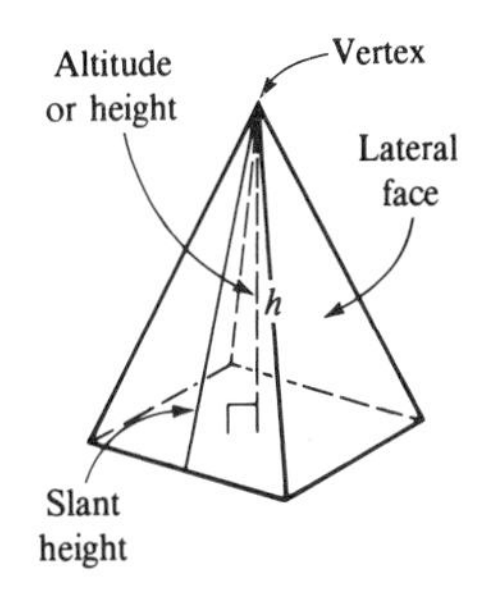

Figure 14.104

Regular pyramid
Slant height

Our discussion of volumes and lateral surface areas will be restricted to regular pyramids. A **regular pyramid** is a pyramid whose base is a regular polygon and whose lateral faces (Figure 14.104) are congruent isosceles triangles. The altitude of each of these triangles is called the **slant height** of the pyramid.

The volume of a pyramid is only one-third of the volume of a prism with the same base and altitude. (Figure 14.105.)

Volume of pyramid: $V = \frac{1}{3}Bh$

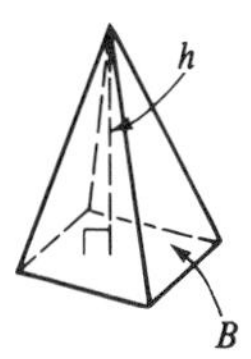

Figure 14.105

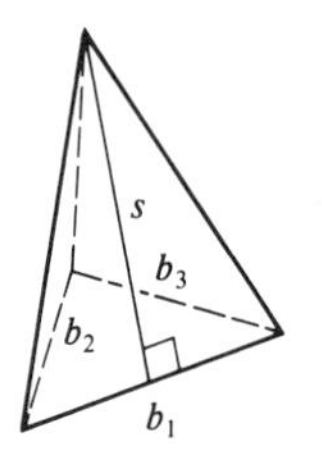

Figure 14.106

The lateral surface area is the sum of the areas of the triangles. Suppose the base is a triangle with sides b_1, b_2, and b_3, respectively. Then the lateral surface area is $\frac{1}{2}b_1s + \frac{1}{2}b_2s + \frac{1}{2}b_3s$, where s is the slant height (Figure 14.106). So if S is the lateral surface area, then

$$S = \frac{1}{2}b_1s + \frac{1}{2}b_2s + \frac{1}{2}b_3s$$

$$= \frac{1}{2}s(b_1 + b_2 + b_3) \qquad \text{Common factor } \frac{1}{2}s$$

$$= \frac{1}{2}sp = \frac{1}{2}ps \qquad p = b_1 + b_2 + b_3$$

where p is the perimeter of the base. The formula $S = \frac{1}{2}ps$ holds for any regular pyramid.

Lateral surface area of regular pyramid: $S = \dfrac{1}{2}ps$

where p is the perimeter of the base and s is the slant height.

Example 6 Find the volume and lateral surface area of the pyramid in Figure 14.107.

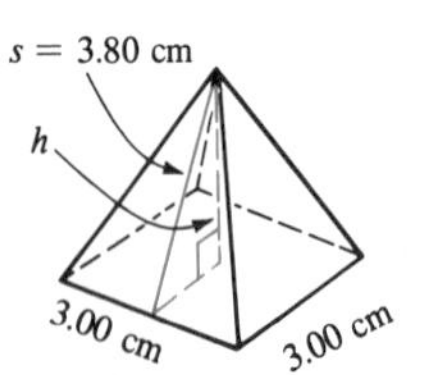

Figure 14.107

Solution. Note that the slant height is $s = 3.80$ cm. To find the height h, we use the Pythagorean theorem:

$$h^2 + (1.50)^2 = (3.80)^2 \qquad \text{or} \qquad h = 3.491 \text{ cm}$$

Since

$$B = (3.00 \text{ cm})(3.00 \text{ cm}) = 9.00 \text{ cm}^2$$

we get for the volume

$$V = \frac{1}{3}Bh = \frac{1}{3}(9.00 \text{ cm}^2)(3.491 \text{ cm})$$

$$= 10.5 \text{ cm}^3$$

To find the lateral surface area, note that the perimeter of the base is

$$p = 4(3.00 \text{ cm}) = 12.0 \text{ cm}$$

It follows that

$$S = \frac{1}{2}(12.0\text{ cm})(3.80\text{ cm}) \qquad S = \frac{1}{2}ps$$
$$= 22.8\text{ cm}^2$$

Vertex

A *cone* is similar to a pyramid, but its base is a circle rather than a polygon. More precisely, if a line through a fixed point (the **vertex**) is rotated so that it follows a circular path, we obtain a *conical surface*. If a plane cuts this surface, we obtain a solid called a *circular cone* (Figure 14.108). The line from the vertex to the center of the circle is called the *axis* of the cone. If the intersecting plane is perpendicular to the axis, the resulting solid is called a **right circular cone** (Figure 14.108).

Right circular cone

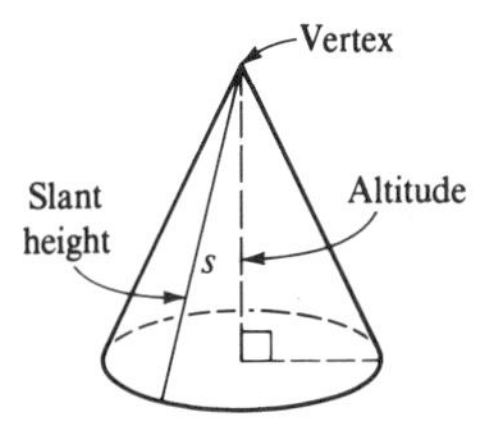

Figure 14.108

Altitude
Slant height

The **altitude** or **height** of a right circular cone is perpendicular to the circular base. The **slant height** s is drawn from the vertex to the base, as shown in Figure 14.108. (From now on, *cone* will refer to a right circular cone.)

The formulas for the volume and lateral surface area of a cone follow the same pattern as for a pyramid. In particular, $S = \frac{1}{2}ps = \frac{1}{2}(2\pi r)s = \pi rs$. (See Figure 14.109.) Also, $V = \frac{1}{3}Bh = \frac{1}{3}\pi r^2 h$.

Volume of cone: $V = \frac{1}{3}\pi r^2 h$

where h is the altitude of the cone and r the radius of the base.

Lateral surface area of cone: $S = \pi rs$

where s is the slant height of the cone and r the radius of the base.

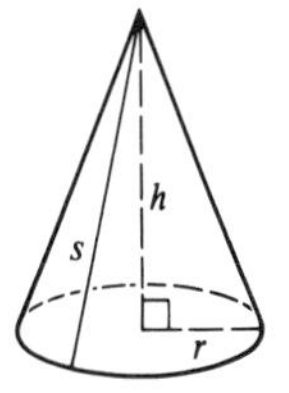

Figure 14.109

Example 7 Find the volume and lateral surface area of the cone in Figure 14.110.

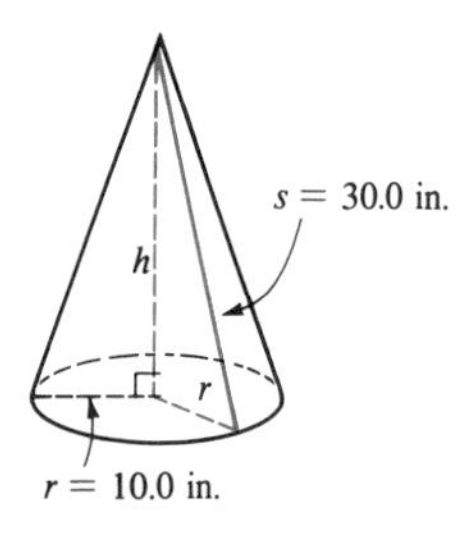

Figure 14.110

Solution. Note that the altitude, slant height, and radius form a right triangle (Figure 14.110). We can therefore use the Pythagorean theorem to find h.

$$h^2 + (10.0)^2 = (30.0)^2 \qquad \text{or} \qquad h = 28.28 \text{ in.}$$

Volume: $V = \frac{1}{3}\pi r^2 h = \frac{1}{3}\pi(10.0 \text{ in.})^2(28.28 \text{ in.}) = 2960 \text{ in.}^3$

Lateral surface area: $S = \pi r s = \pi(10.0 \text{ in.})(30.0 \text{ in.}) = 942 \text{ in.}^2$

Spheres

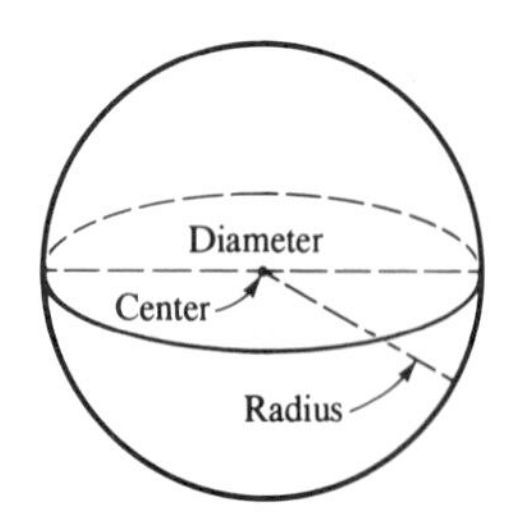

Figure 14.111

Our last geometric solid is the sphere. A **sphere** is a curved surface all points of which are equally distant from a fixed point called the **center** of the sphere. A line segment through the center with endpoints on the sphere is called a **diameter** of the sphere. A line segment from the center to the point on the sphere is called a **radius** (Figure 14.111).

A sphere can also be defined as a solid formed by rotating a circle about a diameter. Because of this connection with circles, the formulas for the volume and surface area of a sphere are expressed in terms of π (Figure 14.112).

Volume and surface area of sphere of radius r:

$$V = \frac{4}{3}\pi r^3$$

$$S = 4\pi r^2$$

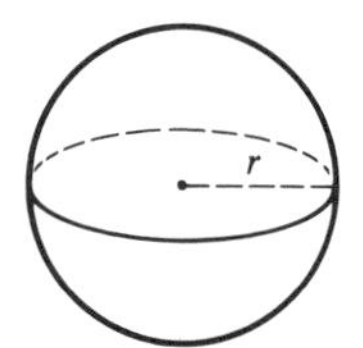

Figure 14.112

Example 8 Find the volume and surface area of a sphere of radius 2.00 m.

Solution. $$V = \frac{4}{3}\pi r^3 = \frac{4}{3}\pi(2.00\text{ m})^3 = 33.5\text{ m}^3$$

$$S = 4\pi r^2 = 4\pi(2.00\text{ m})^2 = 50.3\text{ m}^2$$

Hemisphere

The storage tank in Figure 14.113 has the shape of a cylinder with a hemisphere at each end. A **hemisphere** is half a sphere.

Example 9 Find the volume of the storage tank in Figure 14.113.

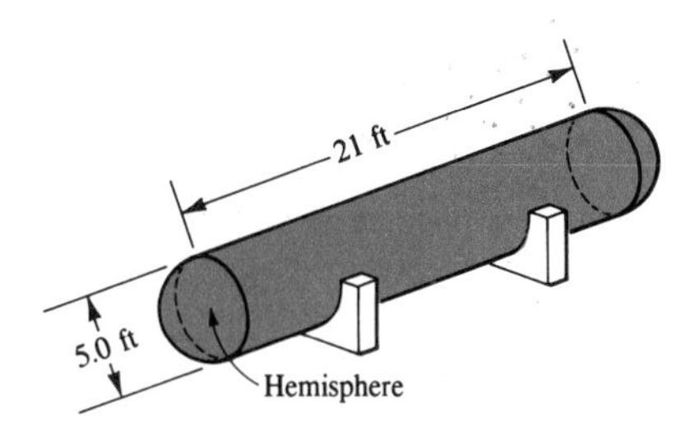

Figure 14.113

Solution. Each end of the tank is a hemisphere of radius 2.5 ft. The total volume of the two hemispheres is equal to the volume of a sphere of radius 2.5 ft, or

$$\frac{4}{3}\pi(2.5\text{ ft})^3$$

The center portion of the tank is a cylinder of radius 2.5 ft and height 21 ft. Its volume is

$$\pi(2.5\text{ ft})^2(21\text{ ft})$$

The total volume V is

$$V = \frac{4}{3}\pi(2.5\text{ ft})^3 + \pi(2.5\text{ ft})^2(21\text{ ft})$$

$$= \frac{4}{3}\pi(2.5)^3\text{ft}^3 + \pi(2.5)^2\text{ft}^2(21\text{ ft}) \qquad (ab)^n = a^n b^n$$

$$= \left[\frac{4}{3}\pi(2.5)^3 + \pi(2.5)^2(21)\right]\text{ft}^3 \qquad \text{Factoring ft}^3$$

$$= \pi(2.5)^2\left[\frac{4}{3}(2.5) + 21\right]\text{ft}^3 \qquad \text{Common factor: } \pi(2.5)^2$$

The sequence is

4 [÷] 3 [×] 2.5 [+] 21 [=] [×] 2.5 [x^2] [×] [π] [=]
→ 477.78388

So the volume of the tank is 480 ft^3.

Exercises / Section 14.5

(Use a calculator in this exercise set.)

In Exercises 1–8, find the volumes of the prisms with the given dimensions.

1. Base: rectangle with length 15 cm and width 20 cm; $h = 10$ cm
2. Base: square with side 8 cm; $h = 10$ cm
3. Base: triangle with base 3.2 in. and height 4.7 in.; $h = 1.6$ in.
4. Base: triangle with base 3.6 m and height 9.4 m; $h = 5.0$ m
5. Base: parallelogram with base 7.83 ft and height 5.52 ft; $h = 1.36$ ft
6. Base: parallelogram with base 0.36 in. and height 0.076 in.; $h = 1.34$ in.
7. Cube with side 5.041 cm
8. Cube with side 10.30 in.

In Exercises 9–14, find the lateral surface area and the total surface area of the prisms with the given dimensions.

9. Cube with edge 11.6 ft
10. Cube with edge 1.4 cm
11. Rectangular solid with length 4.5 m, width 3.4 m, and height 5.6 m
12. Rectangular solid: $l = 20$ ft, $w = 25$ ft, $h = 30$ ft
13. Base: right triangle with base 8.6 m and height 5.3 m; $h = 1.2$ m
14. Base: right triangle with base 14.5 cm and height 18.4 cm; $h = 10.3$ cm

In Exercises 15–20, find in each case the volume of the cylinder with the given radius (or diameter) and height.

15. $r = 3.5$ cm, $h = 7.6$ cm
16. $r = 10.0$ ft, $h = 3.00$ ft
17. $r = 25.4$ in., $h = 11.9$ in.
18. $r = 0.34$ yd, $h = 0.76$ yd
19. $D = 0.8640$ yd, $h = 0.9604$ yd
20. $D = 2.680$ cm, $h = 2.368$ cm

In Exercises 21–26, find in each case the lateral surface area and the total surface area of the cylinder with the given radius and height.

21. $r = 15.0$ mm, $h = 11.0$ mm
22. $r = 0.96$ yd, $h = 0.43$ yd
23. $r = 1.30$ in., $h = 2.40$ in.

24. $r = 3.76$ m, $h = 2.13$ m

25. $r = 20.0$ m, $h = 15.6$ m

26. $r = 3.460$ cm, $h = 7.340$ cm

In Exercises 27–32, find the volumes of the pyramids with the given dimensions.

27. Base: square with side 3.60 in.; $h = 1.26$ in.

28. Base: rectangle with length 10.3 in. and width 8.00 in.; $h = 2.30$ in.

29. Base: quadrilateral with $B = 20.61$ cm^2; $h = 1.306$ cm

30. Base: polygon with $B = 76.46$ in.^{2}; $h = 10.30$ in.

31. Base: triangle with base 4.76 ft and height 4.12 ft; $h = 2.36$ ft

32. Base: triangle with base 16.4 in. and height 14.2 in., $h = 12.0$ in.

In Exercises 33–36, find the lateral surface area of each given pyramid.

33. Base: polygon with $p = 16.0$ cm; $s = 9.00$ cm

34. Base: polygon with $p = 20.0$ in., $s = 12.3$ in.

35. Base: square with side 4.00 in.; $s = 2.30$ in.

36. Base: square with side 10.5 cm; $s = 11.6$ cm

In Exercises 37–42, find in each case the volume of the cone with the given radius (or diameter) of the base and the given height.

37. $r = 5.179$ ft, $h = 3.647$ ft

38. $r = 12.62$ in., $h = 13.40$ in.

39. $r = 7.6$ cm, $h = 5.2$ cm

40. $r = 3.4$ cm, $h = 2.4$ cm

41. $D = 20.6$ m, $h = 5.00$ m

42. $D = 1.52$ yd, $h = 0.210$ yd

In Exercises 43–46, find the lateral surface area of each cone.

43. $r = 10.6$ in., $s = 12.4$ in.

44. $r = 20.4$ cm, $s = 14.6$ cm

45. $r = 45.2$ cm, $h = 30.4$ cm

46. $r = 28.4$ in., $h = 5.03$ in.

In Exercises 47–52, find the volume and surface area of each sphere with the given radius (or diameter).

47. $r = 0.32$ ft

48. $r = 0.47$ yd

49. $D = 7.60$ cm

50. $D = 12.0$ cm

51. $r = 0.3680$ m

52. $r = 1.749$ in.

53. A typewriter is shipped in a box 2.1 ft long, 1.9 ft wide, and 1.2 ft high. Determine the volume.

54. How many cubic feet of concrete are needed to build a concrete patio 11 ft long, 8.0 ft wide, and 0.75 ft thick?

55. A cylindrical tank with a 1.2-m radius and 2.3-m height is filled with brine to a level of 1.5 m. Find the volume of the brine.

56. If it costs $18.50/m^2 to polish the side of the tank in Exercise 55, find the cost of polishing the side.

57. It costs $0.32/cm^2 to finish the surface of a metal bar. Determine the cost of finishing a metal bar in the shape of a rectangular solid 45.0 cm long, 2.50 cm wide, and 1.80 cm thick.

58. The density of copper is 8.90 g/cm^3. Find the mass of a copper wire 0.400 cm in diameter and 3.20 m long.

59. A paperweight made of copper has the shape of a pyramid with a square base 8.13 cm on the side and 10.1 cm high. Determine its mass. (Density of copper: 8.90 g/cm^3)

60. Find the weight of a steel bar with a radius of 0.750 in. and a length of 20.0 in. (Steel weighs 0.283 lb/in.^{3}.)

61. A glass prism, used in studying the refraction of light, has a length of 10.0 cm and a base in the shape of an equilateral triangle 3.50 cm on the side. Find the volume of the prism.

62. The Great Pyramid of Cheops has a square base 755 ft on the side and a height of 481 ft. Determine the lateral surface area.

63. The Great Pyramid of Egypt has a square base 762 ft on the side and a height of 484 ft. What is its volume?

64. A cone has a base diameter of 4.62 cm and a slant height of 3.00 cm. Find the volume.

65. A tank full of water is a cone with vertex down. The tank measures 6.0 ft across the top and is 5.0 ft deep in the center. Find the weight of the water in the tank (62.4 lb/ft^3).

66. A cylindrical tank has a diameter of 8.3 ft. It is filled with water to a level of 3.6 ft. Determine the weight of the water (62.4 lb/ft^3).

67. A conical tank with vertex down measures 6.4 ft across the top. Its slant height is 4.6 ft. Find the cost of rust-proofing the outside surface at $0.85/ft^2.

68. A paper cup in the shape of a cone measures 3.6 in. across the top and is 3.4 in. high. What is the surface area of the cup?

69. Find the mass of a gold cone, given that the radius of the base is 8.00 mm and the height 1.00 cm. (Density of gold: 19.3 g/cm^3)

70. Find the weight of a gold sphere of radius 0.340 in. (Gold weighs 0.697 lb/in.^{3}.)

71. Given that the radius of the earth is 3960 mi, determine the volume and surface area of the earth.

72. The diameter of the moon is 1080 km. Find the volume and surface area of the moon.

73. A storage tank has the shape of a cylinder with a hemisphere at each end. (See Example 9.) If the diameter of the cylinder is 6.0 ft and the total length of the tank is 15 ft, find the volume and surface area.

74. A cylindrical silo with a hemispherical top has a diameter of 12 ft and a height of 36 ft. Find the volume and surface area.

75. A hemispherical tank with 5.60-ft diameter is full of water. Find the weight of the water (62.4 lb/ft^3).

76. A cast-iron ball has a diameter of 2.0 ft. Determine the weight of the ball, given that cast iron weighs 0.26 lb/in^3.

77. Assume that it costs $0.85/ft^2 to finish a metal surface. Determine the cost of finishing the surface of a metal ball with a 3.0-ft radius.

Review Exercises / Chapter 14

In Exercises 1–6, refer to Figure 14.114.

Identify the following:

1. An acute angle
2. An obtuse angle
3. A right angle
4. A pair of adjacent angles
5. A pair of complementary angles
6. A pair of supplementary angles

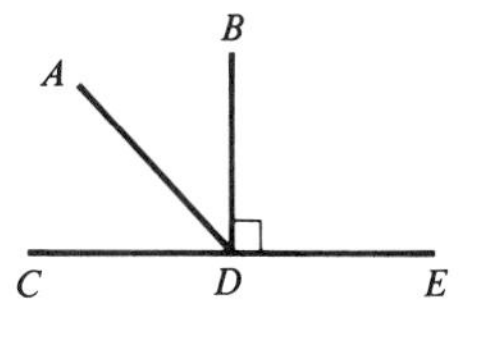

Figure 14.114

In Exercises 7–12, refer to Figure 14.115.

Identify or evaluate the following:

7. A pair of alternate interior angles
8. A pair of alternate exterior angles
9. A pair of corresponding angles
10. A pair of vertical angles
11. $\angle 1 + \angle 3$
12. $\angle 2 + \angle 3$

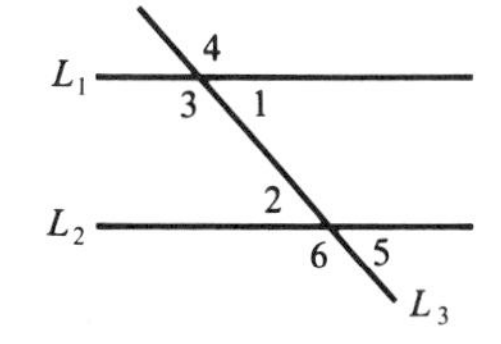

Figure 14.115

13. Explain why the quadrilateral in Figure 14.116 is a parallelogram.

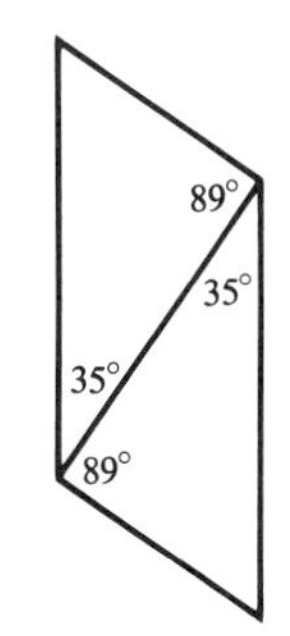

Figure 14.116

14. In Figure 14.117, $DE \parallel AC$. Find $\angle D$ and $\angle E$.

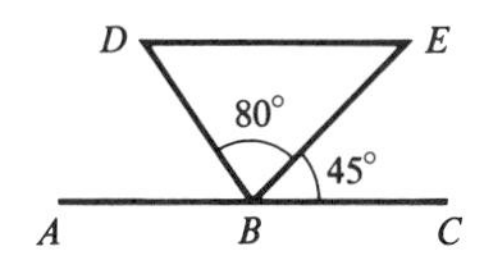

Figure 14.117

15. Find all the angles of the parallelogram in Figure 14.118.

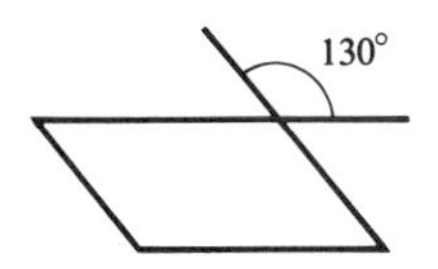

Figure 14.118

16. In Figure 14.119, determine:
a. The degree measure of $\widehat{AB}$
b. $\angle C$

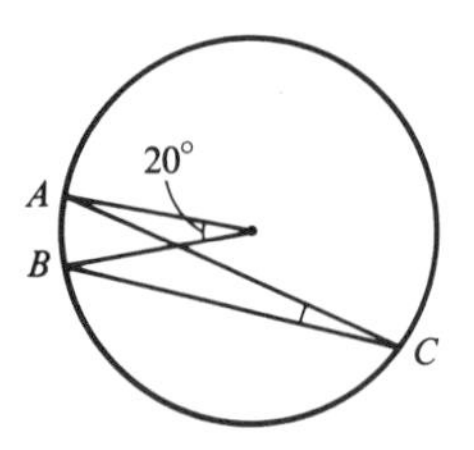

Figure 14.119

17. The circle in Figure 14.120 has a diameter of 10 cm. Given that $AC = BC$, find AC.

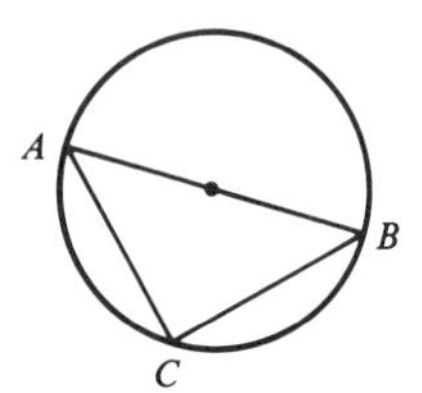

Figure 14.120

18. The circle in Figure 14.121 has a radius of 5.00 cm. If $AB = 9.00$ cm, find OB.

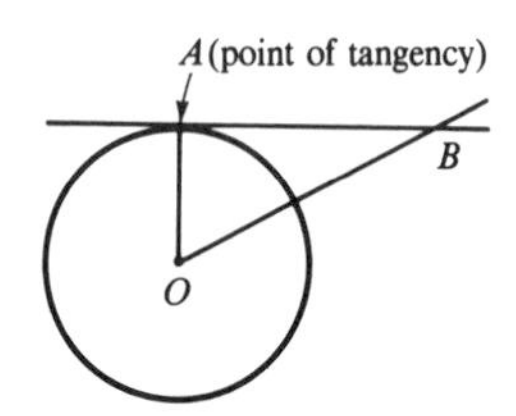

Figure 14.121

19. In Figure 14.122, ΔABC is an isosceles triangle. Show that $\Delta ACD \cong \Delta BCD$.

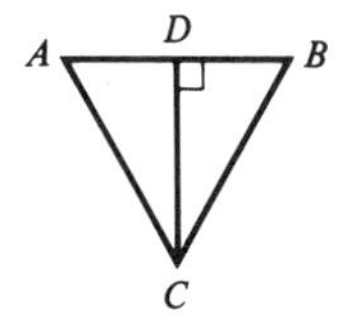

Figure 14.122

20. Find x in Figure 14.123.

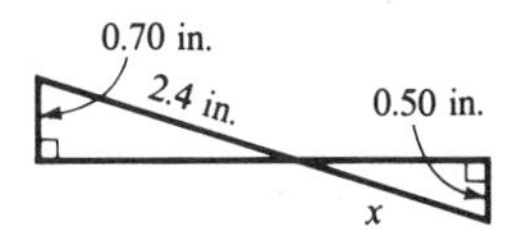

Figure 14.123

In Exercises 21–27, find the volumes of the indicated solids.

21. Cube of side 0.96 m

22. Prism: base of prism is a parallelogram with base 0.34 yd and height 0.79 yd; altitude of prism: 1.2 yd

23. Pyramid: $B = 10.4\text{ cm}^2$, $h = 5.76$ cm

24. Cylinder: diameter 14.6 ft, height 8.86 ft

25. Cone: radius of base 3.9 cm, height 0.64 cm

26. Sphere: radius 8.4 in.

27. Rectangular solid: square base 15.6 cm on the side and height 11.4 cm

In Exercises 28–30, find the total surface areas of the indicated solids.

28. Rectangular solid: base 4.20 in. × 3.80 in., height 2.75 in.

29. Prism: base of prism is a right triangle with base 3.4 cm and height 4.8 cm; altitude of prism: 5.4 cm

30. Sphere: radius 3.5 cm

In Exercises 31–33, find the lateral surface area of each solid.

31. Cylinder: radius 10.4 ft, altitude 3.80 ft

32. Cone: diameter of base 15.4 m, altitude 6.40 m

33. Pyramid: base of pyramid is a square 2.0 m on the side; slant height of pyramid: 12 m

34. $\angle A$ and $\angle B$ are complementary angles that differ by 15°. Find the angles.

35. One angle of a quadrilateral measures 63°. Find the sum of the remaining angles.

36. Find the diagonal distance from two opposite vertices of a rectangular machine part measuring 3.40 cm × 2.79 cm.

37. A ship sails 30.0 km due west. How far north must it travel to be 60.0 km from the starting point?

38. A roof truss has a 13.0-ft rafter and a span of 20.0 ft. Find the height of the truss in the center.

39. The base of a 6.0-m ladder is 2.8 m from the wall. How far does the ladder reach up on the wall?

40. A blueprint of a machine has a scale of $1\frac{1}{4}$ in. = 2 ft. Find the actual length of a shaft whose length on the blueprint is $\frac{7}{8}$ in.

41. A tree casts a shadow 22.0 ft long. At the same time, a stick 2.00 ft in length casts a shadow 1.85 ft long. How high is the tree?

42. Find the length x of the support in Figure 14.124.

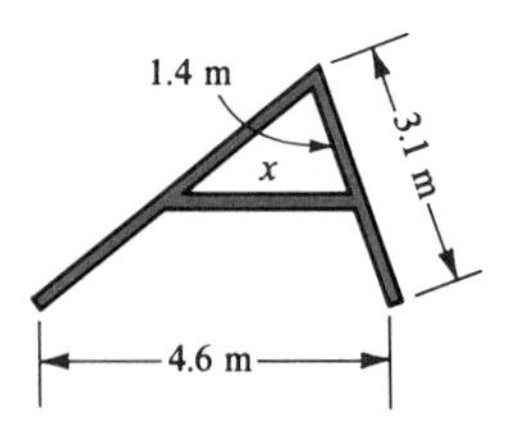

Figure 14.124

43. A sandpile in the shape of a cone measures 39.6 cm across the base and is 10.3 cm high. What is its volume?

44. A paperweight in the shape of a cylinder is made from copper (0.32 lb/in.^{3}). Determine its weight, given that the altitude of the cylinder is 1.1 in. and the diameter of the base is 2.5 in.

45. Find the weight of a steel ball of diameter 16.0 in. (Steel weighs 0.283 lb/in.^{3}.)

46. Determine the weight of the water in a full conical tank 3.0 ft across the top and 3.7 ft deep in the center. (Water weighs 62.4 lb/ft^3.)

47. The Great Pyramid of Egypt has a square base 762 ft on the side and a height of 484 ft. Determine its surface area.

48. A silo is a cylinder with a conical top. Find the volume and surface area of the silo, given that the radius is 2.5 m and the total height 15 m. The top 2.5 m is the height of the cone.

49. The diameter of the sun is 860,000 mi. Find the volume and surface area of the sun.

50. Determine the cost of rust-proofing the outside surface of a conical tank measuring 3.0 m across the top and having a slant height of 2.6 m, given that the cost per square meter is $5.25.

51. A cylindrical water heater has a diameter of 0.85 m and a height of 2.1 m. How many square meters of insulation are required for the side and top?

52. Determine the weight of a hollow cast-iron sphere with outer radius 19 in. and inner radius 18 in. (Cast iron weighs 0.26 lb/in.^{3}.)

CHAPTER 15

Trigonometry

The literal meaning of *trigonometry* is "triangle measurement." Many problems in science and technology involve triangles. Trigonometry is a powerful tool for solving such problems.

15.1 Angles

Before we can introduce the basic concepts of trigonometry, we need to recall the definition of an angle and then combine the concept of an angle with that of a rectangular coordinate system.

Standard Position of an Angle

In Section 6.1 we defined an **angle** as a geometric figure consisting of two rays with a common endpoint. In trigonometry, angle measurement is of primary importance. So an angle is better defined in terms of rotation: We start with a single ray and rotate this ray about its endpoint. The original position is called the **initial side** of the angle and the new position is called the **terminal side,** as shown in Figure 15.1. The common endpoint is called the **vertex.**

Initial side
Terminal side
Vertex

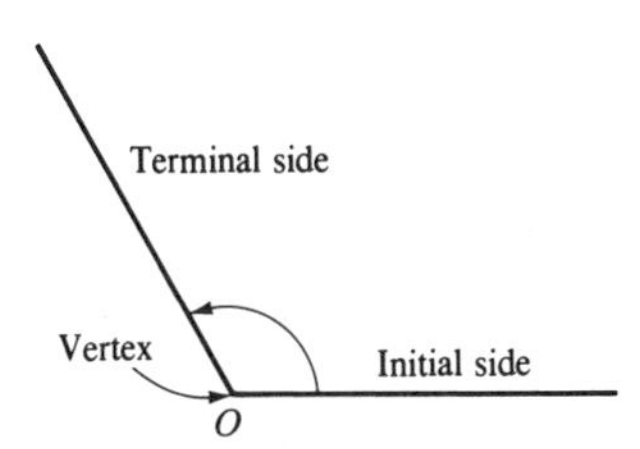

Figure 15.1

Standard position

Using the rectangular coordinate system, an angle can be placed in **standard position.** (Figure 15.2).

Standard position of an angle

An angle is said to be in **standard position** if its vertex is at the origin and its initial side is on the positive x-axis. (The angle is designated by a curved arrow.)

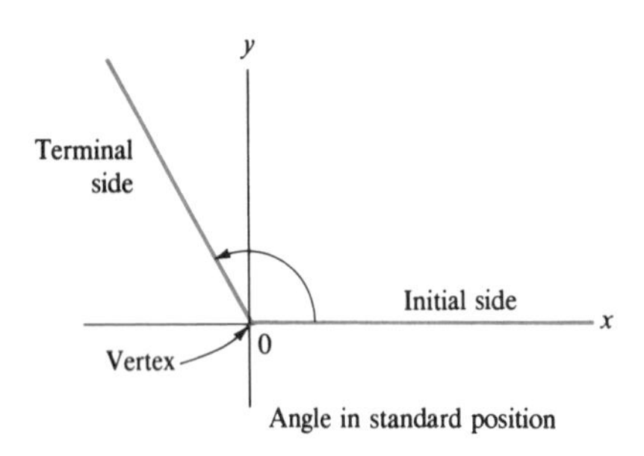

Figure 15.2

Once in standard position, an angle can be described by specifying one point on the terminal side.

Angles are frequently named by use of Greek letters. Particularly common are θ (theta), α (alpha), and β (beta).

Example 1 Draw the angle θ in standard position, given that the point $(-2, 1)$ lies on its terminal side.

Solution. We plot the point $(-2, 1)$ on the rectangular coordinate system. The terminal side of θ is a ray from the origin through $(-2, 1)$, as shown in Figure 15.3. Note that the letter θ is placed next to the curved arrow.

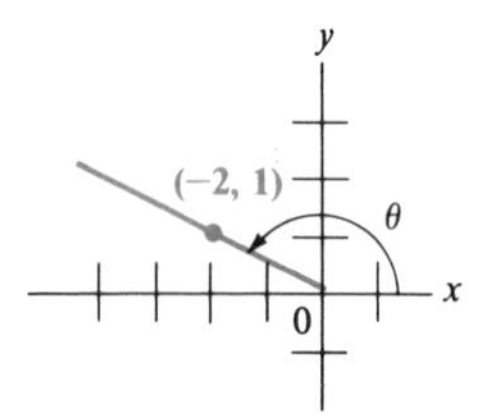

Figure 15.3

Since an angle is defined in terms of rotation, there is no limit to its size. For example, $\theta = 380°$ is $20°$ more than a complete rotation ($360°$).

Example 2 Draw the angle $\theta = 400°$ in standard position.

Solution. Note that $\theta = 400°$ is $40°$ more than one complete rotation ($360°$). This angle is indicated by the curved arrow in Figure 15.4.

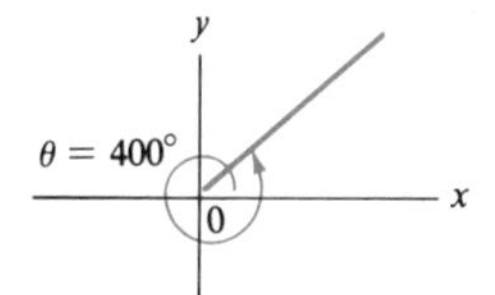

Figure 15.4

Quadrantal angle

If the terminal side of an angle lies on one of the coordinate axes, it is called a **quadrantal angle.** For example, the angle θ whose terminal side passes through $(0, -1)$ is quadrantal. (See Figure 15.5.) Note that $\theta = 270°$.

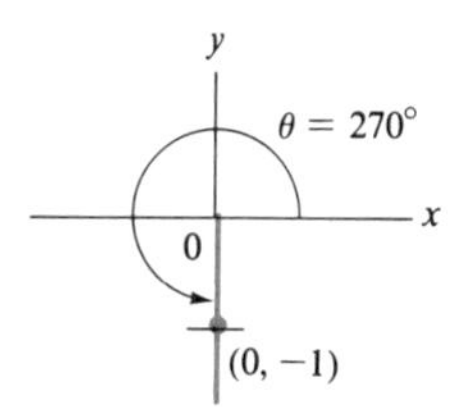

Figure 15.5

Similarly, the angle θ whose terminal side passes through $(-2, 0)$ is quadrantal (Figure 15.6).

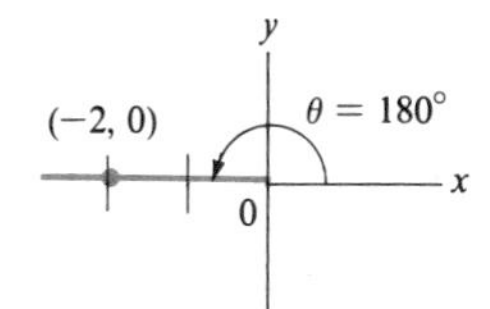

Figure 15.6

Positive and Negative Angles

Positive and negative angles

Angles in standard position have algebraic signs: If the rotation is **counterclockwise,** the angle is considered **positive.** If the rotation is **clockwise,** the angle is considered **negative.** Study the next example.

Example 3 Draw the following angles in standard position:

a. $\theta = 240°$ **b.** $\alpha = -120°$

Solution.
a. Since $\theta = 240°$ is positive, the rotation (from the initial side) is counterclockwise (Figure 15.7).
b. Since $\alpha = -120°$ is negative, the rotation is clockwise (Figure 15.8).

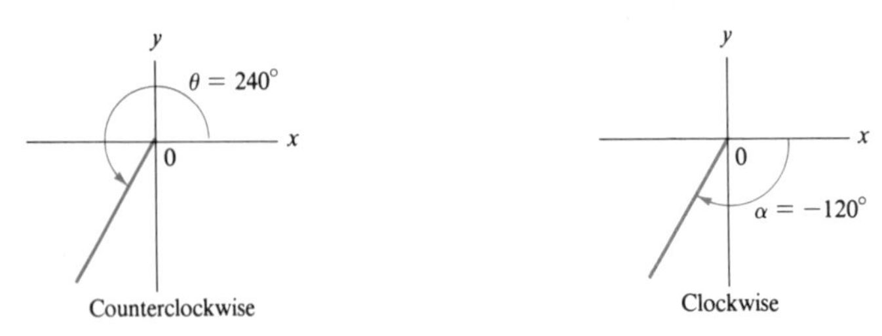

Figure 15.7 **Figure 15.8**

Coterminal Angles

In Example 3 we considered two angles, $\theta = 240°$ and $\alpha = -120°$ (shown in Figures 15.7 and 15.8, respectively). Now, since a positive number cannot be equal to a negative number, the angles are completely different. Yet the terminal sides of the angles *are* the same. Such angles are said to be **coterminal.**

> Two angles in standard position are **coterminal** if they have the same terminal side.

Example 4 Find the smallest positive angle α so that α and $\theta = -32°51'$ are coterminal.

Solution. To obtain α, we subtract $32°51'$ from $360°$. To be able to do so, we write

$$360° \quad \text{as} \quad 359°60'$$

We now get

$$\begin{array}{rl} 359°60' & \\ -\ \ 32°51' & \quad \theta = 32°51' \\ \hline \alpha = \mathbf{327°\ 9'} & \quad \text{Subtracting} \end{array}$$

The angles are shown in Figure 15.9.

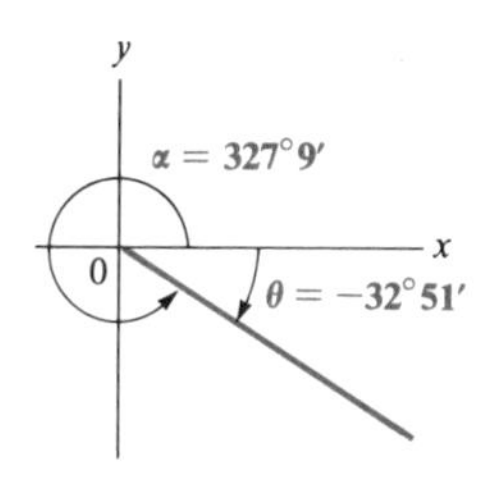

Figure 15.9

Exercises / Section 15.1

In Exercises 1–8, draw the angle θ in standard position so that its terminal side passes through the given point. (See Example 1.)

1. (1, 3) **2.** (−2, 4) **3.** (−3, −1) **4.** (2, −3) **5.** (−6, 2)

6. (−7, −2) **7.** (2, −5) **8.** (3, 2)

In Exercises 9–20, draw the given angles in standard position. (See Examples 2 and 3.)

9. 370° **10.** 410° **11.** 512° **12.** 670° **13.** −45°

14. −60° **15.** −150° **16.** −270° **17.** −90° **18.** −180°

19. −350° **20.** −370°

In Exercises 21–50, find the smallest positive angle coterminal with the given angle. (See Example 4.)

21. 421°16′ **22.** 542°33′ **23.** 624°41′ **24.** 605°52′ **25.** 830°20′

26. 790°24′ **27.** −150° **28.** −250° **29.** −20°40′ **30.** −40°19′

31. −50°10′ **32.** −35°8′ **33.** −92°48′ **34.** −100°11′ **35.** −115°37′

36. −119°22′ **37.** −172°34′ **38.** −210°42′ **39.** −230°5′ **40.** −310°14′

41. −340°45′ **42.** −328°38′ **43.** −365°1′ **44.** −380°56′ **45.** −182°55′

46. −170°26′ **47.** −280°23′ **48.** −18°50′ **49.** −600° **50.** −800°

15.2 The Trigonometric Functions

The study of trigonometry, as well as the applications of trigonometry, are based on the ratios of two sides of a right triangle. These ratios are used to define the trigonometric functions of an angle.

The Trigonometric Ratios

The definitions of the trigonometric ratios are based on the properties of similar triangles. Recall from Section 14.4 that the corresponding sides of two similar triangles are proportional. For example, given that the triangles in Figure 15.10 are similar, we have

$$\frac{AB}{A'B'} = \frac{AC}{A'C'} = \frac{BC}{B'C'}$$

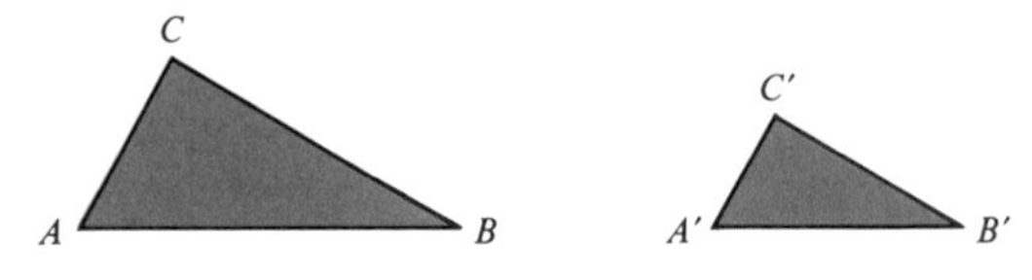

Figure 15.10

Now consider the right triangles in Figure 15.11. Since the corresponding sides are proportional, the triangles are similar. As a result, the angles denoted by θ in the two triangles are equal. Consider next the ratio of the side opposite angle θ to the hypotenuse in each triangle:

$$\frac{3 \text{ in.}}{5 \text{ in.}} = \frac{3}{5} \quad \text{and} \quad \frac{6 \text{ in.}}{10 \text{ in.}} = \frac{6}{10} = \frac{3}{5}$$

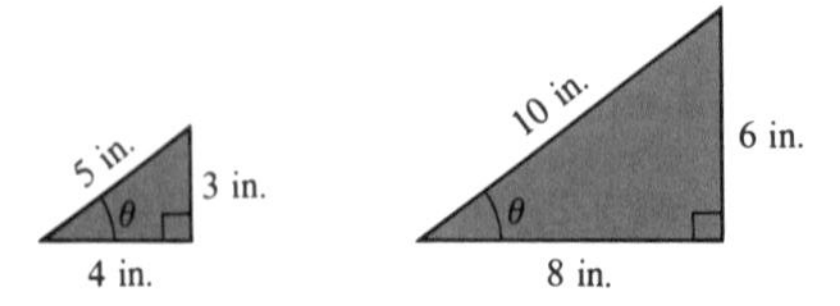

Figure 15.11

The ratios are equal.

Similarly, the respective ratios of the opposite side to the bottom side are

$$\frac{3 \text{ in.}}{4 \text{ in.}} = \frac{3}{4} \quad \text{and} \quad \frac{6 \text{ in.}}{8 \text{ in.}} = \frac{6}{8} = \frac{3}{4}$$

Again, the ratios are equal.

It is true in general that *the ratio of two sides of a triangle is equal to the ratio of the corresponding sides of any similar triangle.*

Now start with an acute angle θ and construct the right triangles in Figure 15.12. Since any two such triangles are similar, the ratio of the side opposite angle θ to the hypotenuse is the same regardless of the triangle chosen; that is,

$$\frac{BC}{AC} = \frac{DE}{AE} = \frac{FG}{AG}$$

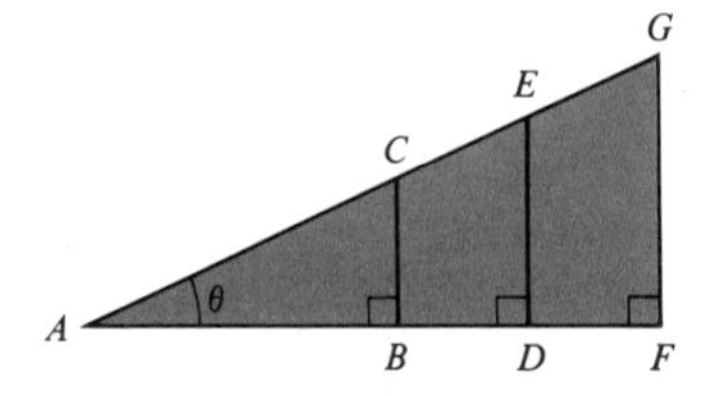

Figure 15.12

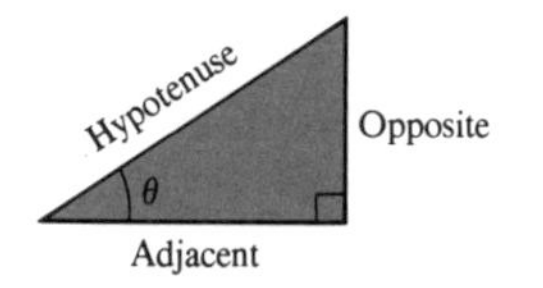

Figure 15.13

So for any given angle θ (Figure 15.13), the ratio of the side opposite θ to the hypotenuse is a unique number. The number does *not* depend on the size of the triangle—it depends only on the angle. The ratio is called the **sine of θ**, abbreviated **sin θ**. We may now say

$$\sin \theta = \frac{\text{side opposite angle } \theta}{\text{hypotenuse}}$$

The sine of θ is only one of six possible ratios. Referring to Figure 15.13, we get another unique number by taking the ratio of the side *opposite* θ to the side *adjacent* to θ. (See Figure 15.13.) This ratio is called the **tangent of θ**, abbreviated **tan θ**. Thus

$$\tan \theta = \frac{\text{side opposite angle } \theta}{\text{side adjacent to angle } \theta}$$

The names of the remaining four ratios are given next.

The Trigonometric Functions

Using the descriptive labels in Figure 15.14, we can list six possible ratios. These ratios are called the six **trigonometric functions.** They are given in the following table:

Trigonometric functions

Name	*Abbreviation*	*Value*
sine of θ	$\sin \theta$	$\sin \theta = \dfrac{\text{opposite}}{\text{hypotenuse}}$
cosine of θ	$\cos \theta$	$\cos \theta = \dfrac{\text{adjacent}}{\text{hypotenuse}}$
tangent of θ	$\tan \theta$	$\tan \theta = \dfrac{\text{opposite}}{\text{adjacent}}$
cosecant of θ	$\csc \theta$	$\csc \theta = \dfrac{\text{hypotenuse}}{\text{opposite}}$
secant of θ	$\sec \theta$	$\sec \theta = \dfrac{\text{hypotenuse}}{\text{adjacent}}$
cotangent of θ	$\cot \theta$	$\cot \theta = \dfrac{\text{adjacent}}{\text{opposite}}$

Hypotenuse
Opposite
θ
Adjacent

Figure 15.14

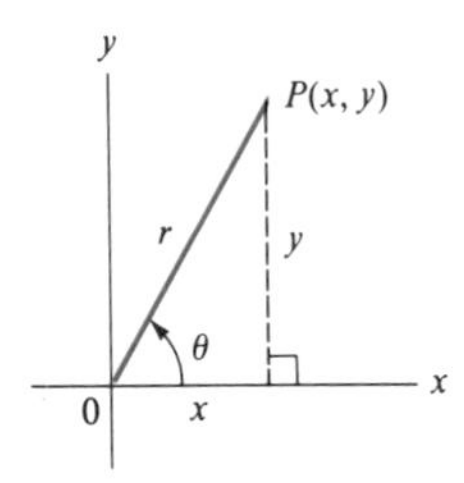

Figure 15.15

(The word *function* is used because for every angle θ the corresponding ratio is a unique number. For example, for $y = \sin \theta$, every value of the independent variable θ yields a unique value of the dependent variable y; so $y = \sin \theta$ is a function. We will study the graphs of the sine and cosine functions in Chapter 17.)

The trigonometric functions can also be stated for angles in standard position. (See Figure 15.15.) Consider the acute angle θ whose terminal side passes through the point $P(x, y)$. Drop a perpendicular from $P(x, y)$ to the x-axis. To find the six ratios, we need to compute the length r of the hypotenuse. By the Pythagorean theorem, $r^2 = x^2 + y^2$, or

Radius vector

$$r = \sqrt{x^2 + y^2}$$

The distance r from the origin to P is called the **radius vector.**

Using the point $P(x, y)$ and the radius vector r (Figure 15.16), the trigonometric functions can also be stated in the form given next.

Trigonometric functions

$$\sin \theta = \frac{y}{r} \qquad \csc \theta = \frac{r}{y}$$

$$\cos \theta = \frac{x}{r} \qquad \sec \theta = \frac{r}{x}$$

$$\tan \theta = \frac{y}{x} \qquad \cot \theta = \frac{x}{y}$$

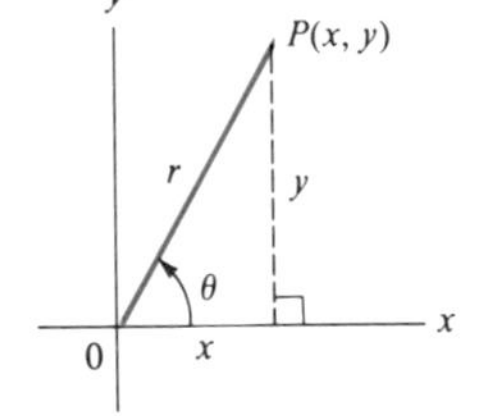

Figure 15.16

In the next section we will learn how to find the values of the functions of a given angle. In this section we merely familiarize ourselves with these definitions by means of some appropriate examples and exercises.

Example 1 Find $\sin \theta$ and $\cos \theta$, given that θ is an angle in standard position whose terminal side passes through the point (3, 4).

Solution. Drop a perpendicular from the point **(3, 4)** to the x-axis, as shown in Figure 15.17. Placing the numbers **3** and **4** along the sides of the resulting triangle, we get from the Pythagorean theorem

$$r^2 = 3^2 + 4^2 = 9 + 16 = 25$$

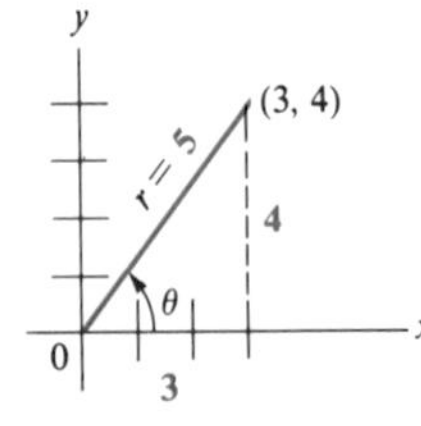

Figure 15.17

Taking the positive square root, we get $r = \sqrt{25} = 5$. It now follows from the definitions that

$$\sin \theta = \frac{y}{r} = \frac{4}{5} \quad \text{and} \quad \cos \theta = \frac{x}{r} = \frac{3}{5}$$

or

$$\sin\theta = \frac{\text{opposite}}{\text{hypotenuse}} = \frac{4}{5}$$

and

$$\cos\theta = \frac{\text{adjacent}}{\text{hypotenuse}} = \frac{3}{5}$$

We saw in our study of the Pythagorean theorem that the lengths of the sides of a triangle often involve square roots. For problems containing approximate numbers, such as physical measurements, the square roots are converted to decimals and rounded off to the proper number of significant digits. If the given lengths are exact numbers, then the results are left in radical form.

Example 2 Find the values of the six trigonometric functions of the angle θ in standard position, given that its terminal side passes through the point $(2, \sqrt{5})$.

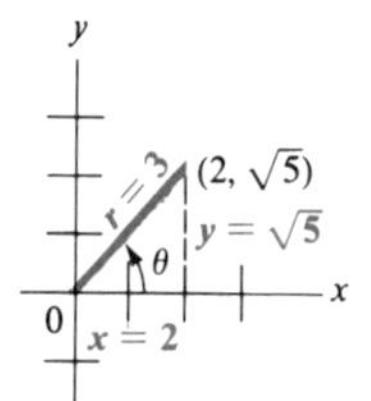

Figure 15.18

Solution. Draw Figure 15.18. By the Pythagorean theorem,

$$r = \sqrt{2^2 + (\sqrt{5})^2} = \sqrt{4+5} = \sqrt{9} = 3$$

We now get

$$\sin\theta = \frac{y}{r} = \frac{\sqrt{5}}{3} \qquad \csc\theta = \frac{r}{y} = \frac{3}{\sqrt{5}}$$

$$\cos\theta = \frac{x}{r} = \frac{2}{3} \qquad \sec\theta = \frac{r}{x} = \frac{3}{2}$$

$$\tan\theta = \frac{y}{x} = \frac{\sqrt{5}}{2} \qquad \cot\theta = \frac{x}{y} = \frac{2}{\sqrt{5}}$$

Since the given numbers are exact, the answers should be left in radical form. Moreover, as we saw in Chapter 10, it is customary to rationalize denominators. Thus

$$\csc\theta = \frac{3}{\sqrt{5}} = \frac{3}{\sqrt{5}} \cdot \frac{\sqrt{5}}{\sqrt{5}} = \frac{3\sqrt{5}}{(\sqrt{5})^2} = \frac{3\sqrt{5}}{5}$$

and

$$\cot\theta = \frac{2}{\sqrt{5}} = \frac{2}{\sqrt{5}} \cdot \frac{\sqrt{5}}{\sqrt{5}} = \frac{2\sqrt{5}}{(\sqrt{5})^2} = \frac{2\sqrt{5}}{5}$$

The next example involves approximate numbers, so the answers should be given in decimal form.

Example 3 Find $\cos\theta$ and $\csc\theta$ for θ in Figure 15.19.

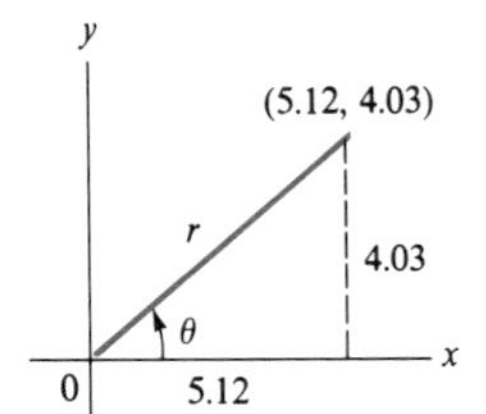

Figure 15.19

Solution. The radius vector r is

$$r = \sqrt{(4.03)^2 + (5.12)^2}$$

The sequence is

4.03 [x²] [+] 5.12 [x²] [=] [√] → 6.5157732

Carrying one extra digit, let

$$r = 6.516 \qquad \text{Four significant digits}$$

Then

$$\cos\theta = \frac{5.12}{6.516} = 0.786 \qquad \text{Rounding off to three significant digits}$$

$$\csc\theta = \frac{6.516}{4.03} = 1.62$$

Remark. To avoid rounding off in the intermediate calculations, it is best to store the value of r from the calculator sequence in the memory and press [MR] whenever the value is needed.

$\cos\theta$: 5.12 [÷] [MR] [=] → 0.7857854
$\csc\theta$: [MR] [÷] 4.03 [=] → 1.6168172

If the value of one trigonometric function of an angle is known, the values of the remaining functions of this angle can be found, as illustrated in the next example.

Example 4 Given that $\sec\theta = \frac{7}{3}$, find

$$\cot\theta \qquad \text{and} \qquad \sin\theta$$

Solution. Since

$$\sec\theta = \frac{\text{hypotenuse}}{\text{adjacent}}$$

we need to draw our figure with 7 along the hypotenuse and 3 along the adjacent side (Figure 15.20). Now denote the opposite side by y and use the Pythagorean theorem:

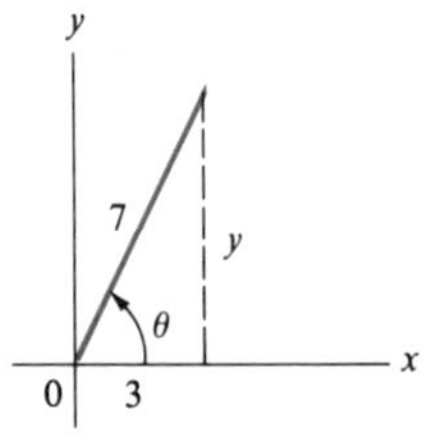

Figure 15.20

$$y^2 + 3^2 = 7^2 \qquad x = 3,\ r = 7$$

$$y^2 + 9 = 49$$

$$y^2 = 49 - 9 \qquad \text{Subtracting 9}$$

$$y^2 = 40$$

$$y = \sqrt{40} = \sqrt{4 \cdot 10} = 2\sqrt{10}$$

We get

$$\cot\theta = \frac{\text{adjacent}}{\text{opposite}}$$
$$= \frac{3}{2\sqrt{10}} = \frac{3}{2\sqrt{10}}\frac{\sqrt{10}}{\sqrt{10}}$$
$$= \frac{3\sqrt{10}}{2(\sqrt{10})^2} = \frac{3\sqrt{10}}{2 \cdot 10} = \frac{3\sqrt{10}}{20}$$

and

$$\sin\theta = \frac{\text{opposite}}{\text{hypotenuse}} = \frac{2\sqrt{10}}{7}$$

Reciprocal Relations

It follows directly from the definitions of the trigonometric functions that certain ratios are reciprocals. For example, since

$$\sin\theta = \frac{y}{r} \qquad \text{and} \qquad \csc\theta = \frac{r}{y}$$

it follows that $\sin\theta = 1/\csc\theta$. The reciprocal relations are stated next.

Reciprocal relations

$$\sin\theta = \frac{1}{\csc\theta} \qquad \cos\theta = \frac{1}{\sec\theta} \qquad \tan\theta = \frac{1}{\cot\theta}$$

Example 5

If $\sin\theta = \frac{1}{3}$, then $\csc\theta = \frac{1}{1/3} = 3$.

If $\cos\theta = \frac{2}{5}$, then $\sec\theta = \frac{1}{2/5} = \frac{5}{2}$.

If $\cot\theta = \frac{7}{4}$, then $\tan\theta = \frac{1}{7/4} = \frac{4}{7}$.

Exercises / Section 15.2

In Exercises 1–10, find the values of the indicated trigonometric functions for angles in standard position and with the given point on the terminal side. Express the answer in exact (nondecimal) form. (See Example 1.)

1. (4, 3); find $\sin\theta$ and $\tan\theta$

2. (12, 5); find $\cos\theta$ and $\cot\theta$

3. (5, 12); find $\sec\theta$ and $\cot\theta$

4. (3, 4); find $\csc\theta$ and $\tan\theta$

5. (2, 1); find $\csc\theta$ and $\sec\theta$

6. (3, 2); find $\cot\theta$ and $\sec\theta$

7. (4, 1); find $\sin\theta$ and $\cos\theta$

8. (3, 5); find $\sin\theta$ and $\cos\theta$

9. $(\sqrt{3}, \sqrt{6})$; find $\sec\theta$ and $\csc\theta$

10. $(\sqrt{5}, 2)$; find $\sec\theta$ and $\tan\theta$

In Exercises 11–16, find the values of the indicated trigonometric functions for angles in standard position and with the given point on the terminal side. Express answers in decimal form. (See Example 3.)

11. (1.0, 3.0); find $\sin\theta$, $\cos\theta$

12. (2.0, 1.0); find $\cos\theta$, $\cot\theta$

13. (3.20, 1.60); find $\sec\theta$, $\cot\theta$

14. (4.60, 3.90); find $\sin\theta$, $\sec\theta$

15. (4.23, 7.00); find $\csc\theta$, $\cos\theta$

16. (2.60, 6.49); find $\tan\theta$, $\csc\theta$

In Exercises 17–24, find the values of the six trigonometric functions for angles in standard position and with the given point on the terminal side. Express answers in exact (nondecimal) form. (See Example 2.)

17. (4, 2)

18. (8, 2)

19. $(\sqrt{2}, 4)$

20. $(5, \sqrt{3})$

21. $(\sqrt{3}, 3)$

22. $(\sqrt{11}, 1)$

23. $(\sqrt{5}, \sqrt{7})$

24. $(\sqrt{3}, \sqrt{5})$

In Exercises 25–30, find the values of the six trigonometric functions for angles in standard position and with the given point on the terminal side. Express answers in decimal form.

25. (3.2, 4.8)

26. (5.9, 3.6)

27. (7.50, 3.90)

28. (10.0, 5.37)

29. (7.43, 11.4)

30. (12.1, 14.6)

In Exercises 31–42, write the answers in exact (nondecimal) form. (See Example 4.)

31. Given $\sin\theta = \frac{1}{2}$, find $\cos\theta$.

32. Given $\cos\theta = \frac{1}{2}$, find $\sin\theta$.

33. Given $\tan\theta = \frac{3}{2}$, find $\sec\theta$.

34. Given $\cot\theta = \frac{3}{2}$, find $\csc\theta$.

35. Given $\tan\theta = 3$, find $\sec\theta$. $\left(\textit{Note: } \tan\theta = \frac{3}{1}.\right)$

36. Given $\sec\theta = 3$, find $\tan\theta$.

37. Given $\csc\theta = 3$, find $\cos\theta$.

38. Given $\cos\theta = \frac{\sqrt{3}}{3}$, find $\tan\theta$.

39. Given $\sin\theta = \frac{\sqrt{6}}{4}$, find $\sec\theta$.

40. Given $\cot\theta = \frac{\sqrt{5}}{2}$, find $\sec\theta$.

41. Given $\cos\theta = \frac{\sqrt{13}}{4}$, find $\csc\theta$.

42. Given $\sec\theta = 2\sqrt{2}$, find $\tan\theta$.

In Exercises 43–52, write the answers in decimal form.

43. Given $\sin\theta = 0.236$, find $\tan\theta$.

44. Given $\cos\theta = 0.786$, find $\sin\theta$.

45. Given $\tan\theta = 3.16$, find $\cos\theta$.

46. Given $\cot\theta = 4.00$, find $\sec\theta$.

47. Given $\csc\theta = 3.04$, find $\tan\theta$.

48. Given $\tan\theta = 5.36$, find $\cos\theta$.

49. Given $\cos\theta = 0.559$, find $\cot\theta$.

50. Given $\csc\theta = 3.64$, find $\cot\theta$.

51. Given $\sec\theta = 4.03$, find $\sin\theta$.

52. Given $\sin\theta = 0.642$, find $\tan\theta$.

In Exercises 53–60, find the values of the indicated reciprocal functions. (See Example 5.)

53. Given $\sin\theta = \frac{1}{4}$, find $\csc\theta$.

54. Given $\cos\theta = \frac{3}{4}$, find $\sec\theta$.

55. Given $\sec\theta = \frac{7}{6}$, find $\cos\theta$.

56. Given $\csc\theta = \frac{5}{2}$, find $\sin\theta$.

57. Given $\cot\theta = 5$, find $\tan\theta$.

58. Given $\tan\theta = 4$, find $\cot\theta$.

59. Given $\csc\theta = 10$, find $\sin\theta$.

60. Given $\cot\theta = 2$, find $\tan\theta$.

15.3 Values of Trigonometric Functions

So far we have concentrated only on the definitions of the trigonometric functions. In this section we will see how to obtain the values of the trigonometric functions of a given angle. We will start with the function values of certain special angles and then continue with the function values of arbitrary angles.

Special Angles

For certain special angles the values of the trigonometric functions can be obtained from a diagram. These angles are 0°, 30°, 45°, 60°, and 90°.

For the functions of 30° and 60°, we need the following special property of a 30°-60° right triangle:

> In a 30°-60° right triangle, the side opposite the 30° angle is one-half the hypotenuse.

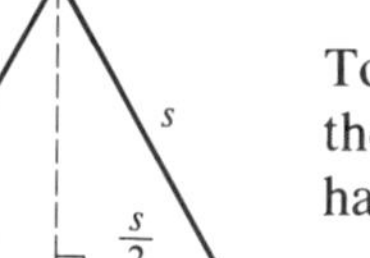

Figure 15.21

To see why, consider the equilateral triangle ABC in Figure 15.21. Note that the altitude CD divides ΔABC into two 30°-60° right triangles, each of which has the property claimed.

Now consider the 30°-60° right triangle in Figure 15.22. If we assign a value of 1 unit to the side opposite the 30° angle, then the hypotenuse is 2 units long. If x denotes the length of the adjacent side, then

$$x^2 + 1^2 = 2^2 \qquad \text{or} \qquad x = \sqrt{3}$$

For the 60° angle, the adjacent side is 1 unit long and the opposite side is $\sqrt{3}$ (Figure 15.22).

Special angles

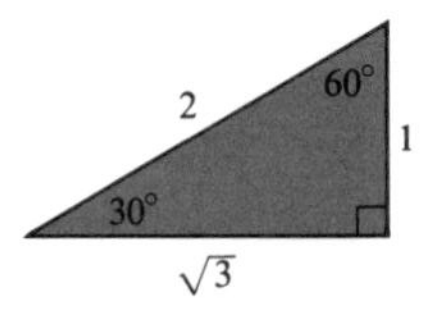

Figure 15.22

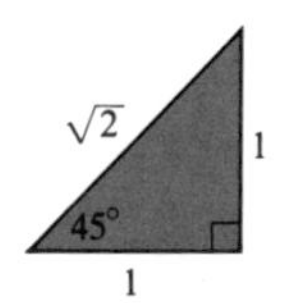

Figure 15.23

For the 45° angle (Figure 15.23), the side opposite and the side adjacent are each 1 unit long and the hypotenuse is

$$\sqrt{1^2 + 1^2} = \sqrt{2}$$

Example 1 Find sin 30° and cos 30°.

Solution. We place the 30° angle in standard position, as shown in Figure 15.24. We now have

$$\sin 30° = \frac{1}{2} \qquad \text{and} \qquad \cos 30° = \frac{\sqrt{3}}{2}$$

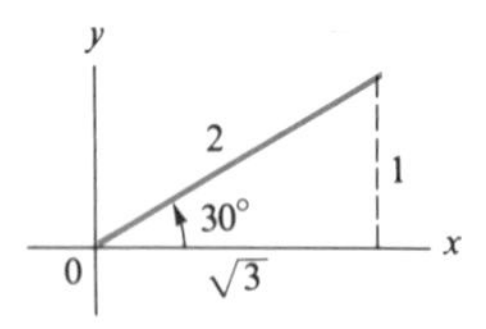

Figure 15.24

Example 2 Find sin 60°, sec 60°, and cot 60°.

Solution. First we put the 60° angle in standard position (Figure 15.25). By Figure 15.22, the adjacent side is 1 unit long and the opposite side $\sqrt{3}$. So

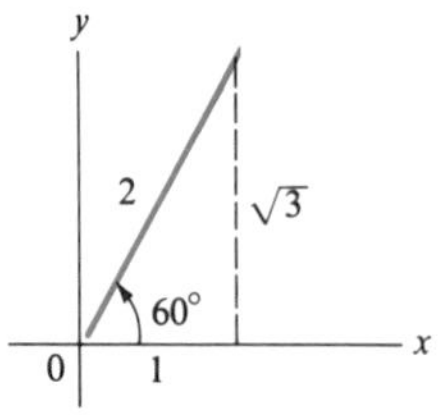

$$\sin 60° = \frac{\sqrt{3}}{2}$$

$$\sec 60° = \frac{2}{1} = 2$$

$$\cot 60° = \frac{1}{\sqrt{3}} = \frac{1 \cdot \sqrt{3}}{\sqrt{3} \cdot \sqrt{3}} = \frac{\sqrt{3}}{3}$$

Figure 15.25

Example 3 Using the information from Figure 15.23 (45° triangle), we obtain the angle (in standard position) shown in Figure 15.26. It follows that

$$\cos 45° = \frac{1}{\sqrt{2}} = \frac{1}{\sqrt{2}} \cdot \frac{\sqrt{2}}{\sqrt{2}} = \frac{\sqrt{2}}{(\sqrt{2})^2} = \frac{\sqrt{2}}{2}$$

$$\csc 45° = \frac{\sqrt{2}}{1} = \sqrt{2}$$

$$\tan 45° = \frac{1}{1} = 1$$

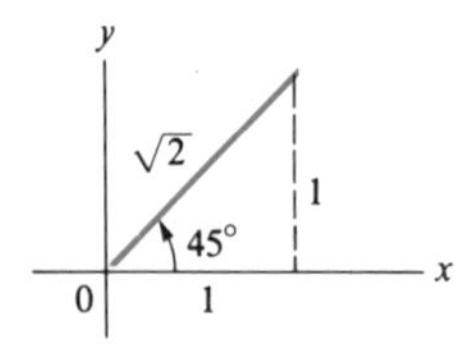

Figure 15.26

For the quadrantal angles 0° and 90°, we pick a point on the terminal side and apply the definitions.

Example 4 Evaluate sin 0°, cos 0°, tan 0°, and cot 0°.

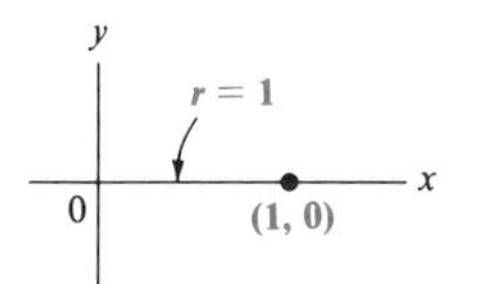

Figure 15.27

Solution. If the angle is 0°, the terminal side coincides with the initial side, and no triangle is formed. (See Figure 15.27.) However, by picking a point, say (1,0), on the terminal side, we can use the definitions:

$$\sin 0° = \frac{y}{r} = \frac{0}{1} = 0$$

$$\cos 0° = \frac{x}{r} = \frac{1}{1} = 1$$

$$\tan 0° = \frac{y}{x} = \frac{0}{1} = 0$$

$$\cot 0° = \frac{x}{y} = \frac{1}{0} \qquad \text{(undefined)}$$

(Recall that we cannot divide by 0; so 1/0 is not defined.)

Example 5 Evaluate sin 90°, cot 90°, cos 90°, and sec 90°.

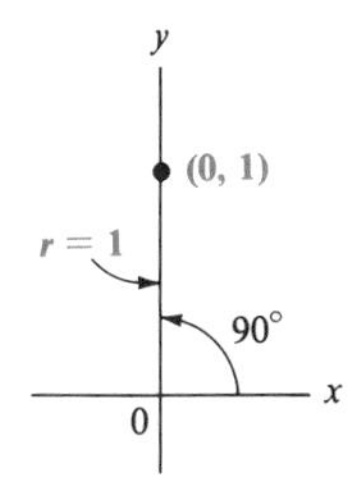

Figure 15.28

Solution. Since 90° is a quadrantal angle, we do not get a triangle. (See Figure 15.28.) As in the case of the 0° angle, we pick a point on the terminal side, say (0, 1), and use the definitions:

$$\sin 90° = \frac{y}{r} = \frac{1}{1} = 1$$

$$\cot 90° = \frac{x}{y} = \frac{0}{1} = 0$$

$$\cos 90° = \frac{x}{r} = \frac{0}{1} = 0$$

$$\sec 90° = \frac{r}{x} = \frac{1}{0} \qquad \text{(undefined)}$$

The values of all the special angles just discussed are listed in the table on the next page.

Although the table lists the values of all the special angles, it is pointless to memorize the whole set. Later we are going to study special angles in other quadrants. (For example, 120° is also a special angle.) So even if you could remember them, it is much better in the long run to construct a diagram and read off the values required.

Special angles

θ	$0°$	$30°$	$45°$	$60°$	$90°$
$\sin\theta$	0	$\frac{1}{2}$	$\frac{\sqrt{2}}{2}$	$\frac{\sqrt{3}}{2}$	1
$\cos\theta$	1	$\frac{\sqrt{3}}{2}$	$\frac{\sqrt{2}}{2}$	$\frac{1}{2}$	0
$\tan\theta$	0	$\frac{\sqrt{3}}{3}$	1	$\sqrt{3}$	undefined
$\csc\theta$	undefined	2	$\sqrt{2}$	$\frac{2\sqrt{3}}{3}$	1
$\sec\theta$	1	$\frac{2\sqrt{3}}{3}$	$\sqrt{2}$	2	undefined
$\cot\theta$	undefined	$\sqrt{3}$	1	$\frac{\sqrt{3}}{3}$	0

For special angles, diagrams can also be used to find θ, given the value of a trigonometric function.

Example 6 Find θ, given that $\tan\theta = \frac{\sqrt{3}}{3}$.

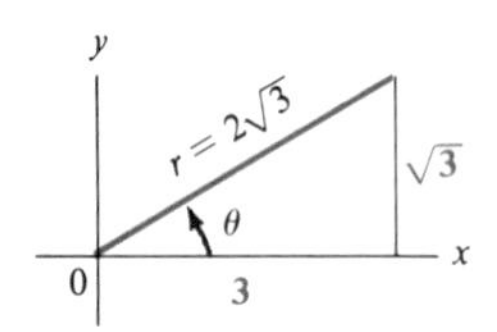

Figure 15.29

Solution. We draw the diagram in Figure 15.29, placing $\sqrt{3}$ on the opposite side and 3 on the adjacent side. The radius vector is found to be

$$r = \sqrt{(\sqrt{3})^2 + 3^2} = \sqrt{3+9} = \sqrt{12} = \sqrt{4\cdot 3}$$
$$= \sqrt{4}\sqrt{3} = 2\sqrt{3}$$

Since the hypotenuse ($r = 2\sqrt{3}$) is twice as long as the opposite side, we conclude that $\theta = 30°$.

Arbitrary Angles—Using a Table

For angles other than special angles, the values of the trigonometric functions can be obtained only by more advanced methods. For that reason, tables of values have been constructed. One such table is given in Appendix B, and a portion is shown in Table 15.1.

In the table, the angles from 0° to 45° are listed in the left column, while the names of the functions are given on top (first row). Angles from 45° to 90° are listed in the right column with the names of the functions given at the bottom (last row).

Table 15.1

Degrees	Sin θ	Cos θ	Tan θ	Cot θ	Sec θ	Csc θ	
24°00′	0.4067	0.9135	0.4452	2.246	1.095	2.459	66°00′
10	0.4094	0.9124	0.4487	2.229	1.096	2.443	50
20	0.4120	0.9112	0.4522	2.211	1.097	2.427	40
30	0.4147	0.9100	0.4557	2.194	1.099	2.411	30
40	0.4173	0.9088	0.4592	2.177	1.100	2.396	20
50	0.4200	0.9075	0.4628	2.161	1.102	2.381	10
25°00′	0.4226	0.9063	0.4663	2.145	1.103	2.366	65°00′
	Cos θ	Sin θ	Cot θ	Tan θ	Csc θ	Sec θ	Degrees

Example 7 Find cos 24°30′.

Solution. Since the angle is listed in the left column, we need to refer to the names on top. Reading from the cosine column, we get

$$\cos 24°30' = 0.9100$$

Example 8 Find tan 65°20′.

Solution. Since the angle is listed in the right column, we refer to the names at the bottom of the table. In locating the angle, note that 65°20′ is *above* 65°:

$$\tan 65°20' = 2.177$$

It is often necessary to reverse the process: Find the angle, given a function value of the angle.

Example 9 Find θ, given that sec θ = 2.443.

Solution. We locate the value 2.443 in the secant column. Since the name of the function is given at the bottom of the table, we read off the angle in the right column to obtain

$$\theta = 65°50'$$

Since the angles are listed in the table in intervals of 10′, function values for angles between must be estimated by a process called **interpolation**. Space does not permit a detailed discussion of this process, but the next example illustrates the procedure.

Example 10 Find $\sin 24°28'$.

Solution. The desired value is between

$$\sin 24°20' = 0.4120 \qquad \text{and} \qquad \sin 24°30' = 0.4147$$

Since $28'$ is $\frac{8}{10}$ of the way from $20'$ to $30'$, we would expect $\sin 24°28'$ to be about $\frac{8}{10}$ of the way from 0.4120 to 0.4147. Ignoring the decimal point for now, as well as the digits immediately following, we see that 27 is the *tabular difference*. The calculations can be done systematically as follows:

$$10\left[\begin{array}{l} 8\left[\begin{array}{l}\sin 24°20' = 0.4120 \\ \\ \sin 24°28' = \;.\;.\;. \end{array}\right]x \\ \\ \sin 24°30' = 0.4147 \end{array}\right]27$$

From the diagram, we set up the following proportion:

$$\frac{x}{27} = \frac{8}{10} \qquad \text{or} \qquad x = 21.6$$

Since we cannot obtain more than four-decimal-place accuracy, 21.6 is rounded off to 22. So 22 is the tabular difference between 0.4120 and the desired value. It follows that

$$\begin{aligned} \sin 24°28' &= 0.0022 + 0.4120 \\ &= 0.4142 \end{aligned}$$

Interpolation may also be necessary for obtaining the angle from a given function value.

Example 11 Find θ, given that $\cot \theta = 0.4536$.

Solution. The nearest values are $\cot 65°30' = 0.4557$ and $\cot 65°40' = 0.4522$. We now set up the following diagram:

$$10\left[\begin{array}{l} x\left[\begin{array}{l}\cot 65°30' = 0.4557 \\ \\ \cot \theta = 0.4536 \end{array}\right]21 \\ \\ \cot 65°40' = 0.4522 \end{array}\right]35$$

$$\frac{x}{10} = \frac{21}{35} \quad \text{or} \quad x = 6$$

It follows that $\theta = 65°36'$.

Arbitrary Angles—Using a Calculator

Scientific calculators are programmed to evaluate the basic trigonometric functions sine, cosine, and tangent: Put the calculator in **degree mode**, enter the degree measure of the angle, and press

[SIN], [COS], or [TAN]

To find csc θ, sec θ, or cot θ, we use the reciprocal relations

$$\csc\theta = \frac{1}{\sin\theta}, \quad \sec\theta = \frac{1}{\cos\theta}, \quad \cot\theta = \frac{1}{\tan\theta}$$

Reciprocals are found by pressing [1/x].

Example 12 Use a calculator to find

a. sin 20° **b.** cos 70.6°

Solution.

a. To find sin 20°, enter 20 and press [SIN]. The sequence is

20 [SIN] → 0.3420201

So sin 20° = 0.3420 to four decimal places.

b. For cos 70.6°, the sequence is

70.6 [COS] → 0.3321611

Thus cos 70.6° = 0.3322 to four decimal places.

Example 13 Find (a) csc 29.4°, (b) cot 65.8°.

Solution.

a. To find csc 29.4°, we use the reciprocal relation $\csc\theta = 1/\sin\theta$:

29.4 [SIN] [1/x] → 2.0370592

b. Since $\cot\theta = 1/\tan\theta$, the sequence is

65.8 [TAN] [1/x] → 0.4494178

Some calculators will not accept angles expressed in degrees and minutes. In that case, the angle must be changed to decimal form. For example, to change 30°15′ to decimal form, recall that 1° = 60′. Thus

$$30°15' = 30° + \left(\frac{15}{60}\right)^{\circ} = \left(30 + \frac{15}{60}\right)^{\circ} = 30.25°$$

Example 14 Find (a) tan 30°16′, (b) sec 52°49′.

Solution.

a. Since $30°16' = \left(30 + \frac{16}{60}\right)^{\circ}$, the sequence is

30 [+] 16 [÷] 60 [=] [TAN] → 0.5835726

Thus tan 30°16′ = 0.5836.

b. Since $52°49' = \left(52 + \frac{49}{60}\right)^{\circ}$, the sequence is

52 [+] 49 [÷] 60 [=] [COS] [1/x] → 1.6546227

So sec 52°49′ = 1.6546.

To obtain the angle θ, given the value of a trigonometric function, use

[INV] [SIN], [ARCSIN], or [SIN^{-1}]
[INV] [COS], [ARCCOS], or [COS^{-1}]
[INV] [TAN], [ARCTAN], or [TAN^{-1}]

or similar keys. For the remaining functions, we enter the function value, press [1/x], followed by one of the above operations.

Example 15 Find θ, given that tan θ = 1.347.

Solution. The sequence is

1.347 [INV] [TAN] → 53.410158

So θ = 53.41° to the nearest hundredth of a degree.

Example 16 Find θ, given that sec θ = 3.843.

Solution. Since sec θ = 1/cos θ, we use the sequence

3.843 [1/x] [INV] [COS] → 74.917277

So θ = 74.92° to the nearest hundredth of a degree.

If the angle is to be expressed in degrees and minutes, we need to change the decimal form, as shown next.

Example 17 Find θ in degrees and minutes, given that $\csc \theta = 2.107$.

Solution. Since $\csc \theta = 1/\sin \theta$, we use the sequence

$$2.107 \boxed{1/x} \boxed{\text{INV}} \boxed{\text{SIN}} \rightarrow 28.333859$$

Thus $\theta = 28.33°$ in decimal form.

To change the fractional part (0.33) to minutes, we set up the following proportion:

$$\frac{33}{100} = \frac{x}{60} \qquad 0.33 = \frac{33}{100}$$

or

$$x = \frac{(33)(60)}{100} = 19.8 \approx 20$$

So $\theta = 28°20'$.

Exercises / Section 15.3

In Exercises 1–12, draw a diagram to determine the exact value of each trigonometric function. (See Examples 1–3.)

1. $\sin 60°$ **2.** $\sin 45°$ **3.** $\cos 45°$ **4.** $\tan 30°$

5. $\tan 60°$ **6.** $\sec 60°$ **7.** $\csc 30°$ **8.** $\csc 45°$

9. $\cot 45°$ **10.** $\cot 30°$ **11.** $\cot 60°$ **12.** $\tan 45°$

In Exercises 13–24, find the values of the functions of the given quadrantal angles. (See Examples 4 and 5.)

13. $\cos 0°$ **14.** $\cos 90°$ **15.** $\tan 90°$ **16.** $\tan 0°$

17. $\sin 0°$ **18.** $\sin 90°$ **19.** $\sec 0°$ **20.** $\sec 90°$

21. $\csc 0°$ **22.** $\csc 90°$ **23.** $\cot 90°$ **24.** $\cot 0°$

In Exercises 25–38, use a diagram to find θ. (See Example 6.)

25. $\tan \theta = 1$ **26.** $\tan \theta = \sqrt{3}$ **27.** $\sin \theta = \frac{1}{2}$ **28.** $\sin \theta = \frac{\sqrt{3}}{2}$

29. $\cos \theta = \frac{\sqrt{3}}{2}$ **30.** $\cos \theta = \frac{1}{2}$ **31.** $\csc \theta = \frac{2\sqrt{3}}{3}$ **32.** $\sec \theta = \frac{2\sqrt{3}}{3}$

33. $\cot \theta = \frac{\sqrt{3}}{3}$ **34.** $\cot \theta = 1$ **35.** $\cos \theta = \frac{\sqrt{2}}{2}$ **36.** $\sin \theta = \frac{\sqrt{2}}{2}$

37. $\sec \theta = 2$ **38.** $\csc \theta = 2$

39. For what value of θ is $\csc \theta$ undefined?

40. For what value of θ is $\sec \theta$ undefined?

In Exercises 41–50, use Table 1 of Appendix B to find the value of each trigonometric function.

41. sin 23°40′ **42.** cos 34°10′ **43.** sec 29°50′ **44.** tan 42°30′

45. csc 53°20′ **46.** sin 64°0′ **47.** tan 47°26′ **48.** sin 9°37′

49. cos 31°4′ **50.** cot 35°48′

In Exercises 51–56, use Table 1 of Appendix B to find θ.

51. $\sec\theta = 1.195$ **52.** $\csc\theta = 1.216$ **53.** $\sin\theta = 0.8025$ **54.** $\tan\theta = 1.260$

55. $\cos\theta = 0.9031$ **56.** $\cot\theta = 1.414$

In Exercises 57–80, use a calculator to determine the values of the given trigonometric functions. Give answers to four decimal places.

57. sin 26.5° **58.** sin 38.7° **59.** cos 5.4° **60.** cos 8.0°

61. tan 76.0° **62.** tan 71.2° **63.** csc 55.6° **64.** csc 41.8°

65. sec 20.4° **66.** sec 32.9° **67.** cot 73.5° **68.** cot 23.7°

69. sin 30°16′ **70.** sin 39°10′ **71.** cos 76°44′ **72.** cos 59°18′

73. tan 28°6′ **74.** tan 6°55′ **75.** csc 4°6′ **76.** csc 84°38′

77. sec 43°29′ **78.** sec 52°47′ **79.** cot 20°22′ **80.** cot 40°11′

In Exercises 81–90, use a calculator to find θ, accurate to the nearest hundredth of a degree.

81. $\sin\theta = 0.1492$ **82.** $\cos\theta = 0.1776$ **83.** $\tan\theta = 0.1865$ **84.** $\cot\theta = 0.1812$

85. $\csc\theta = 2.172$ **86.** $\sec\theta = 3.031$ **87.** $\cos\theta = 0.8453$ **88.** $\sin\theta = 0.5730$

89. $\cot\theta = 1.506$ **90.** $\tan\theta = 2.937$

In Exercises 91–100, use a calculator to find θ, expressed in degrees and minutes.

91. $\sin\theta = 0.1945$ **92.** $\sin\theta = 0.1918$ **93.** $\sec\theta = 2.793$ **94.** $\csc\theta = 1.812$

95. $\cot\theta = 0.5441$ **96.** $\tan\theta = 1.386$ **97.** $\sin\theta = 0.7005$ **98.** $\cos\theta = 0.6172$

99. $\csc\theta = 2.006$ **100.** $\cot\theta = 3.109$

15.4 Solving Right Triangles

The main purpose of this section is to develop a method for solving right triangles. To **solve a triangle** means to find the unknown parts of the triangle. First, however, we need to make some additional observations about trigonometric ratios for angles not in standard position.

More on Trigonometric Ratios

So far we have concentrated on the trigonometric functions of angles in standard position. For such angles the values of the trigonometric functions can be given in terms of the coordinates of a point on the terminal side and the radius vector. In some applications of trigonometry, it is more convenient to use the descriptive definitions in terms of opposite side, adjacent side, and hypotenuse, a concept that was first introduced in Section 15.2. (In particular, see Figure 15.14 and the corresponding definitions.)

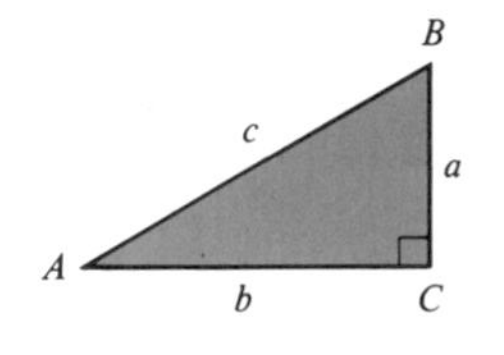

Figure 15.30

Now, whenever we use the descriptive definitions, we must remember that the terms *opposite side* and *adjacent side* are **relative.** For example, in Figure 15.30, side a is **opposite** $\angle A$. But as far as $\angle B$ is concerned, a is the **adjacent** side. Similarly, side b is *opposite* $\angle B$, but *adjacent* to $\angle A$. Thus

$$\tan A = \frac{a}{b} \quad \text{and} \quad \tan B = \frac{b}{a}$$

The hypotenuse, however, is always the same, being opposite the right angle.

Example 1 Find sin A and cos B for the angles in Figure 15.31.

Solution. As far as $\angle A$ is concerned, BC is the opposite side. So

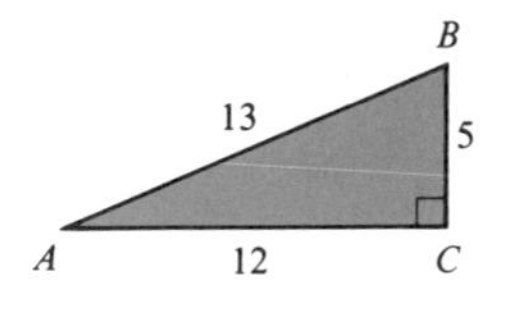

Figure 15.31

$$\sin A = \frac{5}{13}$$

As far as $\angle B$ is concerned, BC is the adjacent side. So

$$\cos B = \frac{5}{13}$$

The next example generalizes the ideas in Example 1.

Example 2 Referring to Figure 15.32, note that

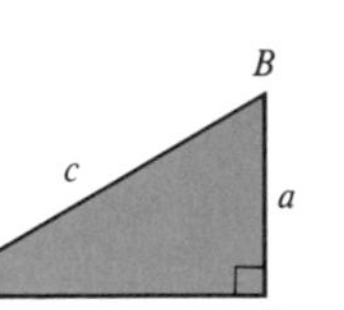

Figure 15.32

1. $\sin A = \dfrac{a}{c}$ and $\cos B = \dfrac{a}{c}$
2. $\tan A = \dfrac{a}{b}$ and $\cot B = \dfrac{a}{b}$
3. $\sec A = \dfrac{c}{b}$ and $\csc B = \dfrac{c}{b}$

We see that $\sin A = \cos B$, $\tan A = \cot B$, and $\sec A = \csc B$.

Since the sum of the angles of a triangle is 180°, the sum of the acute angles A and B is 90° (Figure 15.32). By Example 2, if angles A and B are complementary, then $\sin A = \cos B$. The sine and cosine functions are therefore called **cofunctions** (*cosine* means "complementary sine"). Also, the tangent and cotangent functions are cofunctions ($\tan A = \cot B$), as are the secant and cosecant functions ($\sec A = \csc B$). In summary: **Any trigonometric function of an acute angle is equal to the corresponding cofunction of its complement.**

Cofunctions

Solving Right Triangles

We will now use the trigonometric functions to solve a given right triangle. In this section we concentrate on the technique; physical applications are taken up in the next section.

For any right triangle, if we know the length of two sides, or the length of one side and an acute angle, we can find the remaining parts. The next example illustrates the technique.

Example 3 Find x in the right triangle in Figure 15.33.

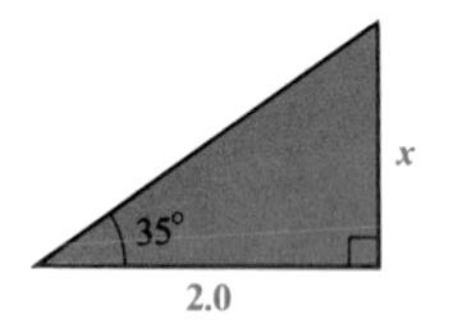

Figure 15.33

Solution. As far as the 35° angle is concerned, x is the **opposite** side and **2.0** is the **adjacent** side. So

$$\frac{x}{2.0} = \tan 35^\circ \qquad \frac{\text{opposite side}}{\text{adjacent side}} = \tan\theta$$

Since tan 35° is known (obtained from Table 1 or a calculator), the last equation can be solved for x:

$$\frac{x}{2.0} = 0.7002$$

Multiplying both sides by 2.0, we get

$$x = (0.7002)(2.0) = 1.4$$

Example 3, which is quite typical, gives us a plan of attack:

> To find the unknown side of a triangle:
>
> **1.** Form a ratio involving the unknown side and a known side.
> **2.** Set this ratio equal to the corresponding function of the known angle of the triangle.
> **3.** Solve the resulting equation for the unknown.

Remark on significant digits. Two significant digits in the side measurements corresponds to the nearest degree in the angle measurements. Three significant digits corresponds to the nearest multiple of 10′ or tenth of a degree. Four significant digits corresponds to the nearest minute or hundredth of a degree. These rules are summarized in Table 15.2.

Table 15.2

Accuracy of degree measurement		Significant digits in side measurements
1°	1°	2
10′	0.1°	3
1′	0.01°	4

The remaining examples refer to the labels in Figure 15.34. Note that the capital letters are used to denote the angles and the same lowercase letters, the respective opposite sides. Angle C is the right angle. (In this section, A, B, and C represent the degree measures of angles A, B, and C.)

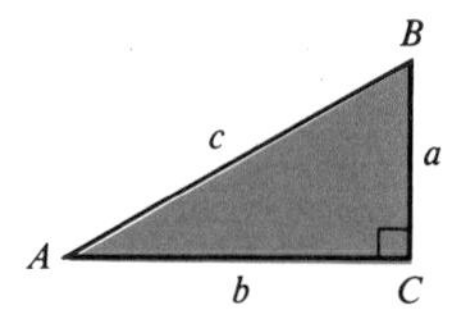

Figure 15.34

Example 4 Solve the right triangle in Figure 15.35, given that $B = 36.7°$ and $c = 1.34$.

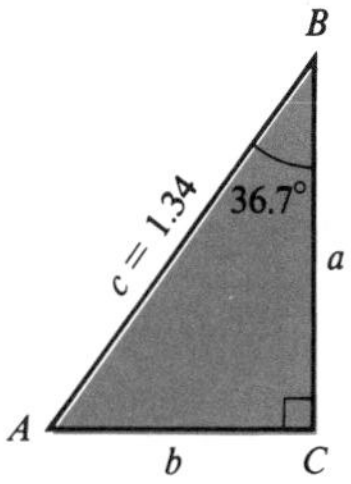

Figure 15.35

Solution. Let us find side a first. Observe that for $\angle B$, a is the **adjacent** side (and b the opposite side). Since c is the **hypotenuse,** we have by the definition of cosine that

$$\cos 36.7° = \frac{a}{c} = \frac{a}{1.34} \qquad \frac{\text{adjacent side}}{\text{hypotenuse}} = \cos\theta$$

It follows that

$$\frac{a}{1.34} = 0.8018$$

$$a = (0.8018)(1.34) \qquad \text{Multiplying by 1.34}$$

$$= 1.07 \qquad \text{Three significant digits}$$

To find side b, note that b is **opposite** $\angle B$. Again, c is the **hypotenuse,** so that

$$\sin 36.7° = \frac{b}{c} = \frac{b}{1.34} \qquad \frac{\text{opposite side}}{\text{hypotenuse}} = \sin\theta$$

Solving for b, we get

$$\frac{b}{1.34} = 0.5976$$

$$b = (0.5976)(1.34) = 0.801$$

Finally, to obtain A, recall that the sum of the angles of a triangle is 180°. So

$$A + 36.7° + 90° = 180°$$

$$A + 126.7° = 180°$$

$$A = 180° - 126.7° = 53.3°$$

Example 5 If $A = 58°38'$ and $a = 5.379$, find b (Figure 15.36).

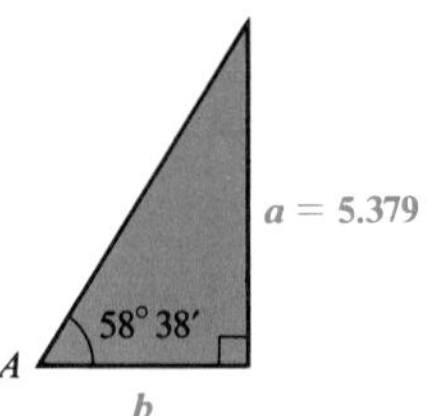

Figure 15.36

Solution. The two sides involved are a and b, so that the appropriate ratio is either tan A or cot A. Using cot A, we get

$$\frac{b}{5.379} = \cot 58°38' \qquad \frac{\text{adjacent side}}{\text{opposite side}} = \cot \theta$$

or

$$b = 5.379 \cot 58°38'$$

The sequence is

58 [+] 38 [÷] 60 [=] [TAN] [1/x] [×] 5.379 [=] → 3.2790618

So $b = 3.279$ to four significant digits.

The next example illustrates how to find an unknown angle, given two sides.

Example 6 If $a = 7.56$ and $b = 10.6$, find B (Figure 15.37).

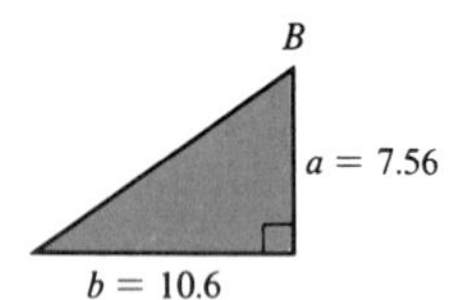

Figure 15.37

Solution. Since b is opposite $\angle B$ and a is adjacent to $\angle B$, we get

$$\tan B = \frac{b}{a} = \frac{10.6}{7.56} \qquad \frac{\text{opposite side}}{\text{adjacent side}} = \tan \theta$$

The sequence is

10.6 [÷] 7.56 [=] [INV] [TAN] → 54.503248

So $B = 54.5°$ to the nearest tenth of a degree.

Exercises / Section 15.4

The exercises below refer to Figure 15.38.

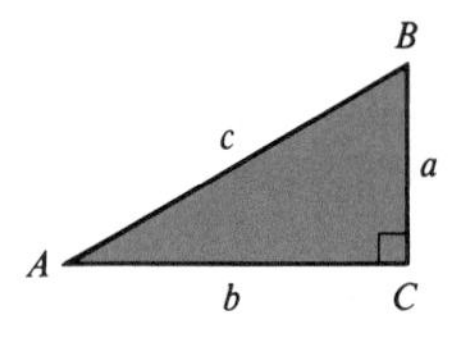

Figure 15.38

In Exercises 1–12, find the parts indicated.

1. $A = 32.3°$, $a = 1.46$; find b

2. $A = 64.7°$, $b = 4.60$; find c

3. $B = 53.6°$, $b = 5.36$; find c

4. $B = 47.4°$, $a = 5.07$; find b

5. $A = 36°10'$, $b = 10.0$; find a

6. $A = 47°20'$, $c = 12.6$; find a

7. $B = 29°44'$, $c = 13.64$; find b

8. $B = 48°19'$, $c = 0.8492$; find b

9. $B = 23.64°$, $a = 17.64$; find b

10. $B = 39.06°$, $b = 1.943$; find a

11. $a = 1.84$, $b = 2.76$; find A

12. $a = 3.67$, $b = 2.18$; find B

In Exercises 13–20, solve each of the given triangles.

13. $A = 72.6°$, $a = 9.60$

14. $B = 63.7°$, $b = 6.42$

15. $B = 40°$, $b = 7.0$

16. $A = 50°$, $a = 8.0$

17. $A = 53°24'$, $c = 3.200$

18. $B = 32°56'$, $c = 5.760$

19. $B = 72.76°$, $a = 10.30$

20. $A = 76.47°$, $b = 12.71$

15.5 Applications of Right Triangles

The method for solving right triangles we studied in the last section has numerous applications in technology.

Many of these applications involve indirect measurement. For example, suppose we wish to find the width of the river shown in Figure 15.39. Assuming the bank to be straight, a distance of 106 ft is measured along a line perpendicular to the line directly across. The angle on the right is 50.1°. Denoting the distance across by x, we form the ratio $x/106$. We now have

River
x
50.1°
106 ft

Figure 15.39

$$\frac{x}{106} = \tan 50.1° \qquad \frac{\text{opposite side}}{\text{adjacent side}} = \tan\theta$$

or

$$x = 106 \tan 50.1° = 127 \text{ ft}$$

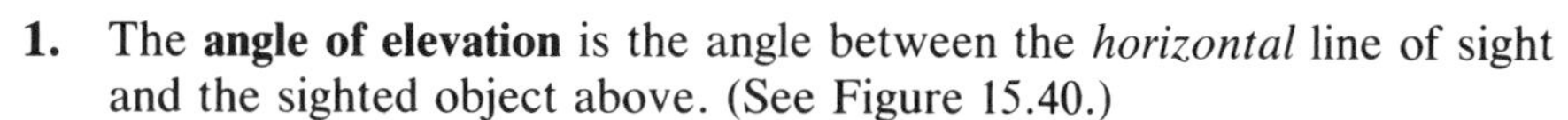

To make some of the applied problems easier to state, we need to introduce the following terms:

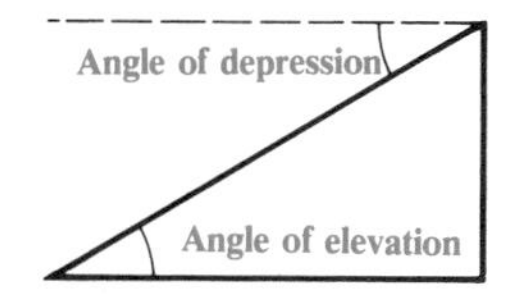

Figure 15.40

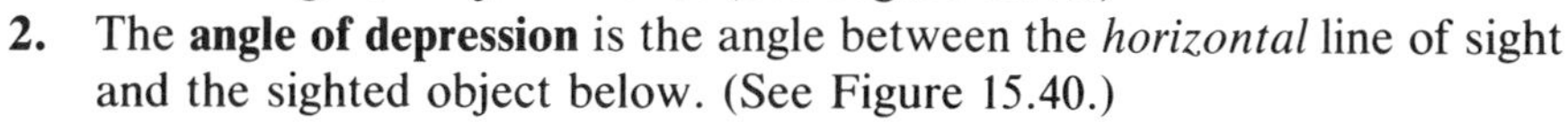

1. The **angle of elevation** is the angle between the *horizontal* line of sight and the sighted object above. (See Figure 15.40.)
2. The **angle of depression** is the angle between the *horizontal* line of sight and the sighted object below. (See Figure 15.40.)

While some problems involve the angle of elevation and some the angle of depression, whenever we have a particular triangle (as in Figure 15.40), the two angles, being alternate interior angles, are equal.

Example 1 The angle of elevation of the sun is 29° at the time when the shadow of a tree is 41 ft long. Find the height of the tree.

Solution. The situation is illustrated in Figure 15.41. If we denote the unknown height by x, we can set up a ratio involving the opposite and adjacent sides. Thus

$$\frac{x}{41} = \tan 29° \qquad \frac{\text{opposite side}}{\text{adjacent side}} = \tan\theta$$

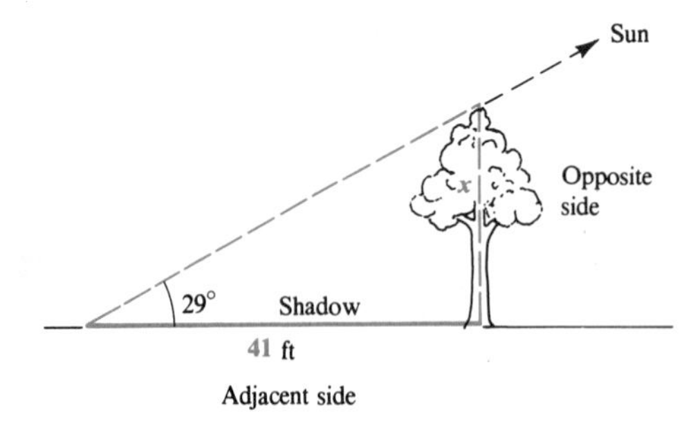

Figure 15.41

Solving for x:

$$x = 41 \tan 29° = 23 \text{ ft}$$

rounded off to two significant digits.

Example 2 From the top of a cliff 118.0 ft above the surface of the water, the angle of depression of a boat was measured to be 12.42°. Find the distance from the boat to the base of the cliff.

Solution. In Figure 15.42, we denote the unknown distance by x. Since the angle of depression is 12.42°, the angle on the right is also 12.42°. The ratio involves the adjacent and opposite sides, so that x can be found by using the cotangent:

$$\frac{x}{118.0} = \cot 12.42° \qquad \frac{\text{adjacent side}}{\text{opposite side}} = \cot \theta$$

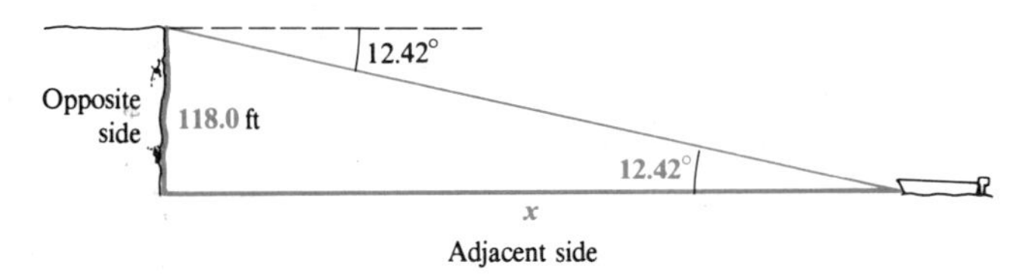

Figure 15.42

Solving for x:

$$x = 118.0 \cot 12.42°$$

The sequence is

118.0 [×] 12.42 [TAN] [1/x] [=] → 535.80293

Rounding off to four significant digits, the distance is 535.8 ft.

Example 3 From a point 15.2 m from the base of a building, the angle of elevation of the bottom of a flagpole mounted on top of the building is 60.9°, and the angle of elevation of the top of the flagpole is 63.4°. Find the height of the flagpole.

Solution. The situation is pictured in Figure 15.43.

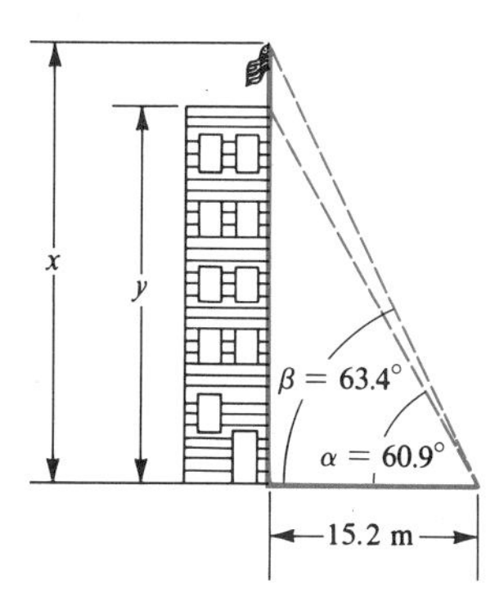

Figure 15.43

Let us denote the two angles by $\alpha = 60.9°$ and $\beta = 63.4°$. Also, let x and y denote the respective distances from the top and bottom of the flagpole to the ground. (See Figure 15.43.) Calculating x and y separately, we get

$$\frac{x}{15.2} = \tan \beta \qquad \frac{y}{15.2} = \tan \alpha$$

$$\frac{x}{15.2} = \tan 63.4° \qquad \frac{y}{15.2} = \tan 60.9°$$

$$x = 15.2 \tan 63.4° \qquad y = 15.2 \tan 60.9°$$

$$x = 30.4 \text{ m} \qquad y = 27.3 \text{ m}$$

It follows that the height of the flagpole is

$$x - y = 30.4 \text{ m} - 27.3 \text{ m} = 3.1 \text{ m}$$

Example 4 Five screws are equally spaced on a circle of radius 10.2 cm. Find the center-to-center distance between two adjacent screws.

Solution. The arrangement of screws is shown in Figure 15.44. Let x be the length of half the chord determined by the two adjacent screws on top in the figure. The central angle determined by two adjacent screws is $360° \div 5 = 72°$. So for the triangle in the figure, the angle is $\frac{1}{2}(72°) = 36°$. We now get

$$\frac{x}{10.2} = \sin 36° \qquad \frac{\text{opposite side}}{\text{hypotenuse}} = \sin \theta$$

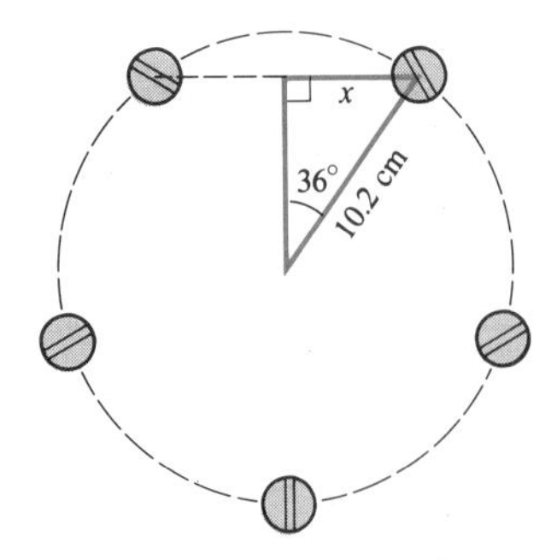

Figure 15.44

Solving for x:

$$x = 10.2 \sin 36° = 6.00 \text{ cm}$$

It follows that the center-to-center distance between two adjacent screws is 2(6.00 cm) = 12.0 cm.

Exercises / Section 15.5

1. A tree casts a shadow 11.2 m long. If the angle of elevation of the sun is 42.8°, how tall is the tree?
2. From a point 39.7 ft from the base of a flagpole, the angle of elevation of the top is 53.1°. Find the height of the flagpole.
3. A 9.0-ft ladder is leaning against a wall. If the ladder makes an angle of 62° with the ground, how high does the ladder reach up on the wall?
4. A guy wire attached to the top of a tower makes an angle of 61.2° with the ground. If the tower is known to be 40.2 m high, find the length of the wire.
5. From a point 223 ft from the base of the Empire State Building, the angle of elevation of the top is 79.9°. Find the height.
6. From a point 229 m from the base of the World Trade Center Tower, the angle of elevation of the top is 60.9°. Find the height of the tower.
7. The Great Pyramid of Egypt has a base of 762 ft and a height of 484 ft. What angle do the sides make with the ground?
8. A 12-ft ladder reaches 10 ft on a vertical wall. What angle does the ladder make with the ground?
9. Find the angle α between the vertical bar and the support in Figure 15.45.

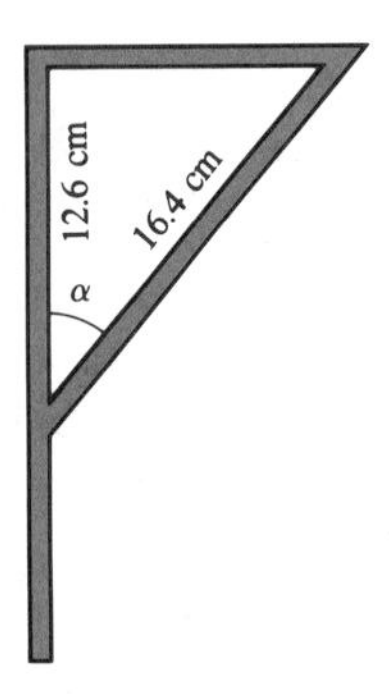

Figure 15.45

10. Find the lateral surface area of the conical tank in Figure 15.46.

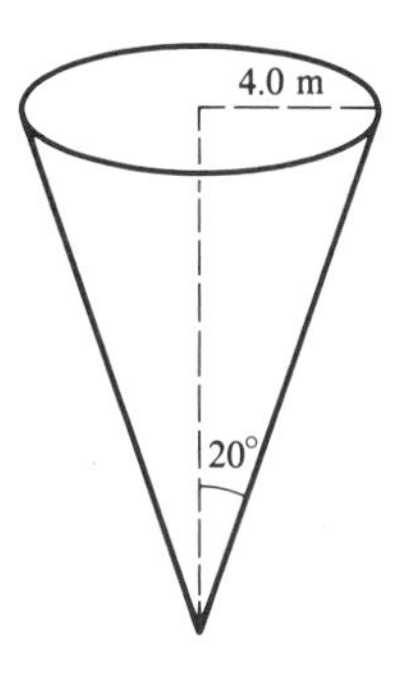

Figure 15.46

11. Find the center-to-center distance d between the pulleys in Figure 15.47.

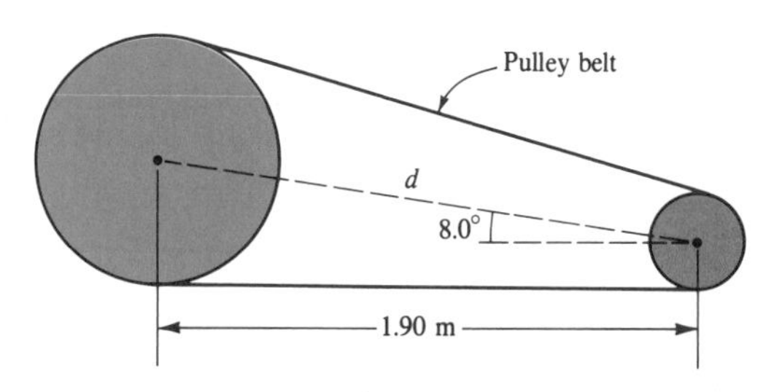

Figure 15.47

12. The metal plate in Figure 15.48 has the shape of a regular pentagon. Find the perimeter.

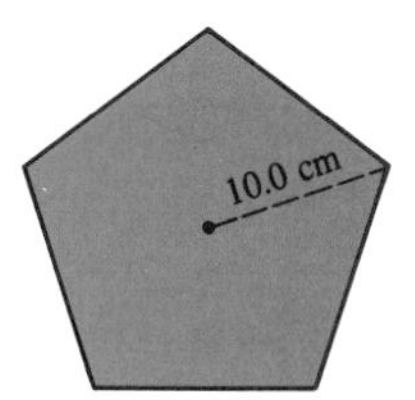

Figure 15.48

13. For the hex nut in Figure 15.49, find the distance d from the center to a side.

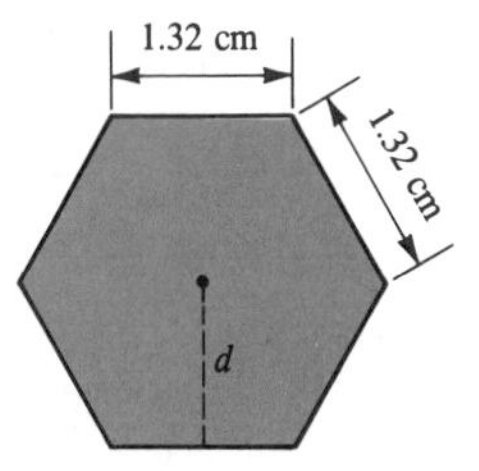

Figure 15.49

14. A rectangular computer chip measures 1.70 mm by 1.06 mm. Find the angle that the diagonal makes with the longer side.

15. Nine rivets are equally spaced on the circumference of a circle with radius 1.80 in. Find the center-to-center distance between two adjacent rivets.

16. Ten holes are equally spaced on the circumference of a circle. The distance from center to center between two adjacent holes is 10.0 cm. Find the radius of the circle.

17. The angle of depression from the top of a building 30.4 ft high to a park bench is 19.0°. Find the distance from the park bench to the base of the building.

18. From a weather balloon 684 ft above the ground the angle of depression to a distant building is 21.4°. Find the direct-line distance from the building to the balloon.

19. From the top of a lighthouse 40.6 m above the surface of a lake the angle of depression of a sailboat is 12°40′. Find the direct-line distance from the boat to the top of the lighthouse.

20. The angle of depression from a look-out tower 112 ft high to an intersection is 28°20′. Find the distance from the intersection to the base of the tower.

21. If a pendulum 26.40 cm long swings through an arc of 24°43′, how high will the bob rise above its lowest point?

22. A swimming pool measures 15.0 ft by 30.0 ft. Assuming that the bottom is flat and inclined 15.0° with the horizontal, determine the total area of the vertical walls. (See Figure 15.50.)

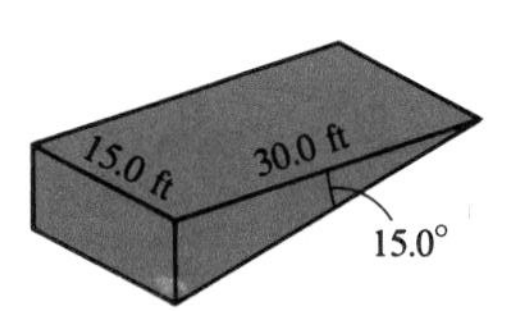

Figure 15.50

23. In an alternating-current circuit containing a resistance R and an inductive reactance X_L, the *phase angle* ϕ is the angle between the impedance Z and the resistance. (See Figure 15.51.) Find the phase angle if $R = 15.6\ \Omega$ and $X_L = 8.74\ \Omega$.

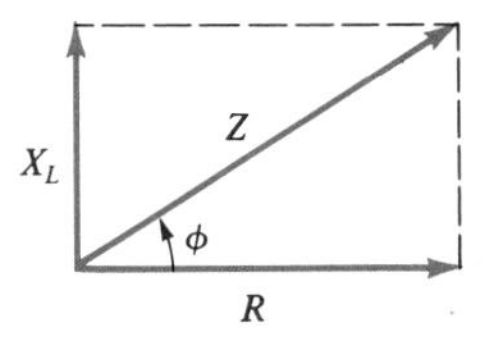

Figure 15.51

24. A platform 12.5 ft high is 46.5 ft from a building. From the top of the platform, the angle of elevation of the top of the building is 57°40′. Find the height of the building.

25. A building is located 655.0 ft from a transmission tower. From the top of the building the angle of depression of the base of the tower is 9°42′, and the angle of elevation of the top of the tower is 10°13′. Determine the height of the tower.

26. A war memorial features the figure of a soldier on a pedestal. From a point 6.00 m from the pedestal, the angle of elevation of the head of the figure is 56.4° and the angle of elevation of the base of the figure is 43.9°. Find the height of the figure.

27. From the top of a lighthouse 30.0 m above the surface of the water, the angles of depression of two boats due east are 12°20′ and 15°30′, respectively. Find the distance between the boats.

28. An observer in a balloon 86.40 m high measures the respective angles of depression of two houses to be 15°11′ and 18°42′. Assuming that the houses are on the same side of the balloon and in the same vertical plane as the balloon, find the distance between the houses.

15.6 Signs of Trigonometric Functions

So far we have dealt only with trigonometric functions of acute angles. In this section we will define the trigonometric functions of any angle.

If an angle is in standard position, then our definitions from Section 15.2 carry over directly. (Refer to Figure 15.52.)

Trigonometric functions

$$\sin\theta = \frac{y}{r} \qquad \csc\theta = \frac{r}{y}$$

$$\cos\theta = \frac{x}{r} \qquad \sec\theta = \frac{r}{x}$$

$$\tan\theta = \frac{y}{x} \qquad \cot\theta = \frac{x}{y}$$

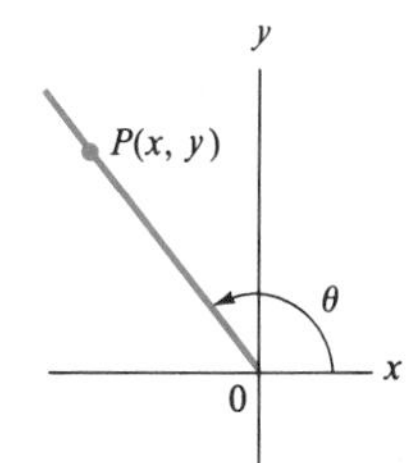

Figure 15.52

Note that the point $P(x, y)$ can be located in any quadrant. We will say that an angle is in the second quadrant if its terminal side is in the second quadrant. An angle is in the third quadrant if its terminal side is in the third quadrant, and so on.

While the definitions appear to be identical to our earlier definitions, some important differences should be noted. Since x and y can be either positive or negative, the values of the trigonometric functions are either positive or negative, depending on the quadrant in which the angle lies. However, **the radius vector r is always positive**:

Radius vector

$$r = \sqrt{x^2 + y^2}$$

Example 1 Find the values of the six trigonometric functions of the angle θ in standard position, given that its terminal side passes through the point $(-3, -4)$.

Solution. The angle θ is shown in Figure 15.53. The radius vector is

$$r = \sqrt{x^2 + y^2} = \sqrt{(-3)^2 + (-4)^2} = \sqrt{9 + 16} = \sqrt{25} = 5$$

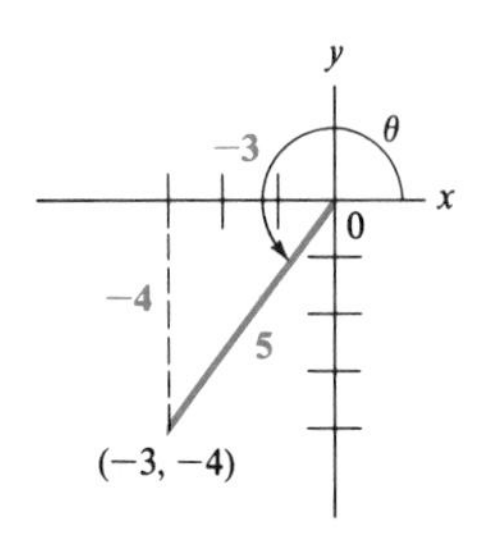

Figure 15.53

It follows that

$$\sin\theta = \frac{y}{r} = \frac{-4}{5} = -\frac{4}{5} \qquad \csc\theta = \frac{r}{y} = \frac{5}{-4} = -\frac{5}{4}$$

$$\cos\theta = \frac{x}{r} = \frac{-3}{5} = -\frac{3}{5} \qquad \sec\theta = \frac{r}{x} = \frac{5}{-3} = -\frac{5}{3}$$

$$\tan\theta = \frac{y}{x} = \frac{-4}{-3} = \frac{4}{3} \qquad \cot\theta = \frac{x}{y} = \frac{-3}{-4} = \frac{3}{4}$$

Example 2 Find the values of the six trigonometric functions of the angle θ in standard position, given that its terminal side passes through the point $(4.36, -7.84)$.

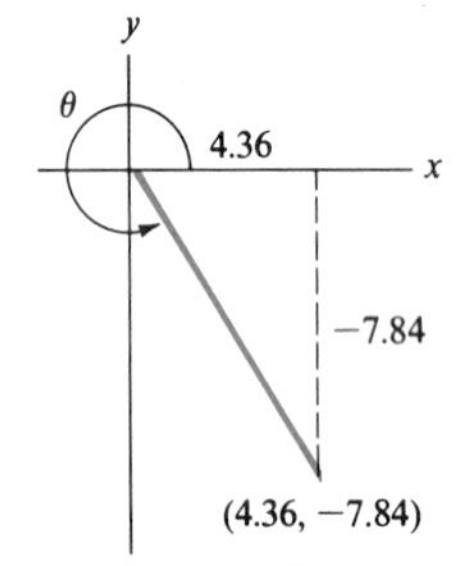

Figure 15.54

Solution. The angle θ is shown in Figure 15.54.

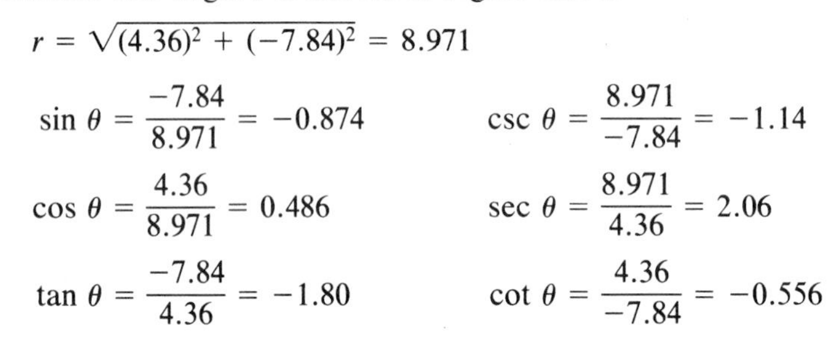

$$r = \sqrt{(4.36)^2 + (-7.84)^2} = 8.971$$

$$\sin\theta = \frac{-7.84}{8.971} = -0.874 \qquad \csc\theta = \frac{8.971}{-7.84} = -1.14$$

$$\cos\theta = \frac{4.36}{8.971} = 0.486 \qquad \sec\theta = \frac{8.971}{4.36} = 2.06$$

$$\tan\theta = \frac{-7.84}{4.36} = -1.80 \qquad \cot\theta = \frac{4.36}{-7.84} = -0.556$$

Remark. To minimize round-off errors in the intermediate calculations, r was written with an extra digit. It is even better to store the calculated value for r in the memory and press [MR] whenever the value is needed:

4.36 [x^2] [+] 7.84 [+/−] [x^2] [=] [√] [STO]

Now $\sin\theta$ is found by the sequence

7.84 [+/−] [÷] [MR] [=] → −0.8739468

So $\sin\theta = -0.874$, as before.

To find the **sign of the value of a trigonometric function**, we locate the quadrant of the angle and determine the sign from the definition.

Example 3 Determine the sign of cos 137°.

Solution. The angle, shown in Figure 15.55, is in quadrant II. For any point (x, y) in the second quadrant,

$$x < 0 \qquad \text{and} \qquad y > 0$$

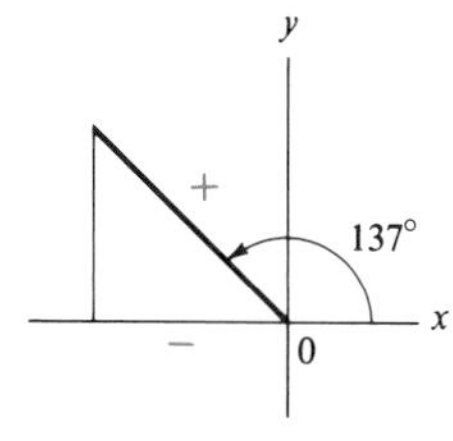

Figure 15.55

Since r is always positive,

$$\cos \theta = \frac{x}{r} \text{ is negative}$$

The conclusion also follows from Figure 15.55. By placing the signs of x and r right on the triangle, we get

$$\cos \theta = \frac{-}{+} = -$$

Example 4 Determine the sign of tan 310°.

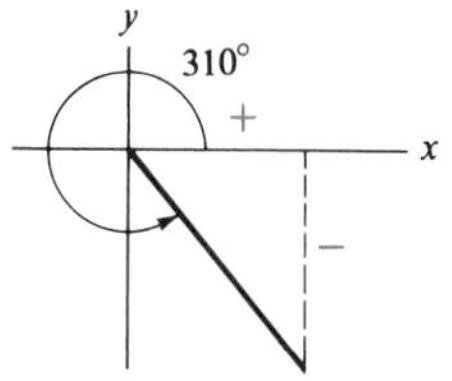

Solution. The angle, shown in Figure 15.56, is in quadrant IV. For any point (x, y) in the fourth quadrant,

$$x > 0 \qquad \text{and} \qquad y < 0$$

So

$$\tan \theta = \frac{y}{x} \text{ is negative}$$

Figure 15.56

Using Figure 15.56, we also have

$$\tan \theta = \frac{-}{+} = -$$

We can see from these examples that all function values are positive in the first quadrant. The sine and cosecant functions are positive in quadrants I and II. The tangent and cotangent functions are positive in quadrants I and III. Finally, the cosine and secant functions are positive in quadrants I and IV.

Example 5 Determine the quadrant in which θ lies, given that $\cot \theta > 0$ and $\sec \theta < 0$.

Solution. *Using the definitions:*

$$\cot \theta = \frac{x}{y} > 0 \quad \text{in I and III}$$

and

$$\sec \theta = \frac{r}{x} < 0 \quad \text{in II and III}$$

It follows that θ is in quadrant III.

Using diagrams:

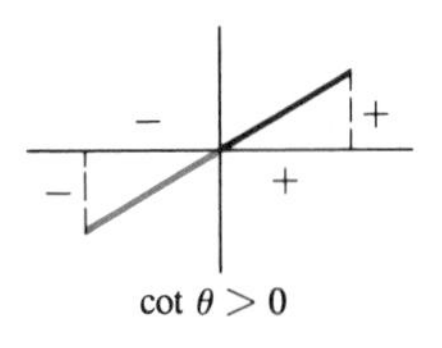

Figure 15.57

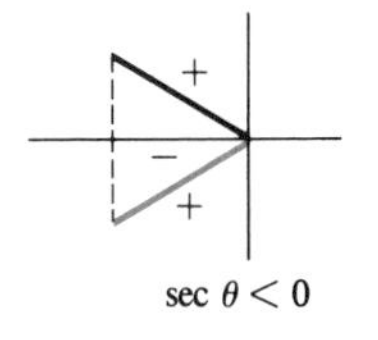

Figure 15.58

By Figure 15.57, $\cot\theta > 0$ in I and III. By Figure 15.58, $\sec\theta < 0$ in II and III. So θ is in quadrant III.

Exercises / Section 15.6

In Exercises 1–10, find the values of the indicated trigonometric functions for the angle θ in standard position whose terminal side passes through the given point. Express answers in exact form. (See Example 1.)

1. $(5, 12)$ $\cos\theta$, $\tan\theta$
2. $(4, 3)$ $\tan\theta$, $\sec\theta$
3. $(-4, 3)$ $\sin\theta$, $\cot\theta$
4. $(-12, -5)$ $\cot\theta$, $\csc\theta$
5. $(3, -2)$ $\csc\theta$, $\tan\theta$
6. $(-3, 5)$ $\sec\theta$, $\tan\theta$
7. $(-\sqrt{3}, -\sqrt{6})$ $\sin\theta$, $\cos\theta$
8. $(\sqrt{5}, -2)$ $\sec\theta$, $\sin\theta$
9. $(-\sqrt{2}, 4)$ $\sec\theta$, $\tan\theta$
10. $(-3, -\sqrt{3})$ $\cot\theta$, $\csc\theta$

In Exercises 11–20, find the values of the indicated trigonometric functions for the angle θ in standard position whose terminal side passes through the given point. Express answers in decimal form. (See Example 2.)

11. $(-2.34, 1.07)$ $\tan\theta$, $\sec\theta$
12. $(2.06, -3.60)$ $\sin\theta$, $\cot\theta$
13. $(-7.6, -3.9)$ $\cot\theta$, $\csc\theta$
14. $(-5.8, -2.0)$ $\csc\theta$, $\tan\theta$
15. $(11.0, -15.6)$ $\cos\theta$, $\tan\theta$
16. $(-14.6, 20.1)$ $\sin\theta$, $\cos\theta$
17. $(2.05, 3.67)$ $\sec\theta$, $\sin\theta$
18. $(-1.48, -6.70)$ $\csc\theta$, $\cot\theta$
19. $(-6.60, -10.3)$ $\cos\theta$, $\csc\theta$
20. $(5.90, -8.70)$ $\cos\theta$, $\tan\theta$

In Exercises 21–40, determine the algebraic sign of each trigonometric function. (See Examples 3 and 4.)

21. $\sin 80°$
22. $\cos 85°$
23. $\tan 110°$
24. $\cot 100°$
25. $\sec 160°$
26. $\sec 175°$
27. $\csc 210°$
28. $\tan 230°$
29. $\cos 285°$
30. $\cot 280°$
31. $\tan 320°$
32. $\sec 330°$
33. $\sin 350°$
34. $\tan 354°$
35. $\cos(-35°)$
36. $\cos(-60°)$
37. $\csc 144°$
38. $\csc 170°$
39. $\sec 260°$
40. $\cos 220°$

In Exercises 41–56, determine the quadrant in which θ must lie to satisfy the given conditions. (See Example 5.)

41. $\sin\theta > 0$, $\cos\theta < 0$
42. $\sin\theta < 0$, $\cos\theta > 0$
43. $\tan\theta > 0$, $\sec\theta < 0$
44. $\tan\theta < 0$, $\sec\theta > 0$
45. $\csc\theta > 0$, $\sec\theta < 0$
46. $\csc\theta > 0$, $\cos\theta > 0$

47. $\cos\theta > 0$, $\cot\theta < 0$
48. $\cot\theta < 0$, $\sin\theta > 0$
49. $\sec\theta < 0$, $\sin\theta < 0$
50. $\tan\theta > 0$, $\csc\theta < 0$
51. $\csc\theta > 0$, $\cot\theta < 0$
52. $\sin\theta < 0$, $\sec\theta < 0$
53. $\cot\theta > 0$, $\sin\theta < 0$
54. $\tan\theta < 0$, $\cos\theta < 0$
55. $\cos\theta > 0$, $\cot\theta > 0$
56. $\sin\theta > 0$, $\sec\theta < 0$

15.7 Values of Trigonometric Functions

In this section we will determine the values of trigonometric functions of angles in any quadrant.

Special Angles

To find the values of trigonometric functions of special angles, we first need to recall the 30°-60° right triangle and the 45° right triangle (Figure 15.59).

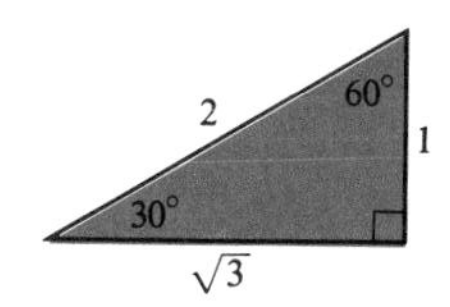

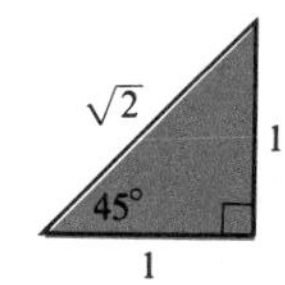

Figure 15.59

Now consider the special angle 150° shown in Figure 15.60. Since the angle $\alpha = 180° - 150° = 30°$, the point $(-\sqrt{3}, 1)$ lies on the terminal side. So $\sin 150° = \frac{1}{2}$ and $\cos 150° = -\sqrt{3}/2$. The angle α is called the **reference angle**. The reference angle is the acute angle formed by the terminal side and the x-axis.

Reference angle

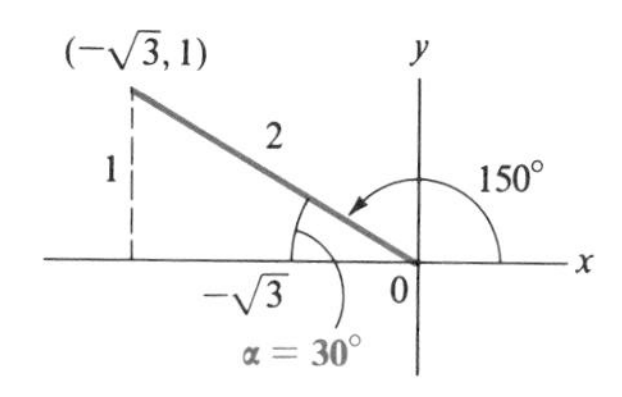

Figure 15.60

The values of the trigonometric functions can be given in terms of the reference angles: If α is the reference angle of θ, then the value of the trigonometric function of θ is numerically equal to the value of the trigonometric function of α. The sign depends on the quadrant in which θ lies. Consider the next example.

Example 1 Find $\tan 210°$ and $\sec 210°$.

Solution. The angle is shown in Figure 15.61.

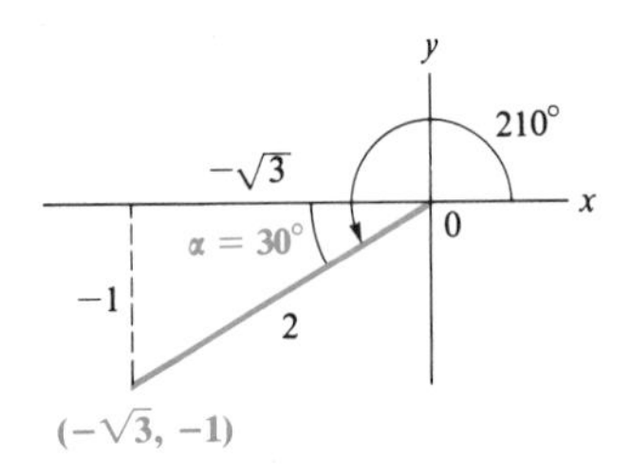

Figure 15.61

Since $210° - 180° = 30°$, we see that the reference angle α is 30°. So the point $(-\sqrt{3}, -1)$ lies on the terminal side. It follows that

$$\tan 210° = \frac{-1}{-\sqrt{3}} = \frac{1}{\sqrt{3}} = \frac{1}{\sqrt{3}} \cdot \frac{\sqrt{3}}{\sqrt{3}} = \frac{\sqrt{3}}{3}$$

and

$$\sec 210° = \frac{2}{-\sqrt{3}} = -\frac{2}{\sqrt{3}} = -\frac{2}{\sqrt{3}} \cdot \frac{\sqrt{3}}{\sqrt{3}} = -\frac{2\sqrt{3}}{3}$$

Example 2 Find sin 315° and cot 315°.

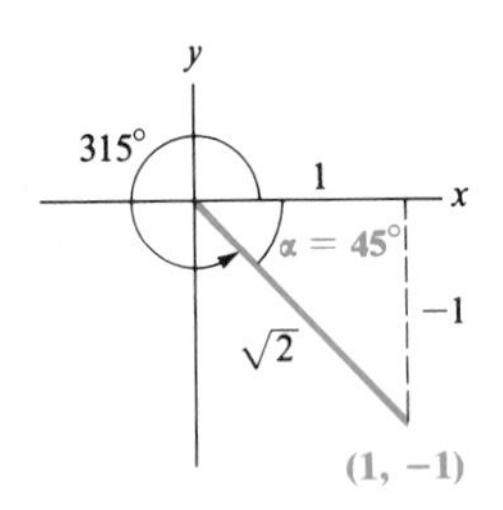

Figure 15.62

Solution. The angle is shown in Figure 15.62.

Since $360° - 315° = 45°$, the reference angle α is 45°. So the point $(1, -1)$ lies on the terminal side. Thus

$$\sin 135° = \frac{-1}{\sqrt{2}} = -\frac{1}{\sqrt{2}} = -\frac{1}{\sqrt{2}} \cdot \frac{\sqrt{2}}{\sqrt{2}} = -\frac{\sqrt{2}}{2}$$

and

$$\cot 135° = \frac{1}{-1} = -1$$

To find the values of quadrantal angles, we return to the definitions.

Example 3 Find cos 180° and tan 180°.

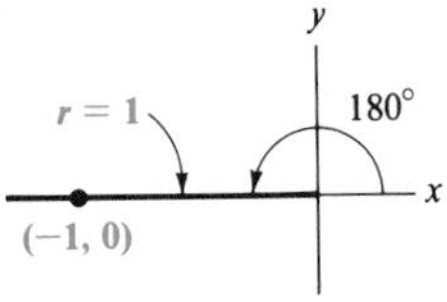

Solution. To find the functions of 180°, let $(-1, 0)$ be a point on the terminal side (Figure 15.63). Since $r = 1$, we have

$$\cos 180° = \frac{x}{r} = \frac{-1}{1} = -1$$

and

$$\tan 180° = \frac{y}{x} = \frac{0}{-1} = 0$$

Figure 15.63

Example 4 Find sin 270° and sec 270°.

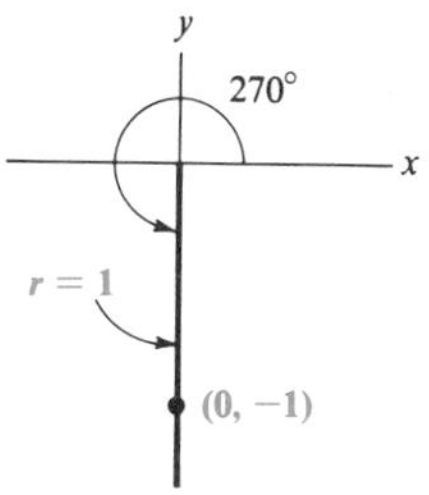

Solution. To find the functions of 270°, let (0, −1) be a point on the terminal side (Figure 15.64). Since $r = 1$, we get

$$\sin 270^\circ = \frac{y}{r} = \frac{-1}{1} = -1$$

and

$$\sec 270^\circ = \frac{r}{x} = \frac{1}{0} \qquad \text{(undefined)}$$

y
270°
x
r = 1
(0, −1)

Figure 15.64

Tables

We will discuss the use of tables only briefly. (Ordinarily, we find the values of trigonometric functions with a calculator.)

Recall that Table 1 in Appendix B lists the values of the trigonometric functions only for angles from 0° to 90°. Outside this range, the angles in the table serve only as reference angles.

Example 5 Find sin 170°.

Solution. Since 180° − 170° = 10°, the reference angle is 10°. Since $\sin \theta > 0$ in II,

$$\sin 170^\circ = \sin 10^\circ = 0.1736$$

Example 6 Find cot 347°40′.

Solution. The reference angle is found as follows:

$$\begin{array}{r} 359^\circ 60' \\ \underline{347^\circ 40'} \\ 12^\circ 20' \end{array}$$

Since $\cot \theta < 0$ in IV,

$$\cot 347^\circ 40' = -\cot 12^\circ 20' = -4.574$$

Calculators

To find the values of trigonometric functions with a calculator, we use the procedure discussed in Section 15.3.

Example 7 Find

a. $\sin 263.4°$ **b.** $\sec 228°14'$

Solution.

a. The sequence is

263.4 [SIN] → −0.9933727

Using four decimal places,

$$\sin 263.4° = -0.9934$$

b. Here we change the angle to decimal form first:

$$228°14' = \left(228 + \frac{14}{60}\right)^{\circ}$$

The sequence is

228 [+] 14 [÷] 60 [=] [COS] [1/x] → −1.5012791

Using four decimal places,

$$\sec 228°14' = -1.5013$$

A special problem arises when finding an angle, given the value of a trigonometric function. Since the calculator has no way of knowing what quadrant you have in mind, the angles are always given in the following range:

function	*range*
$\sin\theta$	$-90° \le \theta \le 90°$
$\tan\theta$	$-90° < \theta < 90°$
$\cos\theta$	$0° \le \theta \le 180°$

Example 8 Find θ, given that $\sin\theta = -0.6039$.

Solution. The sequence is

0.6039 [+/−] [INV] [SIN] → −37.149728

So $\theta = -37.15°$ to the nearest hundredth of a degree.

In many cases the angle obtained with a calculator will serve only as a reference angle. Consider the next example.

Example 9 Find θ, given that $\sin \theta = -0.2986$, θ in quadrant III.

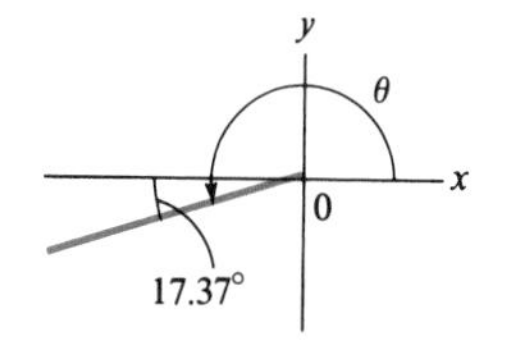

Figure 15.65

Solution. The sequence is

0.2986 [+/−] [INV] [SIN] → −17.373535

To find the angle in the third quadrant, we use the fact that 17.37° is the reference angle. (See Figure 15.65.) So

$$\theta = 180° + 17.37° = 197.37°$$

Note: The reference angle can also be found by ignoring the minus sign in the given function value and using the sequence

0.2986 [INV] [SIN] → 17.373535

Example 10 Find θ, given that $\cot \theta = -2.164$, in quadrant II.

Figure 15.66

Solution. The sequence is

2.164 [+/−] [1/x] [INV] [TAN] → −24.801999

or

2.164 [1/x] [INV] [TAN] → 24.801999

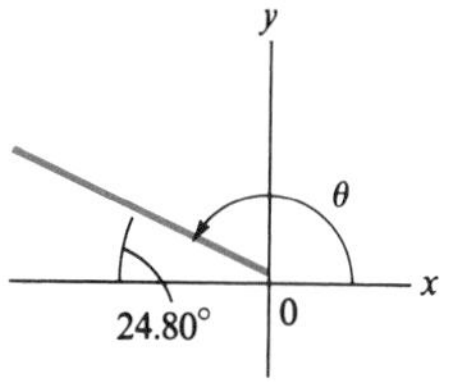

So the reference is 24.80°. To find θ in quadrant II, we subtract the value of the reference angle from 180° (Figure 15.66):

$$\theta = 180° - 24.80° = 155.20°$$

Trigonometric functions in BASIC require the use of radian measure. We will discuss these topics in Section 17.1.

Example 11 If a projectile is hurled with velocity v (in feet per second) along an inclined plane making a constant angle α with the horizontal, then the range R of the projectile up the plane is

$$R = \frac{2v^2 \cos \theta \sin(\theta - \alpha)}{32(\cos \alpha)^2}$$

where θ is the angle with the horizontal at which the projectile is aimed. Find the range if $v = 92.5$ ft/s, $\alpha = 11.5°$, and $\theta = 22.0°$.

Solution. By the given formula,

$$R = \frac{2(92.5)^2 \cos 22.0° \sin(22.0° - 11.5°)}{32(\cos 11.5°)^2}$$

The sequence is

2 [×] 92.5 [x^2] [×] 22.0 [COS] [×] [(] 22.0 [−] 11.5 [)] [SIN] [÷] 32 [÷] 11.5 [COS] [x^2] [=] → 94.097259

So the projectile lands 94.1 ft away along the plane.

Exercises / Section 15.7

In Exercises 1–34, find the exact values of the given functions of special angles. (See Examples 1–4.)

1. sin 45° **2.** cos 60° **3.** tan 120°
4. cos 120° **5.** sec 135° **6.** csc 135°
7. cos 135° **8.** sin 150° **9.** tan 150°
10. cot 150° **11.** csc 210° **12.** cos 210°
13. sec 240° **14.** cot 240° **15.** cot 225°
16. tan 225° **17.** csc 300° **18.** sin 300°
19. cot 300° **20.** sin 315° **21.** csc 315°
22. sec 315° **23.** sin 330° **24.** tan 330°
25. sin 0° **26.** cos 90° **27.** tan 270°
28. sec 180° **29.** sin 270° **30.** sin 180°
31. cos 270° **32.** tan 0° **33.** cot 270°
34. sec 90°

In Exercises 35–42, find the values of the given trigonometric functions by using Table 1 in Appendix B.

35. sin 132°40′ **36.** cos 232°10′ **37.** sec 100°10′
38. tan 140°20′ **39.** cot 215°30′ **40.** csc 264°50′
41. cos 204°20′ **42.** sec 320°10′

In Exercises 43–70, use a calculator to find the values of the given trigonometric functions to four decimal places.

43. sin 110.4° **44.** cos 131.6° **45.** tan 159.6°
46. cot 213.4° **47.** sec 183.7° **48.** csc 284.7°
49. cos 326.0° **50.** sin 76.73° **51.** cot 234.1°
52. tan 349.2° **53.** csc 119.3° **54.** sec 100.5°
55. sin 316.9° **56.** cos 116.8° **57.** tan 341.5°
58. cot 229.2° **59.** sec(−16.4°) **60.** csc(−26.8°)
61. cos(−39.3°) **62.** sec(−75.7°) **63.** sin 386.0°
64. cos 406.1° **65.** sin 124°16′ **66.** cos 284°37′
67. tan 293°26′ **68.** cot 117°42′ **69.** sec 236°55′
70. csc 331°18′

In Exercises 71–84, find θ to the nearest hundredth of a degree.

71. $\sin\theta = 0.2842$, θ in II
72. $\cos\theta = 0.3173$, θ in IV
73. $\tan\theta = -2.856$, θ in IV
74. $\cot\theta = -0.9577$, θ in II
75. $\sec\theta = 2.749$, θ in IV
76. $\csc\theta = 2.865$, θ in II
77. $\cos\theta = -0.1782$, θ in II
78. $\sin\theta = -0.3227$, θ in III
79. $\cot\theta = -1.836$, θ in IV
80. $\tan\theta = -0.4917$, θ in II
81. $\csc\theta = 2.614$, θ in I
82. $\sec\theta = 3.715$, θ in I
83. $\cos\theta = -0.1612$, θ in II
84. $\sin\theta = -0.8163$, θ in III

85. The *index of refraction* μ of a medium is given by

$$\mu = \frac{\sin\theta}{\sin\alpha} \qquad \mu = \text{mu}$$

where θ is the angle of incidence and α the angle of refraction. Find μ if $\theta = 38.16°$ and $\alpha = 26.74°$.

86. The instantaneous voltage in a coil rotating in a magnetic field is given by $V = V_m \cos\theta$, where V_m is the maximum voltage and θ is the angle that the coil makes with the magnetic field. Given that $V_m = 115$ V, find V at the instant when $\theta = -36.4°$.

87. The range R (in meters) along the ground of a projectile fired at velocity v (in meters per second) at an angle θ with the horizontal is

$$R = \frac{v^2}{9.8}\sin 2\theta$$

Find the range if $\theta = 28.4°$ and $v = 18.6$ m/s.

88. The largest weight that can be pulled up a plane inclined at an angle θ with the horizontal by a force F is

$$W = \frac{F}{\mu}(\cos\theta + \mu\sin\theta)$$

where μ is the coefficient of friction. Find W if $\theta = 25.3°$, $F = 65.4$ lb, and $\mu = 0.320$.

Review Exercises / Chapter 15

In Exercises 1–4, find the smallest positive angle coterminal with the given angle.

1. 394°17′
2. −14°39′
3. −27°44′
4. −193°54′

In Exercises 5–8, find the values of the six trigonometric functions for angles in standard position and with the given point on the terminal side. Express answers in exact form.

5. $(1, \sqrt{3})$
6. $(\sqrt{2}, \sqrt{7})$
7. $(-\sqrt{3}, \sqrt{6})$
8. $(-1, -\sqrt{15})$

In Exercises 9–12, find the values of the six trigonometric functions for angles in standard position and with the given point on the terminal side. Express answers in decimal form.

9. (1.0, 4.3)
10. (2.6, −3.7)
11. (−7.640, −10.91)
12. (−4.81, 2.75)

In Exercises 13–16, θ is in quadrant I.

13. Given $\sin\theta = \frac{1}{3}$, find $\cos\theta$ in exact form.

14. Given $\cot\theta = \frac{5}{2}$, find $\csc\theta$ in exact form.

15. Given $\tan\theta = 1.87$, find $\sec\theta$.

16. Given $\cos\theta = \frac{5}{9}$, find $\sec\theta$.

In Exercises 17–20, determine the algebraic sign of each trigonometric function.

17. $\sin 165°$

18. $\sec 223°$

19. $\cot 263°$

20. $\tan 347°$

In Exercises 21–24, determine the quadrant in which θ must lie to satisfy the given conditions.

21. $\cos\theta < 0$, $\tan\theta > 0$

22. $\csc\theta < 0$, $\cot\theta < 0$

23. $\sin\theta > 0$, $\sec\theta < 0$

24. $\tan\theta > 0$, $\sin\theta < 0$

In Exercises 25–34, find the exact values of the indicated trigonometric functions. (Do not use a calculator.)

25. $\tan 60°$

26. $\sin 90°$

27. $\sec 120°$

28. $\cot 150°$

29. $\tan 180°$

30. $\csc 225°$

31. $\sin 240°$

32. $\cos 270°$

33. $\sec 315°$

34. $\tan 330°$

In Exercises 35–38, use a diagram to find θ, θ in quadrant I.

35. $\cos\theta = \frac{\sqrt{2}}{2}$

36. $\tan\theta = \frac{\sqrt{3}}{3}$

37. $\sin\theta = \frac{\sqrt{3}}{2}$

38. $\csc\theta = \sqrt{2}$

In Exercises 39–42, use Table 1 in Appendix B to find the value of each trigonometric function.

39. $\sin 46°14'$

40. $\cos 132°10'$

41. $\tan 192°40'$

42. $\sec 316°50'$

In Exercises 43–50, use a calculator to determine the values of the given trigonometric functions. Give answers to four decimal places.

43. $\sin 14.7°$

44. $\cos 97.8°$

45. $\tan 117.4°$

46. $\csc 152.7°$

47. $\sec 210.8°$

48. $\cot 263.2°$

49. $\sin 318°14'$

50. $\cos 348°35'$

51. If $\cos\theta = 0.2813$, θ in quadrant I, find θ in degrees and minutes.

In Exercises 52–57, use a calculator to find θ to the nearest hundredth of a degree.

52. $\sin\theta = 0.7218$, θ in II

53. $\cos\theta = -0.5816$, θ in II

54. $\tan\theta = -3.8147$, θ in IV

55. $\cot\theta = 2.015$, θ in III

56. $\csc\theta = -1.8041$, θ in IV

57. $\sec\theta = 3.120$, θ in I

58. A weight W is to be dragged along a horizontal plane by a force whose line of action makes an angle θ with the plane. The force required to move the weight is given by

$$F = \frac{\mu W}{\mu \sin \theta + \cos \theta}$$

where μ is the coefficient of friction. Find F, if $W = 87.3$ lb, $\theta = 25.6°$, and $\mu = 0.250$.

59. From a point 231 ft from the base of the Washington Monument, the angle of elevation of the top is 67.4°. Find the height of the monument.

60. A flagpole casts a shadow 43.0 m long. Given that the angle of elevation of the sun is 15.4°, determine the height of the flagpole.

61. A straight highway rises 152 m for every 915 m along the road. Find the angle of inclination of the highway.

62. The conical tank in Figure 15.67 measures 4.00 m across the top, and the side makes an angle of 21.0° with the vertical. Find the volume of the tank.

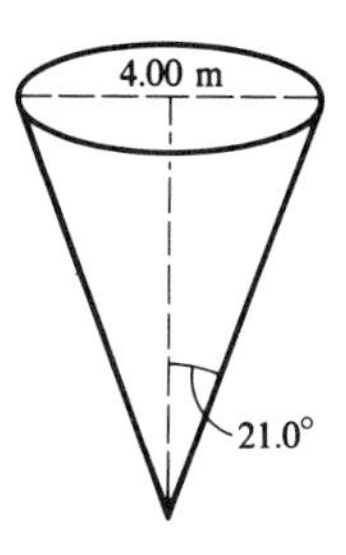

Figure 15.67

63. In an alternating-current circuit containing a resistance R and an inductive reactance X_L, the phase angle ϕ is the angle between the impedance Z and the resistance. Find the phase angle if $R = 20.8\ \Omega$ and $X_L = 50.4\ \Omega$. (See Figure 15.51.)

64. Twelve bolts are equally spaced on the rim of a circular metal plate 28.60 cm in diameter. Find the center-to-center distance between any two adjacent bolts.

65. A regular pentagon is inscribed in a circle of radius 4.735 cm. Find the perimeter of the pentagon.

66. The angle of depression from a window to a fire hydrant 65.8 ft from the base of the building is 52°50′. How high is the window above the ground?

67. A tower is standing on top of a cliff 165 ft high. From a distant point, the angle of elevation of the bottom of the tower is 40°30′ and the angle of elevation of the top is 51°10′. How tall is the tower?

68. From the top of a lighthouse 41.0 m above the surface of the water, the angles of depression of two boats due west are 11°40′ and 13°10′, respectively. Determine the distance between the boats.

Cumulative Review Exercises / Chapters 13–15

1. Solve the following system of equations graphically. Estimate the solution to the nearest tenth of a unit.

$$3x + 2y = 6$$
$$x - y = 1$$

2. Solve the system

$$3x + 2y = 8$$
$$x - y = 1$$

a. By addition or subtraction
b. By substitution

3. A rectangular machine part must meet the following specifications: The length is 2.5 cm more than twice the width, and the perimeter is 25.4 cm. Find the dimensions.

4. The acute angles of a right triangle differ by 7.4°. Find the angles.

5. In the figure, $AC \| DE$. Determine $\angle 1$ and $\angle 2$.

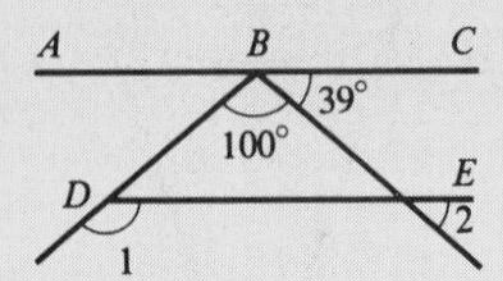

6. Find x in the figure.

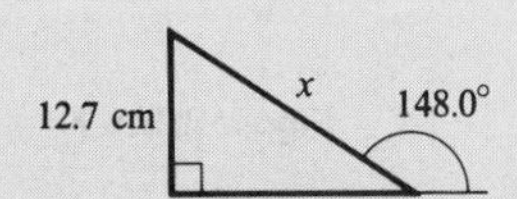

7. The radius of the circle in the figure is 10.0 in. Find x.

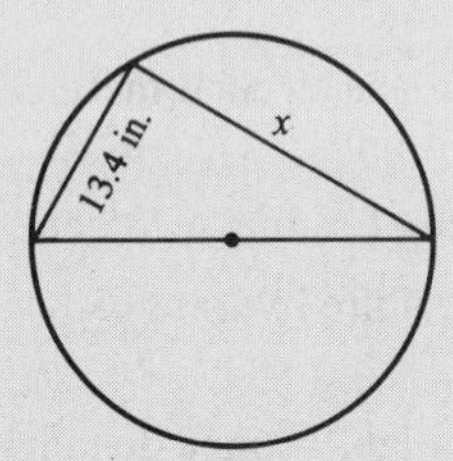

8. In the figure, $BC = 6.00$ m. Find AB.

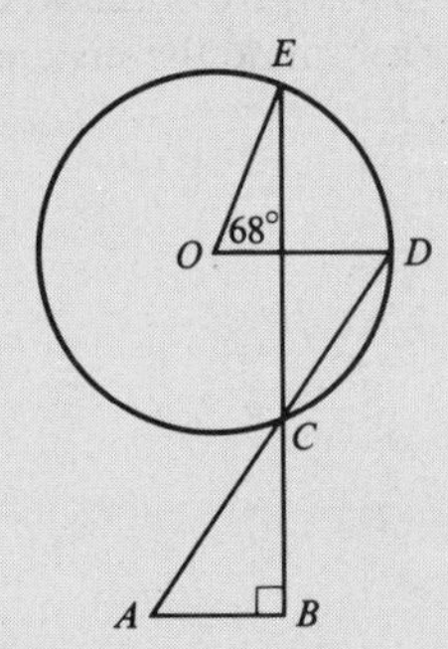

9. Find x in the figure.

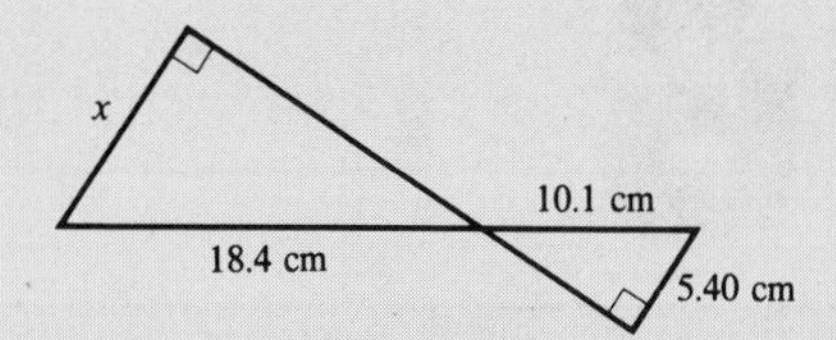

10. One angle of a parallelogram is 115°. What is the sum of the remaining angles?

In Exercises 11 and 12, find the values of the six trigonometric functions for the angles in standard position and with the given point on the terminal side.

11. $(-\sqrt{2}, -\sqrt{7})$; give answers in exact form.

12. $(-5.85, 4.32)$; give answers in decimal form.

13. If $\cot \theta > 0$ and $\sec \theta < 0$, in what quadrant must θ lie?

14. Show that
a. $(\cos 150°)^2 + (\sin 150°)^2 = 1$
b. $1 + (\tan 315°)^2 = (\sec 315°)^2$

15. Find θ, given that $\csc \theta = -2.8471$, θ in quadrant III.

16. For what values of θ $(0° \leq \theta < 360°)$ is $\sec \theta$ undefined?

17. Find the volume and lateral surface area of the following pyramid: The base is a square 16.4 cm on the side, and the altitude is 11.8 cm.

18. Find the volume and lateral surface area of the following cylinder: diameter of base = 8.6 in., height = 9.7 in.

19. Find the volume and surface area of a sphere of radius 6.074 cm.

20. Find the volume and lateral surface area of the conical tank in the figure.

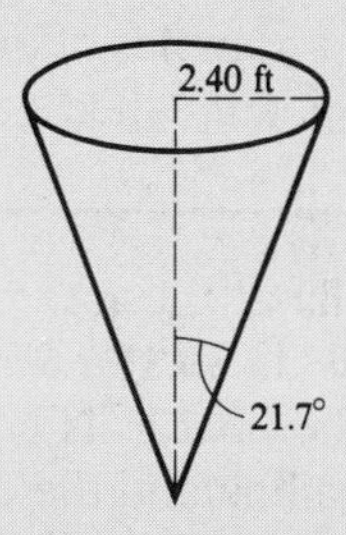

21. The glass prism in the figure is used in the study of refraction of light. Determine its volume.

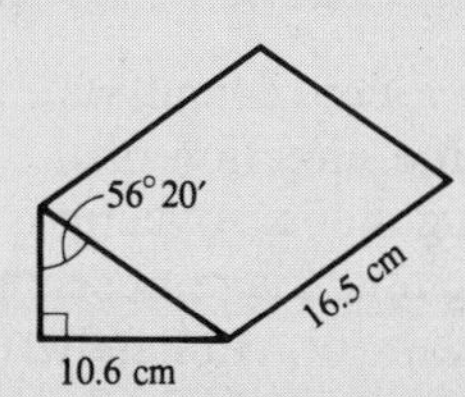

22. A tank full of water has the shape of a rectangular solid with dimensions 3.40 ft by 1.82 ft by 2.19 ft. Given that water weighs 62.4 lb/ft^3, find the weight of the water.

CHAPTER **16**

Vectors and Oblique Triangles

16.1 Vectors

Introduction

Most quantities we have discussed so far can be fully described by a number. Examples are areas, volumes, and temperature. Quantities that can be fully described by magnitude are called **scalar** quantities.

Scalar

Other physical phenomena require, for a full description, both **magnitude and direction.** Examples of these types of phenomena are velocities and forces. Such entities are called **vectors.**

Vector

Example 1 Suppose a boat sails at the rate of 18 mi/h. From this statement we know only the rate traveled. So 18 mi/h is a *scalar* quantity.

If we say, "The boat sails east at 18 mi/h," we know both the rate at which the boat is sailing and the direction. In particular, we can now determine its position at any time. So the velocity is a *vector*.

An arrow is a convenient way to represent a vector graphically. The arrow points in the direction of the vector, while the length of the arrow represents the magnitude. A vector is denoted by a boldface letter such as **A**, **B**, or **C**. The magnitude is represented by the same letter in italics, *A*, *B*, or *C*, or by $|\mathbf{A}|$, $|\mathbf{B}|$, or $|\mathbf{C}|$. (In handwriting, a common notation for **A** is $\vec{A}$.)

Initial and terminal points

The base of the arrow is called the **initial point.** The tip of the arrow is the **terminal point.** (See Figure 16.1.)

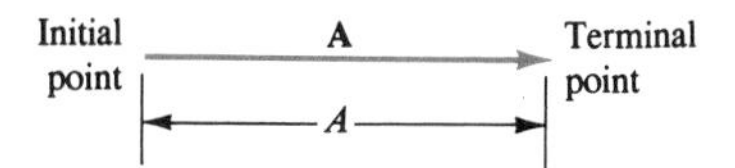

Figure 16.1

Addition of Vectors

If two vectors are to be added, both magnitude and direction must be taken into account. For example, if a plane travels north at 200 km/h and if the wind is from the south at 45 km/h, we expect the velocity to be 245 km/h. But what if the wind is from the east? Then we have the situation pictured in Figure 16.2. The resulting velocity vector, called the *resultant,* is the hypotenuse of the triangle determined by the given velocity vectors.

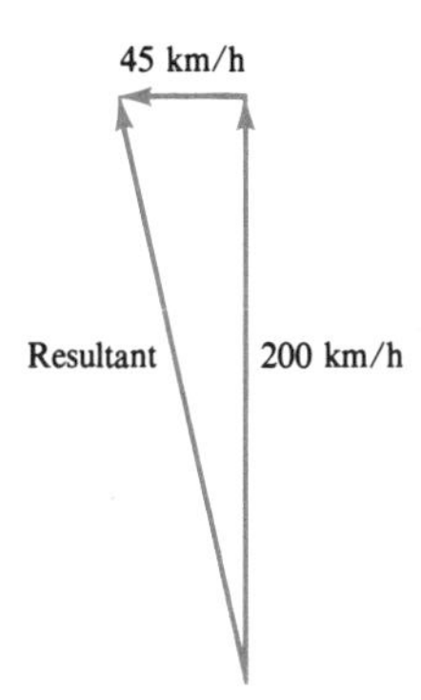

Figure 16.2

Resultant

To generalize, let **A** and **B** represent two vectors with the same initial point O (Figure 16.3). In the parallelogram determined by **A** and **B**, the vector sum **A** + **B**, called the **resultant,** is the vector with initial point O coinciding with the diagonal of the parallelogram. The initial point is introduced here only for convenience, for the initial point of a vector can be placed anywhere in the plane. For that reason the sum is also given by the vector in Figure 16.4.

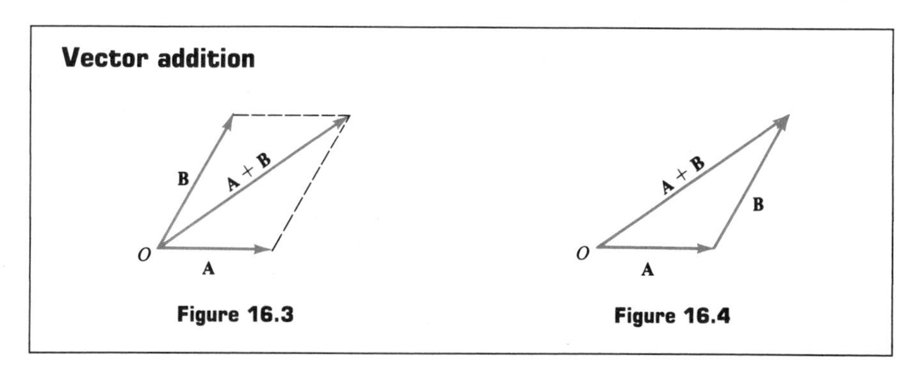

Figure 16.3

Figure 16.4

If more than two vectors are added, then the procedure shown in Figure 16.4 is by far the most convenient.

Example 2 Graphically construct the sum of the vectors in Figure 16.5.

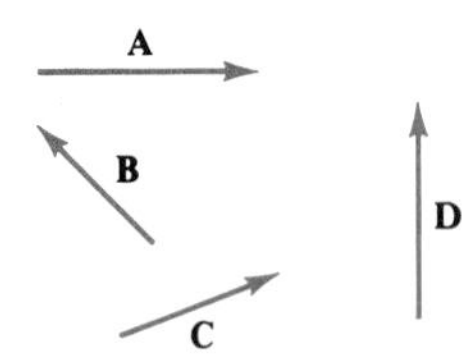

Figure 16.5

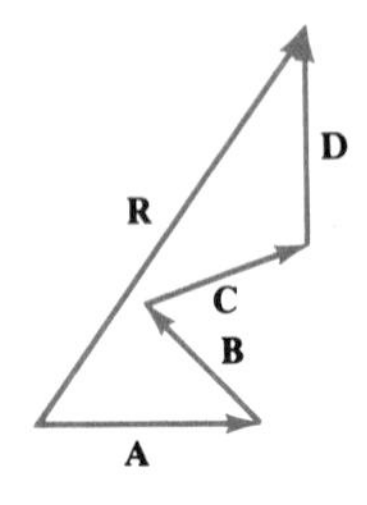

Figure 16.6

Solution. Draw vector **A**. Place the initial point of **B** at the terminal point of **A**. Now place the initial point of **C** at the terminal point of **B**, and the initial point of **D** at the terminal point of **C**. The final resultant **R** is the vector from the initial point of **A** to the terminal point of **D**, as shown in Figure 16.6.

While the vectors in Figure 16.6 were added in alphabetical order, they can actually be added in any order. For example, the same resultant **R** can be obtained by placing the vectors **A**, **C**, **D**, and **B** together as shown in Figure 16.7.

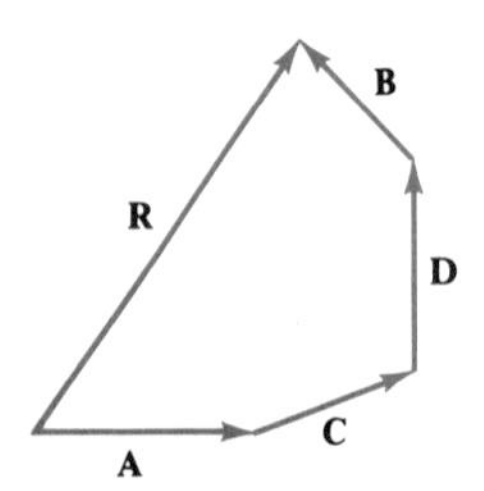

Figure 16.7

Scalar Multiplication

Scalar product

If a vector **A** is multiplied by a number c, the resulting vector $c\mathbf{A}$ is called the **scalar product.** The scalar product $c\mathbf{A}$ has the same direction as **A**, but the magnitude is multiplied by c. The resulting magnitude is therefore cA or $c|\mathbf{A}|$. For example, the vector $3\mathbf{A}$ is three times as long as **A** and has the same direction as **A**.

Example 3 Given the vector **A** in Figure 16.8, draw $2\mathbf{A}$ and $\frac{1}{2}\mathbf{A}$.

Figure 16.8

Solution. Both vectors, $2\mathbf{A}$ and $\frac{1}{2}\mathbf{A}$, have the same direction as **A**. The vector $2\mathbf{A}$ is twice as long as **A** (Figure 16.9) and $\frac{1}{2}\mathbf{A}$ is half as long as **A** (Figure 16.10).

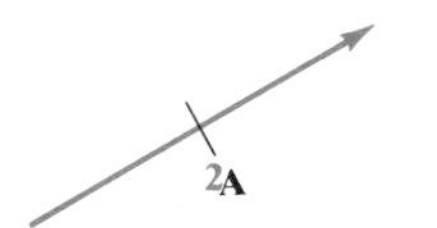

Figure 16.9

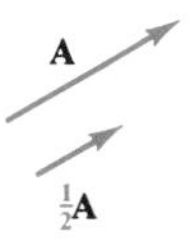

Figure 16.10

Example 4 Given the vectors **A** and **B** in Figure 16.11, draw $3\mathbf{A} + 2\mathbf{B}$.

Figure 16.11

Solution. From vector **A**, construct the vector $3\mathbf{A}$ (see Figure 16.12). From vector **B**, construct the vector $2\mathbf{B}$. The resultant $\mathbf{R} = 3\mathbf{A} + 2\mathbf{B}$ is shown in the figure.

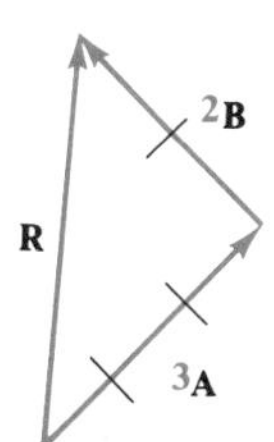

Figure 16.12

Subtraction of Vectors

To subtract vectors, we define $-\mathbf{B}$ to be a vector that has the same magnitude as **B**, but whose direction is opposite that of **B**. So

$$\mathbf{A} - \mathbf{B} = \mathbf{A} + (-\mathbf{B})$$

which can be added in the usual way. (See Figure 16.13.)

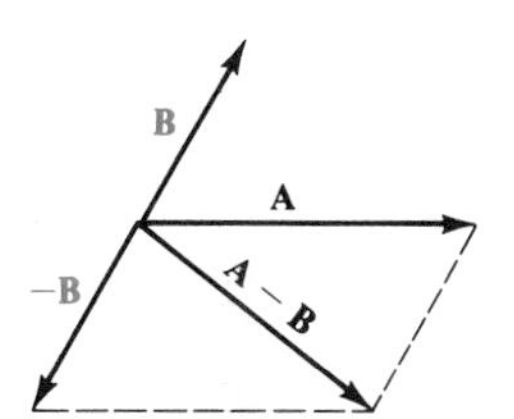

Figure 16.13

It can be seen from Figure 16.13 that **A** − **B** is the second diagonal in the parallelogram determined by **A** and **B**, as shown in Figure 16.14.

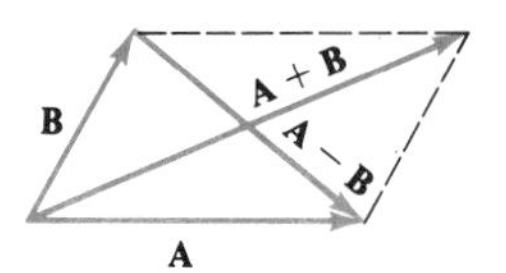

Figure 16.14

Exercises / Section 16.1

Carry out the indicated operations graphically with the vectors in Figure 16.15.

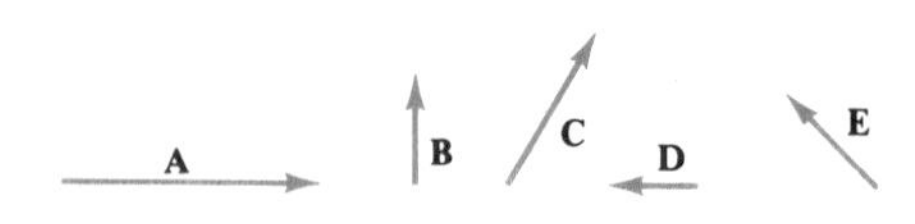

Figure 16.15

1. **A** + **B**
2. **A** − **B**
3. **A** − **C**
4. **A** + **C**
5. **A** + **B** + **D**
6. **A** + **C** + **E**
7. **A** + **B** + **C** + **D** + **E**
8. **A** + **B** + **C** + 2**D**
9. 2**A** + **E**
10. **B** + 2**C**
11. 2**B** + 3**C**
12. 2**C** + 3**D**
13. 2**A** − 3**B** + 2**D**
14. 2**D** − 3**E**
15. 2**B** − 3**C** − 2**D**
16. **A** − 3**C** + 2**E**
17. **B** + 2**C** − 3**D**
18. 2**B** + **C** − 4**D** − 2**E**
19. **B** + **C** − **D** − 2**E**
20. **A** + **B** − **D** + 3**E**

16.2 Vector Components

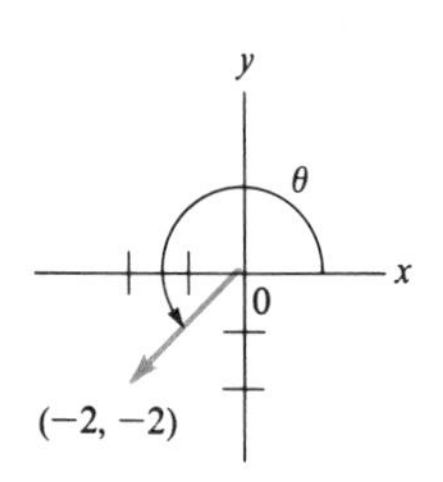

Figure 16.16

In the last section we considered some of the operations with vectors by means of diagrams. While graphical methods can teach us a lot about vectors, diagrams are not very practical for computational work. In this section we will see how the magnitude and direction of the resultant can be determined exactly.

To this end we need to combine the vector concept with that of a coordinate system: We place the initial point of a vector at the origin, so that the terminal point uniquely determines the vector. For example, the vector in Figure 16.16 can be specified by the point $(-2, -2)$. The magnitude can be computed directly using the Pythagorean theorem and turns out to be $\sqrt{(-2)^2 + (-2)^2} = \sqrt{4 + 4} = \sqrt{8} = \sqrt{4 \cdot 2} = 2\sqrt{2}$. The direction is $\theta = 225°$.

Components

In general, a vector **A** is denoted by (A_x, A_y), the coordinates of the terminal point. The numbers A_x and A_y are called the **components** of the vector.

Magnitude and direction of vector $\mathbf{A} = (A_x, A_y)$
Magnitude: $|\mathbf{A}| = A = \sqrt{A_x^2 + A_y^2}$
Direction: The direction is obtained from the relation

$$\tan \theta = \frac{A_y}{A_x}$$

A_x and A_y are the **components.**

Example 1

Find the magnitude and direction of the vector $\mathbf{A} = (A_x, A_y) = (-3, 5)$.

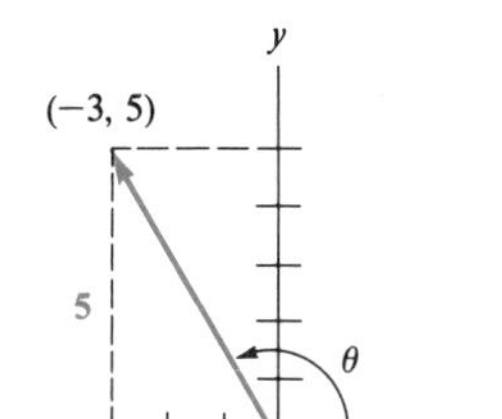

Figure 16.17

Solution. The vector is shown in Figure 16.17.
Since the terminal point is $(-3, 5)$, the magnitude is

$$A = \sqrt{(-3)^2 + 5^2} = \sqrt{9 + 25} = \sqrt{34}$$

To find the direction, note that

$$\tan \theta = \frac{5}{-3} = -\frac{5}{3}$$

From the sequence

5 [÷] 3 [+/−] [=] [INV] [TAN] → −59.036243

we see that the reference angle is 59.0°. Since θ is in the second quadrant, the direction is

$$\theta = 180° - 59.0° = 121.0°$$

Example 2

A vector $\mathbf{A}$ has magnitude 3 and direction 215°. Express $\mathbf{A}$ in the form (A_x, A_y), called *resolving the vector into its components.*

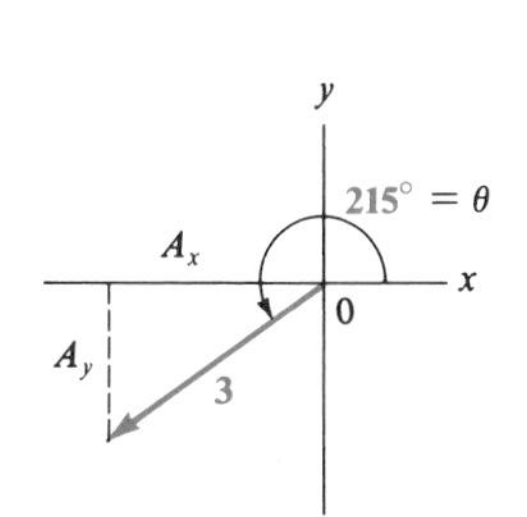

Figure 16.18

Solution. The vector is shown in Figure 16.18.
To find A_x, note that

$$\frac{A_x}{3} = \cos 215° \qquad \cos \theta = \frac{\text{adjacent side}}{\text{hypotenuse}}$$

so that

$$A_x = 3 \cos 215° = -2.46$$

Similarly,

$$\frac{A_y}{3} = \sin 215° \qquad \sin \theta = \frac{\text{opposite side}}{\text{hypotenuse}}$$

or

$$A_y = 3 \sin 215° = -1.72$$

It follows that $\mathbf{A} = (-2.46, -1.72)$.

In general, if vector **A** has direction θ, then

$$A_x = A\cos\theta \qquad \text{and} \qquad A_y = A\sin\theta$$

or

$$\mathbf{A} = (A\cos\theta, A\sin\theta)$$

Vector Addition by Components

To add two or more vectors by components, we proceed as follows:

1. Resolve each vector into its components.
2. Add all x-components and all y-components to obtain the components of the resultant.
3. Find the magnitude and direction of the resultant from the components obtained in Step 2.

Example 3 Add the vectors **A** and **B** in Figure 16.19. Obtain the components (R_x, R_y) of the resultant **R** to three significant digits. Find $|\mathbf{R}|$ to two significant digits and the direction of **R** to the nearest tenth of a degree. (**A** has magnitude 2 and direction 56°; **B** has magnitude 4 and direction 162°.)

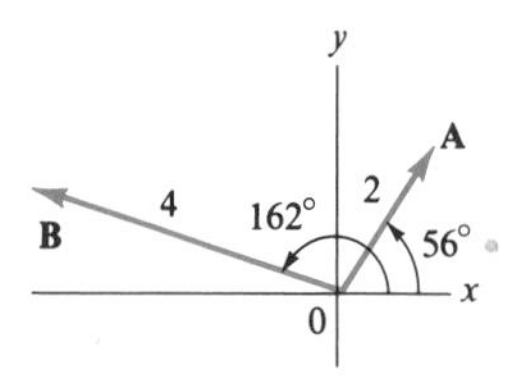

Figure 16.19

Solution. We resolve the vectors **A** and **B** into their components. We then add the corresponding components to obtain the components of **R**:

$$\begin{array}{llll} A_x = 2\cos 56^\circ = 1.12 & A_y = 2\sin 56^\circ = 1.66 & \\ B_x = 4\cos 162^\circ = -3.80 & B_y = 4\sin 162^\circ = 1.24 & \\ \hline R_x = \mathbf{-2.68} & R_y = \mathbf{2.90} & \text{Adding} \end{array}$$

It follows that

$$\mathbf{R} = (R_x, R_y) = (\mathbf{-2.68, 2.90})$$

and

$$R = \sqrt{(-2.68)^2 + (2.90)^2} = 3.9$$

To find θ, observe that

$$\tan\theta = \frac{2.90}{-2.68}, \ \theta \text{ in quadrant II}$$

(θ is in quadrant II because x is negative and y positive.) So $\theta = 132.7°$. The vectors **A** and **B** and the resultant **R** are shown in Figure 16.20.

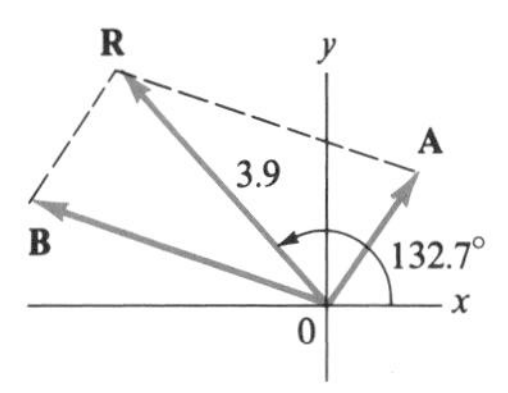

Figure 16.20

Example 4 Add the vectors **A** and **B** in Figure 16.21.

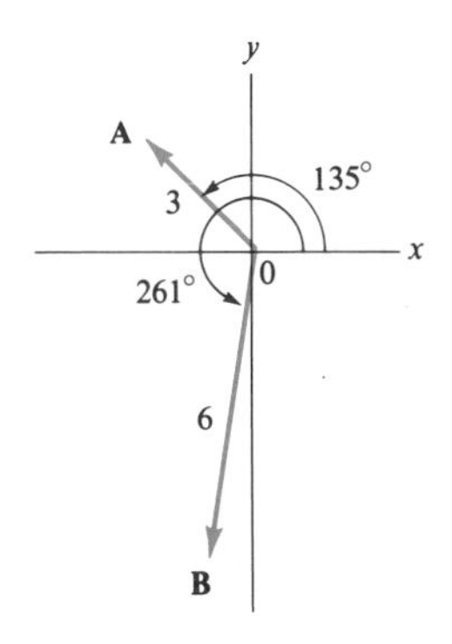

Figure 16.21

Solution.

$A_x = 3 \cos 135° = -2.12$	$A_y = 3 \sin 135° = 2.12$
$B_x = 6 \cos 261° = -0.939$	$B_y = 6 \sin 261° = -5.93$
$R_x = -3.06$	$R_y = -3.81$ Adding

$$\mathbf{R} = (-3.06, -3.81) \quad \text{Three significant digits}$$

$$R = \sqrt{(-3.06)^2 + (-3.81)^2} = 4.9 \quad \text{Two significant digits}$$

$$\tan \theta = \frac{-3.81}{-3.06}, \ \theta \text{ in quadrant III}$$

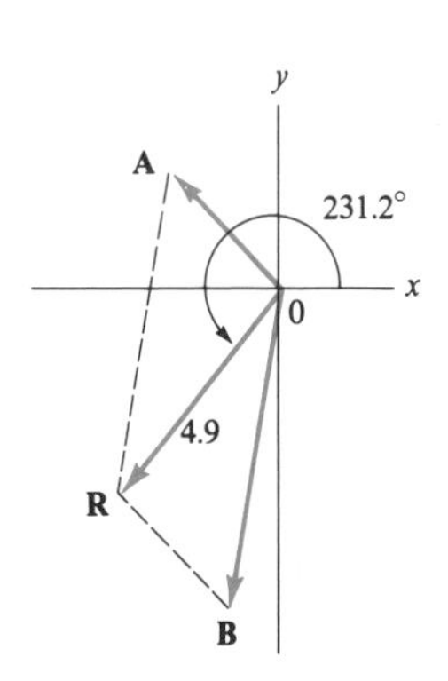

Figure 16.22

(θ is in quadrant III because x and y are both negative.) The reference angle is 51.2°. So

$$\theta = 180° + 51.2° = 231.2°$$

The vectors **A** and **B** and the resultant **R** are shown in Figure 16.22.

The next example illustrates addition of three vectors by components.

Example 5 Add the vectors **A**, **B**, and **C** in Figure 16.23.

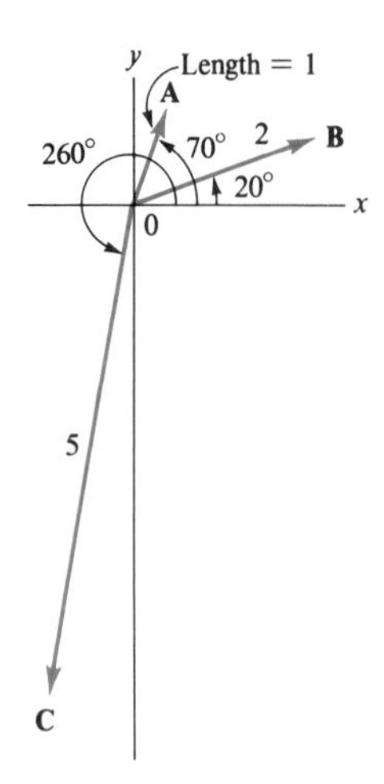

Figure 16.23

Solution.

$$\begin{array}{ll} A_x = 1\cos 70° = 0.342 & A_y = 1\sin 70° = 0.940 \\ B_x = 2\cos 20° = 1.88 & B_y = 2\sin 20° = 0.684 \\ C_x = 5\cos 260° = -0.868 & C_y = 5\sin 260° = -4.92 \\ \hline R_x = 1.35 & R_y = -3.30 \end{array} \quad \text{Adding}$$

$$\mathbf{R} = (1.35, -3.30) \qquad \text{Three significant digits}$$

$$R = \sqrt{(1.35)^2 + (-3.30)^2} = 3.6 \qquad \text{Two significant digits}$$

$$\tan\theta = \frac{-3.30}{1.35}, \ \theta \text{ in quadrant IV}$$

The reference angle is 67.8°. So

$$\theta = 360° - 67.8° = 292.2°$$

The vectors are shown in Figure 16.24.

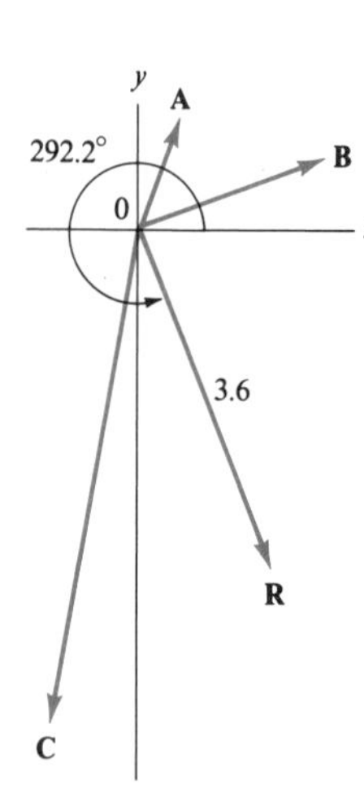

Figure 16.24

Remark on rounding off. To show the steps in this example, R_x was calculated by adding the components A_x, B_x, and C_x, each expressed with an accuracy of three significant digits. (R_y was obtained similarly.) R_x can also be obtained directly with a calculator by using the following sequence:

70 [COS] [+] 2 [×] 20 [COS] [+] 5 [×] 260 [COS] [=] → 1.3531645

This operation leads to $R_x = 1.35$, as before. In some cases, however, the resulting value may differ slightly from the value obtained by the other method, since no rounding off is done in the intermediate steps.

For consistency, in Exercises 31–40, the answers were obtained by the first method. (See also Examples 3 and 4.)

Exercises / Section 16.2

In Exercises 1–20, find the magnitude and direction of each vector (A_x, A_y). (See Example 1.)

1. $(1, 1)$
2. $(\sqrt{3}, 1)$
3. $(-1, \sqrt{3})$
4. $(-1, -1)$
5. $(-2, -2\sqrt{3})$
6. $(3\sqrt{3}, -3)$
7. $(2, 3)$
8. $(-1, 3)$
9. $(-4, 7)$
10. $(-3, -7)$
11. $(-2, -6)$
12. $(4, -8)$
13. $(6, -4)$
14. $(6, -8)$
15. $(-\sqrt{2}, \sqrt{7})$
16. $(-1, -\sqrt{11})$
17. $(-1.23, 2.65)$
18. $(-3.78, -5.36)$
19. $(-10.0, -6.84)$
20. $(15.8, -16.4)$

In Exercises 21–30, resolve each vector into its components. Round off the answer to three significant digits. (See Example 2.)

21. $A = 3, \theta = 80°$
22. $A = 2, \theta = 72°$
23. $A = 8, \theta = 120°$
24. $A = 10, \theta = 140°$
25. $A = 5, \theta = 217°$
26. $A = 7, \theta = 245°$
27. $A = 6, \theta = 280.6°$
28. $A = 11, \theta = 295.1°$
29. $A = 13, \theta = 312.5°$
30. $A = 15, \theta = 348.7°$

In Exercises 31–40, add the given vectors by components. Find the components of **R** to three significant digits, $|\mathbf{R}|$ to two significant digits, and θ to the nearest tenth of a degree. (See Examples 3–5.)

31. **A**: $A = 2$, direction: 10°
B: $B = 1$, direction: 100°

32. **A**: $A = 4$, direction: 20°
B: $B = 5$, direction: 110°

33. **A**: $A = 6$, direction: 116°
B: $B = 4$, direction: 171°

34. **A**: $A = 6$, direction: 25°
B: $B = 8$, direction: 280°

35. **A**: $A = 6$, direction: 300°
B: $B = 9$, direction: 350°

36. **A**: $A = 5$, direction: 190°
B: $B = 8$, direction: 290°

37. **A**: $A = 1$, direction: 10°
B: $B = 2$, direction: 15°
C: $C = 6$, direction: 300°

38. **A**: $A = 2$, direction: 66°
B: $B = 2$, direction: 80°
C: $C = 6$, direction: 170°

39. **A**: $A = 4$, direction: 160°
B: $B = 2$, direction: 200°
C: $C = 5$, direction: 290°

40. **A**: $A = 8$, direction: 100°
B: $B = 3$, direction: 110°
C: $C = 2$, direction: 231°

16.3 Basic Applications of Vectors

In this section we will apply our vector concepts to physical problems. But first we need to recall the rule for significant digits.

Degree measurement to nearest	Significant digits for measurement of side
1°	2
10′ or 0.1°	3
1′ or 0.01°	4

Vectors at Right Angles

In the first part of this section we will deal with vectors at right angles to each other. The first two examples illustrate the problem of finding the resultant from two given vectors.

Example 1 A boat, whose velocity in still water is 20.1 mi/h, is heading directly across a river. Given that the river flows at 4.60 mi/h, determine the direction of the boat and its velocity with respect to the bank.

Solution. The situation is pictured in Figure 16.25.

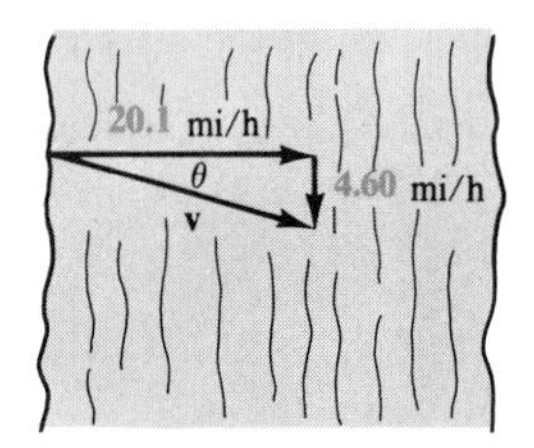

Figure 16.25

Denoting the resultant velocity by **v**, we get for the magnitude

$$v = \sqrt{(20.1)^2 + (4.60)^2} = 20.6\,\frac{\text{mi}}{\text{h}}$$

The direction is found from

$$\tan\theta = \frac{4.60}{20.1}$$

so that $\theta = 12.9°$.

We conclude that the boat travels at 20.6 mi/h with respect to the bank, and its direction is 12.9° with the line directly across.

The next example features a force acting on a body hanging on a rope. In this and similar problems, we assume the system to be in *equilibrium*. A system is in equilibrium if there are no net forces acting on it in any direction. In particular, in the next example the tension on the rope is the resultant of the two forces acting on it.

Example 2 A 44.5-lb weight hanging on a rope is pulled sideways by a force of 15.0 lb. Determine the resulting tension on the rope and the angle that the rope makes with the vertical.

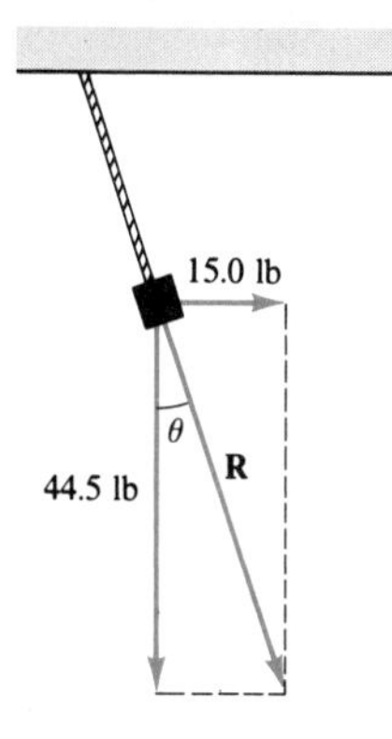

Figure 16.26

Solution. The tension on the rope must be equal to the resultant **R** of two forces, the 44.5-lb downward force due to the weight and the 15.0-lb pull to the side, as shown in Figure 16.26. So the tension is

$$|\mathbf{R}| = \sqrt{(44.5)^2 + (15.0)^2} = 47.0 \text{ lb}$$

The angle made with the vertical is found from $\tan\theta = 15.0/44.5$, or $\theta = 18.6°$.

The next example shows why resolving a vector into its components may be of interest for physical reasons.

Example 3 A cart weighing 125 lb is positioned on a ramp inclined 18.0° with the horizontal. Determine the force required to keep the cart from rolling down the ramp. (See Figure 16.27; the weight of an object always acts vertically downward.)

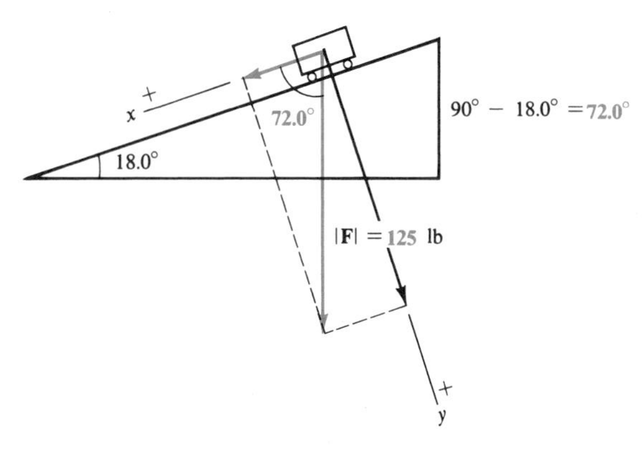

Figure 16.27

Solution. In this problem the downward force **F**, where $|\mathbf{F}| = 125$ lb, has to be split into two components, one directed down the plane and the other perpendicular to the plane. (The components must obey the parallelogram law.) The component along the plane represents the tendency of the cart to roll down the ramp. In other words,

$$F_x = 125 \cos 72.0° = 38.6 \text{ lb}$$

So a force of 38.6 lb is required to keep the cart on the ramp.

Vectors Not at Right Angles

The next example involves vectors not positioned at right angles to each other. In this case, the given vectors have to be resolved into their components first.

Example 4 A small plane is headed 12.34° *south of east* (Figure 16.28) at an air speed of 150.0 km/h. The wind is from the north at 25.75 km/h. Find the resulting direction and velocity with respect to the ground.

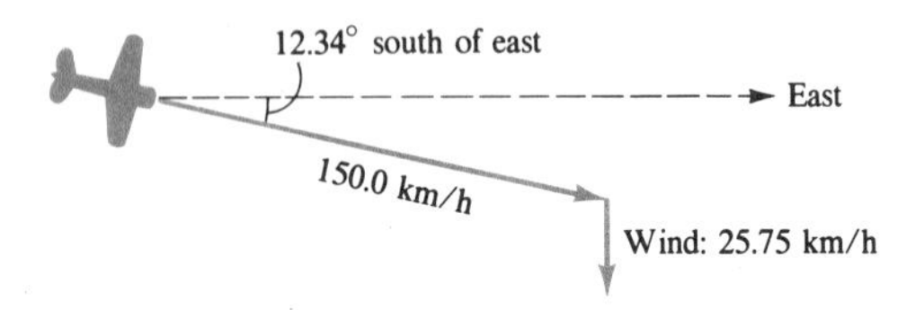

Figure 16.28

Solution. First we place the vectors with initial points at the origin, as shown in Figure 16.29. Let **P** represent the velocity of the plane and **W** the velocity of the wind.

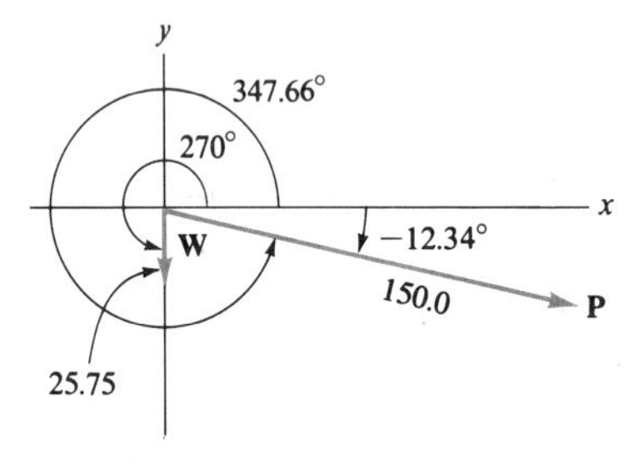

Figure 16.29

The direction of the plane (12.34° south of east) is now represented by 360° − 12.34° = 347.66° (or by −12.34°).

To find the resulting magnitude and direction, we resolve the vectors **P** and **W** into their components:

$$\begin{aligned} P_x &= 150.0 \cos 347.66° &&= 146.53 \\ W_x &= 25.75 \cos 270° &&= 0 \\ \hline R_x & &&= 146.53 \end{aligned}$$

Adding

$$P_y = 150.0 \sin 347.66° = -32.057$$
$$W_y = 25.75 \sin 270° = -25.75$$
$$R_y = -57.807 \quad \text{Adding}$$
$$R = \sqrt{(146.53)^2 + (-57.807)^2} = 157.5 \quad \text{Magnitude}$$
$$\tan \theta = \frac{-57.807}{146.53}, \ \theta \text{ in quadrant IV}$$
$$\theta = -21.53° \quad \text{Direction}$$

So the direction is 21.53° south of east and the speed with respect to the ground is 157.5 km/h.

Exercises / Section 16.3

In Exercises 1–4, two forces with respective magnitudes F_1 and F_2 are acting on the same object at right angles. Find the magnitude of the resultant and its direction with respect to the larger force.

1. $F_1 = 9.8$ N, $F_2 = 6.4$ N

2. $F_1 = 3.76$ N, $F_2 = 2.60$ N

3. $F_1 = 7.300$ lb, $F_2 = 15.89$ lb

4. $F_1 = 15.00$ kg, $F_2 = 21.00$ kg

5. Two boys are pulling on a cart in mutually perpendicular directions with forces of 28 lb and 39 lb, respectively. What single force will have the same effect?

6. Two tractors are pulling on a tree stump in mutually perpendicular directions with respective forces of 13,000 lb and 16,000 lb. What single force will have the same effect?

7. A boat heads directly across a river. Its velocity in still water is 12.0 mi/h. If the river flows at 3.60 mi/h, what is the direction of the boat and its speed relative to the bank?

8. A boat heads directly across a river that is flowing at 5.70 km/h. If the speed of the boat in still water is 16.4 km/h, find its direction.

9. A jet is heading due north with an air speed of 412 km/h. The wind is from the east at 38.0 km/h. Find the resulting direction and velocity with respect to the ground.

10. A small plane is heading east with an air speed of 125 mi/h, while the wind is from the south at 21.6 mi/h. Find the resulting velocity and direction.

11. A boat crosses a river that is flowing at 3.80 km/h and lands at a point directly across. If the velocity of the boat is 12.5 km/h in still water, find the direction in which the boat must head, as well as the velocity with respect to the bank.

12. A boat crossing a river lands at a point directly across from where it started. Due to the flow of the river, it must head in a direction 15° upstream from the line directly across. Given that the speed of the boat in still water is 25 mi/h, determine the rate of flow of the river.

13. A pilot wants to fly directly south. His air speed is 245 km/h, and the wind is from the west at 35.5 km/h. In what direction should he orient his plane, and what is the resulting velocity relative to the ground?

14. With the sun directly overhead, a jet is taking off at 120 mi/h at an angle of 39°. How fast is the shadow moving along the ground?

15. A girl is pulling a cart with a force of 58 lb at an angle of 42°. What single force parallel to the floor would have the same effect?

16. Find the force required to keep a 3250-lb car on a hill that makes an angle of 12.5° with the horizontal.

17. Determine the force required to keep an 80.0-lb block of ice from slipping down a plane inclined 15.0° with the horizontal.

18. What is the force required to pull a 250-lb wagon up a ramp inclined 8° with the horizontal?

19. An 85-lb weight is supported as shown in Figure 16.30. Determine the tension on the rope.

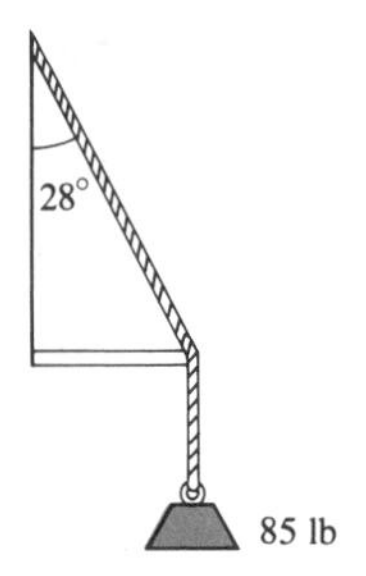

Figure 16.30

20. Determine the force against the horizontal support AB in Figure 16.31.

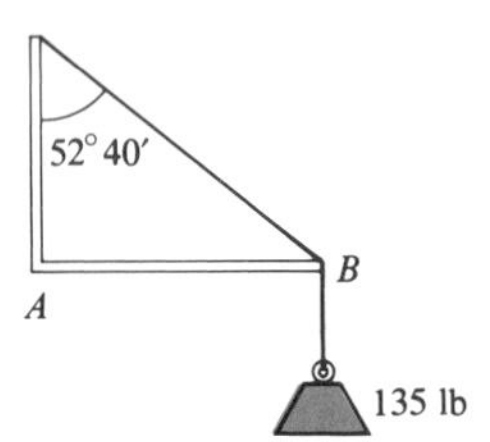

Figure 16.31

21. A 21-N weight hanging on a rope is pulled sideways by a force of 9.0 N. If the system is in equilibrium, determine the tension on the rope and the angle that the rope makes with the vertical.

22. A 45.0-lb weight hanging on a rope is pulled sideways by a force of 23.5 lb. What is the resulting tension on the rope?

The remaining exercises involve vectors not at right angles.

23. A small plane is flying 13.4° north of west at 145 mi/h. The wind is from the north at 26.5 mi/h. Find the resulting velocity and direction.

24. A plane heading 11.6° west of south at 426 km/h encounters a wind from the east at 19.5 km/h. Find the resulting velocity and direction.

25. A boy and his younger sister are pulling a cart with forces of 35.0 lb and 25.0 lb, respectively. If the forces are 45.0° apart, determine what single force will have the same effect.

26. Two forces of 45 N and 75 N, respectively, are acting on an object. If the forces are 50° apart, what single force will have the same effect?

27. An object is resting on the floor. Two forces are pulling on the object from the right. A force of 50.0 lb acts at an angle of 25.0° with the horizontal and a force of 40.0 lb acts at an angle of 40.0° with the horizontal. Find the resultant.

28. An object resting on the floor is pulled by two forces acting on opposite sides. One force of 75.0 N is acting on the right at a 26.5° angle with respect to the floor. Another force of 112 N is acting on the left at a 42.5° angle with respect to the floor. Determine the resultant of the two forces.

16.4 The Law of Sines

So far we have solved only right triangles. We will now solve **oblique triangles** (triangles not containing a right angle), and thereby greatly extend our previous methods. We will study the *law of sines* in this section and the *law of cosines* in the next. Both methods can be applied to vector problems.

To derive the law of sines, consider the triangles in Figure 16.32. Pick an arbitrary vertex such as B and drop a perpendicular to the side opposite, or the side opposite extended. (The triangles in the figure illustrate the two possibilities.)

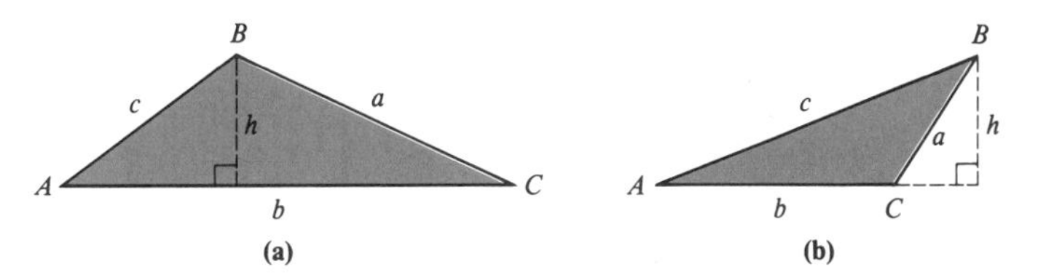

Figure 16.32

Note that in Figure 16.32(a)

$$\sin A = \frac{h}{c} \qquad \text{and} \qquad \sin C = \frac{h}{a}$$

or

$$h = c \sin A \qquad \text{and} \qquad h = a \sin C$$

From Figure 16.32(b),

$$\sin A = \frac{h}{c} \qquad \text{and} \qquad \sin(180° - C) = \frac{h}{a}$$

But for any angle C between 0° and 180°,

$$\sin C = \sin(180° - C)$$

So

$$h = c \sin A \qquad \text{and} \qquad h = a \sin C$$

In both cases

$$c \sin A = a \sin C$$

and

$$\frac{c \cancel{\sin A}}{\cancel{\sin A} \sin C} = \frac{a \cancel{\sin C}}{\sin A \cancel{\sin C}} \qquad \text{Dividing by } \sin A \sin C$$

$$\frac{c}{\sin C} = \frac{a}{\sin A} \tag{16.1}$$

If we drop a perpendicular from C to c, we get, by a similar argument,

$$\frac{b}{\sin B} = \frac{a}{\sin A} \tag{16.2}$$

Combining formulas (16.1) and (16.2), we get

$$\frac{a}{\sin A} = \frac{b}{\sin B} = \frac{c}{\sin C}$$

called the **law of sines**, or simply the **sine law**.

Law of sines

$$\frac{a}{\sin A} = \frac{b}{\sin B} = \frac{c}{\sin C}$$

or

$$\frac{\sin A}{a} = \frac{\sin B}{b} = \frac{\sin C}{c}$$

The law of sines enables us to solve many oblique triangles. Many of the problems in this and the next section refer to the triangle in Figure 16.33. Note that capital letters are used to denote the angles and the corresponding lowercase letters, the respective opposite sides.

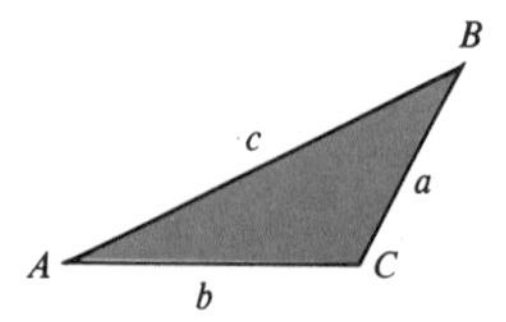

Figure 16.33

Example 1 Solve the triangle in Figure 16.34. Note that $A = 46.4°$, $C = 115.7°$, and $b = 5.76$.

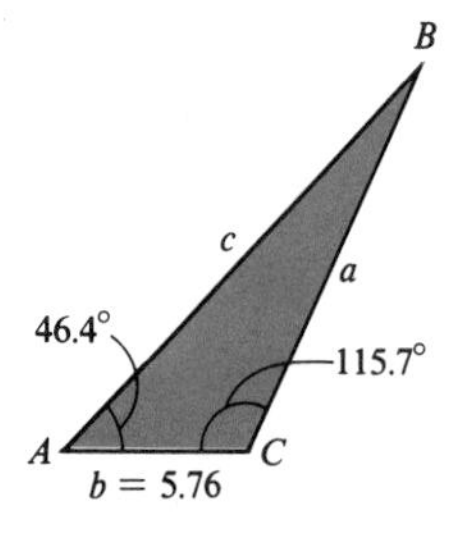

Figure 16.34

Solution. Since the sum of the angles of a triangle is 180°,

$$B = 180° - (46.4° + 115.7°) = 17.9°$$

By the law of sines, we now have

$$\frac{a}{\sin 46.4°} = \frac{5.76}{\sin 17.9°} \qquad \frac{a}{\sin A} = \frac{b}{\sin B}$$

or

$$a = \frac{5.76 \sin 46.4°}{\sin 17.9°} \qquad \text{Multiplying by } \sin 46.4°$$

The sequence is

5.76 [×] 46.4 [SIN] [÷] 17.9 [SIN] [=] → 13.571303

So $a = 13.6$. The side c is found similarly:

$$\frac{c}{\sin 115.7°} = \frac{5.76}{\sin 17.9°} \qquad \frac{c}{\sin C} = \frac{b}{\sin B}$$

or

$$c = \frac{5.76 \sin 115.7°}{\sin 17.9°} = 16.9$$

Thus $B = 17.9°$, $a = 13.6$, and $c = 16.9$.

Remark. Side c could also have been obtained by using the calculated value of a and its opposite angle:

$$\frac{c}{\sin 115.7°} = \frac{13.6}{\sin 46.4°}$$

However, it is better to use the given numbers rather than the calculated values. In this way, any calculation error or round-off error will not be carried over to the rest of the calculations.

The next example shows how the sine law can be used to find an unknown angle.

Example 2 Given that $B = 53.40°$, $b = 15.10$, and $c = 11.30$, find the remaining parts. (See Figure 16.35.)

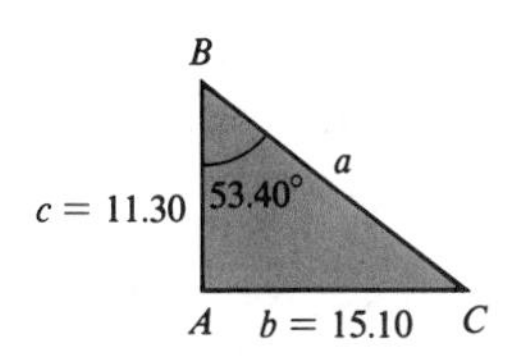

Figure 16.35

Solution. Since only one angle is known, one of the ratios must be

$$\frac{\sin 53.40^\circ}{15.10} \quad \text{or} \quad \frac{15.10}{\sin 53.40^\circ}$$

We must therefore find angle C first:

$$\frac{\sin C}{11.30} = \frac{\sin 53.40^\circ}{15.10} \qquad \frac{\sin C}{c} = \frac{\sin B}{b}$$

or

$$\sin C = \frac{11.30 \sin 53.40^\circ}{15.10} \qquad \text{Multiplying by 11.30}$$

The sequence is

11.30 [×] 53.40 [SIN] [÷] 15.10 [=] [INV] [SIN] → 36.926064

So $C = 36.93^\circ$ to the nearest hundredth of a degree. It now follows that

$$A = 180^\circ - (53.40^\circ + 36.93^\circ) = 89.67^\circ$$

Using angle A, we now find side a:

$$\frac{a}{\sin 89.67^\circ} = \frac{15.10}{\sin 53.40^\circ} \qquad \frac{a}{\sin A} = \frac{b}{\sin B}$$

or

$$a = \frac{15.10 \sin 89.67^\circ}{\sin 53.40^\circ} = 18.81$$

Thus $A = 89.67^\circ$, $C = 36.93^\circ$, and $a = 18.81$.

If three sides of a triangle are known, the sine law is of no help: All ratios involve an angle, at least one of which has to be known. We run into the same problem if $a = 5$, $b = 6$, and $C = 20^\circ$ (Figure 16.36):

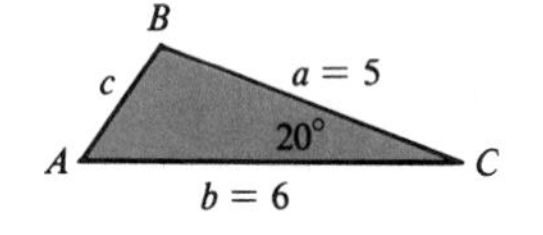

Figure 16.36

$$\frac{c}{\sin 20^\circ} = \frac{6}{\sin B} \quad \text{or} \quad \frac{c}{\sin 20^\circ} = \frac{5}{\sin A}$$

Here one angle is known, yet both equations contain two unknowns. These cases will be discussed in the next section. The cases to which the sine law does apply are given next.

> To solve an oblique triangle by the **law of sines**, we need either:
>
> **1.** Two angles and one side, or
> **2.** Two sides and the angle opposite one of them.

The sine law can be applied to certain problems involving vectors, as shown in the next example.

Example 3 A plane has an air speed of 515 km/h and wants to fly on a course of 25°10′ east of north. If the wind is from the north at 60.1 km/h, determine the direction in which the plane must head to stay on the proper course. What is the resulting velocity with respect to the ground?

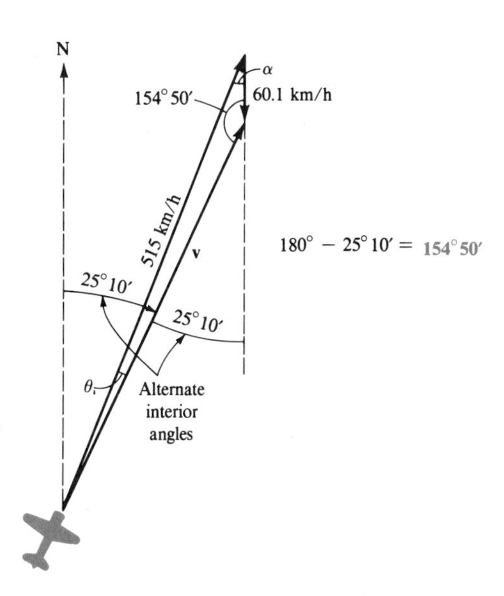

Figure 16.37

Solution. Let us denote the resulting velocity vector by **v** (Figure 16.37). To determine the proper course, we need to find angle θ first. Applying the law of sines, we get

$$\frac{\sin\theta}{60.1} = \frac{\sin 154°50'}{515}$$

Solving for $\sin\theta$,

$$\sin\theta = \frac{60.1 \sin 154°50'}{515} = \frac{60.1 \sin\left(154 + \frac{50}{60}\right)^\circ}{515}$$

The sequence is

154 [+] 50 [÷] 60 [=] [SIN] [×] 60.1 [÷] 515 [=]
[INV] [SIN] → 2.8445626

Thus $\theta = 2°50'$ $\quad \frac{x}{60} = \frac{84}{100}$ or $x = 50$

To find v, we need the angle opposite **v** (denoted by α in Figure 16.37). Since $\theta = 2°50'$, $\alpha = 180° - (2°50' + 154°50') = 180° - (156°100') = 180°$

$-\ (157°40') = 22°20'$. Applying the law of sines,

$$\frac{v}{\sin 22°20'} = \frac{515}{\sin 154°50'}$$

Solving for v,

$$v = \frac{515 \sin 22°20'}{\sin 154°50'}$$

and $v = 460$ km/h.

The direction is $25°10' - \theta = 25°10' - 2°50' = 24°70' - 2°50' = 22°20'$ east of north.

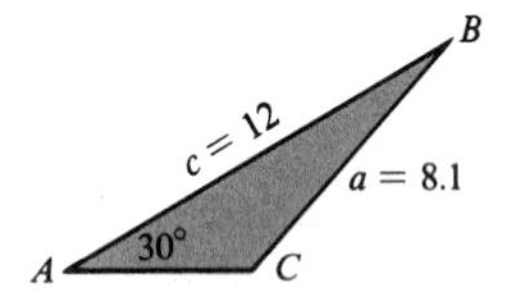

Figure 16.38

Caution. Suppose we want to find the obtuse angle C in Figure 16.38. By the sine law

$$\frac{\sin C}{12} = \frac{\sin 30°}{8.1} \qquad \text{or} \qquad \sin C = 0.74$$

If we now use a calculator to find C, we might conclude from

0.74 [INV] [SIN]

that $C = 48°$. However, since C is obtuse, the correct value is $C = 180° - 48° = 132°$. In other words, *we have to know in advance whether a given angle is obtuse or acute*. If this information is not available, then there may exist two solutions to the triangle.

The case just discussed is called the *ambiguous case* (since two solutions are possible). The different cases are summarized next (see Figures 16.39–16.41).

Number of possible triangles	Sketch	Condition
0	Figure 16.39	$a < h$ or $a < b \sin A$
1	Figure 16.40	$a = h$ or $a = b \sin A$

Number of possible triangles	Sketch	Condition
2	Figure 16.41	$b > a > h$ or $b > a > b \sin A$

In the exercises that follow, the conditions in the problems are clearly stated, so the ambiguous case does not arise.

Exercises / Section 16.4

In Exercises 1–20, refer to Figure 16.42.

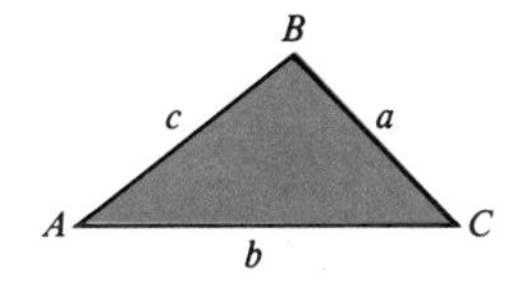

Figure 16.42

Solve each of the triangles from the given information.

1. $A = 52.7°$, $C = 105.3°$, $b = 8.35$

2. $A = 115.2°$, $C = 43.1°$, $b = 13.2$

3. $B = 23.74°$, $C = 49.00°$, $a = 18.56$

4. $B = 39.18°$, $C = 95.63°$, $a = 25.75$

5. $A = 120.3°$, $B = 32.4°$, $c = 12.7$

6. $A = 15.4°$, $B = 65.5°$, $c = 15.8$

7. $A = 48.3°$, $B = 69.6°$, $a = 6.34$

8. $B = 57.6°$, $C = 72.4°$, $a = 8.04$

9. $A = 20°$, $C = 118°$, $c = 16$

10. $A = 115°$, $C = 17°$, $a = 25$

11. $B = 106°10'$, $C = 16°40'$, $b = 12.4$

12. $B = 45°30'$, $C = 53°50'$, $c = 14.8$

13. $A = 21.7°$, $a = 11.3$, $b = 5.20$

14. $C = 51.2°$, $a = 2.90$, $c = 5.36$

15. $A = 28.4°$, $a = 0.180$, $c = 0.234$, angle C acute

16. Same as Exercise 15 with angle C obtuse

17. $C = 62.8°$, $a = 14.6$, $c = 13.5$, angle A obtuse

18. Same as Exercise 17 with angle A acute

19. $A = 34°$, $a = 0.25$, $b = 0.36$, angle B acute

20. $A = 56°$, $a = 24$, $b = 28$, angle B obtuse

21. A small plane has an air speed of 160 km/h and wants to fly east. The wind is from the northwest at 44 km/h. In what direction should the pilot orient the plane, and what is the resulting velocity with respect to the ground?

22. A plane maintaining an air speed of 585 mi/h is heading 10.5° south of west. A wind out of the west causes the actual course to be 12.4° south of west. Find the velocity of the plane with respect to the ground.

23. A surveyor wants to find the width of a river from a certain point on the bank. Since no other points on the bank are accessible, he takes the measurements shown in Figure 16.43. Determine the width of the river.

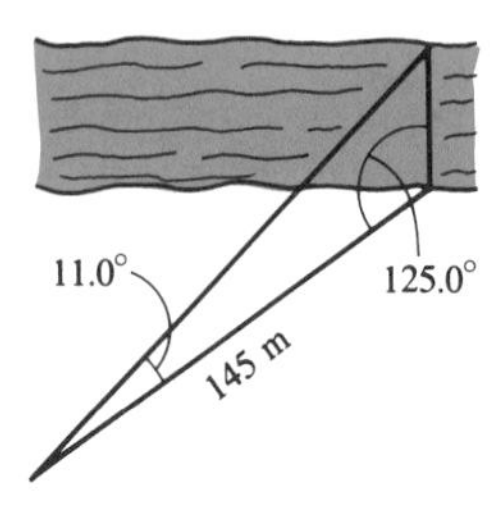

Figure 16.43

24. From a point on the ground, the angle of elevation of a balloon is 49.0°. From a second point 1250 ft away on the opposite side of the balloon and in the same vertical plane as the balloon and the first point, the angle of elevation is 33.0°. Find the distance from the second point to the balloon.

25. Two forces act on the same object in directions that are 40.0° apart. If one force is 48.0 kg, what must the other force be so that the combined effect is equivalent to a force of 65.1 kg?

26. Two forces, 25.65° apart, act on the same object. If one force is 51.36 lb, what must the other force be so that the combined effect is a force of 100.0 lb?

27. A guy wire supporting a tower is attached to a point 15.0 ft from the top of the tower. At the point where the wire is anchored to the ground, the angle of elevation of the top of the tower is 59.8°. Given that the guy wire makes an angle of 51.3° with the ground, determine the length of the wire.

28. Find the perimeter of the triangular metal plate in Figure 16.44.

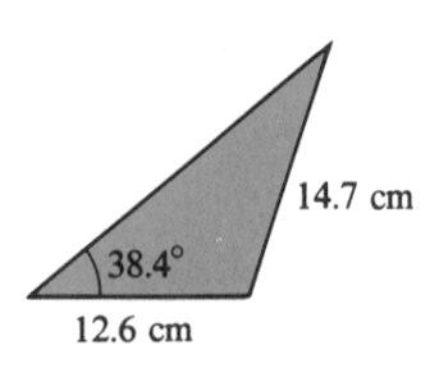

Figure 16.44

16.5 The Law of Cosines

We saw in the last section that some triangles cannot be solved by the law of sines. In this section we will consider the solution of such triangles by the *law of cosines*.

> Use the **law of cosines** to solve an oblique triangle given:
>
> **1.** Two sides and the included angle, or
> **2.** Three sides.

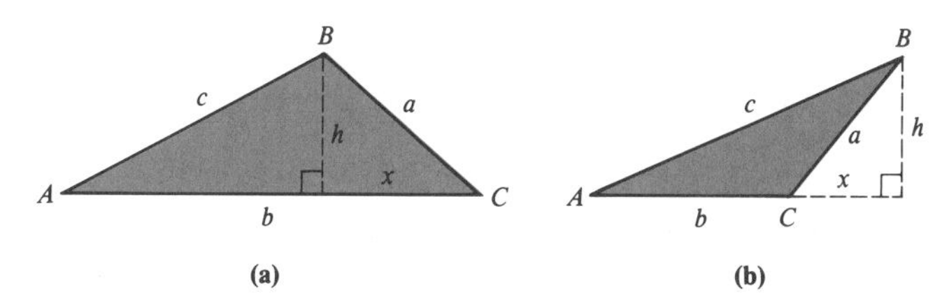

Figure 16.45

To derive the law of cosines, consider the triangles in Figure 16.45. Drop a perpendicular from B to the opposite side (or the opposite side extended) and let x denote the distance from C to the perpendicular. Since the sine of an angle is the opposite side over the hypotenuse, we have

$$\sin A = \frac{h}{c}$$

Solving for h,

$$h = c \sin A$$

We can use this expression in the Pythagorean theorem as follows:

$$a^2 = h^2 + x^2$$

Substituting for h,

$$a^2 = c^2(\sin A)^2 + x^2 \qquad (16.3)$$

In Figure 16.45(a), $b - x = c \cos A$, or $x = b - c \cos A$. In Figure 16.45(b), $x + b = c \cos A$, or $x = c \cos A - b$. Substituting the respective expressions into equation (16.3), we get

$$a^2 = c^2(\sin A)^2 + (b - c \cos A)^2$$

and

$$a^2 = c^2(\sin A)^2 + (c \cos A - b)^2$$

If we multiply the expressions on the right, we get in either case

$$a^2 = c^2(\sin A)^2 + c^2(\cos A)^2 - 2bc \cos A + b^2$$

Factoring c^2 from the first two terms, we obtain

$$a^2 = c^2[(\sin A)^2 + (\cos A)^2] - 2bc \cos A + b^2 \qquad (16.4)$$

It can be shown that $(\sin A)^2 + (\cos A)^2 = 1$. Recall that for an angle in standard position, $\sin \theta = y/r$ and $\cos \theta = x/r$. So

$$(\sin \theta)^2 + (\cos \theta)^2 = \frac{y^2}{r^2} + \frac{x^2}{r^2} = \frac{x^2 + y^2}{r^2} = \frac{r^2}{r^2} = 1$$

It follows that equation (16.4) reduces to

$$a^2 = c^2(1) - 2bc \cos A + b^2$$

or

$$a^2 = b^2 + c^2 - 2bc \cos A \qquad (16.5)$$

This formula is known as the **law of cosines**, or simply the **cosine law**.

By dropping perpendiculars from the other vertices, we get the other forms of the cosine law, stated next.

Law of cosines

$$a^2 = b^2 + c^2 - 2bc \cos A$$
$$b^2 = a^2 + c^2 - 2ac \cos B$$
$$c^2 = a^2 + b^2 - 2ab \cos C$$

The law of cosines can also be stated verbally.

Law of cosines (verbal form): The square of any side of a triangle equals the sum of the squares of the other two sides minus twice the product of those two sides and the cosine of the angle between them.

Note especially that if $A = 90°$, then formula (16.5) becomes

$$a^2 = b^2 + c^2 - 2bc \cdot 0 = b^2 + c^2$$

or $a^2 = b^2 + c^2$. It follows that the cosine law reduces to the Pythagorean theorem if the triangle is a right triangle.

We will now see how the cosine law is used to solve triangles.

Example 1 Given that $B = 117.5°$, $a = 7.50$, and $c = 3.90$, find the remaining parts (Figure 16.46).

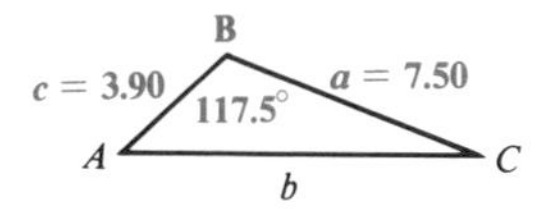

Figure 16.46

Solution. Note first that this triangle cannot be solved by the sine law. By the cosine law, however, we can find side b:

$$b^2 = a^2 + c^2 - 2ac \cos B$$
$$b^2 = (7.50)^2 + (3.90)^2 - 2(7.50)(3.90) \cos 117.5°$$

The sequence is

7.50 [x²] [+] 3.90 [x²] [−] 2 [×] 7.50 [×] 3.90
[×] 117.5 [COS] [=] [√] → 9.9233207

So $b = 9.92$.

Now that the side opposite B is known, we can find angle A (or angle C) by the sine law:

$$\frac{\sin A}{7.50} = \frac{\sin 117.5°}{9.92} \quad \text{or} \quad A = 42.1° \qquad \frac{\sin A}{a} = \frac{\sin B}{b}$$

Finally, $C = 180° - (42.1° + 117.5°) = 20.4°$. So

$$b = 9.92,\ A = 42.1°, \text{ and } C = 20.4°$$

Note. The angle A can also be found by the cosine law:

$$a^2 = b^2 + c^2 - 2bc \cos A$$
$$(7.50)^2 = (9.92)^2 + (3.90)^2 - 2(9.92)(3.90) \cos A$$

Solving for $\cos A$, we get

$$\cos A = \frac{(7.50)^2 - (9.92)^2 - (3.90)^2}{-2(9.92)(3.90)}$$

The sequence is

7.50 [x²] [−] 9.92 [x²] [−] 3.90 [x²] [=] [÷] 2 [+/−]
[÷] 9.92 [÷] 3.90 [=] [INV] [COS] → 42.149373

However, the calculation using the sine law is easier.

Example 1 shows that even though the cosine law may be needed to find the first of the unknown parts, it is easier to find the remaining parts by the sine law. As we saw in the last section, however, to find an angle by the sine law, we have to know in advance whether the angle is acute or obtuse. We can handle this difficulty in one of two ways: We can find all angles by the cosine law (if $\cos \theta$ is negative, a calculator yields an obtuse angle automatically). A simpler alternative is the following:

> First use the cosine law to find the angle opposite the longest side. Then use the sine law to find the remaining angles.

This procedure is illustrated in the next example.

Example 2 Given that $a = 4.3$, $b = 5.2$, and $c = 8.2$, find the three angles (Figure 16.47).

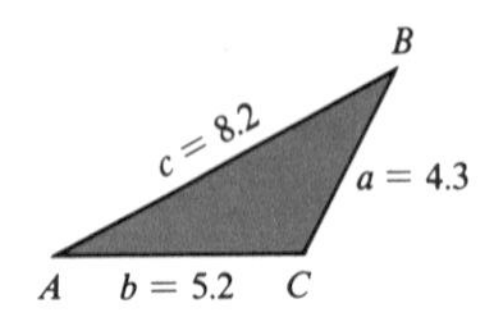

Figure 16.47

Solution. Since angle C is the angle opposite the longest side, we find angle C by the cosine law:

$$c^2 = a^2 + b^2 - 2ab\cos C$$

$$(8.2)^2 = (4.3)^2 + (5.2)^2 - 2(4.3)(5.2)\cos C$$

$$\cos C = \frac{(8.2)^2 - (4.3)^2 - (5.2)^2}{-2(4.3)(5.2)}$$

The sequence is

8.2 [x²] [−] 4.3 [x²] [−] 5.2 [x²] [=] [÷] 2 [+/−]
[÷] 4.3 [÷] 5.2 [=] [INV] [COS] → 119.04295

So $C = 119°$. We can now safely apply the sine law to obtain angle A:

$$\frac{\sin A}{4.3} = \frac{\sin 119°}{8.2} \quad \text{or} \quad A = 27° \qquad \frac{\sin A}{a} = \frac{\sin C}{c}$$

Finally $B = 180° - (27° + 119°) = 34°$.

Example 3 Given $A = 20.4°$, $b = 10.0$, and $c = 7.60$, find the remaining parts (Figure 16.48).

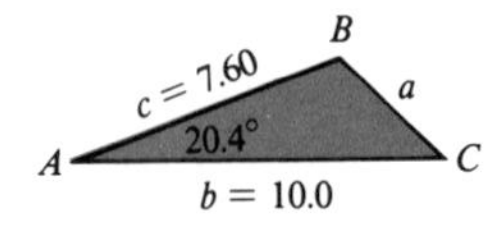

Figure 16.48

Solution. First we need to find side a by the cosine law:

$$a^2 = b^2 + c^2 - 2bc \cos A$$
$$a^2 = (10.0)^2 + (7.60)^2 - 2(10.0)(7.60) \cos 20.4°$$

which yields $a = 3.91$.

Since B is the angle opposite the longest side, it may be an obtuse angle. (Judging from the figure, it is.) So the safest procedure is to use the cosine law again:

$$b^2 = a^2 + c^2 - 2ac \cos B$$
$$(10.0)^2 = (3.91)^2 + (7.60)^2 - 2(3.91)(7.60) \cos B$$
$$\cos B = \frac{(10.0)^2 - (3.91)^2 - (7.60)^2}{-2(3.91)(7.60)}$$

or

$$B = 117.0°$$

Finally, $C = 180° - (117.0° + 20.4°) = 42.6°$.

Caution. Occasionally, using the sine law (instead of the cosine law) to find an angle leads to a slightly different result due to round-off errors. For example, if angle B in Example 3 is found by the sine law, we have

$$\frac{\sin B}{10.0} = \frac{\sin 20.4°}{3.91}$$

Assuming B to be obtuse, this leads to 116.9°, instead of 117.0° from the cosine law.

We saw in Sections 16.2 and 16.3 that the sum of two vectors not mutually perpendicular can be found by adding components. Using the cosine law, some of these problems can be solved in a very simple way. Let us rework Example 4, Section 16.3.

Example 4 (Same as Example 4, Section 16.3.) A small plane is headed 12.34° south of east at an air speed of 150.0 km/h. The wind is from the north at 25.75 km/h. Find the resulting direction and velocity with respect to the ground. (See Figure 16.49.)

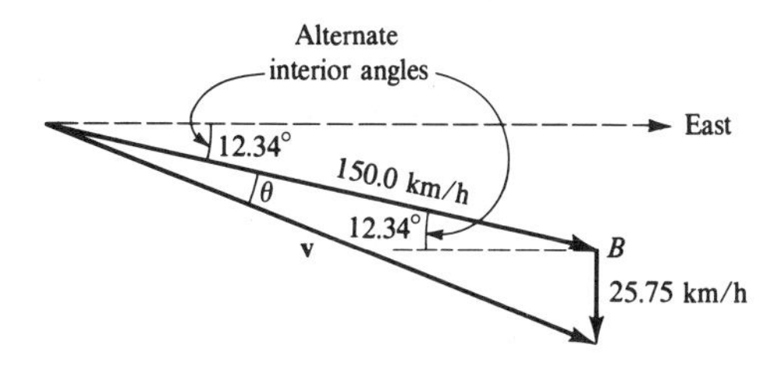

Figure 16.49

Solution. Let us denote the desired velocity vector by $\mathbf{v}$. Also note that the angle labeled B in the figure is

$$B = 12.34° + 90° = 102.34°$$

To find $v = |\mathbf{v}|$, we use the cosine law:

$$v^2 = (150.0)^2 + (25.75)^2 - 2(150.0)(25.75) \cos 102.34°$$

or

$$v = 157.5 \text{ km/h}$$

To find the direction, we need angle θ in Figure 16.49. Since B is the angle opposite the longest side, we can safely apply the sine law to find θ:

$$\frac{\sin \theta}{25.75} = \frac{\sin 102.34°}{157.5} \quad \text{or} \quad \theta = 9.19°$$

So the direction is $12.34° + 9.19° = 21.53°$ south of east.

The next example shows the power of the cosine law particularly well.

Example 5 Two forces of 55.0 N and 37.0 N, respectively, are acting on the same object. The angle between the forces is 23.4°. (See Figure 16.50.) What single force will have the same effect?

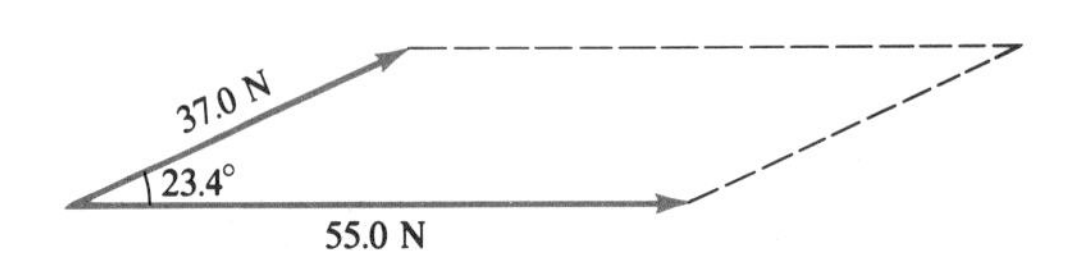

Figure 16.50

Solution. To be able to use the cosine law, we draw the vectors as shown in Figure 16.51. Note that the 23.4° angle between the forces appears on the right in the figure, so the angle inside the triangle is $180° - 23.4° = 156.6°$.

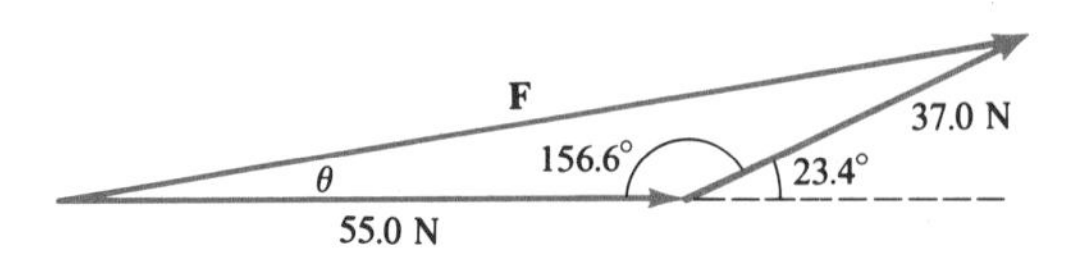

Figure 16.51

By the cosine law,

$$|\mathbf{F}|^2 = (55.0)^2 + (37.0)^2 - 2(55.0)(37.0) \cos 156.6°$$

or

$$|\mathbf{F}| = 90.2 \text{ N}$$

Since the angle opposite the longest side is already known, we can safely apply the sine law to find angle θ, the direction of **F** relative to the 55.0-N force. From

$$\frac{\sin \theta}{37.0} = \frac{\sin 156.6^\circ}{90.2}$$

we get $\theta = 9.4^\circ$. (If the cosine law is used, the calculated value turns out to be 9.3°.)

Exercises / Section 16.5

Exercises 1–14 refer to the triangle in Figure 16.52. Solve each of the given triangles.

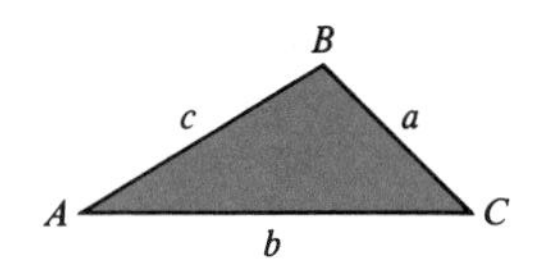

Figure 16.52

1. $a = 6.39$, $b = 8.47$, $c = 7.63$

2. $a = 15.4$, $b = 13.1$, $c = 12.6$

3. $a = 25.7$, $b = 15.3$, $c = 17.6$

4. $a = 4.72$, $b = 2.16$, $c = 6.83$

5. $a = 0.3182$, $b = 0.4641$, $c = 0.2916$

6. $a = 234.6$, $b = 142.1$, $c = 172.4$

7. $C = 73.4^\circ$, $a = 59.5$, $b = 84.2$

8. $A = 37.6^\circ$, $b = 7.63$, $c = 8.72$

9. $A = 135^\circ 20'$, $b = 4.38$, $c = 3.74$

10. $B = 129^\circ 40'$, $a = 15.7$, $c = 12.1$

11. $C = 58^\circ$, $a = 72$, $b = 88$

12. $B = 82^\circ$, $a = 5.8$, $c = 4.6$

13. $B = 99.0^\circ$, $a = 24.3$, $c = 17.0$

14. $C = 61.5^\circ$, $a = 7.56$, $b = 8.94$

15. Two forces of 64.2 lb and 38.6 lb, respectively, are acting on the same object. If the angle between the directions is 21.7°, what single force would have the same effect?

16. Two forces 15.0° apart are acting on an object. If their respective magnitudes are 145 N and 114 N, find the resultant.

17. A ship sails 21.3 km due west, turns 10.4° south of west, and then continues for another 30.0 km. Find the distance from the starting point.

18. A plane travels 130 mi due north, turns 25° east of north, and continues for another 150 mi. Find the distance from the starting point.

19. To find the length of a proposed tunnel, a surveyor observes that from a distant point the distances to the ends of the tunnel are 465 m and 529 m, respectively. If the angle between the lines of sight is 38.6°, find the length of the proposed tunnel.

20. To find the distance from A to B in Figure 16.53, a surveyor takes the measurements given in the figure. Find the distance.

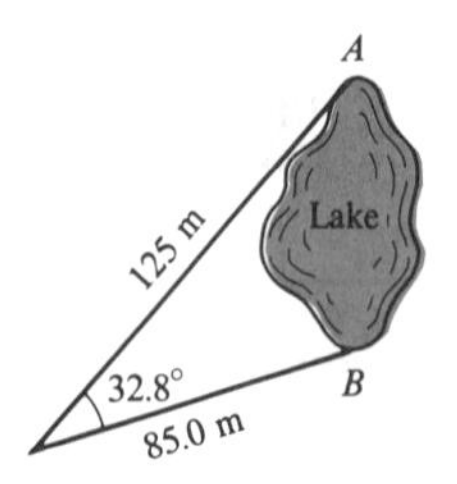

Figure 16.53

21. A small plane is heading 6.0° west of north with an air speed of 165 mi/h. The wind is from the east at 31.0 mi/h. Find the actual course and the velocity with respect to the ground.

22. A small plane is heading 5.0° east of north with an air speed of 192 mi/h. If the wind is from the south at 35.0 mi/h, determine the actual course and the velocity with respect to the ground.

23. A boat is heading 19° north of east when it runs into a strong current of 8.0 knots from the north. If the velocity of the boat in still water is 22 knots, find the resulting speed and direction.

Review Exercises / Chapter 16

1. Given the vectors in Figure 16.54, graphically determine $\mathbf{A} + 2\mathbf{B} - 3\mathbf{C}$.

Figure 16.54

In Exercises 2–7, find the magnitude and direction of each vector (A_x, A_y).

2. $(-1, 1)$

3. $(-3\sqrt{3}, -3)$

4. $(1, -2)$

5. $(2, 5)$

6. $(-2.4, 3.6)$

7. $(-5.82, -10.3)$

In Exercises 8–10, resolve each vector into its components. Round off the answers to three significant digits.

8. $A = 7$, $\theta = 125.6°$

9. $A = 6$, $\theta = 314.4°$

10. $A = 9$, $\theta = 233.6°$

In Exercises 11 and 12, add the given vectors by components. Find R_x and R_y to three significant digits, $|\mathbf{R}|$ to two significant digits, and θ to the nearest tenth of a degree.

11. **A**: $A = 6$, direction: 162°
B: $B = 9$, direction: 285°

12. **A**: $A = 4$, direction: 10°
B: $B = 5$, direction: 342°
C: $C = 2$, direction: 260°

In Exercises 13–22, solve each triangle from the given information.

13. $A = 22.5°$, $C = 57.8°$, $b = 10.0$

14. $A = 125.6°$, $B = 20.4°$, $a = 37.7$

15. $A = 34.50°$, $C = 100.00°$, $c = 0.8461$

16. $B = 110.00°$, $C = 38.46°$, $b = 6.469$

17. $A = 42.3°$, $a = 12.8$, $c = 16.4$, angle C acute

18. Same as Exercise 17 with angle C obtuse

19. $a = 58$, $b = 32$, $c = 36$

20. $a = 1.846$, $b = 2.017$, $c = 2.739$

21. $C = 112°40'$, $a = 6.72$, $b = 4.34$

22. $B = 95.7°$, $a = 27.6$, $c = 25.4$

23. A boat heads directly across a river. Its velocity in still water is 16.0 km/h. Given that the river flows at 4.10 km/h, what is the direction of the boat and its velocity relative to the bank?

24. A railroad car weighing 12.4 tons is on a track inclined at 5.6° with the horizontal. What is the force required to keep the car from rolling downhill?

25. A 25-kg weight hanging on a rope is pulled sideways by a force of 11 kg. Determine the tension on the rope, and the angle that the rope makes with the vertical.

26. Use addition of components to solve the following problem: A small plane flying with an air speed of 156 mi/h is heading 14.3° south of west and encounters a wind out of the north at 21.4 mi/h. Find the resulting direction and velocity with respect to the ground.

27. Work Exercise 26 by the law of cosines.

28. A surveyor wants to determine the width of a river. She can reach one point on the bank, but no other points nearby are accessible. So she takes the measurements in Figure 16.55. Determine the width of the river.

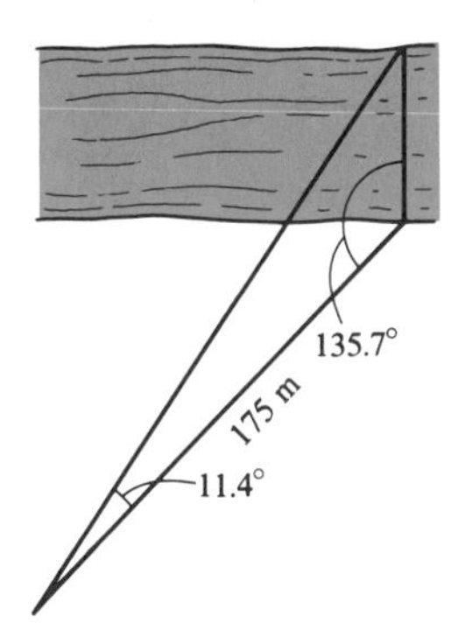

Figure 16.55

29. Two forces act on the same object in directions that are 36.5° apart. If one force is 28.4 lb, what must the other force be so that the combined effect is a force of 54.0 lb?

30. A force of 36.7 kg and a force of 43.8 kg produce a resultant force of 72.4 kg. Find the angle between the forces.

31. From a point on the ground the angle of elevation of the top of a tower is 26.3°. From a second point 48.3 ft closer to the tower, the angle of elevation of the top is 37.4°. Find the height of the tower.

32. From a certain point on the ground, the angle of elevation of the top of a pillar is 24°30′. From a point 10.0 m further back, the angle of elevation of the top is 16°20′. Find the height of the pillar.

33. Find the perimeter of the metal plate in Figure 16.56.

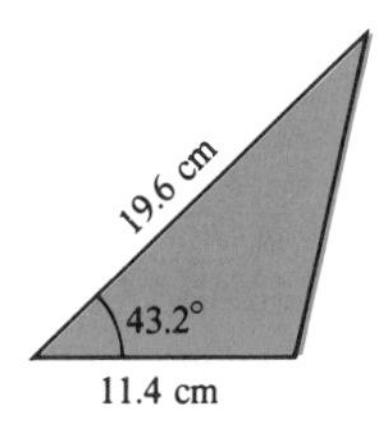

Figure 16.56

CHAPTER **17**

Radian Measure and Graphs of Sinusoidal Functions

So far all our angle measurements have been made in degrees. In the first part of this chapter we introduce another unit of measure, the **radian**. Later in this chapter we will use radian measure to study the graphs of the sine and cosine functions.

17.1 Radian Measure

Radian measure is highly useful in technical applications. The reason is that radian measure is based on certain properties of the circle. Consider, for example, the circle of radius r in Figure 17.1. If we measure off r units along the circumference, the central angle formed has a measure of 1 **radian** (rad).

Definition of radian measure:

One **radian** is the size of an angle whose vertex is at the center of a circle and whose intercepted arc on the circumference is equal in length to the radius of the circle.

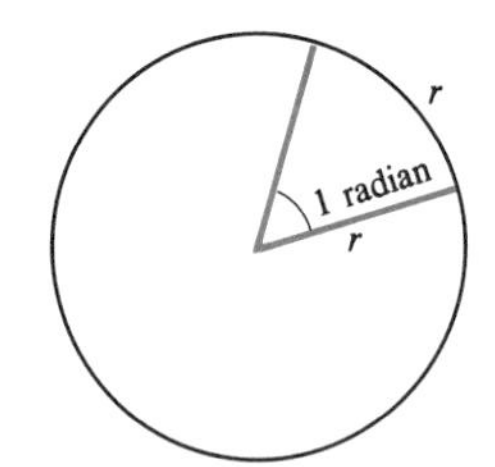

Figure 17.1

To see the relationship between radian and degree measure, consider the angles of measure 2 rad, 3 rad, and 6 rad, respectively, shown in Figure 17.2.

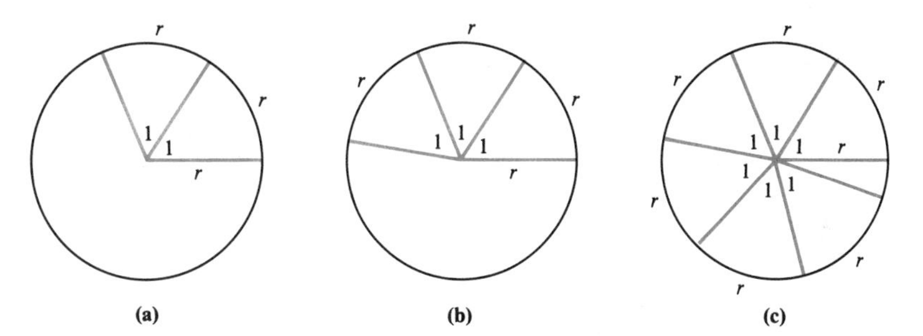

Figure 17.2

To obtain the radian measure of one complete rotation (360°), we need to determine how many times r can be measured off along the circumference. According to Figure 17.2(c), this appears to be just over six times. So 360° must be approximately equal to 6 rad. Can we find the exact number? The answer is yes, since this number is given to us by the formula for the circumference of a circle: $C = 2\pi r$. So r can be measured off exactly 2π times along the circumference. (Note that $2\pi \approx 6.28$.) It follows that 2π rad = 360°, or

$$\pi \text{ rad} = 180°$$

From this relationship we can obtain all our special angles in radian measure. For example, since π rad = 180°, we get $\pi/2 = 90°$, $\pi/3 = 60°$, and so on. Some of these angles are listed next.

Degrees:	0°	30°	45°	60°	90°	180°	270°	360°
Radians:	0	$\frac{\pi}{6}$	$\frac{\pi}{4}$	$\frac{\pi}{3}$	$\frac{\pi}{2}$	π	$\frac{3\pi}{2}$	2π

Of particular interest are the measures corresponding to 1° and 1 rad. From the relationship

$$\pi \text{ rad} = 180°$$

we get

$$\frac{\pi \text{ rad}}{180} = \frac{180°}{180} = 1° \qquad \text{Dividing by 180}$$

and

$$\frac{180°}{\pi} = \frac{\pi \text{ rad}}{\pi} = 1 \text{ rad} \qquad \text{Dividing by } \pi$$

It follows that

$$1^\circ = \frac{\pi}{180}\text{ rad} = 0.017453\text{ rad}$$

$$1\text{ rad} = \frac{180^\circ}{\pi} = 57^\circ 17' 45'' = 57.296^\circ$$

From these relationships we obtain a simple method for converting an angle measure from one system to the other. For example,

$$15^\circ = 15 \cdot 1^\circ = 15 \cdot \frac{\pi}{180}\text{ rad} = \frac{\pi}{12}\text{ rad}$$

Conversely,

$$\frac{4\pi}{5}\text{ rad} = \frac{4\pi}{5} \cdot 1\text{ rad} = \frac{4\pi}{5}\left(\frac{180^\circ}{\pi}\right) = 144^\circ$$

The general rule is given next.

To convert an angle measure from

1. Degree measure to radian measure, multiply by $\pi/180^\circ$
2. Radian measure to degree measure, multiply by $180^\circ/\pi$

Example 1 Change 64° to radians.

Solution. Multiplying by $\pi/180^\circ$, we get

$$64^\circ = 64 \cdot 1^\circ = 64\left(\frac{\pi}{180}\text{ rad}\right) = \frac{16\pi}{45}\text{ rad}$$

Example 2 Change $5\pi/6$ rad to degrees.

Solution. Multiplying by $180^\circ/\pi$, we have

$$\frac{5\pi}{6}\text{ rad} = \frac{5\pi}{6} \cdot 1\text{ rad} = \frac{5\pi}{6}\left(\frac{180^\circ}{\pi}\right) = 150^\circ$$

Before proceeding with any further examples, let us take another look at the radian as a unit of measure. Consider the circle in Figure 17.3, with a radius of 2 cm. As noted earlier, the central angle in radians is equal to the number of times that the radius can be measured off along the intercepted arc. In Figure 17.3, the arc length is $2k$ cm for $k \geq 0$. So k is equal to the number of times that $r = 2$ cm can be measured off along the arc. It follows

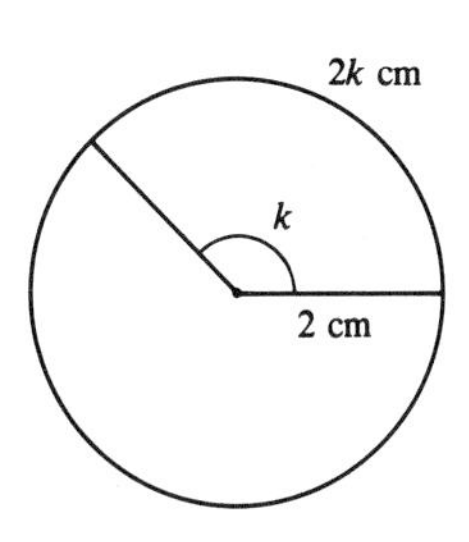

Figure 17.3

that k is the measure of the central angle in radians. Expressed in another way,

$$\frac{2k \text{ cm}}{2 \text{ cm}} = \frac{\not{2}k \not{\text{cm}}}{\not{2} \not{\text{cm}}} = k$$

Since the units cancel, radian measure is free of units.

> Whenever an angle is expressed in radians, no units are indicated.

For example, 30° is expressed simply as $\pi/6$.

In general, if s is the arc length and θ the central angle (Figure 17.4), then $\theta = s/r$.

Radian measure of central angles

$$\theta = \frac{s}{r}$$

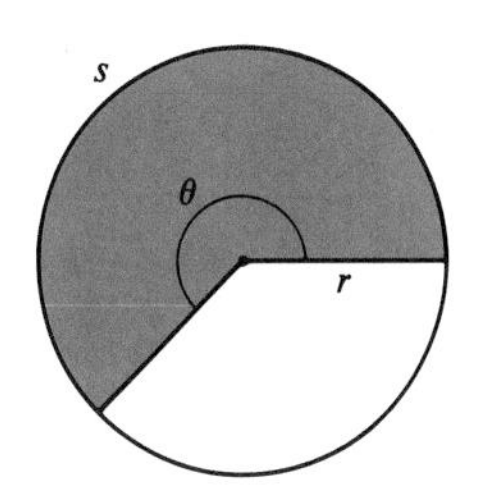

Figure 17.4

We now consider some additional examples of conversions from one system to the other.

Example 3 Convert 12° and 240° to radian measure.

Solution. Multiplying each angle measure by $\pi/180°$, we get

$$12° = 12° \cdot \frac{\pi}{180°} = \frac{\pi}{15} \qquad \text{Degrees cancel}$$

$$240° = 240° \cdot \frac{\pi}{180°} = \frac{4\pi}{3}$$

Example 4 Convert $\pi/9$ and $7\pi/6$ to degree measure.

Solution. We multiply by $180°/\pi$ in each case:

$$\frac{\pi}{9} = \frac{\pi}{9} \cdot \frac{180°}{\pi} = 20°$$

$$\frac{7\pi}{6} = \frac{7\pi}{6} \cdot \frac{180°}{\pi} = 210°$$

Using a Calculator

If the angle measures are expressed in decimal form, the conversions should be carried out with a calculator.

Example 5 Convert 36.74° to radians.

Solution. We multiply by $\pi/180°$:

$$36.74° = 36.74° \left(\frac{\pi}{180°}\right)$$

The sequence is

$$36.74 \boxed{\times} \boxed{\pi} \boxed{\div} 180 \boxed{=} \rightarrow 0.6412339$$

So 36.74° = 0.6412.

Example 6 Convert 2.86 to degrees.

Solution. We multiply by $180°/\pi$:

$$2.86 = 2.86 \left(\frac{180°}{\pi}\right)$$

The sequence is

$$2.86 \boxed{\times} 180 \boxed{\div} \boxed{\pi} \boxed{=} \rightarrow 163.86593$$

So 2.86 = 163.9°.

Radian mode

When finding the value of a trigonometric function of an angle expressed in radians, take care to set your calculator in the **radian mode**. (Once put into the radian mode, most calculators will remain in this mode until changed back to degree mode or turned off.)

Example 7 Evaluate sin 1.683 and cos 2.904.

Solution. To find sin 1.683, set your calculator in radian mode and use the sequence

$$1.683 \boxed{\text{SIN}} \rightarrow 0.9937117$$

or sin 1.683 = 0.9937.

Similarly,

$$\cos 2.904 = -0.9719$$

Trigonometric Functions in BASIC

To find the values of the trigonometric functions in BASIC, use the library functions

SIN(X), COS(X), and TAN(X)

In most systems, the angles must be expressed in radians.

Example 8 Evaluate

a. sin 2 **b.** tan 3 **c.** cos(−1)

Solution.

Function	*Result*
a. SIN(2)	0.9092974268
b. TAN(3)	−0.142546543
c. COS(−1)	0.5403023059

Exercises / Section 17.1

In Exercises 1–24, convert each degree measure to radian measure. Express the radian measures in terms of π. (See Example 3.)

1. 30° **2.** 45° **3.** 60° **4.** 20°

5. −45° **6.** −60° **7.** 24° **8.** 72°

9. 44° **10.** 135° **11.** 150° **12.** 210°

13. 315° **14.** 140° **15.** 160° **16.** 232°

17. 144° **18.** 96° **19.** 156° **20.** 108°

21. 66° **22.** 126° **23.** 192° **24.** 228°

In Exercises 25–48, convert each radian measure to degree measure. (See Example 4.)

25. $\frac{\pi}{4}$ **26.** $\frac{\pi}{3}$ **27.** $\frac{5\pi}{4}$ **28.** $\frac{5\pi}{3}$

29. $-\frac{5\pi}{6}$ **30.** $-\frac{11\pi}{6}$ **31.** $\frac{5\pi}{12}$ **32.** $\frac{7\pi}{15}$

33. $\frac{21\pi}{10}$ **34.** $\frac{16\pi}{9}$ **35.** $\frac{29\pi}{18}$ **36.** $\frac{41\pi}{36}$

37. $\frac{26\pi}{15}$ **38.** $\frac{27\pi}{10}$ **39.** $\frac{19\pi}{9}$ **40.** $\frac{7\pi}{18}$

41. $\frac{17\pi}{36}$ **42.** $\frac{7\pi}{60}$ **43.** $\frac{13\pi}{90}$ **44.** $\frac{11\pi}{4}$

45. $\frac{17\pi}{6}$ **46.** $\frac{25\pi}{18}$ **47.** $-\frac{19\pi}{12}$ **48.** $-\frac{26\pi}{9}$

In Exercises 49–64, use a calculator to change each degree measure to radian measure. Use four decimal places in the result.

49. $62°$ **50.** $43°$ **51.** $93°$ **52.** $85°$

53. $39°$ **54.** $47°$ **55.** $120.6°$ **56.** $163.7°$

57. $271.4°$ **58.** $342.1°$ **59.** $45°20'$ **60.** $58°40'$

61. $125°30'$ **62.** $160°50'$ **63.** $280°10'$ **64.** $342°40'$

In Exercises 65–76, use a calculator to change each radian measure to degree measure, accurate to the nearest tenth of a degree.

65. 0.5642 **66.** 0.8458 **67.** 0.7763 **68.** 0.2896

69. 0.3968 **70.** 0.5504 **71.** 1.732 **72.** 1.207

73. 2.960 **74.** 2.407 **75.** 1.871 **76.** 2.884

In Exercises 77–90, use a calculator to find the values of the given trigonometric functions. Set your calculator in the **radian mode**.

77. $\sin(0.5079)$ **78.** $\cos(0.1390)$ **79.** $\tan(1.369)$ **80.** $\sin(1.106)$

81. $\cos(1.943)$ **82.** $\tan(2.092)$ **83.** $\sec(0.4451)$ **84.** $\csc(3.6015)$

85. $\cot(1.357)$ **86.** $\sec(2.394)$ **87.** $\csc(3.468)$ **88.** $\cot(0.3407)$

89. $\tan(2.268)$ **90.** $\sin(1.114)$

91. The current i in an alternating-series circuit as a function of time t (in seconds) is $i = 2.00 \cos 277t$. Find i when $t = 2.70$ ms.

92. The voltage across a certain inductor as a function of time t (in seconds) is $V = 140 \sin 350t$. Find V when $t = 1.80$ ms.

93. The displacement x (in centimeters) of a weight on a spring as a function of time is given by $x = 4.0 \cos 3.0t + 5.0 \sin 3.0t$. Find x when $t = 2.1$ s.

94. The velocity (in feet per second) of a mass on a certain spring is $v = 0.42 \cos 2(t - 1.5)$. Find v when $t = 5.2$ s.

17.2 Applications of Radian Measure

The basic applications of radian measure that we will consider are the determination of the length of a circular arc, the area of a circular sector, and linear and angular velocity.

Arc Length

To obtain the formula for the length of a circular arc, we refer back to the definition of radian measure:

$$\theta = \frac{s}{r}$$

Solving for s, we get $s = r\theta$, where θ is the central angle shown in Figure 17.5. With this formula we can compute the length of a circular arc, provided that θ is expressed in radians.

Length of circular arc

$$s = r\theta, \qquad \theta \text{ in radians}$$

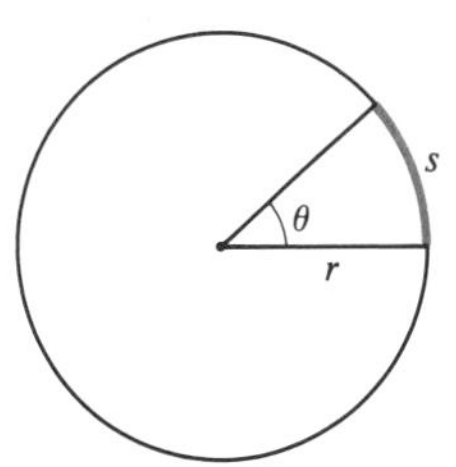

Figure 17.5

Example 1 Find the arc length s in Figure 17.6.

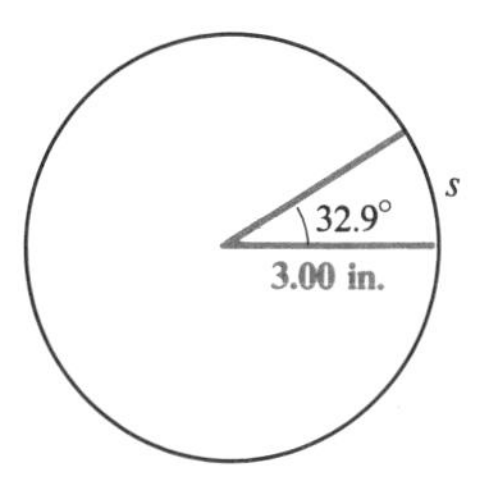

Figure 17.6

Solution. First we change 32.9° to radians:

$$32.9° = 32.9°\left(\frac{\pi}{180°}\right) = 0.5742$$

Then by the formula $s = r\theta$, we get

$$s = (3.00 \text{ in.})(0.5742) = 1.72 \text{ in.}$$

Note that since radian measure is free of units, no radians are included in the final result. The answer is therefore expressed as 1.72 in.

A particularly interesting application of the arc length formula is finding the approximate height of a distant object, as shown in the next example.

Example 2 A tree 206 ft from an observer on the ground intercepts an angle of 7.35°. (See Figure 17.7.) Find the approximate height of the tree.

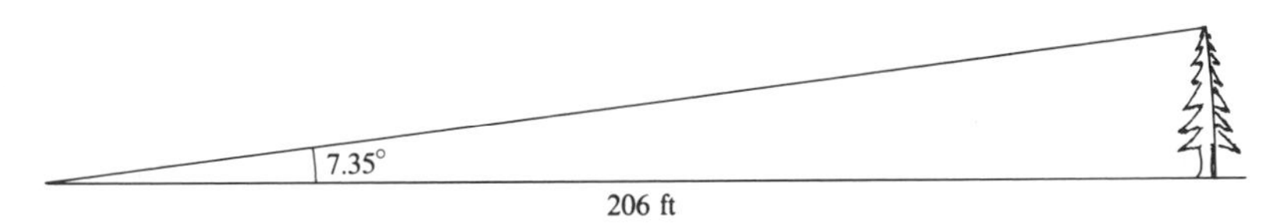

Figure 17.7

Solution. The height of the tree is approximately equal to the length of the intercepted arc shown in Figure 17.7.

To find the length of the arc, we use the formula $s = r\theta$. Since

$$\theta = 7.35° = \frac{7.35\pi}{180} = 0.1283$$

it follows that the height is approximately equal to

$$(206 \text{ ft})(0.1283) = 26.4 \text{ ft}$$

(This method of approximation works only for small angles.)

Area of Circular Sector

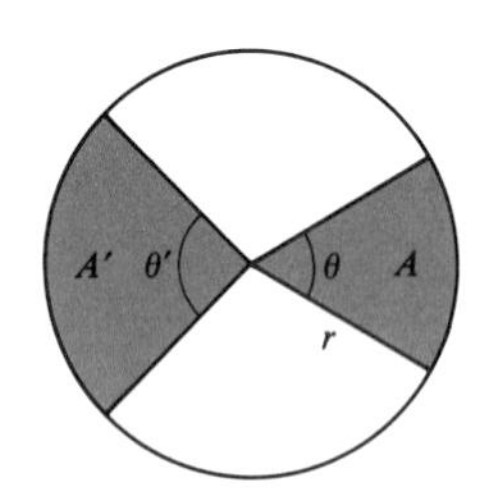

Figure 17.8

Radian measure also enables us to find the area of the sector of a circle. The formula can be obtained by using the fact that the area of a circular sector is proportional to the central angle. This statement means (referring to Figure 17.8)

$$\frac{A}{\theta} = \frac{A'}{\theta'}$$

Now suppose that sector A' consists of the entire circle. Then $A' = \pi r^2$ and $\theta' = 2\pi$, and the equation becomes

$$\frac{A}{\theta} = \frac{\pi r^2}{2\pi}$$

or

$$A = \frac{1}{2} r^2\theta$$

The sector is shown in Figure 17.9.

Area of circular sector

$$A = \frac{1}{2} r^2\theta, \qquad \theta \text{ in radians}$$

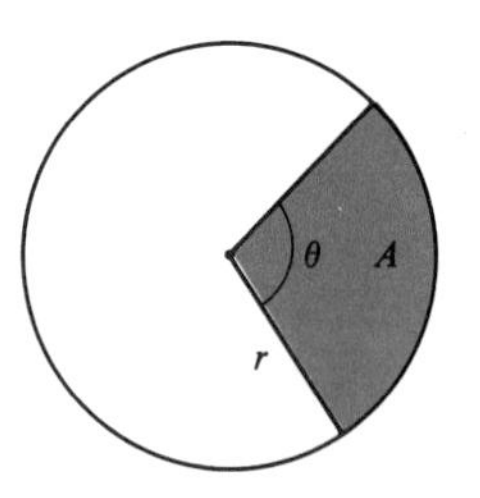

Figure 17.9

Example 3 Find the area of the sector in Figure 17.10.

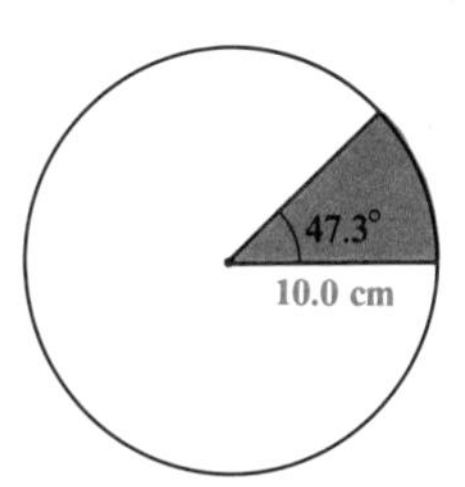

Figure 17.10

Solution. First we convert the angle to radian measure:

$$47.3° = \frac{47.3\pi}{180} = 0.8255$$

From the formula $A = \frac{1}{2}r^2\theta$, we get

$$A = \frac{1}{2}(10.0 \text{ cm})^2(0.8255) = 41.3 \text{ cm}^2$$

Common error Forgetting to change the angle θ to radians when finding the length of a circular arc or the area of a circular sector.

Linear and Angular Velocity

Another interesting application of radian measure involves rotational motion. Recall that for motion along a path, distance = rate × time, where the rate is assumed constant. Suppose a particle moves around a circle at a constant rate. If we denote the distance along the circle by s and the rate by v (for velocity), then v = distance ÷ time, or $v = s/t$. Since $s = r\theta$, we have $v = (r\theta)/t = r(\theta/t)$. The ratio θ/t, measured in radians per unit time, is called the **angular velocity** and is denoted by the Greek letter ω (omega); v is called the **linear velocity.** Letting $\omega = \theta/t$, $v = r(\theta/t)$ becomes $v = r\omega$ or $v = \omega r$.

Angular velocity
Linear velocity

> Relationship between angular and linear velocity:
>
> $v = \omega r$
>
> Units for ω: radians per unit time

Example 4 An object is moving about a circle of radius 4.0 cm with angular velocity of 2.0 rad/s. Find the linear velocity v.

Solution. By the formula $v = \omega r$

$$v = \left(2.0\ \frac{\text{rad}}{\text{s}}\right) \cdot 4.0 \text{ cm} = 8.0\ \frac{\text{cm}}{\text{s}}$$

(Note that "rad" is not included in the final result.) So the object is moving about the circle at the rate of 8.0 cm/s.

In many problems ω is given in revolutions per unit time. In such a case ω has to be converted to radians per unit time. Consider the next example.

Example 5 A computer disk rotates at the rate of 325 rev/min (revolutions per minute). If the radius of the disk is 6.25 cm, find the linear velocity of a point on the rim in meters per minute.

Solution. To convert the angular velocity ω to radians per minute, we use the fact that

$$1 \text{ rev} = 2\pi \text{ rad}$$

So

$$325 \frac{\text{rev}}{\text{min}} = 325 \frac{\cancel{\text{rev}}}{\text{min}} \frac{2\pi \text{ rad}}{1 \cancel{\text{rev}}} = 325 \cdot 2\pi \frac{\text{rad}}{\text{min}}$$

and

$$v = \omega r = (325 \cdot 2\pi) \frac{\text{rad}}{\text{min}} \cdot 6.25 \text{ cm}$$

$$= 12{,}762.72 \frac{\text{cm}}{\text{min}}$$

Converting to meters per minute, we get

$$v = 12{,}762.72 \frac{\cancel{\text{cm}}}{\text{min}} \cdot \frac{1 \text{ m}}{100 \cancel{\text{cm}}} = 128 \frac{\text{m}}{\text{min}}$$

Example 6 Find the velocity in miles per second of a communications satellite that remains 22,300 mi above a point on the equator at all times. (The radius of the earth is about 3960 mi.)

Solution. First we need to express the angular velocity ω in radians per second. To do so, note that the satellite makes one revolution every 24 h, that is, $\omega = (1/24)$rev/h. So

$$\omega = \frac{1 \cancel{\text{rev}}}{24 \cancel{\text{h}}} \frac{2\pi \text{ rad}}{1 \cancel{\text{rev}}} \frac{1 \cancel{\text{h}}}{60 \cancel{\text{min}}} \frac{1 \cancel{\text{min}}}{60 \text{ s}} \qquad 1 \text{ rev} = 2\pi \text{ rad}$$

$$= 7.2722 \times 10^{-5} \frac{\text{rad}}{\text{s}}$$

Since the radius of the earth is 3960 mi, the satellite moves about a circle of radius

$$3960 \text{ mi} + 22{,}300 \text{ mi} = 26{,}260 \text{ mi}$$

It follows that

$$v = \omega r = \left(7.2722 \times 10^{-5} \frac{\text{rad}}{\text{s}}\right)(26{,}260 \text{ mi})$$

$$= 1.91 \frac{\text{mi}}{\text{s}}$$

Exercises / Section 17.2

In Exercises 1–10, find the length of the circular arc intercepted by the given central angle.

1. $\theta = 2.64$, $r = 10.0$ cm

2. $\theta = 1.04$, $r = 3.46$ in.

3. $\theta = 30.4°$, $r = 6.34$ m

4. $\theta = 43.64°$, $r = 7.642$ cm

5. $\theta = 39°30'$, $r = 2.64$ ft

6. $\theta = 53°20'$, $r = 1.25$ m

7. $\theta = 10.76°$, $r = 1.732$ m

8. $\theta = 130°$, $r = 4.8$ ft

9. $\theta = 140°$, $r = 5.8$ ft

10. $\theta = 145.0°$, $r = 5.76$ in.

In Exercises 11–20, find the area of the circular sector satisfying the given conditions.

11. $\theta = 1.32$, $r = 12.0$ cm

12. $\theta = 2.16$, $r = 3.76$ in.

13. $\theta = 32.6°$, $r = 7.84$ m

14. $\theta = 45.75°$, $r = 9.020$ cm

15. $\theta = 35°40'$, $r = 3.72$ ft

16. $\theta = 55°10'$, $r = 1.79$ cm

17. $\theta = 12.40°$, $r = 2.394$ m

18. $\theta = 125.0°$, $r = 5.38$ ft

19. $\theta = 132.0°$, $r = 4.05$ ft

20. $\theta = 150.0°$, $r = 4.59$ in.

21. A tree 530 ft away intercepts an angle of 1.6°. Find the approximate height of the tree. (See Example 2.)

22. A building 990 ft away intercepts an angle of 2.8°. Find the approximate height of the building. (See Example 2.)

23. Find the degree measure of the central angle that intercepts an arc length of 7.26 cm on a circle of radius 2.47 cm.

24. A central angle of a circle intercepts an arc length of 10.4 cm. If the radius of the circle is 3.41 cm, find the degree measure of the central angle.

25. The full moon intercepts an angle of 0.518°. Find the diameter of the moon, given that the distance from the earth to the moon is about 239,000 mi.

26. The mean distance from the earth to the sun is 93 million miles. The diameter of the sun is 866,000 mi. Find its angular size (the intercepted angle). Why do the sun and moon appear to be the same size? (Refer to Exercise 25.)

27. Find the area swept out by a pendulum of length 16.0 in. that swings through an arc of 10.3°.

28. A pendulum of length 35.60 cm swings through an arc of 15°46′. Find the distance covered by the end of the pendulum as it swings from one end of the arc to the other.

29. An object is moving about a circle of radius 12.0 cm with an angular velocity of 0.64 rad/s. Find its linear velocity in centimeters per second.

30. A circular disk 8.0 in. in diameter rotates at the rate of 4.7 rad/min. Find the linear velocity of a point on the rim in inches per minute.

31. A wheel with radius 5.75 in. rotates at the rate of 10.0 rev/s. Find the linear velocity of a point on the rim in feet per second.

32. A flywheel spins at the rate of 295 rev/min. If the radius of the flywheel is 6.30 cm, find the velocity of a point on the rim in meters per minute.

33. A floppy disk has a radius of 2.75 in. and rotates at the rate of 467 rev/min. Find the linear velocity of a point on the rim in feet per second.

34. The wheel driving the piston in Figure 17.11 rotates at the rate of 256 rev/min. Find the linear velocity of the point of connection in feet per second.

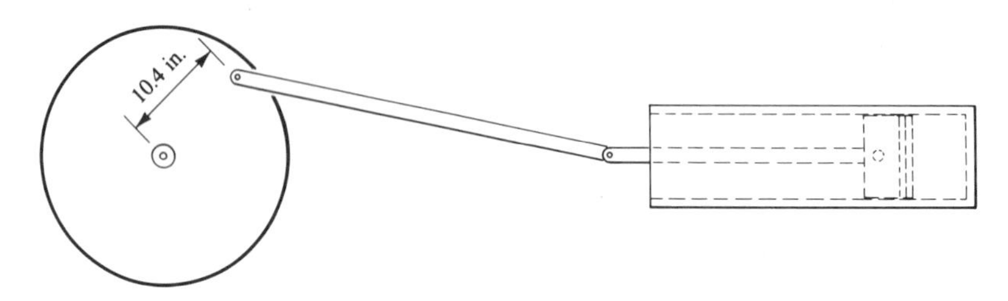

Figure 17.11

35. Find the velocity in miles per hour of a point on the equator, due to the earth's rotation. (The radius of the earth is 3960 mi.)

36. Determine the velocity in miles per hour of the earth relative to the sun. (The distance from the earth to the sun is about 93 million miles.)

37. The radius of the tires on a truck is 16.5 in. If the angular velocity of each tire is 9.39 rev/s, find the velocity of the truck in feet per minute.

38. Find the velocity of a truck in miles per hour if the tires are 32.0 in. in diameter and if the angular velocity is 8.51 rev/s. (Recall that 60 mi/h = 88 ft/s.)

39. A bicycle is traveling at the rate of 1500 ft/min. Find the angular velocity of the wheels in radians per second if each wheel has a diameter of 32 in.

40. Determine the angular velocity of the wheels of a car in revolutions per second if each wheel has a radius of 14 in. and the car is traveling at 30 mi/h.

41. A pulley belt is 10 ft long and takes 45 s to make a complete circuit. If the radius of the pulley is 12 in., determine its angular velocity in radians per second.

42. A pulley belt 40.0 cm long takes 3.64 s to make one complete circuit. Determine the angular velocity of the pulley in revolutions per minute, given that the radius of the pulley is 6.10 cm.

17.3 Graphs of Sine and Cosine Functions

Our discussion in Chapter 12 showed that the graph of a function gives a revealing picture of the behavior of the function. In this section we will study the graphs of $y = a \sin x$ and $y = a \cos x$.

When graphing trigonometric functions, we usually use radian measure. To be able to construct a table of values readily, let us recall the radian measure of certain special angles. From the relationship

$$\pi = 180°$$

we get the following basic special angles:

$$\frac{\pi}{2} = 90°, \quad \frac{\pi}{3} = 60°, \quad \frac{\pi}{4} = 45°, \quad \text{and } \frac{\pi}{6} = 30°$$

These relationships, in turn, yield the special angles in other quadrants:

$$150° = 5(30°) = \frac{5\pi}{6}$$

$$210° = 7(30°) = \frac{7\pi}{6}$$

$$330° = 11(30°) = \frac{11\pi}{6}$$

and so on. The multiples of 45° are obtained from $\pi/4 = 45°$. Thus

$$135° = 3(45°) = \frac{3\pi}{4}$$

$$225° = 5(45°) = \frac{5\pi}{4}$$

and so on. Finally, the multiples of 60° are obtained from $\pi/3 = 60°$:

$$120° = 2(60°) = \frac{2\pi}{3}$$

$$240° = 4(60°) = \frac{4\pi}{3}$$

$$300° = 5(60°) = \frac{5\pi}{3}$$

and so on.

Using these special angles, we can construct a table of values for the function $y = \sin x$.

$y = \sin x$

x:	0	$\frac{\pi}{6}$	$\frac{\pi}{4}$	$\frac{\pi}{3}$	$\frac{\pi}{2}$	$\frac{2\pi}{3}$	$\frac{5\pi}{6}$	π	$\frac{7\pi}{6}$	$\frac{4\pi}{3}$	$\frac{3\pi}{2}$	$\frac{11\pi}{6}$	2π	$2\pi + \frac{\pi}{6}$
y (*exact*):	0	$\frac{1}{2}$	$\frac{1}{\sqrt{2}}$	$\frac{\sqrt{3}}{2}$	1	$\frac{\sqrt{3}}{2}$	$\frac{1}{2}$	0	$-\frac{1}{2}$	$-\frac{\sqrt{3}}{2}$	-1	$-\frac{1}{2}$	0	$\frac{1}{2}$
y (*decimal*):	0	0.5	0.7	0.87	1	0.87	0.5	0	−0.5	−0.87	−1	−0.5	0	0.5

Plotting these points on the rectangular coordinate system, we obtain the graph shown in Figure 17.12.

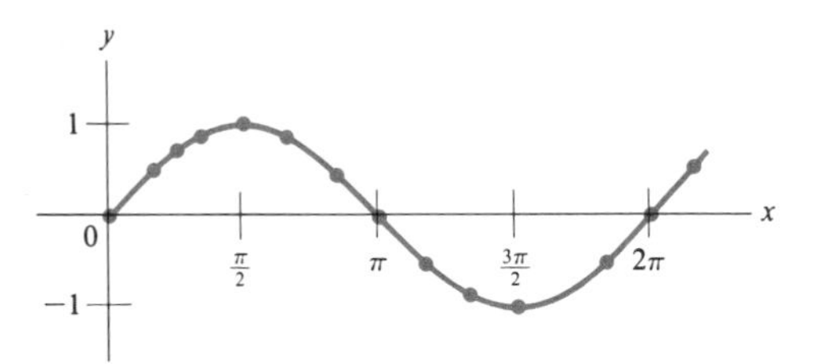

Figure 17.12

Observe that as x increases from 0 to $\pi/2$, the values of $\sin x$ increase from 0 to 1. As x continues to increase from $\pi/2$ to π, the values of $\sin x$ repeat in reverse order from 1 to 0. As x increases from π to $3\pi/2$, the values of $\sin x$ become negative, decreasing from 0 to -1. Finally, as x increases from $3\pi/2$ to 2π, the values of $\sin x$ increase once again, from -1 back to 0. *Starting at 2π, the values of $\sin x$ repeat.*

Note especially the zero values of $\sin x$, as well as the largest and smallest values:

x:	0	$\frac{1}{4}(2\pi) = \frac{\pi}{2}$	$\frac{1}{2}(2\pi) = \pi$	$\frac{3}{4}(2\pi) = \frac{3\pi}{2}$	$1 \cdot (2\pi) = 2\pi$
$\sin x$:	0	1	0	-1	0

To obtain the graph of $y = \cos x$, we construct the following table:

$y = \cos x$

x:	0	$\frac{\pi}{6}$	$\frac{\pi}{4}$	$\frac{\pi}{3}$	$\frac{\pi}{2}$	$\frac{2\pi}{3}$	$\frac{5\pi}{6}$	π	$\frac{7\pi}{6}$	$\frac{4\pi}{3}$	$\frac{3\pi}{2}$	$\frac{5\pi}{3}$	$\frac{11\pi}{6}$	2π
y (*exact*):	1	$\frac{\sqrt{3}}{2}$	$\frac{1}{\sqrt{2}}$	$\frac{1}{2}$	0	$-\frac{1}{2}$	$-\frac{\sqrt{3}}{2}$	-1	$-\frac{\sqrt{3}}{2}$	$-\frac{1}{2}$	0	$\frac{1}{2}$	$\frac{\sqrt{3}}{2}$	1
y (*decimal*):	1	0.87	0.7	0.5	0	-0.5	-0.87	-1	-0.87	-0.5	0	0.5	0.87	1

Plotting these points, we get the graph shown in Figure 17.13.

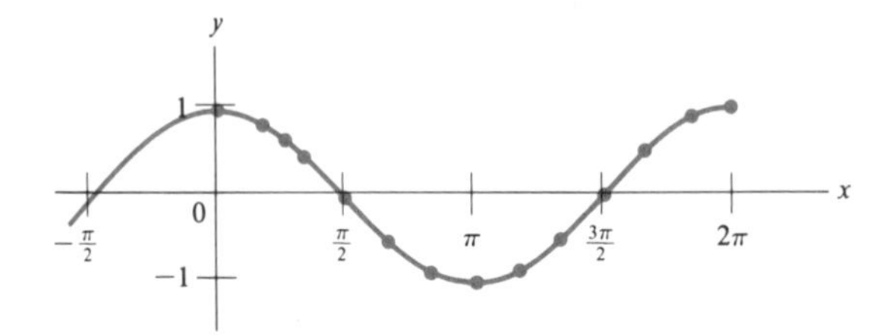

Figure 17.13

Note especially the zero values of $\cos x$, as well as the largest and smallest values:

x:	0	$\frac{1}{4}(2\pi) = \frac{\pi}{2}$	$\frac{1}{2}(2\pi) = \pi$	$\frac{3}{4}(2\pi) = \frac{3\pi}{2}$	$1 \cdot (2\pi) = 2\pi$
$\cos x$:	1	0	-1	0	1

A closer inspection of the tables for $y = \sin x$ and $y = \cos x$ suggests that the values of the sine and cosine functions, and hence the shapes of their graphs, are identical except for their positions. Indeed, if the graph of the cosine function is moved $\pi/2$ units to the right, it becomes the graph of the sine function.

The graphs shown in Figures 17.12 and 17.13 not only continue indefinitely in both directions, but the function values repeat every 2π radians. A function that repeats regularly is called a **periodic function**. So the sine and cosine functions are periodic with **period** 2π. Note also that the largest value is 1 and the smallest value -1. The sine and cosine functions are said to have an **amplitude** of 1 unit.

The remainder of this section and the next two sections will be devoted to certain variations on these basic forms.

The Graphs of $y = a \sin x$ and $y = a \cos x$

To see the difference between the graph of $y = \sin x$ and that of $y = a \sin x$, recall that the amplitude of $y = \sin x$ is 1. Since $a \sin x$ is the product of a and $\sin x$, the largest value of $a \sin x$ is a (and the smallest value is $-a$). The number $|a|$ is the **amplitude** of $y = a \sin x$. Similarly, $|a|$ is the amplitude of $y = a \cos x$. The amplitude will be denoted by the letter A. Thus $A = |a|$ for $y = a \sin x$ and $y = a \cos x$.

The multiplication by a does not affect the period, however. The function values still repeat every 2π radians.

Period and **amplitude** of $y = a \sin x$ and $y = a \cos x$:

Period: $P = 2\pi$ **Amplitude:** $A = |a|$

Example 1 Sketch the graph of $y = 2 \cos x$.

Solution. We first construct a table of values. Note that every value of $y = 2 \cos x$ is twice the corresponding value in the table of $y = \cos x$.

x:	0	$\frac{\pi}{6}$	$\frac{\pi}{4}$	$\frac{\pi}{3}$	$\frac{\pi}{2}$	$\frac{2\pi}{3}$	$\frac{5\pi}{6}$	π	$\frac{7\pi}{6}$	$\frac{4\pi}{3}$	$\frac{3\pi}{2}$	$\frac{5\pi}{3}$	$\frac{11\pi}{6}$	2π
y:	2	$\sqrt{3}$	$\frac{2}{\sqrt{2}}$	1	0	-1	$-\sqrt{3}$	-2	$-\sqrt{3}$	-1	0	1	$\sqrt{3}$	2

The graph, shown in Figure 17.14, looks like a tall version of the graph in Figure 17.13.

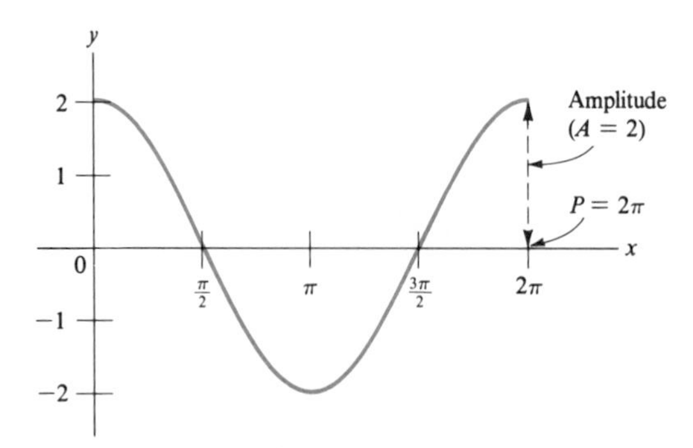

Figure 17.14

Returning to Figure 17.14 for a moment, since the amplitude of $y = 2\cos x$ is 2 (and the period 2π), we could have obtained the graph from Figure 17.13 by simply increasing the amplitude to 2. The resulting graph is similar, although a bit taller.

> To sketch the graph of $y = a \sin x$ and $y = a \cos x$, mark the x-intercepts and the highest and lowest points on the curve, and obtain the sketch from the basic shapes.

Note: Recall that the x-intercepts are the points where the graph crosses the x-axis.

Example 2 Sketch the graph of $y = \frac{1}{2} \sin x$.

Solution. Note that

$$A = \frac{1}{2} \quad \text{and} \quad P = 2\pi$$

(See Figure 17.15.) From the basic function, $y = \sin x$, we know that the x-intercepts are $x = 0, \pi$, and 2π. (See Figure 17.12.) To obtain the intercepts and the highest and lowest points, we need to plot only the values in the following table:

x:	0	$\frac{\pi}{2}$	π	$\frac{3\pi}{2}$	2π
y:	0	$\frac{1}{2}$	0	$-\frac{1}{2}$	0

The graph, shown in Figure 17.15, has a smaller amplitude than the graph in Figure 17.12.

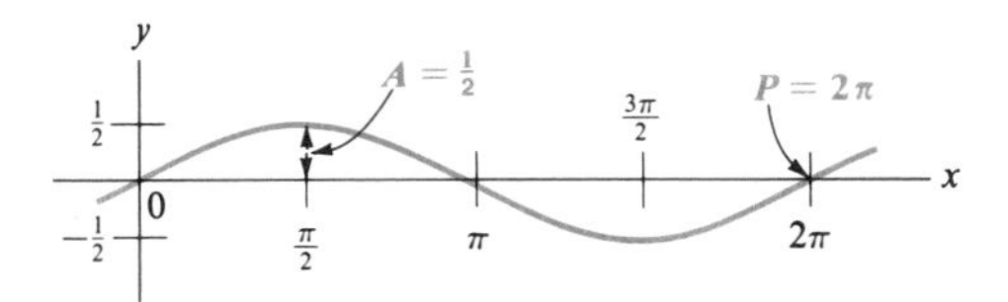

Figure 17.15

Example 3 Sketch the graph of $y = 3 \cos x$.

Solution. Since

$$A = 3 \quad \text{and} \quad P = 2\pi$$

we need to plot only the values in the next table. (For the x-intercepts, see Figure 17.13.)

x:	0	$\frac{\pi}{2}$	π	$\frac{3\pi}{2}$	2π
y:	3	0	-3	0	3

The graph is shown in Figure 17.16 over two periods.

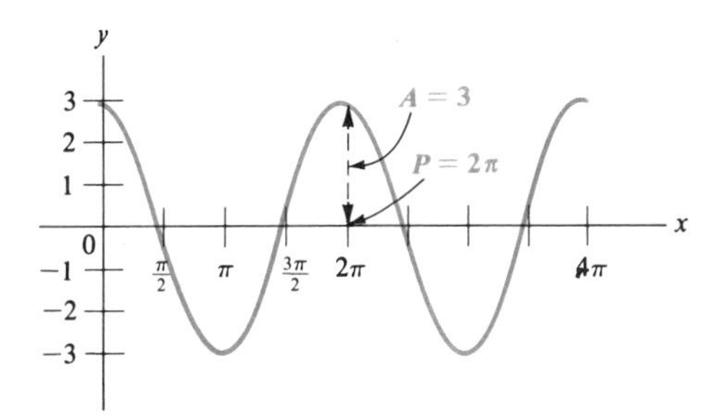

Figure 17.16

Example 4 Sketch the graph of $y = -3 \sin x$.

Solution. The effect of the coefficient -3 is twofold: It multiplies the values of the basic sine function by 3 and changes the sign of each value. The curve of $y = -3 \sin x$ is, therefore, the "mirror image" of the curve of $y = 3 \sin x$. For $y = 3 \sin x$,

$$A = 3 \quad \text{and} \quad P = 2\pi$$

The curve is shown as the dashed curve in Figure 17.17.

The reflection of the dashed curve is the graph of $y = -3 \sin x$, shown as the solid curve in Figure 17.17.

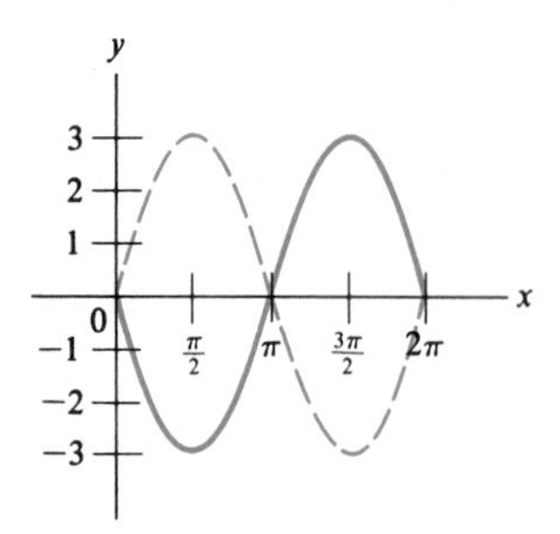

Figure 17.17

Instead of using the mirror image, the graph of $y = -3 \sin x$ can also be sketched from the following short table (Figure 17.18):

x:	0	$\frac{\pi}{2}$	π	$\frac{3\pi}{2}$	2π
y:	0	-3	0	3	0

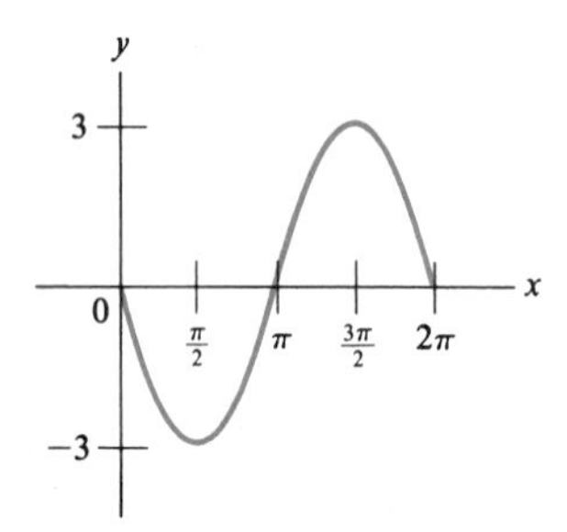

Figure 17.18

Exercises / Section 17.3

State the amplitude and period of each of the given functions. Plot the values corresponding to $x = 0, \frac{\pi}{2}, \frac{3\pi}{2}$, and 2π, and sketch the curves. (See Examples 2–4.)

1. $y = 2 \sin x$

2. $y = \frac{3}{2} \sin x$

3. $y = \frac{1}{2} \cos x$

4. $y = 4 \cos x$

5. $y = 3 \sin x$

6. $y = \frac{1}{3} \cos x$

7. $y = -\cos x$

8. $y = -2 \sin x$

9. $y = -2 \cos x$ **10.** $y = -\frac{1}{2} \sin x$ **11.** $y = 2.5 \sin x$ **12.** $y = 1.5 \cos x$

13. $y = 1.8 \cos x$ **14.** $y = 2.3 \sin x$ **15.** $y = -1.5 \sin x$ **16.** $y = -2.2 \cos x$

17.4 More on Graphs of Sinusoidal Functions

In this section we will consider another variation of the functions discussed in the previous section: $y = a \sin bx$ and $y = a \cos bx$. Together, these functions are called **sinusoidal functions** and their graphs, **sinusoidal curves**.

Since the maximum value of $\sin \theta$ and $\cos \theta$ is 1, the amplitude of $y = a \sin bx$ and $y = a \cos bx$ is $A = |a|$, as before. The period, however, is $P = 2\pi/b$, as we will see shortly.

> **Period** and **Amplitude** of $y = a \sin bx$ and $y = a \cos bx$:
>
> **Period:** $P = \dfrac{2\pi}{b}$ **Amplitude:** $A = |a|$

To see why the period is $2\pi/b$, note first that the graph of $y = \sin x$ passes through the origin and crosses the x-axis at

$$x = \pi, \quad 2\pi, \quad 3\pi, \quad \text{and so on}$$

The second intercept, 2π, is the period. To obtain the period of $y = a \sin bx$, observe that

$$a \sin bx = 0$$

at the origin and whenever

$$bx = \pi, \quad 2\pi, \quad 3\pi, \quad \text{and so on}$$

or

$$x = \frac{\pi}{b}, \quad \frac{2\pi}{b}, \quad \frac{3\pi}{b}, \quad \text{and so on}$$

As before, the second intercept, $2\pi/b$, is the period.

Example 1 Find the amplitude and period of $y = 2 \cos 3x$ and sketch the curve.

Solution. From $A = |a|$, we get $A = 2$. The period is

$$P = \frac{2\pi}{b} = \frac{2\pi}{3}$$

To sketch the curve, we need only mark the highest and lowest points on the curve and the end of the period $2\pi/3$ rad from the starting point (origin). The sketch is shown in Figure 17.19 over two periods.

x:	0	$1 \cdot \frac{\pi}{6}$	$2 \cdot \frac{\pi}{6} = \frac{\pi}{3}$	$3 \cdot \frac{\pi}{6} = \frac{\pi}{2}$	$4 \cdot \frac{\pi}{6} = \frac{2\pi}{3}$
y:	2	0	-2	0	2

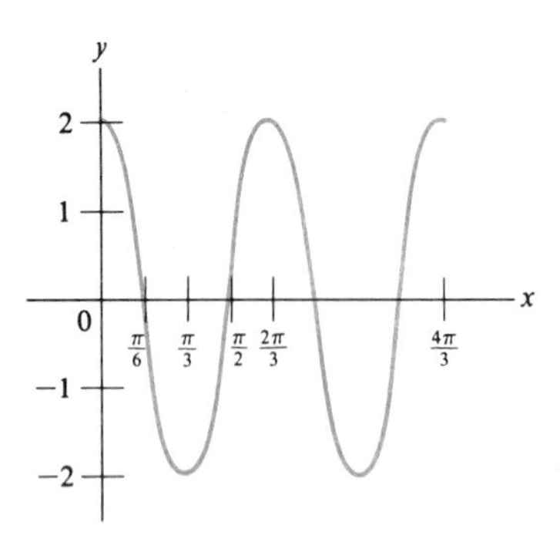

Figure 17.19

Given the basic shapes, we can sketch the graph of any sinusoidal function using only the period and amplitude.

To sketch a sinusoidal curve: $y = a \sin bx$ or $y = a \cos bx$:

1. Determine the period $P = 2\pi/b$. Start at $x = 0$ and mark off the distance $2\pi/b$.
2. Divide the interval from 0 to $2\pi/b$ into *four* equal parts.
3. Locate the x-intercepts.
   a. For $y = a \sin bx$, the intercepts are the beginning, midpoint, and end of the interval.
   b. For $y = a \cos bx$, the intercepts are one-fourth and three-fourths of the way through the interval.
4. Determine the amplitude and mark off the highest and lowest points on the curve.
5. Sketch the curve over one period.
6. Extend the graph over additional periods, if desired.

Example 2 Sketch the graph of $y = \frac{3}{4} \sin 2x$.

Solution. $y = \frac{3}{4} \sin 2x$

Step 1. $P = \frac{2\pi}{2} = \pi$. Mark off $x = \pi$ (Figure 17.20).

Step 2. The four equally spaced points (and $x = 0$) are

$$x = 0, \quad \frac{\pi}{4}, \quad \frac{\pi}{2}, \quad \frac{3\pi}{4}, \quad \pi$$

Step 3. The intercepts are

$x = 0$ Beginning of interval

$x = \frac{\pi}{2}$ Midpoint

$x = \pi$ End of interval

Step 4. The highest and lowest points occur at $x = \pi/4$ and $3\pi/4$, the points midway between the intercepts. (These are the points in Step 2 that are not in Step 3.) Note that

$$A = \frac{3}{4} \qquad \text{(Figure 17.20)}$$

Step 5. The graph is shown in Figure 17.20.
Step 6. The graph is extended to the right over one additional period.

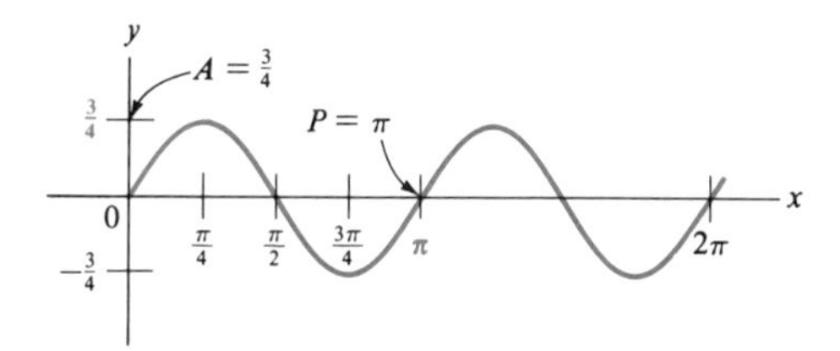

Figure 17.20

Example 3 Sketch the graph of $y = 4 \cos \frac{1}{2} x$.

Solution. $y = 4 \cos \frac{1}{2} x$

Step 1. $P = \dfrac{2\pi}{1/2} = 4\pi$. Mark off $x = 4\pi$ (Figure 17.21).

Step 2. The four equally spaced points (and $x = 0$) are

$$x = 0, \quad \pi, \quad 2\pi, \quad 3\pi, \quad 4\pi$$

Step 3. The intercepts are

$x = \pi$ One-fourth of the way

$x = 3\pi$ Three-fourths of the way

Step 4. The highest and lowest points occur at $x = 0, 2\pi, 4\pi$ (the points in Step 2 that are not in Step 3). Also note that

$A = 4$ (Figure 17.21)

Step 5. The graph is shown in Figure 17.21.

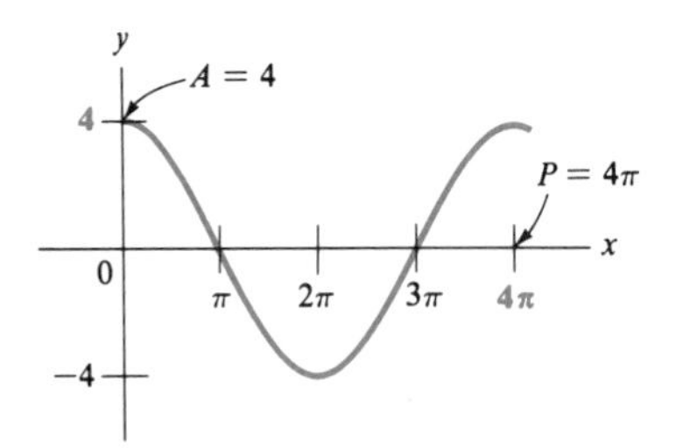

Figure 17.21

An important application of sinusoidal functions is the study of **simple harmonic motion**. Simple harmonic motion can be described by

$$y = r \sin \omega t \quad \text{or} \quad y = r \cos \omega t \qquad \omega = \text{omega}$$

where t represents time. Many natural occurrences, such as a mass oscillating on a spring, water waves, sound waves, and alternating current, are simple harmonic.

Example 4 A weight hanging on a spring is allowed to come to rest at the origin (Figure 17.22). Once set in motion, the displacement x is the distance from the origin as a function of time. For a certain weight, x is given by

$$x = 3.4 \sin 4\pi t$$

where x is measured in centimeters and t in seconds. Graph this function.

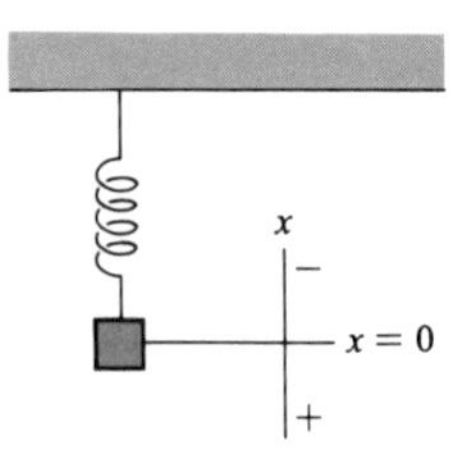

Figure 17.22

Solution. $x = 3.4 \sin 4\pi t$

$$A = 3.4 \qquad P = \frac{2\pi}{4\pi} = \frac{1}{2}$$

The graph is shown in Figure 17.23 over one period.

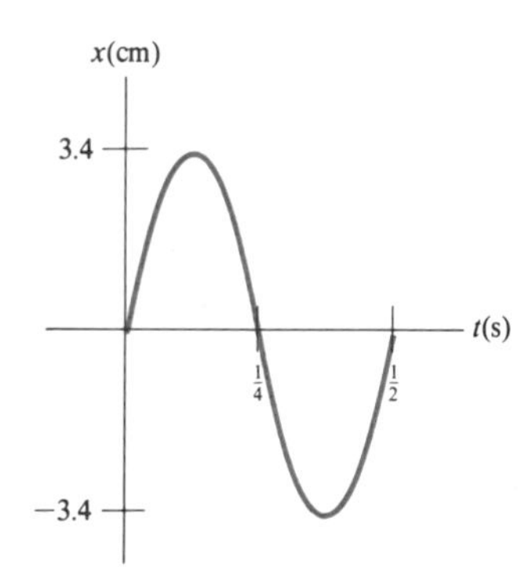

Figure 17.23

Exercises / Section 17.4

In Exercises 1–20, state the amplitude and period of each function; sketch the curve.

1. $y = \sin 2x$ **2.** $y = \cos \frac{1}{2}x$ **3.** $y = \cos 2x$ **4.** $y = \sin 3x$

5. $y = 2 \sin 3x$ **6.** $y = 3 \cos 2x$ **7.** $y = \frac{1}{2} \cos 2x$ **8.** $y = \frac{1}{2} \sin 2x$

9. $y = 3 \sin \frac{1}{2}x$ **10.** $y = 3 \cos \frac{1}{2}x$ **11.** $y = 5.4 \cos \frac{1}{3}x$ **12.** $y = 3.2 \sin \frac{1}{3}x$

13. $y = -\sin 3x$ **14.** $y = -\cos 2x$ **15.** $y = -\frac{1}{4} \cos 4x$ **16.** $y = -\frac{1}{5} \sin 6x$

17. $y = \cos \pi x$ **18.** $y = \sin \pi x$ **19.** $y = \frac{1}{3} \sin 3\pi x$ **20.** $y = \frac{1}{3} \cos 2\pi x$

21. A weight hanging on a spring is oscillating vertically (see Example 4). The displacement x as a function of time is given by

$$x = 2.1 \sin 3\pi t$$

where x is in centimeters and t in seconds. Sketch the curve of this motion.

22. The displacement of the weight on a spring is given by

$$x = 3.4 \cos \frac{2}{3} \pi t$$

where x is in feet and t in seconds. Sketch the graph of this motion.

23. The sound of a tuning fork may be expressed in the form $y = A \sin 2\pi ft$, where f, called the *frequency,* is equal to the reciprocal of the period ($f = 1/P$). If $y = 0.002 \sin 2\pi(150)t$ for a certain tuning fork, graph y as a function of time t. (Assume that y is measured in inches and t in seconds.)

24. Suppose an object is moving about a circle of radius 4.0 cm centered at the origin with a constant angular velocity of 1.5 rad/s. If the motion starts at (4.0, 0), then the distance y from the horizontal axis as a function of time is $y = 4.0 \sin 1.5t$. Sketch the graph.

17.5 Phase Shifts

The sinusoidal curves discussed in the last section can be shifted to the left or right by a certain amount called the *phase shift*.

To see what this means, consider the function

$$y = \sin(x - 1)$$

and compare its graph to the graph of $y = \sin x$. The function $y = \sin x$ has amplitude 1 and period 2π. The y-values of $y = \sin(x - 1)$ and $y = \sin x$ are identical, but they correspond to different x-values. In particular, to get $\sin 0$, we must let $x = 1$ in $y = \sin(x - 1)$, so that

$$y = \sin(1 - 1) = \sin 0$$

It follows that the point $(1, 0)$ corresponds to the origin. The other points are shifted similarly (see Figure 17.24). In particular, the y-values start repeating at $x = 2\pi + 1$.

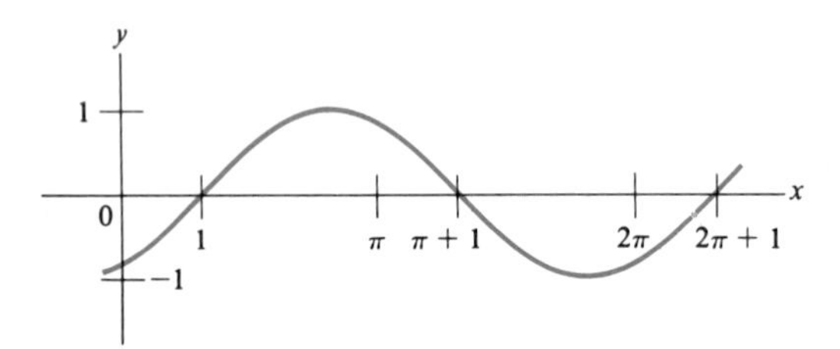

Figure 17.24

Example 1 Sketch the graph of $y = 2 \sin(x + 1)$.

Solution. The graph of this function is similar to the graph in Figure 17.24, but this time we need to let $x = -1$ to obtain $\sin 0$. So the point $(-1, 0)$ corresponds to the origin. It follows that the graph of $y = 2 \sin(x + 1)$ is the graph of $y = 2 \sin x$ shifted one unit to the left (Figure 17.25.)

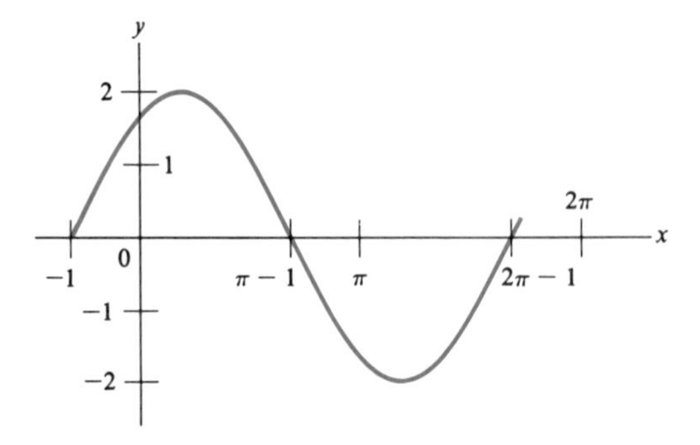

Figure 17.25

To sketch the graph of $y = a \sin(bx + c)$ or $y = a \cos(bx + c)$, we factor b and write

$$y = a \sin\left[b\left(x + \frac{c}{b}\right)\right] \quad \text{or} \quad y = a \cos\left[b\left(x + \frac{c}{b}\right)\right]$$

These forms show that the shift is given by c/b. The number c/b is called the **phase shift**.

Period, amplitude, and phase shift of $y = a \sin(bx - c)$ and $y = a \cos(bx - c)$, $(b, c > 0)$:

Period: $P = \dfrac{2\pi}{b}$ **Amplitude:** $A = |a|$

Phase shift: $\dfrac{c}{b}$ units to *right*

Period, amplitude, and phase shift of $y = a \sin(bx + c)$ and $y = a \cos(bx + c)$, $(b, c > 0)$:

Period: $P = \dfrac{2\pi}{b}$ **Amplitude:** $A = |a|$

Phase shift: $\dfrac{c}{b}$ units to *left*

Rule: To sketch the graph of $y = a \sin(bx \pm c)$, we first sketch the graph of $y = a \sin bx$ by the technique discussed in Section 17.4. We then shift this curve by c/b units to obtain the graph of $y = a \sin(bx \pm c)$. The procedure for graphing $y = a \cos(bx \pm c)$ is similar.

Example 2 Sketch the graph of

$$y = \frac{1}{2} \sin\left(2x - \frac{\pi}{2}\right)$$

Solution. We first factor 2 and write the function in the form

$$y = \frac{1}{2} \sin 2\left(x - \frac{\pi}{4}\right)$$

This form shows that the phase shift is $\pi/4$. The phase shift can also be obtained from

$$\frac{c}{b} = \frac{\pi/2}{2} = \frac{\pi}{4} \quad \text{(shifted right)} \qquad \text{phase shift}$$

It follows that the graph of $y = \frac{1}{2} \sin 2[x - (\pi/4)]$ is the graph of $y = \frac{1}{2} \sin 2x$ shifted **$\pi/4$ units to the right**

For the function $y = \frac{1}{2}\sin 2x$, we have

$$A = \frac{1}{2} \qquad \text{and} \qquad P = \frac{2\pi}{2} = \pi$$

The graph of $y = \frac{1}{2}\sin 2x$ is the dashed curve in Figure 17.26. This graph is now shifted $\pi/4$ units to the right, shown as the solid curve in the figure.

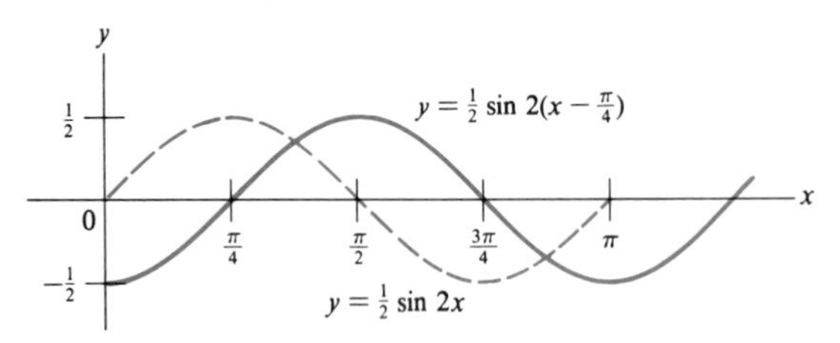

Figure 17.26

Example 3 Sketch the graph of

$$y = 3\cos\left(\frac{1}{2}x + \frac{\pi}{4}\right)$$

Solution. First we write the function in the form

$$y = 3\cos\frac{1}{2}\left(x + \frac{\pi}{2}\right)$$

It follows that this graph is the graph of $y = 3\cos\frac{1}{2}x$ shifted **$\pi/2$ units to the left**. This also follows from

$$\frac{c}{b} = \frac{\pi/4}{1/2} = \frac{\pi}{2} \quad \text{(shifted left)} \qquad \text{phase shift}$$

For the function $y = 3\cos\frac{1}{2}x$, we have

$$A = 3 \qquad \text{and} \qquad P = \frac{2\pi}{1/2} = 4\pi$$

The graph of $y = 3\cos\frac{1}{2}x$ is the dashed curve in Figure 17.27. Shifting this curve $\pi/2$ units to the left, we obtain the solid curve in the same figure.

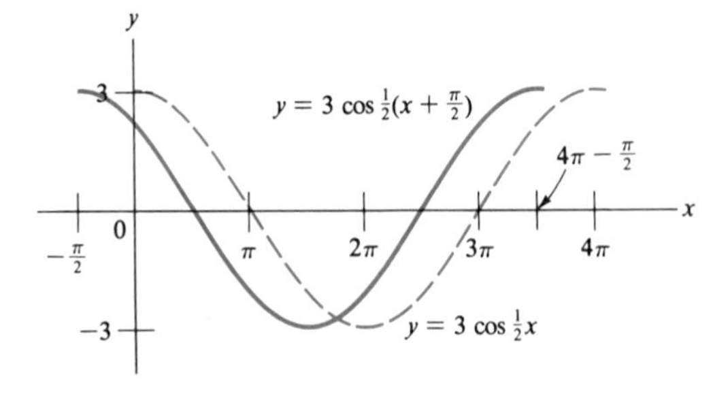

Figure 17.27

Phase shifts commonly occur in the study of alternating current. The current and voltage produced by an alternating-current generator can be described by sinusoidal functions, often subject to a displacement. The current i typically has the form

$$i = I \sin(\omega t + \alpha) \tag{17.1}$$

and the voltage v has the form

$$v = E \sin(\omega t + \alpha) \tag{17.2}$$

Example 4 If $I = 4.00$ A, $\omega = 120\pi$ rad/s, and $\alpha = \pi/2$, sketch the graph of the current.

Solution. By formula (17.1),

$$i = 4.00 \sin\left(120\pi t + \frac{\pi}{2}\right)$$

Factoring 120π, we obtain the form

$$i = 4.00 \sin 120\pi\left(t + \frac{1}{240}\right)$$

So

$$A = 4.00 \text{ A}, P = \frac{2\pi}{120\pi} = \frac{1}{60} \text{ s}$$

and the shift is $\frac{1}{240}$ s to the left. The graph is shown in Figure 17.28.

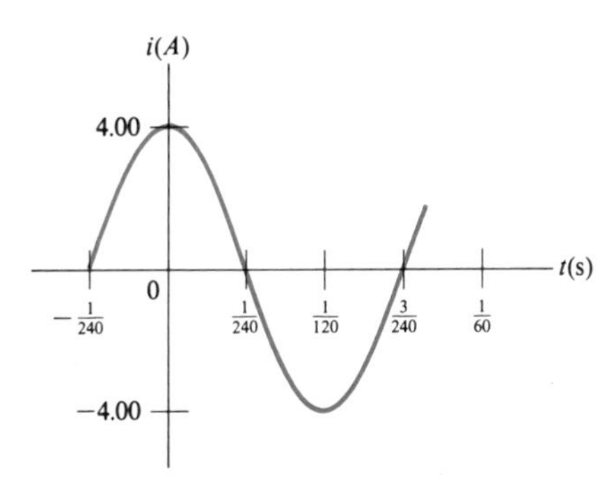

Figure 17.28

Note that the graph crosses the t-axis at

$$\frac{1}{120} - \frac{1}{240} = \frac{1}{240} \text{ s} \qquad \text{phase shift} = \frac{1}{240} \text{ s (left)}$$

At that instant, i becomes negative, indicating that the current reverses direction. The current is said to be **alternating**.

The graph crosses the t-axis again at

$$\frac{1}{60} - \frac{1}{240} = \frac{3}{240}\text{ s}$$

At $t = \frac{3}{240}$ s, the current becomes positive.

Exercises / Section 17.5

In Exercises 1–20, state the amplitude, period, and phase shift. Sketch each curve over one period.

1. $y = 2\sin(x - 1)$

2. $y = 2\sin\left(x + \frac{1}{2}\right)$

3. $y = 2\cos(x + 1)$

4. $y = 2\cos(x - 1)$

5. $y = \frac{3}{4}\sin\left(x - \frac{\pi}{4}\right)$

6. $y = \frac{4}{3}\cos\left(x - \frac{\pi}{4}\right)$

7. $y = \frac{1}{2}\cos\left(2x + \frac{\pi}{4}\right)$

8. $y = \frac{1}{3}\sin\left(2x + \frac{\pi}{4}\right)$

9. $y = -0.8\sin\left(2x - \frac{\pi}{8}\right)$

10. $y = -0.6\cos\left(2x - \frac{\pi}{8}\right)$

11. $y = 4\sin\left(\frac{1}{2}x + \frac{\pi}{2}\right)$

12. $y = 3\cos\left(\frac{1}{2}x + \frac{\pi}{2}\right)$

13. $y = 2\cos(\pi x - \pi)$

14. $y = 2\sin(\pi x - \pi)$

15. $y = 2\sin\left(2x + \frac{\pi}{2}\right)$

16. $y = \frac{1}{2}\cos\left(3x + \frac{3\pi}{4}\right)$

17. $y = 0.5\cos(2x - 1)$

18. $y = 1.2\sin(3x - 1)$

19. $y = 3\sin\left(\frac{1}{2}x - 2\right)$

20. $y = 0.2\sin(2\pi x + 2)$

21. If $I = 5.00$ A, $\omega = 120\pi$ rad/s, and $\alpha = \pi/4$, sketch the graph of i as a function of time. (See Example 4.)

22. If $I = 4.50$ A, $\omega = 100\pi$ rad/s, and $\alpha = \pi/3$, sketch the graph of i as a function of time.

23. If $E = 4.50$ V, $\omega = 80\pi$ rad/s, and $\alpha = \pi/4$, sketch the graph of v as a function of time. [See formula (17.2).]

24. If $E = 12.0$ V, $\omega = 120\pi$ rad/s, and $\alpha = \pi/6$, sketch the graph of v as a function of time.

Review Exercises / Chapter 17

In Exercises 1–6, convert each degree measure to radian measure expressed in terms of π.

1. 40°

2. 84°

3. 112°

4. 155°

5. 235°

6. 320°

In Exercises 7–12, convert each radian measure to degree measure.

7. $\frac{29\pi}{15}$

8. $\frac{67\pi}{36}$

9. $\frac{19\pi}{18}$

10. $\frac{11\pi}{9}$

11. $\frac{5\pi}{12}$

12. $\frac{2\pi}{15}$

In Exercises 13–16, use a calculator to change each degree measure to radian measure. Use four-decimal-place accuracy.

13. 112°

14. 284°

15. 73.7°

16. 72°40′

In Exercises 17–20, use a calculator to change each radian measure to degree measure, accurate to the nearest tenth of a degree.

17. 0.1360
18. 0.8943
19. 2.6341
20. 4.0372

In Exercises 21–24, use a calculator to find the values of the given trigonometric functions. (Set calculator in radian mode.)

21. sin(2.549)
22. tan(5.037)
23. sec(3.604)
24. cos(−0.6781)

In Exercises 25–28, find the length of the circular arc intercepted by the given central angle.

25. $\theta = 2.19$, $r = 10.0$ cm
26. $\theta = 24.3°$, $r = 8.40$ m
27. $\theta = 95.47°$, $r = 20.50$ in.
28. $\theta = 162°40'$, $r = 17.4$ ft

In Exercises 29–32, find the area of each circular sector satisfying the given conditions.

29. $\theta = 1.94$, $r = 18.0$ cm
30. $\theta = 29.46°$, $r = 12.75$ m
31. $\theta = 74.3°$, $r = 1.60$ ft
32. $\theta = 134°10'$, $r = 4.25$ yd

In Exercises 33–40, state the amplitude, period, and phase shift of each function, and sketch the curve.

33. $y = 3.5 \sin x$
34. $y = -2.5 \cos x$
35. $y = 4 \cos \frac{1}{3} x$
36. $y = \frac{1}{4} \sin 3\pi x$
37. $y = -\frac{1}{2} \sin 2x$
38. $y = \frac{1}{3} \cos\left(2x - \frac{\pi}{4}\right)$
39. $y = -1.25 \sin\left(2x + \frac{\pi}{8}\right)$
40. $y = 4 \cos\left(\frac{1}{2} x - \frac{\pi}{2}\right)$

41. A tower 740 ft away intercepts an angle of 2.9°. Find the approximate height of the tower.

42. Find the degree measure of the central angle that intercepts an arc length of 11.4 m on a circle of radius 16.3 m.

43. Find the area swept out by a pendulum of length 23.8 cm that swings through an arc of 12.3°.

44. A flywheel of diameter 16.8 in. does 3120 revolutions per minute. Find the linear velocity of a point on the rim in feet per second.

45. The wheels on a truck have a diameter of 33.0 in. and an angular velocity of 512 rev/min. Find the speed of the truck in miles per hour.

46. A phonograph record with a 6.0-in. radius rotates at the rate of 33.3 rev/min. Find the linear velocity of a point on the rim in feet per second.

47. The wheels of a bicycle have a radius of 16.0 in. and rotate at the rate of 2.20 rev/s. Find the speed of the bicycle in miles per hour.

48. An artificial satellite travels around the earth in a circular orbit at an altitude of 300 mi. Find its velocity (in miles per hour) if it makes one revolution every 90 min. (The radius of the earth is approximately 4000 mi.)

49. A pulley belt 85.0 cm long takes 4.52 s to make one complete circuit. Determine the angular velocity of the pulley in revolutions per minute, given that the diameter of the pulley is 16.0 cm.

50. A weight hanging on a spring is oscillating vertically. The displacement x as a function of time is given by

$$x = 4.2 \sin 2\pi t$$

where x is in centimeters and t in seconds. Sketch the curve of this motion.

CHAPTER 18

Logarithms

Logarithms were first introduced into mathematics as a tool for carrying out lengthy arithmetic computations. This aspect of logarithms has lost its importance with the availability of scientific calculators. However, logarithms still play an important role in technology.

18.1 The Definition of Logarithm

To see how logarithms are defined, let us consider a typical expression containing an exponent:

$$3^2 = 9$$

Logarithm

Recall that 3 is called the *base* and 2 the *exponent,* and we say "3 squared equals 9." If we start with the number 9, we could also say "9 is written as a power of 3," or "The exponent corresponding to 9 is 2, provided that the base is 3." Now, a **logarithm** is merely an exponent. Using the word logarithm instead of exponent, we could say "The logarithm of 9 is 2, provided that the base is 3." More simply,

"The logarithm of 9 to the base 3 is 2"

This statement is abbreviated

$$\log_3 9 = 2$$

where *log* stands for *logarithm*. In summary,

$$3^2 = 9 \quad \text{means the same as} \quad \log_3 9 = 2$$

In general,

$$x = b^y \quad \text{means the same as} \quad y = \log_b x$$

Definition of logarithm

$y = \log_b x$ means the same as $b^y = x$, $b > 0, b \neq 1$

The equation $y = \log_b x$ is read *y is equal to the logarithm of x to the base b*, or *y equals log x to the base b*.

Example 1

$\log_2 8 = 3$ means $2^3 = 8$

$\log_2 16 = 4$ means $2^4 = 16$

$\log_4 2 = \frac{1}{2}$ means $4^{1/2} = 2$

$\log_8 \frac{1}{2} = -\frac{1}{3}$ means $8^{-1/3} = \frac{1}{2}$

Conversely, an equation of the form $b^y = x$ can be written as a logarithm.

Example 2 Write $2^{-3} = \frac{1}{8}$ in logarithmic form.

Solution. Note that 2 is the base and -3 the exponent. Since the logarithm of a number is an exponent, we write

$$\log_2 \frac{1}{8} = -3$$

Example 3 Write $\log_9 3 = \frac{1}{2}$ in exponential form.

Solution. The equation says that the logarithm of 3 (base 9) is $\frac{1}{2}$. So $\frac{1}{2}$ is the exponent. We get

$$9^{1/2} = 3$$

If a logarithmic equation contains an unknown, we can find the unknown by writing the equation in exponential form.

Example 4 Find x in the equation

$$\log_{125} x = -\frac{1}{3}$$

Solution. To find x, we write the equation in exponential form. Thus

$$125^{-1/3} = x$$

or

$$x = \frac{1}{125^{1/3}} = \frac{1}{\sqrt[3]{125}} = \frac{1}{5}$$

Example 5 Find b in the equation

$$\log_b \frac{1}{16} = -2$$

Solution. Written in exponential form, the equation is

$$b^{-2} = \frac{1}{16}$$

Since $b^{-2} = 1/b^2$, we get

$$\frac{1}{b^2} = \frac{1}{16} \qquad b^{-2} = \frac{1}{b^2}$$

$$b^2 = 16 \qquad \text{Taking reciprocals}$$

$$b = 4 \qquad \text{Positive square root}$$

(Since the base has to be positive, we take the positive square root.)

Examples 4 and 5 and the corresponding exercises are only intended to illustrate the definition of logarithm. More complicated equations will be taken up in Section 18.5.

Exercises / Section 18.1

In Exercises 1–16, change each exponential form to logarithmic form.

1. $8^2 = 64$
2. $2^5 = 32$
3. $10^2 = 100$
4. $6^2 = 36$
5. $3^4 = 81$
6. $3^{-2} = \frac{1}{9}$
7. $2^{-4} = \frac{1}{16}$
8. $81^{1/2} = 9$
9. $4^0 = 1$
10. $9^0 = 1$
11. $\left(\frac{2}{3}\right)^2 = \frac{4}{9}$
12. $\left(\frac{1}{3}\right)^{-2} = 9$
13. $\left(\frac{4}{3}\right)^{-2} = \frac{9}{16}$
14. $10^{-3} = 0.001$
15. $\left(\frac{1}{2}\right)^{-3} = 8$
16. $\left(\frac{1}{7}\right)^{-2} = 49$

In Exercises 17–32, change each logarithmic form to exponential form.

17. $\log_4 64 = 3$
18. $\log_{10} 1000 = 3$
19. $\log_5 1 = 0$
20. $\log_7 7 = 1$
21. $\log_3 3 = 1$
22. $\log_2 \frac{1}{32} = -5$

23. $\log_3 \frac{1}{27} = -3$

24. $\log_5 \frac{1}{125} = -3$

25. $\log_{1/6} \frac{1}{36} = 2$

26. $\log_{1/3} 81 = -4$

27. $\log_{1/4} 64 = -3$

28. $\log_5 \frac{1}{25} = -2$

29. $\log_{14} 196 = 2$

30. $\log_{11} \frac{1}{121} = -2$

31. $\log_{9/4} \frac{4}{9} = -1$

32. $\log_{3/10} \frac{10}{3} = -1$

In Exercises 33–60, find the value of the unknown. (See Examples 4 and 5.)

33. $\log_3 x = 3$

34. $\log_4 x = 3$

35. $\log_3 x = -2$

36. $\log_2 x = -3$

37. $\log_4 x = \frac{1}{2}$

38. $\log_9 x = \frac{1}{2}$

39. $\log_{16} x = -\frac{1}{2}$

40. $\log_{64} x = -\frac{1}{3}$

41. $\log_{1/3} x = -2$

42. $\log_{1/2} x = -3$

43. $\log_b 49 = 2$

44. $\log_b 32 = 5$

45. $\log_b 64 = 3$

46. $\log_b 27 = 3$

47. $\log_b 16 = -2$

48. $\log_b \frac{27}{8} = -3$

49. $\log_b \frac{8}{27} = -3$

50. $\log_b \frac{1}{64} = -6$

51. $\log_3 27 = a$

52. $\log_5 125 = a$

53. $\log_4 16 = a$

54. $\log_3 81 = a$

55. $\log_{49} \frac{1}{7} = a$

56. $\log_{27} \frac{1}{3} = a$

57. $\log_5 \frac{1}{125} = a$

58. $\log_4 \frac{1}{16} = a$

59. $\log_{3/2} \frac{2}{3} = a$

60. $\log_{3/4} \frac{16}{9} = a$

18.2 Properties of Logarithms

To be able to use logarithms effectively, we need the following five properties:

Properties of logarithms: For any positive real number b ($b \neq 1$), for any *positive* real numbers x and y, and for any real number r:

1. $\log_b xy = \log_b x + \log_b y$ (18.1)
2. $\log_b \frac{x}{y} = \log_b x - \log_b y$ (18.2)
3. $\log_b x^r = r \log_b x$ (18.3)
4. $\log_b 1 = 0$ (18.4)
5. $\log_b b = 1$ (18.5)

To check law (18.1), let

$$M = \log_b x \quad \text{and} \quad N = \log_b y$$

Then

$$b^M = x \quad \text{and} \quad b^N = y$$

and

$$xy = b^M b^N = b^{M+N} \qquad \text{Adding exponents}$$

Converting back to logarithmic form, we get

$$\log_b xy = M + N$$

Since $M = \log_b x$ and $N = \log_b y$, it follows that

$$\log_b xy = \log_b x + \log_b y$$

This is rule (18.1).

To check rule (18.2), take the quotient of x and y:

$$\frac{x}{y} = \frac{b^M}{b^N} = b^{M-N} \qquad \text{Subtracting exponents}$$

In logarithmic form,

$$\log_b \frac{x}{y} = M - N$$

Since $M = \log_b x$ and $N = \log_b y$, we get

$$\log_b \frac{x}{y} = \log_b x - \log_b y$$

which is rule (18.2).

To check rule (18.3), let $M = \log_b x$, so that $b^M = x$. The rth power of x is

$$x^r = (b^M)^r = b^{Mr} \qquad \text{Multiplying exponents}$$

Converting to logarithms again,

$$\log_b x^r = Mr = r \log_b x$$

and rule (18.3) is verified.

To check properties (18.4) and (18.5), we simply observe that

$$\log_b 1 = 0 \quad \text{means} \quad b^0 = 1$$

and

$$\log_b b = 1 \quad \text{means} \quad b^1 = b$$

The properties are illustrated in the next example.

Example 1

1. $\log_2 21 = \log_2 3 \cdot 7$
   $= \log_2 3 + \log_2 7$ By property (18.1)

2. $\log_3 \frac{5}{7} = \log_3 5 - \log_3 7$ By property (18.2)

3. $\log_{10} 8 = \log_{10} 2^3 = 3 \log_{10} 2$ By property (18.3)

4. $\log_5 1 = 0$ By property (18.4)

5. $\log_6 6 = 1$ By property (18.5)

Certain logarithmic expressions can be evaluated by using the properties of logarithms.

Example 2 Evaluate

a. $\log_5 \frac{1}{5}$ **b.** $\log_7 49$

Solution.

a. $\log_5 \frac{1}{5} = \log_5 1 - \log_5 5$ By property (18.2)

$= 0 - \log_5 5$ $\log_b 1 = 0$

$= 0 - 1 = -1$ $\log_b b = 1$

b. Since $49 = 7^2$, we can write $\log_7 49$ as

$$\log_7 7^2$$

Since $\log_b x^r = r \log_b x$,

$\log_7 7^2 = 2 \log_7 7$ Property (18.3)

$= 2 \cdot 1 = 2$ $\log_b b = 1$

The properties of logarithms can be used to break up complicated logarithmic expressions into sums, differences, and multiples of logarithms. Consider the next example.

Example 3 Write the given logarithms as sums, differences, and multiples of logarithms. If possible, use the properties $\log_b 1 = 0$ and $\log_b b = 1$:

a. $\log_2 2y^3$ **b.** $\log_6 \frac{1}{x^4}$ **c.** $\log_4 \sqrt[3]{4}$

Solution.

a. $\log_2 2y^3 = \log_2 2 + \log_2 y^3$ log of a product

$= 1 + \log_2 y^3$ $\log_b b = 1$

$= 1 + 3 \log_2 y$ $\log_b x^r = r \log_b x$

$= 1 + 3 \log_2 y$

b. $\log_6 \frac{1}{x^4} = \log_6 1 - \log_6 x^4$ log of a quotient

$= 0 - \log_6 x^4$ $\log_b 1 = 0$

$= -4 \log_6 x$ $\log_b x^r = r \log_b x$

c. $\log_4 \sqrt[3]{4} = \log_4 4^{1/3}$ $\sqrt[3]{4} = 4^{1/3}$

$= \frac{1}{3} \log_4 4$ $\log_b x^r = r \log_b x$

$= \frac{1}{3} \cdot 1 = \frac{1}{3}$ $\log_b b = 1$

The properties of logarithms can also be used to combine expressions involving logarithms, as shown in the next example.

Example 4 Combine $\frac{1}{2} \log_3 x - 3 \log_3 y$ into a single logarithm.

Solution.

$$\frac{1}{2} \log_3 x - 3 \log_3 y = \log_3 x^{1/2} - \log_3 y^3 \qquad r \log_b x = \log_b x^r$$

$$= \log_3 \sqrt{x} - \log_3 y^3$$

$$= \log_3 \frac{\sqrt{x}}{y^3} \qquad \text{log of a quotient}$$

Common error Writing

$$\log_b(x + y) \quad \text{as} \quad \log_b x + \log_b y$$

or

$$\log_b(x - y) \quad \text{as} \quad \log_b x - \log_b y$$

By the first property of logarithms,

$$\log_b x + \log_b y = \log_b xy$$

and by the second property,

$$\log_b x - \log_b y = \log_b \frac{x}{y}$$

The expressions $\log_b(x + y)$ and $\log_b(x - y)$ cannot be written as a sum or difference.

Example 5 The fallout from a nuclear explosion can be written in the form $\log_{10} R - \log_{10} R_0 = kt$, where R is the amount of radiation after the explosion, R_0 the amount before the explosion, and k a constant. Write R as a function of time.

Solution.

$\log_{10}R - \log_{10}R_0 = kt$	Given formula
$\log_{10}\dfrac{R}{R_0} = kt$	Log of a quotient
$\dfrac{R}{R_0} = 10^{kt}$	Exponential form
$R = R_0(10^{kt})$	Multiplying by R_0

Exercises / Section 18.2

In Exercises 1–30, write each expression as a sum, difference, or multiple of logarithms. Whenever possible, use $\log_b 1 = 0$ and $\log_b b = 1$. (See Example 3.)

1. $\log_6 25$
2. $\log_5 49$
3. $\log_{10} 36$
4. $\log_2 121$
5. $\log_5 16$
6. $\log_7 81$
7. $\log_7 64$
8. $\log_2 16$
9. $\log_3 27$
10. $\log_5 125$
11. $\log_7 \sqrt{7}$
12. $\log_9 \sqrt[3]{9}$
13. $\log_6 36x^2$
14. $\log_{10} 100y^3$
15. $\log_5 \dfrac{1}{3y^2}$
16. $\log_3 \dfrac{1}{2V^3}$
17. $\log_7 \dfrac{1}{7S^2}$
18. $\log_{10} \dfrac{1}{10x^4}$
19. $\log_2 \dfrac{1}{\sqrt{2P}}$
20. $\log_7 \dfrac{1}{\sqrt{7Q}}$
21. $\log_3 \dfrac{1}{9x^2y^3}$
22. $\log_4 \dfrac{1}{4x^3z}$
23. $\log_a a^2$
24. $\log_b b^3$
25. $\log_a \dfrac{1}{a^3}$
26. $\log_b \dfrac{1}{b^4}$
27. $\log_e \sqrt{2e}$
28. $\log_e \sqrt[3]{e}$
29. $\log_e \dfrac{1}{49\sqrt{e}}$
30. $\log_e \dfrac{16}{\sqrt[3]{e}}$

In Exercises 31–44, write each expression as a single logarithm. (See Example 4.)

31. $\log_3 5 + \log_3 7$
32. $\log_5 2 + \log_5 x$
33. $\log_9 10 - \log_9 5$
34. $\log_6 20 - \log_6 4$
35. $2\log_{10} 5$
36. $3\log_3 2$
37. $2\log_4 x + \log_4 y$
38. $5\log_3 y - \log_3 z$
39. $\dfrac{1}{2}\log_7 x - 2\log_7 y$
40. $\dfrac{1}{3}\log_{10} P - 2\log_{10} Q$
41. $\dfrac{3}{2}\log_{10} R - 3\log_{10} L$
42. $\log_{10} m + \dfrac{1}{2}\log_{10} v$
43. $2\log_5 x + 3\log_5 y - 4\log_5 z$
44. $\dfrac{1}{2}\log_6 x - 2\log_6 y - 3\log_6 z$

45. The loudness β of sound (in decibels) is given by

$$\beta = 10(\log_{10} I - \log_{10} I_0)$$

where I is the intensity of the sound in watts per meter and I_0 the intensity of the faintest audible sound. Write β as a single logarithm.

46. If p_0 is the pressure at sea level and p the pressure at the top of a column of air h meters in height and with a uniform temperature T (in Kelvin), then

$$\log_{10}p - \log_{10}p_0 = -0.0149\,\frac{k}{T}$$

where k is a constant. Simplify this equation by combining the terms on the left side.

47. In a certain chemical reaction, a substance is converted into another substance. Starting with 1 kg of unconverted substance, the time required for the amount of unconverted substance to shrink to N $(0 < N \leq 1)$ is given by

$$t = -10\log_{10}N, \qquad 0 < N \leq 1$$

where t is in seconds. Write this equation in exponential form. (See Example 5.)

48. The tension T (in pounds) on a certain rope around a cylindrical post at a point P is related to the central angle θ determined by P by the formula $\log_{10}T = 0.5\theta$, where $\theta \geq 0$ is measured in radians. Write this formula in exponential form. (See Example 5.)

49. The number of bacteria in a certain culture doubles every hour. If there are 1000 bacteria initially, then the number N of bacteria after t hours can be determined from the formula

$$\log_{10}N = 3 + t\log_{10}2$$

Write N as a function of t.

50. If p (in millimeters of mercury) is the vapor pressure of carbon tetrachloride, then

$$\log_{10}p + \log_{10}(5.97 \times 10^7) = -\frac{1706.4}{T}$$

where T is the temperature in Kelvin. Write this equation in exponential form.

18.3 Common Logarithms

Although logarithms are defined for any base b $(b > 0,\ b \neq 1)$, in practice only two bases are generally used. Logarithms to base 10, called **common logarithms,** have traditionally been employed for numerical computations, but they also arise in certain applications, as we will see. Logarithms to a base designated by e $(e \approx 2.71828)$ are called **natural logarithms** and will be discussed in the next section. In this section we discuss only common logarithms.

Before calculators came to be widely used, common logarithms of numbers were obtained from a table. (Table 2 in Appendix B is a table of common logarithms.) The table lists the logarithms only for numbers between 1 and 10. For example, to find $\log_{10}3.79$, we look under n and find 3.7 (the first two significant digits). In this row, under the 9 (the third significant digit), we find 0.5786. The answer is

$$\log_{10}3.79 = 0.5786$$

To see the practical importance of base 10, let us find $\log_{10}379$. In scientific notation,

$$379 = 3.79 \times 10^2$$

As a result,

$$\begin{aligned}
\log_{10}379 &= \log_{10}(3.79 \times 10^2) \\
&= \log_{10}3.79 + \log_{10}10^2 \\
&= \log_{10}3.79 + 2\log_{10}10 && \log_b x^r = r\log_b x \\
&= \log_{10}3.79 + 2 \cdot 1 && \log_b b = 1 \\
&= 0.5786 + 2 && \log_{10}3.79 = 0.5786 \\
&= 2.5786
\end{aligned}$$

In the statement

$$\log_{10}379 = 0.5786 + 2$$

the number 0.5786 is the *mantissa* and the number 2 the *characteristic*.

In general, for the number

$$M = N \times 10^k \qquad 1 \le N < 10$$

$\log_{10}N$ is the **mantissa** and k the **characteristic.** In summary, to find the common logarithm of a number using Table 2, write the number in scientific notation, obtain the mantissa from the table, and add the characteristic to the mantissa.

Example 1 Use Table 2 in Appendix B to find $\log_{10}43800$.

Solution. In scientific notation,

$$43800 = \mathbf{4.38} \times 10^4$$

From the table,

$$\log_{10}4.38 = \mathbf{0.6415} \qquad \text{(mantissa)}$$

Since the characteristic is **4**, we get

$$\log_{10}43800 = \mathbf{0.6415 + 4} = 4.6415$$

Example 2 Use Table 2 to find $\log_{10}0.0000716$.

Solution. In scientific notation,

$$0.0000716 = 7.16 \times 10^{-5}$$

From the table, $\log_{10}7.16 = 0.8549$. Since the characteristic is -5, we get

$$\begin{aligned}
\log_{10}0.0000716 &= 0.8549 + (-5) \\
&= -4.1451
\end{aligned}$$

When writing a common logarithm, it is customary to omit the number 10 indicating the base:

log N means $\log_{10} N$

Using a Calculator

Common logarithms can be found with a scientific calculator as well: Enter the number and press [LOG].

Example 3 Use a calculator to find

a. log 9328 **b.** log 0.0001496

Solution.

a. The sequence is

9328 [LOG] → 3.9697885

or log 9328 = 3.9698 to four decimal places.

b. The sequence is

0.0001496 [LOG] → −3.8250684

So log 0.0001496 = −3.8251 to four decimal places.

If log N is known, we can find N. The number N is called the **antilogarithm.** The sequence for finding an antilogarithm is

[INV] [LOG] or [10^x]

Example 4 Given that $\log N = -5.894782$, find N.

Solution. The sequence is

5.894782 [+/−] [INV] [LOG] → 1.2741425×10^{-6}

So $N = 1.274 \times 10^{-6}$.

Example 5 The atmospheric pressure P (in pounds per square inch) can be obtained from the equation

$$\log_{10} P = 0.434(2.69 - 0.21h)$$

where h is the altitude in miles above sea level. Find the atmospheric pressure at an altitude of 34 mi.

Solution. From the given formula,

$$\log_{10}P = 0.434(2.69 - 0.21 \cdot 34)$$

The sequence is

2.69 [−] 0.21 [×] 34 [=] [×] 0.434 [=] [INV] [LOG] → 0.0117138

So the atmospheric pressure is 0.012 lb/in.^{2}.

Exercises / Section 18.3

In Exercises 1–6, use Table 2 in Appendix B to find each common logarithm. (See Examples 1 and 2.)

1. log 2.86 **2.** log 5.73 **3.** log 6.79 **4.** log 8.43

5. log 3.04 **6.** log 7.12

In Exercises 7–16, use a calculator to find each common logarithm to four decimal places.

7. log 8.35 **8.** log 5.41 **9.** log 25.36 **10.** log 36.34

11. log 0.1572 **12.** log 0.2309 **13.** log 0.005124 **14.** log 0.00736

15. log 26,472 **16.** log 15,803

In Exercises 17–26, use a calculator to find the antilogarithm in each case. Use four significant digits.

17. 0.8472 **18.** 0.3016 **19.** 1.5263 **20.** 2.8752

21. 3.0914 **22.** 2.8117 **23.** −0.0931 **24.** −1.1273

25. 2.1915 **26.** 3.9078

27. The acidity of a chemical solution is determined by the concentration of the hydrogen ion H^+, written $[H^+]$, and is measured in moles per liter (mol/L). The hydrogen potential pH is defined by

$$ph = -\log_{10}[H^+]$$

Given that the acid concentration of water is 10^{-7} mol/L, determine the pH of water.

28. The acid concentration of blood is 3.98×10^{-8} mol/L. Determine the pH value.

29. A certain satellite has a power supply whose output P (in watts) can be determined from the equation $\log_{10}P = -t/895$, where t is the number of days that the battery has operated. What will the output be after 365 days?

30. A record company estimates that its monthly profit (in thousands of dollars) from a new hit record is $P = 12 - 15\log_{10}(t + 1)$, where t is the number of months after release. Determine when the record should be withdrawn.

18.4 Natural Logarithms

In the last section we discussed logarithms to base 10. We saw that 10 is a convenient choice for the base because our number system is based on 10.

Another number base, which arises in the study of calculus, is particularly useful for theoretical work. This number, denoted by e, is an irrational number whose approximate value is 2.71828. Logarithms to base e are called **natural logarithms**. Natural logarithms are denoted by ln (instead of log, which is used for common logarithms). (Table 3 in Appendix B is a table of natural logarithms.)

> **Notation for natural logarithms**
>
> $\log_e x$ is denoted by $\ln x$
>
> where $e \approx 2.71828$.

The examples and exercises in this section will illustrate the use of natural logarithms in technology.

Natural logarithms can be found with a scientific calculator by entering the number and pressing [LN].

Antilogarithms are found by using the sequence [INV] [LN] or by pressing [e^x].

Example 1 Find

a. ln 376.4 **b.** ln 0.06178

Solution. The sequences are

a. 376.4 [LN] → 5.9306524

b. 0.06178 [LN] → −2.7841756

Example 2 Given $\ln N = -4.12736$, find N, the antilogarithm.

Solution. The sequence is

4.12736 [+/−] [INV] [LN] → 0.0161253

or

4.12736 [+/−] [e^x] → 0.0161253

Example 3 The *half-life* of a radioactive element is the time required for half of a given amount of the element to decay. If a radioactive substance has a half-life H, then the time taken for the initial amount N_0 to shrink to N is given by

$$t = \frac{H}{\ln 2} \ln \frac{N_0}{N} \tag{18.6}$$

The half-life of radium is 1590 years. How long will it take for 25.0 g of radium to shrink to 15.0 g?

Solution. Substituting the given values into formula (18.6), we have

$$t = \frac{1590}{\ln 2} \ln \frac{25.0}{15.0}$$

The sequence is

1590 [÷] 2 [LN] [×] [(] 25.0 [÷] 15.0 [)] [LN] [=] → 1171.7753

The time required is 1170 years.

Since e is the base for natural logarithms, $\ln N = x$ is equivalent to $N = e^x$. Because of the close relationship between logarithmic and exponential forms, the **exponential function** $f(x) = e^x$ also occurs frequently in applications. The function values can be obtained with a calculator by pressing [e^x] or [INV] [LN].

Exponential function

To find natural logarithms in BASIC, use the library function LOG(X) [that is, LOG(X) is the function $\ln x$]. The library function EXP(X) is the exponential function e^x.

Example 4 An electric circuit consists of a resistor and inductor. When the current source is removed, the current dies out quickly according to the equation

$$i = I_0 e^{-Rt/L}$$

If $R = 20.0\ \Omega$, $L = 0.0986$ H, and $I_0 = 2.50$ A, find i after 3.50 ms.

Solution. If $t = 3.50$ ms $= 0.00350$ s, we get

$$i = 2.50e^{(-20.0)(0.00350)/0.0986}$$

The sequence is

20.0 [+/−] [×] 0.00350 [÷] 0.0986 [=] [e^x] [×] 2.50 [=] → 1.2291853

So $i = 1.23$ A after 3.50 ms.

Exercises / Section 18.4

In Exercises 1–10, use a calculator to find the indicated natural logarithms to four decimal places.

1. ln 0.1528 **2.** ln 0.7091 **3.** ln 2.664 **4.** ln 2.179 **5.** ln 37.34
6. ln 59.01 **7.** ln 326.4 **8.** ln 719.9 **9.** ln 0.001726 **10.** ln 0.09881

In Exercises 11–16, use a calculator to find the (natural) antilogarithm in each case. Use four significant digits.

11. 1.8926 **12.** 2.5402 **13.** 0.01724 **14.** 0.6819 **15.** 3.2483
16. 4.3376

In Exercises 17–24, use a calculator to find the exponential values to four decimal places.

17. $e^{2.64}$ **18.** $e^{0.179}$ **19.** $e^{-1.28}$ **20.** $e^{-2.83}$ **21.** $e^{1.65}$

22. $e^{2.39}$ **23.** $e^{0.0146}$ **24.** $e^{0.0812}$

25. The radioactive element polonium has a half-life of 140 days. Use formula (18.6) to find the time required for 35 g of polonium to shrink to 26 g.

26. A radioactive isotope has a half-life of 85 years. How long will it take for 6.4 kg to shrink to 5.1 kg?

27. Just as interest can be compounded quarterly or daily, it can also be compounded continuously. The formula is

$$P = P_0e^{rt}$$

where P_0 is the amount invested, t the time in years, and r the interest rate in decimal form. To what amount will \$5000 accumulate after 5.5 years at 8.25% interest compounded continuously?

28. A woman deposits \$3500 in a savings account that pays 9.75% interest compounded continuously. How much will she have after 8.5 years?

29. For a certain gas, an enclosed volume of 10 cm^3 is gradually increased to v. The average pressure (in atmospheres) is

$$P_{av} = \frac{4}{v - 10} \ln \frac{v}{10}$$

Find the average pressure if the volume is increased to 50 cm^3.

30. The distance traveled by a motorboat after the engine is shut off is

$$x = \frac{1}{k} \ln(v_0kt + 1)$$

where v_0 is the velocity when the motor is running and k a constant. If $k = 0.00300$, a typical value, determine how far a motorboat traveling at 20.0 ft/s will continue in the first 15.0 s after the motor is shut off.

31. A certain satellite has a power supply whose output in watts is given by $P = 20.0e^{-t/655}$, where t is the number of days that the battery has operated. What is the output after 180 days?

32. An object cools according to the formula $T = 15(1 + e^{-0.10t})$, where T is in degrees Celsius and t in minutes. Since $T = 30°C$ when $t = 0$ min, the initial temperature is 30°C. What will the temperature be after 6 min?

18.5 Exponential Equations

One of the most interesting applications of logarithms is the solution of equations in which the unknown is an exponent. To solve such an equation, we write the equation in the form $a^x = b$ and then take the logarithm of both sides:

$$a^x = b$$
$$\log a^x = \log b$$

By the third law of logarithms,

$$x \log a = \log b$$

and

$$x = \frac{\log b}{\log a}$$

Example 1 Solve the equation

$$5^x = 16$$

Solution. First we take the common logarithm of both sides:

$$\log 5^x = \log 16$$

By the third law of logarithms, we get

$$x \log 5 = \log 16$$

Dividing both sides by log 5, we get

$$x \log 5 = \log 16$$

$$x = \frac{\log 16}{\log 5}$$

The sequence is

16 [LOG] [÷] 5 [LOG] [=] → 1.7227062

or $x = 1.7227$ to four decimal places.

Natural logarithms can also be used to solve exponential equations, as shown in the next example.

Example 2 Solve the equation

$$2(4^x) = 35$$

Solution.

$$2(4^x) = 35 \qquad \text{Given equation}$$

$$4^x = \frac{35}{2} \qquad \text{Dividing by 2}$$

$$4^x = \frac{35}{2}$$

$$\ln 4^x = \ln \frac{35}{2} \qquad \text{Taking logarithms}$$

$$x \ln 4 = \ln \frac{35}{2} \qquad \log_b x^r = r \log_b x$$

$$x = \frac{\ln \frac{35}{2}}{\ln 4} \qquad \text{Dividing by } \ln 4$$

The sequence is

35 [÷] 2 [=] LN [÷] 4 [LN] [=] → 2.0646415

So $x = 2.0646$ to four decimal places.

Example 3 Logarithms are used to determine the age of a fossil in a procedure called *carbon dating*. The method is based on the fact that carbon-14 is found in all organisms in a fixed percentage. When an organism dies, the carbon-14 decays according to the formula

$$P = P_0(2^{-t/5580}) \tag{18.7}$$

where P_0 is the initial amount. Suppose a fossil contains 32.0% of the original amount. How long has the organism been dead?

Solution. We are given that 32.0% of the original amount P_0 is left. Since 32.0% of P_0 is $0.320\ P_0$, the problem is to find t such that

$$P = 0.320\ P_0$$

Substituting in formula (18.7), we get

$$\begin{aligned} 0.320\ P_0 &= P_0(2^{-t/5580}) \\ 0.320 &= 2^{-t/5580} && \text{Dividing by } P_0 \\ \ln 0.320 &= \ln 2^{-t/5580} && \text{Taking logarithms} \\ \ln 0.320 &= -\frac{t}{5580}\ln 2 && \ln x^r = r \ln x \\ t &= -\frac{5580 \ln 0.320}{\ln 2} \\ &= 9170 \text{ years} \end{aligned}$$

Exercises / Section 18.5

In Exercises 1–14, solve each equation for x, accurate to four decimal places.

1. $3^x = 5$ **2.** $2^x = 6$ **3.** $4^x = 15$ **4.** $3^x = 10$ **5.** $10^x = 3$

6. $12^x = 7$ **7.** $2(4^x) = 23$ **8.** $3(2^x) = 10$ **9.** $7(6^x) = 50$ **10.** $4(7^x) = 25$

11. $10(8^x) = 7$ **12.** $5(7^x) = 2$ **13.** $12(15^x) = 5$ **14.** $15(20^x) = 6$

15. Solve for t: $i = \frac{E}{R} e^{-t/(RC)}$

16. Solve for V_1: $W = p_0 V_0 \log \frac{V_1}{V_0}$

17. The subjective impression of sound, measured in decibels, is related to the intensity of sound (measured in watts per square meter) by the formula

$$\beta = 10 \log \frac{I}{I_0}$$

where I_0 is the intensity level of the faintest audible sound (10^{-12} W/m^2). Rustling leaves have a loudness of 10 decibels. What is the intensity level?

18. An elevated train has a loudness of 90 decibels. What is the intensity level? (Refer to Exercise 17.)

19. A bacteria culture doubles every hour. If there are 1000 bacteria initially, then the number of bacteria after t hours is $N = 1000(2^t)$.
a. Determine the bacteria population after 4.5 h.
b. Determine when the population will reach 15,000.

20. Suppose a fossil contains 20% of the original amount of carbon-14. How long has the organism been dead? (See Example 3.)

21. In a body of water, the light intensity I diminishes with the depth. If I_0 is the intensity level at the surface, then at depth d (in feet)

$$I = I_0 e^{-kd}$$

where k is a constant. For a chlorinated swimming pool, a typical value for k is 0.0080. At what depth is the intensity 95% of the intensity at the surface? (See Example 3.)

22. Uranium-238, which has a half-life of 4.5 billion years, decays according to the formula

$$N = N_0 e^{-(1.54 \times 10^{-10})t}$$

Determine how long 0.1% of the initial amount N_0 takes to decay.

23. From the formula $\text{pH} = -\log_{10}[\text{H}^+]$, find the concentration of hydrogen ions (in moles per liter) of milk whose hydrogen potential is 6.39. (Refer to Exercise 27 in Section 18.3.)

24. If P dollars is compounded n times a year, then the value of the investment at the end of t years is $P(1 + r/n)^{nt}$, where r is the interest rate. If \$1000 is invested at 10% compounded quarterly (four times a year), how long will it take for the invested amount to double?

Review Exercises / Chapter 18

In Exercises 1–2, change each exponential form to logarithmic form.

1. $4^3 = 64$ **2.** $2^{-5} = \dfrac{1}{32}$

In Exercises 3–4, change each logarithmic form to exponential form.

3. $\log_{1/2} 16 = -4$ **4.** $\log_{27} \dfrac{1}{3} = -\dfrac{1}{3}$

In Exercises 5–8, find the value of the unknown.

5. $\log_{1/6} x = -2$ **6.** $\log_b 81 = -4$ **7.** $\log_b \dfrac{1}{32} = 5$ **8.** $\log_{64} 8 = a$

In Exercises 9–18, write each expression as a sum, difference, or multiple of logarithms. Whenever possible, use $\log_b 1 = 0$ and $\log_b b = 1$.

9. $\log_5 81$ **10.** $\log 16$ **11.** $\log_3 \sqrt{11}$ **12.** $\log_{11} \sqrt{11}$

13. $\log_6 \sqrt{5x}$ **14.** $\log_6 \sqrt{6x}$ **15.** $\log 10x^2$ **16.** $\log_5 \dfrac{1}{5V}$

17. $\ln \dfrac{1}{\sqrt{2e}}$ **18.** $\ln \dfrac{1}{6\sqrt[3]{e}}$

In Exercises 19–22, write each expression as a single logarithm.

19. $\log_5 36 - \log_5 12$

20. $2 \log_5 x + \log_5 2$

21. $\frac{1}{3} \ln P - 2 \ln Q$

22. $\log V + \frac{1}{2} \log W$

23. Use Table 2 to find log 7.84.

In Exercises 24–32, use a calculator to find the values to four significant digits.

24. log 8.406

25. log 26.72

26. log 0.1893

27. log 0.001682

28. ln 0.09243

29. ln 376.7

30. ln 26,350

31. $e^{2.362}$

32. $e^{-0.01740}$

In Exercises 33–36, solve the given equations for x, accurate to four decimal places.

33. $4^x = 17$

34. $11^x = 5$

35. $3(5^x) = 19$

36. $34(10^x) = 6$

37. The time t required for a bacteria culture to grow from N_0 to N is given by

$$t = \frac{1}{k}(\ln N - \ln N_0)$$

Write the equation in exponential form.

38. In a certain circuit the current i as a function of time is $i = 2.5e^{-0.12t}$. Find i after 1.3 s.

39. An object of weight w moving around a circle of radius a is subject to a retarding force. If the initial velocity is v_0, then the angular displacement as a function of time is given by

$$\theta = \frac{w}{kga} \ln\left(1 + \frac{kgv_0 t}{w}\right)$$

where k is a constant. Suppose $k = 0.0040$. Determine the angular displacement of an object after 8.0 s if $v_0 = 25$ ft/s, $w = 10$ lb, and $a = 50$ ft ($g = 32$ ft/s^2).

40. The radioactive element strontium-90 decays according to the formula

$$N = N_0 e^{-0.025t}$$

where N_0 is the initial amount. Determine the half-life—that is, the time required for half of a given amount to decay.

41. Recall that carbon-14 decays according to the formula

$$P = P_0(2^{-t/5580})$$

If a fossil contains 10.0% of the original carbon, how long has the organism been dead?

42. Newton's law of cooling states that the rate of change of the temperature of a body is directly proportional to the difference of the temperature of the body and the temperature of the surrounding medium. If T_0 is the initial temperature and M_T the temperature of the medium, then the temperature of the body as a function of time is

$$T = M_T + (T_0 - M_T)e^{-kt}$$

where k is a constant. Solve this formula for t.

43. The atmospheric pressure (in millimeters of mercury) is given by

$$P = 760e^{-0.00013h}$$

where h is the height (in meters) above sea level. At what height is the atmospheric pressure equal to 550 mm of mercury?

Cumulative Review Exercises / Chapters 16–18

1. Solve the following triangle: $a = 15.62$, $b = 25.76$, $c = 17.14$
2. Change 124° to radians (π form).
3. Change 75°40′ to radians (decimal form).
4. Change 2.397 to degrees.
5. Combine into a single logarithm:

$$\frac{1}{3} \log R + 2 \log L - 3 \log C$$

6. Simplify $\ln \frac{1}{\sqrt{3e}}$.
7. Write the equation $\log_3 x = a - \log_3 4$ in exponential form.
8. A 36.4-N weight hanging on a rope is pulled sideways by a force of 14.3 N. Determine the tension on the rope and the angle that the rope makes with the vertical.
9. Use addition of components to solve the following problem: Two forces 35.3° apart are pulling on an object. Given that the respective forces have magnitudes 46.4 lb and 63.8 lb, what single force will have the same effect?
10. Work Exercise 9 by using the cosine law.
11. A cart weighing 145 lb is resting on a ramp inclined 11.0° with the horizontal. What force is required to keep the cart from rolling down the ramp?
12. Two tractors are pulling a heavy machine in directions that are 25.0° apart. If one force is 2560 lb, what must the other force be so that the combined effect is a force of 5380 lb?
13. A pendulum has a length of 35.0 cm and swings through an arc of 17.5°. Find (a) the area swept out by the pendulum and (b) the distance covered by the tip as it swings from one end of the arc to the other.
14. A floppy disk has a diameter of 5.50 in. Given that the linear velocity of a point on the rim is 10.9 ft/s, determine the angular velocity in revolutions per minute.
15. Determine the amplitude and period, and sketch the curve:

$$y = -5 \sin \frac{1}{2} x$$

16. Determine the amplitude, period, and phase shift, and sketch the curve:

$$y = 2 \cos \left(2x - \frac{\pi}{2}\right)$$

17. Solve the following formula for t:

$$v = v_1 - v_2 e^{-kt}$$

18. If \$10,000 is invested at 9.25% compounded continuously, how much will there be after 6.5 years?
19. An object at room temperature (20°C) is placed in an oven kept at 100°C. If, after 1 min, the temperature has risen to 40°, it can be shown that the temperature at any time t is

$$T = 100 - 80 \left(\frac{3}{4}\right)^t, \qquad t \geq 0$$

How long will it take for the temperature to reach 60°C?

APPENDIX A

Introduction to Basic

The language of BASIC (*B*eginners *A*ll-Purpose *S*ymbolic *I*nstruction *C*ode) is the most widely used computer language today. The purpose of this appendix is to give a brief introduction to BASIC.

Algebraic Expressions, Operations, and Functions

Algebraic expressions and operations are discussed in Section 4.5 and in Section 7.6. Functions are discussed in Section 12.5. For trigonometric functions, see Section 17.1, and for logarithmic and exponential functions, see Section 18.4.

Format

A computer program is a list of step-by-step instructions that a computer can follow. These instructions consist of a list of numbered statements. Although the statements may be numbered 1, 2, 3, 4, . . . , it is more practical to use a sequence such as 10, 20, 30, 40, . . . This numbering scheme allows you to insert additional lines, if necessary, without retyping the entire program. A typical program, then, has the following form:

```
1Ø Statement 1
2Ø Statement 2
3Ø Statement 3
and so on
```

(Note that the symbol Ø is used for zero.) The lines do not have to be typed in order. The computer automatically orders the statements when entered.

To correct a line, it is sufficient to enter the corrected line—the last typed line replaces the previous one.

Example 1 Suppose the following lines are part of a program:

```
10 LET A = 1
20 LET B = 2
30 LET C = 6
20 LET B = 4
```

Since the second line 20 replaces the first line 20, the computer reads these lines as if they had been typed as follows:

```
10 LET A = 1
20 LET B = 4
30 LET C = 6
```

System Commands

A **system command** is a general command to a computer. It is not part of a program. The most important system commands are LIST, DELETE, and RUN.

LIST: Lists an entered program. (This command is usually used to check or correct an entered program.)

DELETE: Erases a designated line.

RUN: Executes the program.

LET Statement

The LET statement is a program statement that assigns a value to a variable.

Example 2 The program statement

```
30 LET X = 2.71
```

assigns the value 2.71 to the memory location named X. It also assigns 2.71 to X in all statements that follow.

Example 3 The statement

```
50 LET A = A + 1
```

replaces the value of A stored in the memory by A + 1. (Note that in BASIC, the symbol = means "replaces.")

INPUT Statement

The INPUT statement causes the computer to pause until a set of data is entered. (See Example 6.)

PRINT Statement

The PRINT statement instructs the computer to print or display an expression or the value of a variable or expression. For example, the statement

```
50 PRINT X
```

causes the computer to print the *value* of the variable X. On the other hand, the statement

```
50 PRINT "X"
```

(X in quotation marks) causes the *letter* X to be printed, not the value of X.

It is possible to print or display more than one variable. For example, depending on the system, the statement

```
60 PRINT X, X + 1
```

causes the values of X and X + 1 to be printed

1. On two separate lines, or
2. On the same line, separated by a wide space.

If you want the results printed closer together on the same line, separate the variables by a semicolon:

```
60 PRINT X; X + 1
```

(See Example 4.)

END Statement

The END statement tells the computer that the program is finished.

Examples of Simple Programs

Given the information we have so far, it is possible to write some simple programs.

Example 4 Consider the program

```
10 LET X = 4
20 LET Y = X + 1
30 PRINT X; Y
40 END
```

Note that the last line consists of an END statement.

The command RUN will result in the following display:

```
4  5
```

(The numbers are printed close together on the same line because a semicolon is used in line 30.)

Example 5 A circuit contains a resistor and inductor. If R is the resistance and X_L the inductive reactance, then the impedance Z is given by $Z = \sqrt{R^2 + (X_L)^2}$. Write a BASIC program for computing Z, if $R = 32.4\ \Omega$ and $X_L = 27.6\ \Omega$.

Solution. In the program, X1 represents X_L:

```
10 PRINT "IMPEDANCE"
20 LET R = 32.4
30 LET X1 = 27.6
40 LET Z = SQR(R↑2 + X1↑2)
50 PRINT "Z="; Z
60 END
```

(Line 10 merely identifies the program.)

The RUN command will result in the following display:

```
IMPEDANCE
Z = 42.56195484
```

In the next program, the descriptive labels are used to identify not only the program but also the data to be entered.

Example 6 Write a program for the solution of the equation

$$ax + b = c$$

in terms of a, b, and c.

Solution. Note that the solution is

$$x = \frac{c - b}{a}, \qquad a \neq 0$$

Program:

```
 10 PRINT "SOLUTION OF LINEAR EQUATION"
 20 PRINT "A="
 30 INPUT A
 40 PRINT "B="
 50 INPUT B
 60 PRINT "C="
 70 INPUT C
 80 LET X = (C - B)/A
 90 PRINT "X="; X
100 END
```

On some systems the INPUT statement may contain a reminder in quotation marks. On such a system, the program can be written as follows:

```
10 PRINT "SOLUTION OF LINEAR EQUATION"
20 INPUT "A="; A
30 INPUT "B="; B
40 INPUT "C ="; C
50 LET X = (C - B)/A
60 PRINT "X="; X
70 END
```

The values of A, B, and C are entered while the program is running. The computer prints or displays a question mark on lines 20, 30, and 40, allowing the constants to be "input."

When solving the equation $2x + 3 = 12$, the RUN command will result in the following display:

```
SOLUTION OF LINEAR EQUATION
A = ?        2        Input 2
B = ?        3        Input 3
C = ?       12        Input 12
X = 4.5
```

GO TO and IF-THEN Statements

The GO TO statement directs the computer to go to a specific line.

The IF-THEN statement tells the computer to go to a specific line if a certain condition is met. Consider, for example, the following programming steps:

```
100 IF N = 8 THEN 120
110 GO TO 80
```

When the computer reads line 100, it jumps to line 120 **if N = 8**. Otherwise it reads line 110 and follows the instruction to go to line 80.

Example 7 Write a program to print the squares of the first 12 positive integers.

Solution.

```
10 LET N = 1
20 PRINT N↑2
30 LET N = N + 1
40 IF N < = 12 THEN 20
50 END
```

Line 20 instructs the computer to print N^2 (starting with N = 1). On line 30 the index is increased by 1. Line 40 says that if N is *less than or equal to 12*, the computer is to jump to line 20 to repeat the cycle. When N *exceeds* 12, the condition in line 40 is not met, so the computer goes to line 50, thereby terminating the program.

In the next program, the sum of the squares of the first 12 positive integers is found. Note how the equality symbol is used to perform the addition.

Example 8 Write a program for finding the sum of the squares of the first 12 positive integers.

Solution.

```
10 LET M = 0
20 LET N = 1
30 LET M = M + N↑2
40 LET N = N + 1
50 IF N < = 12 THEN 30
60 PRINT "SUM OF SQUARES="; M
70 END
```

Note that in line 30, M is replaced by the original value plus N^2, the next term in the sum.

The RUN command will cause the following result to be printed or displayed:

```
SUM OF SQUARES = 650
```

FOR-NEXT Statement

The FOR-NEXT statement provides a convenient method for programming simple loops of the type in Example 8. The FOR-NEXT statement has the following form:

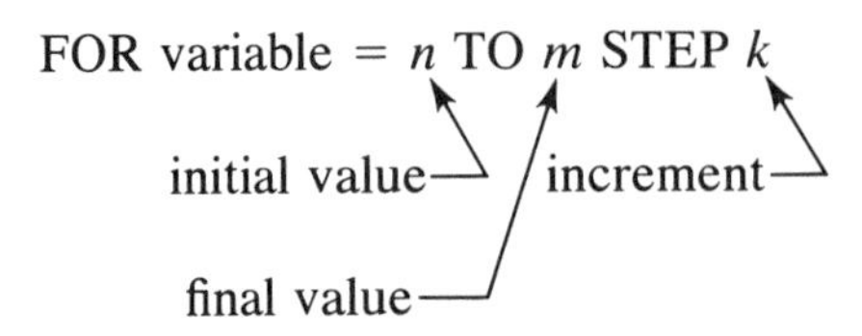

NEXT variable

For example,

```
FOR I = 1 TO 1Ø STEP 1
    . . .
NEXT I
```

repeatedly executes a specified command while the value of I changes from 1 to 10 in increments of 1 unit. (If the increment is 1, as in this case, STEP 1 can be omitted.)

Example 9 Write a program for constructing a table of values for the function $y = x^2 + 1$ from $x = -2$ to $x = 2$ in increments of 0.1. (See also Section 12.3 on plotting points and Section 12.5.)

Solution.

```
1Ø FOR X = -2 TO 2 STEP Ø.1
2Ø LET Y = X↑2 + 1
3Ø PRINT X; Y
4Ø NEXT X
5Ø END
```

Exercises / Appendix A

All the programs for these exercises are given in the answer section.

1. Write a program entitled "Pythagorean theorem" for finding the hypotenuse of a right triangle, given the legs.
2. Write a program for solving a pure quadratic equation of the form $ax^2 - b = 0$ $(a, b > 0)$.
3. Write a program to print the sum of the cubes of the first eight positive integers, using the IF-THEN statement. (See Example 8.)
4. Write the program in Exercise 3 using the FOR-NEXT statement. (See Example 9.)
5. Write a program for constructing a table of values for the function $f(x) = \sin 2x$ for $x = 0$ to $x = 2$ (radians) in increments of 0.05.

6. Repeat Exercise 5 for $f(x) = \ln x^2$ for $x = 1$ to $x = 2$ in increments of 0.1.
7. Write a program to print the volume and surface area of a sphere of radius 2.47 in.
8. Write a program for finding the length of a side of a triangle, given the lengths of the other two sides and the angle between them. Use the law of cosines.
9. Write a program for solving the quadratic equation

$$ax^2 + bx + c = 0, \; a \neq 0$$

valid for real roots.

10. Write a program for finding the area of a circular sector of radius r and central angle θ, measured in degrees.

APPENDIX **B**

Tables

Table 1 *Trigonometric functions*

Degrees	Sin θ	Cos θ	Tan θ	Cot θ	Sec θ	Csc θ	
0°00′	0.0000	1.0000	0.0000	—	1.000	—	90°00′
10	0.0029	1.0000	0.0029	343.8	1.000	343.8	50
20	0.0058	1.0000	0.0058	171.9	1.000	171.9	40
30	0.0087	1.0000	0.0087	114.6	1.000	114.6	30
40	0.0116	0.9999	0.0116	85.94	1.000	85.95	20
50	0.0145	0.9999	0.0145	68.75	1.000	68.76	10
1°00′	0.0175	0.9998	0.0175	57.29	1.000	57.30	89°00′
10	0.0204	0.9998	0.0204	49.10	1.000	49.11	50
20	0.0233	0.9997	0.0233	42.96	1.000	42.98	40
30	0.0262	0.9997	0.0262	38.19	1.000	38.20	30
40	0.0291	0.9996	0.0291	34.37	1.000	34.38	20
50	0.0320	0.9995	0.0320	31.24	1.001	31.26	10
2°00′	0.0349	0.9994	0.0349	28.64	1.001	28.65	88°00′
10	0.0378	0.9993	0.0378	26.43	1.001	26.45	50
20	0.0407	0.9992	0.0407	24.54	1.001	24.56	40
30	0.0436	0.9990	0.0437	22.90	1.001	22.93	30
40	0.0465	0.9989	0.0466	21.47	1.001	21.49	20
50	0.0494	0.9988	0.0495	20.21	1.001	20.23	10
3°00′	0.0523	0.9986	0.0524	19.08	1.001	19.11	87°00′
10	0.0552	0.9985	0.0553	18.07	1.002	18.10	50
20	0.0581	0.9983	0.0582	17.17	1.002	17.20	40
30	0.0610	0.9981	0.0612	16.35	1.002	16.38	30
40	0.0640	0.9980	0.0641	15.60	1.002	15.64	20
50	0.0669	0.9978	0.0670	14.92	1.002	14.96	10
4°00′	0.0698	0.9976	0.0699	14.30	1.002	14.34	86°00′
10	0.0727	0.9974	0.0729	13.73	1.003	13.76	50
20	0.0756	0.9971	0.0758	13.20	1.003	13.23	40
30	0.0785	0.9969	0.0787	12.71	1.003	12.75	30
40	0.0814	0.9967	0.0816	12.25	1.003	12.29	20
50	0.0843	0.9964	0.0846	11.83	1.004	11.87	10
5°00′	0.0872	0.9962	0.0875	11.43	1.004	11.47	85°00′
	Cos θ	Sin θ	Cot θ	Tan θ	Csc θ	Sec θ	Degrees

Table 1 *Trigonometric functions* (continued)

Degrees	Sin θ	Cos θ	Tan θ	Cot θ	Sec θ	Csc θ	
5°00′	0.0872	0.9962	0.0875	11.43	1.004	11.47	85°00′
10	0.0901	0.9959	0.0904	11.06	1.004	11.10	50
20	0.0929	0.9957	0.0934	10.71	1.004	10.76	40
30	0.0958	0.9954	0.0963	10.39	1.005	10.43	30
40	0.0987	0.9951	0.0992	10.08	1.005	10.13	20
50	0.1016	0.9948	0.1022	9.788	1.005	9.839	10
6°00′	0.1045	0.9945	0.1051	9.514	1.006	9.567	84°00′
10	0.1074	0.9942	0.1080	9.255	1.006	9.309	50
20	0.1103	0.9939	0.1110	9.010	1.006	9.065	40
30	0.1132	0.9936	0.1139	8.777	1.006	8.834	30
40	0.1161	0.9932	0.1169	8.556	1.007	8.614	20
50	0.1190	0.9929	0.1198	8.345	1.007	8.405	10
7°00′	0.1219	0.9925	0.1228	8.144	1.008	8.206	83°00′
10	0.1248	0.9922	0.1257	7.953	1.008	8.016	50
20	0.1276	0.9918	0.1287	7.770	1.008	7.834	40
30	0.1305	0.9914	0.1317	7.596	1.009	7.661	30
40	0.1334	0.9911	0.1346	7.429	1.009	7.496	20
50	0.1363	0.9907	0.1376	7.269	1.009	7.337	10
8°00′	0.1392	0.9903	0.1405	7.115	1.010	7.185	82°00′
10	0.1421	0.9899	0.1435	6.968	1.010	7.040	50
20	0.1449	0.9894	0.1465	6.827	1.011	6.900	40
30	0.1478	0.9890	0.1495	6.691	1.011	6.765	30
40	0.1507	0.9886	0.1524	6.561	1.012	6.636	20
50	0.1536	0.9881	0.1554	6.435	1.012	6.512	10
9°00′	0.1564	0.9877	0.1584	6.314	1.012	6.392	81°00′
10	0.1593	0.9872	0.1614	6.197	1.013	6.277	50
20	0.1622	0.9868	0.1644	6.084	1.013	6.166	40
30	0.1650	0.9863	0.1673	5.976	1.014	6.059	30
40	0.1679	0.9858	0.1703	5.871	1.014	5.955	20
50	0.1708	0.9853	0.1733	5.769	1.015	5.855	10
10°00′	0.1736	0.9848	0.1763	5.671	1.015	5.759	80°00′
10	0.1765	0.9843	0.1793	5.576	1.016	5.665	50
20	0.1794	0.9838	0.1823	5.485	1.016	5.575	40
30	0.1822	0.9833	0.1853	5.396	1.017	5.487	30
40	0.1851	0.9827	0.1883	5.309	1.018	5.403	20
50	0.1880	0.9822	0.1914	5.226	1.018	5.320	10
11°00′	0.1908	0.9816	0.1944	5.145	1.019	5.241	79°00′
10	0.1937	0.9811	0.1974	5.066	1.019	5.164	50
20	0.1965	0.9805	0.2004	4.989	1.020	5.089	40
30	0.1994	0.9799	0.2035	4.915	1.020	5.016	30
40	0.2022	0.9793	0.2065	4.843	1.021	4.945	20
50	0.2051	0.9787	0.2095	4.773	1.022	4.876	10
12°00′	0.2079	0.9781	0.2126	4.705	1.022	4.810	78°00′
10	0.2108	0.9775	0.2156	4.638	1.023	4.745	50
20	0.2136	0.9769	0.2186	4.574	1.024	4.682	40
30	0.2164	0.9763	0.2217	4.511	1.024	4.620	30
40	0.2193	0.9757	0.2247	4.449	1.025	4.560	20
50	0.2221	0.9750	0.2278	4.390	1.026	4.502	10
13°00′	0.2250	0.9744	0.2309	4.331	1.026	4.445	77°00′
	Cos θ	Sin θ	Cot θ	Tan θ	Csc θ	Sec θ	Degrees

Table 1 *Trigonometric functions* (continued)

Degrees	Sin θ	Cos θ	Tan θ	Cot θ	Sec θ	Csc θ	
13°00′	0.2250	0.9744	0.2309	4.331	1.026	4.445	77°00′
10	0.2278	0.9737	0.2339	4.275	1.027	4.390	50
20	0.2306	0.9730	0.2370	4.219	1.028	4.336	40
30	0.2334	0.9724	0.2401	4.165	1.028	4.284	30
40	0.2363	0.9717	0.2432	4.113	1.029	4.232	20
50	0.2391	0.9710	0.2462	4.061	1.030	4.182	10
14°00′	0.2419	0.9703	0.2493	4.011	1.031	4.134	76°00′
10	0.2447	0.9696	0.2524	3.962	1.031	4.086	50
20	0.2476	0.9689	0.2555	3.914	1.032	4.039	40
30	0.2504	0.9681	0.2586	3.867	1.033	3.994	30
40	0.2532	0.9674	0.2617	3.821	1.034	3.950	20
50	0.2560	0.9667	0.2648	3.776	1.034	3.906	10
15°00′	0.2588	0.9659	0.2679	3.732	1.035	3.864	75°00′
10	0.2616	0.9652	0.2711	3.689	1.036	3.822	50
20	0.2644	0.9644	0.2742	3.647	1.037	3.782	40
30	0.2672	0.9636	0.2773	3.606	1.038	3.742	30
40	0.2700	0.9628	0.2805	3.566	1.039	3.703	20
50	0.2728	0.9621	0.2836	3.526	1.039	3.665	10
16°00′	0.2756	0.9613	0.2867	3.487	1.040	3.628	74°00′
10	0.2784	0.9605	0.2899	3.450	1.041	3.592	50
20	0.2812	0.9596	0.2931	3.412	1.042	3.556	40
30	0.2840	0.9588	0.2962	3.376	1.043	3.521	30
40	0.2868	0.9580	0.2994	3.340	1.044	3.487	20
50	0.2896	0.9572	0.3026	3.305	1.045	3.453	10
17°00′	0.2924	0.9563	0.3057	3.271	1.046	3.420	73°00′
10	0.2952	0.9555	0.3089	3.237	1.047	3.388	50
20	0.2979	0.9546	0.3121	3.204	1.048	3.356	40
30	0.3007	0.9537	0.3153	3.172	1.049	3.326	30
40	0.3035	0.9528	0.3185	3.140	1.049	3.295	20
50	0.3062	0.9520	0.3217	3.108	1.050	3.265	10
18°00′	0.3090	0.9511	0.3249	3.078	1.051	3.236	72°00′
10	0.3118	0.9502	0.3281	3.047	1.052	3.207	50
20	0.3145	0.9492	0.3314	3.018	1.053	3.179	40
30	0.3173	0.9483	0.3346	2.989	1.054	3.152	30
40	0.3201	0.9474	0.3378	2.960	1.056	3.124	20
50	0.3228	0.9465	0.3411	2.932	1.057	3.098	10
19°00′	0.3256	0.9455	0.3443	2.904	1.058	3.072	71°00′
10	0.3283	0.9446	0.3476	2.877	1.059	3.046	50
20	0.3311	0.9436	0.3508	2.850	1.060	3.021	40
30	0.3338	0.9426	0.3541	2.824	1.061	2.996	30
40	0.3365	0.9417	0.3574	2.798	1.062	2.971	20
50	0.3393	0.9407	0.3607	2.773	1.063	2.947	10
20°00′	0.3420	0.9397	0.3640	2.747	1.064	2.924	70°00′
10	0.3448	0.9387	0.3673	2.723	1.065	2.901	50
20	0.3475	0.9377	0.3706	2.699	1.066	2.878	40
30	0.3502	0.9367	0.3739	2.675	1.068	2.855	30
40	0.3529	0.9356	0.3772	2.651	1.069	2.833	20
50	0.3557	0.9346	0.3805	2.628	1.070	2.812	10
21°00′	0.3584	0.9336	0.3839	2.605	1.071	2.790	69°00′
	Cos θ	Sin θ	Cot θ	Tan θ	Csc θ	Sec θ	Degrees

Table 1 *Trigonometric functions* (continued)

Degrees	Sin θ	Cos θ	Tan θ	Cot θ	Sec θ	Csc θ	
21°00′	0.3584	0.9336	0.3839	2.605	1.071	2.790	69°00′
10	0.3611	0.9325	0.3872	2.583	1.072	2.769	50
20	0.3638	0.9315	0.3906	2.560	1.074	2.749	40
30	0.3665	0.9304	0.3939	2.539	1.075	2.729	30
40	0.3692	0.9293	0.3973	2.517	1.076	2.709	20
50	0.3719	0.9283	0.4006	2.496	1.077	2.689	10
22°00′	0.3746	0.9272	0.4040	2.475	1.079	2.669	68°00′
10	0.3773	0.9261	0.4074	2.455	1.080	2.650	50
20	0.3800	0.9250	0.4108	2.434	1.081	2.632	40
30	0.3827	0.9239	0.4142	2.414	1.082	2.613	30
40	0.3854	0.9228	0.4176	2.394	1.084	2.595	20
50	0.3881	0.9216	0.4210	2.375	1.085	2.577	10
23°00′	0.3907	0.9205	0.4245	2.356	1.086	2.559	67°00′
10	0.3934	0.9194	0.4279	2.337	1.088	2.542	50
20	0.3961	0.9182	0.4314	2.318	1.089	2.525	40
30	0.3987	0.9171	0.4348	2.300	1.090	2.508	30
40	0.4014	0.9159	0.4383	2.282	1.092	2.491	20
50	0.4041	0.9147	0.4417	2.264	1.093	2.475	10
24°00′	0.4067	0.9135	0.4452	2.246	1.095	2.459	66°00′
10	0.4094	0.9124	0.4487	2.229	1.096	2.443	50
20	0.4120	0.9112	0.4522	2.211	1.097	2.427	40
30	0.4147	0.9100	0.4557	2.194	1.099	2.411	30
40	0.4173	0.9088	0.4592	2.177	1.100	2.396	20
50	0.4200	0.9075	0.4628	2.161	1.102	2.381	10
25°00′	0.4226	0.9063	0.4663	2.145	1.103	2.366	65°00′
10	0.4253	0.9051	0.4699	2.128	1.105	2.352	50
20	0.4279	0.9038	0.4734	2.112	1.106	2.337	40
30	0.4305	0.9026	0.4770	2.097	1.108	2.323	30
40	0.4331	0.9013	0.4806	2.081	1.109	2.309	20
50	0.4358	0.9001	0.4841	2.066	1.111	2.295	10
26°00′	0.4384	0.8988	0.4877	2.050	1.113	2.281	64°00′
10	0.4410	0.8975	0.4913	2.035	1.114	2.268	50
20	0.4436	0.8962	0.4950	2.020	1.116	2.254	40
30	0.4462	0.8949	0.4986	2.006	1.117	2.241	30
40	0.4488	0.8936	0.5022	1.991	1.119	2.228	20
50	0.4514	0.8923	0.5059	1.977	1.121	2.215	10
27°00′	0.4540	0.8910	0.5095	1.963	1.122	2.203	63°00′
10	0.4566	0.8897	0.5132	1.949	1.124	2.190	50
20	0.4592	0.8884	0.5169	1.935	1.126	2.178	40
30	0.4617	0.8870	0.5206	1.921	1.127	2.166	30
40	0.4643	0.8857	0.5243	1.907	1.129	2.154	20
50	0.4669	0.8843	0.5280	1.894	1.131	2.142	10
28°00′	0.4695	0.8829	0.5317	1.881	1.133	2.130	62°00′
10	0.4720	0.8816	0.5354	1.868	1.134	2.118	50
20	0.4746	0.8802	0.5392	1.855	1.136	2.107	40
30	0.4772	0.8788	0.5430	1.842	1.138	2.096	30
40	0.4797	0.8774	0.5467	1.829	1.140	2.085	20
50	0.4823	0.8760	0.5505	1.816	1.142	2.074	10
29°00′	0.4848	0.8746	0.5543	1.804	1.143	2.063	61°00′
	Cos θ	Sin θ	Cot θ	Tan θ	Csc θ	Sec θ	Degrees

Table 1 *Trigonometric functions* (continued)

Degrees	Sin θ	Cos θ	Tan θ	Cot θ	Sec θ	Csc θ	
29°00′	0.4848	0.8746	0.5543	1.804	1.143	2.063	61°00′
10	0.4874	0.8732	0.5581	1.792	1.145	2.052	50
20	0.4899	0.8718	0.5619	1.780	1.147	2.041	40
30	0.4924	0.8704	0.5658	1.767	1.149	2.031	30
40	0.4950	0.8689	0.5696	1.756	1.151	2.020	20
50	0.4975	0.8675	0.5735	1.744	1.153	2.010	10
30°00′	0.5000	0.8660	0.5774	1.732	1.155	2.000	60°00′
10	0.5025	0.8646	0.5812	1.720	1.157	1.990	50
20	0.5050	0.8631	0.5851	1.709	1.159	1.980	40
30	0.5075	0.8616	0.5890	1.698	1.161	1.970	30
40	0.5100	0.8601	0.5930	1.686	1.163	1.961	20
50	0.5125	0.8587	0.5969	1.675	1.165	1.951	10
31°00′	0.5150	0.8572	0.6009	1.664	1.167	1.942	59°00′
10	0.5175	0.8557	0.6048	1.653	1.169	1.932	50
20	0.5200	0.8542	0.6088	1.643	1.171	1.923	40
30	0.5225	0.8526	0.6128	1.632	1.173	1.914	30
40	0.5250	0.8511	0.6168	1.621	1.175	1.905	20
50	0.5275	0.8496	0.6208	1.611	1.177	1.896	10
32°00′	0.5299	0.8480	0.6249	1.600	1.179	1.887	58°00′
10	0.5324	0.8465	0.6289	1.590	1.181	1.878	50
20	0.5348	0.8450	0.6330	1.580	1.184	1.870	40
30	0.5373	0.8434	0.6371	1.570	1.186	1.861	30
40	0.5398	0.8418	0.6412	1.560	1.188	1.853	20
50	0.5422	0.8403	0.6453	1.550	1.190	1.844	10
33°00′	0.5446	0.8387	0.6494	1.540	1.192	1.836	57°00′
10	0.5471	0.8371	0.6536	1.530	1.195	1.828	50
20	0.5495	0.8355	0.6577	1.520	1.197	1.820	40
30	0.5519	0.8339	0.6619	1.511	1.199	1.812	30
40	0.5544	0.8323	0.6661	1.501	1.202	1.804	20
50	0.5568	0.8307	0.6703	1.492	1.204	1.796	10
34°00′	0.5592	0.8290	0.6745	1.483	1.206	1.788	56°00′
10	0.5616	0.8274	0.6787	1.473	1.209	1.781	50
20	0.5640	0.8258	0.6830	1.464	1.211	1.773	40
30	0.5664	0.8241	0.6873	1.455	1.213	1.766	30
40	0.5688	0.8225	0.6916	1.446	1.216	1.758	20
50	0.5712	0.8208	0.6959	1.437	1.218	1.751	10
35°00′	0.5736	0.8192	0.7002	1.428	1.221	1.743	55°00′
10	0.5760	0.8175	0.7046	1.419	1.223	1.736	50
20	0.5783	0.8158	0.7089	1.411	1.226	1.729	40
30	0.5807	0.8141	0.7133	1.402	1.228	1.722	30
40	0.5831	0.8124	0.7177	1.393	1.231	1.715	20
50	0.5854	0.8107	0.7221	1.385	1.233	1.708	10
36°00′	0.5878	0.8090	0.7265	1.376	1.236	1.701	54°00′
10	0.5901	0.8073	0.7310	1.368	1.239	1.695	50
20	0.5925	0.8056	0.7355	1.360	1.241	1.688	40
30	0.5948	0.8039	0.7400	1.351	1.244	1.681	30
40	0.5972	0.8021	0.7445	1.343	1.247	1.675	20
50	0.5995	0.8004	0.7490	1.335	1.249	1.668	10
37°00′	0.6018	0.7986	0.7536	1.327	1.252	1.662	53°00′
	Cos θ	Sin θ	Cot θ	Tan θ	Csc θ	Sec θ	Degrees

Table 1 *Trigonometric functions* (continued)

Degrees	Sin θ	Cos θ	Tan θ	Cot θ	Sec θ	Csc θ	
37°00′	0.6018	0.7986	0.7536	1.327	1.252	1.662	53°00′
10	0.6041	0.7969	0.7581	1.319	1.255	1.655	50
20	0.6065	0.7951	0.7627	1.311	1.258	1.649	40
30	0.6088	0.7934	0.7673	1.303	1.260	1.643	30
40	0.6111	0.7916	0.7720	1.295	1.263	1.636	20
50	0.6134	0.7898	0.7766	1.288	1.266	1.630	10
38°00′	0.6157	0.7880	0.7813	1.280	1.269	1.624	52°00′
10	0.6180	0.7862	0.7860	1.272	1.272	1.618	50
20	0.6202	0.7844	0.7907	1.265	1.275	1.612	40
30	0.6225	0.7826	0.7954	1.257	1.278	1.606	30
40	0.6248	0.7808	0.8002	1.250	1.281	1.601	20
50	0.6271	0.7790	0.8050	1.242	1.284	1.595	10
39°00′	0.6293	0.7771	0.8098	1.235	1.287	1.589	51°00′
10	0.6316	0.7753	0.8146	1.228	1.290	1.583	50
20	0.6338	0.7735	0.8195	1.220	1.293	1.578	40
30	0.6361	0.7716	0.8243	1.213	1.296	1.572	30
40	0.6383	0.7698	0.8292	1.206	1.299	1.567	20
50	0.6406	0.7679	0.8342	1.199	1.302	1.561	10
40°00′	0.6428	0.7660	0.8391	1.192	1.305	1.556	50°00′
10	0.6450	0.7642	0.8441	1.185	1.309	1.550	50
20	0.6472	0.7623	0.8491	1.178	1.312	1.545	40
30	0.6494	0.7604	0.8541	1.171	1.315	1.540	30
40	0.6517	0.7585	0.8591	1.164	1.318	1.535	20
50	0.6539	0.7566	0.8642	1.157	1.322	1.529	10
41°00′	0.6561	0.7547	0.8693	1.150	1.325	1.524	49°00′
10	0.6583	0.7528	0.8744	1.144	1.328	1.519	50
20	0.6604	0.7509	0.8796	1.137	1.332	1.514	40
30	0.6626	0.7490	0.8847	1.130	1.335	1.509	30
40	0.6648	0.7470	0.8899	1.124	1.339	1.504	20
50	0.6670	0.7451	0.8952	1.117	1.342	1.499	10
42°00′	0.6691	0.7431	0.9004	1.111	1.346	1.494	48°00′
10	0.6713	0.7412	0.9057	1.104	1.349	1.490	50
20	0.6734	0.7392	0.9110	1.098	1.353	1.485	40
30	0.6756	0.7373	0.9163	1.091	1.356	1.480	30
40	0.6777	0.7353	0.9217	1.085	1.360	1.476	20
50	0.6799	0.7333	0.9271	1.079	1.364	1.471	10
43°00′	0.6820	0.7314	0.9325	1.072	1.367	1.466	47°00′
10	0.6841	0.7294	0.9380	1.066	1.371	1.462	50
20	0.6862	0.7274	0.9435	1.060	1.375	1.457	40
30	0.6884	0.7254	0.9490	1.054	1.379	1.453	30
40	0.6905	0.7234	0.9545	1.048	1.382	1.448	20
50	0.6926	0.7214	0.9601	1.042	1.386	1.444	10
44°00′	0.6947	0.7193	0.9657	1.036	1.390	1.440	46°00′
10	0.6967	0.7173	0.9713	1.030	1.394	1.435	50
20	0.6988	0.7153	0.9770	1.024	1.398	1.431	40
30	0.7009	0.7133	0.9827	1.018	1.402	1.427	30
40	0.7030	0.7112	0.9884	1.012	1.406	1.423	20
50	0.7050	0.7092	0.9942	1.006	1.410	1.418	10
45°00′	0.7071	0.7071	1.000	1.000	1.414	1.414	45°00′
	Cos θ	Sin θ	Cot θ	Tan θ	Csc θ	Sec θ	Degrees

Table 2 *Common Logarithms*

n	0	1	2	3	4	5	6	7	8	9
1.0	.0000	.0043	.0086	.0128	.0170	.0212	.0253	.0294	.0334	.0374
1.1	.0414	.0453	.0492	.0531	.0569	.0607	.0645	.0682	.0719	.0755
1.2	.0792	.0828	.0864	.0899	.0934	.0969	.1004	.1038	.1072	.1106
1.3	.1139	.1173	.1206	.1239	.1271	.1303	.1335	.1367	.1399	.1430
1.4	.1461	.1492	.1523	.1553	.1584	.1614	.1644	.1673	.1703	.1732
1.5	.1761	.1790	.1818	.1847	.1875	.1903	.1931	.1959	.1987	.2014
1.6	.2041	.2068	.2095	.2122	.2148	.2175	.2201	.2227	.2253	.2279
1.7	.2304	.2330	.2355	.2380	.2405	.2430	.2455	.2480	.2504	.2529
1.8	.2553	.2577	.2601	.2625	.2648	.2672	.2695	.2718	.2742	.2765
1.9	.2788	.2810	.2833	.2856	.2878	.2900	.2923	.2945	.2967	.2989
2.0	.3010	.3032	.3054	.3075	.3096	.3118	.3139	.3160	.3181	.3201
2.1	.3222	.3243	.3263	.3284	.3304	.3324	.3345	.3365	.3385	.3404
2.2	.3424	.3444	.3464	.3483	.3502	.3522	.3541	.3560	.3579	.3598
2.3	.3617	.3636	.3655	.3674	.3692	.3711	.3729	.3747	.3766	.3784
2.4	.3802	.3820	.3838	.3856	.3874	.3892	.3909	.3927	.3945	.3962
2.5	.3979	.3997	.4014	.4031	.4048	.4065	.4082	.4099	.4116	.4133
2.6	.4150	.4166	.4183	.4200	.4216	.4232	.4249	.4265	.4281	.4298
2.7	.4314	.4330	.4346	.4362	.4378	.4393	.4409	.4425	.4440	.4456
2.8	.4472	.4487	.4502	.4518	.4533	.4548	.4564	.4579	.4594	.4609
2.9	.4624	.4639	.4654	.4669	.4683	.4698	.4713	.4728	.4742	.4757
3.0	.4771	.4786	.4800	.4814	.4829	.4843	.4857	.4871	.4886	.4900
3.1	.4914	.4928	.4942	.4955	.4969	.4983	.4997	.5011	.5024	.5038
3.2	.5051	.5065	.5079	.5092	.5105	.5119	.5132	.5145	.5159	.5172
3.3	.5185	.5198	.5211	.5224	.5237	.5250	.5263	.5276	.5289	.5302
3.4	.5315	.5328	.5340	.5353	.5366	.5378	.5391	.5403	.5416	.5428
3.5	.5441	.5453	.5465	.5478	.5490	.5502	.5514	.5527	.5539	.5551
3.6	.5563	.5575	.5587	.5599	.5611	.5623	.5635	.5647	.5658	.5670
3.7	.5682	.5694	.5705	.5717	.5729	.5740	.5752	.5763	.5775	.5786
3.8	.5798	.5809	.5821	.5832	.5843	.5855	.5866	.5877	.5888	.5899
3.9	.5911	.5922	.5933	.5944	.5955	.5966	.5977	.5988	.5999	.6010
4.0	.6021	.6031	.6042	.6053	.6064	.6075	.6085	.6096	.6107	.6117
4.1	.6128	.6138	.6149	.6160	.6170	.6180	.6191	.6201	.6212	.6222
4.2	.6232	.6243	.6253	.6263	.6274	.6284	.6294	.6304	.6314	.6325
4.3	.6335	.6345	.6355	.6365	.6375	.6385	.6395	.6405	.6415	.6425
4.4	.6435	.6444	.6454	.6464	.6474	.6484	.6493	.6503	.6513	.6522
4.5	.6532	.6542	.6551	.6561	.6571	.6580	.6590	.6599	.6609	.6618
4.6	.6628	.6637	.6646	.6656	.6665	.6675	.6684	.6693	.6702	.6712
4.7	.6721	.6730	.6739	.6749	.6758	.6767	.6776	.6785	.6794	.6803
4.8	.6812	.6821	.6830	.6839	.6848	.6857	.6866	.6875	.6884	.6893
4.9	.6902	.6911	.6920	.6928	.6937	.6946	.6955	.6964	.6972	.6981
5.0	.6990	.6998	.7007	.7016	.7024	.7033	.7042	.7050	.7059	.7067
5.1	.7076	.7084	.7093	.7101	.7110	.7118	.7126	.7135	.7143	.7152
5.2	.7160	.7168	.7177	.7185	.7193	.7202	.7210	.7218	.7226	.7235
5.3	.7243	.7251	.7259	.7267	.7275	.7284	.7292	.7300	.7308	.7316
5.4	.7324	.7332	.7340	.7348	.7356	.7364	.7372	.7380	.7388	.7396
5.5	.7404	.7412	.7419	.7427	.7435	.7443	.7451	.7459	.7466	.7474
5.6	.7482	.7490	.7497	.7505	.7513	.7520	.7528	.7536	.7543	.7551
5.7	.7559	.7566	.7574	.7582	.7589	.7597	.7604	.7612	.7619	.7627
5.8	.7634	.7642	.7649	.7657	.7664	.7672	.7679	.7686	.7694	.7701
5.9	.7709	.7716	.7723	.7731	.7738	.7745	.7752	.7760	.7767	.7774

Table 2 *Common Logarithms* (continued)

n	0	1	2	3	4	5	6	7	8	9
6.0	.7782	.7789	.7796	.7803	.7810	.7818	.7825	.7832	.7839	.7846
6.1	.7853	.7860	.7868	.7875	.7882	.7889	.7896	.7903	.7910	.7917
6.2	.7924	.7931	.7938	.7945	.7952	.7959	.7966	.7973	.7980	.7987
6.3	.7993	.8000	.8007	.8014	.8021	.8028	.8035	.8041	.8048	.8055
6.4	.8062	.8069	.8075	.8082	.8089	.8096	.8102	.8109	.8116	.8122
6.5	.8129	.8136	.8142	.8149	.8156	.8162	.8169	.8176	.8182	.8189
6.6	.8195	.8202	.8209	.8215	.8222	.8228	.8235	.8241	.8248	.8254
6.7	.8261	.8267	.8274	.8280	.8287	.8293	.8299	.8306	.8312	.8319
6.8	.8325	.8331	.8338	.8344	.8351	.8357	.8363	.8370	.8376	.8382
6.9	.8388	.8395	.8401	.8407	.8414	.8420	.8426	.8432	.8439	.8445
7.0	.8451	.8457	.8463	.8470	.8476	.8482	.8488	.8494	.8500	.8506
7.1	.8513	.8519	.8525	.8531	.8537	.8543	.8549	.8555	.8561	.8567
7.2	.8573	.8579	.8585	.8591	.8597	.8603	.8609	.8615	.8621	.8627
7.3	.8633	.8639	.8645	.8651	.8657	.8663	.8669	.8675	.8681	.8686
7.4	.8692	.8698	.8704	.8710	.8716	.8722	.8727	.8733	.8739	.8745
7.5	.8751	.8756	.8762	.8768	.8774	.8779	.8785	.8791	.8797	.8802
7.6	.8808	.8814	.8820	.8825	.8831	.8837	.8842	.8848	.8854	.8859
7.7	.8865	.8871	.8876	.8882	.8887	.8893	.8899	.8904	.8910	.8915
7.8	.8921	.8927	.8932	.8938	.8943	.8949	.8954	.8960	.8965	.8971
7.9	.8976	.8982	.8987	.8993	.8998	.9004	.9009	.9015	.9020	.9025
8.0	.9031	.9036	.9042	.9047	.9053	.9058	.9063	.9069	.9074	.9079
8.1	.9085	.9090	.9096	.9101	.9106	.9112	.9117	.9122	.9128	.9133
8.2	.9138	.9143	.9149	.9154	.9159	.9165	.9170	.9175	.9180	.9186
8.3	.9191	.9196	.9201	.9206	.9212	.9217	.9222	.9227	.9232	.9238
8.4	.9243	.9248	.9253	.9258	.9263	.9269	.9274	.9279	.9284	.9289
8.5	.9294	.9299	.9304	.9309	.9315	.9320	.9325	.9330	.9335	.9340
8.6	.9345	.9350	.9355	.9360	.9365	.9370	.9375	.9380	.9385	.9390
8.7	.9395	.9400	.9405	.9410	.9415	.9420	.9425	.9430	.9435	.9440
8.8	.9445	.9450	.9455	.9460	.9465	.9469	.9474	.9479	.9484	.9489
8.9	.9494	.9499	.9504	.9509	.9513	.9518	.9523	.9528	.9533	.9538
9.0	.9542	.9547	.9552	.9557	.9562	.9566	.9571	.9576	.9581	.9586
9.1	.9590	.9595	.9600	.9605	.9609	.9614	.9619	.9624	.9628	.9633
9.2	.9638	.9643	.9647	.9652	.9657	.9661	.9666	.9671	.9675	.9680
9.3	.9685	.9689	.9694	.9699	.9703	.9708	.9713	.9717	.9722	.9727
9.4	.9731	.9736	.9741	.9745	.9750	.9754	.9759	.9763	.9768	.9773
9.5	.9777	.9782	.9786	.9791	.9795	.9800	.9805	.9809	.9814	.9818
9.6	.9823	.9827	.9832	.9836	.9841	.9845	.9850	.9854	.9859	.9863
9.7	.9868	.9872	.9877	.9881	.9886	.9890	.9894	.9899	.9903	.9908
9.8	.9912	.9917	.9921	.9926	.9930	.9934	.9939	.9943	.9948	.9952
9.9	.9956	.9961	.9965	.9969	.9974	.9978	.9983	.9987	.9991	.9996

Table 3 *Natural Logarithms*

n	$\log_e n$	n	$\log_e n$	n	$\log_e n$	n	$\log_e n$
0.0	—	3.5	1.2528	7.0	1.9459	15	2.7081
0.1	−2.3026	3.6	1.2809	7.1	1.9601	16	2.7726
0.2	−1.6094	3.7	1.3083	7.2	1.9741	17	2.8332
0.3	−1.2040	3.8	1.3350	7.3	1.9879	18	2.8904
0.4	−0.9163	3.9	1.3610	7.4	2.0015	19	2.9444
0.5	−0.6931	4.0	1.3863	7.5	2.0149	20	2.9957
0.6	−0.5108	4.1	1.4110	7.6	2.0281	25	3.2189
0.7	−0.3567	4.2	1.4351	7.7	2.0412	30	3.4012
0.8	−0.2231	4.3	1.4586	7.8	2.0541	35	3.5553
0.9	−0.1054	4.4	1.4816	7.9	2.0669	40	3.6889
1.0	0.0000	4.5	1.5041	8.0	2.0794	45	3.8067
1.1	0.0953	4.6	1.5261	8.1	2.0919	50	3.9120
1.2	0.1823	4.7	1.5476	8.2	2.1041	55	4.0073
1.3	0.2624	4.8	1.5686	8.3	2.1163	60	4.0943
1.4	0.3365	4.9	1.5892	8.4	2.1282	65	4.1744
1.5	0.4055	5.0	1.6094	8.5	2.1401	70	4.2485
1.6	0.4700	5.1	1.6292	8.6	2.1518	75	4.3175
1.7	0.5306	5.2	1.6487	8.7	2.1633	80	4.3820
1.8	0.5878	5.3	1.6677	8.8	2.1748	85	4.4427
1.9	0.6419	5.4	1.6864	8.9	2.1861	90	4.4998
2.0	0.6931	5.5	1.7047	9.0	2.1972	95	4.5539
2.1	0.7419	5.6	1.7228	9.1	2.2083	100	4.6052
2.2	0.7885	5.7	1.7405	9.2	2.2192	200	5.2983
2.3	0.8329	5.8	1.7579	9.3	2.2300	300	5.7038
2.4	0.8755	5.9	1.7750	9.4	2.2407	400	5.9915
2.5	0.9163	6.0	1.7918	9.5	2.2513	500	6.2146
2.6	0.9555	6.1	1.8083	9.6	2.2618	600	6.3969
2.7	0.9933	6.2	1.8245	9.7	2.2721	700	6.5511
2.8	1.0296	6.3	1.8405	9.8	2.2824	800	6.6846
2.9	1.0647	6.4	1.8563	9.9	2.2925	900	6.8024
3.0	1.0986	6.5	1.8718	10	2.3026		
3.1	1.1314	6.6	1.8871	11	2.3979		
3.2	1.1632	6.7	1.9021	12	2.4849		
3.3	1.1939	6.8	1.9169	13	2.5649		
3.4	1.2238	6.9	1.9315	14	2.6391		

APPENDIX C

Answers to Selected Exercises

Chapter 1

Section 1.1 (page 8)

1. 1266 **3.** 1046 **5.** 1342 **7.** 1857 **9.** 1587 **11.** 2015 **13.** 28,280 **15.** 13,365
17. 176,304 **19.** 47,155 **21.** 17,924 **23.** 12,928 **25.** 1269 **27.** 149 **29.** 6537
31. 698 **33.** 42,489 **35.** 2659 **37.** 20,211 **39.** 14,937 **41.** 32 in. **43.** 487 lb
45. 1066 **47.** 18 Ω **49.** \$85 **51.** \$1670 **53.** 2 ft 10 in. **55.** 4 ft 6 in. **57.** 36 cm

Section 1.2 (page 13)

1. 9728 **3.** 21,483 **5.** 334,946 **7.** 800,280 **9.** 233,728 **11.** 240,720 **13.** 1,672,512
15. 20,527,648 **17.** 5,535,660 **19.** 53,751,049 **21.** 1596 cm^2 **23.** \$348 **25.** 1092 gal
27. 239,000 mi (approximately) **29.** 434 V **31.** 12,250 ft

Section 1.3 (page 17)

1. 42 **3.** 61 **5.** 498 **7.** 529 **9.** 527 **11.** 836 **13.** 107 **15.** 17,365
17. 12,327, remainder 15 **19.** 1027, remainder 6 **21.** 16 **23.** 11 **25.** 16 **27.** 0
29. 35 **31.** 24 **33.** 29 in. **35.** \$2334 **37.** 18 **39.** 48 lb **41.** yes

Section 1.4 (page 19)

1. $2 \times 2 \times 3$ **3.** $2 \times 2 \times 2 \times 2 \times 3$ **5.** $2 \times 2 \times 3 \times 3$ **7.** $2 \times 2 \times 7$ **9.** $2 \times 2 \times 3 \times 7$
11. $2 \times 3 \times 3 \times 7$ **13.** $3 \times 3 \times 11$ **15.** $2 \times 5 \times 5 \times 11$ **17.** $3 \times 3 \times 3 \times 7$
19. $3 \times 11 \times 11$ **21.** $3 \times 3 \times 3 \times 7 \times 13$ **23.** $2 \times 2 \times 2 \times 3 \times 11 \times 13$
25. $2 \times 2 \times 7 \times 7 \times 13$ **27.** $3 \times 3 \times 13$ **29.** $2 \times 2 \times 3 \times 17$ **31.** 19×29

Section 1.5 (page 24)

1. 9 is the numerator, 16 is the denominator **3.** 1 is the numerator, 10 is the denominator
5. 25 is the numerator, 31 is the denominator **7.** $\frac{3}{7}$ **9.** $\frac{7}{6}$ **11.** $\frac{7}{9}$ **13.** $\frac{2}{3}$ **15.** $\frac{5}{12}$

17. $\frac{3}{7}$ **19.** $\frac{9}{20}$ **21.** $\frac{33}{13}$ **23.** $\frac{5}{6}$ **25.** $\frac{28}{27}$ **27.** $\frac{4}{9}$ **29.** $\frac{3}{4}$ **31.** $\frac{51}{38}$ **33.** $\frac{29}{31}$ **35.** $\frac{3}{4}$
37. $\frac{4}{11}$ **39.** $\frac{4}{9}$ **41.** $\frac{3}{13}$ **43.** $\frac{1}{3}$

Section 1.6 (page 30)

1. $\frac{2}{5}$ **3.** $\frac{3}{4}$ **5.** $\frac{2}{3}$ **7.** 3 **9.** 12 **11.** $\frac{1}{7}$ **13.** $\frac{1}{2}$ **15.** $\frac{2}{3}$ **17.** $\frac{6}{77}$ **19.** $\frac{8}{5}$
21. $\frac{54}{77}$ **23.** $\frac{13}{10}$ **25.** $\frac{1}{5}$ **27.** $\frac{1}{15}$ **29.** $\frac{2}{147}$ **31.** 3 **33.** 10 lb **35.** $\frac{9}{20}$ s **37.** $\frac{5}{3}$ s
39. $\frac{6}{11}$ ms **41.** 34 gal **43.** $\frac{6}{5}$ ft^2 **45.** 6 h

Section 1.7 (page 38)

1. 1 **3.** $\frac{1}{3}$ **5.** $\frac{3}{4}$ **7.** $\frac{1}{4}$ **9.** $\frac{8}{15}$ **11.** $\frac{1}{3}$ **13.** $\frac{1}{12}$ **15.** $\frac{3}{14}$ **17.** $\frac{19}{4}$ **19.** $\frac{3}{8}$
21. $\frac{11}{24}$ **23.** $\frac{41}{56}$ **25.** $\frac{17}{105}$ **27.** $\frac{15}{36}$ **29.** $\frac{9}{16}$ **31.** $\frac{11}{7}$ **33.** $\frac{53}{140}$ **35.** $\frac{67}{168}$ **37.** 2
39. 10 **41.** $\frac{10}{3}$ **43.** $\frac{7}{6}$ in. **45.** $\frac{13}{35}$ A **47.** $\frac{19}{72}$ **49.** $\frac{29}{60}$ lb **51.** $\frac{149}{240}$ lb **53.** $\frac{55}{112}$ in.

Section 1.8 (page 43)

1. $\frac{21}{5}$ **3.** $\frac{25}{11}$ **5.** $\frac{81}{8}$ **7.** $\frac{120}{13}$ **9.** $6\frac{2}{3}$ **11.** $3\frac{1}{13}$ **13.** $4\frac{2}{7}$ **15.** $6\frac{5}{6}$ **17.** $3\frac{1}{2}$
19. $3\frac{1}{3}$ **21.** $6\frac{3}{4}$ **23.** $12\frac{7}{12}$ **25.** $14\frac{5}{8}$ **27.** $10\frac{13}{48}$ **29.** $1\frac{17}{26}$ **31.** $4\frac{1}{3}$ **33.** $4\frac{2}{9}$ in.
35. $1\frac{19}{45}$ cm^2 **37.** $21\frac{20}{63}$ cm **39.** $16\frac{1}{24}$ in. **41.** $1510\frac{3}{4}$ lb **43.** 4 pieces, $1\frac{3}{5}$ in. left over
45. $1\frac{2}{3}$ h **47.** $6\frac{7}{60}$ ft

Section 1.9 (page 52)

1. 0.8 **3.** 0.16 **5.** 0.52 **7.** 0.09 **9.** 0.43 **11.** 0.47 **13.** 0.65 **15.** 0.33
17. $\frac{3}{4}$ **19.** $\frac{2}{25}$ **21.** $\frac{1}{400}$ **23.** $\frac{3}{8}$ **25.** 16.655 **27.** 38.66 **29.** 2.471 **31.** 14.16
33. 2.611 **35.** 93.24 **37.** 65.664 **39.** 2.5029 **41.** 0.0000065 **43.** 7.2 **45.** 0.097
47. 0.723 **49.** 0.857 **51.** 0.152 **53.** 0.0179 **55.** 9.81 mm^2 **57.** 9.79 m/s^2
59. 63.4 Ω **61.** 12.25 in. **63.** 25 lb **65.** 628.4 lb **67.** 0.14 h

Section 1.10 (page 56)

1. 0.29 **3.** 0.862 **5.** 0.0134 **7.** 0.5736 **9.** 36% **11.** 42.19% **13.** 162% **15.** 6
17. 15 **19.** 9.6 **21.** 44.4% **23.** 80% **25.** 38.5% **27.** 160 **29.** 142.5 **31.** 170
33. $31.24 **35.** 80% **37.** 26,275 **39.** 54 lb copper, 5.4 lb lead, 30.6 lb zinc **41.** 6.3 lb
43. 120 hp **45.** No, $8.62 per hour **47.** 10 yd^3

Section 1.11 (page 58)

1. 9 **3.** 125 **5.** 49 **7.** 64 **9.** 144 **11.** 576 **13.** 288 **15.** 12,544 **17.** 23,328
19. 72,000

Section 1.12 (page 66)

1. 170 **3.** 151 **5.** 7636 **7.** 53 **9.** 4 **11.** 9409 **13.** 153.76 **15.** 26 **17.** 46.3 **19.** 22.12 **21.** 2030 **23.** 87.81 **25.** 5.46 **27.** 36.59 **29.** 1.04 **31.** 2.34 **33.** 8.17 **35.** 51,529 **37.** 529 **39.** 961 **41.** 3.8 s **43.** 18.6 Ω **45.** 67.8 Ω

Review Exercises/Chapter 1 (page 67)

1. 23,300 **2.** 2495 **3.** 5734 **4.** 5203 **5.** 4,508,672 **6.** 11,832,444 **7.** 429 **8.** 3048 **9.** $2 \times 2 \times 3 \times 3 \times 7$ **10.** $2 \times 3 \times 3 \times 3 \times 5 \times 5$ **11.** $\frac{3}{10}$ **12.** $\frac{21}{22}$ **13.** $\frac{2}{33}$ **14.** $\frac{1}{48}$ **15.** $\frac{5}{54}$ **16.** $\frac{1}{6}$ **17.** $\frac{1}{4}$ **18.** $\frac{57}{100}$ **19.** $\frac{359}{560}$ **20.** $\frac{73}{140}$ **21.** $8\frac{7}{8}$ **22.** $4\frac{3}{4}$ **23.** $12\frac{8}{9}$ **24.** $1\frac{5}{7}$ **25.** $1\frac{31}{44}$ **26.** $2\frac{1}{4}$ **27.** 86.197 **28.** 23.98 **29.** 6.716 **30.** 446.814 **31.** 0.0000595 **32.** 0.01092 **33.** 14.4 **34.** 0.96 **35.** 0.26 **36.** 0.24 **37.** 0.04 **38.** 0.28 **39.** 0.072 **40.** 0.052 **41.** $\frac{1}{40}$ **42.** $\frac{19}{2000}$ **43.** $\frac{3}{25}$ **44.** $\frac{3}{125}$ **45.** 0.463 **46.** 7.2% **47.** 0.075 **48.** 22.5 **49.** 31.25% **50.** 40% **51.** 188.5 **52.** 182.6 **53.** 288 **54.** 43,200 **55.** 26 lb 14 oz **56.** 228 V **57.** 129.3 ft **58.** $\frac{7}{38}$ **59.** 40 s **60.** 40 days **61.** $19\frac{5}{12}$ V **62.** 41.05°C **63.** 62.4 lb/ft^3 **64.** 25% **65.** 15.4% **66.** 244 acres **67.** $1543 **68.** 70.28 **69.** 8.02 **70.** 0.014 **71.** 2.09 **72.** 256 **73.** 289 **74.** 8.68

Chapter 2
Section 2.1 (page 74)

1. approximate **3.** approximate **5.** exact **7.** approximate **9.** 4 **11.** 3 **13.** 3 **15.** 3 **17.** 5 **19.** 5 **21.** 6 **23.** 2 **25.** 4 **27.** 3 **29.** 60,500 **31.** 28,000 **33.** 19,000 **35.** 2.59 **37.** 3.55 **39.** 0.00550 **41.** **a.** 5.230 **b.** 5.230 **43.** **a.** 38.7 **b.** 0.76 **45.** **a.** 3.764 **b.** 0.0042 **47.** **a.** same accuracy **b.** 0.0960 **49.** 8.55 ft, 8.65 ft **51.** 5.37 in. **53.** same accuracy; 0.938 cm is more precise

Section 2.2 (page 78)

1. 12.4 **3.** 104.64 **5.** 88.98 **7.** 73.50 **9.** 0.021 **11.** 0.647 **13.** 0.019 **15.** 1.7 **17.** 36.2 **19.** 34.70 **21.** 2.924 **23.** 6.9126 **25.** 0.1152 **27.** 36.6 lb **29.** 10.7 cm **31.** 38 V **33.** 3.39 hours **35.** 78.5 ft/s

Section 2.3 (page 84)

1. mm **3.** mg **5.** cL **7.** kW **9.** μF **11.** mΩ **13.** μs **15.** mA **17.** millisecond **19.** megaohm **21.** centiliter **23.** milligram **25.** centimeter **27.** microsecond **29.** 340 mm **31.** 264 mL **33.** 0.023 g **35.** 4060 W **37.** 0.102 s **39.** 890 μs **41.** 0.025 mΩ **43.** 0.0364 mF **45.** 76,000,000 mΩ **47.** 72.6 cW **49.** 9600 mW **51.** 0.00023 F **53.** 0.02 s **55.** 0.000036 ms

Section 2.4 (page 90)

1. 1.60 ft **3.** 82.8 in. **5.** 131 ms **7.** 238 μA **9.** 28,000 μA **11.** 0.047 mF **13.** 5.1 cm **15.** 3.94 in. **17.** 26 cm^2 **19.** 76 cm **21.** 17.2 km **23.** 16.7 N **25.** 1.99 lb **27.** $7.6\ \frac{\text{cm}}{\text{s}}$ **29.** $0.0740\ \frac{\text{km}}{\text{min}}$ **31.** $2.70\ \frac{\text{N}}{\text{cm}^2}$ **33.** $0.01050\ \frac{\text{lb}}{\text{in.}^2}$

35. $0.000217\ \frac{lb}{in.^3}$ **37.** $193\ \frac{kg}{m^2}$ **39.** $2{,}807{,}000\ \frac{N}{m^3}$ **41.** 0.79 L **43.** 7110 L
45. 385,000 km **47.** $9800\ \frac{N}{m^3}$ **49.** $1240\ \frac{km}{h}$ **51.** $170\ \frac{lb}{ft^3}$ **53.** $150\ \frac{kg}{cm^2}$

Review Exercises/Chapter 2 (page 91)

1. approximate **2.** exact by definition **3.** 2 **4.** 3 **5.** 3 **6.** 4 **7.** 27,700
8. 99,630 **9.** 0.0576 **10.** 0.0030 **11.** 14.68 **12.** 44.81 **13.** 455 **14.** 0.465
15. 25.0 **16.** 0.276 **17.** 1.112 **18.** 0.50526 **19.** microfarad **20.** millisecond
21. megaohm **22.** milliohm **23.** milliliter **24.** microsecond **25.** 37.2 ms **26.** 27.9 μA
27. 2.7 MΩ **28.** 0.000017 F **29.** 54.4 cm^2 **30.** 17.8 in.^{2} **31.** $1536\ \frac{cm}{min}$ **32.** $34.3\ \frac{km}{s}$
33. $0.0399\ \frac{lb}{in.^2}$ **34.** $12\ \frac{N}{m^2}$ **35.** $8.54\ \frac{kg}{m^2}$ **36.** $0.118\ \frac{lb}{in.^2}$ **37.** **a.** same accuracy **b.** 0.86 in.
38. 181.8 Ω **39.** 2.38 A **40.** 0.000051 A **41.** 0.0023 mm **42.** $1200\ \frac{lb}{ft^3}$ **43.** $107{,}000\ \frac{km}{h}$
44. 960 L

Chapter 3

Section 3.1 (page 97)

1. $2 < 4$ **3.** $-4 < -2$ **5.** $4 > -4$ **7.** $0 > -\frac{1}{2}$ **9.** $-\frac{3}{4} < -\frac{1}{4}$ **11.** $-\frac{1}{3} < 4$
13. $-2.5 > -5.4$ **15.** $0 > -2.7$ **17.** 2 **19.** 4 **21.** 7 **23.** 15 **25.** 2 **27.** 7
29. 1.8 **31.** a decrease of $1.60 **33.** -2 ft **35.** -40 lb **37.** 6 s after takeoff
39. the largest loss

Section 3.2 (page 103)

1. 3 **3.** -3 **5.** -5 **7.** 0 **9.** -11 **11.** 9 **13.** -12 **15.** 5 **17.** -32
19. 0 **21.** -9 **23.** -16 **25.** 18 **27.** -12 **29.** -9.9 **31.** -2.36 **33.** -6°C
35. $6 **37.** 5¢ gain **39.** $I_2 = 1.4$ A

Section 3.3 (page 108)

1. -14 **3.** 9 **5.** -3 **7.** -7 **9.** -24 **11.** 6 **13.** 12 **15.** -7 **17.** -8
19. 0 **21.** 11.0 **23.** 6.4 **25.** $\frac{7}{12}$ **27.** 14 V **29.** $-5\ \frac{lb}{in.^2}$ **31.** $60\ \frac{ft}{s}$
33. $100 - (-50) = 150$ ft

Section 3.4 (page 113)

1. 72 **3.** -45 **5.** 0 **7.** 64 **9.** 240 **11.** -36 **13.** 432 **15.** 192 **17.** -5040
19. 4800 **21.** -3 **23.** -3 **25.** $\frac{3}{5}$ **27.** 6 **29.** 2 **31.** -144 **33.** $\frac{35}{2}$ **35.** $-\frac{2}{3}$
37. $\frac{1}{90}$ **39.** 4 **41.** -72 **43.** -500 **45.** 256 **47.** 144 **49.** -1 **51.** 1 **53.** -1
55. -114°C **57.** -181.2°C **59.** 0.6 Ω **61.** 3.0 V (left), -3.0 V (right)

Section 3.5 (page 116)

1. -10 **3.** 10 **5.** -20 **7.** 4 **9.** -7 **11.** 19 **13.** 29 **15.** -3 **17.** 2
19. 6 **21.** -23 **23.** 15 **25.** 27 **27.** -11 **29.** 36

Section 3.6 (page 118)

1. 6.7 **3.** −13.3 **5.** −0.4207 **7.** 82.7 **9.** 9.820 **11.** 8.77 **13.** −0.00203
15. −19.28 **17.** −39.9 **19.** −0.823 **21.** 0.189 **23.** 525 **25.** −0.7616 **27.** 1.94
29. 0.525 A **31.** $-8.06\ \frac{\text{cm}}{\text{s}}$

Review Exercises/Chapter 3 (page 119)

1. −2 **2.** 9 **3.** −13 **4.** −12 **5.** 19 **6.** −13 **7.** 23 **8.** 4 **9.** −1.4
10. −16.1 **11.** 5.8 **12.** −7.2 **13.** $\frac{3}{4}$ **14.** $\frac{1}{6}$ **15.** 72 **16.** −384 **17.** −630
18. 0 **19.** 18 **20.** −14 **21.** $-\frac{3}{2}$ **22.** $\frac{8}{3}$ **23.** 2 **24.** 14 **25.** 28 **26.** 3
27. $-\frac{1}{90}$ **28.** $\frac{3}{20}$ **29.** $\frac{9}{10}$ **30.** $-\frac{56}{45}$ **31.** −108 **32.** 144 **33.** 144
34. −5 **35.** 17.5 V **36.** 2.1, −3.4 A **37.** \$(−2150), \$(−5375) **38.** 5.52 A **39.** −0.1 A
40. −14°C **41.** 65,229 ft **42.** −2.28 N **43.** −138.84°C **44.** $-21.3\ \frac{\text{m}}{\text{s}}$ **45.** $I_4 = -1.1$ A
46. 7.8 V (left), −9.6 V (right)

Cumulative Review Exercises/Chapter 1–3 (page 121)

1. 17,440 **2.** 5215 **3. a.** 290,772 **b.** 290.772 **4. a.** 358 **b.** 35.8 **5.** 12.8794 **6.** 90
7. $\frac{2}{3}$ **8.** $\frac{4}{9}$ **9.** $12\frac{3}{5}$ **10.** $\frac{25}{33}$ **11.** $\frac{1}{252}$ **12.** 17 **13.** −48 **14.** 0 **15.** 20
16. −428 **17.** 0.402 kg **18.** $4860\ \frac{\text{km}}{\text{h}}$ **19.** 26.9 lb **20.** 235 hp **21.** 25.0 Ω **22.** 1.3 s
23. 21.4 lb **24.** 1260 L **25.** 1.26 m^3 **26.** −1.2 A **27.** $15\ \frac{\text{ft}}{\text{s}}$

Chapter 4
Section 4.1 (page 125)

1. constant: 2; variables: P, b, c **3.** constants: 2, 3; variables: Q, s_1, s_2
5. constant: 2; variables: L, a_1, a_2 **7.** constants: 4, 6; variables: Z, F_1, F_2 **9.** $\frac{3}{5}$ **11.** 0.67
13. $F = 4m$ **15.** $l = 4s$ **17.** \$80; $D = 8r$ **19.** $C = 0.75\,lw$ **21.** $i = Prt$; $i = \$276.25$
23. $C = C_1 + C_2$; $C = 5.43\ \mu\text{F}$ **25.** $V_1 = V_2\frac{N_1}{N_2}$; $V_1 = 15$ V

Section 4.2 (page 129)

1. 3 **3.** −6 **5.** −0.4 **7.** 2.30 **9.** coefficient: 2; factors: a and b
11. coefficient: −4; factors: a and b **13.** coefficient: 10; factors: p and q **15.** monomial
17. monomial **19.** trinomial **21.** trinomial **23.** binomial **25.** binomial **27.** monomial
29. trinomial **31.** 12 **33.** 1.0 **35.** 7.8 **37.** 1.95 m^3 **39.** 2.01 μF
41. total cost of contents **43.** $98\ \frac{\text{ft}}{\text{min}}$ **45.** $L - x$ **47.** total monthly wages

Section 4.3 (page 136)

1. $-3x + 9y$ **3.** $-5m - 5n$ **5.** $10V_a - 9V_b$ **7.** $5R^2$ **9.** 0 **11.** $5mv - 8ab$
13. $4\sqrt{z} + 5xz$ **15.** $3ef$ **17.** $-y - z$ **19.** $4x - 4x^2$ **21.** $3V + 2\left(\frac{1}{C}\right)$

23. $V_a + 6V_b - 2V_c$ **25.** $12m_1 - 8m_2$ **27.** $x - \frac{1}{3}w - \frac{3}{2}z$ **29.** $-3t_1 - t_2 - 5t_3$
31. $-12x^2y + 3xy^2 + 7x^2y^2$ **33.** $30mv - 16p$ **35.** $15x^2 - 4m^2 - 3n^2 + 10p^2$ **37.** $-1.86x - 6.25$
39. $6.3x + 12.2y$ **41.** $-52.91V + 20.77$ **43.** $(4y + 2a)$ m **45.** $43R$ ohms **47.** $(6x + 2)$ cm
49. $F_n = 5f_1 + 8f_2$ **51.** $0.00069T^2 + 0.0116T + 18.1$

Section 4.4 (page 141)

1. $2x - y$ **3.** $-a - b + c$ **5.** $2R - C$ **7.** $4x - 2a$ **9.** $2R + 4V$ **11.** $2x$ **13.** $-a$
15. $2x - 10z$ **17.** $-x - 8y + 8z$ **19.** $2m_1 + 6m_2 + 5m_3$ **21.** $-9x - 11w + 11y$ **23.** $6C_2$
25. $C = \frac{5}{9}F - \frac{160}{9}$ **27.** $E_b - E$ **29.** $\frac{1}{4}mv^2 + \frac{1}{6}$

Section 4.5 (page 143)

1. 2 * X + 4 * Y **3.** 2 * B↑2 **5.** −4 * C↑3 + 6 * D **7.** (2 * X − 4)/A
9. 2 * (X − 2 * Y)↑4 **11.** SQR(B − 2) **13.** (S − 4)↑3/SQR(T) **15.** (2 * V1 − 3 * V2)/V3
17. SQR(I2 − I1)/(2 − 3 * A) **19.** 2/SQR(V1 − 2 * V2) **21.** $3x^2 - 4$ **23.** $\frac{v_2 - v_1}{1 - a}$
25. $\frac{2x^2 - y}{a - 3b}$ **27.** $\frac{\sqrt{t_1 - t_2}}{s_1 + s_2}$ **29.** $\frac{2 + \sqrt{a}}{y}$

Review Exercises/Chapter 4 (page 143)

1. constants: 3, −4; variables: L, C, D **2.** constants: 3, −6; variables: S, t_1, t_2
3. constants: $\frac{1}{4}$, −2; variables: N, T_a, T_b **4.** constant: 4; variables: A, x, b **5.** −4.7 **6.** 6.46
7. −2.3 **8.** 28 **9.** monomial **10.** binomial **11.** trinomial **12.** trinomial
13. $6x^2 - 8y^2 - 6z^2$ **14.** $-10P_a - 4P_b + 4P$ **15.** $-12a^2 - b^2$ **16.** $7x^2y - 8xy^2 + 2xy$
17. $a^2b + 2ab^2 + a^2b^2$ **18.** $4X_1 - 14X_2 - 17X_3$ **19.** $13R^2 + 6R - 12$ **20.** $3r^2 - 6s^2$
21. $-6d + \sqrt{e} + f$ **22.** $-3xy + 3w - 7z$ **23.** $5a^2 - c^2 - 4d^2$ **24.** $5D_1 + D_2 + 4D_3$
25. $-2C_a - 14C_b + 18C$ **26.** $-15T_1 - T_2 + 2T_3$ **27.** $-3A + B$ **28.** $6a - 2b$
29. $-x + 2y - 7z$ **30.** $3x^2 + 4y^2 - 5$ **31.** $-2C$ **32.** $-5p + q$ **33.** 4.2 **34.** 5.7
35. 22.7 **36.** 42.0 **37.** $1.895x - 3.785$ **38.** $-3.68y + 61.83z$ **39.** $C = 0.0145t + 0.0048$
40. $20,000x **41.** total weight of shipment **42.** total volume **43.** $F = -m_Am_B + m_Cm_D$
44. $l = 12 - x$ **45.** $E = 1 - \frac{O}{I}$ **46.** $C = 10.50ab$ **47.** 18.3 Ω **48.** $2m_1v_1 + 2m_2v_2$
49. $3.21R - 4.37$ **50.** 74 W **51.** $W = mgy_2 - mgy_1$ **52.** $\frac{3}{2}mv^2 - 2.0$

Chapter 5
Section 5.1 (page 151)

1. x^6 **3.** a^{12} **5.** $-8a^5$ **7.** $12s^5$ **9.** $20V^7$ **11.** $-8x^3y^3$ **13.** $-20T_1^3T_2^3$ **15.** a^2b^2
17. $-a^3b^3$ **19.** $-m^3v^6$ **21.** $t_1^8t_2^{12}$ **23.** $4a^2b^4c^2$ **25.** $-25P^8Q^7$ **27.** $42L_1^3L_2^3L_3^4$
29. $-4a^7b^9$ **31.** $-s^8t^9$ **33.** $8v_1^9v_2^{10}$ **35.** $-128R^{13}C^{17}$ **37.** $-27.27V_1^4V_2^4$ **39.** $0.0016r^6s^7$
41. $7.18x^4y^6$ **43.** $-73.21R^{10}V^{15}$

Section 5.2 (page 155)

1. $-x^2 - xy$ **3.** $8x^2 - 6x$ **5.** $6P - 3Q$ **7.** $10x^3 - 4x^4 + 14x^5$ **9.** $6a^2b^3 - 8a^3b^4$
11. $35T_1^3T_2^2 - 21T_1^2T_2^3$ **13.** $6x$ **15.** $9a - 22$ **17.** $-5a$ **19.** $-2v - 10w$ **21.** $x^2 - 1$
23. $4x^2 - 9$ **25.** $2x^2 + 2x - 12$ **27.** $x^2 + xy - 20y^2$ **29.** $P^2 + PV - 6V^2$
31. $S^2 - 2ST + T^2$ **33.** $T_a^2 - 4T_aT_b + 4T_b^2$ **35.** $x^3 + 4x - 5$ **37.** $2s^3 - 5s^2 - 31s - 15$
39. $x^3 + xy^2 - 2y^3$ **41.** $2s^3 - 7s^2t + 5st^2 - t^3$ **43.** $6R^3 - R^2V - 4RV^2 - V^3$

45. $2x^4 - 6x^3 - 2x^2 + 7x - 3$ **47.** $7x^4 - 46x^3 - 21x^2 + 4x - 28$ **49.** $2x^4 + 2x^3 - 24x^2 - 14x - 2$
51. $6v^4 + v^3 + 3v - 2$ **53.** $2x^4 + 5x^3y - 6x^2y^2 - 7xy^3 + 6y^4$ **55.** $3.33x + 15.9$
57. $-75.2 + 93.1V$ **59.** $3.0x^2 - 9.7$ **61.** $70a + 115b$ **63.** $\frac{1}{2}mv_2^2 - \frac{1}{2}mv_1^2$
65. $y = 24kx - kx^2$ **67.** $A = x^2 - x - 6$ **69.** $K = \frac{1}{2}m_0v^2c^2$ **71.** $R = kab - kay - kby + ky^2$

Section 5.3 (page 160)

1. x^3 **3.** $2a^2$ **5.** $-2C^2D^3$ **7.** $\frac{2a^4}{3b^3}$ **9.** $3a^2$ **11.** $-\frac{2}{T_1^2T_2^2}$ **13.** $-\frac{5y^2}{3z}$ **15.** $\frac{C_1}{3C_2}$
17. b **19.** $-\frac{1}{2x}$ **21.** $-2RC$ **23.** $\frac{1}{3}x^2z^3$ **25.** $-\frac{64R^2}{V}$ **27.** $-4xz^5$ **29.** $\frac{L_3^4}{81L_2^4}$
31. $\frac{x^6}{y^3}$ **33.** $\frac{8x^6}{y^9}$ **35.** $\frac{4V^4}{R^6}$

Section 5.4 (page 168)

1. $1 - \frac{y}{x}$ **3.** $2y - 3$ **5.** $3 + 6s$ **7.** $-2y + 3x$ **9.** $5 - x$ **11.** $1 - b + 2a$
13. $3 + C_1 - 2C_2$ **15.** $2s - 3 + 4m_0s$ **17.** $1.54x + 3.54$ **19.** $2.02 - 0.798t$
21. $1.414V - 0.3502$ **23.** $x - 3$ **25.** $x - 4$ **27.** $3y - 4$ **29.** $s + 3$ **31.** $z + 3$
33. $5r - 2$ **35.** $3n - 5$ **37.** $5x + y$ **39.** $s + t$ **41.** $3c - 4d$ **43.** $x^2 - 4x - 1$
45. $2y^2 - 3y - 1$ **47.** $R + 4$ **49.** $2x - 3z$ **51.** $3S - 2T$ **53.** $R + V$
55. $3a^2 + 3ab + b^2$ **57.** $2s^2 - 3st - t^2$ **59.** $x^2 + 4xy + 8y^2$ **61.** $x^2 + xy + y^2$
63. $x^2 - y^2 + \frac{y^4}{x^2 - y^2}$ **65.** $2a^2 - 6b^2 + \frac{8b^4}{a^2 + 2b^2}$ **67.** $V^2 + W^2$ **69.** $u = 4k\left(\frac{1}{4} - \frac{1}{n^2}\right)$
71. $w = x^2 + 2x + 1$ **73.** $30\ \frac{\text{mi}}{\text{gal}}$; $(x - 3)\ \frac{\text{mi}}{\text{gal}}$

Review Exercises/Chapter 5 (page 170)

1. $32s^3t^4$ **2.** $-15M^7T^3$ **3.** $-8C_1^6C_2^9$ **4.** $16C_1^8C_2^{12}$ **5.** $-72v^{13}w^8$ **6.** $9.7a^{14}b^4$
7. $35a^4c - 25a^3c^2 + 30a^5c^3$ **8.** $6p^3q^2 - 12p^4q + 10p^4q^2$ **9.** $8x^2 - 10xy - 3y^2$
10. $6a^2 + 11ab + 3b^2$ **11.** $L^2 - 4v^2$ **12.** $9m^2 - 4n^2$ **13.** $6s^3 + s^2t - 4st^2 + t^3$
14. $6R^3 - 22R^2V + 15RV^2 - 2V^3$ **15.** $4C_1^2 + 12C_1C_2 + 9C_2^2$ **16.** $T_1^2 - 12T_1T_2 + 36T_2^2$
17. $7.41x^2 - 19.0xy + 9.84y^2$ **18.** $6.34x^2 - 2.59xy - 9.95y^2$
19. $4R^4 + 6R^3C - 4R^2C^2 - 4RC^3 + 2C^4$ **20.** $6M^4 - M^3N - 3MN^3 - 2N^4$ **21.** $-2x^2y^5$
22. $-3a^7b^2$ **23.** $\frac{3R^2}{NT^4}$ **24.** $\frac{3T_b}{2T_a}$ **25.** $-2M$ **26.** $-\frac{16}{27z^4}$ **27.** $-2abc$ **28.** $-\frac{1}{cde}$
29. $-\frac{8x^3}{y^9}$ **30.** $\frac{16x^{12}}{y^{16}}$ **31.** $4.34c - 4.75$ **32.** $1.957p + 3.754q$ **33.** $2 - 3F$ **34.** $-1 - 2D$
35. $-a + 1 - 2b$ **36.** $-2m^2v - 1 - 4mv^2$ **37.** $x + 2y$ **38.** $x + 5y$ **39.** $4x^2 - 3xy + y^2$
40. $3R^2 - 2RC + C^2$ **41.** $3a^2 + 4ab - 2b^2$ **42.** $2a^2 - ab + 2b^2$ **43.** $2m - 3n$
44. $2M - N$ **45.** $2x - 2 + \frac{1}{2x + 1}$ **46.** $2R^2 - 2R + 1 - \frac{1}{R - 1}$ **47.** $W = mgy_2 - mgy_1$
48. $T = 100ax - ax^3$ **49.** $v = 4lx - 4x^2$ **50.** $m_1t_2 + m_2ct_2 - m_1t_1 - m_2ct_1$
51. $\frac{PV}{R} + \frac{a}{RV} - \frac{bP}{R} - \frac{ab}{RV^2}$ **52.** $L = 0.28t + 0.03$ **53.** $P = 57.6t^2 - 76.8t + 25.6$
54. $1 - 2mv^2$ **55.** $\frac{1}{R_1} + \frac{1}{R_2} + \frac{1}{R_3} + \frac{1}{R_4}$ **56.** $x^2 + 2x + 1$ (pounds) **57.** $80n + 110$

Chapter 6
Section 6.1 (page 179)

1. 15′ **3.** 24′ **5.** 38′ **7.** 2°11′ **9.** 12°41′ **11.** 48°9′ **13.** 60°54′ **15.** 41°28′ **17.** 0.23° **19.** 0.80° **21.** 0.67° **23.** 10.90° **25.** 70.53° **27.** 64.20° **29.** 21.48° **31.** 19.47° **33.** 130° **35.** 122.1° **37.** 149°10′ **39.** 128°24′ **41.** 137°53′ **43.** 122°42′ **45.** $\angle BCA$, $\angle DCA$ **47.** $\angle ABC$, $\angle DBE$ **49.** ED and AC **51.** isosceles **53.** right **55.** scalene **57.** right **59.** isosceles

Section 6.2 (page 184)

1. $AD \parallel BC$; $AB \parallel DC$ **3.** $DC = 8$ cm; $BC = 5$ cm **5.** 12 in. **7.** $AD = 3$ ft; $DC = 8$ ft **9.** $DC \parallel AB$ **11. a.** $BCEF$ **b.** $ACDF$ **c.** $ACEF$, $BCDF$ **13.** 20 ft **15.** 11.2 cm **17.** 1.18 ft **19.** parallelogram **21.** 239,000 mi

Section 6.3 (page 189)

1. 15.0 cm **3.** 22.0 ft **5.** 48.0 m **7.** 31.1 cm **9.** 1.92 cm **11.** 18.4 cm **13.** 33.2 in. **15.** 8.0 ft **17.** 13 in. **19.** 20.1 ft **21.** 17 cm **23.** 19.31 in. **25.** 117.3 m **27.** 25 cm **29.** 11.3 cm; 16.3 cm **31.** 14.0 mm **33.** 165,000 mi **35.** 7920 mi **37.** 461 m

Section 6.4 (page 199)

1. 100 cm^2 **3.** 882 ft^2 **5.** 33.4 $in.^2$ **7.** 16.47 cm^2 **9.** 19.7 ft^2 **11.** 909 cm^2 **13.** 220 ft^2 **15.** 14.5 m^2 **17.** 131.5 cm^3 **19.** 21.0 $in.^3$ **21.** 34 cm^2 **23.** 50.8 m^2 **25.** 358 $in.^2$ **27.** 37.6 cm^2 **29.** 31 ft^2 **31.** 1260 yd^2 **33.** 149 ft^2 **35.** 1804 cm^2 **37.** 25 $in.^2$ **39.** 595 cm^3 **41.** $30 **43.** 38.0 $in.^2$ **45.** 127,000 lb **47.** 24.3 cm^2

Review Exercises/Chapter 6 (page 204)

1. 21′ **2.** 41′ **3.** 2°2′ **4.** 7°22′ **5.** 12°47′ **6.** 10°29′ **7.** 0.17° **8.** 0.47° **9.** 0.73° **10.** 0.53° **11.** 60.80° **12.** 45.92° **13.** 48°57′ **14.** 88°50′ **15.** 145°50′ **16.** 59°10′ **17.** 37.6 ft **18.** 26.4 cm **19.** 81.6 in. **20.** 84.3 cm **21.** 31 cm^2 **22.** 39.2 cm^2 **23.** 8.84 ft^2 **24.** 645 $in.^2$ **25.** 44.4 cm^2 **26.** 84.9 $in.^2$ **27.** 51.3 cm, 136 cm^2 **28.** 15.6 in., 13.3 $in.^2$ **29.** 44.0 m, 90.7 m^2 **30.** 15 cm, 17 cm^2 **31.** 30.2 in., 72.7 $in.^2$ **32.** 85.0 cm, 458 cm^2 **33.** 82.7 m, 394 m^2 **34.** 18.7 in., 9.80 $in.^2$ **35.** 36.1 cm, 30.7 cm^2 **36.** 14.5 mm^2 **37.** 6.69 $in.^2$ **38.** 10.93 ft **39.** 8100 ft^2 **40.** 0.25 $in.^3$ **41.** 2500 $in.^3$ **42.** 40 in. **43.** 8.5 lb **44.** 2.2 kg **45.** 20 ft **46.** 24,900 mi **47.** 3480 km **48.** 26.9 cm^2 **49.** 84¢ **50.** 16.1 ft

Cumulative Review Exercises/Chapters 4–6 (page 208)

1. $(wx^2 - 2wx)$ lb **2. a.** trinomial **b.** binomial **3.** $-5a\sqrt{b}$ **4.** $22.3R - 41.9$ **5.** $-3C_1 + 2C_2$ **6.** $-4a - 2b$ **7.** $-50L^5M^5$ **8.** $-128a^7b^8$ **9.** $-24.1a^2 + 16.9a^3$ **10.** $-3R - 26C$ **11.** $2x^4 + 8x^3 + 3x^2 + 10x - 8$ **12.** $6x^4 + 5x^3 - x^2 + 14x - 4$ **13.** $\frac{n^6}{m^2}$ **14.** $-2PV + 3P^2V^3 + 1$ **15.** $2w - 4z$ **16.** $5x^2 - 3x + 4$ **17.** 42.63° **18.** 17.8 cm^2 **19.** 124.7 $in.^2$ **20.** 13 $in.^2$ **21.** 68.6 m **22.** $12.3C + 2.6$ **23.** $4x^2 - 5xy - 6y^2$ **24.** $\$(x - 2)$ **25.** 21,900 lb

Chapter 7
Section 7.1 (page 214)

1. -4 **3.** 15 **5.** 2 **7.** -3 **9.** $\frac{2}{5}$ **11.** 5 **13.** 5 **15.** $\frac{3}{2}$ **17.** $\frac{8}{7}$ **19.** -2 **21.** 2 **23.** -6 **25.** $-\frac{2}{5}$ **27.** $\frac{7}{4}$ **29.** 0 **31.** $\frac{2}{5}$ **33.** -3.02 **35.** 2.12 **37.** 1.491 **39.** -0.90

Section 7.2 (page 220)

1. $x = 10$ **3.** $x = \frac{3}{2}$ **5.** $x = 6$ **7.** $x = -9$ **9.** $x = -1$ **11.** $x = 4$ **13.** $y = -1$
15. $z = 1$ **17.** $w = -\frac{1}{12}$ **19.** $m = \frac{1}{4}$ **21.** $x = -10$ **23.** $x = -\frac{1}{2}$ **25.** $y = -\frac{3}{4}$
27. $t = 2$ **29.** $T = -1$ **31.** $m = \frac{3}{2}$ **33.** $x = \frac{2}{3}$ **35.** $z = -\frac{1}{8}$ **37.** $w = \frac{4}{3}$
39. $n = \frac{20}{13}$ **41.** $\frac{1}{b}$ **43.** $-\frac{3}{a}$ **45.** $\frac{b}{a}$ **47.** $2 - b$ **49.** $2a + b$ **51.** $\frac{1-b}{2}$ **53.** $\frac{3}{a}$
55. $\frac{d-6}{c}$ **57.** $\frac{a+m}{b}$ **59.** $\frac{t+2}{a}$ **61.** $\frac{c+3}{b}$ **63.** $-\frac{3}{a}$ **65.** $\frac{2c}{a}$ **67.** $\frac{11}{a}$ **69.** $\frac{10}{7c}$
71. $\frac{3b}{7a}$ **73.** $\frac{6}{b}$ **75.** $-\frac{1}{3a}$ **77.** $\frac{1}{c}$ **79.** $\frac{5}{3a}$

Section 7.3 (page 225)

1. $a = \frac{b}{3}$ **3.** $p = \frac{s}{q}$ **5.** $R_2 = \frac{R_1}{3T}$ **7.** $D = \frac{Q+R}{A}$ **9.** $m_1 = \frac{2m_3 + 8}{m_2}$
11. $x = \frac{1-2y}{2y}$ **13.** $b = \frac{c-2a}{2}$ **15.** $a = \frac{2bP+3}{P}$ **17.** $a = \frac{2b+3Q}{3}$ **19.** $v = \frac{aA+4}{3}$
21. $P_2 = \frac{aP_0 - 2P_1}{4}$ **23.** $s = \frac{Mt + 4v - 12}{M}$ **25.** $P_1 = \frac{P_0P_2 + 2P_2}{2}$ **27.** $P = \frac{aV - ab}{r}$
29. $T_0 = \frac{bT_2 - bT_1}{a}$ **31.** $V = \frac{k}{P}$ **33.** $t = \frac{A-P}{Pr}$ **35.** $K = \frac{E_dI - H^2}{H^2}$ **37.** $C = \frac{5}{9}(F - 32)$
39. $V = \frac{P + RI^2}{I}$ **41.** $t = \frac{v - v_0}{g}$ **43.** $r_1^2 = \frac{\pi r_2^2 - LT}{\pi}$ **45.** $v^2 = \frac{rT + mgr}{m}$
47. $k_2 = AC - k_1A^2$ **49.** $T = \frac{V - V_0 + bT_0V_0}{bV_0}$ **51.** $p_1 = \frac{Pn_1 + Pn_2 - n_2p_2}{n_1}$

Section 7.4 (page 235)

1. 12, 14 **3.** 17, 34 **5.** 9, 29 **7.** \$15, \$23 **9.** \$2.39, \$4.78
11. 0.2316 F, 0.5872 F **13.** 10 yr **15.** 15 cm, 12 cm **17.** 2.1 mm, 1.4 mm
19. 11.6 cm, 5.8 cm **21.** 31.3 Ω, 45.5 Ω **23.** 10.2 Ω, 14.7 Ω, 16.4 Ω **25.** 40 lb
27. 12 gal of 20% solution; 8 gal of 15% solution **29.** 38 mL **31.** 10 mL **33.** 1.5 h
35. 2 mi/h **37.** 2 km/h **39.** 5:30 P.M. **41.** 6 km/h **43.** 36 h **45.** 2 h **47.** 2.5 h

Section 7.5 (page 242)

1. $0 < 1$ **3.** $-6 < 30$ **5.** $-1 > -5$ **7.** $x < 8$ **9.** $x < -2$ **11.** $x < 2$ **13.** $x < \frac{5}{9}$
15. $x < \frac{4}{3}$ **17.** $x \le 0$ **19.** $x \ge \frac{1}{2}$ **21.** $x < \frac{11}{7}$ **23.** $x \le \frac{8}{9}$ **25.** $x \ge 2$ **27.** $t \ge \frac{3}{2}$ s
29. $T \ge 48$°C

Section 7.6 (page 243)

1. Y = X + 4 * X ↑ 2 **3.** P = K/V **5.** E = (K + 1) * H ↑ 2/I **7.** P = V * I − R * I ↑ 2
9. V = K * R ↑ 2 * (A − R) **11.** 1/R = 1/R1 + 1/R2 **13.** Q = K * A * (T1 − T2)/L
15. X − 2 <= 2 * X + 3 **17.** Z − 2 < 4 − Z **19.** 1 − 3 * W < 3 **21.** 2 − 3 * W >= 7

Review Exercises/Chapter 7 (page 244)

1. $x = 2$ **2.** $x = 6$ **3.** $x = \frac{9}{5}$ **4.** $x = \frac{10}{7}$ **5.** $x = -\frac{3}{2}$ **6.** $x = -\frac{1}{2}$ **7.** $y = \frac{1}{8}$
8. $z = \frac{27}{16}$ **9.** $w = -\frac{1}{4}$ **10.** $v = -\frac{1}{2}$ **11.** $u = 1$ **12.** $y = -\frac{1}{3}$ **13.** $s = \frac{5}{2}$ **14.** $t = \frac{1}{2}$
15. $1 - 2b$ **16.** $3a + 2$ **17.** $-\frac{d}{c}$ **18.** $\frac{n+2}{b}$ **19.** $\frac{b+1}{c}$ **20.** $\frac{b+c}{n}$ **21.** $\frac{b}{2}$
22. $-a$ **23.** $\frac{24}{5a}$ **24.** $-\frac{1}{6a}$ **25.** $T_2 = \frac{T_1}{2N}$ **26.** $t = \frac{w+4}{s}$ **27.** $s_1 = \frac{2s_3 + 2}{s_2}$
28. $A = \frac{2B - 2}{B}$ **29.** $a = \frac{cP + 4b}{2}$ **30.** $n = \frac{dQ - 4m}{3}$ **31.** $P_1 = \frac{bP_0 - aP_2}{a}$
32. $t = s - abL$ **33.** $x > -\frac{1}{2}$ **34.** $x < -\frac{1}{4}$ **35.** $x < \frac{5}{3}$ **36.** $x > \frac{3}{2}$ **37.** $x \le \frac{11}{10}$
38. $x \le 5$ **39.** $x \ge \frac{4}{3}$ **40.** $x < 1$ **41.** $x > 2$ **42.** $x < \frac{22}{19}$ **43.** $m_0 = \frac{c^2m - T}{c^2}$
44. $m_0 = \frac{E_m - T_m}{c^2}$ **45.** $v_0 = \frac{2s - 2s_0 - gt^2}{2t}$ **46.** $t_2 = \frac{HL + KAt_1}{KA}$ **47.** $p = \frac{fq}{q - f}$
48. $m_1 = \frac{m_2T - m_2T'}{T'}$ **49.** $T = \frac{R - 2.5}{0.0000082} = 120{,}000(R - 2.5)$ **50.** $h = \frac{A - 2\pi r^2}{2\pi r}$
51. \$5.26; \$13.15 **52.** 26.50 Ω; 34.45 Ω **53.** 25 lb **54.** 17.5 mL of each **55.** 4:40 P.M.
56. 12 km/h **57.** 18 h **58.** 1:18 P.M. **59.** $x \ge 21$ ft

Chapter 8
Section 8.1 (page 249)

1. $3x - 12y$ **3.** $4ax + 3ay$ **5.** $-3x^3 - x^2$ **7.** $A^2B^2 - AB$ **9.** $2T_1^2T_2 - 4T_1^3T_2$
11. $-4R^3S^3 - 8R^4S^3$ **13.** $-21s^5t^3 + 28s^2t^2$ **15.** $3(x - 2y)$ **17.** $3(3a - 5b)$ **19.** $2xy(x + 2y)$
21. $2ax^2(1 - 2x)$ **23.** $5p^3q(3q^3 + 5)$ **25.** $2x^2(x^2 - 2x + 5)$ **27.** $ay^2(y^2 - 2y - 1)$
29. $AB(AB + 1)$ **31.** $T_aT_b^2(T_a - 1)$ **33.** $3x(1 + 2x^3 - 4x^4)$ **35.** $7L^2V(3L^2 - 2V^2 + 1)$
37. $3T_1T_2^2(1 - 3T_1 - 6T_1^2)$ **39.** $11vw(1 - v + 2w)$ **41.** $M = m(v_2 - v_1)$ **43.** $l = l_0(1 + aT)$
45. $W = mg(y_2 - y_1)$ **47.** $E = I(R_1 + R_2 + R_3)$ **49.** $ax(1 - 2x + 3x^2)$

Section 8.2 (page 254)

1. $x^2 - 9$ **3.** $4x^2 - b^2$ **5.** $25V^2 - 1$ **7.** $\frac{1}{4}W^2 - 4$ **9.** $1 - a^2y^2$ **11.** $4L^2V^2 - 16$
13. $a^2w^2 - z^2$ **15.** $a^2V_0^2 - 4b^2$ **17.** $(x - 4)(x + 4)$ or $(x + 4)(x - 4)$
19. $(3y - 2)(3y + 2)$ or $(3y + 2)(3y - 2)$ **21.** $(2W - 1)(2W + 1)$ **23.** $(2 - 3V_0)(2 + 3V_0)$
25. $(z - 2b)(z + 2b)$ **27.** $\left(\frac{1}{2}x - y\right)\left(\frac{1}{2}x + y\right)$ **29.** $\left(\frac{1}{2}L - \frac{1}{3}M\right)\left(\frac{1}{2}L + \frac{1}{3}M\right)$
31. $\left(\frac{x}{a} - b\right)\left(\frac{x}{a} + b\right)$ **33.** $2(2V_0 - V_1)(2V_0 + V_1)$ **35.** $a(z - 2w)(z + 2w)$
37. $V_a(V_b - V_c)(V_b + V_c)$ **39.** $(s - t)(s + t)(s^2 + t^2)$ **41.** $k(V_1 - V_2)(V_1 + V_2)(V_1^2 + V_2^2)$
43. $a(2R - C)(2R + C)(4R^2 + C^2)$ **45.** $k(v_2 - v_1)(v_2 + v_1)$ **47.** $V = \pi l(r_2 - r_1)(r_2 + r_1)$
49. $\frac{1}{2}m(v_2 - v_1)(v_2 + v_1) = \frac{1}{2}M(\omega_2 - \omega_1)(\omega_2 + \omega_1)$ **51.** $v^2 = \frac{k}{m}(A - x)(A + x)$
53. $k\left(\frac{1}{d_1} - \frac{1}{d_2}\right)\left(\frac{1}{d_1} + \frac{1}{d_2}\right)$

Section 8.3 (page 258)

1. $x^2 + 2x - 8$ **3.** $2x^2 - 7x + 3$ **5.** $2y^2 - 5y - 7$ **7.** $6x^2 - 10xy - 4y^2$
9. $12x^2 + 10xy + 2y^2$ **11.** $12V^2 - 2VW - 4W^2$ **13.** $30R_1^2 - 16R_1R_2 + 2R_2^2$

15. $35w^2 + 11wz - 6z^2$ **17.** $15P^2 + 7PQ - 2Q^2$ **19.** $15r^2 + 23rs + 4s^2$ **21.** $4x^2 - 20x + 25$
23. $4V^2 - 16V + 16$ **25.** $25R_1^2 + 20R_1R_2 + 4R_2^2$ **27.** $16v_1^2 - 24v_1v_2 + 9v_2^2$ **29.** $1 - 12L + 36L^2$

Section 8.4 (page 263)

1. $(x - 1)(x + 2)$ **3.** $(y + 2)(y - 3)$ **5.** $(L + 4)(L - 1)$ **7.** $(x - 4y)(x + y)$
9. $(x + y)(x + 6y)$ **11.** $(x - 4y)(x - 6y)$ **13.** $(p - 5q)(p + 6q)$ **15.** $(2x - 3y)(x - 4y)$
17. $(2V - b)(V + 6b)$ **19.** $(2v_1 - v_2)(2v_1 + 3v_2)$ **21.** $(4s_1 - s_2)(s_1 - s_2)$ **23.** $(5T - n)(T + 2n)$
25. $(2x - w)(3x + 2w)$ **27.** $3(2k - a)(k + 6a)$ **29.** not factorable **31.** $(x + 3y)^2$
33. $(2m - n)^2$ **35.** $2(x - 4y)(x - 2y)$ **37.** $3(a - 3b)(4a - b)$ **39.** $a(4x - 3y)(x + y)$
41. $4(3x - y)(2x - y)$ **43.** $2(5x - 2y)(x + 3y)$ **45.** $2(y - 2z)(y + 8z)$ **47.** $3(w + 2z)(w + 4z)$
49. $2(a - 3b)^2$ **51.** not factorable **53.** $a(2V - 3)^2$ **55.** $(3T_a + 2T_b)^2$
57. $d = (ax - 2L)(2ax - 3L)$ **59.** $s = 8(2t + 1)(t - 4)$ **61.** $(R - 12)(R - 6) = 0$

Review Exercises/Chapter 8 (page 264)

1. $as - 2at$ **2.** $2bx - 3by$ **3.** $-10a^3b^3 + 4a^3b^2 - 6a^5b^3$ **4.** $3x^2y^4 + 2x^3y^6 - x^4y^5$
5. $2V_a^3V_b^2 - 4V_a^2V_b^2 - V_aV_b$ **6.** $T_1^4T_2^2 - T_1^2T_2^3 + T_1T_2$ **7.** $y^2 - 25$ **8.** $P^2Q^2 - 1$ **9.** $\frac{1}{4}x^2 - 1$
10. $\frac{1}{16}a^2 - \frac{1}{4}b^2$ **11.** $16S_0^2 - T_0^2$ **12.** $9t_1^2 - 4t_2^2$ **13.** $x^2 + x - 20$ **14.** $y^2 + 3y - 18$
15. $2z^2 - 7zw + 3w^2$ **16.** $6t^2 + 7t + 2$ **17.** $8R_1^2 - 2R_1R_2 - R_2^2$ **18.** $3T_a^2 + 11T_aT_b - 4T_b^2$
19. $16S^2 - 8ST + T^2$ **20.** $V^2 - 8VR + 16R^2$ **21.** $a(b^2 - 2)$ **22.** $d(c - d)$
23. $3PQ(PQ - 2)$ **24.** $4ST(1 - 2ST)$ **25.** $xw(3xw + 1)$ **26.** $np(4np - 1)$
27. $7LM(5L - 4M - 2LM)$ **28.** $4S_aS_b(7S_a^2S_b^2 - 4S_aS_b - 3)$ **29.** $13RV(1 + 2RV^2 - 3R^2V)$
30. $14st(t^2 - 2s^2 - 3st)$ **31.** $(P - 4)(P + 4)$ **32.** $(3V - 2)(3V + 2)$ **33.** $(1 - 4V_0)(1 + 4V_0)$
34. $(1 - 5t_0)(1 + 5t_0)$ **35.** $\left(\frac{1}{2}s - 1\right)\left(\frac{1}{2}s + 1\right)$ **36.** $\left(\frac{1}{4}p - 1\right)\left(\frac{1}{4}p + 1\right)$ **37.** $a(x - y)(x + y)$
38. $2(V_a - V_b)(V_a + V_b)$ **39.** $3(S - T)(S + T)(S^2 + T^2)$ **40.** $c(N_1 - N_2)(N_1 + N_2)(N_1^2 + N_2^2)$
41. $(y - 1)(2y + 1)$ **42.** $(w - 2)(w + 5)$ **43.** $(2P - 1)(2P - 3)$ **44.** $(4L - P)(L - P)$
45. $(2s + t)(s - 6t)$ **46.** $(5x + y)(x - 2y)$ **47.** $2(a - 2b)(a - b)$ **48.** $2(2m - 3n)(m - 2n)$
49. $3(5y - 2w)(y + 3w)$ **50.** $3(s_1 - 2s_2)(s_1 + 8s_2)$ **51.** $2(3P - V)^2$ **52.** $3(1 - 2L)^2$
53. $4(3a - b)^2$ **54.** not factorable **55.** $E = c^2(m - m_0)$ **56.** $V = V_0(1 + bt)$
57. $T_{\max} = k(\nu - \nu_0)$ **58.** $E_d = k(T_1 - T_2)(T_1 + T_2)(T_1^2 + T_2^2)$ **59.** $P_d = R(i_2 - i_1)(i_2 + i_1)$
60. $R = 3.00(3.00t + 4.00)(2.00t - 5.00)$ **61.** $P = 3(m + 1)(m - 8)$ **62.** $T = K\left(1 - \frac{2a}{h}\right)\left(1 + \frac{2a}{h}\right)$

Chapter 9
Section 9.1 (page 272)

1. $\frac{2}{3x^2}$ **3.** $\frac{8}{7w}$ **5.** $-\frac{y^3}{4}$ **7.** $\frac{2a^2}{LW}$ **9.** $\frac{1}{x}$ **11.** s^2 **13.** $-\frac{1}{2}$ **15.** $\frac{x - 4}{y}$ **17.** $x - y$
19. $-V - W$ **21.** $-2T - T_0$ **23.** $v_1 + v_2$ **25.** $\frac{m - 2n}{m + n}$ **27.** $-\frac{f + g}{2f + g}$ **29.** $\frac{2T_1 - T_2}{3T_1 + 4T_2}$
31. $\frac{s_1 + s_2}{2s_1 + s_2}$ **33.** $\frac{1}{4x + 1}$ **35.** $\frac{1}{2a + 1}$ **37.** $\frac{1}{3x + 1}$ **39.** $\frac{m + M}{m}$
41. $i = 2 + 5t$ (amperes) **43.** $w = 2a + 5b$

Section 9.2 (page 277)

1. $\frac{a^2}{x}$ **3.** $\frac{2}{m}$ **5.** $\frac{a}{z^2}$ **7.** $\frac{x}{2}$ **9.** $\frac{2}{3b}$ **11.** $\frac{1}{2(x + w)}$ **13.** $\frac{P + Q}{(P - Q)(x - 2)}$
15. $\frac{a + 3b}{(a + 2b)(c - 4)}$ **17.** $\frac{(v + 2s)(2p - 1)}{2v + s}$ **19.** $-\frac{(1 + 2n)(3R - 2C)}{2R + C}$ **21.** $-\frac{2x + y}{x + 2y}$

23. $\frac{3}{2(1-s)}$ 25. $-\frac{2}{3p+7}$ 27. $\frac{R_1R_2}{R_1+R_2}$ 29. $F = \frac{2am_1m_2}{m_1-m_2}$ 31. $\frac{b^2}{c}$ 33. $i = \frac{t+3}{2t+1}$
35. $\frac{M}{M+m}$

Section 9.3 (page 285)

1. $8x^2y^2$ 3. $30R^3C^3$ 5. $y(x-y)$ 7. $(2P-Q)(2P+Q)$ 9. $(x-1)(x+2)$
11. $2(r-s)(2r+s)$ 13. $(x-y)(x+y)(2x+3y)$ 15. $(x+2y)(x-y)(2x+3y)$ 17. $\frac{2}{x}$
19. $\frac{x-2}{2x}$ 21. $\frac{a+1}{a^2}$ 23. $\frac{5}{4V}$ 25. $\frac{x+1}{2x^2}$ 27. $\frac{6y-1}{4xy^2}$ 29. $\frac{16C^2+15R}{36R^3C^4}$
31. $\frac{c^2+2ab-3a^2c}{a^3b^2c^3}$ 33. $-\frac{6}{(x+4)(x-2)}$ 35. $\frac{6(L-5)}{(L-4)(L-7)}$ 37. $\frac{1}{a}$ 39. $-\frac{1}{w}$
41. $\frac{2}{s+2t}$ 43. -1 45. 1 47. $\frac{-x+8}{(x-3)(x-4)}$ 49. $\frac{3x-9y+4}{2(x+y)(x-3y)}$ 51. $\frac{43}{48}$ in.
53. $E = \frac{T_1-T_2}{T_1}$ 55. $E = K\frac{a^2W^2+b^2L^2}{L^2W^2}$ 57. $R = \frac{R_1R_2+R_1R_3+R_2R_3}{R_2+R_3}$
59. $I = \frac{a(L-x)^2+bx^2}{x^2(L-x)^2}$

Section 9.4 (page 293)

1. $x = 2$ 3. $x = -3$ 5. $x = 1$ 7. $x = -12$ 9. $x = -1$ 11. no solution
13. $x = -2$ 15. $x = -8$ 17. $x = \frac{1}{2}$ 19. $x = -3$ 21. $x = \frac{1}{3}$ 23. $x = \frac{1}{2}$
25. $a = \frac{b}{2b-1}$ 27. $R = \frac{CL}{2(L-C)}$ 29. $N = \frac{R-Q}{Q}$ 31. $z = \frac{a^2-b^2}{b}$ 33. $v_1 = \frac{v_2^2}{1-v_2}$
35. $m_1 = -\frac{m_2(s+1)}{s-1} = \frac{m_2(1+s)}{1-s}$ 37. $a = \frac{b-bc-1}{1-c}$ 39. $T_1 = \frac{P_1V_1T_2}{P_2V_2}$
41. $p_1 = \frac{aE-E+p_2V_2}{V_1}$ 43. $f = \frac{f_1f_2}{f_1+f_2}$ 45. $C = \frac{C_1(C_2+C_3)}{C_1+C_2+C_3}$ 47. $I_M = -\frac{I_m(P+1)}{P-1}$
49. $N = \frac{2D}{D_0-D}$

Review Exercises/Chapter 9 (page 295)

1. $-\frac{W^2}{3V^2}$ 2. $\frac{C_2^2}{bC_1^3}$ 3. P_0 4. $v_1 - v_2$ 5. $\frac{1}{4m_1+m_2}$ 6. $\frac{2s+t}{2}$ 7. $\frac{p+q}{3p-5q}$
8. $\frac{2w+y}{3w-y}$ 9. $\frac{1}{b^2y^4}$ 10. $\frac{12}{C^2}$ 11. $\frac{2av^3}{9b^4w}$ 12. $\frac{10m^2nP^2}{3Q^2}$ 13. $\frac{2}{a^2b}$ 14. $\frac{3(2V-L)}{M^2}$
15. $\frac{4}{a(s+3t)}$ 16. $\frac{c(m-4n)}{9}$ 17. $-\frac{(bV+1)(c+d)}{c+4d}$ 18. $-\frac{(1+dv)(s_1-s_2)}{s_1+2s_2}$
19. $\frac{a-b}{(x-2y)(2a+b)}$ 20. $\frac{x-1}{8}$ 21. $-\frac{1}{4V}$ 22. $\frac{2y-x}{x^2y^2}$ 23. $\frac{15ac+2}{18a^2bc^2}$ 24. $\frac{2R_2^2+4R_1}{R_1^2R_2^3}$
25. $\frac{-x+11}{(x-4)(x+3)}$ 26. $\frac{2V-11}{(V-6)(V-5)}$ 27. $\frac{1}{2(s_1-s_2)}$ 28. $\frac{8}{3(V_1+V_2)}$ 29. $\frac{6}{s+3}$
30. $-\frac{8}{v+4}$ 31. $\frac{1}{V_a-2V_b}$ 32. $\frac{3s+4t}{2(s-t)(s+2t)}$ 33. $\frac{5y-4z}{(y-2z)(y+3z)(y+z)}$
34. $\frac{9R+5S}{(R+S)(R-3S)(2R+S)}$ 35. $x = 2$ 36. $x = \frac{1}{2}$ 37. $x = 3$ 38. $x = 0$ 39. $x = -2$

40. $x = -5$ **41.** no solution **42.** $x = 3$ **43.** $P = \frac{12V}{4 - V}$ **44.** $N = \frac{P(S^2 - T^2)}{2S}$

45. $P_1 = \frac{P_0 P_2}{P_0 - 1}$ **46.** $l_2 = \frac{l_1(L + 1)}{L - 1}$ **47.** $\frac{m_2}{m_1 + m_2}$ **48.** $\frac{2}{N^2}$ (dollars) **49.** $\frac{m_1 - m_2}{ab}$

50. $D = k\frac{4x^3 - 3x^2}{48}$ **51.** $\frac{C_1C_2 + C_1C_3 + C_2C_3}{C_2 + C_3}$ **52.** $\frac{m_0 c(\lambda' - \lambda)}{h\lambda\lambda'}$ **53.** $k = \frac{k_1 k_2}{k_1 + k_2}$

54. $C = \frac{C_1C_2}{C_1 + C_2}$ **55.** $q = \frac{pf}{p - f}$ **56.** $\frac{264{,}800\, v}{(331)^2 - v^2}$ **57.** $P = \frac{KV^2 - aV + ab}{V^2(V - b)}$

Cumulative Review Exercises/Chapters 7–9 (page 299)

1. $y = \frac{13}{11}$ **2.** $x = \frac{3b - 1}{4a}$ **3.** $x \le -\frac{1}{2}$ **4.** $17S_aS_b(S_a^3S_b - 3S_aS_b^2 + 1)$ **5.** $3(1 - T_0)(1 + T_0)$

6. $2(a - 2)(a - 7)$ **7.** $2(t_1 + 6t_2)(3t_1 - 4t_2)$ **8.** $-\frac{2(P + V)}{3P + 8V}$ **9.** $-2a(a + b)$

10. $\frac{(3x - 4)(x + 6)}{(2x - 3)(2x + 1)}$ **11.** $\frac{w^2 + w - 1}{w(w - 1)}$ **12.** $\frac{2A + 1}{A + 3}$ **13.** $m_0 = \frac{mc^2 - E}{c^2}$ **14.** $b_1 = \frac{2A - hb_2}{h}$

15. $P = 11$ **16.** $V_1 = \frac{3V_2}{a - 1}$ **17.** $V = I(R_1 + R_2 + R_3)$ **18.** $A = \pi(r_2 - r_1)(r_2 + r_1)$

19. $a(T_0 - T_1)(T_0 + T_1)(T_0^2 + T_1^2)$ **20.** $r_1 = \frac{rr_2}{r_2 - r}$ **21.** $i = 2(t + 2)$A **22.** 40.5 cm × 16.2 cm
23. 60 L of 10% solution, 40 L of 15% solution

Chapter 10
Section 10.1 (page 308)

1. $\frac{1}{9}$ **3.** $\frac{1}{7}$ **5.** $\frac{1}{16}$ **7.** 4 **9.** 36 **11.** 25 **13.** $\frac{1}{9}$ **15.** 12 **17.** $3x^2$ **19.** $\frac{3}{2V}$

21. $16x^4$ **23.** $-a^4b^2$ **25.** x^4 **27.** $\frac{b^5}{a^3}$ **29.** $\frac{1}{m^5n}$ **31.** $\frac{1}{2d^2e^3}$ **33.** $\frac{V^8}{V_0^4}$ **35.** L^8M^8

37. $\frac{b^4}{c^6}$ **39.** $\frac{4}{y^4z^6}$ **41.** $\frac{y^5}{6x^5}$ **43.** $\frac{2s_2^4}{s_1^2}$ **45.** $\frac{9A}{B^5}$ **47.** $\frac{x^2z^2}{2y^5}$ **49.** $\frac{16v_2^8v_3^7}{v_1}$ **51.** $-\frac{e^5}{d^5}$

53. $9s^2t^2$ **55.** $\frac{T^5}{18V^3}$ **57.** $\frac{n}{n - 1}$ **59.** $\frac{1 - 2b^4}{b^5}$ **61.** $\frac{a + b}{b}$ **63.** $\frac{P^2 - 1}{P^2 + 1}$ **65.** $\frac{2R + 2}{3R + 2}$

67. $\frac{1 + a^4}{a^4}$ **69.** $P = \frac{P_0}{2^{t/5580}}$ **71.** $V = \frac{k}{P}$ **73.** $R = \frac{R_1R_2}{R_1 + R_2}$ **75.** 4 cm

Section 10.2 (page 315)

1. 3.2×10^5 **3.** 5.70×10^{-5} **5.** 9.7612×10^8 **7.** 1.00×10^{-6} **9.** 3×10^6 **11.** 32,000
13. 0.000000319 **15.** 176,000,000 **17.** 0.000637 **19.** 0.0000502 **21.** 2.11×10^{10}
23. 3.27×10^{11} **25.** 1.60×10^{-14} **27.** 5.000×10^{10} **29.** 7.1×10^{-6} **31.** 5.9×10^{-8}
33. 1.5×10^7 **35.** 8.10×10^{-5} **37.** 1.50×10^9 **39.** 6.16×10^{-20} **41.** 1.929×10^{-14}
43. 3.71×10^{-23} **45.** 7.8×10^{19} **47.** 7.6×10^{-17} **49.** 9.1×10^{-19} **51.** 7.9×10^{18}

53. 8.3×10^{-27} **55.** 2.23×10^4 mi **57.** 264,000 **59.** $0.000000000001\ \frac{\text{W}}{\text{m}^2}$

61. $30{,}000{,}000{,}000\ \frac{\text{cm}}{\text{s}}$ **63.** 0.00000003 cm **65.** $6.670 \times 10^{-11}\ \frac{\text{N} \cdot \text{m}^2}{\text{kg}^2}$ **67.** 8.3 min

69. 9×10^{12} km **71.** 4.7×10^{11} **73.** $5540\ \frac{\text{kg}}{\text{m}^3}$

Section 10.3 (page 319)

1. 1 **3.** 9 **5.** −5 **7.** 7 **9.** $-\frac{1}{3}$ **11.** 2 **13.** 3 **15.** $\frac{3}{8}$ **17.** 0.3 **19.** 0.5 **21.** 2.236 **23.** 3.317 **25.** 4.359 **27.** r **29.** i **31.** r **33.** i **35.** r **37.** i **39.** r **41.** i **43.** i

Section 10.4 (page 324)

1. $\sqrt{V}$ **3.** $\sqrt[4]{z}$ **5.** $\sqrt[3]{s^2}$ **7.** $\sqrt[3]{m^5}$ **9.** 3 **11.** $\frac{1}{3}$ **13.** 3 **15.** $\frac{3}{2}$ **17.** −2 **19.** −8 **21.** 1 **23.** 12 **25.** 2 **27.** $\frac{9}{125}$ **29.** $-\frac{5}{9}$ **31.** $\frac{1}{2}$ **33.** x **35.** $a^{3/4}$ **37.** $b^{1/6}$ **39.** $-\frac{2}{V^{1/6}}$ **41.** $\frac{1}{p^{1/4}}$ **43.** $\frac{s_1^{3/2}s_2^{11/6}}{2^{1/3}}$ **45.** $\frac{1}{2s^{5/3}t^{10/3}}$ **47.** $\frac{R_a^{1/2}}{2R_b^{7/8}}$ **49.** $\frac{x^4}{y}$ **51.** t^4w^2 **53.** $\frac{p^3}{3q^{1/2}}$ **55.** $\frac{R}{2^{1/2}S^{1/4}}$ **57.** $\frac{m^2}{2n}$ **59.** $\frac{N^4}{M}$ **61.** 28,561 **63.** 1.40 **65.** 0.118 **67.** 39.9 **69.** −0.525 **71.** 25.9 **73.** 0.00499 **77.** $\sqrt[3]{x^2} + \sqrt[3]{y^2} = \sqrt[3]{a^2}$ **79.** $P = 0.75$ atmosphere **81.** $w = 9.0$ lb **83.** $i = 0.0664$ A **85.** 27.6 g

Section 10.5 (page 331)

1. $2\sqrt{2}$ **3.** $4\sqrt{3}$ **5.** $5\sqrt{5}$ **7.** $2\sqrt{3}$ **9.** $6\sqrt{2}$ **11.** $2\sqrt{7}$ **13.** $3\sqrt[3]{2}$ **15.** $2\sqrt[3]{4}$ **17.** $2\sqrt[5]{2}$ **19.** $2\sqrt[4]{3}$ **21.** $x^2\sqrt{x}$ **23.** $2V\sqrt{2V}$ **25.** $s_1\sqrt{s_1s_2}$ **27.** $4mn\sqrt{n}$ **29.** $5v_1v_2\sqrt{2v_2v_3}$ **31.** $2st\sqrt[3]{s}$ **33.** $2T_aT_b\sqrt[4]{2T_a}$ **35.** $2L_1^2L_2^3\sqrt[5]{L_2}$ **37.** $3x^3y^4z^5\sqrt[3]{z^2}$ **39.** $2a^2b^2c^3\sqrt[4]{2b^2c^2}$ **41.** $4j$ **43.** $8j$ **45.** $4\sqrt{2}j$ **47.** $6\sqrt{2}j$ **49.** $2\sqrt{3}j$ **51.** $2\sqrt{5}j$ **53.** $\frac{\sqrt{6}}{6}$ **55.** $\frac{a\sqrt{b}}{b}$ **57.** $\sqrt{2}$ **59.** $\frac{4\sqrt{t}}{st}$ **61.** $\frac{\sqrt{T_0T_1}}{T_0T_1}$ **63.** $\frac{L\sqrt{3RC}}{3RC}$ **65.** $\frac{\sqrt{2ab}}{2b}$ **67.** $\sqrt{C_2}$ **69.** $\frac{2\sqrt{vz}}{z}$ **71.** $\frac{\sqrt{pq}}{q}$ **73.** $\frac{\sqrt{35P}}{5}$ **75.** $\frac{2p^2\sqrt{s_1}}{s_1s_2}$ **77.** $\frac{\sqrt{6t_2}}{2}$ **79.** $\frac{\sqrt{6V_1V_2}}{3V_2}$ **81.** $f = \frac{\sqrt{LC}}{2\pi LC}$ **83.** $\frac{\sqrt{mv_0}}{mv}$ **85.** $\frac{3\sqrt{np(1-p)}}{n}$ **87.** $v = \frac{r\sqrt{2gmy(mr^2 + I)}}{mr^2 + I}$ **89.** $\frac{2\sqrt{3}}{3}$ kg

Section 10.6 (page 338)

1. $-\sqrt{5}$ **3.** $6\sqrt{2}$ **5.** $7\sqrt{2}$ **7.** $8\sqrt{2} - 7\sqrt{3}$ **9.** $5\sqrt{a}$ **11.** $-5\sqrt{V}$ **13.** $\sqrt{3} + 7\sqrt{V}$ **15.** $7a\sqrt{b}$ **17.** $11\sqrt{N} - R\sqrt{C}$ **19.** $\sqrt{2} + (1 + 2b)\sqrt[3]{a}$ **21.** $2\sqrt{ab}$ **23.** $2y\sqrt{3z}$ **25.** $2t\sqrt[3]{t}$ **27.** $3LS\sqrt[4]{4L}$ **29.** $2S_aS_b\sqrt[4]{S_aS_c^3}$ **31.** $\sqrt{3} + 3$ **33.** −2 **35.** $a - b^2$ **37.** $-1 - 2\sqrt{2}$ **39.** $-10 - 3\sqrt{6}$ **41.** $6 - 2\sqrt{3}$ **43.** $\frac{\sqrt{5a}}{5}$ **45.** $\frac{\sqrt{3t}}{t}$ **47.** $2\sqrt{a}$ **49.** $\frac{\sqrt{14a}}{7}$ **51.** $\frac{\sqrt{5C_1C_2}}{C_2}$ **53.** $-\frac{1}{4}(\sqrt{3} + \sqrt{15})$ **55.** $-\sqrt{7} - \sqrt{14}$ **57.** $\frac{3}{2}(\sqrt{5} - \sqrt{3})$ **59.** $\frac{1}{4}(\sqrt{15} + \sqrt{3} + \sqrt{5} + 1)$ **61.** $\frac{a(\sqrt{a} - b)}{a - b^2}$ **63.** $\frac{\sqrt{R}(1 + \sqrt{C})}{1 - C}$ **65.** $\frac{3 - 2\sqrt{3t} + t}{3 - t}$ **67.** $\frac{\sqrt{mv} - v}{m - v}$ **69.** $R = \frac{\mu_1 - 2\sqrt{\mu_1\mu_2} + \mu_2}{\mu_1 - \mu_2}$

Review Exercises/Chapter 10 (page 339)

1. $\sqrt[4]{a}$ **2.** $\sqrt[3]{T^2}$ **3.** $\frac{1}{36}$ **4.** $\frac{1}{16}$ **5.** $\frac{1}{6}$ **6.** 5 **7.** 36 **8.** $\frac{1}{8}$ **9.** $\frac{1}{9}$

10. $\frac{5}{2}$ **11.** 16 **12.** $-\frac{125}{2}$ **13.** $-8c^3d^6$ **14.** $x^{10}y$ **15.** $\frac{2A^2}{3B^5}$ **16.** $3x^8z^3$ **17.** $\frac{T_1^9}{T_2^5}$
18. $\frac{V_a^3}{V_b^4}$ **19.** $\frac{R + 3R^2}{1 + 2R^2}$ **20.** $\frac{S^4 - 1}{S^4}$ **21.** $\frac{s_2^{1/2}}{2^{3/4}s_1^{5/2}}$ **22.** $\frac{2q^{1/8}}{p^{5/9}}$ **23.** V^3R **24.** $\frac{M^{2/9}N^4}{4}$
25. 4.3×10^7 **26.** 9.83×10^8 **27.** 1.2×10^{-7} **28.** 9.30×10^{-11} **29.** 0.0000000360
30. 0.0000001400 **31.** 2,300,000 **32.** 63,400,000,000 **33.** 4.4×10^{-21} **34.** 7.10×10^{-14}
35. 6.50×10^7 **36.** 8.4×10^{-17} **37.** 1.20×10^{-8} **38.** -3.67×10^{-6} **39.** 7.7×10^{-19}
40. 1.1×10^{12} **41.** 1.14×10^7 **42.** 2.058 **43.** 7.40 **44.** 1.23 **45.** −4 **46.** $\frac{1}{2}$
47. 0.6 **48.** 0.3 **49.** r **50.** i **51.** r **52.** i **53.** $4ab^3\sqrt{2a}$ **54.** $2P_1P_2^2\sqrt[4]{P_1^2P_2}$
55. $2mn^2\sqrt[5]{2mn^2}$ **56.** $\frac{\sqrt{5}}{10}$ **57.** $\frac{P\sqrt{2Q}}{2Q}$ **58.** $\frac{\sqrt{2LV}}{10}$ **59.** $\frac{\sqrt{3s_1}}{3}$ **60.** $\frac{A\sqrt{10}}{4}$ **61.** $4\sqrt{3}j$
62. $6\sqrt{3}j$ **63.** $\sqrt{2}$ **64.** $2\sqrt{2} + 7C\sqrt{V}$ **65.** $3\sqrt{2ab}$ **66.** $3PV\sqrt[3]{2P}$ **67.** $2MN\sqrt[4]{4M}$
68. $A^2 - B$ **69.** $-28 - \sqrt{5}$ **70.** $\sqrt{15} - 9$ **71.** $\frac{\sqrt{3}}{3}$ **72.** $\sqrt{5}$ **73.** $\frac{a\sqrt{b} - b}{a^2 - b}$
74. $\frac{1}{3}(7 - 2\sqrt{10})$ **75.** $\frac{1}{3}(\sqrt{7} + 2\sqrt{5} - 2 - \sqrt{35})$ **76.** $\frac{T_1 + \sqrt{T_1T_2}}{T_1 - T_2}$ **77.** $1.86 \times 10^5 \frac{\text{mi}}{\text{s}}$
78. 9.3×10^7 mi **79.** 0.0000036 F **80.** $60{,}000 \frac{\text{lb}}{\text{in.}^2}$ **81.** $6.3 \times 10^{-5} \frac{\Omega}{\text{in.}}$ **82.** $\frac{n}{n-1}$
83. $\lambda = \frac{9n^2}{R(n^2 - 9)}$ **84.** 2,000,000 years **86.** 0.0644 A **87.** $T = \frac{2\pi\sqrt{Lg}}{g}$
88. $m = \frac{m_0c\sqrt{c^2 - v^2}}{c^2 - v^2}$

Chapter 11
Section 11.1 (page 349)

1. ±3 **3.** ±7 **5.** $\pm\sqrt{5}$ **7.** $\pm\frac{1}{3}$ **9.** $\pm\frac{\sqrt{3}}{4}$ **11.** $\pm\frac{\sqrt{6}}{3}$ **13.** $\pm 2j$ **15.** $\pm\sqrt{7}j$
17. $\pm\frac{\sqrt{7}}{2}j$ **19.** $\pm\frac{\sqrt{7}}{7}j$ **21.** 0, 1 **23.** $0, -\frac{1}{3}$ **25.** $0, \frac{7}{4}$ **27.** 1, 2 **29.** 1, −4
31. −3, −4 **33.** $\frac{1}{2}, 1$ **35.** $\frac{2}{3}, -1$ **37.** $\frac{3}{2}, 2$ **39.** $-\frac{2}{3}, 3$ **41.** $-\frac{2}{3}, -2$ **43.** $\frac{1}{4}, -2$
45. $-\frac{2}{5}, 3$ **47.** −3, −3 **49.** $\frac{1}{2}, \frac{1}{2}$ **51.** $\frac{1}{3}, \frac{1}{3}$ **53.** $\frac{3}{2}, -\frac{1}{2}$ **55.** $-\frac{2}{3}, -\frac{5}{2}$

Section 11.2 (page 356)

1. 3, 4 **3.** $\frac{1}{2}, -3$ **5.** $\frac{1}{2} \pm \frac{1}{2}\sqrt{5}$ **7.** $-1 \pm \sqrt{2}$ **9.** $1 \pm \sqrt{5}$ **11.** 3, 3 **13.** $-\frac{1}{3}, -\frac{1}{3}$
15. $\frac{5}{2}, \frac{5}{2}$ **17.** $\pm\frac{1}{2}\sqrt{10}$ **19.** 0, 2 **21.** $0, \frac{5}{3}$ **23.** $\frac{1}{2} \pm \frac{1}{2}\sqrt{3}$ **25.** $1 \pm \frac{1}{2}\sqrt{6}$
27. $\frac{1}{4} \pm \frac{1}{4}\sqrt{5}$ **29.** $-\frac{1}{3} \pm \frac{1}{3}\sqrt{13}$ **31.** $-\frac{1}{6} \pm \frac{1}{6}\sqrt{13}$ **33.** $\frac{2}{7} \pm \frac{3}{7}\sqrt{2}$ **35.** $1 \pm j$
37. $1 \pm \sqrt{5}j$ **39.** $-\frac{1}{2} \pm \frac{1}{2}\sqrt{7}j$ **41.** $1 \pm \sqrt{3}j$ **43.** $-\frac{1}{3} \pm \frac{1}{3}\sqrt{5}j$ **45.** $\frac{3}{5} \pm \frac{1}{5}\sqrt{6}j$
47. 3.02, −1.02 **49.** 0.552, −1.02 **51.** 1.68, −0.865 **53.** −0.18, −6.8 **55.** $\frac{1}{2}, -2$
57. 3, −4 **59.** −4, 6 **61.** $8, \frac{4}{3}$

Section 11.3 (page 362)

1. $t = 3$ s **3.** 0 A, 2 A **5.** 0 m, 3.0 m **7.** 0.789 A or 0.401 A **9.** 7, 13 **11.** 8.92, 6.38 **13.** 7.0 in. × 12 in. **15.** 10 ft × 15 ft **17.** 16 cm × 7.5 cm or 15 cm × 8 cm **19.** 4 in. × 6 in. **21.** 3.0 in. **23.** 6 Ω, 12 Ω **25.** 20.4 μF, 28.6 μF **27.** 36 min, 45 min **29.** 6 h **31.** 4 km/h **33.** 30 mi/h, 45 mi/h **35.** 30 mi/h, 40 mi/h **37.** 19.8 mi/h **39.** 6.49 mi/h, 4.49 mi/h

Review Exercises/Chapter 11 (page 365)

1. ± 7 **2.** $\pm 7j$ **3.** $\pm \frac{3}{2}$ **4.** $\pm \frac{3}{2}j$ **5.** $\pm \frac{\sqrt{2}}{3}$ **6.** $\pm \frac{\sqrt{15}}{3}j$ **7.** $0, \frac{1}{2}$ **8.** $0, -5$

9. $3, -1$ **10.** $3, 4$ **11.** $\frac{1}{2}, -1$ **12.** $-\frac{2}{3}, -1$ **13.** $\frac{1}{2}, 2$ **14.** $-\frac{1}{4}, 2$ **15.** $-\frac{1}{2}, -\frac{1}{2}$

16. $\frac{3}{2}, \frac{3}{2}$ **17.** $0, 7$ **18.** $2, 2$ **19.** $1 \pm \sqrt{2}$ **20.** $-1 \pm j$ **21.** $5, -3$ **22.** $\frac{3}{2}, -5$

23. $-2 \pm \sqrt{10}$ **24.** $-1 \pm \sqrt{3}j$ **25.** $\frac{3}{2}, -\frac{5}{3}$ **26.** $\frac{1}{3} \pm \frac{1}{3}\sqrt{5}j$ **27.** $-\frac{1}{2}, 2$ **28.** $-\frac{2}{3}, -1$

29. $-\frac{1}{4} \pm \frac{1}{4}\sqrt{13}$ **30.** $-\frac{2}{5}, 1$ **31.** $\frac{3}{4} \pm \frac{1}{4}\sqrt{17}$ **32.** $-\frac{1}{6} \pm \frac{1}{6}\sqrt{13}$ **33.** $-\frac{3}{10} \pm \frac{1}{10}\sqrt{29}$

34. $-\frac{7}{6} \pm \frac{1}{6}\sqrt{73}$ **35.** 3.32, 0.257 **36.** 0.073, −1.0 **37.** 8, −24 **38.** 4, −6 **39.** $x = \frac{1}{2}L$

40. $i = 2.54$ A or $i = 0.116$ A **41.** $t = 4$ s **42.** $b = 10.0$ cm, $h = 8.00$ cm **43.** 12 μF, 24 μF

44. 6 h **45.** $40 \frac{\text{mi}}{\text{h}}$ **46.** $25 \frac{\text{mi}}{\text{gal}}$ **47.** 50 shares

Chapter 12
Section 12.1 (page 371)

1. independent variable: x; dependent variable: y **3.** independent variable: R; dependent variable: E **5.** independent variable: r; dependent variable: A **7.** independent variable: A; dependent variable: C

9. $f(x) = 3x - 2$ **11.** $f(v) = \sqrt{7v + 1}$ **13.** $f(t) = -7t^2 + 3$ **15.** $f(s) = \frac{1}{\sqrt[3]{s + 2}}$

17. $0, 2, -2$ **19.** $\frac{1}{2}, \frac{1}{3}, \frac{1}{6}$ **21.** $0, \sqrt{3}, 2$ **23.** $1, \frac{1}{2}, \frac{1}{3}$ **25.** $-2, 2, 23$

27. $1 + a, 1 + b, a$ **29.** $\sqrt{2a^2 - 1}, \sqrt{2a^2 - 4a + 1}$ **31.** all x except $x = -1$ **33.** all t except $t = 1$ and $t = 2$ **35.** $x \leq 1$ **37.** $t \geq -3$ **39.** $r \geq 3$ **41.** $A = s^2$

43. $A = \frac{\pi}{4}D^2$ **45.** $C = 5A$ **47.** $E = kT^4$

Section 12.2 (page 375)

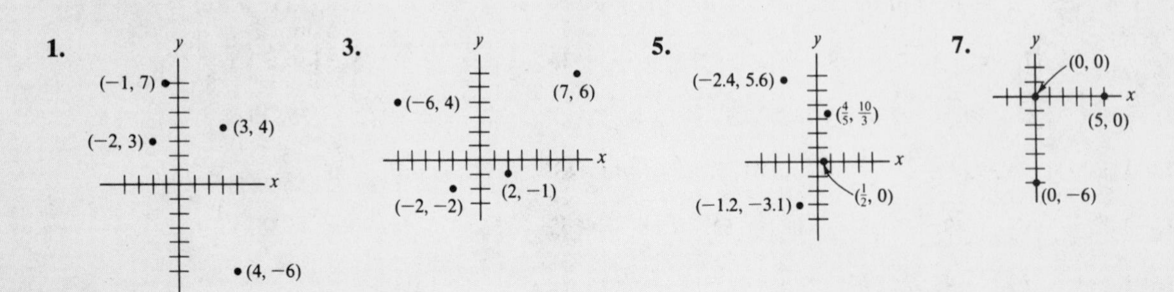

9. **a.** y-axis **b.** x-axis

11.

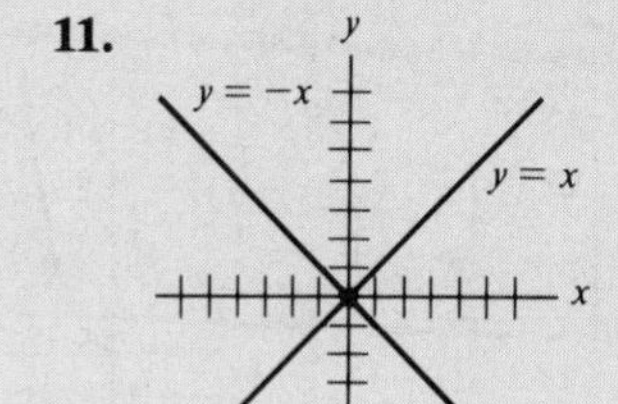

13.

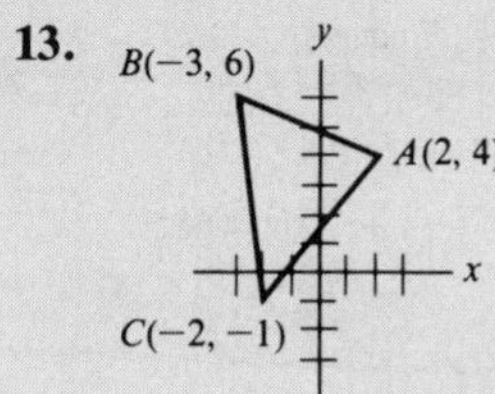

15.

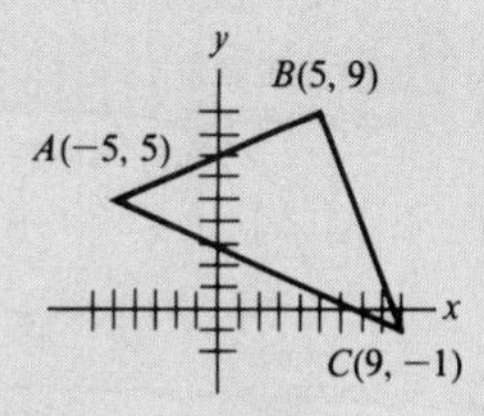

17.

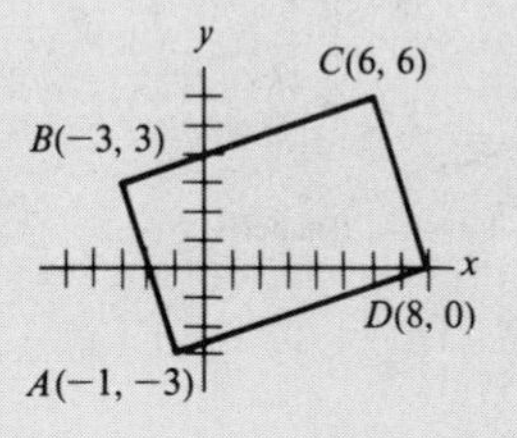

Section 12.3 (page 382)

1.

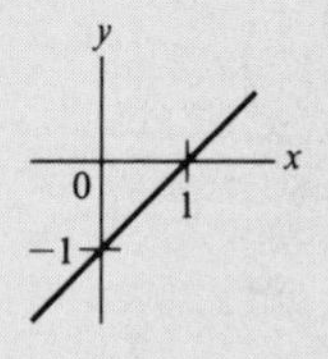

3.

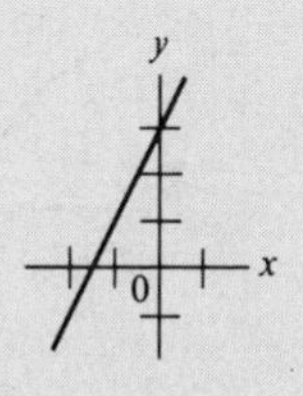

5.

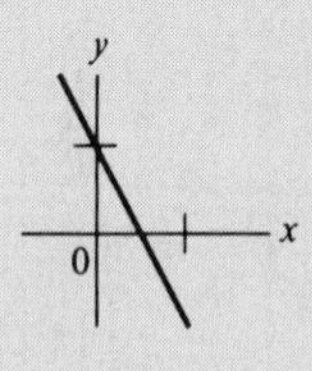

7.

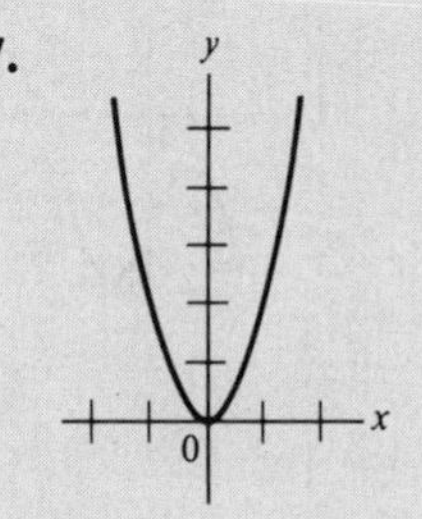

9.

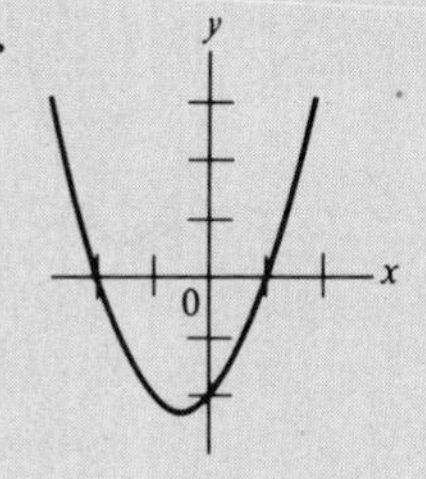

11.

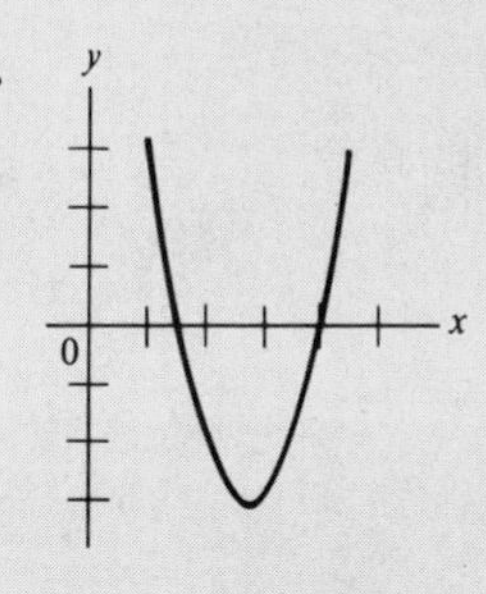

13.

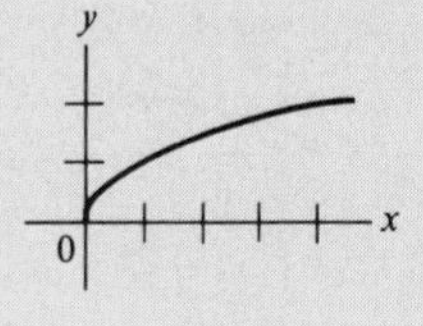

15.

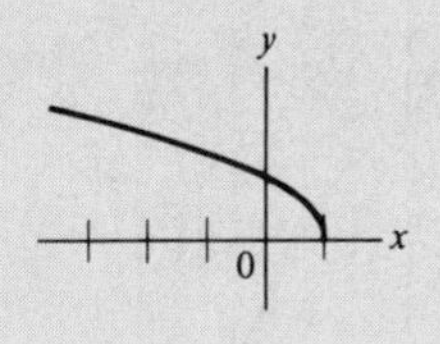

17.

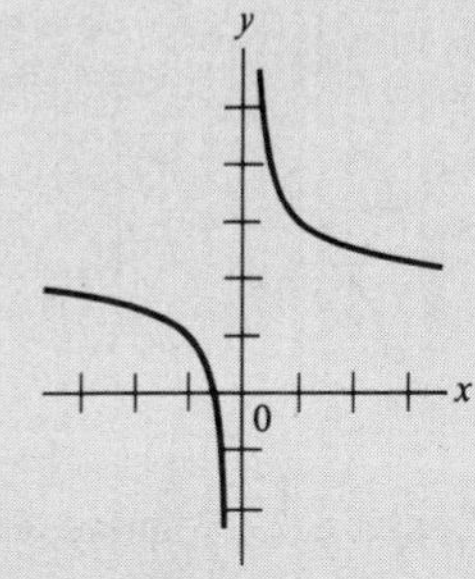

19.

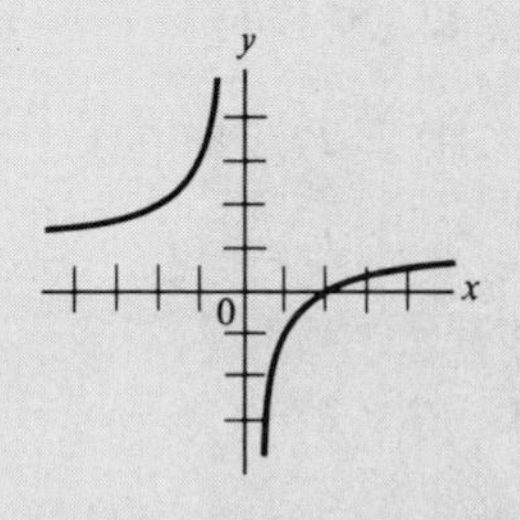

21.

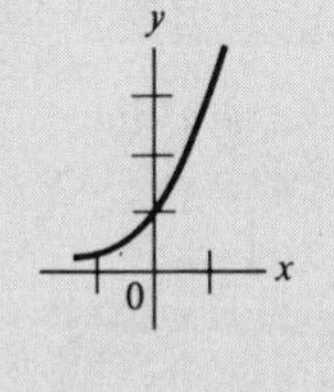

23.

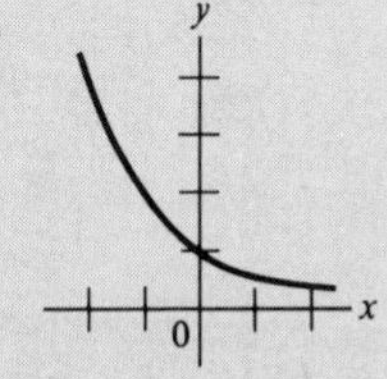

25.

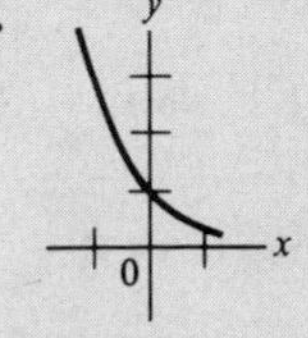

27. 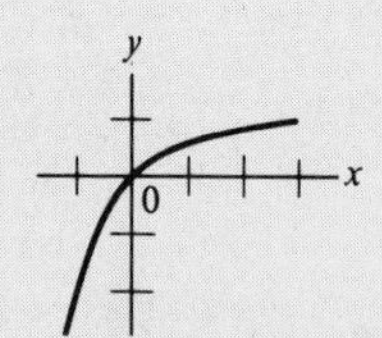

29.

31. 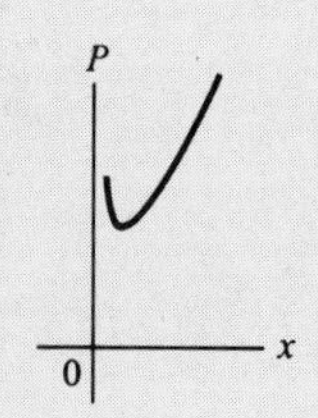

33. 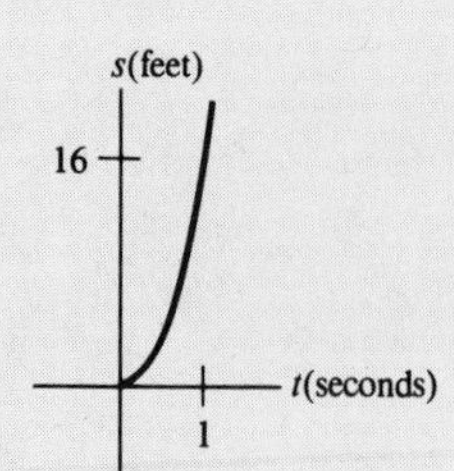

35. 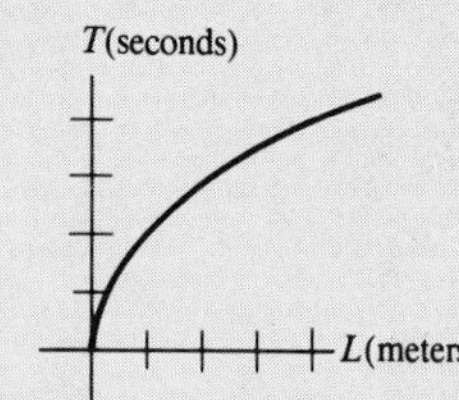

37. 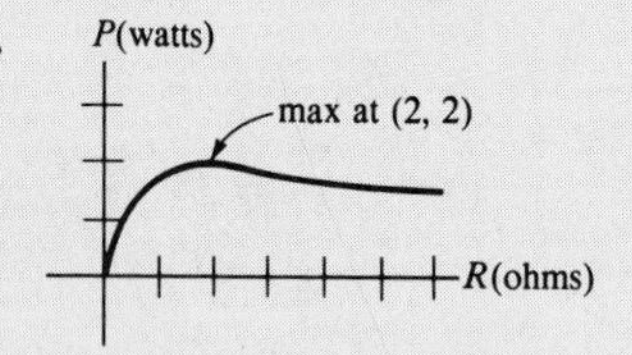

39. 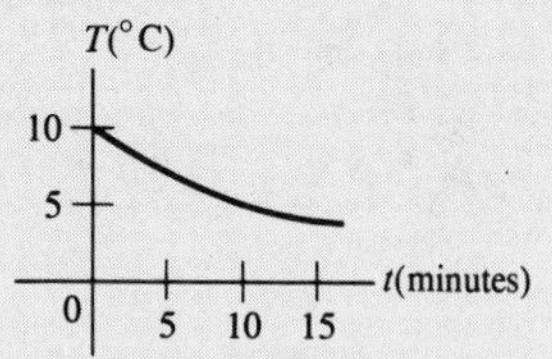

Section 12.4 (page 385)

1. 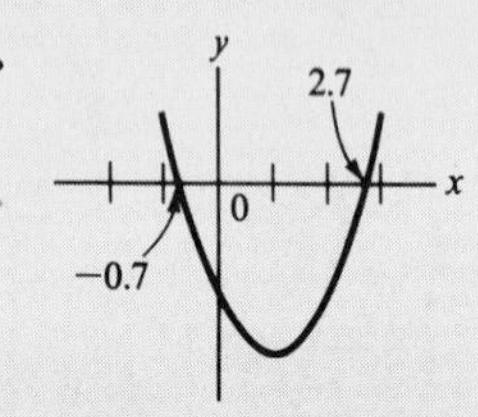

3. 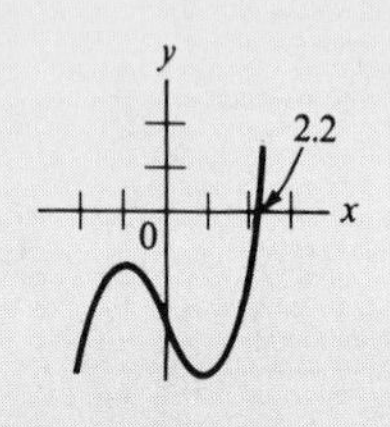

5. 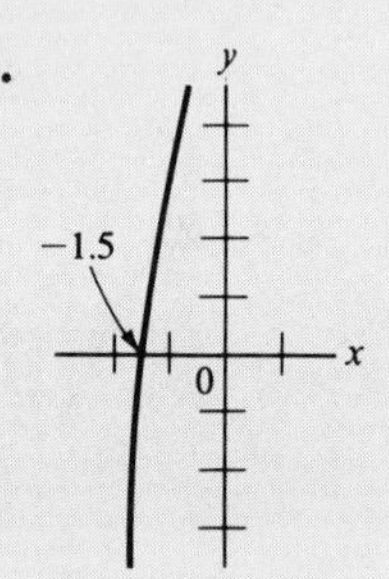

7. 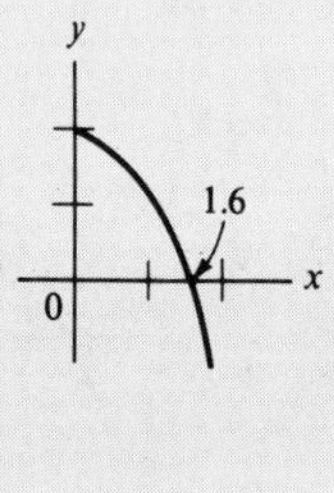

9. 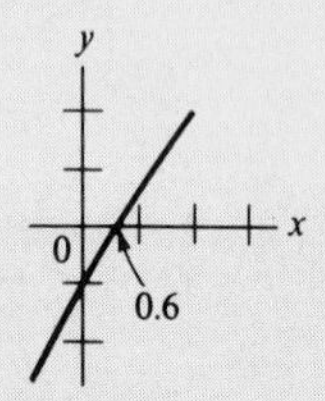

11. 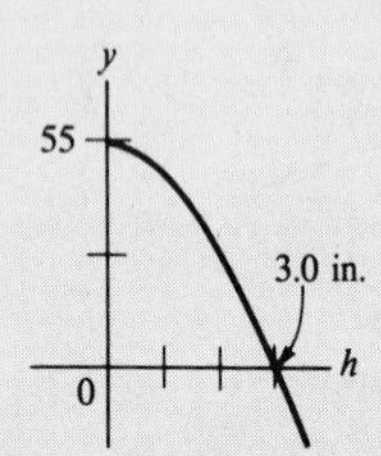

13. 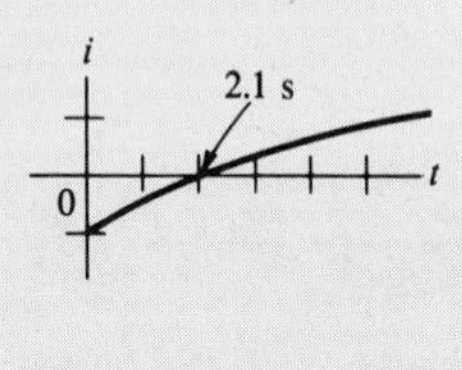

Section 12.6 (page 392)

1. 7.6 **3.** $816 \frac{\text{lb}}{\text{ft}^2}$ **5.** 0.82 **7.** $81 \frac{\text{ft}}{\text{s}}$ **9.** $17.2 \frac{\text{mi}}{\text{h}}$ **11.** 8.03 cm **13.** 0.17 Ω

15. 2.3 qt **17.** 48 ft **19.** 30 ft **21.** $3.2 \frac{\text{rev}}{\text{s}}$ **23.** 46.7 lb **25.** $1.41 **27.** 86.5 Ω

Section 12.7 (page 397)

1. $z = ky$ **3.** $w = kz^2$ **5.** $P = kS^3$ **7.** $s = \frac{k}{\sqrt{t}}$ **9.** $A = k\frac{b}{c}$ **11.** $Q = k\frac{pq}{s}$

13. $E = k\dfrac{vw}{\sqrt{u}}$ **15.** $z = 3x$ **17.** $P = \dfrac{1}{V}$ **19.** $n = \dfrac{5}{3}st$ **21.** $F = \dfrac{1}{2}\dfrac{ab}{c}$ **23.** 7.5 lb

25. $S = kwd^2$ **27.** 256 ft **29.** $60\ \dfrac{\text{rev}}{\text{min}}$ **31.** $6.10\ \dfrac{\text{lb}}{\text{ft}^2}$

33. **a.** 4 times as large **b.** $\dfrac{1}{4}$ as large **35.** 4.70 Ω

Review Exercises/Chapter 12 (page 399)

1. independent variable: t; dependent variable: s **2.** $-1, \dfrac{1}{3}, \dfrac{1}{5}$ **3.** $0, 1, \sqrt{5}, 3$ **4.** $t \geq -2$

5. $A = \dfrac{5}{2}h$ **6.** **a.** y-axis **b.** quadrants III and IV **7.**

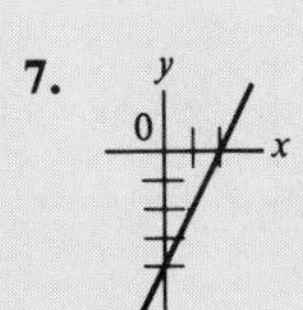

8.

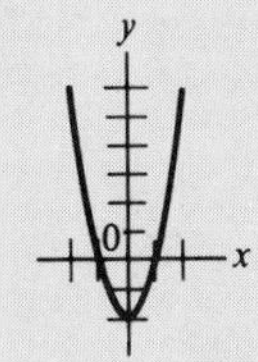

9.

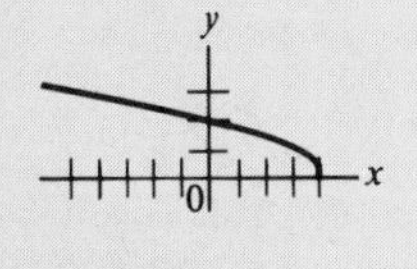

10.

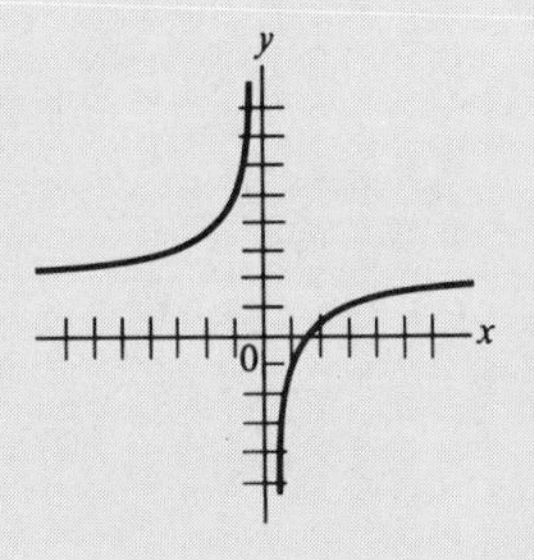

11.

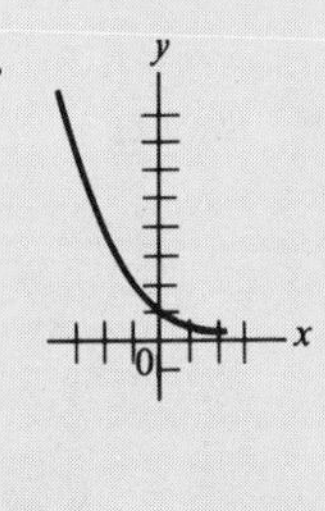

12.

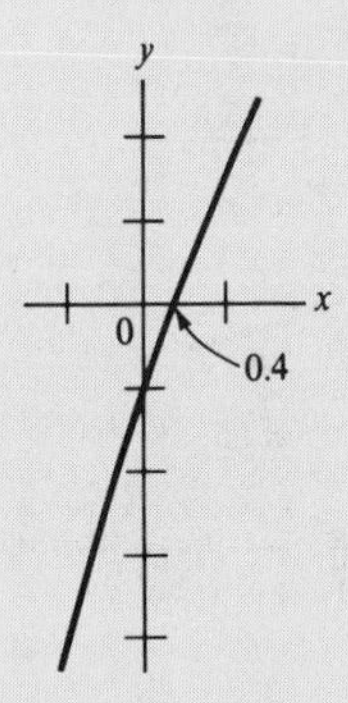

13. $y = k\dfrac{x}{z}$ **14.** $P = k\dfrac{a}{\sqrt{b}}$ **15.** $Z = kst$ **16.** $V = kRC^2$ **17.** $S = 20\dfrac{p}{q}$ **18.** $T = k\dfrac{v_1 v_2}{v_3}$

19.

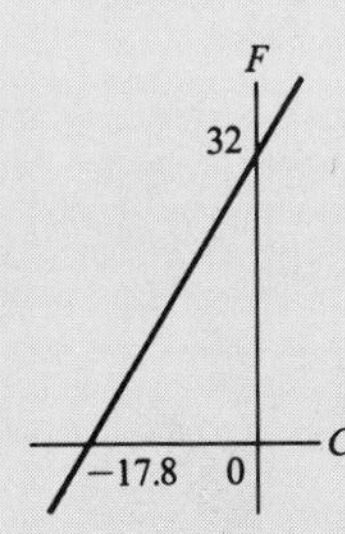

20.

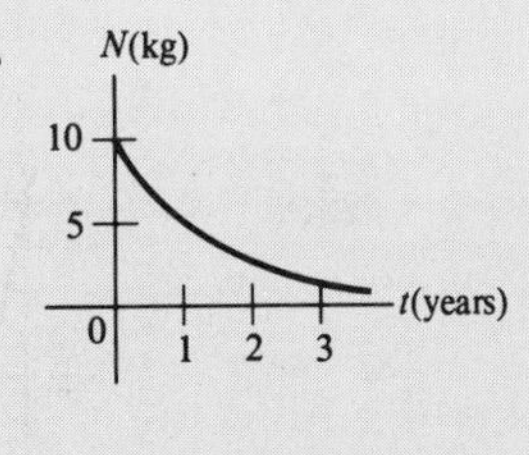

21.

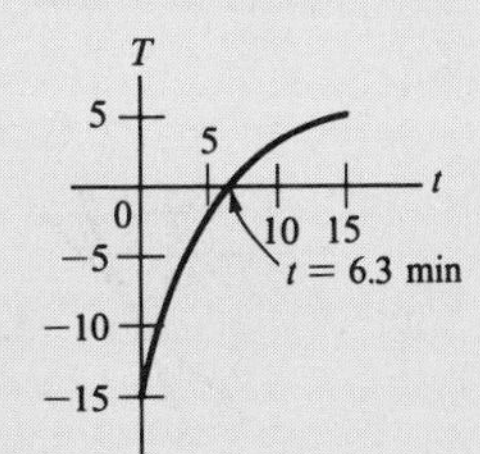

22. 66% **23.** 1.7%

24. 30 lb of tin, 18 lb of lead **25.** 137.5 min **26.** $C = \dfrac{k}{d}$ **27.** 29 V **28.** 52.2 N

29. $F = k\dfrac{q_1 q_2}{d^2}$ **30.** $R = k(T - T_0)$

Cumulative Review Exercises/Chapters 10–12 (page 401)

1. **a.** $-\frac{1}{2}$ **b.** 1 **2.** $3a^4b^6$ **3.** T_a **4.** $\frac{9p^2}{4q^{1/3}}$ **5.** $\frac{C^2 - C}{C + 1}$ **6.** 1.24×10^{-19}

7. $2P_1P_2^2\sqrt[4]{2P_1^2P_2^2}$ **8.** $\frac{V\sqrt{2Q}}{4Q}$ **9.** $\frac{\sqrt{2C}}{2}$ **10.** $3m\sqrt{2n}$ **11.** $3M\sqrt[3]{2N^2}$ **12.** $-7\sqrt{x}$

13. $3s^2 - s\sqrt{t} - 2t$ **14.** $4 + \sqrt{15}$ **15.** $\pm\frac{\sqrt{5}}{2}j$ **16.** $\frac{3}{2}, -2$ **17.** $-1 \pm \sqrt{3}$

18. $\frac{1}{3} \pm \frac{1}{3}\sqrt{5}j$ **19.** $t \leq \frac{1}{2}$; $\sqrt{2}, 3\sqrt{2}$ **20.** **21.** $V = k\frac{vw}{\sqrt{t}}$ **22.** $v = \sqrt{2gh}$

23. $f = \frac{\sqrt{T\rho}}{2L\rho}$ **24.** $r = 1.37\sqrt{h^5}$ **25.** $35.6\,\frac{\text{m}^3}{\text{s}}$ **27.** 7.66 h **28.** 3.142

29.

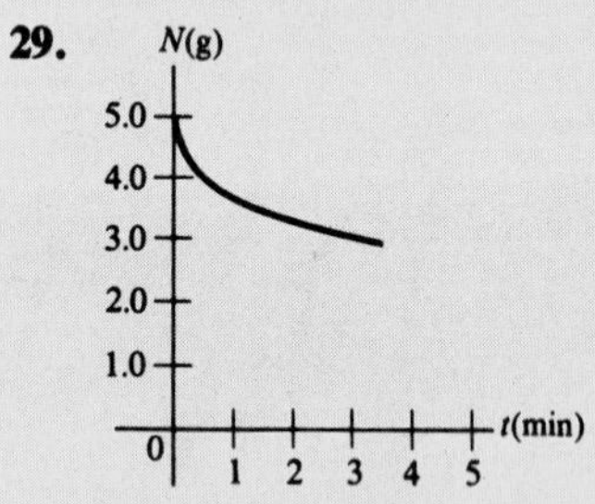

30. 82.9 Ω

Chapter 13
Section 13.1 (page 409)

1. (2, 0)

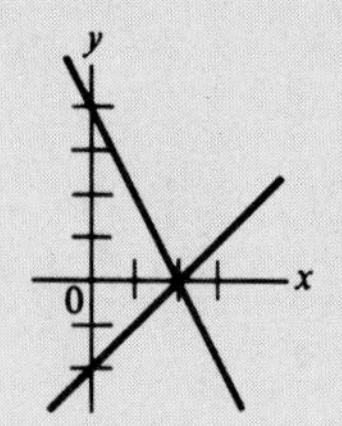

3. (1.0, −1.0)

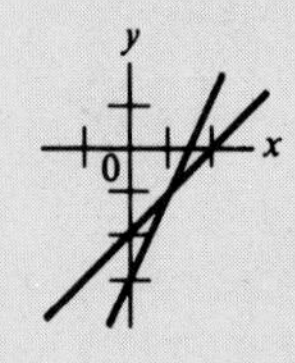

5. (1.4, −1.7)

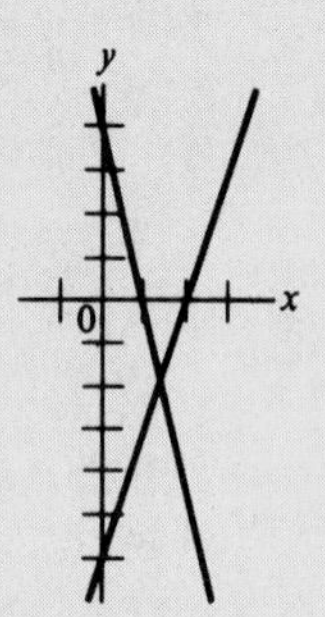

7. (8.0, 3.3)

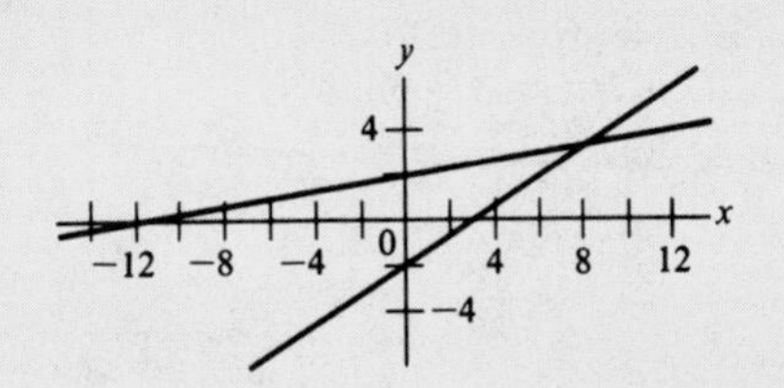

9. (−2.4, −5.6)

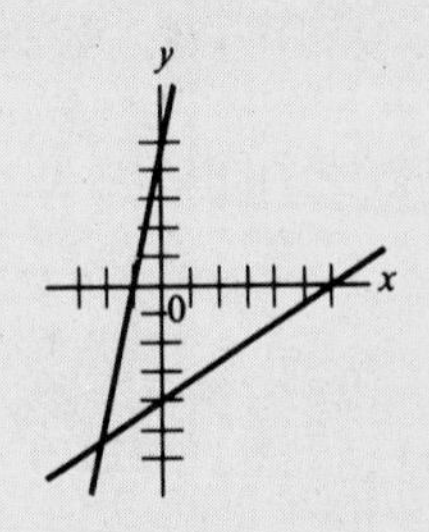

11. (5.8, −3.3)

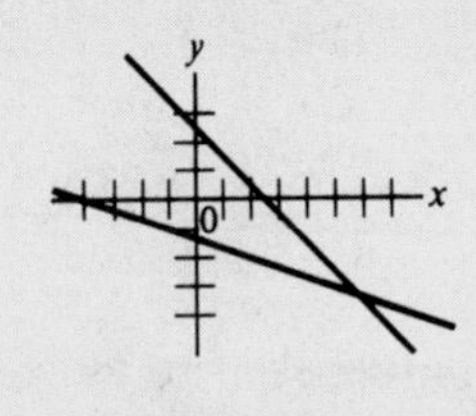

13. (−0.2, 0.8)

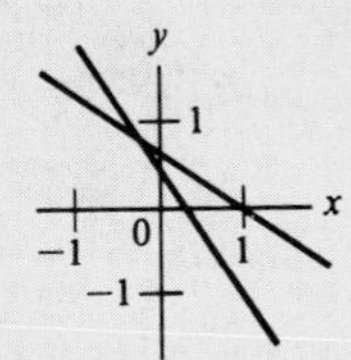

15. (2.7, 2.3)

17. inconsistent

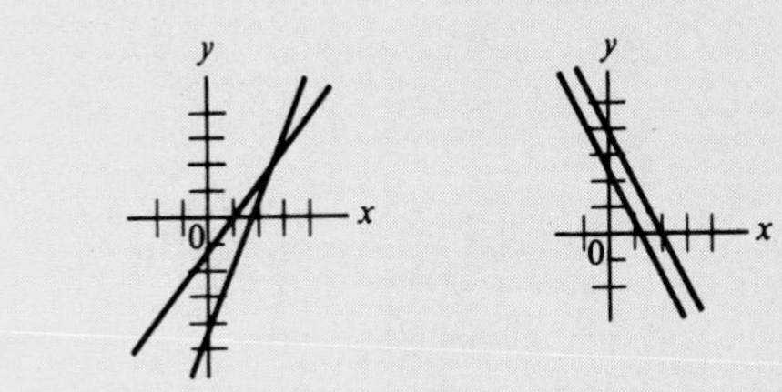

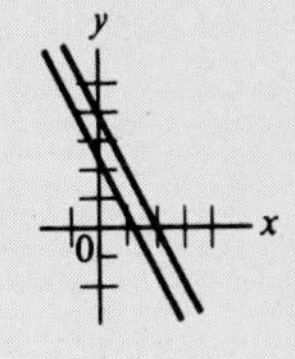

19. inconsistent

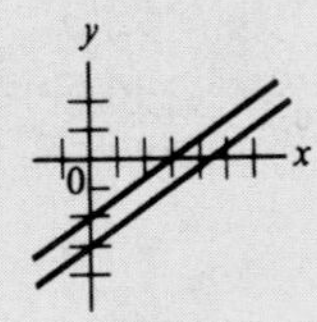

21. dependent

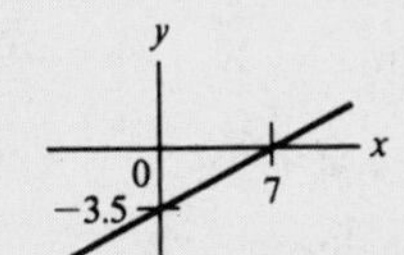

23. dependent

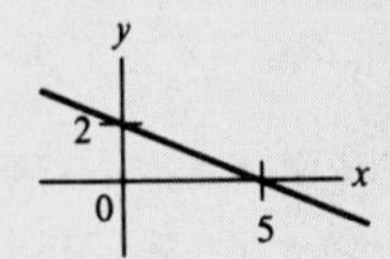

Section 13.2 (page 415)

1. (3, 0) **3.** (−1, 1) **5.** (−5, −8) **7.** (3, −6) **9.** $\left(\frac{7}{20}, \frac{3}{20}\right)$ **11.** inconsistent **13.** dependent **15.** (−13, 10) **17.** (22, 31) **19.** (−44, −38) **21.** (20, 10) **23.** $\left(\frac{7}{3}, \frac{4}{3}\right)$ **25.** $\left(2, \frac{7}{2}\right)$ **27.** $\left(-2, \frac{1}{4}\right)$ **29.** $\left(\frac{39}{5}, \frac{13}{5}\right)$ **31.** (0.68, 0.085) **33.** (11.2, 6.48)

Section 13.3 (page 419)

1. (−2, 1) **3.** (−1, 4) **5.** $\left(-\frac{1}{2}, 3\right)$ **7.** (1, −1) **9.** (3, 3) **11.** (−1, 3) **13.** $\left(-1, \frac{1}{2}\right)$ **15.** $\left(\frac{1}{2}, -1\right)$ **17.** $\left(-\frac{3}{2}, -4\right)$ **19.** $\left(\frac{3}{4}, 2\right)$ **21.** $\left(4, -\frac{3}{4}\right)$ **23.** (1, 1) **25.** $\left(1, \frac{1}{2}\right)$ **27.** $\left(-\frac{5}{2}, -\frac{3}{2}\right)$ **29.** $\left(\frac{9}{5}, -\frac{1}{5}\right)$ **31.** $\left(2, \frac{1}{3}\right)$ **33.** inconsistent **35.** dependent **37.** (2, −1) **39.** $\left(1, \frac{1}{2}\right)$ **41.** (2, 4) **43.** $\left(\frac{1}{2}, 4\right)$

Section 13.4 (page 425)

1. $v_1 = 10$ cm/s, $v_2 = -5$ cm/s **3.** $s = 566 - 0.00200T$ **5.** $w_1 = 0.50$ N, $w_2 = 3.0$ N **7.** 14, 18
9. 3.6, 7.2 **11.** 13 ft, 16 ft **13.** 7 m, 22 m **15.** 47, 15 **17.** 80 Ω, 100 Ω
19. 2.7 A, 1.2 A **21.** 6.8 cm × 2.4 cm **23.** 20 ft × 30 ft **25.** 1.2 mm × 3.6 mm
27. 40 mL of 10% solution **29.** $33\frac{1}{3}$ lb of 12% alloy **31.** $I_1 = \frac{24}{11}$ A; $I_2 = \frac{14}{11}$ A **33.** 190, 25
35. \$6000 at 10%, \$5000 at 9%

Review Exercises/Chapter 13 (page 428)

1. (2.6, −1.1) **2.** (−7.2, −2.4) **3.** (−3, 2) **4.** (3, 4) **5.** $\left(-\frac{1}{2}, 3\right)$

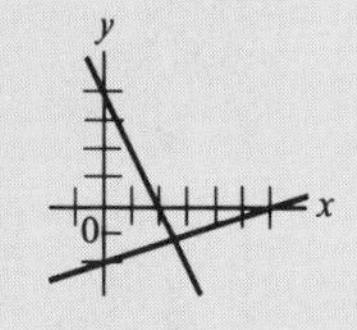

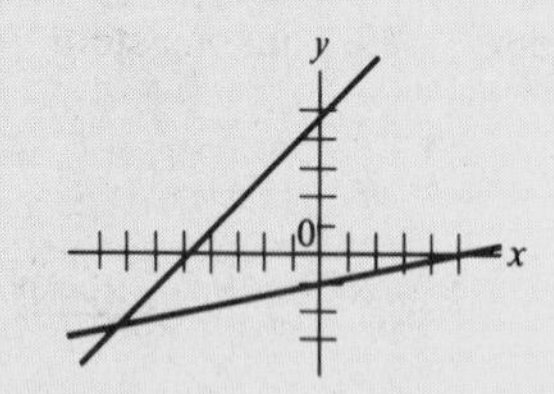

6. $\left(2, -\frac{1}{3}\right)$ **7.** $\left(-\frac{1}{3}, -\frac{1}{4}\right)$ **8.** $\left(-\frac{1}{4}, -2\right)$ **9.** (−3, 2) **10.** (2, −3) **11.** $\left(-\frac{1}{3}, 4\right)$
12. $\left(\frac{2}{3}, -2\right)$ **13.** $\left(4, \frac{3}{4}\right)$ **14.** $\left(-\frac{1}{4}, -3\right)$ **15.** $\left(-\frac{4}{5}, 3\right)$ **16.** $\left(\frac{1}{6}, 5\right)$ **17.** (−3, 4)
18. $\left(\frac{1}{3}, -2\right)$ **19.** $\left(\frac{2}{5}, -\frac{1}{3}\right)$ **20.** $\left(\frac{3}{4}, \frac{1}{3}\right)$ **21.** $\left(\frac{5}{6}, \frac{5}{8}\right)$ **22.** (2, 3) **23.** (1, −2)
24. (−2, 5) **25.** $\left(\frac{1}{2}, -3\right)$ **26.** (−2, −3) **27.** dependent **28.** inconsistent **29.** (2, −5)
30. (4, 5) **31.** $\left(-\frac{1}{3}, -4\right)$ **32.** $\left(1, \frac{5}{6}\right)$ **33.** $R = 0.012T + 1.4$ **34.** 5.8, 26.4
35. 50 Ω, 80 Ω **36.** 3.4 cm × 11.2 cm **37.** 75 m × 100 m **38.** 18 lb of 6% alloy
39. 8 smaller offices, 12 larger offices **40.** 7 days at \$250, 6 days at \$200

Chapter 14
Section 14.1 (page 434)

1. ∠*CBE* **3.** ∠*ABD* and ∠*DBE*, ∠*DBE* and ∠*EBC* **5.** ∠*ABE*
7. ∠*ABD* and ∠*CBD*, ∠*ABE* and ∠*CBE* **9.** ∠*DBE* **11.** 60°26′ **13.** 50°, 40°
15. vertical angles **17.** alternate interior **19.** corresponding **21.** alternate exterior **23.** 75°
25. 105° **27.** 75° **29.** 105° **31.** 50° **33.** 50° **35.** 130°
37. Alternate interior and corresponding angles are equal (all 90°).
39. Since alternate interior angles are equal, $AD \parallel BC$ and $AB \parallel DC$.

Section 14.2 (page 443)

1. quadrilateral, pentagon, hexagon **3.** 30° **5.** 58° **7.** 109° **9.** 80° **11.** 100°
13. ∠*ABC* = 100°, ∠*A* = 80°, ∠*C* = 80°, ∠*D* = 100° **15.** *EF* **17.** *BD* **19.** ∠*ACB*, ∠*CBD*
21. 120° **23.** 30° **25.** 17.5° **27.** 65°

Section 14.3 (page 449)

1. 3 **3.** 5 **5.** 10 **7.** 8.09 **9.** 52.4 **11.** 1.20 **13.** 10.3 **15.** 109 **17.** 518
19. 0.268 **21.** 6.64 cm **23.** 34.0 Ω **25.** 23.5 ft **27.** 5.9 ft **29.** 286 m **31.** 30.54 mi
33. 5.8 ft **35.** 73.8 in.

Section 14.4 (page 459)

1. $A \leftrightarrow E, B \leftrightarrow F, C \leftrightarrow D, AB \leftrightarrow EF$ **3.** $A \leftrightarrow D, C \leftrightarrow E, B \leftrightarrow F, AB \leftrightarrow DF$ **9.** 7.0 cm
11. 14.4 in. **13.** 18.0 cm **15.** 8.85 in. **17.** 26 ft **19.** 11 m **21.** $\frac{3}{10}$ ft **23.** 46.8 m
25. 5130 cm^2 = 0.513 m^2

Section 14.5 (page 472)

1. 3000 cm^3 **3.** 12 in.^{3} **5.** 58.8 ft^3 **7.** 128.1 cm^3 **9.** 538 ft^2, 807 ft^2 **11.** 88 m^2, 119 m^2
13. 29 m^2, 74 m^2 **15.** 290 cm^3 **17.** 24,100 in.^{3} **19.** 0.5631 yd^3 **21.** 1040 mm^2, 2450 mm^2
23. 19.6 in.^{2}, 30.2 in.^{2} **25.** 1960 m^2, 4470 m^2 **27.** 5.44 in.^{3} **29.** 8.972 cm^3 **31.** 7.71 ft^3
33. 72.0 cm^2 **35.** 18.4 in.^{2} **37.** 102.4 ft^3 **39.** 310 cm^3 **41.** 555 m^3 **43.** 413 in.^{2}
45. 7740 cm^2 **47.** 0.14 ft^3, 1.3 ft^2 **49.** 230 cm^3, 181 cm^2 **51.** 0.2088 m^3, 1.702 m^2
53. 4.8 ft^3 **55.** 6.8 m^3 **57.** \$126.72 **59.** 1980 g **61.** 53.0 cm^3 **63.** 93,700,000 ft^3
65. 2900 lb **67.** \$39 **69.** 12.9 g **71.** volume: 2.60×10^{11} mi^3; surface area: 1.97×10^8 mi^2
73. volume: 370 ft^3; surface area: 280 ft^2 **75.** 2870 lb **77.** \$96

Review Exercises/Chapter 14 (page 474)

1. $\angle ADC, \angle ADB$ **2.** $\angle ADE$ **3.** $\angle BDC, \angle BDE$ **4.** $\angle ADC$ and $\angle ADB$, $\angle ADB$ and $\angle BDE$
5. $\angle ADC$ and $\angle ADB$ **6.** $\angle ADC$ and $\angle ADE$, $\angle BDC$ and $\angle BDE$ **7.** $\angle 1$ and $\angle 2$ **8.** $\angle 4$ and $\angle 6$
9. $\angle 1$ and $\angle 5$, $\angle 3$ and $\angle 6$ **10.** $\angle 3$ and $\angle 4$, $\angle 2$ and $\angle 5$ **11.** 180° **12.** 180°
13. Since alternate interior angles are equal, opposite sides are parallel. **14.** $\angle D = 55°$, $\angle E = 45°$
15. 50°, 130° **16.** **a.** 20° **b.** 10° **17.** $5\sqrt{2}$ cm **18.** 10.3 cm **20.** 1.7 in. **21.** 0.88 m^3
22. 0.32 yd^3 **23.** 20.0 cm^3 **24.** 1480 ft^3 **25.** 10 cm^3 **26.** 2500 in.^{3} **27.** 2770 cm^3
28. 75.9 in.^{2} **29.** 92 cm^2 **30.** 150 cm^2 **31.** 248 ft^2 **32.** 242 m^2 **33.** 48 m^2
34. 37.5°, 52.5° **35.** 297° **36.** 4.40 cm **37.** 52.0 km **38.** 8.31 ft **39.** 5.3 m
40. $1\frac{2}{5}$ ft **41.** 23.8 ft **42.** 2.1 m **43.** 4230 cm^3 **44.** 1.7 lb **45.** 607 lb
46. 544 lb **47.** 939,000 ft^2 **48.** volume: 260 m^3; surface area: 224 m^2
49. volume: 3.3×10^{17} mi^3; surface area: 2.3×10^{12} mi^2 **50.** \$64 **51.** 6.2 m^2 **52.** 1100 lb

Chapter 15
Section 15.1 (page 483)

1.
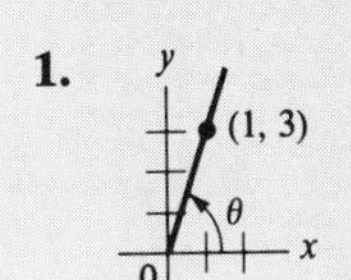

3.
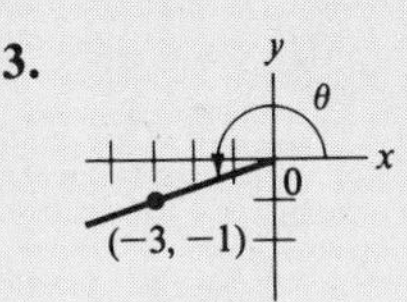

5.
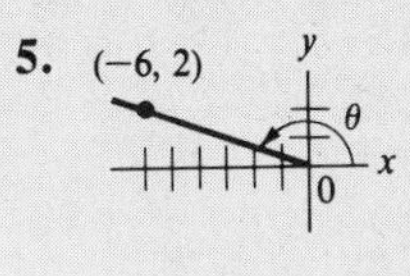

7.
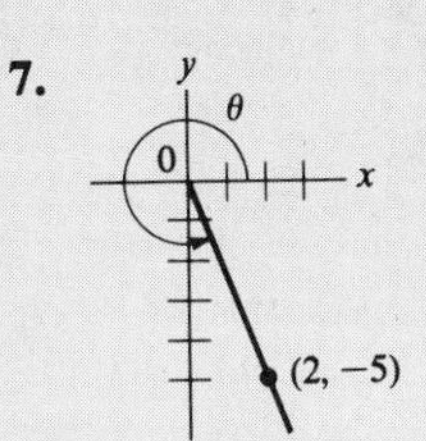

9.
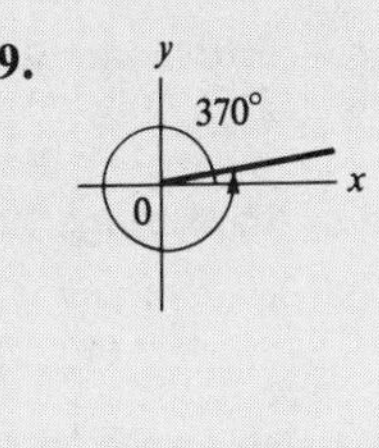

11.
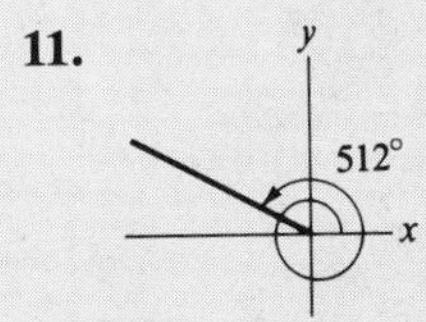

13.
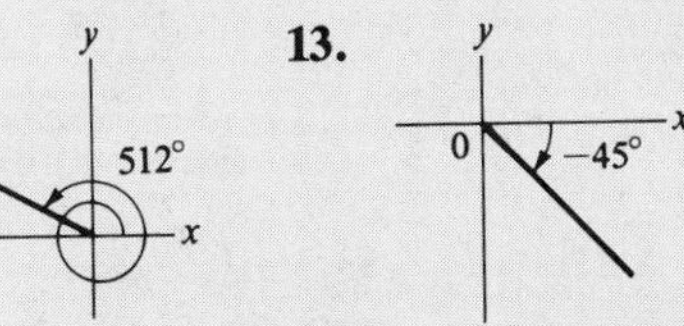

15.
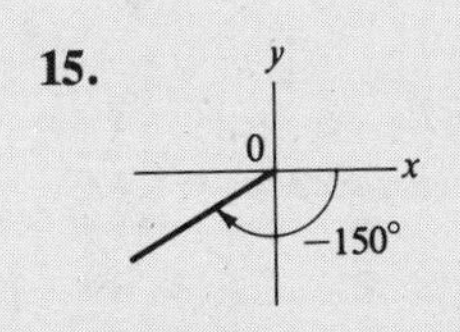

17.
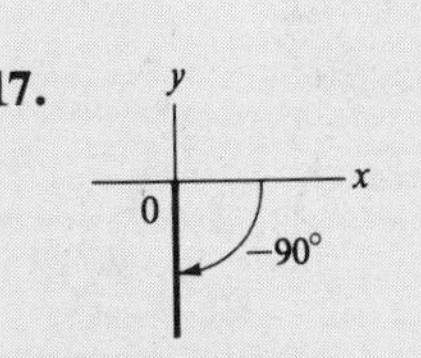

19.
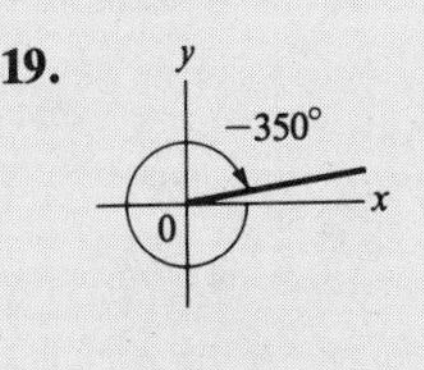

21. 61°16′ **23.** 264°41′ **25.** 110°20′ **27.** 210° **29.** 339°20′ **31.** 309°50′ **33.** 267°12′
35. 244°23′ **37.** 187°26′ **39.** 129°55′ **41.** 19°15′ **43.** 354°59′ **45.** 177°5′ **47.** 79°37′
49. 120°

Section 15.2 (page 489)

1. $\frac{3}{5}, \frac{3}{4}$ **3.** $\frac{13}{5}, \frac{5}{12}$ **5.** $\sqrt{5}, \frac{\sqrt{5}}{2}$ **7.** $\frac{\sqrt{17}}{17}, \frac{4\sqrt{17}}{17}$ **9.** $\sqrt{3}, \frac{\sqrt{6}}{2}$ **11.** 0.95, 0.32
13. 1.12, 2.00 **15.** 1.17, 0.517

In Exercises 17–29, the trigonometric functions are given in the following order: sin θ, cos θ, tan θ, csc θ, sec θ, and cot θ.

17. $\frac{1}{\sqrt{5}} = \frac{\sqrt{5}}{5}, \frac{2}{\sqrt{5}} = \frac{2\sqrt{5}}{5}, \frac{1}{2}, \sqrt{5}, \frac{\sqrt{5}}{2}, 2$ **19.** $\frac{4}{3\sqrt{2}} = \frac{2\sqrt{2}}{3}, \frac{1}{3}, \frac{4}{\sqrt{2}} = 2\sqrt{2}, \frac{3\sqrt{2}}{4}, 3, \frac{\sqrt{2}}{4}$
21. $\frac{3}{2\sqrt{3}} = \frac{\sqrt{3}}{2}, \frac{1}{2}, \frac{3}{\sqrt{3}} = \sqrt{3}, \frac{2\sqrt{3}}{3}, 2, \frac{\sqrt{3}}{3}$ **23.** $\frac{\sqrt{21}}{6}, \frac{\sqrt{15}}{6}, \frac{\sqrt{35}}{5}, \frac{2\sqrt{21}}{7}, \frac{2\sqrt{15}}{5}, \frac{\sqrt{35}}{7}$
25. 0.83, 0.55, 1.5, 1.2, 1.8, 0.67 **27.** 0.461, 0.887, 0.520, 2.17, 1.13, 1.92
29. 0.838, 0.546, 1.53, 1.19, 1.83, 0.652 **31.** $\frac{\sqrt{3}}{2}$ **33.** $\frac{\sqrt{13}}{2}$ **35.** $\sqrt{10}$ **37.** $\frac{2\sqrt{2}}{3}$
39. $\frac{2\sqrt{10}}{5}$ **41.** $\frac{4\sqrt{3}}{3}$ **43.** 0.243 **45.** 0.302 **47.** 0.348 **49.** 0.674 **51.** 0.969
53. 4 **55.** $\frac{6}{7}$ **57.** $\frac{1}{5}$ **59.** $\frac{1}{10}$

Section 15.3 (page 499)

1. $\frac{\sqrt{3}}{2}$ **3.** $\frac{\sqrt{2}}{2}$ **5.** $\sqrt{3}$ **7.** 2 **9.** 1 **11.** $\frac{\sqrt{3}}{3}$ **13.** 1 **15.** undefined **17.** 0
19. 1 **21.** undefined **23.** 0 **25.** 45° **27.** 30° **29.** 30° **31.** 60° **33.** 60°
35. 45° **37.** 60° **39.** 0° **41.** 0.4014 **43.** 1.153 **45.** 1.247 **47.** 1.089 **49.** 0.8566
51. 33°10′ **53.** 53°22′ **55.** 25°26′ **57.** 0.4462 **59.** 0.9956 **61.** 4.0108 **63.** 1.2120
65. 1.0669 **67.** 0.2962 **69.** 0.5040 **71.** 0.2295 **73.** 0.5340 **75.** 13.9865 **77.** 1.3782
79. 2.6937 **81.** 8.58° **83.** 10.56° **85.** 27.41° **87.** 32.30° **89.** 33.58° **91.** 11°13′
93. 69°1′ **95.** 61°27′ **97.** 44°28′ **99.** 29°54′

Section 15.4 (page 504)

1. 2.31 **3.** 6.66 **5.** 7.31 **7.** 6.765 **9.** 7.721 **11.** 33.7°
13. $b = 3.01$, $c = 10.1$, $B = 17.4°$ **15.** $a = 8.3$, $c = 11$, $A = 50°$
17. $a = 2.569$, $b = 1.908$, $B = 36°36'$ **19.** $b = 33.19$, $c = 34.75$, $A = 17.24°$

Section 15.5 (page 508)

1. 10.4 m **3.** 7.9 ft **5.** 1250 ft **7.** 51.8° **9.** 39.8° **11.** 1.92 m **13.** 1.14 cm
15. 1.23 in. **17.** 88.3 ft **19.** 185 m **21.** 0.61 cm **23.** 29.3° **25.** 230.0 ft **27.** 29.0 m

Section 15.6 (page 514)

1. $\frac{5}{13}, \frac{12}{5}$ **3.** $\frac{3}{5}, -\frac{4}{3}$ **5.** $-\frac{\sqrt{13}}{2}, -\frac{2}{3}$ **7.** $-\frac{\sqrt{6}}{3}, -\frac{\sqrt{3}}{3}$ **9.** $-3, -2\sqrt{2}$
11. −0.457, −1.10 **13.** 1.9, −2.2 **15.** 0.576, −1.42 **17.** 2.05, 0.873 **19.** −0.540, −1.19
21. + **23.** − **25.** − **27.** − **29.** + **31.** − **33.** − **35.** + **37.** +
39. − **41.** II **43.** III **45.** II **47.** IV **49.** III **51.** II **53.** III **55.** I

Section 15.7 (page 520)

1. $\frac{\sqrt{2}}{2}$ **3.** $-\sqrt{3}$ **5.** $-\sqrt{2}$ **7.** $-\frac{\sqrt{2}}{2}$ **9.** $-\frac{\sqrt{3}}{3}$ **11.** −2 **13.** −2 **15.** 1

17. $-\frac{2\sqrt{3}}{3}$ **19.** $-\frac{\sqrt{3}}{3}$ **21.** $-\sqrt{2}$ **23.** $-\frac{1}{2}$ **25.** 0 **27.** undefined **29.** -1
31. 0 **33.** 0 **35.** 0.7353 **37.** -5.665 **39.** 1.402 **41.** -0.9112 **43.** 0.9373
45. -0.3719 **47.** -1.0021 **49.** 0.8290 **51.** 0.7239 **53.** 1.1467 **55.** -0.6833
57. -0.3346 **59.** 1.0424 **61.** 0.7738 **63.** 0.4384 **65.** 0.8264 **67.** -2.3072
69. -1.8320 **71.** 163.49° **73.** 289.30° **75.** 291.33° **77.** 100.26° **79.** 331.42°
81. 22.49° **83.** 99.28° **85.** 1.373 **87.** 29.5 m

Review Exercises/Chapter 15 (page 521)

1. 34°17′ **2.** 345°21′ **3.** 332°16′ **4.** 166°6′

In Exercises 5–12, the trigonometric functions are given in the following order: sin θ, cos θ, tan θ, csc θ, sec θ, and cot θ.

5. $\frac{\sqrt{3}}{2}, \frac{1}{2}, \sqrt{3}, \frac{2\sqrt{3}}{3}, 2, \frac{\sqrt{3}}{3}$ **6.** $\frac{\sqrt{7}}{3}, \frac{\sqrt{2}}{3}, \frac{\sqrt{14}}{2}, \frac{3\sqrt{7}}{7}, \frac{3\sqrt{2}}{2}, \frac{\sqrt{14}}{7}$
7. $\frac{\sqrt{6}}{3}, -\frac{\sqrt{3}}{3}, -\sqrt{2}, \frac{\sqrt{6}}{2}, -\sqrt{3}, -\frac{\sqrt{2}}{2}$ **8.** $-\frac{\sqrt{15}}{4}, -\frac{1}{4}, \sqrt{15}, -\frac{4\sqrt{15}}{15}, -4, \frac{\sqrt{15}}{15}$
9. 0.97, 0.23, 4.3, 1.0, 4.4, 0.23 **10.** -0.82, 0.57, -1.4, -1.2, 1.7, -0.70
11. -0.8191, -0.5736, 1.428, -1.221, -1.743, 0.7003 **12.** 0.496, -0.868, -0.572, 2.01, -1.15, -1.75
13. $\frac{2\sqrt{2}}{3}$ **14.** $\frac{\sqrt{29}}{2}$ **15.** 2.12 **16.** $\frac{9}{5}$ **17.** + **18.** − **19.** + **20.** − **21.** III
22. IV **23.** II **24.** III **25.** $\sqrt{3}$ **26.** 1 **27.** -2 **28.** $-\sqrt{3}$ **29.** 0 **30.** $-\sqrt{2}$
31. $-\frac{\sqrt{3}}{2}$ **32.** 0 **33.** $\sqrt{2}$ **34.** $-\frac{\sqrt{3}}{3}$ **35.** 45° **36.** 30° **37.** 60° **38.** 45°
39. 0.7222 **40.** -0.6713 **41.** 0.2247 **42.** 1.371 **43.** 0.2538 **44.** -0.1357
45. -1.9292 **46.** 2.1803 **47.** -1.1642 **48.** 0.1192 **49.** -0.6661 **50.** 0.9802
51. 73°40′ **52.** 133.80° **53.** 125.56° **54.** 284.69° **55.** 206.39° **56.** 326.34°
57. 71.31° **58.** 21.6 lb **59.** 555 ft **60.** 11.8 m **61.** 9.6° **62.** 21.8 m^3 **63.** 67.6°
64. 7.402 cm **65.** 27.83 cm **66.** 86.8 ft **67.** 75.0 ft **68.** 23.3 m

Cumulative Review Exercises/Chapters 13–15 (page 524)

1. (1.6, 0.6) **2.** (2, 1) **3.** 9.3 cm by 3.4 cm **4.** 48.7°, 41.3° **5.** 139°, 39° **6.** 24.0 cm
7. 14.8 in. **8.** 4.05 m **9.** 9.84 cm **10.** 245°
11. $\sin\theta = -\frac{\sqrt{7}}{3}$, $\cos\theta = -\frac{\sqrt{2}}{3}$, $\tan\theta = \frac{\sqrt{14}}{2}$, $\csc\theta = -\frac{3\sqrt{7}}{7}$, $\sec\theta = -\frac{3\sqrt{2}}{2}$, $\cot\theta = \frac{\sqrt{14}}{7}$
12. $\sin\theta = 0.594$, $\cos\theta = -0.804$, $\tan\theta = -0.738$, $\csc\theta = 1.68$, $\sec\theta = -1.24$, $\cot\theta = -1.35$
13. III **15.** 200.56° **16.** 90°, 270° **17.** 1060 cm^3, 471 cm^2 **18.** 560 $in.^3$, 260 $in.^2$
19. 938.7 cm^3, 463.6 cm^2 **20.** 36.4 ft^3, 48.9 ft^2 **21.** 617 cm^3 **22.** 846 lb

Chapter 16
Section 16.1 (page 530)

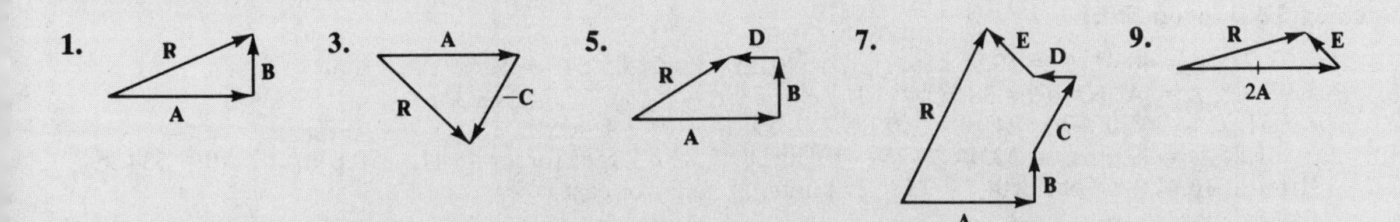

11.

13.

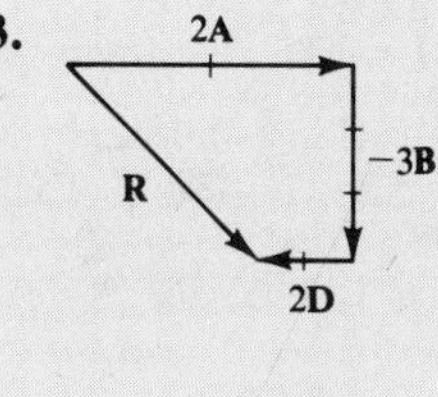

15.

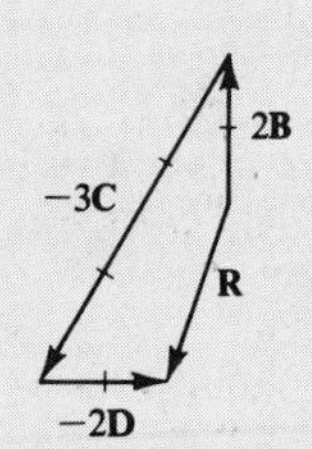

17.

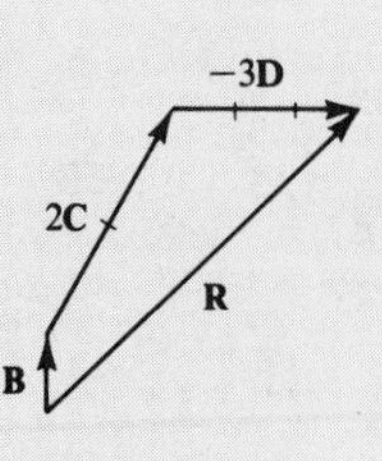

19.

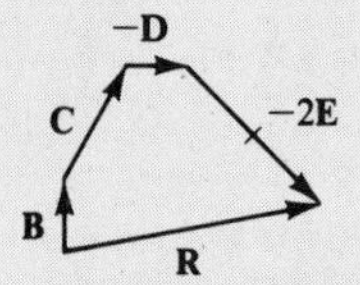

Section 16.2 (page 535)

1. $\sqrt{2}$, 45° **3.** 2, 120° **5.** 4, 240° **7.** $\sqrt{13}$, 56.3° **9.** $\sqrt{65}$, 119.7° **11.** $2\sqrt{10}$, 251.6° **13.** $2\sqrt{13}$, 326.3° **15.** 3, 118.1° **17.** 2.92, 114.9° **19.** 12.1, 214.4° **21.** (0.521, 2.95) **23.** (−4.00, 6.93) **25.** (−3.99, −3.01) **27.** (1.10, −5.90) **29.** (8.78, −9.58) **31.** (1.80, 1.33), 2.2, 36.5° **33.** (−6.58, 6.02), 8.9, 137.5° **35.** (11.9, −6.76), 14, 330.4° **37.** (5.92, −4.51), 7.4, 322.7° **39.** (−3.93, −4.01), 5.6, 225.6°

Section 16.3 (page 539)

1. 12 N, 33° **3.** 17.49 lb, 24.67° **5.** 48 lb, 36° with larger force **7.** 12.5 mi/h, 16.7° with line across **9.** 414 km/h, 5.3° west of north **11.** 11.9 km/h, 17.7° upstream with line across **13.** 242 km/h, 8.3° west of south **15.** 43 lb **17.** 20.7 lb **19.** 96 lb **21.** 23 N, 23° **23.** 141 mi/h, 2.9° north of west **25.** 55.6 lb, 18.6° with larger force **27.** 89.2 lb, 31.7° with respect to horizontal

Section 16.4 (page 547)

1. $B = 22.0°$, $a = 17.7$, $c = 21.5$ **3.** $A = 107.26°$, $b = 7.824$, $c = 14.67$ **5.** $C = 27.3°$, $a = 23.9$, $b = 14.8$ **7.** $C = 62.1°$, $b = 7.96$, $c = 7.50$ **9.** $B = 42°$, $a = 6.2$, $b = 12$ **11.** $A = 57°10'$, $a = 10.8$, $c = 3.70$ **13.** $B = 9.8°$, $C = 148.5°$, $c = 16.0$ **15.** $C = 38.2°$, $B = 113.4°$, $b = 0.347$ **17.** $A = 105.9°$, $B = 11.3°$, $b = 2.97$ **19.** $B = 54°$, $C = 92°$, $c = 0.45$ **21.** 188 km/h, 11° north of east **23.** 39.8 m **25.** 20.5 kg **27.** 51.0 ft

Section 16.5 (page 555)

1. $B = 73.7°$, $A = 46.4°$, $C = 59.9°$ **3.** $A = 102.5°$, $B = 35.5°$, $C = 42.0°$ **5.** $B = 99.02°$, $A = 42.62°$, $C = 38.36°$ **7.** $c = 88.1$, $A = 40.3°$, $B = 66.3°$ **9.** $a = 7.51$, $C = 20°30'$, $B = 24°10'$ **11.** $c = 79$, $B = 71°$, $A = 51°$ **13.** $b = 31.8$, $A = 49.0°$, $C = 32.0°$ **15.** 101 lb, 8.1° with larger force **17.** 51.1 km **19.** 334 m **21.** 171 mi/h, 16.4° west of north **23.** 21 knots, 2° south of east

Review Exercises/Chapter 16 (page 556)

1. −3C R 2B A **2.** $\sqrt{2}$, 135° **3.** 6, 210° **4.** $\sqrt{5}$, 296.6° **5.** $\sqrt{29}$, 68.2° **6.** 4.3, 124°

7. 11.8, 240.5° **8.** (−4.07, 5.69) **9.** (4.20, −4.29) **10.** (−5.34, −7.24)
11. (−3.38, −6.84), 7.6, 243.7° **12.** (8.35, −2.82), 8.8, 341.3° **13.** $B = 99.7°$, $a = 3.88$, $c = 8.58$
14. $C = 34.0°$, $b = 16.2$, $c = 25.9$ **15.** $B = 45.50°$, $a = 0.4866$, $b = 0.6128$
16. $A = 31.54°$, $a = 3.601$, $c = 4.282$ **17.** $C = 59.6°$, $B = 78.1°$, $b = 18.6$
18. $C = 120.4°$, $B = 17.3°$, $b = 5.66$ **19.** $A = 117°$, $B = 29°$, $C = 34°$
20. $C = 90.20°$, $A = 42.37°$, $B = 47.43°$ **21.** $c = 9.30$, $A = 41°50'$, $B = 25°30'$
22. $b = 39.3$, $A = 44.3°$, $C = 40.0°$ **23.** 16.5 km/h, 14.4° with line across **24.** 1.21 tons
25. 27 kg, 24° **26.** 163 mi/h, 21.6° south of west **28.** 63.7 m **29.** 28.5 lb **30.** 52.1°
31. 67.5 ft **32.** 8.21 m **33.** 44.7 cm

Chapter 17
Section 17.1 (page 563)

1. $\frac{\pi}{6}$ **3.** $\frac{\pi}{3}$ **5.** $-\frac{\pi}{4}$ **7.** $\frac{2\pi}{15}$ **9.** $\frac{11\pi}{45}$ **11.** $\frac{5\pi}{6}$ **13.** $\frac{7\pi}{4}$ **15.** $\frac{8\pi}{9}$ **17.** $\frac{4\pi}{5}$
19. $\frac{13\pi}{15}$ **21.** $\frac{11\pi}{30}$ **23.** $\frac{16\pi}{15}$ **25.** 45° **27.** 225° **29.** −150° **31.** 75° **33.** 378°
35. 290° **37.** 312° **39.** 380° **41.** 85° **43.** 26° **45.** 510° **47.** −285° **49.** 1.0821
51. 1.6232 **53.** 0.6807 **55.** 2.1049 **57.** 4.7368 **59.** 0.7912 **61.** 2.1904 **63.** 4.8898
65. 32.3° **67.** 44.5° **69.** 22.7° **71.** 99.2° **73.** 169.6° **75.** 107.2° **77.** 0.4863
79. 4.8880 **81.** −0.3637 **83.** 1.1080 **85.** 0.2171 **87.** −3.1187 **89.** −1.1940
91. 1.47 A **93.** 4.1 cm

Section 17.2 (page 569)

1. 26.4 cm **3.** 3.36 m **5.** 1.82 ft **7.** 0.3253 m **9.** 14 ft **11.** 95.0 cm² **13.** 17.5 m²
15. 4.31 ft² **17.** 0.6202 m² **19.** 18.9 ft² **21.** 15 ft **23.** 168.4° **25.** 2160 mi
27. 23.0 in.² **29.** 7.7 cm/s **31.** 30.1 ft/s **33.** 11.2 ft/s **35.** 1040 mi/h **37.** 4870 ft/min
39. 19 rad/s **41.** 0.22 rad/s

Section 17.3 (page 576)

1. $A = 2$, $P = 2\pi$

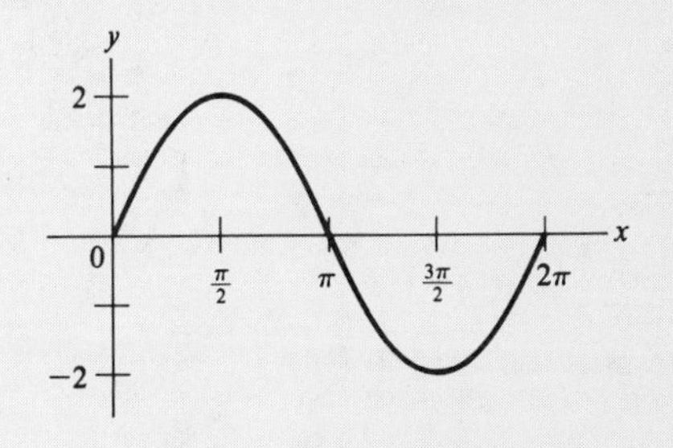

3. $A = \frac{1}{2}$, $P = 2\pi$

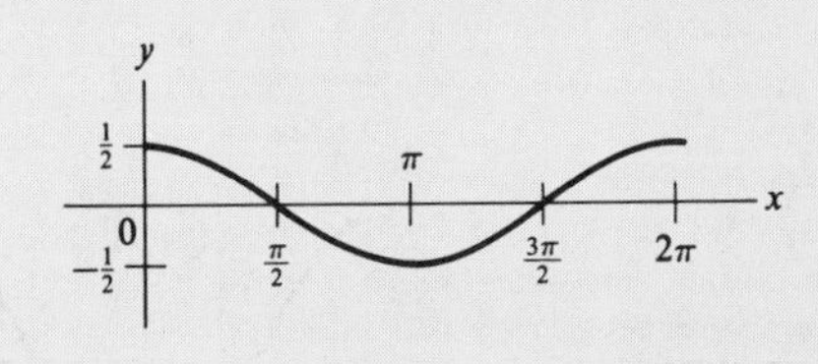

5. $A = 3, P = 2\pi$

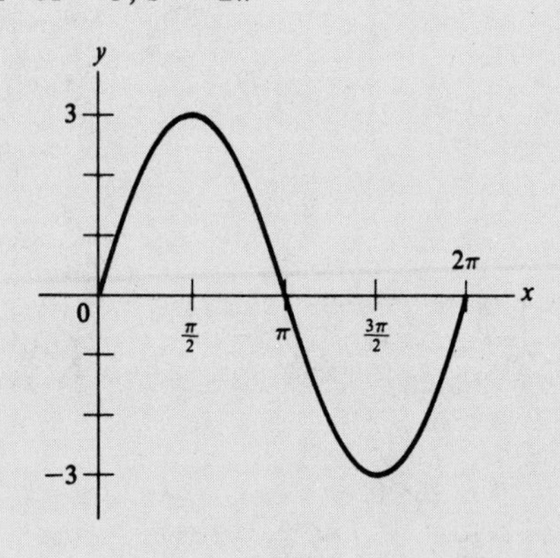

7. $A = 1, P = 2\pi$

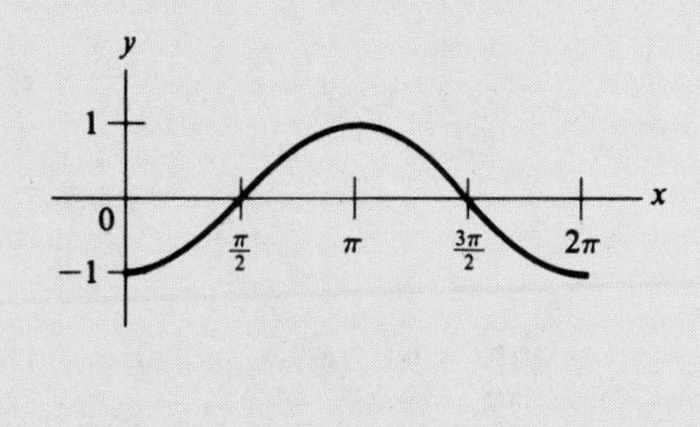

9. $A = 2, P = 2\pi$

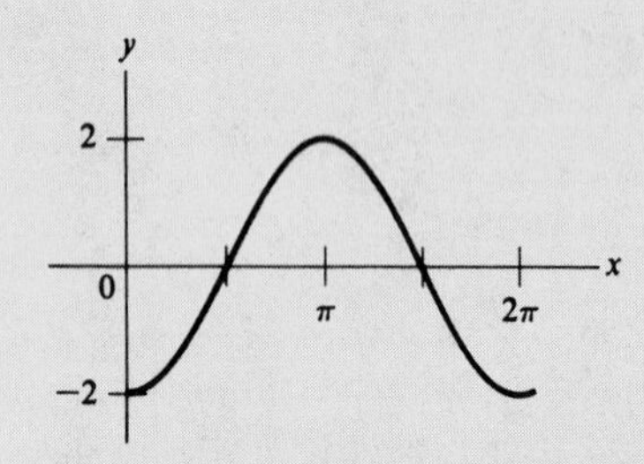

11. $A = 2.5, P = 2\pi$

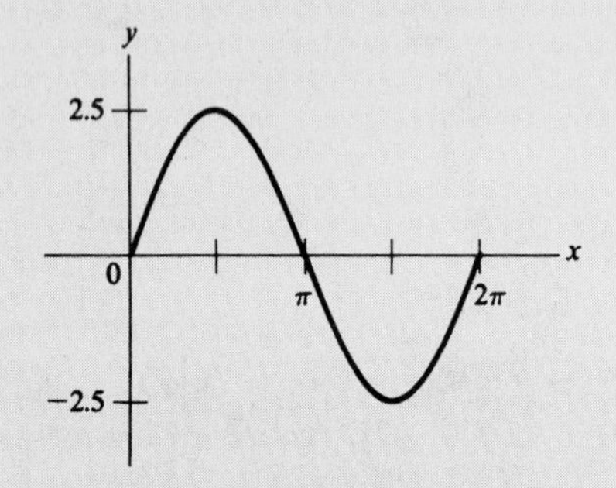

13. $A = 1.8, P = 2\pi$

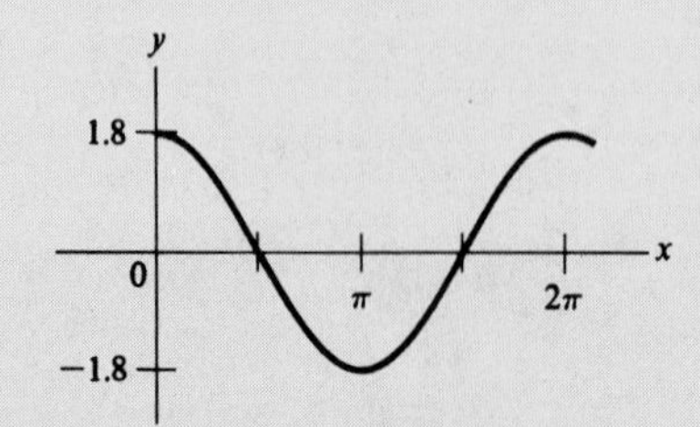

15. $A = 1.5, P = 2\pi$

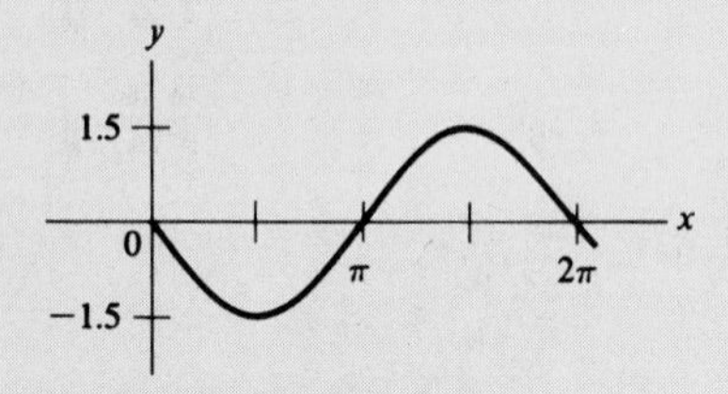

Section 17.4 (page 581)

1. $A = 1, P = \pi$

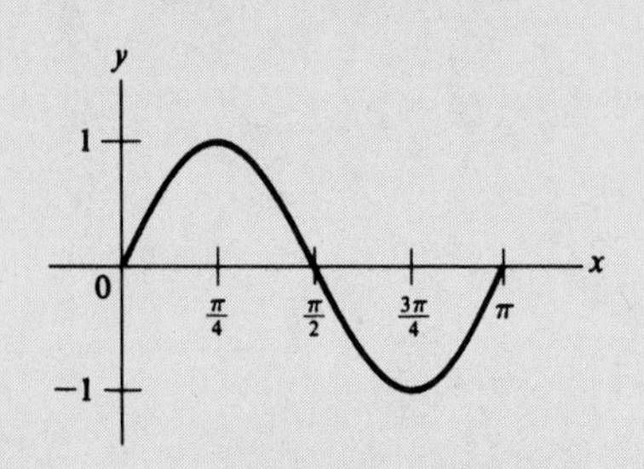

3. $A = 1, P = \pi$

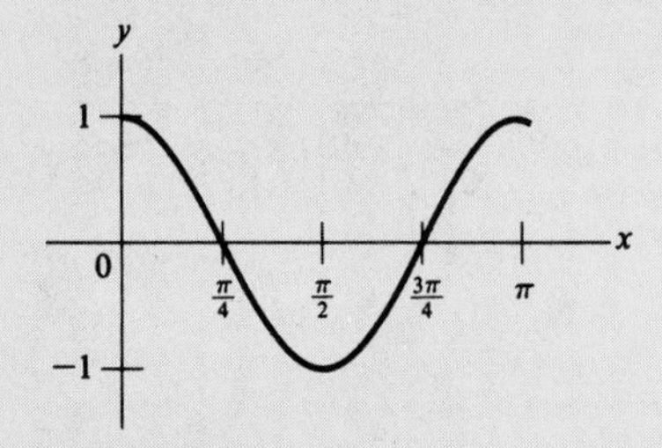

5. $A = 2,\ P = \dfrac{2\pi}{3}$

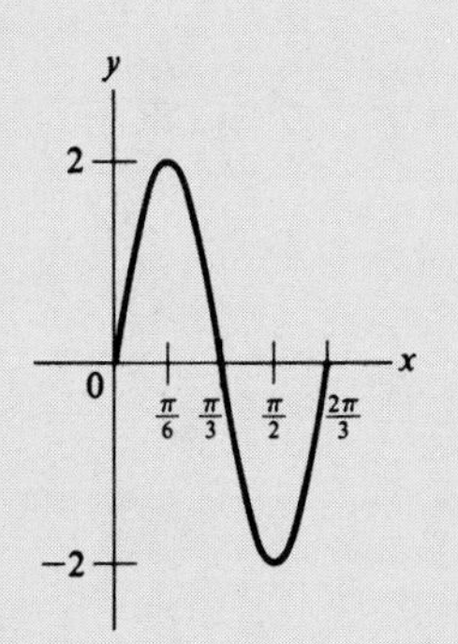

7. $A = \dfrac{1}{2},\ P = \pi$

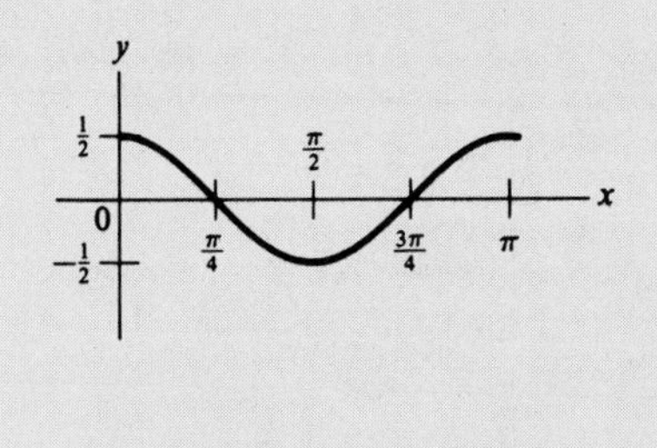

9. $A = 3,\ P = 4\pi$

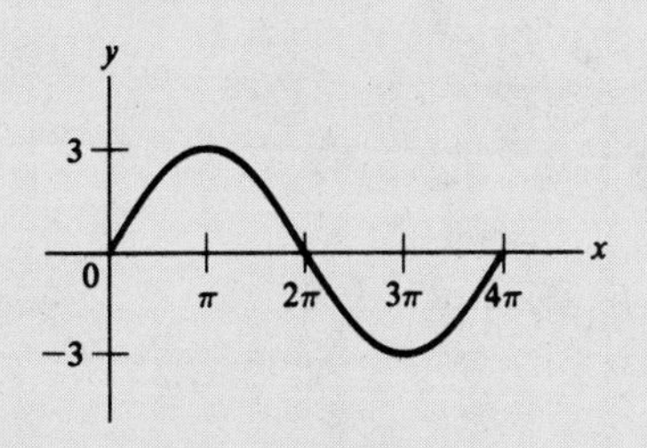

11. $A = 5.4,\ P = 6\pi$

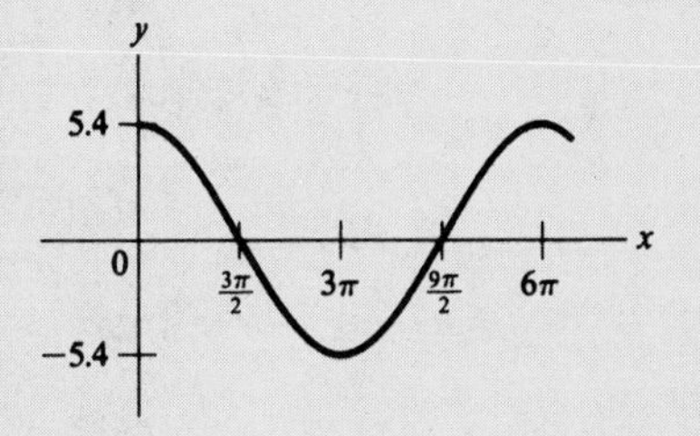

13. $A = 1,\ P = \dfrac{2\pi}{3}$

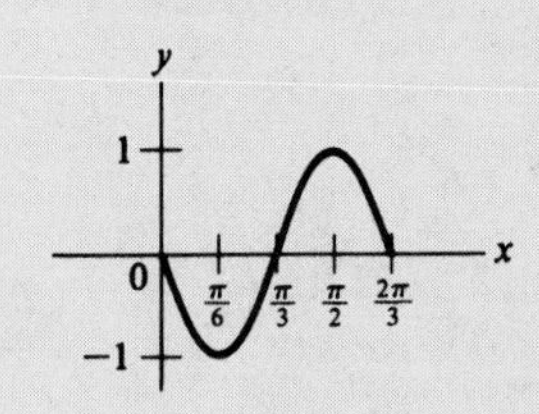

15. $A = \dfrac{1}{4},\ P = \dfrac{\pi}{2}$

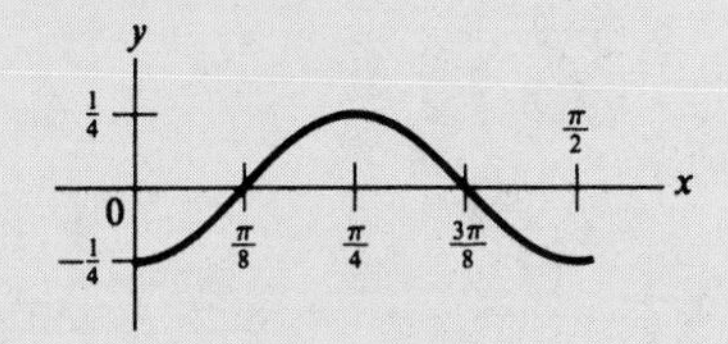

17. $A = 1,\ P = 2$

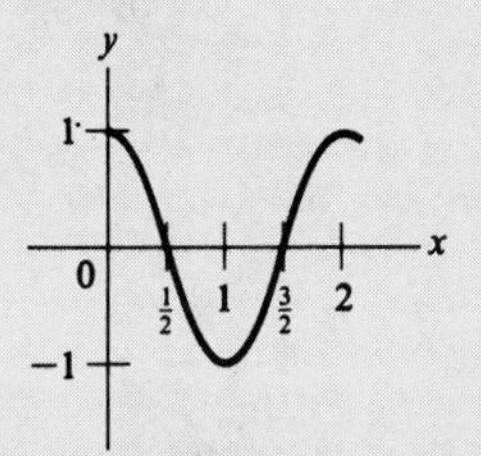

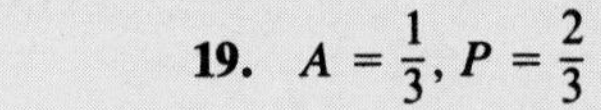

19. $A = \dfrac{1}{3},\ P = \dfrac{2}{3}$

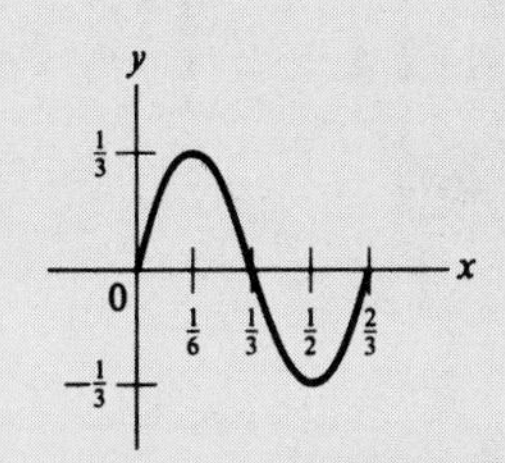

21.

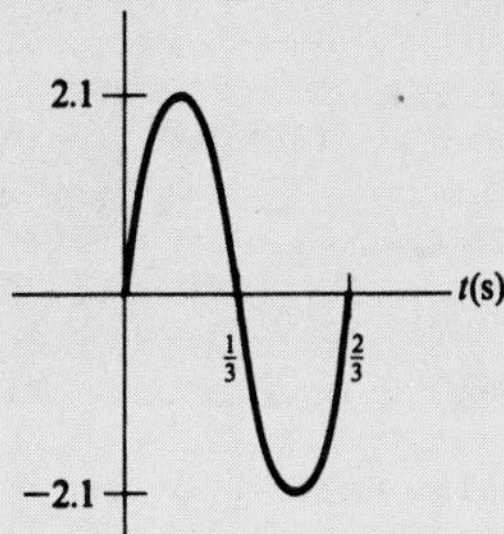

23.

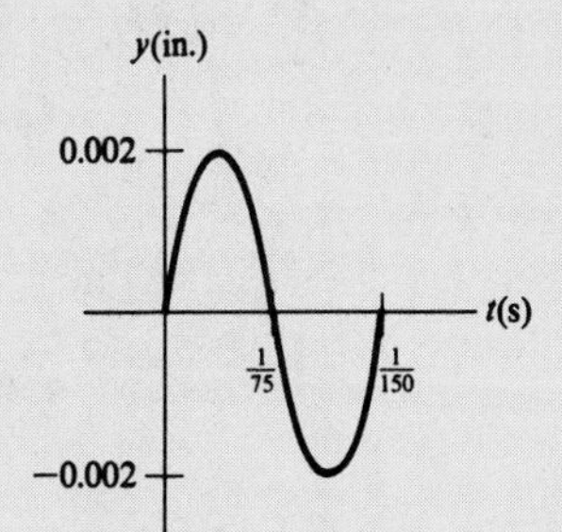

Section 17.5 (page 586)

1. $A = 2$, $P = 2\pi$, shift: 1 to right

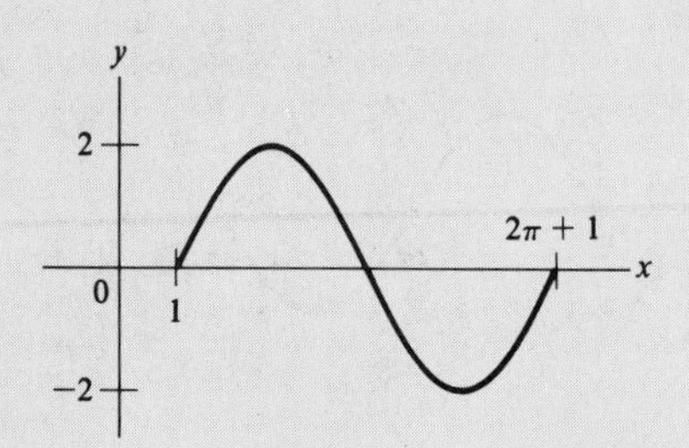

3. $A = 2$, $P = 2\pi$, shift: 1 to left

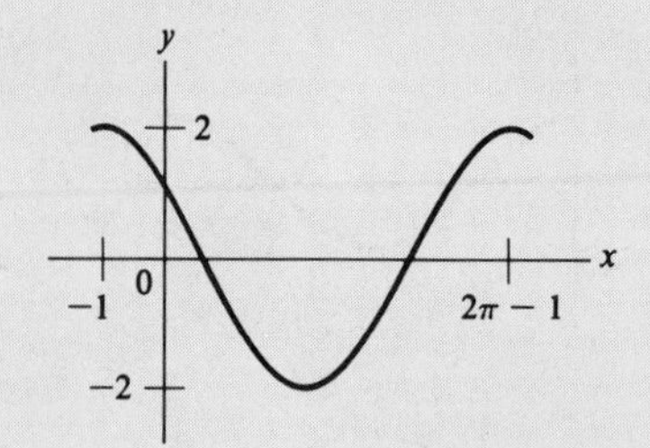

5. $A = \frac{3}{4}$, $P = 2\pi$, shift: $\frac{\pi}{4}$ to right

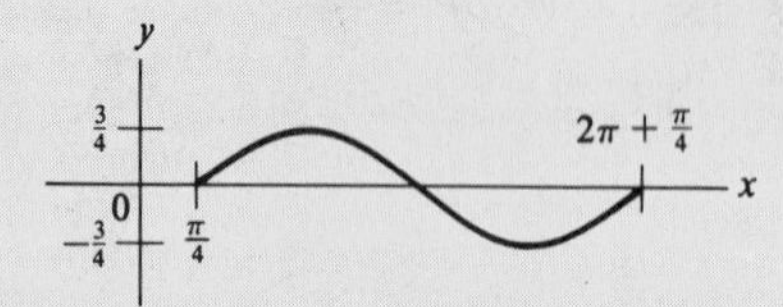

7. $A = \frac{1}{2}$, $P = \pi$, shift: $\frac{\pi}{8}$ to left

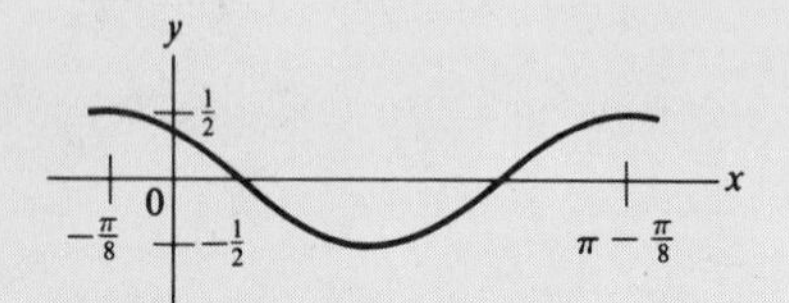

9. $A = 0.8$, $P = \pi$, shift: $\frac{\pi}{16}$ to right

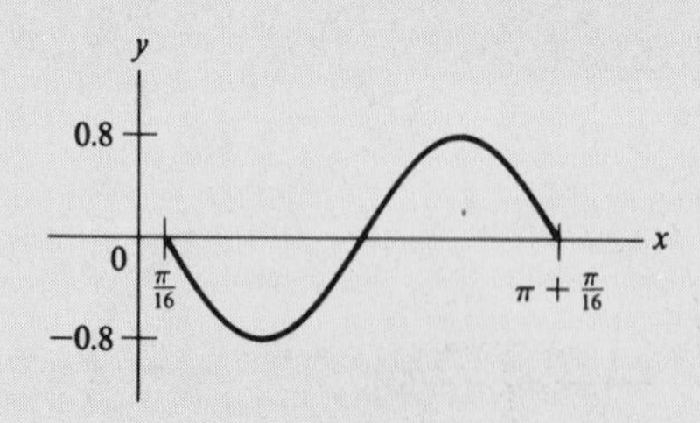

11. $A = 4$, $P = 4\pi$, shift: π to left

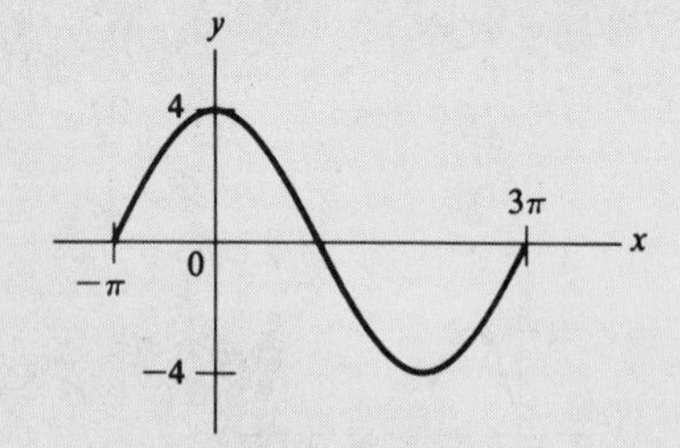

13. $A = 2$, $P = 2$, shift: 1 to right

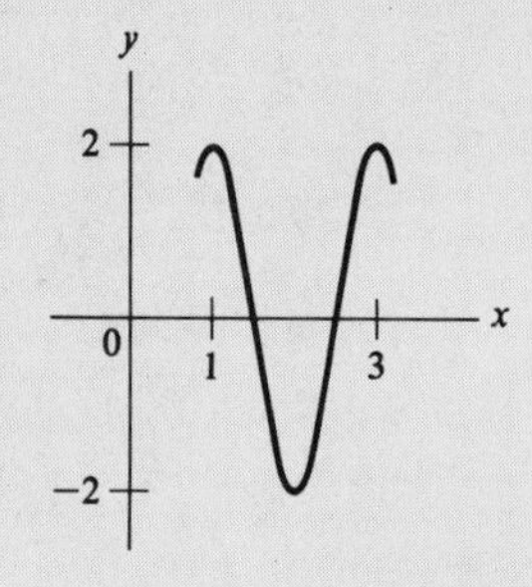

15. $A = 2$, $P = \pi$, shift: $\frac{\pi}{4}$ to left

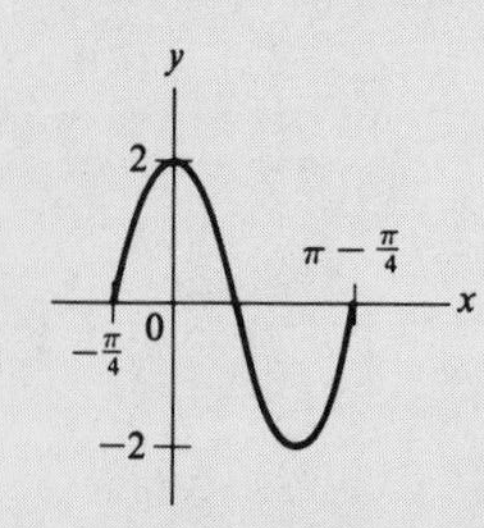

17. $A = 0.5$, $P = \pi$, shift: $\frac{1}{2}$ to right

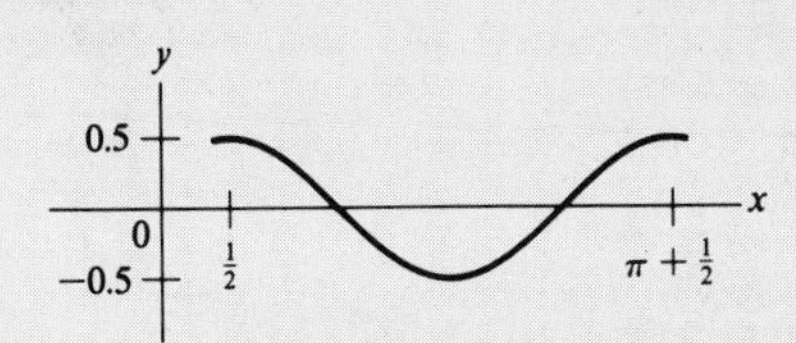

19. $A = 3$, $P = 4\pi$, shift: 4 to right

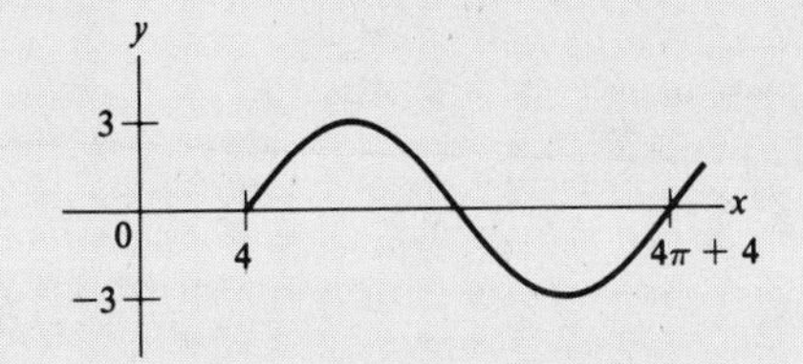

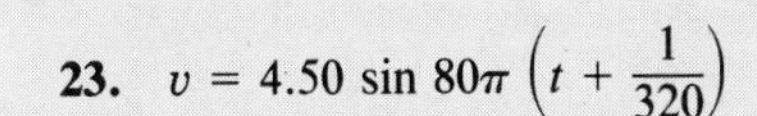

21. $i = 5.00 \sin 120\pi \left(t + \frac{1}{480}\right)$

23. $v = 4.50 \sin 80\pi \left(t + \frac{1}{320}\right)$

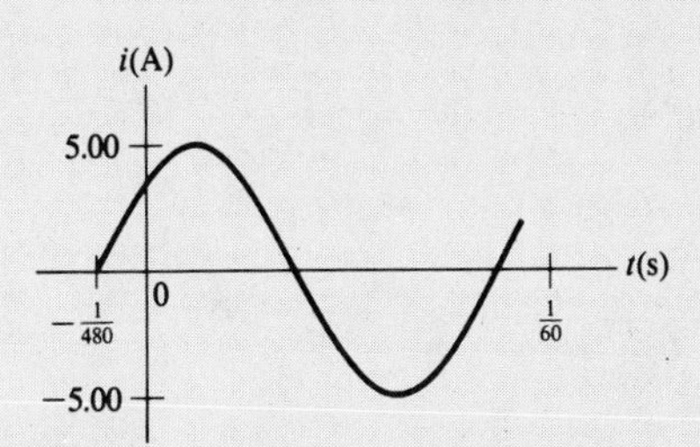

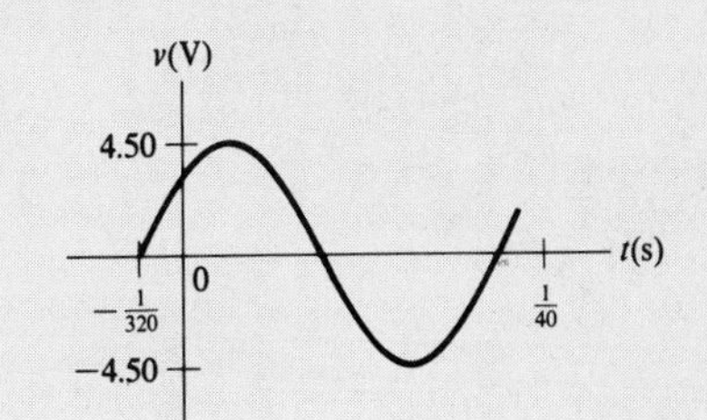

Review Exercises/Chapter 17 (page 586)

1. $\frac{2\pi}{9}$ **2.** $\frac{7\pi}{15}$ **3.** $\frac{28\pi}{45}$ **4.** $\frac{31\pi}{36}$ **5.** $\frac{47\pi}{36}$ **6.** $\frac{16\pi}{9}$ **7.** 348° **8.** 335° **9.** 190° **10.** 220° **11.** 75° **12.** 24° **13.** 1.9548 **14.** 4.9567 **15.** 1.2863 **16.** 1.2683 **17.** 7.8° **18.** 51.2° **19.** 150.9° **20.** 231.3° **21.** 0.5585 **22.** −2.9716 **23.** −1.1173 **24.** 0.7788 **25.** 21.9 cm **26.** 3.56 m **27.** 34.16 in. **28.** 49.4 ft **29.** 314 cm^2 **30.** 41.79 m^2 **31.** 1.66 ft^2 **32.** 21.1 yd^2

33. $A = 3.5$, $P = 2\pi$

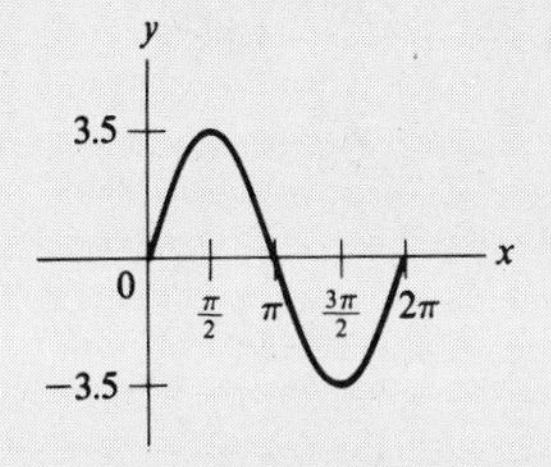

34. $A = 2.5$, $P = 2\pi$

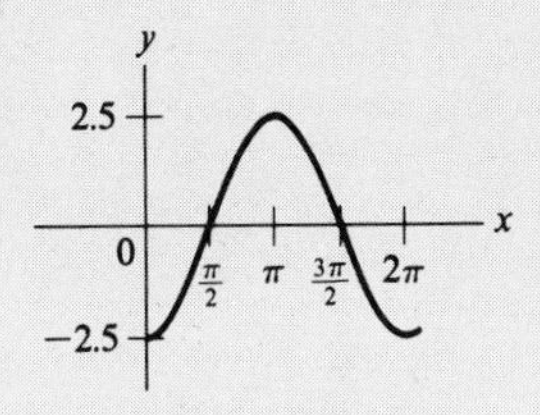

35. $A = 4$, $P = 6\pi$

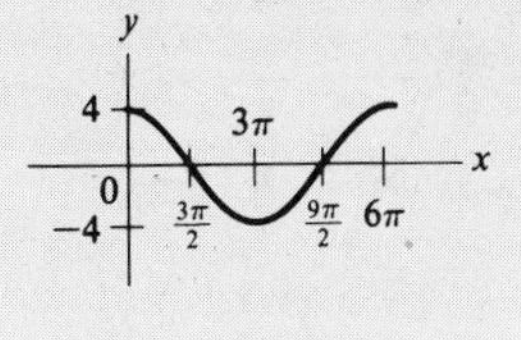

36. $A = \frac{1}{4}$, $P = \frac{2}{3}$

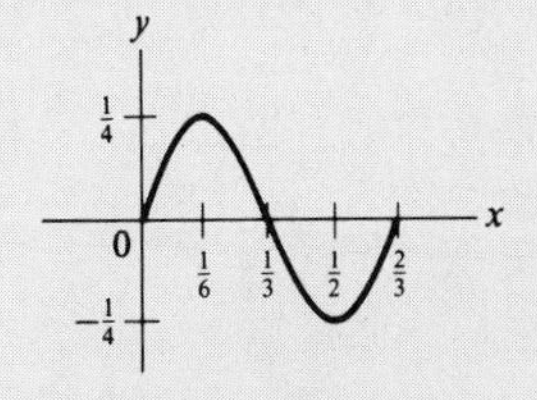

37. $A = \frac{1}{2}$, $P = \pi$

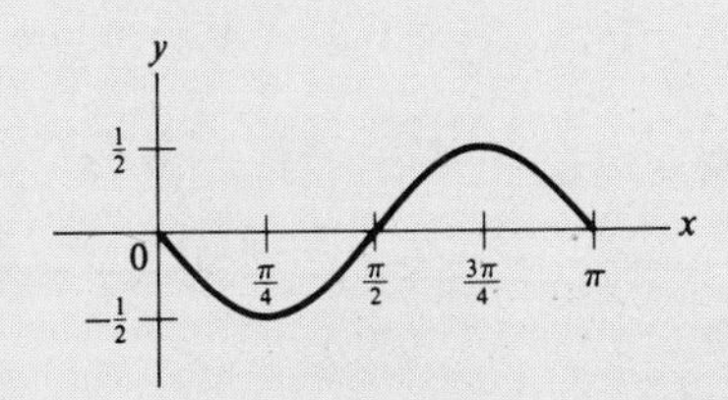

38. $A = \frac{1}{3}$, $P = \pi$, shift: $\frac{\pi}{8}$ to right **39.** $A = 1.25$, $P = \pi$, shift: $\frac{\pi}{16}$ to left

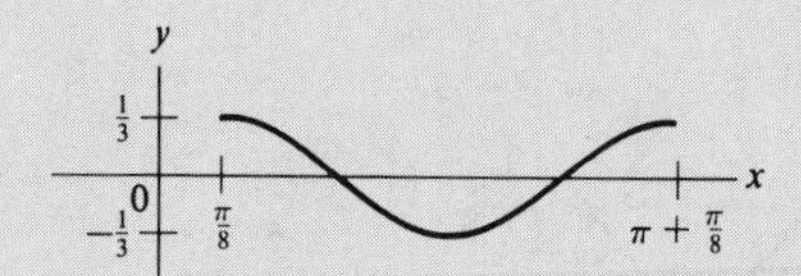

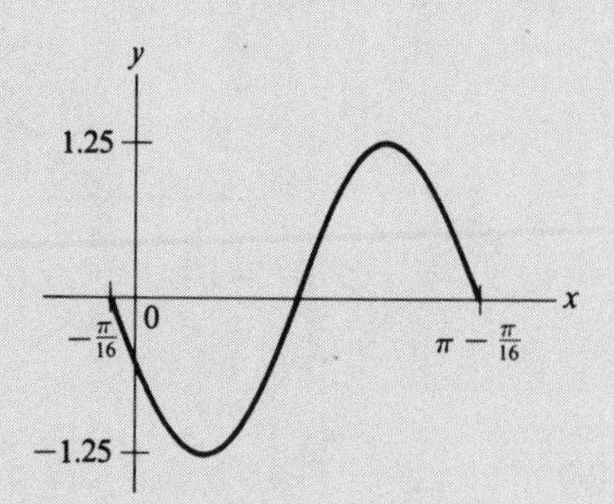

40. $A = 4$, $P = 4\pi$, shift: π to right **41.** 37 ft **42.** 40.1° **43.** 60.8 cm² **44.** 229 ft/s

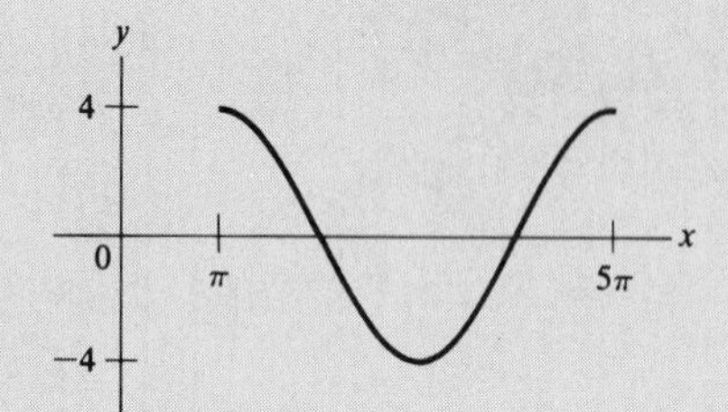

45. 50.3 mi/h **46.** 1.7 ft/s **47.** 12.6 mi/h **48.** 18,000 mi/h **49.** 22.4 rev/min
50. $A = 4.2$ cm, $P = 1$ s

x(cm)
4.2
0
1/2
1
t(s)
−4.2

Chapter 18
Section 18.1 (page 590)

1. $\log_8 64 = 2$ **3.** $\log_{10} 100 = 2$ **5.** $\log_3 81 = 4$ **7.** $\log_2 \frac{1}{16} = -4$ **9.** $\log_4 1 = 0$
11. $\log_{2/3} \frac{4}{9} = 2$ **13.** $\log_{4/3} \frac{9}{16} = -2$ **15.** $\log_{1/2} 8 = -3$ **17.** $4^3 = 64$ **19.** $5^0 = 1$
21. $3^1 = 3$ **23.** $3^{-3} = \frac{1}{27}$ **25.** $\left(\frac{1}{6}\right)^2 = \frac{1}{36}$ **27.** $\left(\frac{1}{4}\right)^{-3} = 64$ **29.** $14^2 = 196$ **31.** $\left(\frac{9}{4}\right)^{-1} = \frac{4}{9}$
33. 27 **35.** $\frac{1}{9}$ **37.** 2 **39.** $\frac{1}{4}$ **41.** 9 **43.** 7 **45.** 4 **47.** $\frac{1}{4}$ **49.** $\frac{3}{2}$ **51.** 3
53. 2 **55.** $-\frac{1}{2}$ **57.** -3 **59.** -1

Section 18.2 (page 595)

1. $2 \log_6 5$ **3.** $2 \log_{10} 6$ **5.** $4 \log_5 2$ **7.** $6 \log_7 2$ **9.** 3 **11.** $\frac{1}{2}$ **13.** $2 + 2 \log_6 x$
15. $-\log_5 3 - 2 \log_5 y$ **17.** $-1 - 2 \log_7 S$ **19.** $-\frac{1}{2}(1 + \log_2 P)$ **21.** $-2 - 2 \log_3 x - 3 \log_3 y$
23. 2 **25.** -3 **27.** $\frac{1}{2}(\log_e 2 + 1)$ **29.** $-2 \log_e 7 - \frac{1}{2}$ **31.** $\log_3 35$ **33.** $\log_9 2$
35. $\log_{10} 25$ **37.** $\log_4 x^2 y$ **39.** $\log_7 \frac{\sqrt{x}}{y^2}$ **41.** $\log_{10} \frac{R^{3/2}}{L^3}$ **43.** $\log_5 \frac{x^2 y^3}{z^4}$ **45.** $\beta = \log_{10}(I/I_0)^{10}$
47. $N = 10^{-t/10}$ **49.** $N = 1000(2^t)$

Section 18.3 (page 599)

1. 0.4564 **3.** 0.8319 **5.** 0.4829 **7.** 0.9217 **9.** 1.4041 **11.** -0.8035 **13.** -2.2904
15. 4.4228 **17.** 7.034 **19.** 33.60 **21.** 1234 **23.** 0.8070 **25.** 155.4 **27.** 7
29. 0.391 W

Section 18.4 (page 601)

1. -1.8786 **3.** 0.9798 **5.** 3.6201 **7.** 5.7881 **9.** -6.3619 **11.** 6.637 **13.** 1.017
15. 25.75 **17.** 14.0132 **19.** 0.2780 **21.** 5.2070 **23.** 1.0147 **25.** 60 days
27. \$7871.02 **29.** 0.16 atmosphere **31.** 15.2 W

Section 18.5 (page 604)

1. 1.4650 **3.** 1.9534 **5.** 0.4771 **7.** 1.7618 **9.** 1.0973 **11.** -0.1715 **13.** -0.3233
15. $t = -RC \ln \frac{Ri}{E}$ **17.** 10^{-11} W/m^2 **19.** **a.** 23,000 **b.** 3.9 h **21.** 6.4 ft
23. 4.1×10^{-7} mol/L

Review Exercises/Chapter 18 (page 605)

1. $\log_4 64 = 3$ **2.** $\log_2 \frac{1}{32} = -5$ **3.** $\left(\frac{1}{2}\right)^{-4} = 16$ **4.** $(27)^{-1/3} = \frac{1}{3}$ **5.** 36 **6.** $\frac{1}{3}$ **7.** $\frac{1}{2}$
8. $\frac{1}{2}$ **9.** $4 \log_5 3$ **10.** $4 \log 2$ **11.** $\frac{1}{2} \log_3 11$ **12.** $\frac{1}{2}$ **13.** $\frac{1}{2}(\log_6 5 + \log_6 x)$
14. $\frac{1}{2}(1 + \log_6 x)$ **15.** $1 + 2 \log x$ **16.** $-1 - \log_5 V$ **17.** $-\frac{1}{2}(\ln 2 + 1)$ **18.** $-\ln 6 - \frac{1}{3}$
19. $\log_5 3$ **20.** $\log_5 2x^2$ **21.** $\ln \frac{\sqrt[3]{P}}{Q^2}$ **22.** $\log V\sqrt{W}$ **23.** 0.8943 **24.** 0.9246 **25.** 1.427
26. -0.7228 **27.** -2.774 **28.** -2.381 **29.** 5.931 **30.** 10.18 **31.** 10.61 **32.** 0.9828
33. 2.0437 **34.** 0.6712 **35.** 1.1469 **36.** -0.7533 **37.** $N = N_0 e^{kt}$ **38.** 2.1 A
39. 2.0 rad **40.** 28 years **41.** 18,500 years **42.** $t = -\frac{1}{k} \ln \frac{T - M_T}{T_0 - M_T}$ **43.** 2500 m

Cumulative Review Exercises/Chapters 16–18 (page 607)

1. $B = 103.59°$, $A = 36.11°$, $C = 40.30°$ **2.** $\frac{31\pi}{45}$ **3.** 1.3206 **4.** 137.34° **5.** $\log \frac{\sqrt[3]{R}\, L^2}{C^3}$
6. $-\frac{1}{2}(1 + \ln 3)$ **7.** $4x = 3^a$ **8.** 39.1 N, 21.4° **9.** 105 lb, 14.8° with larger force
11. 27.7 lb **12.** 2950 lb **13.** **a.** 187 cm^2 **b.** 10.7 cm **14.** 454 rev/min

15. $A = 5,\ P = 4\pi$

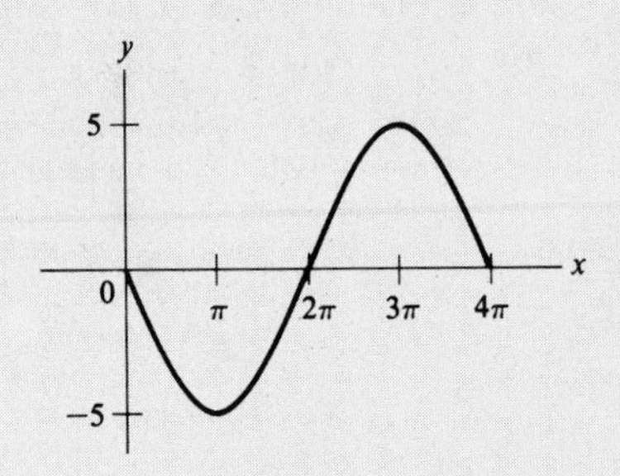

16. $A = 2,\ P = \pi$, shift: $\frac{\pi}{4}$ units to right

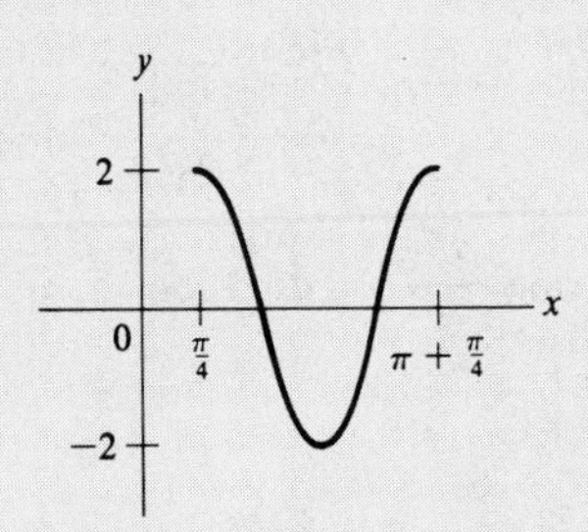

17. $t = -\frac{1}{k} \ln \frac{v_1 - v}{v_2}$ **18.** \$18,243.98 **19.** 2.4 min

Appendix A (page 614)

1.
```
10 PRINT "PYTHAGOREAN THEOREM"
20 INPUT "X="; X
30 INPUT "Y="; Y
40 LET Z = SQR(X↑2 + Y↑2)
50 PRINT "Z="; Z
60 END
```

2.
```
10 INPUT "A="; A
20 INPUT "B="; B
30 LET X = SQR(B/A)
40 PRINT X, −X
50 END
```

or

```
10 PRINT "ENTER THE COEFFICIENT OF THE SQUARE OF X"
20 INPUT A
30 PRINT "ENTER THE CONSTANT"
40 INPUT B
50 LET X = SQR(B/A)
60 PRINT X, −X
70 END
```

3.
```
10 LET M = 0
20 LET N = 1
30 LET M = M + N↑3
40 LET N = N + 1
50 IF N < = 8 THEN 30
60 PRINT M
70 END
```

4.
```
10 LET M = 0
20 FOR N = 1 TO 8
30 LET M = M + N↑3
40 NEXT N
50 PRINT M
60 END
```

5.
```
10 FOR X = 0 TO 2 STEP 0.05
20 LET Y = SIN(2 * X)
30 PRINT X; Y
40 NEXT X
50 END
```

6.
```
10 FOR X = 1 TO 2 STEP 0.1
20 LET Y = LOG(X↑2)
30 PRINT X; Y
40 NEXT X
50 END
```

7.
```
10 LET V = (4/3) * 3.14159265 * 2.47↑3
20 LET S = 4 * 3.14159265 * 2.47↑2
30 PRINT "V="; V
40 PRINT "S="; S
50 END
```

8.
```
10 INPUT "A="; A
20 INPUT "B="; B
30 INPUT "ANGLE="; T
40 LET C = SQR(A↑2 + B↑2 - 2 * A * B * COS(T))
50 PRINT "C="; C
60 END
```

9.
```
10 INPUT "A="; A
20 INPUT "B="; B
30 INPUT "C="; C
40 LET R = SQR(B↑2 - 4 * A * C)
50 LET X1 = (-B + R)/(2 * A)
60 LET X2 = (-B - R)/(2 * A)
70 PRINT X1, X2
80 END
```

10.
```
10 INPUT "R="; R
20 INPUT "ANGLE="; T
30 LET B = T * 3.14159265/180
40 LET A = 0.5 * R↑2 * B
50 PRINT "A="; A
60 END
```

Index

Basic Reduction Values

1 ft = 12 in.

1 yd = 3 ft

1 mi = 5280 ft

1 lb = 16 oz

1 qt = 2 pt

1 gal = 4 qt

1 cm = 10 mm

1 m = 100 cm

1 km = 1000 m

1 kg = 1000 g

1 L = 1000 mL

1 mL = 1 cm^3

Basic Conversion Values

1 in. = 2.54 cm (exact)

1 km = 0.6214 mi

1 mi = 1.609 km

1 lb = 454 g

1 kg = 2.205 lb

1 lb = 4.448 N

1 ft^3 = 28.32 L

Properties of Signed Numbers

$a + b = b + a$

$ab = ba$

$a + (b + c) = (a + b) + c$

$a(bc) = (ab)c$

$a(b \pm c) = ab \pm ac$

$a \pm 0 = a$

$a \cdot 0 = 0$

$\frac{a}{0}$ is undefined

$a + (-a) = 0$

$-(-a) = a$

$-a = (-1)a$

$a = 1 \cdot a$

$a - b = a + (-b)$

$a - (-b) = a + b$

$a - b = -1(b - a)$

$(-a)b = a(-b) = -ab$

$(-a)(-b) = ab$

$\frac{-a}{b} = \frac{a}{-b} = -\frac{a}{b}$

$\frac{-a}{-b} = \frac{a}{b}$